Student Solutions Manual

to accompany

College Algebra

Julie Miller
Daytona State College—Daytona Beach

Prepared by
diacriTech Inc., DBA LaurelTech

The McGraw-Hill Companies

Student Solutions Manual to accompany
COLLEGE ALGEBRA
JULIE MILLER

Published by McGraw-Hill Higher Education, an imprint of The McGraw-Hill Companies, Inc., 1221 Avenue of the Americas, New York, NY 10020.

This book is printed on acid-free paper.

3 4 5 6 7 8 QVS/QVS 19 18 17 16 15

ISBN: 978-0-07-753867-5
MHID: 0-07-753867-6

www.mhhe.com

Contents

Chapter R Review of Prerequisites

Chapter 2 Functions and Graphs

Chapter 3 Polynomial and Rational Functions

Chapter 4 Exponential and Logarithmic Functions

Chapter 5 Systems of Equations and Inequalities

Chapter R Review of Prerequisites

Section R.1 Practice Exercises

1. set

3. natural

5. set-builder

7. rational

9. left

11. $|a-b|$ or $|b-a|$

13. square

15. 3 is an element of the set of natural numbers.

17. -3.1 is not an element of the set of integers.

19. The set of integers is a proper subset of the set of real numbers.

21. a. $\mathbb{N}=\{1,2,3,\ldots\}$ does not include -5. False.

b. $\mathbb{W}=\{0,1,2,3,\ldots\}$ does not include -5. False.

c. $\mathbb{Z}=\{\ldots,-2,-1,0,1,2,\ldots\}$ .This includes -5. True.

d. -5 can be written as $\frac{-5}{1}$, which is the ratio of -5 and 1, so it is an element of $\mathbb{Q}$. True.

23. a. $\mathbb{N}=\{1,2,3,\ldots\}$ does not include $0.\overline{25}$. False.

b. $\mathbb{W}=\{0,1,2,3,\ldots\}$ does not include $0.\overline{25}$. False.

c. $\mathbb{Z}=\{\ldots,-2,-1,0,1,2,\ldots\}$ does not include $0.\overline{25}$. False.

d. $0.\overline{25}$ is a repeating decimal, so it is an element of $\mathbb{Q}$. True.

25. The set given can be written in roster form as $\{\ldots,-10,-5,0,5,10,\ldots\}$. The number 25 is an element of this set. True.

27. The number $-0.\overline{8}$ is a repeating decimal, so it is rational. It can be written as $-0.8\overline{8}$, and $-0.8\overline{8}<-0.8$. False.

29. The set given can be written in roster form as $\{0,2,4,6,\ldots\}$. The number 22 is an element of this set. True.

31. Irrational numbers are real numbers that cannot be represented as a ratio of integers, so a number cannot be both rational and irrational. False.

33. a. The first set is a subset of the second, because all of the elements in the first set are elements of the second set. True.

b. The first set is not a subset of the second, because 0 is in the first set but not in the second set. False.

35. a. The first set is not a subset of the second, because the set of whole numbers does not include negative integers. False.

b. The first set is a subset of the second, because all whole numbers are also integers. True.

37. a. The first set is not a subset of the second, because irrational numbers are real numbers that cannot be represented as a ratio of integers, so a number cannot be both rational and irrational. False.

b. The first set is not a subset of the second, because irrational numbers are real numbers that cannot be represented as a ratio of integers, so a number cannot be both rational and irrational. False.

39. a. 6

b. 6

c. -12, 6

d. $0.\overline{3}, 0.33, -0.9, -12, \frac{11}{4}, 6$

e. $\sqrt{5}, \frac{\pi}{6}$

f. $\sqrt{5}, 0.\overline{3}, 0.33, -0.9, -12, \frac{11}{4}, 6, \frac{\pi}{6}$

41. $a \ge 5$ **43.** $3c \le 9$

45. $m + 4 > 70$

47. $3.14 \overset{?}{<} \pi$
$3.14 \overset{?}{<} 3.141592654\ldots$ True

49. $6.7 \overset{?}{\ge} 6.7$
$6.7 \overset{?}{\ge} 6.7$ True

51. $6.\overline{15} \overset{?}{>} 6.1\overline{5}$
$6.151515\ldots \overset{?}{>} 6.155555\ldots$ False

53. $-\frac{9}{7} \overset{?}{<} -\frac{11}{8}$
$-\frac{9}{7}\cdot\frac{8}{8} \overset{?}{<} -\frac{11}{8}\cdot\frac{7}{7}$
$-\frac{72}{56} \overset{?}{<} -\frac{77}{56}$ False

55. $(7, \infty)$; $\{x \mid x > -7\}$

57. $(-\infty, 4.1]$; $\{x \mid x \le 4.1\}$

59. $[-6, 0)$; $\{x \mid -6 \le x < 0\}$

61. 6 $(-\infty, 6]$

63. $-\frac{7}{6}$ $\frac{1}{3}$ $\left(-\frac{7}{6}, \frac{1}{3}\right]$

65. 4 $(4, \infty)$

67. $\{x \mid -3 < x \le 7\}$ **69.** $\{x \mid x \le 6.7\}$

71. $\left\{x \,\middle|\, x \ge -\frac{3}{5}\right\}$

73. $|-6| = -(-6) = 6$ **75.** $|0| = 0$

77. $\left|\sqrt{2} - 2\right| = -\left(\sqrt{2} - 2\right) = 2 - \sqrt{2}$

79. a. $|\pi - 3| = \pi - 3$

b. $|3 - \pi| = -(3 - \pi) = \pi - 3$

81. a. for $x \ge -2, |x + 2| = x + 2$

b. for $x < -2, |x + 2| = -(x + 2) = -x - 2$

83. a. for $z > 5, \frac{|z - 5|}{z - 5} = \frac{z - 5}{z - 5} = 1$

b. for $z < 5$,
$\frac{|z - 5|}{z - 5} = \frac{-(z - 5)}{z - 5} = -1 \cdot \frac{z - 5}{z - 5} = -1$

85. $|1 - 6| = |-5| = 5$ or $|6 - 1| = |5| = 5$

87. $|3 - (-4)| = |7| = 7$ or $|-4 - 3| = |-7| = 7$

89. $\left|8 - \sqrt{3}\right| = 8 - \sqrt{3}$ or $\left|\sqrt{3} - 8\right| = 8 - \sqrt{3}$

91. $|6 - 2\pi| = 2\pi - 6$ or $|2\pi - 6| = 2\pi - 6$

93. a. $4^2 = 4 \cdot 4 = 16$

b. $(-4)^2 = (-4)\cdot(-4) = 16$

c. $-4^2 = -(4 \cdot 4) = -16$

d. $\sqrt{4} = 2$

e. $-\sqrt{4} = -2$

f. $\sqrt{-4}$ is not a real number.

95. a. $\sqrt[3]{8} = 2$

b. $\sqrt[3]{-8} = -2$

c. $-\sqrt[3]{8}=-\left(\sqrt[3]{8}\right)=-2$

d. $\sqrt{100}=10$

e. $\sqrt{-100}$ is not a real number.

f. $-\sqrt{100}=-10$

97. $\left(\frac{2}{3}\right)^3=\frac{2}{3}\cdot\frac{2}{3}\cdot\frac{2}{3}=\frac{8}{27}$

99. $(-0.2)^4=(-0.2)\cdot(-0.2)\cdot(-0.2)\cdot(-0.2)$
$=0.0016$

101. $\sqrt{\frac{169}{25}}=\frac{13}{5}$

103. $20-12\left(36\div 3^2\div 2\right)=20-12(36\div 9\div 2)$
$=20-12(4\div 2)$
$=20-12(2)$
$=20-24=-4$

105. $6-\left\{-12+3\left[(1-6)^2-18\right]\right\}$
$=6-\left\{-12+3\left[(-5)^2-18\right]\right\}$
$=6-\left[-12+3(25-18)\right]=6-\left[-12+3(7)\right]$
$=6-(-12+21)=6-9=-3$

107. $\sqrt{5^2-3^2}=\sqrt{25-9}=\sqrt{16}=4$

109. $\left(\sqrt{9}+\sqrt{16}\right)^2=(3+4)^2=7^2=49$

111. $-4\cdot\left(\frac{2}{5}-\frac{7}{10}\right)^2=-4\cdot\left(\frac{2}{2}\cdot\frac{2}{5}-\frac{7}{10}\right)^2$
$=-4\cdot\left(\frac{4}{10}-\frac{7}{10}\right)^2$
$=-4\cdot\left(-\frac{3}{10}\right)^2=-4\cdot\frac{9}{100}$
$=-\frac{\cancel{4}^{1}}{1}\cdot\frac{9}{\cancel{100}_{25}}=-\frac{9}{25}$

113. $9-\left(6+\left||3-7|-8\right|\right)\div\sqrt{25}$
$=9-\left(6+\left||-4|-8\right|\right)\div\sqrt{25}$
$=9-\left(6+|4-8|\right)\div\sqrt{25}$
$=9-\left(6+|-4|\right)\div\sqrt{25}$
$=9-(6+4)\div\sqrt{25}$
$=9-10\div 5$
$=9-2$
$=7$

115. $\frac{|11-13|-4\cdot 2}{\sqrt{12^2+5^2}-3-10}=\frac{|-2|-8}{\sqrt{144+25}-3-10}$
$=\frac{2-8}{\sqrt{169}-13}$
$=\frac{-6}{13-13}$
$=\frac{-6}{0}$ Undefined

117. $W_h=(200)\left(\frac{4000}{4000+5.5}\right)^2$
$=(200)\left(\frac{4000}{4005.5}\right)^2$
≈ 119.5 lb

119. $V=\frac{\pi(3.8)^2\cdot 8}{3}=\frac{\pi\cdot 14.44\cdot 8}{3}\approx 121$ cc

121. All terminating decimals can be written as a decimal fraction—that is, a fraction with a denominator that is a power of 10.

123. A parenthesis is used if an endpoint to an interval is not included in the set.

125. $\mathbb{Z}$ represents the set of integers. All integers belong to $\mathbb{Q}$ (the set of rational numbers). Therefore, $\mathbb{Z}$ is a subset of $\mathbb{Q}$. However, $\mathbb{Q}$ contains fractions and decimals, such as $\frac{1}{2}$ and 0.2, that are not integers. Therefore, $\mathbb{Q}\not\subset\mathbb{Z}$.

127. If $n>0$, then $n-|n|=n-n=0$

129. If $n>0$, then $n+|n|=n+n=2n$

131. If $n>0$, then $-|n|=-n$

133. $b>0$

135. $b\geq 0$

137. a is negative, b is positive, and c is negative.

So the sign of $\frac{ab^2}{c^3}$ can be represented by

$\frac{(-)(+)^2}{(-)^3}\to\frac{(-)(+)}{(-)}\to(+)$, which is positive.

139. a is negative, b is positive, and c is negative. The sum of a and c must be negative. So the sign of $\frac{b(a+c)^3}{a^2}$ can be represented by

$\frac{(+)(-)^3}{(-)^2}\to\frac{(+)(-)}{(+)}\to(-)$, which is negative.

141.
```
5000(1+.06/12)^(
12*5)
        6744.250763
```
6744.25

143.
```
(-3+5√(2))/7
         .581581116
```
0.58

145. There are missing parentheses around the denominator.

```
(-3+√(3²-4(-5)(2
))/(2(-5))
                 -.4
```

Section R.2 Models, Algebraic Expressions, and Properties of Real Numbers

1. $[-5,2)$; $\{x\,|\,-5\leq x<2\}$

3. $\left|\sqrt{2}-9\right|=9-\sqrt{2}$ or $\left|9-\sqrt{2}\right|=9-\sqrt{2}$

5. True

7. False

9. $\frac{12-(9-\sqrt{9})^2\div(-2)\cdot 3}{|-7+1|}$

$=\frac{12-(9-3)^2\div(-2)\cdot 3}{|-6|}$

$=\frac{12-(6)^2\div(-2)\cdot 3}{6}=\frac{12-36\div(-2)\cdot 3}{6}$

$=\frac{12-(-18)\cdot 3}{6}=\frac{12-(-54)}{6}=\frac{66}{6}=11$

11. dependent

13. term

15. commutative

17. 1

19. $-x$

21. reciprocal

23. $(a\cdot b)\cdot c$

25. **a.** $m=-0.04x^2+3.6x-49$

$m=-0.04(35)^2+3.6(35)-49$

$m\approx 28$ mpg

b. $m=-0.04x^2+3.6x-49$

$m=-0.04(50)^2+3.6(50)-49$

$m\approx 31$ mpg

c. $m=-0.04x^2+3.6x-49$

$m=-0.04(75)^2+3.6(75)-49$

$m\approx -4$ mpg

The model breaks down. The value of –4 mpg is not possible.

27. a. $E = -27.45t + 704.6$
$E = -27.45(3) + 704.6$
$E \approx \$622.25$

b. $E = -27.45t + 704.6$
$E = -27.45(15) + 704.6$
$E \approx \$292.85$

c. No. Eventually, the model would produce a negative expenditure. This is not possible.

29. $T_n = T_s + 16$

31. $P = W + M$

33. a. $J = C - 1$

b. $C = J + 1$

35. $D \geq 0.25P$

37. $v = \sqrt{2gh}$

39. a. $C = 0.12k + 14.89$

b. $C = 0.12k + 14.89$
$C = 0.12(1200) + 14.89$
$C = \$158.89$

41. a. $C = 640m + 200n + 500$

b. $C = 640m + 200n + 500$
$C = 640(12) + 200(3) + 500$
$C = \$8780$

43. a. $C = 159n + 0.11(159n)$
$C = 159n + 17.49n$
$C = 176.49n$

b. $C = 176.49n$
$C = 176.49(4)$
$C = \$705.96$

45. $12x^2y^5 - xy^4 + 9.2xy^3$
$= 12x^2y^5 + (-xy^4) + 9.2xy^3$
Terms: $12x^2y^5$, $-xy^4$, $9.2xy^3$;
coefficients: 12, -1, 9.2

47. $\frac{5}{m} = 5 \cdot \frac{1}{m}$
Term: $\frac{5}{m}$; coefficient: 5

49. $x|y+z| = 1x|y+z|$
Term: $x|y+z|$; coefficient: 1

51. $7 + x = x + 7$

53. $-3 + w = w + (-3)$

55. $y \cdot \frac{1}{3} = \frac{1}{3}y$

57. $(t+3) + 9 = t + (3+9) = t + 12$

59. $\frac{1}{5}(5w) = \left(\frac{1}{5} \cdot 5\right)w = 1w = w$

61. a. $-8 + 8 = 0$
The additive inverse is 8.

b. $-8 \cdot \left(-\frac{1}{8}\right) = 1$

The multiplicative inverse is $-\frac{1}{8}$.

63. a. $\frac{5}{4} + \left(-\frac{5}{4}\right) = 0$

The additive inverse is $-\frac{5}{4}$.

b. $\frac{5}{4} \cdot \left(\frac{4}{5}\right) = 1$

The multiplicative inverse is $\frac{4}{5}$.

65. a. $0 + 0 = 0$
The additive inverse is 0.

b. $0 \cdot (?) = 1$
The multiplicative inverse is undefined.

67. a. $2.1+(-2.1)=0$

The additive inverse is -2.1.

b. $2.1\cdot\left(\frac{1}{2.1}\right)=1$

The multiplicative inverse is $\frac{1}{2.1}$ or $\frac{10}{21}$.

69. $-14w^3-3w^3+w^3=(-14-3+1)w^3$
$=-16w^3$

71. $3.9x^3y-2.2xy^3+5.1x^3y-4.7xy^3$
$=3.9x^3y+5.1x^3y-2.2xy^3-4.7xy^3$
$=9x^3y-6.9xy^3$

73. $\frac{1}{3}c^7d+\frac{1}{2}cd^7-\frac{2}{5}c^7d-2cd^7$
$=\frac{1}{3}c^7d-\frac{2}{5}c^7d+\frac{1}{2}cd^7-2cd^7$
$=\frac{5}{5}\cdot\frac{1}{3}c^7d-\frac{3}{3}\cdot\frac{2}{5}c^7d+\frac{1}{2}cd^7-\frac{2}{2}\cdot 2cd^7$
$=\frac{5}{15}c^7d-\frac{6}{15}c^7d+\frac{1}{2}cd^7-\frac{4}{2}cd^7$
$=-\frac{1}{15}c^7d-\frac{3}{2}cd^7$

75. $3(4p^3-6.1p^2-8p+2.2)$
$=3\cdot(4p^3)+3\cdot(-6.1p^2)+3(-8p)+3(2.2)$
$=12p^3-18.3p^2-24p+6.6$

77. $-(4x-\pi)=-1[4x+(-\pi)]=-4x+\pi$

79. $-8(3x^2+2x-1)=-8\cdot(3x^2)-8\cdot(2x)-8(-1)$
$=-24x^2-16x+8$

81. $\frac{2}{3}(-6x^2y-18yz^2+2z^3)$
$=\frac{2}{3}\cdot(-6x^2y)+\frac{2}{3}\cdot(-18yz^2)+\frac{2}{3}(2z^3)$
$=-4x^2y-12yz^2+\frac{4}{3}z^3$

83. $2(4w+8)+7(2w-4)+12$
$=8w+16+14w-28+12=22w$

85. $-(4u-8v)-3(7u-2v)+2v$
$=-4u+8v-21u+6v+2v=-25u+16v$

87. $12-4[(8-2v)+5(-3w-4v)]-w$
$=12-4(8-2v-15w-20v)-w$
$=12-4(8-22v-15w)-w$
$=12-32+88v+60w-w=88v+59w-20$

89. $2y^2-\left[13-\frac{2}{3}(6y^2-9)-10\right]+9$
$=2y^2-(13-4y^2+6-10)+9$
$=2y^2-(-4y^2+9)+9$
$=2y^2+4y^2-9+9=6y^2$

91. $\frac{-b}{2a}=\frac{-(-6)}{2(-1)}=\frac{6}{-2}=-3$

93. $\sqrt{(x_2-x_1)^2+(y_2-y_1)^2}$
$=\sqrt{(-1-2)^2+[1-(-4)]^2}$
$=\sqrt{(-3)^2+(5)^2}=\sqrt{9+25}=\sqrt{34}$

95. $\frac{1}{3}\pi r^2h=\frac{1}{3}\left(\frac{22}{7}\right)(7)^2(6)$
$=\frac{22}{21}(49)(6)=\frac{6468}{21}=308$

97. $W=2L-3$

99. $E_{2009}=E_{2008}+489{,}000$

101. $G=\frac{1}{15}c+\frac{1}{25}h$
$G=\frac{1}{15}(240)+\frac{1}{25}(500)$
$G=16+20=36$ gal

103. The commutative property of addition indicates that the order in which two quantities are added does not affect the sum. The associative property of addition indicates that the manner in which quantities are grouped under addition does not affect the sum.

105.
```
((3+0.01)²-3²)/0
.01
                6.01
```

Section R.3 Integer Exponents and Scientific Notation

1. $-\frac{1}{5}(5x+2y-15)-\frac{1}{10}(2y+20)$
$=-x-\frac{2}{5}y+3-\frac{1}{5}y-2=-x-\frac{3}{5}y+1$

3. $|a-6|=-(a-6)=6-a$

5. $t=w+1.93$

7. 1

9. scientific

11. $m+n$

13. a. $8^0=1$

b. $-8^0=-1\cdot 8^0=-1\cdot 1=-1$

c. $8x^0=8\cdot x^0=8\cdot 1=8$

d. $(8x)^0=1$

15. a. $\left(-\frac{2}{3}\right)^0=1$

b. $-\frac{2^0}{3}=-\frac{1}{3}\cdot 2^0=-\frac{1}{3}\cdot 1=-\frac{1}{3}$

c. $-\frac{2}{3}p^0=-\frac{2}{3}\cdot p^0=-\frac{2}{3}\cdot 1=-\frac{2}{3}$

d. $\left(-\frac{2}{3}p\right)^0=1$

17. a. $8^{-2}=\frac{1}{8^2}=\frac{1}{64}$

b. $8x^{-2}=8\cdot\frac{1}{x^2}=\frac{8}{x^2}$

c. $(8x)^{-2}=\frac{1}{(8x)^2}=\frac{1}{64x^2}$

d. $-8^{-2}=-1\cdot 8^{-2}=-1\cdot\frac{1}{8^2}=-1\cdot\frac{1}{64}=-\frac{1}{64}$

19. a. $\frac{1}{q^{-2}}=\frac{1}{\frac{1}{q^2}}=1\cdot\frac{q^2}{1}=q^2$

b. $q^{-2}=\frac{1}{q^2}$

c. $5p^3q^{-2}=5\cdot p^3\cdot q^{-2}=5\cdot p^3\cdot\frac{1}{q^2}=\frac{5p^3}{q^2}$

d. $5p^{-3}q^2=5\cdot p^{-3}\cdot q^2=5\cdot\frac{1}{p^3}\cdot q^2=\frac{5q^2}{p^3}$

21. $2^5\cdot 2^7=2^{5+7}=2^{12}$

23. $x^7\cdot x^6\cdot x^{-2}=x^{7+6+(-2)}=x^{11}$

25. $(-3c^2d^7)(4c^{-5}d)=-3\cdot 4\cdot c^{2-5}\cdot d^{7+1}$
$=-12\cdot c^{-3}\cdot d^8$
$=-12\cdot\frac{1}{c^3}\cdot d^8=-\frac{12d^8}{c^3}$

27. $\frac{y^{-3}y^6}{y^2}=\frac{y^{-3+6}}{y^2}=\frac{y^3}{y^2}=y^{3-2}=y^1=y$

29. $\frac{6^5}{6^8}=6^{5-8}=6^{-3}=\frac{1}{6^3}=\frac{1}{216}$

31. $\dfrac{18k^2p^9}{27k^5p^2}=\dfrac{18}{27}\cdot\dfrac{k^2}{k^5}\cdot\dfrac{p^9}{p^2}$

$=\dfrac{2}{3}\cdot k^{2-5}\cdot p^{9-2}$

$=\dfrac{2}{3}\cdot k^{-3}\cdot p^7$

$=\dfrac{2}{3}\cdot\dfrac{1}{k^3}\cdot p^7=\dfrac{2p^7}{3k^3}$

33. $\dfrac{2m^{-6}n^4}{6m^{-2}n^{-1}}=\dfrac{2}{6}\cdot\dfrac{m^{-6}}{m^{-2}}\cdot\dfrac{n^4}{n^{-1}}$

$=\dfrac{1}{3}\cdot m^{-6-(-2)}\cdot n^{4-(-1)}$

$=\dfrac{1}{3}\cdot m^{-4}\cdot n^5$

$=\dfrac{1}{3}\cdot\dfrac{1}{m^4}\cdot n^5=\dfrac{n^5}{3m^4}$

35. $\left(4^2\right)^3=4^{2\cdot 3}=4^6$

37. $\left(p^{-2}\right)^7=p^{-2\cdot 7}=p^{-14}=\dfrac{1}{p^{14}}$

39. $(-2cd)^3=(-2)^3(c)^3(d)^3=-8c^3d^3$

41. $\left(\dfrac{7a}{b}\right)^2=\dfrac{(7a)^2}{(b)^2}=\dfrac{(7)^2(a)^2}{b^2}=\dfrac{7^2a^2}{b^2}=\dfrac{49a^2}{b^2}$

43. $\left(4x^2y^{-3}\right)^2=4^2\left(x^2\right)^2\left(y^{-3}\right)^2=16x^4y^{-6}=\dfrac{16x^4}{y^6}$

45. $\left(\dfrac{7k}{n^2}\right)^{-2}=\left(7kn^{-2}\right)^{-2}$

$=(7)^{-2}k^{-2}n^4$

$=\dfrac{1}{49}\cdot\dfrac{1}{k^2}\cdot n^4=\dfrac{n^4}{49k^2}$

47. $\left(\dfrac{1}{8}\right)^{-2}-\left(\dfrac{1}{4}\right)^{-3}-\left(\dfrac{1}{2}\right)^{0}=8^2-4^3-1$

$=64-64-1$

$=-1$

49. $\left(\dfrac{-16m^2n^7}{8m^5n^{-2}}\right)^{-2}=\left(-2m^{2-5}n^{7-(-2)}\right)^{-2}$

$=\left(-2m^{-3}n^9\right)^{-2}$

$=(-2)^{-2}\left(m^{-3}\right)^{-2}\left(n^9\right)^{-2}$

$=\dfrac{1}{(-2)^2}\cdot m^6n^{-18}$

$=\dfrac{m^6}{(-2)^2n^{18}}=\dfrac{m^6}{4n^{18}}$

51. $\left(\dfrac{4x^3z^{-5}}{12y^{-2}}\right)^{-3}=\left(\dfrac{1}{3}\cdot x^3y^2z^{-5}\right)^{-3}$

$=\left(\dfrac{1}{3}\right)^{-3}\left(x^3\right)^{-3}\left(y^2\right)^{-3}\left(z^{-5}\right)^{-3}$

$=3^3x^{-9}y^{-6}z^{15}=\dfrac{27z^{15}}{x^9y^6}$

53. $(-2y)^{-3}\left(6y^{-2}z^8\right)^2=(-2)^{-3}y^{-3}\cdot 6^2\left(y^{-2}\right)^2\left(z^8\right)^2$

$=(-2)^{-3}\cdot 6^2\cdot y^{-3}\cdot y^{-4}\cdot z^{16}$

$=(-2)^{-3}\cdot 6^2\cdot y^{-7}\cdot z^{16}$

$=\dfrac{6^2z^{16}}{(-2)^3y^7}=\dfrac{36z^{16}}{-8y^7}$

$=-\dfrac{9z^{16}}{2y^7}$

55. $\left(\dfrac{1}{2}-\dfrac{1}{3}+\dfrac{1}{6}\right)^{-3}$

$=\left[\left(\dfrac{3}{3}\right)\left(\dfrac{1}{2}\right)-\left(\dfrac{2}{2}\right)\left(\dfrac{1}{3}\right)+\left(\dfrac{1}{6}\right)\right]^{-3}$

$=\left(\dfrac{3}{6}-\dfrac{2}{6}+\dfrac{1}{6}\right)^{-3}=\left(\dfrac{2}{6}\right)^{-3}$

$=\left(\dfrac{1}{3}\right)^{-3}=3^3=27$

57. $\left(\dfrac{1}{2}\right)^{-3}-\left(\dfrac{1}{3}\right)^{-3}+\left(\dfrac{1}{6}\right)^{-3}=2^3-3^3+6^3$

$=8-27+216=197$

59. $\left(\dfrac{1}{4x^3y^{-5}}\right)^{-2}\left(\dfrac{8}{x^{-13}y^{14}}\right)^{-1}$

$=\left[(4)^{-1}x^{-3}y^{5}\right]^{-2}\cdot\left(8x^{13}y^{-14}\right)^{-1}$

$=4^2x^6y^{-10}8^{-1}x^{-13}y^{14}$

$=4^2 8^{-1}x^{-7}y^4$

$=\dfrac{16\cdot y^4}{8x^7}=\dfrac{2y^4}{x^7}$

61. $\left(\dfrac{\left(x^{-3}\right)^{-4}x^{-2}}{x^{-6}}\right)^{-1}=\left(x^{12}x^{-2}x^{6}\right)^{-1}$

$=\left(x^{16}\right)^{-1}$

$=x^{-16}$

$=\dfrac{1}{x^{16}}$

63. $\dfrac{\left(4vw^{-3}x^2\right)^2}{\left(2v^2w^3x^{-2}\right)^4}\cdot\left(-v^3w^2x^{-4}\right)^{-5}$

$=\dfrac{4^2v^2w^{-6}x^4}{2^4v^8w^{12}x^{-8}}\cdot(-1)^{-5}\left(v^{-15}w^{-10}x^{20}\right)$

$=4^2v^2w^{-6}x^4\cdot 2^{-4}v^{-8}w^{-12}x^8\cdot(-1)^{-5}\left(v^{-15}w^{-10}x^{20}\right)$

$=4^2\cdot 2^{-4}\cdot(-1)^{-5}\cdot v^{-21}w^{-28}x^{32}$

$=\dfrac{16x^{32}}{16(-1)v^{21}w^{28}}=-\dfrac{x^{32}}{v^{21}w^{28}}$

65. $(3x+5)^{14}(3x+5)^{-2}=(3x+5)^{14+(-2)}=(3x+5)^{12}$

67. $\left[(6v-7)^{10}\right]^9=(6v-7)^{10\cdot 9}=(6v-7)^{90}$

69. $2^{-2}+2^{-1}+2^0+2^1+2^2=\dfrac{1}{2^2}+\dfrac{1}{2^1}+1+2+4$

$=\dfrac{1}{4}+\dfrac{1}{2}+1+2+4$

$=\dfrac{1}{4}+\dfrac{2}{4}+\dfrac{4}{4}+\dfrac{8}{4}+\dfrac{16}{4}$

$=\dfrac{31}{4}$

71. a. Move the decimal point 5 places. The number is greater than 10, so use a positive power of 10.

$350{,}000=3.5\times 10^5$

b. Move the decimal point 5 places. The number is between 0 and 1, so use a negative power of 10.

$0.000035=3.5\times 10^{-5}$

c. Move the decimal point 0 places. The exponent on 10 is 0.

$3.5=3.5\times 10^0$

73. a. Move the decimal point 1 place. The number is between 0 and 1, so use a negative power of 10.

$0.86=8.6\times 10^{-1}$

b. Move the decimal point 0 places. The exponent on 10 is 0.

$8.6=8.6\times 10^0$

c. Move the decimal point 1 place. The number is greater than 10, so use a positive power of 10.

$86=8.6\times 10^1$

75. Move the decimal point 10 places. The number is greater than 10, so use a positive power of 10.

$29{,}980{,}000{,}000\text{ cm/sec}=2.998\times 10^{10}\text{cm/sec}$

77. Move the decimal point 5 places. The number is between 0 and 1, so use a negative power of 10.

$0.00001\text{ cm}=1.0\times 10^{-5}\text{ cm}$

79. Move the decimal point 0 places. The exponent on 10 is 0.

$4.2\text{ L}=4.2\times 10^0\text{ L}$

81. a. 10^{-6} is between 0 and 1. Move the decimal point 6 places to the left.

$2.61\times 10^{-6}=0.00000261$

b. 10^6 is greater than 10. Move the decimal point 6 places to the right.

$2.61\times 10^6=2{,}610{,}000$

c. $10^0 = 1$ The decimal point does not move.
$2.61 \times 10^0 = 2.61$

83. a. 10^{-1} is between 0 and 1. Move the decimal point 1 place to the left.
$6.718 \times 10^{-1} = 0.6718$

b. $10^0 = 1$ The decimal point does not move.
$6.718 \times 10^0 = 6.718$

c. 10^1 is greater than 10. Move the decimal point 1 place to the right.
$6.718 \times 10^1 = 67.18$

85. 1.67×10^{21} molecules
$= 1{,}670{,}000{,}000{,}000{,}000{,}000{,}000$ molecules

87. $7.0 \times 10^{-6}\text{ m} = 0.000007\text{ m}$

89. $$\left(2\times10^{-3}\right)\left(4\times10^{8}\right) = (2)(4)\times\left(10^{-3}\right)\left(10^{8}\right) = 8\times10^{5}$$

91. $$\frac{8.4\times10^{-6}}{2.1\times10^{-2}} = \left(\frac{8.4}{2.1}\right)\times\left(\frac{10^{-6}}{10^{-2}}\right) = 4\times10^{-4}$$

93. $$\begin{aligned}\left(6.2\times10^{11}\right)\left(3\times10^{4}\right) &= (6.2)(3)\times\left(10^{11}\right)\left(10^{4}\right)\\ &= 18.6\times10^{15}\\ &= \left(1.86\times10^{1}\right)\times10^{15}\\ &= 1.86\times10^{16}\end{aligned}$$

95. $$\begin{aligned}\frac{3.6\times10^{-14}}{5\times10^{5}} &= \left(\frac{3.6}{5}\right)\times\left(\frac{10^{-14}}{10^{5}}\right)\\ &= 0.72\times10^{-19}\\ &= \left(7.2\times10^{-1}\right)\times10^{-19}\\ &= 7.2\times10^{-20}\end{aligned}$$

97. $$\begin{aligned}&\frac{\left(6.2\times10^{5}\right)\left(4.4\times10^{22}\right)}{2.2\times10^{17}}\\ &= \left(\frac{6.2\cdot4.4}{2.2}\right)\times\left(\frac{10^{5}\cdot10^{22}}{10^{17}}\right) = 12.4\times10^{10}\\ &= \left(1.24\times10^{1}\right)\times10^{10} = 1.24\times10^{11}\end{aligned}$$

99. a. $$1\text{ yr}\times\frac{365\text{ days}}{1\text{ yr}}\times\frac{24\text{ hr}}{1\text{ day}}\times\frac{3600\text{ sec}}{1\text{ hr}} = 31{,}536{,}000\text{ sec} = 3.1536\times10^{7}\text{ sec}$$

b. $$\begin{aligned}&\left(\frac{1.0\times10^{4}\text{ gal}}{1\text{ sec}}\right)\left(3.1536\times10^{7}\text{ sec}\right)\\ &= (1.0)(3.1536)\times\left(10^{4}\right)\left(10^{7}\right)\text{ gal}\\ &= 3.1536\times10^{11}\text{ gal}\end{aligned}$$

101. $$\frac{8\times10^{10}\text{ bytes}}{4\times10^{6}\ \frac{\text{bytes}}{\text{song}}} = \left(\frac{8}{4}\right)\times\left(\frac{10^{10}}{10^{6}}\right)\text{ songs} = 2\times10^{4}\text{ songs}$$

103. $$\begin{aligned}&5\text{ L}\times\frac{1\times10^{6}\ \mu\text{L}}{1\text{ L}}\times\frac{5\times10^{6}\text{ red blood cells}}{1\ \mu\text{L}}\\ &= 25\times10^{12}\text{ red blood cells}\\ &= 2.5\times10^{13}\text{ red blood cells}\end{aligned}$$

105. In the expression $6x^0$, the exponent 0 applies to x only. In the expression $(6x)^0$, the exponent 0 applies to a base of $(6x)$. The first expression simplifies to 6, and the second expression simplifies to 1.

107. Yes; $(-4)^2 = 16 > 0$

109. No

111. $$\begin{aligned}F &= \frac{Gm_1m_2}{d^2}\\ &= \frac{\left(6.6726\times10^{-11}\right)\left(5.98\times10^{24}\right)\left(1.901\times10^{27}\right)}{\left(7.0\times10^{11}\right)^2}\\ &= \frac{(6.6726)(5.98)(1.901)}{(7.0)^2}\times\frac{\left(10^{-11}\right)\left(10^{24}\right)\left(10^{27}\right)}{\left(10^{11}\right)^2}\\ &\approx 1.55\times10^{18}\text{ N}\end{aligned}$$

113. a. < **b.** >

115. a. > **b.** =

117. $x^m x^4 = x^{m+4}$

119. $x^{m+9}x^{m-2} = x^{m+9+m-2} = x^{2m+7}$

121. $\dfrac{x^m}{x^8} = x^{m-8}$

123. $\dfrac{x^{2m+7}}{x^{m+5}} = x^{2m+7-(m+5)} = x^{2m+7-m-5} = x^{m+2}$

125. $\left(x^{4m}\right)^{3n} = x^{4m\cdot 3n} = x^{12mn}$

127. $\dfrac{x^{4m-3}y^{5n+7}}{x^{m-7}y^{3n+2}} = x^{4m-3-(m-7)}y^{5n+7-(3n+2)}$
$= x^{4m-3-m+7}y^{5n+7-3n-2} = x^{3m+4}y^{2n+5}$

129. $6.284\times 10^6 = 6{,}284{,}000$

131. $2.45\times 10^{-7} = 0.000\,000\,245$

Section R.4 Rational Exponents and Radicals

1. a. $\sqrt{36}$ **b.** $\sqrt{36}, \sqrt{0}$

c. $\sqrt{36}, -\sqrt{4}, \sqrt{0}$

d. $\sqrt{36}, \sqrt{\dfrac{9}{49}}, -\sqrt{4}, \sqrt{0}$

e. $\sqrt{37}$

f. $\sqrt{37}, \sqrt{36}, \sqrt{\dfrac{9}{49}}, -\sqrt{4}, \sqrt{0}$

3. $\dfrac{w^{-4}}{v^{-2}} = w^{-4}v^2 = \dfrac{v^2}{w^4}$

5. $\left(-3x^5y^2\right)\left(2x^{-3}y\right) = -6\cdot x^{5-3}y^{2+1} = -6x^2y^3$

7. n **9.** $\left(\sqrt[n]{a}\right)^m$ or $\sqrt[n]{a^m}$

11. $|x|$ **13.** $\sqrt[n]{a\cdot b}$

15. $\sqrt[4]{81} = 3$ **17.** $\sqrt{\dfrac{4}{49}} = \dfrac{7}{8}$

19. $\sqrt{0.09} = 0.3$

21 $\sqrt[4]{-81}$ is not a real number

23. $-\sqrt[4]{81} = -1\cdot\sqrt[4]{81} = -1\cdot 3 = -3$

25. $\sqrt[3]{-\dfrac{1}{8}} = -\dfrac{1}{2}$

27. a. $25^{1/2} = \sqrt{25} = 5$

b. Undefined

c. $-25^{1/2} = -\sqrt{25} = -(5) = -5$

29. a. $27^{1/3} = \sqrt[3]{27} = 3$

b. $(-27)^{1/3} = \sqrt[3]{-27} = -3$

c. $-27^{1/3} = -\sqrt[3]{27} = -3$

31. a. $\left(\dfrac{121}{169}\right)^{1/2} = \sqrt{\dfrac{121}{169}} = \dfrac{11}{13}$

b. $\left(\dfrac{121}{169}\right)^{-1/2} = \left(\dfrac{169}{121}\right)^{1/2} = \sqrt{\dfrac{169}{121}} = \dfrac{13}{11}$

33. a. $16^{3/4} = \left(\sqrt[4]{16}\right)^3 = (2)^3 = 8$

b. $16^{-3/4} = \dfrac{1}{16^{3/4}} = \dfrac{1}{\left(\sqrt[4]{16}\right)^3} = \dfrac{1}{(2)^3} = \dfrac{1}{8}$

c. $-16^{3/4} = -\left(\sqrt[4]{16}\right)^3 = -(2)^3 = -8$

d. $-16^{-3/4} = -\dfrac{1}{16^{3/4}} = -\dfrac{1}{\left(\sqrt[4]{16}\right)^3} = -\dfrac{1}{(2)^3} = -\dfrac{1}{8}$

e. $(-16)^{3/4}$ is undefined because $\sqrt[4]{-16}$ is not a real number.

f. $(-16)^{-3/4}$ is undefined because $\sqrt[4]{-16}$ is not a real number.

35. a. $64^{2/3}=\left(\sqrt[3]{64}\right)^2=(4)^2=16$

b. $64^{-2/3}=\dfrac{1}{64^{2/3}}=\dfrac{1}{\left(\sqrt[3]{64}\right)^2}=\dfrac{1}{(4)^2}=\dfrac{1}{16}$

c. $-64^{2/3}=-\left(\sqrt[3]{64}\right)^2=-(4)^2=-16$

d. $-64^{-2/3}=-\dfrac{1}{64^{2/3}}=-\dfrac{1}{\left(\sqrt[3]{64}\right)^2}=-\dfrac{1}{(4)^2}=-\dfrac{1}{16}$

e. $(-64)^{2/3}=\left(\sqrt[3]{-64}\right)^2=(-4)^2=16$

f. $(-64)^{-2/3}=\dfrac{1}{(-64)^{2/3}}=\dfrac{1}{\left(\sqrt[3]{-64}\right)^2}=\dfrac{1}{(-4)^2}=\dfrac{1}{16}$

37. a. $y^{4/11}=\sqrt[11]{y^4}$ or $\left(\sqrt[11]{y}\right)^4$

b. $6y^{4/11}=6\sqrt[11]{y^4}$ or $6\left(\sqrt[11]{y}\right)^4$

c. $(6y)^{4/11}=\sqrt[11]{(6y)^4}$ or $6\left(\sqrt[11]{6y}\right)^4$

39. $\sqrt[5]{a^3}=a^{3/5}$

41. $\sqrt{6x}=(6x)^{1/2}$

43. $6\sqrt{x}=6x^{1/2}$

45. $\sqrt[5]{a^5+b^5}=\left(a^5+b^5\right)^{1/5}$

47. $\dfrac{a^{2/3}a^{5/3}}{a^{1/3}}=\dfrac{a^{2/3+5/3}}{a^{1/3}}=\dfrac{a^{7/3}}{a^{1/3}}=a^{7/3-1/3}=a^{6/3}=a^2$

49. $\dfrac{3w^{-2/3}}{y^{-1/3}}=\dfrac{3y^{1/3}}{w^{2/3}}$

51.
$$\begin{aligned}\left(16x^{-8}y^{1/5}\right)^{3/4}&=(16)^{3/4}\left(x^{-8}\right)^{3/4}\left(y^{1/5}\right)^{3/4}\\&=\left(\sqrt[4]{16}\right)^3x^{-8(3/4)}y^{(1/5)(3/4)}\\&=(2)^3x^{-24/4}y^{3/20}\\&=8x^{-6}y^{3/20}=\frac{8y^{3/20}}{x^6}\end{aligned}$$

53.
$$\begin{aligned}\left(\frac{2m^{2/3}}{n^{3/4}}\right)^{12}\left(\frac{n^{1/5}}{2m^{1/2}}\right)^{10}&=\frac{2^{12}m^{(2/3)\cdot12}}{n^{(3/4)\cdot12}}\cdot\frac{n^{(1/5)\cdot10}}{2^{10}m^{(1/2)\cdot10}}\\&=\frac{2^{12}m^8}{n^9}\cdot\frac{n^2}{2^{10}m^5}\\&=\frac{2^2m^3}{n^7}=\frac{4m^3}{n^7}\end{aligned}$$

55.
$$\begin{aligned}\left(\frac{m^2}{m+n}\right)^{-1}\left(\frac{m^2}{m+n}\right)^{1/2}&=\frac{m^{2(-1)}}{(m+n)^{-1}}\cdot\frac{m^{2(1/2)}}{(m+n)^{1/2}}\\&=\frac{m^{-2}}{(m+n)^{-1}}\cdot\frac{m^1}{(m+n)^{1/2}}\\&=\frac{m^{-1}}{(m+n)^{-1/2}}=\frac{(m+n)^{1/2}}{m}\end{aligned}$$

57. a. $\sqrt{t^2}=t$ for $t\ge 0$

b. $\sqrt{t^2}=|t|$ for all real numbers

59. $\sqrt{y^2}=|y|$

61. $\sqrt[3]{y^3}=y$

63. $\sqrt[4]{(2x-5)^4}=|2x-5|$

65. $\sqrt{w^{12}}=w^6$

67. a. $\sqrt{c^7}=\sqrt{c^6\cdot c}=\sqrt{c^6}\cdot\sqrt{c}=c^3\sqrt{c}$

b. $\sqrt[3]{c^7}=\sqrt[3]{c^6\cdot c}=\sqrt[3]{c^6}\cdot\sqrt[3]{c}=c^2\sqrt[3]{c}$

c. $\sqrt[4]{c^7}=\sqrt[4]{c^4\cdot c^3}=\sqrt[4]{c^4}\cdot\sqrt[4]{c^3}=c\sqrt[4]{c^3}$

d. $\sqrt[9]{c^7}=\sqrt[9]{c^7}$

69. a. $\sqrt{24}=\sqrt{2^2\cdot 6}=\sqrt{2^2}\cdot\sqrt{6}=2\sqrt{6}$

b. $\sqrt[3]{24}=\sqrt[3]{2^3\cdot 3}=\sqrt[3]{2^3}\cdot\sqrt[3]{3}=2\sqrt[3]{3}$

71. $\sqrt[3]{250x^2y^6z^{11}}=\sqrt[3]{5^3\cdot 2\cdot x^2y^6z^{11}}$
$=\sqrt[3]{(5^3y^6z^9)(2x^2z^2)}$
$=\sqrt[3]{5^3y^6z^9}\cdot\sqrt[3]{2x^2z^2}$
$=5y^2z^3\sqrt[3]{2x^2z^2}$

73. $\sqrt[4]{96p^{14}q^7}=\sqrt[4]{2^4\cdot 6\cdot p^{14}q^7}$
$=\sqrt[4]{(2^4p^{12}q^4)(6p^2q^3)}$
$=\sqrt[4]{2^4p^{12}q^4}\cdot\sqrt[4]{6p^2q^3}$
$=2p^3q\sqrt[4]{6p^2q^3}$

75. $\sqrt{84(y-2)^3}=\sqrt{2^2\cdot 21(y-2)^3}$
$=\sqrt{2^2(y-2)^2\cdot 21(y-2)}$
$=\sqrt{2^2(y-2)^2}\cdot\sqrt{21(y-2)}$
$=2(y-2)\sqrt{21(y-2)}$

77. $\sqrt{\dfrac{p^7}{36}}=\dfrac{\sqrt{p^7}}{\sqrt{36}}=\dfrac{\sqrt{p^6\cdot p}}{6}=\dfrac{\sqrt{p^6}\cdot\sqrt{p}}{6}=\dfrac{p^3\sqrt{p}}{6}$

79. $4\sqrt[3]{\dfrac{w^3z^5}{8}}=\dfrac{4\sqrt[3]{w^3z^5}}{\sqrt[3]{8}}=\dfrac{4\sqrt[3]{w^3z^3\cdot z^2}}{2}$
$=2\sqrt[3]{w^3z^3}\cdot\sqrt[3]{z^2}=2wz\sqrt[3]{z^2}$

81. $\dfrac{\sqrt[3]{5x^5y}}{\sqrt[3]{625x^2y^4}}=\sqrt[3]{\dfrac{5x^5y}{625x^2y^4}}=\sqrt[3]{\dfrac{x^3}{125y^3}}=\dfrac{x}{5y}$

83. $\sqrt{10}\cdot\sqrt{14}=\sqrt{140}=\sqrt{2^2\cdot 35}$
$=\sqrt{2^2}\cdot\sqrt{35}=2\sqrt{35}$

85. $\sqrt[3]{xy^2}\cdot\sqrt[3]{x^2y}=\sqrt[3]{xy^2\cdot x^2y}=\sqrt[3]{x^3y^3}=xy$

87. $\left(3\sqrt[4]{a^3}\right)\left(-5\sqrt[4]{a^3}\right)=3\cdot(-5)\sqrt[4]{a^3\cdot a^3}$
$=-15\sqrt[4]{a^6}=-15\sqrt[4]{a^4\cdot a^2}$
$=-15\sqrt[4]{a^4}\cdot\sqrt[4]{a^2}=-15a\cdot a^{2/4}$
$=-15a\cdot a^{1/2}=-15a\sqrt{a}$

89. $\left(-\dfrac{1}{2}\sqrt[3]{6a^2b^2c}\right)\left(\dfrac{4}{3}\sqrt[3]{4a^2c^2}\right)$
$=-\dfrac{2}{3}\sqrt[3]{6a^2b^2c\cdot 4a^2c^2}$
$=-\dfrac{2}{3}\sqrt[3]{24a^4b^2c^3}=-\dfrac{2}{3}\sqrt[3]{2^3a^3c^3\cdot 3ab^2}$
$=-\dfrac{2}{3}\sqrt[3]{2^3a^3c^3}\cdot\sqrt[3]{3ab^2}=-\dfrac{4}{3}ac\sqrt[3]{3ab^2}$

91. $\sqrt[5]{x^3y^2}\cdot\sqrt[4]{x}=(x^3y^2)^{1/5}\cdot(x)^{1/4}$
$=x^{3/5}y^{2/5}\cdot x^{1/4}$
$=x^{3/5+1/4}y^{2/5}=x^{12/20+5/20}y^{2/5}$
$=x^{17/20}y^{8/20}=\sqrt[20]{x^{17}y^8}$

93. $\sqrt[6]{m\sqrt[3]{m^2}}=(m\cdot m^{2/3})^{1/6}=(m^{1+2/3})^{1/6}=(m^{5/3})^{1/6}$
$=m^{5/18}=\sqrt[18]{m^5}$

95. $\sqrt{x\sqrt{x\sqrt{x}}}=\left(x\cdot\left(x\cdot x^{1/2}\right)^{1/2}\right)^{1/2}=\left(x\cdot\left(x^{3/2}\right)^{1/2}\right)^{1/2}$
$=(x\cdot x^{3/4})^{1/2}=(x^{7/4})^{1/2}=x^{7/8}=\sqrt[8]{x^7}$

97. $3\sqrt[3]{2y^2}-9\sqrt[3]{2y^2}+\sqrt[3]{2y^2}=(3-9+1)\sqrt[3]{2y^2}$
$=-5\sqrt[3]{2y^2}$

99. $\dfrac{1}{5}\sqrt{50}-\dfrac{7}{3}\sqrt{18}+\dfrac{5}{6}\sqrt{72}$
$=\dfrac{1}{5}\sqrt{5^2\cdot 2}-\dfrac{7}{3}\sqrt{3^2\cdot 2}+\dfrac{5}{6}\sqrt{6^2\cdot 2}$
$=\dfrac{1}{5}\cdot 5\sqrt{2}-\dfrac{7}{3}\cdot 3\sqrt{2}+\dfrac{5}{6}\cdot 6\sqrt{2}$
$=\sqrt{2}-7\sqrt{2}+5\sqrt{2}$
$=(1-7+5)\sqrt{2}=-\sqrt{2}$

101. $-3x\sqrt[3]{16xy^4}+xy\sqrt[3]{54xy}-5\sqrt[3]{250x^4y^4}$
$=-3x\sqrt[3]{2^3y^3\cdot 2xy}+xy\sqrt[3]{3^3\cdot 2xy}$
$-5\sqrt[3]{5^3x^3y^3\cdot 2xy}$
$=-6xy\sqrt[3]{2xy}+3xy\sqrt[3]{2xy}-25xy\sqrt[3]{2xy}$
$=-28xy\sqrt[3]{2xy}$

103. $12\sqrt{2y}+5y\sqrt{2y}=(12+5y)\sqrt{2y}$

105. $-\frac{1}{2}\sqrt{8z^3}+\frac{3}{7}\sqrt{98z}$
$=-\frac{1}{2}\sqrt{2^2z^2\cdot 2z}+\frac{3}{7}\sqrt{7^2\cdot 2z}$
$=-z\sqrt{2z}+3\sqrt{2z}=(-z+3)\sqrt{2z}$

107. $s=\frac{a+b+c}{2}=\frac{6+7+9}{2}=\frac{22}{2}=11$
$A=\sqrt{s(s-a)(s-b)(s-c)}$
$=\sqrt{11(11-6)(11-7)(11-9)}$
$=\sqrt{11(5)(4)(2)}=\sqrt{440}$
$=\sqrt{2^2\cdot 110}=2\sqrt{110}\text{ in.}^2$

109. $b=\sqrt{c^2-a^2}$
$=\sqrt{18^2-12^2}=\sqrt{324-144}$
$=\sqrt{180}=\sqrt{6^2\cdot 5}=6\sqrt{5}\text{ m}$

111. $c=\sqrt{a^2+b^2}$
$=\sqrt{32^2+48^2}=\sqrt{1024+2304}$
$=\sqrt{3328}=\sqrt{16^2\cdot 13}=16\sqrt{13}\text{ in.}\approx 58\text{ in.}$

113. $L=\sqrt{r^2+h^2}$
$=\sqrt{4^2+10^2}=\sqrt{16+100}$
$=\sqrt{116}=\sqrt{2^2\cdot 29}=2\sqrt{29}\text{ in.}\approx 10.8\text{ in.}$

115. $A=\pi r^2$
$=\pi\left(3\sqrt{5}\right)^2=\pi\cdot 3^2\cdot\left(\sqrt{5}\right)^2$
$=\pi\cdot 9\cdot 5=45\pi\text{ in.}^2$

117. $r=1-\left(\frac{S}{C}\right)^{1/n}$
$=1-\left(\frac{11{,}500}{22{,}990}\right)^{1/4}$
$\approx 1-0.841=0.159=15.9\%$

119. **a.** for $n=1$:
$f=440\cdot 2^{n/12}=440\cdot 2^{1/12}\approx 466.2\text{ Hz}$
for $n=2$:
$f=440\cdot 2^{n/12}=440\cdot 2^{2/12}\approx 493.9\text{ Hz}$
for $n=3$:
$f=440\cdot 2^{n/12}=440\cdot 2^{3/12}\approx 523.3\text{ Hz}$

b. for $n=-9$
$f=440\cdot 2^{n/12}=440\cdot 2^{-9/12}\approx 261.6\text{ Hz}$

121. In each case, add like terms or like radicals by using the distributive property.

123. $\sqrt{\frac{8.0\times 10^{12}}{2.0\times 10^4}}=\sqrt{\frac{8.0}{2.0}\times\frac{10^{12}}{10^4}}=\sqrt{\frac{8.0}{2.0}}\times\sqrt{\frac{10^{12}}{10^4}}$
$=\sqrt{4.0}\times\sqrt{10^8}=2.0\times 10^4$

125. $\sqrt{\sqrt[3]{6+\sqrt[4]{16}}+\sqrt{\sqrt{25}+\sqrt{16}}+\sqrt{9}}$
$=\sqrt{\sqrt[3]{6+2}+\sqrt{5+4}+3}=\sqrt{\sqrt[3]{8}+\sqrt{9}+3}$
$=\sqrt{2+3+3}=\sqrt{8}=\sqrt{2^2\cdot 2}=2\sqrt{2}$

127. $T_p=0.7(T_s+273)\left(\frac{r}{d}\right)^{1/2}-273$
$T_p=0.7(7700+273)\left(\frac{1.26\times 10^6}{4.3\times 10^8}\right)^{1/2}-273$
$T_p=0.7(7973)\sqrt{\frac{1.26}{4.3}}\cdot\sqrt{\frac{10^6}{10^8}}-273$
$T_p=0.7(7973)\sqrt{\frac{1.26}{4.3}}\cdot\sqrt{\frac{1}{10^2}}-273$
$T_p=0.7(7973)\sqrt{\frac{1.26}{4.3}}\cdot 0.1-273\approx 29.1$

Yes; The model gives a mean surface temperature of approximately 29.1°C.

Section R.5 Polynomials and Multiplication of Radicals

1. $-3(8-4x)-\frac{1}{3}(15x-6z)$
$=-24+12x-5x+2z$
$=7x+2z-24$

3. $\left(\sqrt{5}\right)^2=5$

5. $\left(3x^2y^{-1}\right)^4\left(-\frac{1}{2}xy^3\right)^3=3^4x^8y^{-4}\cdot\left(-\frac{1}{2}\right)^3x^3y^9$
$=\frac{81x^{11}y^5}{-8}$
$=-\frac{81x^{11}y^5}{8}$

7. polynomial

9. leading

11. binomial; trinomial

13. $(a-b)$

15. a. Yes **b.** Yes
c. No **d.** No

17. $-18x^7+7.2x^3-4.1$;
Leading coefficient -18; Degree 7

19. $-y^2+\frac{1}{3}y$;
Leading coefficient -1; Degree 2

21. 11

23. $\left(-8p^7-4p^4+2p-5\right)+\left(2p^7+6p^4+p^2\right)=-8p^7+2p^7-4p^4+6p^4+p^2+2p-5$
$=-6p^7+2p^4+p^2+2p-5$

25. $\left(0.05c^3b+0.02c^2b^2-0.09cb^3\right)-\left(-0.03c^3b+0.08c^2b^2-0.1cb^3\right)$
$=0.05c^3b+0.03c^3b+0.02c^2b^2-0.08c^2b^2-0.09cb^3+0.1cb^3$
$=0.08c^3b-0.06c^2b^2+0.01cb^3$

27. $\left(\frac{1}{4}x^2+\frac{5}{8}x+5\sqrt{2}\right)-\left(\frac{1}{2}x^2-\frac{3}{4}x-\sqrt{2}\right)$
$=\frac{1}{4}x^2-\frac{1}{2}x^2+\frac{5}{8}x+\frac{3}{4}x+5\sqrt{2}+\sqrt{2}$
$=-\frac{1}{4}x^2+\frac{11}{8}x+6\sqrt{2}$

29. $\left(-6a^5b\right)\left(\frac{1}{3}a^2b^2\right)=-2a^7b^3$

31. $7m^2\left(2m^4-3m+4\right)$
$=7m^2\left(2m^4\right)+7m^2(-3m)+7m^2(4)$
$=14m^6-21m^3+28m^2$

33. $(2x-5)(x+4)$
$=2x(x)+2x(4)+(-5)(x)+(-5)(4)$
$=2x^2+8x-5x-20$
$=2x^2+3x-20$

35. $\left(4u^2-5v^2\right)\left(2u^2+3v^2\right)$
$=4u^2\left(2u^2\right)+4u^2\left(3v^2\right)+\left(-5v^2\right)\left(2u^2\right)$
$+\left(-5v^2\right)\left(3v^2\right)$
$=8u^4+12u^2v^2-10u^2v^2-15v^4$
$=8u^4+2u^2v^2-15v^4$

37. $(3y+6)\left(\frac{1}{3}y^2-5y-4\right)$

$=3y\left(\frac{1}{3}y^2\right)+3y(-5y)+3y(-4)$

$+6\left(\frac{1}{3}y^2\right)+6(-5y)+6(-4)$

$=y^3-15y^2-12y+2y^2-30y-24$

$=y^3-13y^2-42y-24$

39. $(a+b)^2=(a+b)(a+b)$

$=a(a)+a(b)+b(a)+b(b)$

$=a^2+ab+ab+b^2$

$=a^2+2ab+b^2$

41. $(4x-5)(4x+5)=(4x)^2-(5)^2=16x^2-25$

43. $(3w^2-7z)(3w^2+7z)=(3w^2)^2-(7z)^2$

$=9w^4-49z^2$

45. $\left(\frac{1}{5}c-\frac{2}{3}d^3\right)\left(\frac{1}{5}c+\frac{2}{3}d^3\right)=\left(\frac{1}{5}c\right)^2-\left(\frac{2}{3}d^3\right)^2$

$=\frac{1}{25}c^2-\frac{4}{9}d^6$

47. $(5m-3)^2=(5m)^2-2(5m)(3)+(3)^2$

$=25m^2-30m+9$

49. $(4t^2+3p^3)^2=(4t^2)^2+2(4t^2)(3p^3)+(3p^3)^2$

$=16t^4+24t^2p^3+9p^6$

51. $(w+4)^3=(w+4)(w+4)^2$

$=(w+4)\left[w^2+2(w)(4)+4^2\right]$

$=(w+4)(w^2+8w+16)$

$=w(w^2)+w(8w)+w(16)+4(w^2)$

$+4(8w)+4(16)$

$=w^3+8w^2+16w+4w^2+32w+64$

$=w^3+12w^2+48w+64$

53. $[(u+v)-w][(u+v)+w]$

$=(u+v)^2-(w)^2$

$=u^2+2uv+v^2-w^2$

55. $I=45.58x+460.1$

$=45.58(8)+460.1=\$824.74$ billion

57. a. $I+P=45.58x+460.1+10.86x+191.5$

$=56.44x+651.6$

b. The polynomial $I+P$ represents the total amount of health-related expenditures made by individuals with private insurance.

c. $I+P=56.44x+651.6$

$=56.44(6)+651.6=990.24$

\$990.24 billion; In the year 2006, a total of \$990.24 billion was spent on health-related expenses by individuals with private insurance.

59. $P=2(3x+4+x+3+x+4)$

$=2(5x+11)=10x+22$

61. $A=lw$

$=[5y+(2x+3)][5y-(2x+3)]$

$=(5y)^2-(2x+3)^2$

$=25y^2-(4x^2+12x+9)$

$=25y^2-4x^2-12x-9$

63. $V=lwh$

$=(x+4)(x+4)(x+6)$

$=(x+4)^2(x+6)$

$=(x^2+8x+16)(x+6)$

$=x^2(x)+x^2(6)+8x(x)+8x(6)+16(x)$

$+16(6)$

$=x^3+6x^2+8x^2+48x+16x+96$

$=x^3+14x^2+64x+96$

65. $A=lw=(a+b)(c)=ac+bc$

67. $A = l_o w_o - l_i w_i$
$= (2x-3)(x+7) - (x+3)(x+2)$
$= 2x(x) + 2x(7) - 3(x) - 3(7)$
$\quad -[x(x) + x(2) + 3(x) + 3(2)]$
$= 2x^2 + 14x - 3x - 21 - (x^2 + 2x + 3x + 6)$
$= 2x^2 + 11x - 21 - x^2 - 5x - 6$
$= x^2 + 6x - 27$

69. **a.** $x+1$

b. $x + (x+1) = 2x+1$

c. $x(x+1) = x^2 + x$

d. $x^2 + (x+1)^2 = x^2 + x^2 + 2x + 1 = 2x^2 + 2x + 1$

71. $(y+7)^2 - 2(y-3)^2$
$= y^2 + 14y + 49 - 2(y^2 - 6y + 9)$
$= y^2 + 14y + 49 - 2y^2 + 12y - 18$
$= -y^2 + 26y + 31$

73. $(x^n + 3)(x^n - 7)$
$= x^n(x^n) + x^n(-7) + 3(x^n) + 3(-7)$
$= x^{2n} - 7x^n + 3x^n - 21$
$= x^{2n} - 4x^n - 21$

75. $(z^n + w^m)^2 = (z^n)^2 + 2w^m z^n + (w^m)^2$
$= z^{2n} + 2w^m z^n + w^{2m}$

77. $(a^n - 5)(a^n + 5) = (a^n)^2 - 5^2 = a^{2n} - 25$

79. $(6x+5)(6x-5) - (6x+5)^2$
$= (6x)^2 - 5^2 - (36x^2 + 60x + 25)$
$= 36x^2 - 25 - 36x^2 - 60x - 25$
$= -60x - 50$

81. $5\sqrt{2}(2\sqrt{2} + 6\sqrt{3} - 4)$
$= (5\sqrt{2})(2\sqrt{2}) + (5\sqrt{2})(6\sqrt{3}) + (5\sqrt{2})(-4)$
$= 10\sqrt{4} + 30\sqrt{6} - 20\sqrt{2}$
$= 10(2) + 30\sqrt{6} - 20\sqrt{2}$
$= 20 + 30\sqrt{6} - 20\sqrt{2}$

83. $3\sqrt{6}(5\sqrt{3} - 4\sqrt{2} - \sqrt{6})$
$= (3\sqrt{6})(5\sqrt{3}) + (3\sqrt{6})(-4\sqrt{2})$
$\quad + (3\sqrt{6})(-\sqrt{6})$
$= 15\sqrt{18} - 12\sqrt{12} - 3\sqrt{36}$
$= 15\sqrt{3^2 \cdot 2} - 12\sqrt{2^2 \cdot 3} - 3(6)$
$= 45\sqrt{2} - 24\sqrt{3} - 18$

85. $(2\sqrt{y} - 3)(4\sqrt{y} + 5)$
$= (2\sqrt{y})(4\sqrt{y}) + (2\sqrt{y})(5) - (3)(4\sqrt{y})$
$\quad -3(5)$
$= 8\sqrt{y^2} + 10\sqrt{y} - 12\sqrt{y} - 15$
$= 8y - 2\sqrt{y} - 15$

87. $(4\sqrt{3} - 2\sqrt{5})(6\sqrt{3} + 5\sqrt{5})$
$= (4\sqrt{3})(6\sqrt{3}) + (4\sqrt{3})(5\sqrt{5})$
$\quad -(2\sqrt{5})(6\sqrt{3}) - (2\sqrt{5})(5\sqrt{5})$
$= 24\sqrt{9} + 20\sqrt{15} - 12\sqrt{15} - 10\sqrt{25}$
$= 72 + 8\sqrt{15} - 50$
$= 8\sqrt{15} + 22$

89. $(2\sqrt{3} + \sqrt{7})(2\sqrt{3} - \sqrt{7}) = (2\sqrt{3})^2 - (\sqrt{7})^2$
$= 4 \cdot 3 - 7 = 12 - 7 = 5$

91. $(4x\sqrt{y} - 2y\sqrt{x})(4x\sqrt{y} + 2y\sqrt{x})$
$= (4x\sqrt{y})^2 - (2y\sqrt{x})^2$
$= 16x^2 \cdot y - 4y^2 \cdot x = 16x^2 y - 4xy^2$

93. $\left(6z-\sqrt{5}\right)^2=(6z)^2-2(6z)\left(\sqrt{5}\right)+\left(\sqrt{5}\right)^2$
$=36z^2-12\sqrt{5}z+5$

95. $\left(5a^2\sqrt{b}+7b^2\sqrt{a}\right)^2$
$=\left(5a^2\sqrt{b}\right)^2+2\left(5a^2\sqrt{b}\right)\left(7b^2\sqrt{a}\right)+\left(7b^2\sqrt{a}\right)^2$
$=25a^4b+70a^2b^2\sqrt{ab}+49ab^4$

97. $\left(\sqrt{x+1}-5\right)\left(\sqrt{x+1}+5\right)=\left(\sqrt{x+1}\right)^2-5^2$
$=x+1-25=x-24$

99. $\left(\sqrt{x+1}-5\right)^2=\left(\sqrt{x+1}\right)^2-2\left(\sqrt{x+1}\right)(5)+5^2$
$=x+1-10\sqrt{x+1}+25$
$=x-10\sqrt{x+1}+26$

101. $\left(\sqrt{5+2\sqrt{x}}\right)\left(\sqrt{5-2\sqrt{x}}\right)$
$=\sqrt{\left(5+2\sqrt{x}\right)\left(5-2\sqrt{x}\right)}$
$=\sqrt{5^2-\left(2\sqrt{x}\right)^2}=\sqrt{25-4x}$

103. $\left(\sqrt{x+y}-\sqrt{x-y}\right)^2$
$=\left(\sqrt{x+y}\right)^2-2\left(\sqrt{x+y}\right)\left(\sqrt{x-y}\right)$
$+\left(\sqrt{x-y}\right)^2$
$=x+y-2\sqrt{(x+y)(x-y)}+x-y$
$=2x-2\sqrt{x^2-y^2}$

105. $A=\frac{1}{2}bh$
$=\frac{1}{2}\left(2\sqrt{5}+\sqrt{6}\right)\left(2\sqrt{5}-\sqrt{6}\right)$
$=\frac{1}{2}\left[\left(2\sqrt{5}\right)^2-\left(\sqrt{6}\right)^2\right]$
$=\frac{1}{2}\left[4(5)-6\right]$
$=\frac{1}{2}(14)=7\text{ m}^2$

107. A polynomial consists of a finite number of terms in which the coefficient of each term is a real number, and the variable factor x is raised to an exponent that is a whole number.

109. In each case, multiply by using the distributive property.

111. False

113. True

115. $(a+b)^3=(a+b)^2(a+b)$
$=\left(a^2+2ab+b^2\right)(a+b)$
$=\left(a^2\right)(a)+\left(a^2\right)(b)+(2ab)(a)$
$+(2ab)(b)+\left(b^2\right)(a)+\left(b^2\right)(b)$
$=a^3+a^2b+2a^2b+2ab^2+ab^2+b^3$
$=a^3+3a^2b+3ab^2+b^3$

Problem Recognition Exercises: Simplifying Algebraic Expressions

1. a. $64^{1/2}=\sqrt{64}=8$

b. $64^{1/3}=\sqrt[3]{64}=4$

c. $64^{2/3}=\left(\sqrt[3]{64}\right)^2=4^2=16$

d. $64^{-1}=\frac{1}{64}$

e. $-64^{1/2}=-\sqrt{64}=-8$

f. $(-64)^{1/2}=\sqrt{-64}$ Undefined

g. $(-64)^{2/3}=\left(\sqrt[3]{-64}\right)^2=(-4)^2=16$

h. $(-64)^{-2/3}=\left(\sqrt[3]{-64}\right)^{-2}=(-4)^{-2}=\frac{1}{16}$

3. a. $\left(2x^4y\right)^2 = 4x^8y^2$

b. $\left(2x^4 - y\right)^2 = \left(2x^4\right)^2 - 2\left(2x^4\right)(y) + (y)^2$
$= 4x^8 - 4x^4y + y^2$

c. $\left(2x^4y\right)^{-2} = \dfrac{1}{\left(2x^4y\right)^2} = \dfrac{1}{4x^8y^2}$

d. $\left(2x^4 - y\right)^{-2} = \dfrac{1}{\left(2x^4 - y\right)^2}$
$= \dfrac{1}{\left(2x^4\right)^2 - 2\left(2x^4\right)(y) + (y)^2}$
$= \dfrac{1}{4x^8 - 4x^4y + y^2}$

5. a. $\sqrt{x^8} = \left(x^8\right)^{1/2} = x^4$

b. $\sqrt[3]{x^8} = \sqrt[3]{x^6x^2} = x^2\sqrt[3]{x^2}$

c. $\sqrt[5]{x^8} = \sqrt[5]{x^5x^3} = x\sqrt[5]{x^3}$

d. $\sqrt[9]{x^8} = \sqrt[9]{x^8}$

7. a. $\left(a^2 - b^2\right) + \left(a^2 + b^2\right) = a^2 - b^2 + a^2 + b^2$
$= 2a^2$

b. $\left(a^2 - b^2\right)\left(a^2 + b^2\right) = \left(a^2\right)^2 - \left(b^2\right)^2 = a^4 - b^4$

c. $(a-b)^2 - (a+b)^2$
$= a^2 - 2ab + b^2 - \left(a^2 + 2ab + b^2\right)$
$= a^2 - 2ab + b^2 - a^2 - 2ab - b^2 = -4ab$

d. $(a-b)^2(a+b)^2$
$= \left(a^2 - 2ab + b^2\right)\left(a^2 + 2ab + b^2\right)$
$= \left(a^2\right)\left(a^2\right) + \left(a^2\right)(2ab) + \left(a^2\right)\left(b^2\right)$
$+ (-2ab)\left(a^2\right) + (-2ab)(2ab)$
$+ (-2ab)\left(b^2\right) + \left(b^2\right)\left(a^2\right)$
$+ \left(b^2\right)(2ab) + \left(b^2\right)\left(b^2\right)$
$= a^4 + 2a^3b + a^2b^2 - 2a^3b - 4a^2b^2 - 2ab^3$
$+ a^2b^2 + 2ab^3 + b^4$
$= a^4 - 2a^2b^2 + b^4$

9. a. for $x > -2, |x+2| = x + 2$

b. for $x < -2, |x+2| = -(x+2) = -x - 2$

11. a. $\sqrt[3]{2x} \cdot \sqrt[3]{2x} = \sqrt[3]{4x^2}$

b. $\sqrt[3]{2x} + \sqrt[3]{2x} = 2\sqrt[3]{2x^2}$

13. a. $36 \div 12 \div 6 \div 2 = 3 \div 6 \div 2 = \dfrac{1}{2} \div 2 = \dfrac{1}{4}$

b. $36 \div (12 \div 6) \div 2 = 36 \div 2 \div 2 = 18 \div 2 = 9$

c. $36 \div 12 \div (6 \div 2) = 36 \div 12 \div 3 = 3 \div 3 = 1$

d. $(36 \div 12) \div (6 \div 2) = 3 \div 3 = 1$

Section R.6 Factoring

1. a. $\left(2x^4y^2\right)^2 = 4x^8y^4$

b. $\left(2x^4 - y^2\right)^2 = \left(2x^4\right)^2 - 2\left(2x^4\right)\left(y^2\right) + \left(y^2\right)^2$
$= 4x^8 - 4x^4y^2 + y^4$

3. $\left(3x - 4y^2\right)\left(2x + 5y^2\right)$
$= (3x)(2x) + (3x)\left(5y^2\right) + \left(-4y^2\right)(2x)$
$+ \left(-4y^2\right)\left(5y^2\right)$
$= 6x^2 + 15xy^2 - 8xy^2 - 20y^4$
$= 6x^2 + 7xy^2 - 20y^4$

5. $\left(\frac{1}{5}c^4-\frac{3}{8}ab\right)\left(\frac{1}{5}c^4+\frac{3}{8}ab\right)$
$=\left(\frac{1}{5}c^4\right)^2-\left(\frac{3}{8}ab\right)^2$
$=\frac{1}{25}c^8-\frac{9}{64}a^2b^2$

7. $(2x+3)(4x^2-6x+9)$
$=(2x)(4x^2)+(2x)(-6x)+(2x)(9)$
$+(3)(4x^2)+(3)(-6x)+(3)(9)$
$=8x^3-12x^2+18x+12x^2-18x+27$
$=8x^3+27$

9. cubes; $(a+b)(a^2-ab+b^2)$

11. perfect; $(a+b)^2$

13. $15c^5-30c^4+5c^3$
$=5c^3(3c^2)+5c^3(-6c)+5c^3(1)$
$=5c^3(3c^2-6c+1)$

15. $21a^2b^5-14a^3b^4+35a^4b$
$=7a^2b(3b^4)+7a^2b(-2ab^3)+7a^2b(5a^2)$
$=7a^2b(3b^4-2ab^3+5a^2)$

17. $5z(x-6y)+7(x-6y)=(x-6y)(5z+7)$

19. $10k^2(3k^2+7)-5k(3k^2+7)$
$=(3k^2+7)(10k^2-5k)$
$=(3k^2+7)[5k(2k)-5k(1)]$
$=5k(3k^2+7)(2k-1)$

21. a. $-6x^2+12x+9=3(-2x^2+4x+3)$

b. $-6x^2+12x+9=-3(2x^2-4x-3)$

23. $-12x^3y^2-8x^4y^3+4x^2y$
$=-4x^2y(3xy+2x^2y^2-1)$

25. $8ax+18a+20x+45$
$=2a(4x+9)+5(4x+9)$
$=(4x+9)(2a+5)$

27. $12x^3-9x^2-40x+30$
$=3x^2(4x-3)-10(4x-3)$
$=(4x-3)(3x^2-10)$

29. $cd-8d+4c-2d^2=cd+4c-2d^2-8d$
$=c(d+4)-2d(d+4)$
$=(d+4)(c-2d)$

31. $p^2+2p-63=(\square p+\square)(\square p+\square)$
$=(p+9)(p-7)$

Check: $(p+9)(p-7)=p^2+9p-7p-63$
$=p^2+2p-63$

33. $2t^3-28t^2+80t=2t(t^2-14t+40)$
$=2t(\square t+\square)(\square t+\square)$
$=2t(t-4)(t-10)$

Check:
$2t(t-4)(t-10)=2t(t^2-10t-4t+40)$
$=2t(t^2-14t+40)$
$=2t^3-28t^2+80t$

35. $25z+6z^2+14=6z^2+25z+14$
$=(\square z+\square)(\square z+\square)$
$=(2z+7)(3z+2)$

Check:
$(2z+7)(3z+2)=6z^2+4z+21z+14$
$=6z^2+25z+14$

37. $7y^3z - 40y^2z^2 - 12yz^3$
$= yz(7y^2 - 40yz - 12z^2)$
$= yz(\square y + \square z)(\square y + \square z)$
$= yz(7y + 2z)(y - 6z)$

Check:
$yz(7y + 2z)(y - 6z)$
$= yz(7y^2 - 42yz + 2yz - 12z^2)$
$= yz(7y^2 - 40yz - 12z^2)$
$= 7y^3z - 40y^2z^2 - 12yz^3$

39. $t^2 - 18t + 81 = (t)^2 - 2(t)(9) + (9)^2 = (t - 9)^2$

41. $50x^3 + 160x^2y + 128xy^2$
$= 2x(25x^2 + 80xy + 64y^2)$
$= 2x[(5x)^2 + 2(5x)(8y) + (8y)^2]$
$= 2x(5x + 8y)^2$

43. $4c^4 - 20c^2d^3 + 25d^6$
$= (2c^2)^2 - 2(2c^2)(5d^3) + (5d^3)^2 = (2c^2 - 5d^3)^2$

45. $9w^2 - 64 = (3w)^2 - (8)^2 = (3w + 8)(3w - 8)$

47. $200u^4 - 18v^6 = 2(100u^4 - 9v^6)$
$= 2[(10u^2)^2 - (3v^3)^2]$
$= 2(10u^2 + 3v^3)(10u^2 - 3v^3)$

49. $625p^4 - 16 = (25p^2)^2 - (4)^2$
$= (25p^2 + 4)(25p^2 - 4)$
$= (25p^2 + 4)[(5p)^2 - (2)^2]$
$= (25p^2 + 4)(5p + 2)(5p - 2)$

51. $y^3 + 64 = (y)^3 + (4)^3$
$= (y + 4)[(y)^2 - (y)(4) + (4)^2]$
$= (y + 4)(y^2 - 4y + 16)$

53. $c^4 - 27c = c(c^3 - 27)$
$= c[(c)^3 - (3)^3]$
$= c(c - 3)[(c)^2 + (c)(3) + (3)^2]$
$= c(c - 3)(c^2 + 3c + 9)$

55. $8a^6 - 125b^9$
$= (2a^2)^3 - (5b^3)^3$
$= (2a^2 - 5b^3)[(2a^2)^2 + (2a^2)(5b^3) + (5b^3)^2]$
$= (2a^2 - 5b^3)(4a^4 + 10a^2b^3 + 25b^6)$

57. $30x^4 + 70x^3 - 120x^2 - 280x$
$= 10x(3x^3 + 7x^2 - 12x - 28)$
$= 10x[x^2(3x + 7) - 4(3x + 7)]$
$= 10x(3x + 7)(x^2 - 4)$
$= 10x(3x + 7)(x + 2)(x - 2)$

59. $a^2 - y^2 + 10y - 25 = a^2 - (y^2 - 10y + 25)$
$= a^2 - (y - 5)^2$
$= [a + (y - 5)][a - (y - 5)]$
$= (a + y - 5)(a - y + 5)$

61. $30x^3y + 125x^2y + 120xy$
$= 5xy(6x^2 + 25x + 24)$
$= 5xy(3x + 8)(2x + 3)$

63. Let $u = x^2 - 2$.
$(x^2 - 2)^2 - 3(x^2 - 2) - 28$
$= u^2 - 3u - 28$
$= (u + 4)(u - 7)$
$= (x^2 - 2 + 4)(x^2 - 2 - 7)$
$= (x^2 + 2)(x^2 - 9)$
$= (x^2 + 2)(x + 3)(x - 3)$

65. $\left(x^3+12\right)^2-16$

$=\left(x^3+12\right)^2-4^2$

$=\left(x^3+12+4\right)\left(x^3+12-4\right)$

$=\left(x^3+16\right)\left(x^3+8\right)$

$=\left(x^3+16\right)\left(x^3+2^3\right)$

$=\left(x^3+16\right)\left(x+2\right)\left(x^2-2x+4\right)$

67. $(x+y)^2-z^2=(x+y+z)(x+y-z)$

69. $(x+y)^3+z^3$

$=(x+y+z)\left[(x+y)^2-(x+y)(z)+z^2\right]$

$=(x+y+z)\left(x^2+2xy+y^2-xz-yz+z^2\right)$

71. Let $u=3n+1$.

$9m^2+42m(3n+1)+49(3n+1)^2$

$=9m^2+42mu+49u^2$

$=(3m)^2+2(3m)(7u)+(7u)^2$

$=(3m+7u)^2$

$=\left[3m+7(3n+1)\right]^2$

$=(3m+21n+7)^2$

73. $(c-3)^2-(2c-5)^2$

$=\left[(c-3)+(2c-5)\right]\left[(c-3)-(2c-5)\right]$

$=(c-3+2c-5)(c-3-2c+5)$

$=(3c-8)(-c+2)$ or $-(3c-8)(c-2)$

75. $p^{11}-64p^8-p^3+64=p^{11}-p^3-64p^8+64=p^3\left(p^8-1\right)-64\left(p^8-1\right)$

$=\left(p^3-64\right)\left(p^8-1\right)=\left(p^3-4^3\right)\left[\left(p^4\right)^2-1^2\right]$

$=\left(p-4\right)\left(p^2+4p+16\right)\left(p^4+1\right)\left(p^4-1\right)$

$=\left(p-4\right)\left(p^2+4p+16\right)\left(p^4+1\right)\left[\left(p^2\right)^2-1^2\right]$

$=\left(p-4\right)\left(p^2+4p+16\right)\left(p^4+1\right)\left(p^2+1\right)\left(p^2-1\right)$

$=\left(p-4\right)\left(p^2+4p+16\right)\left(p^4+1\right)\left(p^2+1\right)\left(p+1\right)\left(p-1\right)$

77. Let $u=m^3$.

m^6+26m^3-27

$=\left(m^3\right)^2+26m^3-27$

$=u^2+26u-27$

$=(u+27)(u-1)$

$=\left(m^3+27\right)\left(m^3-1\right)$

$=\left(m^3+3^3\right)\left(m^3-1^3\right)$

$=\left(m+3\right)\left(m^2-3m+9\right)\left(m-1\right)\left(m^2+m+1\right)$

79. Let $u=x^3$.

$16x^6z+38x^3z-54z$

$=2z\left(8x^6+19x^3-27\right)$

$=2z\left[8\left(x^3\right)^2+19\left(x^3\right)-27\right]$

$=2z\left(8u^2+19u-27\right)$

$=2z(8u+27)(u-1)$

$=2z\left(8x^3+27\right)\left(x^3-1\right)$

$=2z\left[(2x)^3+3^3\right]\left(x^3-1^3\right)$

$=2x\left(2x+3\right)\left(4x^2-6x+9\right)\left(x-1\right)\left(x^2+x+1\right)$

81. $x^2 - y^2 - x + y = (x+y)(x-y) - 1(x-y)$
$= (x-y)(x+y-1)$

83. $a^2 + ac - 2c^2 - c + a$
$= (a-c)(a+2c) + 1(a-c)$
$= (a-c)(a+2c+1)$

85. $2x^{-4} - 7x^{-3} + x^{-2}$
$= x^{-4}\left(2x^{-4-(-4)} - 7x^{-3-(-4)} + x^{-2-(-4)}\right)$
$= x^{-4}\left(2x^0 - 7x^1 + x^2\right)$
$= x^{-4}\left(2 - 7x + x^2\right)$ or $\dfrac{x^2 - 7x + 2}{x^4}$

87. $y^{-2} - y^{-3} - 12y^{-4}$
$= y^{-4}\left(y^{-2-(-4)} - y^{-3-(-4)} - 12y^{-4-(-4)}\right)$
$= y^{-4}\left(y^2 - y^1 - 12y^0\right)$
$= y^{-4}\left(y^2 - y - 12\right)$
$= y^{-4}(y-4)(y+3)$ or $\dfrac{(y-4)(y+3)}{y^4}$

89. $2c^{7/4} + 4c^{3/4} = 2c^{3/4}\left(c^{7/4-(3/4)} + 2c^{3/4-(3/4)}\right)$
$= 2c^{3/4}\left(c^1 + 2c^0\right)$
$= 2c^{3/4}(c+2)$

91. $5x(3x+1)^{2/3} + (3x+1)^{5/3} = (3x+1)^{2/3}\left[5x(3x+1)^{2/3-(2/3)} + 1(3x+1)^{5/3-(2/3)}\right]$
$= (3x+1)^{2/3}\left[5x(3x+1)^0 + 1(3x+1)^1\right]$
$= (3x+1)^{2/3}(5x+3x+1)$
$= (3x+1)^{2/3}(8x+1)$

93. $x(3x+2)^{-2/3} + (3x+2)^{1/3} = (3x+2)^{-2/3}\left[x(3x+2)^{-2/3-(-2/3)} + (3x+2)^{1/3-(-2/3)}\right]$
$= (3x+2)^{-2/3}\left[x(3x+2)^0 + (3x+2)^1\right]$
$= (3x+2)^{-2/3}(x+3x+2)$
$= (3x+2)^{-2/3}(4x+2)$
$= 2(3x+2)^{-2/3}(2x+1)$ or $\dfrac{2(2x+1)}{(3x+2)^{2/3}}$

95. $A = x^2 - y^2 = (x+y)(x-y)$

97. $2\pi r^2 + 2\pi rh = 2\pi r(r+h)$

99. $\dfrac{4}{3}\pi R^3 - \dfrac{4}{3}\pi r^3 = \dfrac{4}{3}\pi\left(R^3 - r^3\right)$
$= \dfrac{4}{3}\pi(R-r)\left(R^2 + Rr + r^2\right)$

101. $21^2 - 19^2 = (21+19)(21-19) = (40)(2) = 80$

103. Expand the square of any binomial. For example: $(2c+3)^2 = 4c^2 + 12c + 9$.

105. In each case, factor out x to the smallest exponent to which it appears in both terms. That is, $5x^4 + 4x^3 = x^3(5x+4)$ and $5x^{-4} + 4x^{-3} = x^{-4}(5+4x)$.

107. $x^2 - 5 = \left(x+\sqrt{5}\right)\left(x-\sqrt{5}\right)$

109. $z^4 - 36 = (z^2+6)(z^2-6)$
$= (z^2+6)(z+\sqrt{6})(z-\sqrt{6})$

111. $x^2 - 2\sqrt{5}x + 5 = (x)^2 - 2(\sqrt{5})(x) + (\sqrt{5})^2$
$= (x-\sqrt{5})^2$

113. a. $x^5 - 1 = (x-1)(x^4+x^3+x^2+x+1)$

b. $x^n - 1 = (x-1)(x^{n-1}+x^{n-2}+x^{n-3}+\cdots+1)$

115. $36p^2 - 33p - 12$
$b^2 - 4ac = (-33)^2 - 4(36)(-12)$
$= 1089 + 1728 = 2817$
No

117. $8x^2 + 2x - 15$
$b^2 - 4ac = (2)^2 - 4(8)(-15)$
$= 4 + 480$
$= 484$
$= 22^2$
Yes

119. $18y^2 + 45y - 48$
$b^2 - 4ac = (45)^2 - 4(18)(-48)$
$= 2025 + 3456$
$= 5481$
No

Section R.7 Rational Expressions and More Operations on Radicals

1. $30x^3 - 65x^2 - 25x = 5x(6x^2 - 13x - 5)$
$= 5x(3x+1)(2x-5)$

3. $t^4 - t = t(t^3-1) = t(t-1)(t^2+t+1)$

5. $2x^3 + 3x^2 - 32x - 48$
$= 2x^3 - 32x + 3x^2 - 48$
$= 2x(x^2-16) + 3(x^2-16)$
$= (2x+3)(x^2-16) = (2x+3)(x-4)(x+4)$

7. $25w^4 - 40w^2u + 16u^2$
$= (5w^2)^2 - 2(5w^2)(4u) + (4u)^2 = (5w^2-4u)^2$

9. rational

11. $-2; 1$

13. complex (or compound)

15. $\dfrac{x-4}{x+7}$: If -7 were substituted for x, then $-7+7=0$ and the denominator would be 0. So, $x \neq -7$.

17. $\dfrac{a}{a^2-81} = \dfrac{a}{(a+9)(a-9)}$: If -9 or 9 were substituted for a, then $-9+9=0$ or $9-9=0$ and the denominator would be 0. So, $a \neq -9$, $a \neq 9$.

19. There are no values for a that make the denominator 0. So there are no restricted values for this expression.

21. $\dfrac{6c}{7a^3b^2}$: If 0 were substituted for a or b, then the product in the denominator would be 0. So, $a \neq 0$, $b \neq 0$.

23. $-\dfrac{5}{x-3} = \dfrac{-5}{x-3}$ and
$-\dfrac{5}{x-3} = \dfrac{5}{-(x-3)} = \dfrac{5}{-x+3} = \dfrac{5}{3-x}$, so a and b are equal to the original expression

25. $\dfrac{x^2-9}{x^2-4x-21} = \dfrac{\overset{1}{\cancel{(x+3)}}(x-3)}{\underset{1}{\cancel{(x+3)}}(x-7)} = \dfrac{x-3}{x-7}$;
$x \neq -3$, $x \neq 7$

27. $-\dfrac{12a^2bc}{3ab^5} = -\dfrac{\overset{1}{\cancel{3ab}}(4ac)}{\underset{1}{\cancel{3ab}}(b^4)} = -\dfrac{4ac}{b^4}$; $a \neq 0$, $b \neq 0$

29. $\dfrac{10-5\sqrt{6}}{15} = \dfrac{\overset{1}{\cancel{5}}(2-\sqrt{6})}{\underset{1}{\cancel{5}}(3)} = \dfrac{2-\sqrt{6}}{3}$

31. $\dfrac{2y^2-16y}{64-y^2} = \dfrac{2y(y-8)}{(64-y^2)}$

$= \dfrac{2y\,\overset{-1}{\cancel{(y-8)}}}{(8+y)\,\underset{1}{\cancel{(8-y)}}} = -\dfrac{2y}{8+y}$;

$y \neq -8$, $y \neq 8$

33. $\dfrac{4b-4a}{ax-xb-2a+2b} = \dfrac{4(b-a)}{x(a-b)-2(a-b)}$

$= \dfrac{4\,\overset{-1}{\cancel{(b-a)}}}{\underset{1}{\cancel{(a-b)}}(x-2)} = -\dfrac{4}{x-2}$;

$a \neq b$, $x \neq 2$

35. $\dfrac{3a^5b^7}{a-5b} \cdot \dfrac{2a-10b}{12a^4b^{10}} = \dfrac{\cancel{3a^4b^7} \cdot a}{\cancel{a-5b}} \cdot \dfrac{\cancel{2}\,\cancel{(a-5b)}}{\cancel{3a^4b^7} \cdot \underset{2}{\cancel{4}}\,b^3}$

$= \dfrac{a}{2b^3}$

37. $\dfrac{c^2-d^2}{cd^{11}} \div \dfrac{8c^2+4cd-4d^2}{8c^4d^{10}}$

$= \dfrac{c^2-d^2}{cd^{11}} \cdot \dfrac{8c^4d^{10}}{8c^2+4cd-4d^2}$

$= \dfrac{(c+d)(c-d)}{cd^{11}} \cdot \dfrac{8c^4d^{10}}{4(2c^2+cd-d^2)}$

$= \dfrac{\cancel{(c+d)}(c-d)}{\cancel{cd^{10}} \cdot d} \cdot \dfrac{\cancel{cd^{10}} \cdot \overset{2}{\cancel{8}}\,c^3}{\cancel{4}(2c-d)\cancel{(c+d)}}$

$= \dfrac{2c^3(c-d)}{d(2c-d)}$

39. $\dfrac{2a^2b-ab^2}{8b^2+ab} \cdot \dfrac{a^2+16ab+64b^2}{2a^2+15ab-8b^2}$

$= \dfrac{a\cancel{b}\,\cancel{(2a-b)}}{\cancel{b}\,\cancel{(8b+a)}} \cdot \dfrac{\cancel{(a+8b)}\,\cancel{(a+8b)}}{\cancel{(2a-b)}\,\cancel{(a+8b)}}$

$= a$

41. $\dfrac{x^3-64}{16x-x^3} \div \dfrac{2x^2+8x+32}{x^2+2x-8}$

$= \dfrac{x^3-64}{16x-x^3} \div \dfrac{x^2+2x-8}{2x^2+8x+32}$

$= \dfrac{(x-4)(x^2+4x+16)}{x(16-x^2)} \cdot \dfrac{(x+4)(x-2)}{2(x^2+4x+16)}$

$= \dfrac{\overset{-1}{\cancel{(x-4)}}\,\cancel{(x^2+4x+16)}}{x\cancel{(4+x)}\,\cancel{(4-x)}} \cdot \dfrac{\cancel{(x+4)}(x-2)}{2\cancel{(x^2+4x+16)}}$

$= -\dfrac{x-2}{2x}$

43. $\dfrac{7}{6x^5yz^4} = \dfrac{7}{2 \cdot 3x^5yz^4}$, $\dfrac{3}{20xy^2z^3} = \dfrac{3}{5 \cdot 2^2xy^2z^3}$

LCD $= 2^2 \cdot 3 \cdot 5x^5y^2z^4 = 60x^5y^2z^4$

45. $\dfrac{2t+1}{(3t+4)^3(t-2)}$, $\dfrac{4}{t(3t+4)^2(t-2)}$

LCD $= t(3t+4)^3(t-2)$

47. $\dfrac{x+3}{x^2+20x+100} = \dfrac{x+3}{(x+10)^2}$,

$\dfrac{3}{2x^2+20x} = \dfrac{3}{2x(x+10)}$

LCD $= 2x(x+10)^2$

49. $\dfrac{m^2}{m+3} + \dfrac{6m+9}{m+3} = \dfrac{m^2+6m+9}{m+3}$

$= \dfrac{(m+3)\cancel{(m+3)}}{\cancel{m+3}}$

$= m+3$

51. $\frac{2}{9c}+\frac{7}{15c^3}=\frac{2}{3^2c}+\frac{7}{5\cdot 3c^3}$

$=\frac{2\cdot(5c^2)}{3^2c\cdot(5c^2)}+\frac{7\cdot(3)}{5\cdot 3c^3\cdot(3)}$

$=\frac{10c^2}{45c^3}+\frac{21}{45c^3}=\frac{10c^2+21}{45c^3}$

53. $\frac{9}{2x^2y^4}-\frac{11}{xy^5}=\frac{9\cdot(y)}{2x^2y^4\cdot(y)}-\frac{11\cdot(2x)}{xy^5\cdot(2x)}$

$=\frac{9y-22x}{2x^2y^5}$

55. $\frac{1}{x^2+xy}-\frac{2}{x^2-y^2}$

$=\frac{1}{x(x+y)}-\frac{2}{(x+y)(x-y)}$

$=\frac{1\cdot(x-y)}{x(x+y)\cdot(x-y)}-\frac{2\cdot(x)}{(x+y)(x-y)\cdot(x)}$

$=\frac{x-y-2x}{x(x+y)(x-y)}$

$=\frac{-x-y}{x(x+y)(x-y)}$

$=-\frac{\cancel{x+y}}{x\cancel{(x+y)}(x-y)}$

$=-\frac{1}{x(x-y)}$

57. $\frac{5}{y}+\frac{2}{y+1}-\frac{6}{y^2}$

$=\frac{5\cdot(y)(y+1)}{y\cdot(y)(y+1)}+\frac{2\cdot(y^2)}{(y+1)\cdot(y^2)}-\frac{6\cdot(y+1)}{y^2\cdot(y+1)}$

$=\frac{5\cdot(y)(y+1)+2\cdot(y^2)-6\cdot(y+1)}{y^2(y+1)}$

$=\frac{5y^2+5y+2y^2-6y-6}{y^2(y+1)}$

$=\frac{7y^2-y-6}{y^2(y+1)}$

59. $\frac{3w}{w-4}+\frac{2w+4}{4-w}=\frac{3w}{w-4}+\frac{(2w+4)\cdot(-1)}{(4-w)\cdot(-1)}$

$=\frac{3w-2w-4}{w-4}$

$=\frac{\cancel{w-4}}{\cancel{w-4}}$

$=1$

61. $\frac{\frac{1}{27x}+\frac{1}{9}}{\frac{1}{3}+\frac{1}{9x}}=\frac{27x\cdot\left(\frac{1}{27x}+\frac{1}{9}\right)}{27x\cdot\left(\frac{1}{3}+\frac{1}{9x}\right)}$

$=\frac{\frac{27x}{1}\cdot\frac{1}{27x}+\frac{27x}{1}\cdot\frac{1}{9}}{\frac{27x}{1}\cdot\frac{1}{3}+\frac{27x}{1}\cdot\frac{1}{9x}}$

$=\frac{1+3x}{9x+3}$

$=\frac{\cancel{1+3x}}{3\cancel{(3x+1)}}$

$=\frac{1}{3}$

63. $\frac{\frac{x}{6}-\frac{5x+14}{6x}}{\frac{1}{6}-\frac{7}{6x}}=\frac{6x\cdot\left(\frac{x}{6}-\frac{5x+14}{6x}\right)}{6x\cdot\left(\frac{1}{6}-\frac{7}{6x}\right)}$

$=\frac{\frac{6x}{1}\cdot\frac{x}{6}-\frac{6x}{1}\cdot\frac{5x+14}{6x}}{\frac{6x}{1}\cdot\frac{1}{6}-\frac{6x}{1}\cdot\frac{7}{6x}}$

$=\frac{x^2-5x-14}{x-7}$

$=\frac{(x+2)\cancel{(x-7)}}{\cancel{x-7}}$

$=x+2$

65. $\dfrac{2a^{-1}-b^{-1}}{4a^{-2}-b^{-2}}=\dfrac{\frac{2}{a}-\frac{1}{b}}{\frac{4}{a^2}-\frac{1}{b^2}}$

$$=\frac{a^2b^2\cdot\left(\frac{2}{a}-\frac{1}{b}\right)}{a^2b^2\cdot\left(\frac{4}{a^2}-\frac{1}{b^2}\right)}$$

$$=\frac{\frac{a^2b^2}{1}\cdot\frac{2}{a}-\frac{a^2b^2}{1}\cdot\frac{1}{b}}{\frac{a^2b^2}{1}\cdot\frac{4}{a^2}-\frac{a^2b^2}{1}\cdot\frac{1}{b^2}}$$

$$=\frac{2ab^2-a^2b}{4b^2-a^2}$$

$$=\frac{ab\cancel{(2b-a)}}{(2b+a)\cancel{(2b-a)}}$$

$$=\frac{ab}{2b+a}$$

67. $\dfrac{\frac{3}{1+h}-3}{h}=\dfrac{\frac{3}{1+h}-\frac{3}{1}}{h}$

$$=\frac{(1+h)\cdot\left(\frac{3}{1+h}-\frac{3}{1}\right)}{(1+h)\cdot(h)}$$

$$=\frac{\frac{(1+h)}{1}\cdot\left(\frac{3}{1+h}\right)-\frac{(1+h)}{1}\cdot\left(\frac{3}{1}\right)}{(1+h)\cdot(h)}$$

$$=\frac{3-3-3h}{h+h^2}$$

$$=\frac{-3\cancel{h}}{\cancel{h}(1+h)}$$

$$=\frac{-3}{1+h}\text{ or }-\frac{3}{1+h}$$

69. $\dfrac{\frac{7}{x+h}-\frac{7}{x}}{h}=\dfrac{[x(x+h)]\cdot\left(\frac{7}{x+h}-\frac{7}{x}\right)}{[x(x+h)]\cdot(h)}$

$$=\frac{\left[\frac{x(x+h)}{1}\right]\cdot\left(\frac{7}{x+h}-\frac{7}{x}\right)}{[x(x+h)]\cdot(h)}$$

$$=\frac{\left[\frac{x(x+h)}{1}\right]\cdot\left(\frac{7}{x+h}\right)-\left[\frac{x(x+h)}{1}\right]\cdot\left(\frac{7}{x}\right)}{[x(x+h)]\cdot(h)}$$

$$=\frac{7x-7(x+h)}{hx(x+h)}=\frac{7x-7x-7h}{hx(x+h)}$$

$$=\frac{-7\cancel{h}}{\cancel{h}x(x+h)}=\frac{-7}{x(x+h)}\text{ or }-\frac{7}{x(x+h)}$$

71. $\dfrac{4}{\sqrt{y}}=\dfrac{4\cdot\sqrt{y}}{\sqrt{y}\cdot\sqrt{y}}=\dfrac{4\sqrt{y}}{\sqrt{y^2}}=\dfrac{4\sqrt{y}}{y}$

73. $\dfrac{4}{\sqrt[3]{y}}=\dfrac{4\cdot\sqrt[3]{y^2}}{\sqrt[3]{y}\cdot\sqrt[3]{y^2}}=\dfrac{4\sqrt[3]{y^2}}{\sqrt[3]{y^3}}=\dfrac{4\sqrt[3]{y^2}}{y}$

75. $\sqrt[3]{\dfrac{18}{12w}}=\sqrt[3]{\dfrac{3}{2w}}=\dfrac{\sqrt[3]{3}\cdot\sqrt[3]{2^2w^2}}{\sqrt[3]{2w}\cdot\sqrt[3]{2^2w^2}}$

$$=\frac{\sqrt[3]{12w^2}}{\sqrt[3]{2^3w^3}}=\frac{\sqrt[3]{12w^2}}{2w}$$

77. $\dfrac{12}{\sqrt[4]{8wx^2}}=\dfrac{12\cdot\sqrt[4]{2w^3x^2}}{\sqrt[4]{2^3wx^2}\cdot\sqrt[4]{2w^3x^2}}=\dfrac{12\sqrt[4]{2w^3x^2}}{\sqrt[4]{2^4w^4x^4}}$

$$=\frac{12\sqrt[4]{2w^3x^2}}{2wx}=\frac{6\sqrt[4]{2w^3x^2}}{wx}$$

79. $\dfrac{\sqrt{12}}{\sqrt{x+1}}=\dfrac{\sqrt{12}\cdot\sqrt{x+1}}{\sqrt{x+1}\cdot\sqrt{x+1}}=\dfrac{\sqrt{4\cdot3}\cdot\sqrt{x+1}}{\sqrt{(x+1)^2}}$

$$=\frac{2\sqrt{3}\cdot\sqrt{x+1}}{x+1}=\frac{2\sqrt{3x+3}}{x+1}$$

81. $\dfrac{8}{\sqrt{15}-\sqrt{11}}=\dfrac{8\cdot\left(\sqrt{15}+\sqrt{11}\right)}{\left(\sqrt{15}-\sqrt{11}\right)\cdot\left(\sqrt{15}+\sqrt{11}\right)}$

$=\dfrac{8\left(\sqrt{15}+\sqrt{11}\right)}{\left(\sqrt{15}\right)^2-\left(\sqrt{11}\right)^2}$

$=\dfrac{8\left(\sqrt{15}+\sqrt{11}\right)}{15-11}$

$=\dfrac{8\left(\sqrt{15}+\sqrt{11}\right)}{4}$

$=2\left(\sqrt{15}+\sqrt{11}\right)$

83. $\dfrac{x-5}{\sqrt{x}+\sqrt{5}}=\dfrac{(x-5)\cdot\left(\sqrt{x}-\sqrt{5}\right)}{\left(\sqrt{x}+\sqrt{5}\right)\cdot\left(\sqrt{x}-\sqrt{5}\right)}$

$=\dfrac{(x-5)\left(\sqrt{x}-\sqrt{5}\right)}{\left(\sqrt{x}\right)^2-\left(\sqrt{5}\right)^2}$

$=\dfrac{\cancel{(x-5)}\left(\sqrt{x}-\sqrt{5}\right)}{\cancel{x-5}}$

$=\sqrt{x}-\sqrt{5}$

85. $\dfrac{2\sqrt{10}+3\sqrt{5}}{4\sqrt{10}+2\sqrt{5}}$

$=\dfrac{\left(2\sqrt{10}+3\sqrt{5}\right)\cdot\left(2\sqrt{10}-\sqrt{5}\right)}{2\left(2\sqrt{10}+\sqrt{5}\right)\cdot\left(2\sqrt{10}-\sqrt{5}\right)}=\dfrac{\left(2\sqrt{10}\right)\left(2\sqrt{10}\right)+\left(2\sqrt{10}\right)\left(-\sqrt{5}\right)+\left(3\sqrt{5}\right)\left(2\sqrt{10}\right)+\left(3\sqrt{5}\right)\left(-\sqrt{5}\right)}{2\left[\left(2\sqrt{10}\right)^2-\left(\sqrt{5}\right)^2\right]}$

$=\dfrac{4\sqrt{10^2}-2\sqrt{50}+6\sqrt{50}-3\sqrt{5^2}}{2(40-5)}=\dfrac{40+4\sqrt{50}-15}{2(35)}=\dfrac{25+4\sqrt{5^2\cdot 2}}{70}=\dfrac{25+20\sqrt{2}}{70}=\dfrac{5\left(5+4\sqrt{2}\right)}{70}=\dfrac{5+4\sqrt{2}}{14}$

87. $\dfrac{7}{\sqrt{3x}}+\dfrac{\sqrt{3x}}{x}=\dfrac{7\cdot\sqrt{3x}}{\sqrt{3x}\cdot\sqrt{3x}}+\dfrac{\sqrt{3x}}{x}$

$=\dfrac{7\sqrt{3x}}{3x}+\dfrac{\sqrt{3x}\cdot 3}{x\cdot 3}$

$=\dfrac{10\sqrt{3x}}{3x}$

89. $\dfrac{5}{w\sqrt{7}}-\dfrac{\sqrt{7}}{w}=\dfrac{5\cdot\sqrt{7}}{w\sqrt{7}\cdot\sqrt{7}}-\dfrac{\sqrt{7}}{w}$

$=\dfrac{5\sqrt{7}}{7w}-\dfrac{\sqrt{7}\cdot 7}{w\cdot 7}$

$=\dfrac{-2\sqrt{7}}{7w}$ or $-\dfrac{2\sqrt{7}}{7w}$

91. a. $S=\dfrac{2d}{\dfrac{d}{r_1}+\dfrac{d}{r_2}}$

$=\dfrac{r_1r_2\cdot(2d)}{r_1r_2\cdot\left(\dfrac{d}{r_1}+\dfrac{d}{r_2}\right)}$

$=\dfrac{2dr_1r_2}{r_1r_2\cdot\left(\dfrac{d}{r_1}\right)+r_1r_2\cdot\left(\dfrac{d}{r_2}\right)}$

$=\dfrac{2dr_1r_2}{dr_2+dr_1}=\dfrac{2\cancel{d}r_1r_2}{\cancel{d}\left(r_2+r_1\right)}=\dfrac{2r_1r_2}{r_1+r_2}$

b. $S=\dfrac{2r_1r_2}{r_1+r_2}=\dfrac{2(400)(460)}{400+460}=\dfrac{368{,}000}{860}$

≈ 427.9 mph

93. **a.** $C=\dfrac{600t}{t^3+125}$

At 1 hr:

$$C=\frac{600(1)}{(1)^3+125}=\frac{600}{126}\approx 4.8\text{ ng/mL}$$

At 4 hr:

$$C=\frac{600(4)}{(4)^3+125}=\frac{2400}{189}\approx 12.7\text{ ng/mL}$$

At 12 hr:

$$C=\frac{600(12)}{(12)^3+125}=\frac{7200}{1853}\approx 3.9\text{ ng/mL}$$

b. The concentration of the drug appears to be approaching 0 ng/mL for large values of t.

95. $P=\dfrac{5}{x+1}+\dfrac{2}{x+1}+\dfrac{6}{x}=\dfrac{7\cdot x}{(x+1)\cdot x}+\dfrac{6\cdot(x+1)}{x\cdot(x+1)}$

$$=\frac{7x+6x+6}{x(x+1)}=\frac{13x+6}{x(x+1)}\text{ cm}$$

97. $A=\dfrac{1}{2}bh$

$$=\frac{1}{\cancel{2}}\left(\frac{1}{\sqrt{x}}\right)\left(\frac{\overset{2}{\cancel{4}}}{\sqrt{2}}\right)=\frac{2\cdot\sqrt{2x}}{\sqrt{2x}\cdot\sqrt{2x}}$$

$$=\frac{\cancel{2}\sqrt{2x}}{\cancel{2}x}=\frac{\sqrt{2x}}{x}\text{ in.}^2$$

99. $\dfrac{2x^3y}{x^2y+3xy}\cdot\dfrac{x^2+6x+9}{2x+6}\div 5xy^4$

$$=\frac{\cancel{xy}\left(\cancel{2}x^2\right)}{\cancel{xy}\,\cancel{(x+3)}}\cdot\frac{\cancel{(x+3)}\,\cancel{(x+3)}}{\cancel{2}\,\cancel{(x+3)}}\cdot\frac{1}{5xy^4}$$

$$=\frac{x^2}{5xy^4}$$

$$=\frac{x}{5y^4}$$

101. $\left(\dfrac{4}{2t+1}-\dfrac{t}{2t^2+17t+8}\right)(t+8)$

$$=\left(\frac{4}{2t+1}-\frac{t}{(2t+1)(t+8)}\right)(t+8)$$

$$=\left(\frac{4}{2t+1}\right)(t+8)-\left(\frac{t}{(2t+1)\cancel{(t+8)}}\right)\cancel{(t+8)}$$

$$=\frac{4t+32-t}{2t+1}$$

$$=\frac{3t+32}{2t+1}$$

103. $\dfrac{n-2}{n-4}+\dfrac{2n^2-15n+12}{n^2-16}-\dfrac{2n-5}{n+4}=\dfrac{n-2}{n-4}+\dfrac{2n^2-15n+12}{(n+4)(n-4)}-\dfrac{2n-5}{n+4}$

$$=\frac{(n-2)\cdot(n+4)}{(n-4)\cdot(n+4)}+\frac{2n^2-15n+12}{(n+4)(n-4)}-\frac{(2n-5)\cdot(n-4)}{(n+4)\cdot(n-4)}$$

$$=\frac{n^2-2n+4n-8+2n^2-15n+12-2n^2+8n+5n-20}{(n+4)(n-4)}$$

$$=\frac{n^2-16}{n^2-16}=1$$

105. $\dfrac{1-a^{-1}-6a^{-2}}{1-4a^{-1}+3a^{-2}}=\dfrac{1-\frac{1}{a}-\frac{6}{a^2}}{1-\frac{4}{a}+\frac{3}{a^2}}$

$=\dfrac{a^2\cdot\left(1-\frac{1}{a}-\frac{6}{a^2}\right)}{a^2\cdot\left(1-\frac{4}{a}+\frac{3}{a^2}\right)}$

$=\dfrac{a^2-a-6}{a^2-4a+3}$

$=\dfrac{(a+2)\cancel{(a-3)}}{\cancel{(a-3)}(a-1)}$

$=\dfrac{a+2}{a-1}$

107. $\dfrac{34}{2\sqrt{5}-\sqrt{3}}=\dfrac{34\cdot\left(2\sqrt{5}+\sqrt{3}\right)}{\left(2\sqrt{5}-\sqrt{3}\right)\cdot\left(2\sqrt{5}+\sqrt{3}\right)}$

$=\dfrac{34\cdot\left(2\sqrt{5}+\sqrt{3}\right)}{\left(2\sqrt{5}\right)^2-\left(\sqrt{3}\right)^2}$

$=\dfrac{34\cdot\left(2\sqrt{5}+\sqrt{3}\right)}{20-3}$

$=\dfrac{\overset{2}{\cancel{34}}\cdot\left(2\sqrt{5}+\sqrt{3}\right)}{\cancel{17}}$

$=4\sqrt{5}+2\sqrt{3}$

109. $\dfrac{8-\sqrt{48}}{6}=\dfrac{8-\sqrt{4^2\cdot 3}}{6}=\dfrac{8-4\sqrt{3}}{6}$

$=\dfrac{\overset{2}{\cancel{4}}\left(2-\sqrt{3}\right)}{\underset{3}{\cancel{6}}}=\dfrac{4-2\sqrt{3}}{3}$

111. $\dfrac{14}{\sqrt{7x}}-\dfrac{\sqrt{7x}}{x}=\dfrac{14\cdot\sqrt{7x}}{\sqrt{7x}\cdot\sqrt{7x}}-\dfrac{\sqrt{7x}}{x}$

$=\dfrac{14\sqrt{7x}}{7x}-\dfrac{\sqrt{7x}\cdot 7}{x\cdot 7}$

$=\dfrac{\cancel{7}\sqrt{7x}}{\cancel{7}x}=\dfrac{\sqrt{7x}}{x}$

113. $\dfrac{45+9x-5x^2-x^3}{x^3-3x^2-25x+75}$

$=\dfrac{45-5x^2+9x-x^3}{x^3-25x-3x^2+75}$

$=\dfrac{5\left(9-x^2\right)+x\left(9-x^2\right)}{x\left(x^2-25\right)-3\left(x^2-25\right)}$

$=\dfrac{\left(9-x^2\right)(x+5)}{\left(x^2-25\right)(x-3)}=-\dfrac{\left(x^2-9\right)(x+5)}{\left(x^2-25\right)(x-3)}$

$=-\dfrac{(x+3)\cancel{(x-3)}\cancel{(x+5)}}{\cancel{(x+5)}(x-5)\cancel{(x-3)}}=-\dfrac{x+3}{x-5}$

115. $\dfrac{t+6}{1+\frac{2}{t}}-t-4$

$=\dfrac{t+6}{\frac{t+2}{t}}-t-4=(t+6)\cdot\left(\dfrac{t}{t+2}\right)-t-4$

$=\dfrac{t(t+6)}{t+2}-\dfrac{t\cdot(t+2)}{1\cdot(t+2)}-\dfrac{4\cdot(t+2)}{1\cdot(t+2)}$

$=\dfrac{t^2+6t-t^2-2t-4t-8}{t+2}$

$=\dfrac{-8}{t+2}$ or $-\dfrac{8}{t+2}$

117. If $x=y$, then the denominator $x-y$ will equal zero. Division by zero is undefined.

119. For $\dfrac{1}{\sqrt{x}}$, multiply numerator and denominator by $\sqrt{x}$ to make a perfect square in the radicand in the denominator. For $\dfrac{1}{\sqrt[3]{x}}$, multiply numerator and denominator by $\sqrt[3]{x^2}$ to make a perfect cube in the radicand of the denominator.

121. **a.** $\dfrac{14}{3}\cdot\dfrac{30}{7}=\dfrac{420}{21}=20$

b. $\left(5-\sqrt{5}\right)\left(5+\sqrt{5}\right)=(5)^2-\left(\sqrt{5}\right)^2$

$=25-5=20$

123. $\dfrac{w^{3n+1}-w^{3n}z}{w^{n+2}-w^{n}z^{2}}=\dfrac{w^{3n}(w-z)}{w^{n}(w^{2}-z^{2})}$

$=\dfrac{\cancel{w^{n}}\cdot w^{2n}\cancel{(w-z)}}{\cancel{w^{n}}(w+z)\cancel{(w-z)}}=\dfrac{w^{2n}}{w+z}$

125. $\sqrt{\dfrac{x-y}{x+y}}=\dfrac{\sqrt{x-y}\cdot\sqrt{x+y}}{\sqrt{x+y}\cdot\sqrt{x+y}}$

$=\dfrac{\sqrt{(x-y)(x+y)}}{\left(\sqrt{x+y}\right)^{2}}=\dfrac{\sqrt{x^{2}-y^{2}}}{x+y}$

127. $\dfrac{\sqrt{5}}{\sqrt[3]{2}}=\dfrac{\sqrt{5}\cdot\sqrt[3]{2^{2}}}{\sqrt[3]{2}\cdot\sqrt[3]{2^{2}}}=\dfrac{\sqrt{5}\cdot\sqrt[3]{2^{2}}}{\sqrt[3]{2^{3}}}=\dfrac{5^{1/2}\cdot 2^{2/3}}{2}$

$=\dfrac{5^{3/6}\cdot 2^{4/6}}{2}=\dfrac{\left(5^{3}\cdot 2^{4}\right)^{1/6}}{2}=\dfrac{\sqrt[6]{5^{3}2^{4}}}{2}$

129. $\dfrac{a-b}{\sqrt[3]{a}-\sqrt[3]{b}}=\dfrac{\cancel{\left(\sqrt[3]{a}-\sqrt[3]{b}\right)}\left(\sqrt[3]{a^{2}}+\sqrt[3]{ab}+\sqrt[3]{b^{2}}\right)}{\cancel{\sqrt[3]{a}-\sqrt[3]{b}}}$

$=\sqrt[3]{a^{2}}+\sqrt[3]{ab}+\sqrt[3]{b^{2}}$

131. $\dfrac{\sqrt{4+h}-2}{h}=\dfrac{\left(\sqrt{4+h}-2\right)\cdot\left(\sqrt{4+h}+2\right)}{h\cdot\left(\sqrt{4+h}+2\right)}$

$=\dfrac{\left(\sqrt{4+h}\right)^{2}-2^{2}}{h\left(\sqrt{4+h}+2\right)}=\dfrac{4+h-4}{h\left(\sqrt{4+h}+2\right)}$

$=\dfrac{\cancel{h}}{\cancel{h}\left(\sqrt{4+h}+2\right)}=\dfrac{1}{\sqrt{4+h}+2}$

133.

	A	B	C	D	E
1	x	2.9	2.99	2.999	2.9999
2	$\dfrac{x^2-9}{x-3}$	5.9	5.99	5.999	5.9999

	A	B	C	D	E
1	x	3.1	3.01	3.001	3.0001
2	$\dfrac{x^2-9}{x-3}$	6.1	6.01	6.001	6.0001

The expression appears to approach 6 as x gets close to 3.

Chapter R Review Exercises

1. a. $\sqrt{9}$ **b.** $0, \sqrt{9}$

c. $0, -8, \sqrt{9}$

d. $0, -8, 1.\overline{45}, \sqrt{9}, -\dfrac{2}{3}$

e. $\sqrt{6}, 3\pi$

f. $\sqrt{6}, 0, -8, 1.\overline{45}, \sqrt{9}, -\dfrac{2}{3}, 3\pi$

3. $x \geq 4$

5. a. 

$[-3, 7)$; $\{x \mid -3 \leq x < 7\}$

b. 

$(2.1, \infty)$; $\{x \mid x > 2.1\}$

c. 

$(-\infty, 4]$; $\{x \mid 4 \geq x\}$

7. a. $\left|2-\sqrt{5}\right|$ or $\left|\sqrt{5}-2\right|$

b. $\left|2-\sqrt{5}\right| = -\left(2-\sqrt{5}\right) = \sqrt{5}-2$

9. $\sqrt{\dfrac{25}{4}} = \dfrac{5}{2}$

11. $\dfrac{32-\left|-11+3\right|}{36\div 2\div 3-2}=\dfrac{32-(8)}{18\div 3-2}=\dfrac{24}{6-2}=\dfrac{24}{4}=6$

13. a. $J = E + 150$ **b.** $E = J - 150$

15. a. $C = 3.6s + 50p + 250n$

b. $C = 3.6s + 50p + 250n$
$= 3.6(2100) + 50(4) + 250(2)$
$= 7560 + 200 + 500$
$= \$8260$

17. $15.2c^2d - 11.1cd + 8.7c^2d - 5.4cd$
$= 15.2c^2d + 8.7c^2d - 11.1cd - 5.4cd$
$= 23.9c^2d - 16.5cd$

19. d

21. c

23. e

25. a

27. b

29. a. $9^0 = 1$

b. $-9^0 = -1 \cdot 9^0 = -1 \cdot 1 = -1$

c. $9x^0 = 9 \cdot x^0 = 9 \cdot 1 = 9$

d. $(9x)^0 = 1$

31. $p^{-8} \cdot p^{12} \cdot p^{-1} = p^{-8+12-1} = p^3$

33. $(-12a^{-3}b^4)^2 = (-12)^2(a^{-3})^2(b^4)^2$
$= 144a^{-6}b^8 = \dfrac{144b^8}{a^6}$

35. $\left(\dfrac{1}{2u^5v^{-2}}\right)^{-3}\left(\dfrac{4}{u^{-3}v^2}\right) = (2^{-1}u^{-5}v^2)^{-3} \cdot (2^2u^3v^{-2})^{-1}$
$= 2^3u^{15}v^{-6}2^{-2}u^{-3}v^2$
$= 2u^{12}v^{-4} = \dfrac{2u^{12}}{v^4}$

37. a. 10^{-1} is between 0 and 1. Move the decimal point 1 place to the left.
$9.8 \times 10^{-1} = 0.98$

b. $10^0 = 1$ The decimal point does not move.
$9.8 \times 10^0 = 9.8$

c. 10^1 is greater than 10. Move the decimal point 1 place to the right.
$9.8 \times 10^1 = 98$

39. $\dfrac{(8.6\times10^{-3})(4.1\times10^8)}{2.0\times10^{-6}} = \left(\dfrac{8.6\cdot4.1}{2.0}\right)\times\left(\dfrac{10^{-3}\cdot10^8}{10^{-6}}\right)$
$= 17.63\times10^{11}$
$= (1.763\times10^1)\times10^{11}$
$= 1.763\times10^{12}$

41. a. $x^{2/7} = \sqrt[7]{x^2}$ or $\left(\sqrt[7]{x}\right)^2$

b. $9x^{2/7} = 9\sqrt[7]{x^2}$ or $9\left(\sqrt[7]{x}\right)^2$

c. $(9x)^{2/7} = \sqrt[7]{(9x)^2}$ or $\left(\sqrt[7]{9x}\right)^2$

43. a. $-\sqrt[4]{256} = -\sqrt[4]{4^4} = -4$

b. $\sqrt[4]{-256}$ is not a real number.

45. a. $10,000^{3/4} = \left(\sqrt[4]{10,000}\right)^3 = (10)^3 = 1000$

b. $10,000^{-3/4} = \dfrac{1}{10,000^{3/4}} = \dfrac{1}{\left(\sqrt[4]{10,000}\right)^3}$
$= \dfrac{1}{(10)^3} = \dfrac{1}{1000}$

c. $-10,000^{3/4} = -\left(\sqrt[4]{10,000}\right)^3$
$= -(10)^3 = -1000$

d. $-10,000^{-3/4} = -\dfrac{1}{10,000^{3/4}} = -\dfrac{1}{\left(\sqrt[4]{10,000}\right)^3}$
$= -\dfrac{1}{(10)^3} = -\dfrac{1}{1000}$

e. $(-10,000)^{3/4}$ is undefined because $\sqrt[4]{-10,000}$ is not a real number.

f. $(-10,000)^{-3/4}$ is undefined because $\sqrt[4]{-10,000}$ is not a real number.

47. $(9m^{-4}n^{2/3})^{1/2} = (9)^{1/2}(m^{-4})^{1/2}(n^{2/3})^{1/2}$
$= 3m^{-2}n^{1/3} = \dfrac{3n^{1/3}}{m^2}$

49. $\sqrt[3]{54xy^{12}z^{14}} = \sqrt[3]{3^3 \cdot 2 \cdot xy^{12}z^{14}}$
$= \sqrt[3]{\left(3^3 y^{12}z^{12}\right)\left(2xz^2\right)}$
$= \sqrt[3]{3^3 y^{12}z^{12}} \cdot \sqrt[3]{2xz^2}$
$= 3y^4z^4\sqrt[3]{2xz^2}$

51. $\sqrt{\frac{p^{13}}{9}} = \frac{\sqrt{p^{13}}}{\sqrt{9}} = \frac{\sqrt{p^{12} \cdot p}}{3} = \frac{p^6\sqrt{p}}{3}$

53. $\sqrt{10} \cdot \sqrt{35} = \sqrt{350} = \sqrt{5^2 \cdot 14} = \sqrt{5^2} \cdot \sqrt{14} = 5\sqrt{14}$

55. $\sqrt[4]{cd^2} \cdot \sqrt[3]{c^2d} = \left(cd^2\right)^{1/4} \cdot \left(c^2d\right)^{1/3}$
$= c^{1/4}d^{1/2} \cdot c^{2/3}d^{1/3}$
$= c^{1/4+2/3}d^{1/2+1/3} = c^{3/12+8/12}b^{6/12+4/12}$
$= c^{11/12}d^{10/12} = \sqrt[12]{c^{11}d^{10}}$

57. $-2c\sqrt[3]{54c^2d^3} + 5cd\sqrt[3]{2c^2} - 10d\sqrt[3]{250c^5} = -2c\sqrt[3]{3^3d^3 \cdot 2c^2} + 5cd\sqrt[3]{2c^2} - 10d\sqrt[3]{5^3c^3 \cdot 2c^2}$
$= -6cd\sqrt[3]{2c^2} + 5cd\sqrt[3]{2c^2} - 50cd\sqrt[3]{2c^2} = -51cd\sqrt[3]{2c^2}$

59. **a.** Yes **b.** No **c.** Yes

61. 8

63. $\left(-7.2a^2b^3 + 4.1ab^2 - 3.9b\right) - \left(0.8a^2b^3 - 3.2ab^2 - b\right)$
$= -7.2a^2b^3 + 4.1ab^2 - 3.9b - 0.8a^2b^3 + 3.2ab^2 + b$
$= -8a^2b^3 + 7.3ab^2 - 2.9b$

65. $\left(5w^3 + 6y^2\right)\left(2w^3 - y^2\right) = 5w^3\left(2w^3\right) + 5w^3\left(-y^2\right) + 6y^2\left(2w^3\right) + 6y^2\left(-y^2\right)$
$= 10w^6 - 5w^3y^2 + 12w^3y^2 - 6y^4 = 10w^6 + 7w^3y^2 - 6y^4$

67. $(9t-4)(9t+4) = (9t)^2 - (4)^2 = 81t^2 - 16$

69. $(5k-3)^2 = (5k)^2 - 2(5k)(3) + (3)^2 = 25k^2 - 30k + 9$

71. $\left[(2v-1)+w\right]\left[(2v-1)-w\right] = (2v-1)^2 - w^2 = (2v)^2 - 2(2v)(1) + (1)^2 - w^2 = 4v^2 - 4v + 1 - w^2$

73. $\left(6\sqrt{5} - 2\sqrt{3}\right)\left(2\sqrt{5} + 5\sqrt{3}\right) = 6\sqrt{5}\left(2\sqrt{5}\right) + 6\sqrt{5}\left(5\sqrt{3}\right) + \left(-2\sqrt{3}\right)\left(2\sqrt{5}\right) + \left(-2\sqrt{3}\right)\left(5\sqrt{3}\right)$
$= 12\sqrt{25} + 30\sqrt{15} - 4\sqrt{15} - 10\sqrt{9} = 60 + 26\sqrt{15} - 30 = 30 + 26\sqrt{15}$

75. $\left(2c^2\sqrt{d} - 5d^2\sqrt{c}\right)^2 = \left(2c^2\sqrt{d}\right)^2 - 2\left(2c^2\sqrt{d}\right)\left(5d^2\sqrt{c}\right) + \left(5d^2\sqrt{c}\right)^2 = 4c^4d - 20c^2d^2\sqrt{cd} + 25cd^4$

77. $V = lwh = (x+3)(x+2)(2x+5) = (x+3)\left(2x^2 + 5x + 4x + 10\right) = (x+3)\left(2x^2 + 9x + 10\right)$
$= 2x^3 + 9x^2 + 10x + 6x^2 + 27x + 30 = 2x^3 + 15x^2 + 37x + 30$

79. $80m^4n^8 - 48m^5n^3 - 16m^2n = 16m^2n\left(5m^2n^7 - 3m^3n^2 - 1\right)$

81. $15ac-14b-10a+21bc$
$=15ac-10a-14b+21bc$
$=5a(3c-2)-7b(2-3c)$
$=5a(3c-2)+7b(3c-2)$
$=(5a+7b)(3c-2)$

83. $8x^3-40x^2y+50xy^2$
$=2x(4x^2-20xy+25y^2)$
$=2x\left[(2x)^2-2(2x)(5y)+(5y)^2\right]$
$=2x(2x-5y)^2$

85. $3k^4-81k=3k(k^3-27)$
$=3k(k^3-3^3)$
$=3k(k-3)(k^2+3k+9)$

87. $25n^2-m^2-12m-36$
$=25n^2-(m^2+12m+36)$
$=25n^2-(m+6)^2=(5n+m+6)(5n-m-6)$

89. $(2p-5)^2-(4p+1)^2$
$=[(2p-5)+(4p+1)][(2p-5)-(4p+1)]$
$=(6p-4)(-2p-6)$
$=2(3p-2)(-2)(p+3)=-4(3p-2)(p+3)$

91. $x^4+6x^2y+9y^2-x^2-3y$
$=(x^2+3y)(x^2+3y)-(x^2+3y)$
$=(x^2+3y)\left[(x^2+3y)-1\right]$
$=(x^2+3y)(x^2+3y-1)$

93. $12x^{7/2}-4x^{5/2}=4x^{5/2}(3x^{7/2-5/2}-x^{5/2-5/2})$
$=4x^{5/2}(3x^1-x^0)=4x^{5/2}(3x-1)$

95. a. $\dfrac{w-2}{w^2-4}=\dfrac{w-2}{(w+2)(w-2)}$: If -2 or 2 were substituted for w, then $-2+2=0$ or $2-2=0$ and the denominator would be 0. So, $w\neq -2$, $w\neq 2$.

b. There are no values for w that make the denominator 0. So there are no restricted values for this expression.

97. $\dfrac{m^2-16}{m^2-m-12}=\dfrac{(m+4)\cancel{(m-4)}}{(m+3)\cancel{(m-4)}}=\dfrac{m+4}{m+3}$;
$m\neq 4$, $m\neq -3$

99. $\dfrac{4ac}{a^2+4ac}\cdot\dfrac{a^2+8ac+16c^2}{8a+32c}$
$=\dfrac{\cancel{4}\cancel{a}c}{\cancel{a}\cancel{(a+4c)}}\cdot\dfrac{\cancel{(a+4c)}\cancel{(a+4c)}}{\underset{2}{\cancel{8}}\cancel{(a+4c)}}=\dfrac{c}{2}$

101. $\dfrac{7}{20x}+\dfrac{2}{15x^4}=\dfrac{7}{2^2\cdot 5x}+\dfrac{2}{3\cdot 5x^4}$
$=\dfrac{7\cdot(3x^3)}{2^2\cdot 5x\cdot(3x^3)}+\dfrac{2\cdot(4)}{3\cdot 5x^4\cdot(4)}$
$=\dfrac{21x^3+8}{60x^4}$

103. $\dfrac{4}{x^2}+\dfrac{3}{x+3}-\dfrac{2}{x}$
$=\dfrac{4\cdot(x+3)}{x^2\cdot(x+3)}+\dfrac{3\cdot x^2}{(x+3)\cdot x^2}-\dfrac{2\cdot(x)\cdot(x+3)}{x\cdot(x)\cdot(x+3)}$
$=\dfrac{4x+12+3x^2-2x^2-6x}{x^2(x+3)}=\dfrac{x^2-2x+12}{x^2(x+3)}$

105. $\dfrac{\frac{1}{16x}+\frac{1}{8}}{\frac{1}{4}+\frac{1}{8x}}=\dfrac{16x\cdot\left(\frac{1}{16x}+\frac{1}{8}\right)}{16x\cdot\left(\frac{1}{4}+\frac{1}{8x}\right)}$
$=\dfrac{\frac{16x}{1}\cdot\frac{1}{16x}+\frac{16x}{1}\cdot\frac{1}{8}}{\frac{16x}{1}\cdot\frac{1}{4}+\frac{16x}{1}\cdot\frac{1}{8x}}$
$=\dfrac{1+2x}{4x+2}=\dfrac{\cancel{2x+1}}{2\cancel{(2x+1)}}=\dfrac{1}{2}$

107. $\dfrac{5}{\sqrt{k}}=\dfrac{5\cdot\sqrt{k}}{\sqrt{k}\cdot\sqrt{k}}=\dfrac{5\sqrt{k}}{\sqrt{k^2}}=\dfrac{5\sqrt{k}}{k}$

109. $\frac{6}{\sqrt[4]{8x^2y}} = \frac{6}{\sqrt[4]{2^3x^2y}}$

$= \frac{6 \cdot \sqrt[4]{2x^2y^3}}{\sqrt[4]{2^3x^2y} \cdot \sqrt[4]{2x^2y^3}} = \frac{6\sqrt[4]{2x^2y^3}}{\sqrt[4]{2^4x^4y^4}}$

$= \frac{6\sqrt[4]{2x^2y^3}}{2xy} = \frac{3\sqrt[4]{2x^2y^3}}{xy}$

111. $\frac{x-4}{\sqrt{x}+2} = \frac{(x-4)\cdot(\sqrt{x}-2)}{(\sqrt{x}+2)\cdot(\sqrt{x}-2)}$

$= \frac{(x-4)(\sqrt{x}-2)}{(\sqrt{x})^2-(2)^2}$

$= \frac{\cancel{(x-4)}(\sqrt{x}-2)}{\cancel{x-4}} = \sqrt{x}-2$

Chapter R Test

1. a. 8

b. 0, 8

c. 0, 8, − 3

d. $0, 8, -\frac{5}{7}, 2.1, -0.\overline{4}, -3$

e. $\frac{\pi}{6}$

f. $0, \frac{\pi}{6}, 8, -\frac{5}{7}, 2.1, -0.\overline{4}, -3$

3. a. $C = 12A + 40L + 30$

b. $C = 12A + 40L + 30$

$= 12(80) + 40(2) + 30$

$= 960 + 80 + 30$

$= \$1070$

5. a. $|\sqrt{2}-2|$ or $|2-\sqrt{2}|$

b. $|\sqrt{2}-2| = -(\sqrt{2}-2) = 2-\sqrt{2}$

7. $-\{6+4[9x-2(3x-5)]+3\}$

$= -\{6+4[9x-6x+10]+3\}$

$= -[6+4(3x+10)+3]$

$= -(6+12x+40+3)$

$= -(12x+49)$

$= -12x-49$

9. $\frac{7^2-4[8-(-2)]-4^2}{-2|-4+6|} = \frac{7^2-4(10)-4^2}{-2|2|}$

$= \frac{49-40-16}{-4}$

$= \frac{-7}{-4} = \frac{7}{4}$

11. $\left(\frac{1}{2}\right)^0 + \left(\frac{1}{3}\right)^{-3} + \left(\frac{1}{4}\right)^{-1} = 1+3^3+4$

$= 1+27+4$

$= 32$

13. $\sqrt[3]{80k^{15}m^2n^7} = \sqrt[3]{8^3k^{15}n^6 \cdot 10m^2n}$

$= 2k^5n^2\sqrt[3]{10m^2n}$

15. $\sqrt{125ab^3} - 3b\sqrt{20ab}$

$= \sqrt{5^3ab^3} - 3b\sqrt{2^2 \cdot 5ab}$

$= \sqrt{5^2b^2 \cdot 5ab} - 3b \cdot 2\sqrt{5ab}$

$= 5b\sqrt{5ab} - 6b\sqrt{5ab} = -b\sqrt{5ab}$

17. $(12n^2-4)\left(\frac{1}{2}n^2-3n+5\right)$

$= (12n^2)\left(\frac{1}{2}n^2\right) + (12n^2)(-3n) + (12n^2)(5)$

$+(-4)\left(\frac{1}{2}n^2\right) + (-4)(-3n) + (-4)(5)$

$= 6n^4 - 36n^3 + 60n^2 - 2n^2 + 12n - 20$

$= 6n^4 - 36n^3 + 58n^2 + 12n - 20$

19. $\left(\frac{1}{4}\sqrt{z}-p^2\right)\left(\frac{1}{4}\sqrt{z}+p^2\right)=\left(\frac{1}{4}\sqrt{z}\right)^2-\left(p^2\right)^2$
$=\frac{1}{16}z-p^4$

21. $\left(\sqrt{x}-6z\right)^2=\left(\sqrt{x}\right)^2-2\left(\sqrt{x}\right)(6z)+(6z)^2$
$=x-12z\sqrt{x}+36z^2$

23. $\frac{x^2}{x-5}+\frac{10x-25}{5-x}=\frac{x^2}{x-5}-\frac{10x-25}{x-5}$
$=\frac{x^2-10x+25}{x-5}$
$=\frac{(x-5)\cancel{(x-5)}}{\cancel{x-5}}=x-5$

25. $\frac{\frac{x}{4}-\frac{9}{4x}}{\frac{1}{4}-\frac{3}{4x}}=\frac{4x\cdot\left(\frac{x}{4}-\frac{9}{4x}\right)}{4x\cdot\left(\frac{1}{4}-\frac{3}{4x}\right)}$
$=\frac{4x\cdot\frac{x}{4}-4x\cdot\frac{9}{4x}}{4x\cdot\frac{1}{4}-4x\cdot\frac{3}{4x}}=\frac{x^2-9}{x-3}$
$=\frac{(x+3)\cancel{(x-3)}}{\cancel{x-3}}=x+3$

27. $\frac{6}{\sqrt{13}+\sqrt{10}}=\frac{6\cdot\left(\sqrt{13}-\sqrt{10}\right)}{\left(\sqrt{13}+\sqrt{10}\right)\cdot\left(\sqrt{13}-\sqrt{10}\right)}$
$=\frac{6\left(\sqrt{13}-\sqrt{10}\right)}{\left(\sqrt{13}\right)^2-\left(\sqrt{10}\right)^2}$
$=\frac{6\left(\sqrt{13}-\sqrt{10}\right)}{13-10}$
$=\frac{\overset{2}{\cancel{6}}\left(\sqrt{13}-\sqrt{10}\right)}{\cancel{3}}=2\left(\sqrt{13}-\sqrt{10}\right)$

29. $30x^3+2x^2-4x=2x\left(15x^2+x-2\right)$
$=2x(3x-1)(5x+2)$

31. $x^5+2x^4-81x-162$
$=x^5-81x+2x^4-162$
$=x\left(x^4-81\right)+2\left(x^4-81\right)$
$=\left(x^4-81\right)(x+2)=\left(x^2+9\right)\left(x^2-9\right)(x+2)$
$=\left(x^2+9\right)(x-3)(x+3)(x+2)$

33. $27u^3-v^6=(3u)^3-\left(v^2\right)^3$
$=\left(3u-v^2\right)\left(9u^2+3uv^2+v^4\right)$

35. $y(2y-1)^{-3/4}+(2y-1)^{1/4}=(2y-1)^{-3/4}\left[y(2y-1)^{-3/4-(-3/4)}+(2y-1)^{1/4-(-3/4)}\right]$
$=(2y-1)^{-3/4}\left[y(2y-1)^0+(2y-1)^1\right]$
$=(2y-1)^{-3/4}(y+2y-1)$
$=(2y-1)^{-3/4}(3y-1)$ or $\frac{3y-1}{(2y-1)^{3/4}}$

37. $A=(2x-5)(x+6)-(x-2)^2=2x^2+12x-5x-30-\left(x^2-4x+4\right)$
$=2x^2+7x-30-x^2+4x-4=x^2+11x-34$

39. 10^{-7} is between 0 and 1. Move the decimal point 7 places to the left. $8\times10^{-7}=0.0000008$

41. $(2.7,\infty)$

Chapter 1 Equations and Inequalities

Section 1.1 Linear Equations and Rational Equations

1. linear

3. solution

5. equivalent

7. division

9. identity

11. rational

13. a. Linear; $-2x = 8$

$$\frac{-2x}{-2} = \frac{8}{-2}$$
$$x = -4$$

$\{-4\}$

b. Nonlinear

c. Linear; $-\frac{1}{2}x = 8$

$$-2\left(-\frac{1}{2}x\right) = -2(8)$$
$$x = -16$$

$\{-16\}$

d. Nonlinear

e. Linear; $x - 2 = 8$

$$x - 2 + 2 = 8 + 2$$
$$x = 10$$

$\{10\}$

15. $-6x - 4 = 20$

$$-6x = 24$$
$$x = -4$$

$\{-4\}$

17. $4 = 7 - 3(4t + 1)$

$$4 = 7 - 12t - 3$$
$$4 = 4 - 12t$$
$$0 = -12t$$
$$0 = t$$

$\{0\}$

19. $-6(v - 2) + 3 = 9 - (v + 4)$

$$-6v + 12 + 3 = 9 - v - 4$$
$$-6v + 15 = 5 - v$$
$$-5v = -10$$
$$v = 2$$

$\{2\}$

21. $2.3 = 4.5x + 30.2$

$$-27.9 = 4.5x$$
$$-6.2 = x$$

$\{-6.2\}$

23. $0.05y + 0.02(6000 - y) = 270$

$$0.05y + 120 - 0.02y = 270$$
$$0.03y + 120 = 270$$
$$0.03y = 150$$
$$y = 5000$$

$\{5000\}$

25. $2(5x - 6) = 4[x - 3(x - 10)]$

$$10x - 12 = 4(x - 3x + 30)$$
$$10x - 12 = 4(-2x + 30)$$
$$10x - 12 = -8x + 120$$
$$18x = 132$$
$$x = \frac{132}{18} = \frac{22}{3}$$

$\left\{\frac{22}{3}\right\}$

27. $\frac{1}{4}x - \frac{3}{2} = 2$

$4\left(\frac{1}{4}x - \frac{3}{2}\right) = 4(2)$

$x - 6 = 8$

$x = 14$

$\{14\}$

29. $\frac{1}{2}w - \frac{3}{4} = \frac{2}{3}w + 2$

$12\left(\frac{1}{2}w - \frac{3}{4}\right) = 12\left(\frac{2}{3}w + 2\right)$

$6w - 9 = 8w + 24$

$-2w = 33$

$w = -\frac{33}{2}$

$\left\{-\frac{33}{2}\right\}$

31. $\frac{y-1}{5} + \frac{y}{4} = \frac{y+3}{2} + 1$

$20\left(\frac{y-1}{5} + \frac{y}{4}\right) = 20\left(\frac{y+3}{2} + 1\right)$

$4(y-1) + 5y = 10(y+3) + 20$

$4y - 4 + 5y = 10y + 30 + 20$

$9y - 4 = 10y + 50$

$-y = 54$

$y = -54$

$\{-54\}$

33. $\frac{n+3}{4} - \frac{n-2}{5} = \frac{n+1}{10} - 1$

$20\left(\frac{n+3}{4} - \frac{n-2}{5}\right) = 20\left(\frac{n+1}{10} - 1\right)$

$5(n+3) - 4(n-2) = 2(n+1) - 20$

$5n + 15 - 4n + 8 = 2n + 2 - 20$

$n + 23 = 2n - 18$

$-n = -41$

$n = 41$

$\{41\}$

35. a. $V_s = \frac{1}{9}V_t = \frac{1}{9}(50{,}040) = 5560 \text{ m}^3$

b. $V_s = \frac{1}{9}V_t$

$9000 = \frac{1}{9}V_t$

$81{,}000 = V_t$

$V_t = 81{,}000 \text{ m}^3$

37. a. $R = 4.25x + 93.9$

$R = 4.25(6) + 93.9 = \$119.4$ billion

b. $R = 4.25x + 93.9$

$110.9 = 4.25x + 93.9$

$17 = 4.25x$

$4 = x$

$2005 + 4 = 2009$

In 2009

39. a. $T = -1.83a + 212$

$T = -1.83(4) + 212 = 204.68°\text{F}$

b. $T = -1.83a + 212$

$193 = -1.83a + 212$

$-19 = -1.83a$

$10.4 \approx a$

$10.4 \times 10^3 = 10{,}400$

Approximately 10,400 ft

41. $E = 46.2x + 446.2$

$862 = 46.2x + 446.2$

$415.8 = 46.2x$

$9 = x$

$2000 + 9 = 2009$

In 2009

43. $2x - 3 = 4(x-1) - 1 - 2x$

$2x - 3 = 4x - 4 - 1 - 2x$

$2x - 3 = 2x - 5$

$-3 = -5$

Contradiction

45. $-(6-2w)=4(w+1)-2w-10$

$$-6+2w=4w+4-2w-10$$
$$-6+2w=2w-6$$
$$0=0$$

Identity; $\mathbb{R}$

47. $\frac{1}{2}x+3=\frac{1}{4}x+1$

$$4\left(\frac{1}{2}x+3\right)=4\left(\frac{1}{4}x+1\right)$$
$$2x+12=x+4$$
$$x=-8$$

Conditional equation; $\{-8\}$

49. $\frac{3}{x-5}+\frac{2}{x+4}=\frac{5}{7}$

$x\neq 5,\ x\neq -4$

51. $\frac{3}{2x^2+7x-15}-\frac{1}{6}=\frac{1}{x^2-25}$

$$\frac{3}{(2x-3)(x+5)}-\frac{1}{6}=\frac{1}{(x-5)(x+5)}$$
$$2x-3\neq 0$$
$$2x\neq 3$$
$$x\neq\frac{3}{2}, x\neq 5, x\neq -5$$

53. $\frac{1}{2}-\frac{7}{2y}=\frac{5}{y}$

$$2y\left(\frac{1}{2}-\frac{7}{2y}\right)=2y\left(\frac{5}{y}\right)$$
$$y-7=10$$
$$y=17$$

$\{17\}$

55. $\frac{w+3}{4w}+1=\frac{w-5}{w}$

$$4w\left(\frac{w+3}{4w}+1\right)=4w\left(\frac{w-5}{w}\right)$$
$$w+3+4w=4(w-5)$$
$$5w+3=4w-20$$
$$w=-23$$

$\{-23\}$

57. $\frac{c}{c-3}=\frac{3}{c-3}-\frac{3}{4}$

$$4(c-3)\left(\frac{c}{c-3}\right)=4(c-3)\left(\frac{3}{c-3}-\frac{3}{4}\right)$$
$$4c=12-3(c-3)$$
$$4c=12-3c+9$$
$$7c=21$$
$$c=3$$

$\{\ \}$; The value 3 does not check.

59. $\frac{1}{t-1}=\frac{3}{t^2-1}$

$$\frac{1}{t-1}=\frac{3}{(t+1)(t-1)}$$
$$(t+1)(t-1)\left(\frac{1}{t-1}\right)=(t+1)(t-1)\left[\frac{3}{(t+1)(t-1)}\right]$$
$$t+1=3$$
$$t=2$$

$\{2\}$

61.

$$\frac{2}{x-5}-\frac{1}{x+5}=\frac{11}{x^2-25}$$
$$\frac{2}{x-5}-\frac{1}{x+5}=\frac{11}{(x+5)(x-5)}$$
$$(x+5)(x-5)\left(\frac{2}{x-5}-\frac{1}{x+5}\right)=(x+5)(x-5)\left[\frac{11}{(x+5)(x-5)}\right]$$
$$2(x+5)-1(x-5)=11$$
$$2x+10-x+5=11$$
$$x+15=11$$
$$x=-4$$

$\{-4\}$

63.

$$\frac{5}{x^2-x-2}-\frac{2}{x^2-4}=\frac{4}{x^2+3x+2}$$
$$\frac{5}{(x-2)(x+1)}-\frac{2}{(x-2)(x+2)}=\frac{4}{(x+2)(x+1)}$$
$$(x+2)(x-2)(x+1)\left[\frac{5}{(x-2)(x+1)}-\frac{2}{(x-2)(x+2)}\right]=(x+2)(x-2)(x+1)\left[\frac{4}{(x+2)(x+1)}\right]$$
$$5(x+2)-2(x+1)=4(x-2)$$
$$5x+10-2x-2=4x-8$$
$$3x+8=4x-8$$
$$16=x$$

$\{16\}$

65.

$$\frac{5}{m-2}=\frac{3m}{m^2+2m-8}-\frac{2}{m+4}$$
$$\frac{5}{m-2}=\frac{3m}{(m+4)(m-2)}-\frac{2}{m+4}$$
$$(m+4)(m-2)\left(\frac{5}{m-2}\right)=(m+4)(m-2)\left[\frac{3m}{(m+4)(m-2)}-\frac{2}{m+4}\right]$$
$$5(m+4)=3m-2(m-2)$$
$$5m+20=3m-2m+4$$
$$5m+20=m+4$$
$$4m=-16$$
$$m=-4$$

$\{\ \}$; The value -4 does not check.

67.

$$\frac{5x}{3x^2-5x-2}-\frac{1}{3x+1}=\frac{3}{2-x}$$
$$\frac{5x}{(3x+1)(x-2)}-\frac{1}{3x+1}=\frac{-3}{x-2}$$
$$(3x+1)(x-2)\left[\frac{5x}{(3x+1)(x-2)}-\frac{1}{3x+1}\right]=(3x+1)(x-2)\left(\frac{-3}{x-2}\right)$$
$$5x-1(x-2)=-3(3x+1)$$
$$5x-x+2=-9x-3$$
$$4x+2=-9x-3$$
$$13x=-5$$
$$x=-\frac{5}{13}$$

$\left\{-\frac{5}{13}\right\}$

69. $A=lw$ for l

$$\frac{A}{w}=\frac{lw}{w}$$
$$\frac{A}{w}=l \text{ or } l=\frac{A}{w}$$

71. $P=a+b+c$ for c

$$P-a-b=c \text{ or } c=P-a-b$$

73. $\Delta s=s_2-s_1$ for s_1

$$\Delta s-s_2=-s_1$$
$$s_1=s_2-\Delta s$$

75. $7x+2y=8$ for y

$$2y=-7x+8$$
$$\frac{2y}{2}=\frac{-7x+8}{2}$$
$$y=\frac{-7x+8}{2} \text{ or } y=-\frac{7}{2}x+4$$

77. $5x-4y=2$ for y

$$-4y=-5x+2$$
$$\frac{-4y}{-4}=\frac{-5x+2}{-4}$$
$$y=\frac{5x-2}{4} \text{ or } y=\frac{5}{4}x-\frac{1}{2}$$

79. $\frac{1}{2}x+\frac{1}{3}y=1$ for y

$$6\left(\frac{1}{2}x+\frac{1}{3}y\right)=6(1)$$
$$3x+2y=6$$
$$2y=-3x+6$$
$$\frac{2y}{2}=\frac{-3x+6}{2}$$
$$y=\frac{-3x+6}{2} \text{ or } y=-\frac{3}{2}x+3$$

81. $S=\frac{n}{2}(a+d)$ for d

$$2(S)=2\left[\frac{n}{2}(a+d)\right]$$
$$2S=n(a+d)$$
$$2S=na+nd$$
$$2S-na=nd$$
$$\frac{2S-na}{n}=\frac{nd}{n}$$
$$\frac{2S-na}{n}=d$$
$$d=\frac{2S-na}{n} \text{ or } d=\frac{2S}{n}-a$$

83. $V=\frac{1}{3}\pi r^2h$ for h

$$3(V)=3\left(\frac{1}{3}\pi r^2h\right)$$
$$3V=\pi r^2h$$
$$\frac{3V}{\pi r^2}=\frac{\pi r^2h}{\pi r^2}$$
$$\frac{3V}{\pi r^2}=h \text{ or } h=\frac{3V}{\pi r^2}$$

85. $6=4x+tx$ for x

$$6=x(4+t)$$
$$\frac{6}{4+t}=\frac{x(4+t)}{4+t}$$
$$\frac{6}{4+t}=x \text{ or } x=\frac{6}{4+t}$$

87. $6x+ay=bx+5$ for x

$$6x-bx=5-ay$$
$$x(6-b)=5-ay$$
$$\frac{x(6-b)}{6-b}=\frac{5-ay}{6-b}$$
$$x=\frac{5-ay}{6-b} \text{ or } x=\frac{ay-5}{b-6}$$

89. $A=P+Prt$ for P

$$A=P(1+rt)$$
$$\frac{A}{1+rt}=\frac{P(1+rt)}{1+rt}$$
$$\frac{A}{1+rt}=P \text{ or } P=\frac{A}{1+rt}$$

91. $$\frac{5}{2n+1}=\frac{-2}{3n-4}$$

$$(2n+1)(3n-4)\left(\frac{5}{2n+1}\right)=(2n+1)(3n-4)\left(\frac{-2}{3n-4}\right)$$
$$5(3n-4)=-2(2n+1)$$
$$15n-20=-4n-2$$
$$19n=18$$
$$n=\frac{18}{19}$$

$\left\{\frac{18}{19}\right\}$

93. $5-2\{3-[5v+3(v-7)]\}=8v+6(3-4v)-61$

$$5-2[3-(5v+3v-21)]=8v+18-24v-61$$
$$5-2[3-(8v-21)]=-16v-43$$
$$5-2(3-8v+21)=-16v-43$$
$$5-2(24-8v)=-16v-43$$
$$5-48+16v=-16v-43$$
$$16v-43=-16v-43$$
$$32v=0$$
$$v=0$$

$\{0\}$

95. $(x-7)(x+2)=x^2+4x+13$

$$x^2+2x-7x-14=x^2+4x+13$$
$$x^2-5x-14=x^2+4x+13$$
$$-9x=27$$
$$x=-3$$

$\{-3\}$

97.

$$\frac{3}{c^2-4c}-\frac{9}{2c^2+3c}=\frac{2}{2c^2-5c-12}$$

$$\frac{3}{c(c-4)}-\frac{9}{c(2c+3)}=\frac{2}{(2c+3)(c-4)}$$

$$c(2c+3)(c-4)\left[\frac{3}{c(c-4)}-\frac{9}{c(2c+3)}\right]=c(2c+3)(c-4)\left[\frac{2}{(2c+3)(c-4)}\right]$$

$$3(2c+3)-9(c-4)=2c$$

$$6c+9-9c+36=2c$$

$$-3c+45=2c$$

$$-5c=-45$$

$$c=9$$

$\{9\}$

99.

$$\frac{1}{3}x+\frac{1}{2}=\frac{1}{2}(x+1)-\frac{1}{6}x$$

$$6\left(\frac{1}{3}x+\frac{1}{2}\right)=6\left[\frac{1}{2}(x+1)-\frac{1}{6}x\right]$$

$$2x+3=3(x+1)-x$$

$$2x+3=3x+3-x$$

$$2x+3=2x+3$$

$$0=0$$

$\mathbb{R}$

101.

$$(t+2)^2=(t-4)^2$$

$$t^2+4t+4=t^2-8t+16$$

$$12t=12$$

$$t=1$$

$\{1\}$

103.

$$\frac{3}{3a+4}=\frac{5}{5a-1}$$

$$(3a+4)(5a-1)\left(\frac{3}{3a+4}\right)=(3a+4)(5a-1)\left(\frac{5}{5a-1}\right)$$

$$3(5a-1)=5(3a+4)$$

$$15a-3=15a+20$$

$$-3=20$$

$\{\ \}$

105.

$$P=\frac{40+20x}{1+0.05x}$$

$$200=\frac{40+20x}{1+0.05x}$$

$$(1+0.05x)(200)=(1+0.05x)\left(\frac{40+20x}{1+0.05x}\right)$$

$$200+10x=40+20x$$

$$-10x=-160$$

$$x=16\text{ yr}$$

107.

$$A=\frac{1}{22}c+\frac{1}{30}h$$

$$7=\frac{1}{22}c+\frac{1}{30}(165)$$

$$7=\frac{1}{22}c+\frac{11}{2}$$

$$22(7)=22\left(\frac{1}{22}c+\frac{11}{2}\right)$$

$$154=c+121$$

$$c=33\text{ mi}$$

109. The value 5 is not defined within the expressions in the equation. Substituting 5 into the equation would result in division by 0.

111. The equation cannot be written in the form $ax+b=0$. The term $\frac{3}{x}=3x^{-1}$. Therefore, the term $\frac{3}{x}$ is not first degree and the equation is not a first-degree equation.

113. The equation is a contradiction. There is no real number x to which we add 1 that will equal the same real number x to which we add 2.

115.
$$\begin{aligned} ax+6&=4x+14\\ a(4)+6&=4(4)+14\\ 4a+6&=16+14\\ 4a+6&=30\\ 4a&=24\\ a&=6 \end{aligned}$$

117.
$$\begin{aligned} a(2x-5)+6&=5x+7\\ a[2(16)-5]+6&=5(16)+7\\ a(32-5)+6&=80+7\\ 27a+6&=87\\ 27a&=81\\ a&=3 \end{aligned}$$

119. Let $x=0.5\overline{16}$.
$$\begin{aligned} 1000x&=516.\overline{16}\\ -10x&=-5.\overline{16}\\ \hline 990x&=511.00\\ x&=\frac{511}{990} \end{aligned}$$

121. Let $x=0.\overline{534}$.
$$\begin{aligned} 1000x&=534.\overline{534}\\ -x&=-.\overline{534}\\ \hline 999x&=534.000\\ x&=\frac{534}{999}=\frac{178}{333} \end{aligned}$$

Section 1.2 Applications and Modeling with Linear Equations

1. $5x-2$

3. $0.06x$

5. $40-x$

7. $x+1$

9. $P=2l+2w$

11. $90-x$

13. $I=Prt=6000(0.075)(2)=\$900$

15. $\frac{d}{r}$

17. The length of the lot is $l=128+2x$.
The width of the lot is $w=60+2x$.
$$\begin{aligned} P&=2l+2w\\ 440&=2(128+2x)+2(60+2x)\\ 440&=256+4x+120+4x\\ 440&=8x+376\\ 64&=8x\\ 8&=x \end{aligned}$$
The width of the easement is 8 ft.

19. a. The width of the kitchen is w.
The length of the kitchen is $l=w+4$.
$$\begin{aligned} P&=2l+2w\\ 48&=2(w+4)+2w\\ 48&=2w+8+2w\\ 48&=4w+8\\ 40&=4w\\ 10&=w \end{aligned}$$
$l=w+4=10+4=14$
The kitchen is 14 ft by 10 ft.

b. $A=lw+0.1lw=1.1lw$
$$\begin{aligned} A&=1.1(14)(10)\\ &=154\text{ ft}^2 \end{aligned}$$

c. $C=1.06(12)(154)=\$1958.88$

21. $90-x$

23. $180-(131+x)$ or $49-x$

25. The angle between the ladder and the wall is x. The angle between the ladder and the ground is $(x+22)$.

$$x+(x+22)+90=180$$
$$2x+112=180$$
$$2x=68$$
$$x=34$$
$$x+22=34+22=56$$

The angle between the ladder and the ground is $56°$ and the angle that the ladder makes with the wall is $34°$.

27. Let x represent the amount borrowed at 3%. Then, $(5000-x)$ is the amount borrowed at 2.5%.

	3% Interest Loan	2.5% Interest Loan	Total
Principal	x	$5000-x$	
Interest ($I=Prt$)	$x(0.03)(1)$	$(5000-x)(0.025)(1)$	132.50

$$x(0.03)+(5000-x)(0.025)=132.50$$
$$0.03x+125-0.025x=132.50$$
$$0.005x+125=132.50$$
$$0.005x=7.50$$
$$x=1500$$
$$5000-x=5000-1500=3500$$

Rocco borrowed \$1500 at 3% and \$3500 at 2.5%.

29. Let x represent the amount invested in the 3-yr CD. Then, $(x-2000)$ is the amount invested in the 18-month CD.

	3-yr CD	18-month (1.5-yr) CD	Total
Principal	x	$x-2000$	
Interest ($I=Prt$)	$x(0.044)(3)$	$(x-2000)(0.03)(1.5)$	706.50

$$x(0.044)(3)+(x-2000)(0.03)(1.5)=706.50$$
$$0.132x+0.045x-90=706.50$$
$$0.177x-90=706.50$$
$$0.177x=796.50$$
$$x=4500$$
$$x-2000=4500-2000=2500$$

Fernando invested \$4500 in the 3-yr CD and \$2500 in the 18-month CD.

31. Let x represent the amount of the 5% solution (in gallons). 5000 gal is the amount of the 10% solution. Therefore, $(x+5000)$ is the amount of the resulting 9% solution.

	5% Solution	10% Solution	9% Solution
Amount of Solution	x	5000	$x+5000$
Pure Ethanol	$0.05x$	$0.1(5000)$	$0.09(x+5000)$

$$0.05x+0.1(5000)=0.09(x+5000)$$
$$0.05x+500=0.09x+450$$
$$50=0.04x$$
$$1250=x$$

1250 gal of E5 should be mixed with the E10.

33. Let x represent the amount of the pure sand (in cubic feet). 480 ft^2 is the amount of the concrete mix that is 70% sand. Therefore, $(x+480)$ is the amount of the resulting 75% sand mixture.

	100% Sand	70% Sand	75% Sand
Amount of Mixture	x	480	$x+480$
Pure Sand	x	$0.7(480)$	$0.75(x+480)$

$$x+0.7(480)=0.75(x+480)$$
$$x+336=0.75x+360$$
$$0.25x=24$$
$$x=96$$

96 ft^2 of sand should be mixed with the 70% sand mixture.

35. Let x represent the speed of the plane flying to Los Angeles. Then, $(x+60)$ is the speed of the plane flying to New York City.

	Distance	Rate	Time
Los Angeles Flight	$3.4x$	x	3.4
New York City Flight	$2.4(x+60)$	$x+60$	2.4

$$3.4x+2.4(x+60)=2464$$
$$3.4x+2.4x+144=2464$$
$$5.8x=2320$$
$$x=400$$
$$x+60=400+60=460$$

The plane to Los Angeles travels 400 mph and the plane to New York City travels 460 mph.

37. Let x represent the distance from Darren's home to his school.

	Distance	Rate	Time
To School	x	32	$\frac{x}{32}$
To Home	x	48	$\frac{x}{48}$

$$\frac{x}{32}+\frac{x}{48}=1.25$$
$$96\left(\frac{x}{32}+\frac{x}{48}\right)=96(1.25)$$
$$3x+2x=120$$
$$5x=120$$
$$x=24$$

The distance is 24 mi.

39. a. $S_1=45{,}000+2250x$

b. $S_2=48{,}000+2000x$

c.
$$S_1=S_2$$
$$45{,}000+2250x=48{,}000+2000x$$
$$250x=3000$$
$$x=12\text{ yr}$$

41. a. $C=7x$

b. $C=105$
$$7x=105$$
$$x=15$$

The motorist will save money beginning on the 16th working day.

43. Let t represent the time it takes for the runners to cover $\frac{1}{4}$ mile.

$$\frac{1\text{ lap}}{66\text{ sec}}+\frac{1\text{ lap}}{60\text{ sec}}=\frac{1\text{ lap}}{t\text{ sec}}$$
$$660t\left(\frac{1}{66}+\frac{1}{60}\right)=660t\left(\frac{1}{t}\right)$$
$$10t+11t=660$$
$$21t=660$$
$$t=\frac{220}{7}\approx 31.4\text{ sec}$$

45. Let t represent the time it takes the second pump to fill the pool by itself.

$$\frac{1\text{ job}}{10\text{ hr}}+\frac{1\text{ job}}{t\text{ hr}}=\frac{1\text{ job}}{6\text{ hr}}$$
$$30t\left(\frac{1}{10}+\frac{1}{t}\right)=30t\left(\frac{1}{6}\right)$$
$$3t+30=5t$$
$$30=2t$$
$$t=15\text{ hr}$$

47. Let x represent the amount of cement and y represent the amount of gravel.

$$\frac{1}{2.4}=\frac{x}{150}$$
$$2.4x=150$$
$$x=62.5$$
$$\frac{2.4}{3.6}=\frac{150}{y}$$
$$2.4y=540$$
$$y=225$$

62.5 lb of cement and 225 lb of gravel

49. Let x represent the patient's LDL cholesterol level. The HDL cholesterol level is 60 mg/dL, and the total cholesterol is $(x+60)$.

$$\frac{x+60}{60}=3.4$$
$$60\left(\frac{x+60}{60}\right)=60(3.4)$$
$$x+60=204$$
$$x=144$$
$$x+60=144+60=204$$

LDL is 144 mg/dL and the total cholesterol is 204 mg/dL.

51. Let x represent the number of deer in the population.

$$\frac{30}{x}=\frac{5}{80}$$
$$5x=2400$$
$$x=480\text{ deer}$$

53. Let x represent the distance from the epicenter to the station.

	Distance	Rate	Time
P Waves	x	5	$\frac{x}{5}$
S Waves	x	3	$\frac{x}{3}$

$$\frac{x}{3}-\frac{x}{5}=40$$
$$15\left(\frac{x}{3}-\frac{x}{5}\right)=15(40)$$
$$5x-3x=600$$
$$2x=600$$
$$x=300\text{ km}$$

55. Let x represent the price set by the merchant.

$$x-0.25x=180+0.40(180)$$
$$0.75x=180+72$$
$$0.75x=252$$
$$x=\$336$$

57. a. $C=110+60x$

b.
$$C=350$$
$$110+60x=350$$
$$60x=240$$
$$x=4\text{ hr}$$

59. Let x represent the height of the Washington Monument.

$$\frac{5}{4}=\frac{x}{444}$$
$$4x=2220$$
$$x=555\text{ ft}$$

61. Let x represent the height of the pole. Then, $\frac{1}{8}x$ is the length of the pole that is in the ground, and $\frac{2}{3}x$ is the length of the pole that is in the snow.

$$x=1.5+\frac{2}{3}x+\frac{1}{8}x$$
$$x-\frac{2}{3}x-\frac{1}{8}x=1.5$$
$$24\left(x-\frac{2}{3}x-\frac{1}{8}x\right)=24(1.5)$$
$$24x-16x-3x=36$$
$$5x=36$$
$$x=7.2$$
$$\frac{2}{3}x=\frac{2}{3}(7.2)=4.8$$

The pole is 7.2 ft long, and the snow is 4.8 ft deep.

63. Let x represent the amount of 20% fertilizer solution (in liters) to be drained (and therefore the amount of water to be added). 40 L is the amount of the resulting 15% fertilizer solution. Therefore, $(40-x)$ is the amount of 20% fertilizer solution that is not drained.

	0% Solution	20% Solution	15% Solution
Amount of Solution	x	$40-x$	40
Pure fertilizer	$0(x)$	$0.20(40-x)$	$0.15(40)$

$$0(x)+0.20(40-x)=0.15(40)$$
$$8-0.20x=6$$
$$-0.20x=-2$$
$$x=10$$

10 L should be drained and replaced by water.

65. Let x represent the measure of one angle. Then, $(5x+18)$ is the measure of the other angle.

$$x+(5x+18)=180$$
$$6x+18=180$$
$$6x=162$$
$$x=27$$
$$5x+18=5(27)+18=135+18=153$$

The angles measure $153°$ and $27°$.

67. Aliyah had $8000-0.28(8000)=8000-2240=5760$ to invest. Let x represent the amount she invested at 11%. Then, $(5760-x)$ is the amount she invested at 5%.

	11% Investment	5% Investment	Total
Principal	x	$5760-x$	
Interest ($I=Prt$)	$x(0.11)(1)$	$(5760-x)(0.05)(1)$	453.60

$$x(0.11)+(5760-x)(0.05)=453.60$$
$$0.11x+288-0.05x=453.60$$
$$0.06x+288=453.60$$
$$0.06x=165.60$$
$$x=2760$$
$$5760-x=5760-2760=3000$$

Aliyah invested $2760 in the stock returning 11% and $3000 in the stock returning 5%.

69. Let x represent the length of the shortest side. Then, $(x+1)$ is the length of the middle side and $(x+5)$ is the length of the longest side. The perimeter is $(4x)$.

$$P=x+(x+1)+(x+5)$$
$$4x=x+(x+1)+(x+5)$$
$$4x=3x+6$$
$$x=6$$
$$x+1=6+1=7$$
$$x+5=6+5=11$$

The lengths of the sides are 6 ft, 7 ft, and 11 ft.

71. $\frac{7}{8}=\frac{x}{12.8}$

$$8x=89.6$$
$$x=11.2$$

$\frac{8}{y}=\frac{12.8}{12}$

$$12.8y=96$$
$$y=7.5$$

$x=11.2$ ft and $y=7.5$ cm

73. No. If x represents the measure of the smallest angle, then the equation $x+(x+2)+(x+4)=180$ does not result in an odd integer value for x. Instead the measures of the angles would be even integers.

75. No. If x represents the number of each type of bill, then the solution to the equation $20x+10x+5x=100$ is not a whole number.

77. Let x represent the smaller number. Then, $(x+16)$ is the larger number.

$$\frac{x+16}{x}=3+\frac{2}{x}$$
$$x\left(\frac{x+16}{x}\right)=x\left(3+\frac{2}{x}\right)$$
$$x+16=3x+2$$
$$14=2x$$
$$7=x$$
$$x+16=7+16=23$$

The numbers are 7 and 23.

79. Let x represent the tens digit of the number. Then, $(14-x)$ is the ones digit.

$$10(14-x)+1(x)=10(x)+1(14-x)+18$$
$$140-10x+x=10x+14-x+18$$
$$140-9x=9x+32$$
$$108=18x$$
$$6=x$$

$14-x=14-6=8$

The original number is 68.

81.
$$m_1x_1+m_2x_2=0$$
$$(30)(-1.2)+(20)x_2=0$$
$$20x_2=36$$
$$x_2=1.8\text{ m}$$

83.
$$m_1x_1+m_2x_2=0$$
$$(10)(-3.2)+m_2(8)=0$$
$$8m_2=-32$$
$$m_2=4\text{ kg}$$

Section 1.3 Complex Numbers

1. $(3x+2)-(5x+1)=3x+2-5x-1=-2x+1$

3. $(3a-4)(5a+2)=15a^2+6a-20a-8$
$=15a^2-14a-8$

5. $\left(\frac{1}{6}m-\frac{2}{5}n\right)\left(\frac{1}{6}m+\frac{2}{5}n\right)=\left(\frac{1}{6}m\right)^2-\left(\frac{2}{5}n\right)^2$
$=\frac{1}{36}m^2-\frac{4}{25}n^2$

7. $(z+2)^2=z^2+2(2)(z)+2^2=z^2+4z+4$

9. -1

11. real; imaginary

13. pure

15. complex

17. $\sqrt{-121}=i\sqrt{121}=11i$

19. $\sqrt{-98}=i\sqrt{98}=7i\sqrt{2}$

21. $\sqrt{-19}=i\sqrt{19}$

23. $-\sqrt{-16}=-i\sqrt{16}=-4i$

25. $\sqrt{-4}\sqrt{-9}=i\sqrt{4}\cdot i\sqrt{9}$
$=2i\cdot 3i$
$=6i^2$
$=6(-1)$
$=-6$

27. $\sqrt{-10}\sqrt{-5}=i\sqrt{10}\cdot i\sqrt{5}=i^2\sqrt{50}$
$=(-1)\sqrt{5^2\cdot 2}=-5\sqrt{2}$

29. $\sqrt{-6}\sqrt{-14}=i\sqrt{6}\cdot i\sqrt{14}=i^2\sqrt{84}$
$=(-1)\sqrt{2^2\cdot 21}=-2\sqrt{21}$

31. $\frac{\sqrt{-98}}{\sqrt{-2}}=\frac{i\sqrt{98}}{i\sqrt{2}}=\sqrt{\frac{98}{2}}=\sqrt{49}=7$

33. $\frac{\sqrt{-63}}{\sqrt{7}}=\frac{i\sqrt{63}}{\sqrt{7}}=i\sqrt{\frac{63}{7}}=i\sqrt{9}=3i$

35. Real part: 3; Imaginary part: -7

37. Real part: 0; Imaginary part: 19

39. Real part: $-\frac{1}{4}$; Imaginary part: 0

41. a. True **b.** False

43. a. True **b.** True

45. $5=5+0i$

47. $4\sqrt{-4}=4\cdot 2i=8i=0+8i$

49. $2+\sqrt{-12}=2+2\sqrt{3}i$ or $2+2i\sqrt{3}$

51. $\frac{8+3i}{14}=\frac{8}{14}+\frac{3}{14}i=\frac{4}{7}+\frac{3}{14}i$

53. $\dfrac{-18+\sqrt{-48}}{4}=\dfrac{-18+4\sqrt{3}i}{4}$
$=-\dfrac{18}{4}+\dfrac{4\sqrt{3}i}{4}$
$=-\dfrac{9}{2}+\sqrt{3}i$ or $-\dfrac{9}{2}+i\sqrt{3}$

55. a. $i^{20}=1$

b. $i^{29}=i^{28}\cdot i^{1}=(1)\cdot i^{1}=i$

c. $i^{50}=i^{48}\cdot i^{2}=(1)\cdot i^{2}=-1$

d. $i^{-41}=i^{-44}\cdot i^{3}=(1)\cdot i^{3}=-i$

57. a. $i^{37}=i^{36}\cdot i^{1}=(1)\cdot i^{1}=i$

b. $i^{-37}=i^{-40}\cdot i^{3}=(1)\cdot i^{3}=-i$

c. $i^{82}=i^{80}\cdot i^{2}=(1)\cdot i^{2}=-1$

d. $i^{-82}=i^{-84}\cdot i^{2}=(1)\cdot i^{2}=-1$

59. $(2-7i)+(8-3i)=(2+8)+(-7-3)i$
$=10-10i$

61. $(15+21i)-(18-40i)$
$=(15-18)+[21-(-40)]i=-3+61i$

63. $\left(\dfrac{1}{2}+\dfrac{2}{3}i\right)-\left(\dfrac{5}{6}+\dfrac{1}{12}i\right)=\left(\dfrac{1}{2}-\dfrac{5}{6}\right)+\left(\dfrac{2}{3}-\dfrac{1}{12}\right)i$
$=\left(\dfrac{3}{6}-\dfrac{5}{6}\right)+\left(\dfrac{8}{12}-\dfrac{1}{12}\right)i$
$=-\dfrac{2}{6}+\dfrac{7}{12}i=-\dfrac{1}{3}+\dfrac{7}{12}i$

65. $(2.3+4i)-(8.1-2.7i)+(4.6-6.7i)$
$=(2.3-8.1+4.6)+(4+2.7-6.7)i$
$=-1.2+0i$

67. $-\dfrac{1}{8}(16+24i)=-2-3i$

69. $2i(5+i)=10i+2i^2=10i+2(-1)=-2+10i$

71. $\sqrt{-3}\left(\sqrt{11}-\sqrt{-7}\right)=i\sqrt{3}\left(\sqrt{11}-i\sqrt{7}\right)$
$=i\sqrt{33}-i^2\sqrt{21}$
$=i\sqrt{33}-(-1)\sqrt{21}$
$=\sqrt{21}+i\sqrt{33}$

73. $(3-6i)(10+i)$
$=3(10)+3(i)+(-6i)(10)+(-6i)(i)$
$=30+3i-60i-6i^2=30-57i-6(-1)$
$=36-57i$

75. $(3-7i)^2=(3)^2-2(3)(7i)+(7i)^2$
$=9-42i+49i^2=9-42i+49(-1)$
$=9-42i-49=-40-42i$

77. $\left(3-\sqrt{-5}\right)\left(4+\sqrt{-5}\right)$
$=\left(3-i\sqrt{5}\right)\left(4+i\sqrt{5}\right)$
$=3(4)+3\left(i\sqrt{5}\right)+\left(-i\sqrt{5}\right)(4)+\left(-i\sqrt{5}\right)\left(i\sqrt{5}\right)$
$=12+3i\sqrt{5}-4i\sqrt{5}-5i^2$
$=12-i\sqrt{5}-5(-1)=17-i\sqrt{5}$

79. $4(6+2i)-5i(3-7i)=24+8i-15i+35i^2$
$=24-7i+35(-1)$
$=-11-7i$

81. $(2-i)^2+(2+i)^2$
$=(2)^2-2(2)(i)+i^2+(2)^2+2(2)(i)+i^2$
$=4-4i+i^2+4+4i+i^2$
$=8+2i^2=8+2(-1)=6$

83. a. $3+6i$

b. $(3-6i)(3+6i)=(3)^2+(6)^2=9+36=45$

85. a. $0-8i$

b. $(0+8i)(0-8i)=(0)^2+(8)^2=0+64=64$

87. $(10-4i)(10+4i)=(10)^2+(4)^2$
$=100+16=116$

89. $(7i)(-7i)=7^2=49$

91. $\left(\sqrt{2}+\sqrt{3}i\right)\left(\sqrt{2}+\sqrt{3}i\right)=\left(\sqrt{2}\right)^2+\left(\sqrt{3}\right)^2$
$=2+3=5$

93. $\dfrac{6+2i}{3-i}=\dfrac{(6+2i)(3+i)}{(3-i)(3+i)}$
$=\dfrac{18+6i+6i+2i^2}{(3)^2+(1)^2}$
$=\dfrac{18+12i+2(-1)}{9+1}$
$=\dfrac{16+12i}{10}=\dfrac{16}{10}+\dfrac{12}{10}i=\dfrac{8}{5}+\dfrac{6}{5}i$

95. $\dfrac{8-5i}{13+2i}=\dfrac{(8-5i)(13-2i)}{(13+2i)(13-2i)}$
$=\dfrac{104-16i-65i+10i^2}{(13)^2+(2)^2}$
$=\dfrac{104-81i+10(-1)}{169+4}$
$=\dfrac{94-81i}{173}=\dfrac{94}{173}-\dfrac{81}{173}i$

97. $\left(6+\sqrt{5}i\right)^{-1}=\dfrac{1}{6+\sqrt{5}i}$
$=\dfrac{1\left(6-\sqrt{5}i\right)}{\left(6+\sqrt{5}i\right)\left(6-\sqrt{5}i\right)}$
$=\dfrac{6-\sqrt{5}i}{(6)^2+\left(\sqrt{5}\right)^2}$
$=\dfrac{6-\sqrt{5}i}{36+5}$
$=\dfrac{6-\sqrt{5}i}{41}=\dfrac{6}{41}-\dfrac{\sqrt{5}}{41}i$

99. $\dfrac{5}{13i}=\dfrac{5\cdot i}{13i\cdot i}=\dfrac{5i}{13i^2}=\dfrac{5i}{13(-1)}=\dfrac{5i}{-13}=-\dfrac{5}{13}i$
$=0-\dfrac{5}{13}i$

101. $\dfrac{-1}{\sqrt{-3}}=\dfrac{-1}{\sqrt{3}i}=\dfrac{-1\cdot\sqrt{3}i}{\sqrt{3}i\cdot\sqrt{3}i}=\dfrac{-\sqrt{3}i}{3i^2}=\dfrac{-\sqrt{3}i}{3(-1)}$
$=\dfrac{-\sqrt{3}i}{-3}=\dfrac{\sqrt{3}i}{3}=0+\dfrac{\sqrt{3}i}{3}$

103. $\sqrt{b^2-4ac}=\sqrt{(4)^2-4(2)(6)}$
$=\sqrt{16-48}$
$=\sqrt{-32}=i\sqrt{32}=4i\sqrt{2}$

105. $\sqrt{b^2-4ac}=\sqrt{(-6)^2-4(2)(5)}$
$=\sqrt{36-40}$
$=\sqrt{-4}=i\sqrt{4}=2i$

107. a.
$x^2+25=0$
$(5i)^2+25\stackrel{?}{=}0$
$25(-1)+25\stackrel{?}{=}0$
$-25+25\stackrel{?}{=}0$
$0\stackrel{?}{=}0$ ✓

b.
$x^2+25=0$
$(-5i)^2+25\stackrel{?}{=}0$
$25(-1)+25\stackrel{?}{=}0$
$-25+25\stackrel{?}{=}0$
$0\stackrel{?}{=}0$ ✓

109. a.
$x^2-4x+7=0$
$\left(2+i\sqrt{3}\right)^2-4\left(2+i\sqrt{3}\right)+7\stackrel{?}{=}0$
$4+4i\sqrt{3}+3i^2-8-4i\sqrt{3}+7\stackrel{?}{=}0$
$4+3(-1)-8+7\stackrel{?}{=}0$
$4-3-1\stackrel{?}{=}0$
$0\stackrel{?}{=}0$ ✓

b.
$x^2-4x+7=0$
$\left(2-i\sqrt{3}\right)^2-4\left(2-i\sqrt{3}\right)+7\stackrel{?}{=}0$
$4-4i\sqrt{3}+3i^2-8+4i\sqrt{3}+7\stackrel{?}{=}0$
$4+3(-1)-8+7\stackrel{?}{=}0$
$4-3-1\stackrel{?}{=}0$
$0\stackrel{?}{=}0$ ✓

111. $(a+bi)(c+di)=ac+adi+bci+bdi^2$
$=ac+(ad+bc)i+bd(-1)$
$=(ac-bd)+(ad+bc)i$

113. The second step does not follow because the multiplication property of radicals can be applied only if the individual radicals are real numbers. Because $\sqrt{-9}$ and $\sqrt{-4}$ are imaginary numbers, the correct logic for simplification would be
$\sqrt{-9}\cdot\sqrt{-4}=i\sqrt{9}\cdot i\sqrt{4}=i^2\sqrt{36}=-1\cdot 6=-6$

115. Any real number. For example: 5.

117. $z\cdot\overline{z}=(a+bi)(a-bi)=a^2+b^2$

119. a. $x^2-9=(x+3)(x-3)$

b. $x^2+9=(x+3i)(x-3i)$

121. a. $x^2-64=(x+8)(x-8)$

b. $x^2+64=(x+8i)(x-8i)$

123. a. $x^2-3=(x+\sqrt{3})(x-\sqrt{3})$

b. $x^2+3=(x+i\sqrt{3})(x-i\sqrt{3})$

125.
```
√(-16)
                4i
(4-5i)-(2+3i)
              2-8i
(12-15i)(-2+9i)
          111+138i
```

127.
```
(4-9i)²
           -65-72i
7/(2i)▸Frac
             -7/2i
(14+8i)/(3-i)▸Fr
ac
        17/5+19/5i
```

Section 1.4 Quadratic Equations

1. $5t^2+7t-6=(5t-3)(t+2)$

3. $x^2+14x+49=(x+7)(x+7)=(x+7)^2$

5. $\dfrac{8+\sqrt{-44}}{4}=\dfrac{8+2i\sqrt{11}}{4}=\dfrac{8}{4}+\dfrac{2i\sqrt{11}}{4}=2+\dfrac{\sqrt{11}}{2}i$

7. quadratic

9. a; b

11. $\pm\sqrt{k}$

13. $x=\dfrac{-b\pm\sqrt{b^2-4ac}}{2a}$

15.
$n^2+5n=24$
$n^2+5n-24=0$
$(n+8)(n-3)=0$
$n+8=0$ or $n-3=0$
$n=-8$ $\quad$ $n=3$
$\{-8, 3\}$

17.
$8t(t+3)=2t-5$
$8t^2+24t=2t-5$
$8t^2+22t+5=0$
$(2t+5)(4t+1)=0$
$2t+5=0$ or $4t+1=0$
$2t=-5$ $\quad$ $4t=-1$
$t=-\dfrac{5}{2}$ $\quad$ $t=-\dfrac{1}{4}$
$\left\{-\dfrac{5}{2}, -\dfrac{1}{4}\right\}$

19.
$40p^2-90=0$
$10(4p^2-9)=0$
$10(2p-3)(2p+3)=0$
$2p-3=0$ or $2p+3=0$
$2p=3$ $\quad$ $2p=-3$
$p=\dfrac{3}{2}$ $\quad$ $p=-\dfrac{3}{2}$
$\left\{\dfrac{3}{2}, -\dfrac{3}{2}\right\}$

21. $3x^2 = 12x$

$3x^2 - 12x = 0$

$3x(x-4) = 0$

$3x = 0$ or $x - 4 = 0$

$x = 0$ $\quad x = 4$

$\{0, 4\}$

23. $x^2 = 81$

$x = \pm\sqrt{81} = \pm 9$

$\{9, -9\}$

25. $5y^2 - 35 = 0$

$5y^2 = 35$

$y^2 = 7$

$y = \pm\sqrt{7}$

$\{\sqrt{7}, -\sqrt{7}\}$

27. $4u^2 + 64 = 0$

$4u^2 = -64$

$u^2 = -16$

$u = \pm\sqrt{-16} = \pm 4i$

$\{4i, -4i\}$

29. $(k+2)^2 = 28$

$k + 2 = \pm\sqrt{28}$

$k = -2 \pm \sqrt{28} = -2 \pm 2\sqrt{7}$

$\{-2 \pm 2\sqrt{7}\}$

31. $(w-5)^2 = 9$

$w - 5 = \pm\sqrt{9}$

$w = 5 \pm \sqrt{9}$

$w = 5 \pm 3$

$w = 5 + 3$ or $w = 5 - 3$

$w = 8$ $\quad w = 2$

$\{8, 2\}$

33. $\left(t - \frac{1}{2}\right)^2 = -\frac{17}{4}$

$t - \frac{1}{2} = \pm\sqrt{-\frac{17}{4}}$

$t = \frac{1}{2} \pm \sqrt{-\frac{17}{4}} = \frac{1}{2} \pm \frac{i\sqrt{17}}{2} = \frac{1}{2} \pm \frac{\sqrt{17}}{2}i$

$\left\{\frac{1}{2} \pm \frac{\sqrt{17}}{2}i\right\}$

35. $x^2 + 14x + n = x^2 + 14x + \left[\frac{1}{2}(14)\right]^2$

$= x^2 + 14x + (7)^2$

$= x^2 + 14x + 49 = (x+7)^2$

$n = 49; (x+7)^2$

37. $p^2 - 26p + n = p^2 - 26p + \left[\frac{1}{2}(-26)\right]^2$

$= p^2 - 26p + (-13)^2$

$= p^2 - 26p + 169 = (p-13)^2$

$n = 169; (p-13)^2$

39. $w^2 - 3w + n = w^2 - 3w + \left[\frac{1}{2}(-3)\right]^2$

$= w^2 - 3w + \left(-\frac{3}{2}\right)^2$

$= w^2 - 3w + \frac{9}{4} = \left(w - \frac{3}{2}\right)^2$

$n = \frac{9}{4}; \left(w - \frac{3}{2}\right)^2$

41. $m^2 + \frac{2}{9}m + n = m^2 + \frac{2}{9}m + \left[\frac{1}{2}\left(\frac{2}{9}\right)\right]^2$

$= m^2 + \frac{2}{9}m + \left(\frac{1}{9}\right)^2$

$= m^2 + \frac{2}{9}m + \frac{1}{81} = \left(m + \frac{1}{9}\right)^2$

$n = \frac{1}{81}; \left(m + \frac{1}{9}\right)^2$

43.

$$y^2+22y-4=0$$
$$y^2+22y=4$$
$$y^2+22y+\left[\frac{1}{2}(22)\right]^2=4+\left[\frac{1}{2}(22)\right]^2$$
$$y^2+22y+121=4+121$$
$$(y+11)^2=125$$
$$y+11=\pm\sqrt{125}$$
$$y=-11\pm5\sqrt{5}$$

$\{-11\pm5\sqrt{5}\}$

45.

$$t^2-8t=-24$$
$$t^2-8t+\left[\frac{1}{2}(-8)\right]^2=-24+\left[\frac{1}{2}(-8)\right]^2$$
$$t^2-8t+16=-24+16$$
$$(t-4)^2=-8$$
$$t-4=\pm\sqrt{-8}$$
$$t=4\pm2i\sqrt{2}$$

$\{4\pm2i\sqrt{2}\}$

47.

$$4z^2+24z=-160$$
$$\frac{4z^2}{4}+\frac{24z}{4}=\frac{-160}{4}$$
$$z^2+6z=-40$$
$$z^2+6z+\left[\frac{1}{2}(6)\right]^2=-40+\left[\frac{1}{2}(6)\right]^2$$
$$z^2+6z+9=-40+9$$
$$(z+3)^2=-31$$
$$z+3=\pm\sqrt{-31}$$
$$z=-3\pm i\sqrt{31}$$

$\{-3\pm i\sqrt{31}\}$

49.

$$2x(x-3)=4+x$$
$$2x^2-6x=4+x$$
$$2x^2-7x=4$$
$$\frac{2x^2}{2}-\frac{7x}{2}=\frac{4}{2}$$
$$x^2-\frac{7}{2}x=2$$
$$x^2-\frac{7}{2}x+\left[\frac{1}{2}\left(-\frac{7}{2}\right)\right]^2=2+\left[\frac{1}{2}\left(-\frac{7}{2}\right)\right]^2$$
$$x^2-\frac{7}{2}x+\frac{49}{16}=\frac{32}{16}+\frac{49}{16}$$
$$\left(x-\frac{7}{4}\right)^2=\frac{81}{16}$$
$$x-\frac{7}{4}=\pm\sqrt{\frac{81}{16}}$$
$$x=\frac{7}{4}\pm\frac{9}{4}$$
$$x=\frac{7}{4}+\frac{9}{4}\quad\text{or}\quad x=\frac{7}{4}-\frac{9}{4}$$
$$x=\frac{16}{4}=4\qquad x=\frac{-2}{4}=-\frac{1}{2}$$

$\left\{4,-\frac{1}{2}\right\}$

51.

$$-4y^2-12y+5=0$$
$$\frac{-4y^2}{-4}-\frac{12y}{-4}+\frac{5}{-4}=\frac{0}{-4}$$
$$y^2+3y=\frac{5}{4}$$
$$y^2+3y+\left[\frac{1}{2}(3)\right]^2=\frac{5}{4}+\left[\frac{1}{2}(3)\right]^2$$
$$y^2+3y+\frac{9}{4}=\frac{5}{4}+\frac{9}{4}$$
$$\left(y+\frac{3}{2}\right)^2=\frac{7}{2}$$
$$y+\frac{3}{2}=\pm\sqrt{\frac{7}{2}}$$
$$y+\frac{3}{2}=\pm\frac{\sqrt{7}\cdot\sqrt{2}}{\sqrt{2}\cdot\sqrt{2}}$$
$$y+\frac{3}{2}=\pm\frac{\sqrt{14}}{2}$$
$$y=-\frac{3}{2}\pm\frac{\sqrt{14}}{2}$$

$\left\{-\frac{3}{2}\pm\frac{\sqrt{14}}{2}\right\}$

53. False

55. True

57. $x^2 - 3x - 7 = 0$

$a = 1, b = -3, c = -7$

$$x = \frac{-b \pm \sqrt{b^2 - 4ac}}{2a}$$

$$= \frac{-(-3) \pm \sqrt{(-3)^2 - 4(1)(-7)}}{2(1)}$$

$$= \frac{3 \pm \sqrt{9 + 28}}{2} = \frac{3 \pm \sqrt{37}}{2}$$

$$\left\{\frac{3 \pm \sqrt{37}}{2}\right\}$$

59. $y^2 = -4y - 6$

$y^2 + 4y + 6 = 0$

$a = 1, b = 4, c = 6$

$$y = \frac{-b \pm \sqrt{b^2 - 4ac}}{2a}$$

$$= \frac{-(4) \pm \sqrt{(4)^2 - 4(1)(6)}}{2(1)}$$

$$= \frac{-4 \pm \sqrt{16 - 24}}{2}$$

$$= \frac{-4 \pm \sqrt{-8}}{2} = \frac{-4 \pm 2i\sqrt{2}}{2} = -2 \pm i\sqrt{2}$$

$$\left\{-2 \pm i\sqrt{2}\right\}$$

61. $t(t - 6) = -10$

$t^2 - 6t + 10 = 0$

$a = 1, b = -6, c = 10$

$$t = \frac{-b \pm \sqrt{b^2 - 4ac}}{2a}$$

$$= \frac{-(-6) \pm \sqrt{(-6)^2 - 4(1)(10)}}{2(1)}$$

$$= \frac{6 \pm \sqrt{36 - 40}}{2}$$

$$= \frac{6 \pm \sqrt{-4}}{2} = \frac{6 \pm 2i}{2} = 3 \pm i$$

$$\{3 \pm i\}$$

63. $-7c + 3 = -5c^2$

$5c^2 - 7c + 3 = 0$

$a = 5, b = -7, c = 3$

$$c = \frac{-b \pm \sqrt{b^2 - 4ac}}{2a}$$

$$= \frac{-(-7) \pm \sqrt{(-7)^2 - 4(5)(3)}}{2(5)}$$

$$= \frac{7 \pm \sqrt{49 - 60}}{10}$$

$$= \frac{7 \pm \sqrt{-11}}{10}$$

$$= \frac{7 \pm i\sqrt{11}}{10}$$

$$\left\{\frac{7 \pm i\sqrt{11}}{10}\right\}$$

65. $(6x + 5)(x - 3) = -2x(7x + 5) + x - 12$

$6x^2 - 18x + 5x - 15 = -14x^2 - 10x + x - 12$

$6x^2 - 13x - 15 = -14x^2 - 9x - 12$

$20x^2 - 4x - 3 = 0$

$a = 20, b = -4, c = -3$

$$x = \frac{-b \pm \sqrt{b^2 - 4ac}}{2a}$$

$$= \frac{-(-4) \pm \sqrt{(-4)^2 - 4(20)(-3)}}{2(20)}$$

$$= \frac{4 \pm \sqrt{16 + 240}}{40}$$

$$= \frac{4 \pm \sqrt{256}}{40}$$

$$= \frac{4 \pm 16}{40}$$

$x = \frac{4 + 16}{40}$ or $x = \frac{4 - 16}{40}$

$x = \frac{20}{40} = \frac{1}{2}$ $\quad$ $x = \frac{-12}{40} = -\frac{3}{10}$

$$\left\{\frac{1}{2}, -\frac{3}{10}\right\}$$

67. $9x^2 + 49 = 0$

$a = 9, b = 0, c = 49$

$$x = \frac{-b \pm \sqrt{b^2 - 4ac}}{2a}$$

$$= \frac{-(0) \pm \sqrt{(0)^2 - 4(9)(49)}}{2(9)} = \frac{\pm\sqrt{-1764}}{18}$$

$$= \frac{\pm 42i}{18} = \pm\frac{7}{3}i$$

$\left\{\pm\frac{7}{3}i\right\}$

69. $\frac{1}{2}x^2 - \frac{2}{7} = \frac{5}{14}x$

$$14\left(\frac{1}{2}x^2 - \frac{2}{7}\right) = 14\left(\frac{5}{14}x\right)$$

$$7x^2 - 4 = 5x$$

$$7x^2 - 5x - 4 = 0$$

$a = 7, b = -5, c = -4$

$$x = \frac{-b \pm \sqrt{b^2 - 4ac}}{2a}$$

$$= \frac{-(-5) \pm \sqrt{(-5)^2 - 4(7)(-4)}}{2(7)} = \frac{5 \pm \sqrt{137}}{14}$$

$\left\{\frac{5 \pm \sqrt{137}}{14}\right\}$

71. $0.4y^2 = 2y - 2.5$

$$10(0.4y^2) = 10(2y - 2.5)$$

$$4y^2 = 20y - 25$$

$$4y^2 - 20y + 25 = 0$$

$a = 4, b = -20, c = 25$

$$y = \frac{-b \pm \sqrt{b^2 - 4ac}}{2a}$$

$$= \frac{-(-20) \pm \sqrt{(-20)^2 - 4(4)(25)}}{2(4)}$$

$$= \frac{20 \pm \sqrt{0}}{8} = \frac{20}{8} = \frac{5}{2}$$

$\left\{\frac{5}{2}\right\}$

73. Linear

$$2y + 4 = 0$$

$$2y = -4$$

$$y = \frac{-4}{2} = -2$$

$\{-2\}$

75. Quadratic

$$2y^2 + 4y = 0$$

$$\frac{2y^2}{2} + \frac{4y}{2} = \frac{0}{2}$$

$$y^2 + 2y = 0$$

$$y(y + 2) = 0$$

$$y = 0 \quad \text{or} \quad y + 2 = 0$$

$$y = -2$$

$\{0, -2\}$

77. Linear

$$5x(x + 6) = 5x^2 + 27x + 3$$

$$5x^2 + 30x = 5x^2 + 27x + 3$$

$$3x = 3$$

$$x = 1$$

$\{1\}$

79. Neither

81. $(3x - 4)^2 = 0$

$$3x - 4 = 0$$

$$x = \frac{4}{3}$$

$\left\{\frac{4}{3}\right\}$

83. $m^2 + 4m = -2$

$$m^2 + 4m + \left[\frac{1}{2}(4)\right]^2 = -2 + \left[\frac{1}{2}(4)\right]^2$$

$$m^2 + 4m + 4 = -2 + 4$$

$$(m + 2)^2 = 2$$

$$m + 2 = \pm\sqrt{2}$$

$$m = -2 \pm \sqrt{2}$$

$\{-2 \pm \sqrt{2}\}$

85. $\frac{x^2-4x}{6}-\frac{5x}{3}=1$

$6\left(\frac{x^2-4x}{6}-\frac{5x}{3}\right)=6(1)$

$x^2-4x-10x=6$

$x^2-14x=6$

$x^2-14x+\left[\frac{1}{2}(-14)\right]^2=6+\left[\frac{1}{2}(-14)\right]^2$

$x^2-14x+49=6+49$

$(x-7)^2=55$

$x-7=\pm\sqrt{55}$

$x=7\pm\sqrt{55}$

$\{7\pm\sqrt{55}\}$

87. $2(x+4)+x^2=x(x+2)+8$

$2x+8+x^2=x^2+2x+8$

$0=0$

$\mathbb{R}$

89. $\frac{3}{5}x^2-\frac{1}{10}x=\frac{1}{2}$

$10\left(\frac{3}{5}x^2-\frac{1}{10}x\right)=10\left(\frac{1}{2}\right)$

$6x^2-x=5$

$6x^2-x-5=0$

$(x-1)(6x+5)=0$

$x-1=0$ or $6x+5=0$

$x=1$ $\quad 6x=-5$

$x=-\frac{5}{6}$

$\left\{1,-\frac{5}{6}\right\}$

91. $x^2-5x=5x(x-1)-4x^2+1$

$x^2-5x=5x^2-5x-4x^2+1$

$x^2-5x=x^2-5x+1$

$0=1$

$\{\ \}$

93. $(2y+7)(y+1)=2y^2-11$

$2y^2+2y+7y+7=2y^2-11$

$2y^2+9y+7=2y^2-11$

$9y=-18$

$y=-2$

$\{-2\}$

95. $7d^2+5=0$

$7d^2=-5$

$d^2=-\frac{5}{7}$

$d=\pm\sqrt{-\frac{5}{7}}=\pm i\sqrt{\frac{5}{7}}=\pm\frac{\sqrt{5}}{\sqrt{7}}i$

$=\pm\frac{\sqrt{5}\cdot\sqrt{7}}{\sqrt{7}\cdot\sqrt{7}}i=\pm\frac{\sqrt{35}}{7}i$

$\left\{\pm\frac{\sqrt{35}}{7}i\right\}$

97. $x^2-\sqrt{5}=0$

$x^2=\sqrt{5}$

$x=\pm\sqrt[4]{5}$

$\{\pm\sqrt[4]{5}\}$

99. a. $3x^2-4x+6=0$

$b^2-4ac=(-4)^2-4(3)(6)=16-72=-56$

b. $-56<0$; there are two imaginary solutions.

101. a. $-2w^2+8w=3$

$-2w^2+8w-3=0$

$b^2-4ac=(8)^2-4(-2)(-3)=64-24=40$

b. $40>0$; there are two real solutions.

103. a. $3x(x-4)=x-4$

$3x^2-12x=x-4$

$3x^2-13x+4=0$

$b^2-4ac=(-13)^2-4(3)(4)=169-48=121$

b. $121>0$; there are two real solutions.

105. a.

$$-1.4m+0.1=-4.9m^2$$
$$10(-1.4m+0.1)=10(-4.9m^2)$$
$$-14m+1=-49m^2$$
$$49m^2-14m+1=0$$
$$b^2-4ac=(-14)^2-4(49)(1)$$
$$=196-196$$
$$=0$$

b. The discriminant is 0; there is one real solution.

107. $A=\pi r^2$

$$\frac{A}{\pi}=\frac{\pi r^2}{\pi}$$
$$\frac{A}{\pi}=r^2$$
$$r=\sqrt{\frac{A}{\pi}} \text{ or } r=\frac{\sqrt{A\pi}}{\pi}$$

109.

$$s=\frac{1}{2}gt^2$$
$$2(s)=2\left(\frac{1}{2}gt^2\right)$$
$$2s=gt^2$$
$$\frac{2s}{g}=\frac{gt^2}{g}$$
$$\frac{2s}{g}=t^2$$
$$t=\sqrt{\frac{2s}{g}} \text{ or } t=\frac{\sqrt{2sg}}{g}$$

111. $a^2+b^2=c^2$

$$a^2=c^2-b^2$$
$$a=\sqrt{c^2-b^2}$$

113.

$$L=c^2I^2Rt$$
$$\frac{L}{c^2Rt}=\frac{c^2I^2Rt}{c^2Rt}$$
$$\frac{L}{c^2Rt}=I^2$$
$$I=\sqrt{\frac{L}{c^2Rt}}=\frac{1}{c}\sqrt{\frac{L}{Rt}} \text{ or } \frac{\sqrt{LRt}}{cRt}$$

115.

$$kw^2-cw=r$$
$$kw^2-cw-r=0$$
$$a=k, b=-c, c=-r$$
$$w=\frac{-b\pm\sqrt{b^2-4ac}}{2a}$$
$$w=\frac{-(-c)\pm\sqrt{(-c)^2-4(k)(-r)}}{2(k)}$$
$$w=\frac{c\pm\sqrt{c^2+4kr}}{2k}$$

117.

$$s=v_0t+\frac{1}{2}at^2$$
$$2(s)=2\left(v_0t+\frac{1}{2}at^2\right)$$
$$2s=2v_0t+at^2$$
$$at^2+2v_0t-2s=0$$
$$a=a, b=2v_0, c=-2s$$
$$t=\frac{-b\pm\sqrt{b^2-4ac}}{2a}$$
$$t=\frac{-(2v_0)\pm\sqrt{(2v_0)^2-4(a)(-2s)}}{2(a)}$$
$$t=\frac{-2v_0\pm\sqrt{4v_0^{\,2}+8as}}{2a}$$
$$t=\frac{-2v_0\pm2\sqrt{v_0^{\,2}+2as}}{2a}$$
$$t=\frac{-v_0\pm\sqrt{v_0^{\,2}+2as}}{a}$$

119. $LI^2 + RI + \frac{1}{C} = 0$

$C\left(LI^2 + RI + \frac{1}{C}\right) = C(0)$

$CLI^2 + CRI + 1 = 0$

$a = CL, b = CR, c = 1$

$I = \frac{-b \pm \sqrt{b^2 - 4ac}}{2a}$

$I = \frac{-(CR) \pm \sqrt{(CR)^2 - 4(CL)(1)}}{2(CL)}$

$I = \frac{-CR \pm \sqrt{CR^2 - 4CL}}{2CL}$

121. The right side of the equation is not equal to zero.

123. If the discriminant is negative, then the equation has two solutions that are imaginary numbers.

125. $x^2 - xy - 2y^2 = 0$

$(x - 2y)(x + y) = 0$

$x - 2y = 0$ or $x + y = 0$

$x = 2y$ $\quad x = -y$

127. $(x-4)(x+2) = 0$

$x^2 + 2x - 4x - 8 = 0$

$x^2 - 2x - 8 = 0$

129. $\left(x - \frac{2}{3}\right)\left(x - \frac{1}{4}\right) = 0$

$3\left(x - \frac{2}{3}\right) \cdot 4\left(x - \frac{1}{4}\right) = 12 \cdot 0$

$(3x-2)(4x-1) = 0$

$12x^2 - 3x - 8x + 2 = 0$

$12x^2 - 11x + 2 = 0$

131. $\left(x - \sqrt{5}\right)\left(x + \sqrt{5}\right) = 0$

$(x)^2 - \left(\sqrt{5}\right)^2 = 0$

$x^2 - 5 = 0$

133. $(x - 2i)(x + 2i) = 0$

$(x)^2 + (2)^2 = 0$

$x^2 + 4 = 0$

135. $[x - (1+2i)][x - (1-2i)] = 0$

$(x - 1 - 2i)(x - 1 + 2i) = 0$

$[(x-1) - 2i][(x-1) + 2i] = 0$

$(x-1)^2 + (2)^2 = 0$

$x^2 - 2x + 1 + 4 = 0$

$x^2 - 2x + 5 = 0$

137. $x_1 + x_2 = \frac{-b + \sqrt{b^2 - 4ac}}{2a} + \frac{-b - \sqrt{b^2 - 4ac}}{2a}$

$= \frac{-b + \sqrt{b^2 - 4ac} + (-b) - \sqrt{b^2 - 4ac}}{2a}$

$= \frac{-2b}{2a} = -\frac{b}{a}$

139. $x_1 + x_2 = 2 + (-5) = 3$, which is $-\frac{3}{1}$.

$x_1 x_2 = (2)(-5) = -10$, which is $\frac{-10}{1}$.

141. $x = \frac{-b \pm \sqrt{b^2 - 4ac}}{2a}$

$= \frac{-(-5) \pm \sqrt{(-5)^2 - 4(9)(-7)}}{2(9)}$

$= \frac{5 \pm \sqrt{25 + 252}}{18}$

$= \frac{5 \pm \sqrt{277}}{18}$

$= \frac{5 + \sqrt{277}}{18}$ or $\frac{5 - \sqrt{277}}{18}$

≈ 1.2024 or -0.6469

143.

```
(7+√(37))/6→X
            2.180460422
3X²
            14.26322295
7X-1
            14.26322295
```

Problem Recognition Exercises: Simplifying Expressions Versus Solving Equations

1. a. Expression

$$(2x-5)(3x+1)=6x^2+2x-15x-5$$
$$=6x^2-13x-5$$

b. Equation

$$(2x-5)(3x+1)=0$$
$$2x-5=0 \quad \text{or} \quad 3x+1=0$$
$$2x=5 \qquad 3x=-1$$
$$x=\frac{5}{2} \qquad x=-\frac{1}{3}$$
$$\left\{\frac{5}{2},-\frac{1}{3}\right\}$$

3. a. Equation

$$(2x-3)^2=8$$
$$2x-3=\pm\sqrt{8}$$
$$2x-3=\pm2\sqrt{2}$$
$$2x=3\pm\sqrt{2}$$
$$x=\frac{3\pm\sqrt{2}}{2}$$
$$\left\{\frac{3\pm\sqrt{2}}{2}\right\}$$

b. Expression

$$(2x-3)^2-8=(2x)^2-2(2x)(3)+(3)^2-8$$
$$=4x^2-12x+9-8$$
$$=4x^2-12x+1$$

5. a. Equation

$$x^2-11x+28=0$$
$$(x-7)(x-4)=0$$
$$x-7=0 \quad \text{or} \quad x-4=0$$
$$x=7 \qquad x=4$$
$$\{7,4\}$$

b. Equation

$$x^2-11x-28=0$$
$$a=1, b=-11, c=-28$$
$$x=\frac{-b\pm\sqrt{b^2-4ac}}{2a}$$
$$=\frac{-(-11)\pm\sqrt{(-11)^2-4(1)(-28)}}{2(1)}$$
$$=\frac{11\pm\sqrt{121+112}}{2}$$
$$=\frac{11\pm\sqrt{233}}{2}$$
$$\left\{\frac{11\pm\sqrt{233}}{2}\right\}$$

7. a. Equation

$$\frac{35}{x}+12+x=0$$
$$x\left(\frac{35}{x}+12+x\right)=x(0)$$
$$35+12x+x^2=0$$
$$x^2+12x+35=0$$
$$(x+7)(x+5)=0$$
$$x+7=0 \quad \text{or} \quad x+5=0$$
$$x=-7 \qquad x=-5$$
$$\{-7,-5\}$$

b. Expression

$$\frac{35}{x}+12+x=\frac{35}{x}+12\cdot\frac{x}{x}+x\cdot\frac{x}{x}$$
$$=\frac{35+12x+x^2}{x}$$

Section 1.5 Applications of Quadratic Equations

1. $A=\frac{1}{2}bh$

3. $V=lwh$

5. $A=lw$

$629=(2x+3)(x)$

7. $A=\pi r^2$

$88\pi=\pi(x)^2$

9. $A=\frac{1}{2}bh$

$50=\frac{1}{2}x(x-8)$

11. $V=lwh$

$640=x(8)\left(\frac{1}{5}x\right)$

13. $a^2+b^2=c^2$

$(x)^2+(x+2)^2=(2x-2)^2$

15. $x+2$

17. a. $x(x+2)=120$

b.
$$x(x+2)=120$$
$$x^2+2x=120$$
$$x^2+2x-120=0$$
$$(x-10)(x+12)=0$$
$$x-10=0 \quad \text{or} \quad x+12=0$$
$$x=10 \qquad x=-12$$
$$x+2=12 \qquad x+2=-10$$

The integers are 10 and 12 or -10 and -12.

19. a. $x^2+(x+1)^2=113$

b.
$$x^2+(x+1)^2=113$$
$$x^2+x^2+2x+1=113$$
$$2x^2+2x-112=0$$
$$\frac{2x^2}{2}+\frac{2x}{2}-\frac{112}{2}=\frac{0}{2}$$
$$x^2+x-56=0$$
$$(x-7)(x+8)=0$$
$$x-7=0 \quad \text{or} \quad x+8=0$$
$$x=7 \qquad x=-8$$
$$x+1=8 \qquad x+1=-7$$

The integers are 7 and 8 or -7 and -8.

21. Let x represent the width of the cargo space. The length is 12 ft and the height is $(x-1)$ ft.
$$V=lwh$$
$$504=(12)(x)(x-1)$$
$$504=12(x^2-x)$$
$$504=12x^2-12x$$
$$12x^2-12x-504=0$$
$$12(x^2-x-42)=0$$
$$(x-7)(x+6)=0$$
$$x-7=0 \quad \text{or} \quad x+6=0$$
$$x=7 \qquad \cancel{x=-6}$$
$$x-1=7-1=6$$

The dimensions of the cargo space are 6 ft by

23. Let r represent the radius (in yards) of the region watered.
$$A=\pi r^2$$
$$2000=\pi r^2$$
$$\frac{2000}{\pi}=r^2$$
$$r=\pm\sqrt{\frac{2000}{\pi}}\approx\pm25$$

The radius is approximately 25 yd.

25. Let x represent the height of the triangle (in feet) and $(x-3)$ represent the base of the triangle.

$$lw+2\left(\frac{1}{2}bh\right)=A$$

$$(20)(x)+2\left[\frac{1}{2}(x-3)(x)\right]=348$$

$$20x+x^2-3x=348$$

$$x^2+17x-348=0$$

$$(x+29)(x-12)=0$$

$$x+29=0 \quad \text{or} \quad x-12=0$$

$$\cancel{x=-29} \qquad x=12$$

$$x-3=12-3=9$$

The base is 9 ft and the height is 12 ft.

27. Let x represent the distance (in feet) from home plate to second base.

$$a^2+b^2=c^2$$

$$(90)^2+(90)^2=(x)^2$$

$$8100+8100=x^2$$

$$16,200=x^2$$

$$x=\pm\sqrt{16,200}=\pm 90\sqrt{2}\approx \pm 127.3$$

The distance is $90\sqrt{2}$ ft or approximately 127.3 ft.

29. a. Let x represent the length (in feet) of the middle leg.

$$a^2+b^2=c^2$$

$$(x)^2+(x-2)^2=(x+2)^2$$

$$x^2+x^2-4x+4=x^2+4x+4$$

$$2x^2-4x+4=x^2+4x+4$$

$$x^2-8x=0$$

$$x(x-8)=0$$

$$x=0 \quad \text{or} \quad x-8=0$$

$$x=8$$

$$x-2=8-2=6$$

$$x+2=8+2=10$$

The lengths of the sides of the lower triangle are 6 ft, 8 ft, and 10 ft.

b. $A=A_T+A_B$

$$=\frac{1}{2}b_Th_T+=\frac{1}{2}b_Bh_B=\frac{1}{2}(b_Th_T+b_Bh_B)$$

$$=\frac{1}{2}[(10)(4)+(8)(6)]=\frac{1}{2}(40+48)$$

$$=\frac{1}{2}(88)=44$$

The total area is 44 ft^2.

31. a. Let x represent the width (in inches) of the cell phone. Then $1.5x$ is the length of the phone.

$$a^2+b^2=c^2$$

$$(x)^2+(1.5x^2)=(3.5)^2$$

$$x^2+2.25x^2=12.25$$

$$3.25x^2=12.25$$

$$x^2=\frac{12.25}{3.25}$$

$$x=\pm\sqrt{\frac{12.25}{3.25}}\approx \pm 1.94$$

$$1.5x\approx 1.5(1.94)\approx 2.91$$

The length is approximately 2.91 in. and the width is approximately 1.94 in.

b. $2.91(326)\approx 949$

$1.94(326)\approx 632$

Using the rounded values from part (a), the screen is approximately 949 pixels by 632 pixels.

33. Let n represent the number of players.

$$N=\frac{1}{2}n(n-1)$$

$$28=\frac{1}{2}n(n-1)$$

$$56=n^2-n$$

$$n^2-n-56=0$$

$$(n-8)(n+7)=0$$

$$n-8=0 \quad \text{or} \quad n+7=0$$

$$n=8 \qquad \cancel{n=-7}$$

There were 8 players.

35. Let t represent the time(s) at which the population was 600,000.

$$P = -1718t^2 + 82{,}000t + 10{,}000$$
$$600{,}000 = -1718t^2 + 82{,}000t + 10{,}000$$
$$0 = 1718t^2 - 82{,}000t + 590{,}000$$
$$t = \frac{-b \pm \sqrt{b^2 - 4ac}}{2a} = \frac{-(-82{,}000) \pm \sqrt{(-82{,}000)^2 - 4(1718)(590{,}000)}}{2(1718)} = \frac{82{,}000 \pm \sqrt{2{,}669{,}520{,}000}}{3436} \approx 9 \text{ or } 39$$

There were 600,000 organisms approximately 9 hr and 39 hr after the culture was started.

37. a. $d = 0.05v^2 + 2.2v$

$$d = 0.05(50)^2 + 2.2(50)$$
$$= 0.05(2500) + 110$$
$$= 125 + 110 = 235 \text{ ft}$$

b. $d = 0.05v^2 + 2.2v$

$$330 = 0.05v^2 + 2.2v$$
$$d = 0.05v^2 + 2.2v - 330$$
$$v = \frac{-b \pm \sqrt{b^2 - 4ac}}{2a}$$
$$= \frac{-(2.2) \pm \sqrt{(2.2)^2 - 4(0.05)(-330)}}{2(0.05)}$$
$$= \frac{-2.2 \pm \sqrt{70.84}}{0.1} \approx 62 \text{ or } \cancel{-106}$$

The car can travel at 62 mph and stop in time.

39. a. $s = -\frac{1}{2}gt^2 + v_0t + s_0$

$$s = -\frac{1}{2}(32)t^2 + (16)t + 0s = -16t^2 + 16t$$

b. $4 = -16t^2 + 16t$

$$16t^2 - 16t + 4 = 0$$
$$4(4t^2 - 4t + 1) = 0$$
$$4(2t - 1)^2 = 0$$
$$2t - 1 = 0$$
$$2t = 1$$
$$t = \frac{1}{2}$$

It would take Michael Jordan 0.5 sec to reach his maximum height of 4 ft.

41. a. $s = -\frac{1}{2}gt^2 + v_0t + s_0$

$$s = -\frac{1}{2}(32)t^2 + (75)t + 4 = -16t^2 + 75t + 4$$

b. $80 = -16t^2 + 75t + 4$

$$0 = 16t^2 - 75t + 76$$
$$t = \frac{-b \pm \sqrt{b^2 - 4ac}}{2a}$$
$$t = \frac{-(-75) \pm \sqrt{(-75)^2 - 4(16)(76)}}{2(16)}$$
$$t = \frac{75 \pm \sqrt{761}}{32} \approx 1.5 \text{ or } 3.2$$

The ball will be at an 80-ft height 1.5 sec and 3.2 sec after being kicked.

43. a.

$$\frac{L}{W} = \frac{L + W}{L}$$
$$\frac{L}{1} = \frac{L + 1}{L}$$
$$L(L) = L\left(\frac{L + 1}{L}\right)$$
$$L^2 = L + 1$$
$$L^2 - L - 1 = 0$$
$$L = \frac{-b \pm \sqrt{b^2 - 4ac}}{2a}$$
$$L = \frac{-(-1) \pm \sqrt{(-1)^2 - 4(1)(-1)}}{2(1)}$$
$$L = \frac{1 \pm \sqrt{5}}{2} \approx \cancel{-0.62} \text{ or } 1.62$$

b. $\frac{1.62}{1} = \frac{l}{9}$

$$l = 9 \cdot 1.62 \approx 14.6 \text{ ft}$$

45. a. $4x + 6y = 160$

$$6y = 160 - 4x$$

$$y = \frac{160 - 4x}{6} \text{ or } y = \frac{80 - 2x}{3}$$

b. $A = lw$

$$A = x\left(\frac{80 - 2x}{3}\right)$$

c. $A = x\left(\frac{80 - 2x}{3}\right)$

$$250 = x\left(\frac{80 - 2x}{3}\right)$$

$$750 = 80x - 2x^2$$

$$375 = 40x - x^2$$

$$x^2 - 40x + 375 = 0$$

$$(x - 25)(x - 15) = 0$$

$$x - 25 = 0 \quad \text{or} \quad x - 15 = 0$$

$$x = 25 \qquad\qquad x = 15$$

$$y = \frac{80 - 2x}{3}$$

$$y = \frac{80 - 2(25)}{3} = \frac{30}{3} = 10$$

or

$$y = \frac{80 - 2(15)}{3} = \frac{50}{3}$$

Each pen can be 25 yd by 10 yd, or it can be 15 yd by $\frac{50}{3}$ yd.

Section 1.6 More Equations and Applications

1. $x^3 + 27 = (x)^3 + (3)^3$

$$= (x + 3)(x^2 - 3x + 9)$$

3. $\frac{2x + 5}{4x^2 - 25} = \frac{2x + 5}{(2x - 5)(2x + 5)}$

$$2x - 5 \neq 0 \quad \text{and} \quad 2x + 5 \neq 0$$

$$2x \neq 5 \qquad\qquad 2x \neq -5$$

$$x \neq \frac{5}{2} \qquad\qquad x \neq -\frac{5}{2}$$

5. $27^{2/3} = (3^3)^{2/3}$

$$= 3^{3(2/3)}$$

$$= 3^2$$

$$= 9$$

7. radical

9. quadratic; $m^{1/3}$

11. $75y^3 + 100y^2 - 3y - 4 = 0$

$$75y^3 - 3y + 100y^2 - 4 = 0$$

$$3y(25y^2 - 1) + 4(25y^2 - 1) = 0$$

$$(25y^2 - 1)(3y + 4) = 0$$

$$(5y + 1)(5y - 1)(3y + 4) = 0$$

$$5y + 1 = 0 \quad \text{or} \quad 5y - 1 = 0 \quad \text{or} \quad 3y + 4 = 0$$

$$y = -\frac{1}{5} \qquad y = \frac{1}{5} \qquad y = -\frac{4}{3}$$

$$\left\{\pm\frac{1}{5}, -\frac{4}{3}\right\}$$

13. $2x^4 - 32 = 0$

$$2(x^4 - 16) = 0$$

$$2(x^2 - 4)(x^2 + 4) = 0$$

$$2(x - 2)(x + 2)(x^2 + 4) = 0$$

$$x - 2 = 0 \quad \text{or} \quad x + 2 = 0 \quad \text{or} \quad x^2 + 4 = 0$$

$$x = 2 \qquad x = -2 \qquad x^2 = -4$$

$$x = \pm\sqrt{-4}$$

$$x = \pm 2i$$

$$\{\pm 2i, \pm 2\}$$

15.

$$2x^4 = -128x$$
$$2x^4 + 128x = 0$$
$$2x(x^3 + 64) = 0$$
$$2x(x+4)(x^2 - 4x + 16) = 0$$
$$2x = 0 \quad \text{or} \quad x + 4 = 0 \quad \text{or} \quad x^2 - 4x + 16 = 0$$
$$x = 0 \quad \text{or} \quad x = -4 \quad \text{or}$$
$$x = \frac{-(-4) \pm \sqrt{(-4)^2 - 4(1)(16)}}{2(1)}$$
$$= \frac{4 \pm \sqrt{-48}}{2} = \frac{4 \pm 4i\sqrt{3}}{2} = 2 \pm 2i\sqrt{3}$$
$$\{0, -4, 2 \pm 2i\sqrt{3}\}$$

17.

$$3n^2(n^2 + 3) = 20 - 2n^2$$
$$3n^4 + 9n^2 = 20 - 2n^2$$
$$3n^4 + 11n^2 - 20 = 0$$
$$(3n^2 - 4)(n^2 + 5) = 0$$
$$3n^2 - 4 = 0 \quad \text{or} \quad n^2 + 5 = 0$$
$$3n^2 = 4 \qquad n^2 = -5$$
$$n^2 = \frac{4}{3} \qquad n = \pm\sqrt{-5}$$
$$n = \pm\sqrt{\frac{4}{3}} = \pm\frac{2\sqrt{3}}{3} \qquad n = \pm i\sqrt{5}$$
$$\left\{\pm\frac{2\sqrt{3}}{3}, \pm i\sqrt{5}\right\}$$

19.

$$\frac{3x}{x+2} - \frac{5}{x-4} = \frac{2x^2 - 14x}{x^2 - 2x - 8}$$
$$\frac{3x}{x+2} - \frac{5}{x-4} = \frac{2x^2 - 14x}{(x+2)(x-4)}$$
$$(x+2)(x-4)\left(\frac{3x}{x+2} - \frac{5}{x-4}\right) = (x+2)(x-4)\left[\frac{2x^2 - 14x}{(x+2)(x-4)}\right]$$
$$3x(x-4) - 5(x+2) = 2x^2 - 14x$$
$$3x^2 - 12x - 5x - 10 = 2x^2 - 14x$$
$$x^2 - 3x - 10 = 0$$
$$(x+2)(x-5) = 0$$
$$\cancel{x = -2} \quad \text{or} \quad x = 5$$

$\{5\}$; The value -2 does not check.

21.

$$\frac{m}{2m+1} + 1 = \frac{2}{m-3}$$
$$(2m+1)(m-3)\left(\frac{m}{2m+1} + 1\right) = (2m+1)(m-3)\left(\frac{2}{m-3}\right)$$
$$m(m-3) + 1(2m+1)(m-3) = 2(2m+1)$$
$$m^2 - 3m + 2m^2 - 6m + m - 3 = 4m + 2$$
$$3m^2 - 12m - 5 = 0$$
$$m = \frac{-(-12) \pm \sqrt{(-12)^2 - 4(3)(-5)}}{2(3)} = \frac{12 \pm \sqrt{204}}{6} = \frac{12 \pm 2\sqrt{51}}{6} = \frac{6 \pm \sqrt{51}}{3}$$
$$\left\{\frac{6 \pm \sqrt{51}}{3}\right\}$$

23.
$$2-\frac{3}{y}=\frac{5}{y^2}$$
$$y^2\left(2-\frac{3}{y}\right)=y^2\left(\frac{5}{y^2}\right)$$
$$2y^2-3y=5$$
$$2y^2-3y-5=0$$
$$(2y-5)(y+1)=0$$
$$2y-5=0 \quad \text{or} \quad y+1=0$$
$$2y=5 \qquad y=-1$$
$$y=\frac{5}{2}$$
$$\left\{\frac{5}{2};-1\right\}$$

25.
$$\frac{18}{m^2-3m}+2=\frac{6}{m-3}$$
$$\frac{18}{m(m-3)}+2=\frac{6}{m-3}$$
$$m(m-3)\left[\frac{18}{m(m-3)}+2\right]=m(m-3)\left(\frac{6}{m-3}\right)$$
$$18+2m(m-3)=6m$$
$$18+2m^2-6m=6m$$
$$2m^2-12m+18=0$$
$$m^2-6m+9=0$$
$$(m-3)^2=0$$
$$m-3=0$$
$$\cancel{m=3}$$
{ }; The value 3 does not check.

27. Let x represent the speed of the boat in still water.

	Distance (mi)	Rate (mph)	Time (hr)
With current	72	$x+2$	$\frac{72}{x+2}$
Against current	72	$x-2$	$\frac{72}{x-2}$

$$\frac{72}{x-2}-\frac{72}{x+2}=9$$
$$(x-2)(x+2)\left(\frac{72}{x-2}-\frac{72}{x+2}\right)=(x-2)(x+2)(9)$$
$$72(x+2)-72(x-2)=9(x^2-4)$$
$$72x+144-72x+144=9x^2-36$$
$$288=9x^2-36$$
$$324=9x^2$$
$$36=x^2$$
$$x=\pm\sqrt{36}$$
$$=6 \text{ or } \cancel{-6}$$
Jesse travels 6 km/hr in still water.

29. Let x represent the speed at which Jean runs. Then $(x+8)$ is the speed at which she rides.

	Distance (mi)	Rate (mph)	Time (hr)
Running	6	x	$\frac{6}{x}$
Riding	24	$x+8$	$\frac{24}{x+8}$

$$\frac{6}{x}+\frac{24}{x+8}=2.25$$
$$\frac{24}{x}+\frac{96}{x+8}=9$$
$$x(x+8)\left(\frac{24}{x}+\frac{96}{x+8}\right)=x(x+8)(9)$$
$$24(x+8)+96x=9(x^2+8x)$$
$$24x+192+96x=9x^2+72x$$
$$120x+192=9x^2+72x$$
$$0=9x^2-48x-192$$
$$0=3x^2-16x-64$$
$$0=(3x+8)(x-8)$$
$$3x+8=0 \quad \text{or} \quad x-8=0$$
$$3x=-8 \qquad x=8$$
$$\cancel{x=-\frac{8}{3}}$$

$x+8=8+8=16$

Jean runs 8 mph and rides 16 mph.

31. $\sqrt{2x-4}=6$
$$\left(\sqrt{2x-4}\right)^2=(6)^2$$
$$2x-4=36$$
$$2x=40$$
$$x=20$$

$\{20\}$

33. $\sqrt{m+18}+2=m$
$$\sqrt{m+18}=m-2$$
$$\left(\sqrt{m+18}\right)^2=(m-2)^2$$
$$m+18=m^2-4m+4$$
$$0=m^2-5m-14$$
$$0=(m+2)(m-7)$$
$$m=-2 \quad \text{or} \quad m=7$$

Check: $m=-2$
$$\sqrt{m+18}+2=m$$
$$\sqrt{(-2)+18}+2\stackrel{?}{=}(-2)$$
$$\sqrt{16}+2\stackrel{?}{=}-2$$
$$4+2\stackrel{?}{=}-2$$
$$6\stackrel{?}{=}-2 \text{ false}$$

Check: $m=7$
$$\sqrt{m+18}+2=m$$
$$\sqrt{(7)+18}+2\stackrel{?}{=}(7)$$
$$\sqrt{25}+2\stackrel{?}{=}7$$
$$5+2\stackrel{?}{=}7$$
$$7\stackrel{?}{=}7 \checkmark \text{ true}$$

$\{7\}$; The value -2 does not check.

35. $-4\sqrt[3]{2x-5}+6=10$
$$-4\sqrt[3]{2x-5}=4$$
$$\sqrt[3]{2x-5}=-1$$
$$\left(\sqrt[3]{2x-5}\right)^3=(-1)^3$$
$$2x-5=-1$$
$$2x=4$$
$$x=2$$

$\{2\}$

37. $\sqrt[4]{5y-3}-\sqrt[4]{2y+1}=0$
$$\sqrt[4]{5y-3}=-\sqrt[4]{2y+1}$$
$$\left(\sqrt[4]{5y-3}\right)^4=\left(-\sqrt[4]{2y+1}\right)^4$$
$$5y-3=2y+1$$
$$3y=4$$
$$y=\frac{4}{3}$$

Check: $y=\frac{4}{3}$

$$\sqrt[4]{5y-3}-\sqrt[4]{2y+1}=0$$
$$\sqrt[4]{5\left(\frac{4}{3}\right)-3}-\sqrt[4]{2\left(\frac{4}{3}\right)+1}\stackrel{?}{=}0$$
$$\sqrt[4]{\frac{20}{3}-\frac{9}{3}}-\sqrt[4]{\frac{8}{3}+\frac{3}{3}}\stackrel{?}{=}0$$
$$\sqrt[4]{\frac{11}{3}}-\sqrt[4]{\frac{11}{3}}\stackrel{?}{=}0$$
$$0\stackrel{?}{=}0 \checkmark \text{ true}$$

$\left\{\frac{4}{3}\right\}$

39. $\sqrt{8-p}-\sqrt{p+5}=1$

$$\sqrt{8-p}=1+\sqrt{p+5}$$
$$\left(\sqrt{8-p}\right)^2=\left(1+\sqrt{p+5}\right)^2$$
$$8-p=1+2\sqrt{p+5}+p+5$$
$$-2\sqrt{p+5}=2p-2$$
$$\sqrt{p+5}=-p+1$$
$$\left(\sqrt{p+5}\right)^2=(-p+1)^2$$
$$p+5=p^2-2p+1$$
$$0=p^2-3p-4$$
$$0=(p+1)(p-4)$$
$$p=-1 \quad \text{or} \quad p=4$$

Check: $p=-1$

$$\sqrt{8-p}-\sqrt{p+5}=1$$
$$\sqrt{8-(-1)}-\sqrt{(-1)+5}\stackrel{?}{=}1$$
$$\sqrt{9}-\sqrt{4}\stackrel{?}{=}1$$
$$3-2\stackrel{?}{=}1 \checkmark \text{ true}$$

Check: $p=4$

$$\sqrt{8-p}-\sqrt{p+5}=1$$
$$\sqrt{8-(4)}-\sqrt{(4)+5}\stackrel{?}{=}1$$
$$\sqrt{4}-\sqrt{9}\stackrel{?}{=}1$$
$$2-3\stackrel{?}{=}1$$
$$-1\stackrel{?}{=}1 \text{ false}$$

$\{-1\}$; The value 4 does not check.

41.

$$3-\sqrt{y+3}=\sqrt{2-y}$$
$$\left(3-\sqrt{y+3}\right)^2=\left(\sqrt{2-y}\right)^2$$
$$9-6\sqrt{y+3}+y+3=2-y$$
$$-6\sqrt{y+3}=-10-2y$$
$$3\sqrt{y+3}=5+y$$
$$\left(3\sqrt{y+3}\right)^2=(5+y)^2$$
$$9y+27=25+10y+y^2$$
$$0=y^2+y-2$$
$$0=(y+2)(y-1)$$
$$y=-2 \quad \text{or} \quad y=1$$

Check: $y=-2$

$$3-\sqrt{y+3}=\sqrt{2-y}$$
$$3-\sqrt{(-2)+3}\stackrel{?}{=}\sqrt{2-(-2)}$$
$$3-\sqrt{1}\stackrel{?}{=}\sqrt{4}$$
$$3-1\stackrel{?}{=}2$$
$$2\stackrel{?}{=}2 \checkmark \text{ true}$$

Check: $y=1$

$$3-\sqrt{y+3}=\sqrt{2-y}$$
$$3-\sqrt{(1)+3}\stackrel{?}{=}\sqrt{2-(1)}$$
$$3-\sqrt{4}\stackrel{?}{=}\sqrt{1}$$
$$3-2\stackrel{?}{=}1$$
$$1\stackrel{?}{=}1 \checkmark \text{ true}$$

$\{-2,1\}$

43. a. $m^{3/4}=5$

$$\left(m^{3/4}\right)^{4/3}=(5)^{4/3}$$
$$m=5^{4/3}$$

$\left\{5^{4/3}\right\}$

b. $m^{2/3}=5$

$$\left(m^{2/3}\right)^{3/2}=\pm(5)^{3/2}$$
$$m=\pm 5^{3/2}$$

$\left\{\pm 5^{3/2}\right\}$

45.
$$3(t+2)^{5/6}=21$$
$$(t+2)^{5/6}=7$$
$$\left[(t+2)^{5/6}\right]^{6/5}=(7)^{6/5}$$
$$t+2=7^{6/5}$$
$$t=7^{6/5}-2$$
$$\{7^{6/5}-2\}$$

47.
$$2p^{4/5}=\frac{1}{8}$$
$$p^{4/5}=\frac{1}{16}$$
$$(p^{4/5})^{5/4}=\pm\left(\frac{1}{16}\right)^{5/4}$$
$$p=\pm\left(\sqrt[4]{\frac{1}{16}}\right)^5=\pm\left(\frac{1}{2}\right)^5=\pm\frac{1}{32}$$
$$\left\{\pm\frac{1}{32}\right\}$$

49. $(2v+7)^{1/3}-(v-3)^{1/3}=0$
$$(2v+7)^{1/3}=(v-3)^{1/3}$$
$$\left[(2v+7)^{1/3}\right]^3=\left[(v-3)^{1/3}\right]^3$$
$$2v+7=v-3$$
$$v=-10$$
$$\{-10\}$$

51. a. $P=48t^{1/5}$
$$P=48(2)^{1/5}\approx 55\%$$

b.
$$P=48t^{1/5}$$
$$75=48t^{1/5}$$
$$\frac{25}{16}=t^{1/5}$$
$$\left(\frac{25}{16}\right)^5=(t^{1/5})^5$$
$$t\approx 9.3\text{ hr}$$

53. a. $v=\sqrt{2gh}=\sqrt{19.6h}$
$$v=\sqrt{19.6(10)}=14\text{ m/sec}$$

b.
$$v=\sqrt{19.6h}$$
$$26.8=\sqrt{19.6h}$$
$$26.8=\sqrt{19.6}\cdot\sqrt{h}$$
$$\frac{26.8}{\sqrt{19.6}}=\sqrt{h}$$
$$\left(\frac{26.8}{\sqrt{19.6}}\right)^2=\left(\sqrt{h}\right)^2$$
$$h=\frac{26.8^2}{19.6}\approx 36.6\text{ m}$$

55. Let $u=x^2+2$.
$$(x^2+2)^2+(x^2+2)-42=0$$
$$u^2+u-42=0$$
$$(u+7)(u-6)=0$$
$$u=-7 \text{ or } u=6$$
$$x^2+2=-7 \text{ or } x^2+2=6$$
$$x^2=-9 \text{ or } x^2=4$$
$$x=\pm 3i \text{ or } x=\pm 2$$
$$\{\pm 3i, \pm 2\}$$

57. Let $u=2+\frac{3}{t}$.
$$\left(2+\frac{3}{t}\right)^2-\left(2+\frac{3}{t}\right)=12$$
$$\left(2+\frac{3}{t}\right)^2-\left(2+\frac{3}{t}\right)-12=0$$
$$u^2-u-2=0$$
$$(u+3)(u-4)=0$$
$$u=-3 \text{ or } u=4$$
$$2+\frac{3}{t}=-3 \text{ or } 2+\frac{3}{t}=4$$
$$\frac{3}{t}=-5 \text{ or } \frac{3}{t}=2$$
$$t=-\frac{3}{5} \text{ or } t=\frac{3}{2}$$
$$\left\{\frac{3}{2},-\frac{3}{5}\right\}$$

59. Let $u = c^{1/5}$.

$$5c^{2/5} - 11c^{1/5} + 2 = 0$$
$$5u^2 - 11u + 2 = 0$$
$$(5u-1)(u-2) = 0$$
$$u = \frac{1}{5} \quad \text{or} \quad u = 2$$
$$c^{1/5} = \frac{1}{5} \quad \text{or} \quad c^{1/5} = 2$$
$$\left(c^{1/5}\right)^5 = \left(\frac{1}{5}\right)^5 \quad \text{or} \quad \left(c^{1/5}\right)^5 = (2)^5$$
$$c = \frac{1}{3125} \quad \text{or} \quad c = 32$$
$$\left\{\frac{1}{3125}, 32\right\}$$

61. $x^2\left(x^2+5\right) = 7$

$$x^4 + 5x^2 - 7 = 0$$

Let $u = x^2$.

$$u^2 + 5u - 7 = 0$$
$$u = \frac{-(5) \pm \sqrt{(5)^2 - 4(1)(-7)}}{2(1)} = \frac{-5 \pm \sqrt{53}}{2}$$
$$x^2 = \frac{-5 \pm \sqrt{53}}{2}$$
$$x = \pm\sqrt{\frac{-5 \pm \sqrt{53}}{2}}$$
$$\left\{\pm\sqrt{\frac{-5 \pm \sqrt{53}}{2}}\right\}$$

63. Let $u = k^{-1}$.

$$30k^{-2} - 23k^{-1} + 2 = 0$$
$$30u^2 - 23u + 2 = 0$$
$$(10u-1)(3u-2) = 0$$
$$u = \frac{1}{10} \quad \text{or} \quad u = \frac{2}{3}$$
$$k^{-1} = \frac{1}{10} \quad \text{or} \quad k^{-1} = \frac{2}{3}$$
$$k = 10 \quad \text{or} \quad k = \frac{3}{2}$$
$$\left\{\frac{3}{2}, 10\right\}$$

65. a. $y + 4\sqrt{y} = 21$

$$4\sqrt{y} = 21 - y$$
$$\left(4\sqrt{y}\right)^2 = (21-y)^2$$
$$16y = 441 - 42y + y^2$$
$$0 = 441 - 58y + y^2$$
$$0 = (y-9)(y-49)$$
$$y = 9 \quad \text{or} \quad y = 49$$

<u>Check</u>: $y = 9$

$$y + 4\sqrt{y} = 21$$
$$(9) + 4\sqrt{(9)} \stackrel{?}{=} 21$$
$$9 + 12 \stackrel{?}{=} 21$$
$$21 \stackrel{?}{=} 21 \checkmark \text{ true}$$

<u>Check</u>: $y = 49$

$$y + 4\sqrt{y} = 21$$
$$(49) + 4\sqrt{(49)} \stackrel{?}{=} 21$$
$$49 + 28 \stackrel{?}{=} 21$$
$$77 \stackrel{?}{=} 21 \text{ false}$$

$\{9\}$

b. Let $u = \sqrt{y}$.

$$y + 4\sqrt{y} = 21$$
$$y + 4\sqrt{y} - 21 = 0$$
$$u^2 + 4u - 21 = 0$$
$$(u+7)(u-3) = 0$$
$$u = -7 \quad \text{or} \quad u = 3$$
$$\sqrt{y} = -7 \quad \text{or} \quad \sqrt{y} = 3$$
$$\left(\sqrt{y}\right)^2 = (-7)^2 \quad \text{or} \quad \left(\sqrt{y}\right)^2 = (3)^2$$
$$y = 49 \quad \text{or} \quad y = 9$$

$\{9\}$; See checks in part (a).

67.
$$\frac{1}{f}=\frac{1}{p}+\frac{1}{q}$$
$$fpq\left(\frac{1}{f}\right)=fpq\left(\frac{1}{p}+\frac{1}{q}\right)$$
$$pq=fq+fp$$
$$pq-fp=fq$$
$$p(q-f)=fq$$
$$p=\frac{fq}{q-f}$$

69.
$$E=kT^4$$
$$\frac{E}{k}=T^4$$
$$\sqrt[4]{\frac{E}{k}}=\sqrt[4]{T^4}$$
$$T=\sqrt[4]{\frac{E}{k}}$$

71.
$$a=\frac{kF}{m}$$
$$m(a)=m\left(\frac{kF}{m}\right)$$
$$ma=kF$$
$$m=\frac{kF}{a}$$

73.
$$16+\sqrt{x^2-y^2}=z$$
$$\sqrt{x^2-y^2}=z-16$$
$$\left(\sqrt{x^2-y^2}\right)^2=(z-16)^2$$
$$x^2-y^2=(z-16)^2$$
$$x^2=(z-16)^2+y^2$$
$$x=\pm\sqrt{(z-16)^2+y^2}$$

75.
$$\frac{P_1V_1}{T_1}=\frac{P_2V_2}{T_2}$$
$$T_1T_2\left(\frac{P_1V_1}{T_1}\right)=T_1T_2\left(\frac{P_2V_2}{T_2}\right)$$
$$T_2P_1V_1=T_1P_2V_2$$
$$\frac{P_1V_1T_2}{P_2V_2}=T_1$$

77.
$$T=2\pi\sqrt{\frac{L}{g}}$$
$$(T)^2=\left(2\pi\sqrt{\frac{L}{g}}\right)^2$$
$$T^2=\frac{4\pi^2L}{g}$$
$$g(T^2)=g\left(\frac{4\pi^2L}{g}\right)$$
$$gT^2=4\pi^2L$$
$$g=\frac{4\pi^2L}{T^2}$$

79. An equation is in quadratic form if, after a suitable substitution, the equation can be written in the form $au^2+bu+c=0$, where u is a variable expression.

81. When solving a radical equation, if both sides of the equation are raised to an even power, then the potential solutions must be checked. This is because some or all of the solutions may be extraneous solutions.

83. Let t represent the time it takes Joan to fill 100 orders by herself. Then $(t+1)$ is the time it takes Henry to fill 100 orders.
$$\frac{1\text{ job}}{t\text{ hr}}+\frac{1\text{ job}}{(t+1)\text{ hr}}=\frac{1\text{ job}}{3\text{ hr}}$$
$$\frac{1}{t}+\frac{1}{(t+1)}=\frac{1}{3}$$
$$3t(t+1)\left(\frac{1}{t}+\frac{1}{t+1}\right)=3t(t+1)\left(\frac{1}{3}\right)$$
$$3(t+1)+3t=t(t+1)$$
$$3t+3+3t=t^2+t$$
$$0=t^2-5t-3$$

$$t=\frac{-(-5)\pm\sqrt{(-5)^2-4(1)(-3)}}{2(1)}$$
$$=\frac{5\pm\sqrt{37}}{2}$$
≈ 5.5 or ~~0.5~~

$t+1=5.5+1=6.5$

It would take Joan approximately 5.5 hr working alone, and it would take Henry approximately 6.5 hr.

85. Let x represent the distance along the shoreline as shown in the figure.

	Distance	Rate	Time
Row	$\sqrt{400^2+x^2}$	2.5	$\frac{\sqrt{400^2+x^2}}{2.5}$
Walk	$800-x$	5	$\frac{800-x}{5}$

$$\frac{\sqrt{400^2+x^2}}{2.5}+\frac{800-x}{5}=300$$
$$5\left(\frac{\sqrt{400^2+x^2}}{2.5}+\frac{800-x}{5}\right)=5(300)$$
$$2\sqrt{400^2+x^2}+800-x=1500$$
$$2\sqrt{400^2+x^2}=x+700$$
$$(x+700)^2=\left(2\sqrt{400^2+x^2}\right)^2$$
$$x^2+1400x+490{,}000=4\left(400^2+x^2\right)$$
$$x^2+1400x+490{,}000=4x^2+640{,}000$$
$$3x^2-1400x+150{,}000=0$$
$$(3x-500)(x-300)=0$$
$$x=\frac{500}{3}\quad\text{or}\quad x=300$$

Pam can row to a point $166\frac{2}{3}$ ft down the beach or to a point 300 ft down the beach to be home in 5 min.

Section 1.7 Linear Inequalities and Compound Inequalities

1. $(-\infty,-5)$

3. $[4,\infty)$

5. $\left\{x \mid -\frac{5}{6}<x\le 4\right\}$

7. union, $A\cup B$

9. intersection; $A\cap B$

11. $a<x<b$

13. $-2x-5>17$

$-2x>22$

$x<-11$

$\{x \mid x<-11\};(-\infty,-11)$

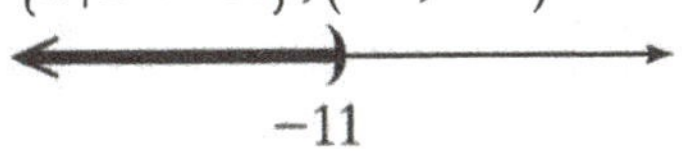

15. $-3\le -\frac{4}{3}w+1$

$-4\le -\frac{4}{3}w$

$3\ge w$ or $w\le 3$

$\{w \mid w\le 3\};(-\infty,3]$

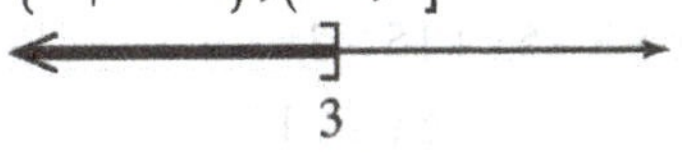

17. $-1.2+0.6a\le 0.4a+0.5$

$0.2a\le 1.7$

$a\le\frac{1.7}{0.2}$

$a\le 8.5$

$\{a \mid a\le 8.5\};(-\infty,8.5]$

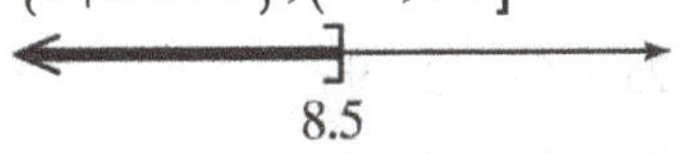

19. $-5 > 6(c-4)+7$
$-5 > 6c-24+7$
$-5 > 6c-17$
$12 > 6c$
$2 > c$ or $c < 2$
$\{c \mid c < 2\}; (-\infty, 2)$

2

21. $\dfrac{4+x}{2} - \dfrac{x-3}{5} < -\dfrac{x}{10}$
$10\left(\dfrac{4+x}{2} - \dfrac{x-3}{5}\right) < 10\left(-\dfrac{x}{10}\right)$
$5(4+x)-2(x-3) < -x$
$20+5x-2x+6 < -x$
$4x < -26$
$x < -\dfrac{26}{4}$
$x < -\dfrac{13}{2}$
$\left\{x \mid x < -\dfrac{13}{2}\right\}; \left(-\infty, -\dfrac{13}{2}\right)$

$-\frac{13}{2}$

23. $\dfrac{1}{3}(x+4) - \dfrac{5}{6}(x-3) \ge \dfrac{1}{2}x+1$
$6\left[\dfrac{1}{3}(x+4) - \dfrac{5}{6}(x-3)\right] \ge 6\left(\dfrac{1}{2}x+1\right)$
$2(x+4)-5(x-3) \ge 3x+6$
$2x+8-5x+15 \ge 3x+6$
$-6x \ge -17$
$x \le \dfrac{17}{6}$
$\left\{x \mid x \le \dfrac{17}{6}\right\}; \left(-\infty, \dfrac{17}{6}\right]$

$\frac{17}{6}$

25. $5(7-x)+2x < 6x-2-9x$
$35-5x+2x < 6x-2-9x$
$35 < -2$ $\{\ \}$

27. $5-3[2-4(x-2)] \ge 6\{2-[4-(x-3)]\}$
$5-3[2-4x+8] \ge 6\{2-[4-x+3]\}$
$5-3[-4x+10] \ge 6\{2-[-x+7]\}$
$5+12x-30 \ge 6\{2+x-7\}$
$12x-25 \ge 6\{x-5\}$
$12x-25 \ge 6x-30$
$6x \ge -5$
$x \ge -\dfrac{5}{6}$
$\left\{x \mid x \ge -\dfrac{5}{6}\right\}; \left[-\dfrac{5}{6}, \infty\right)$

$-\frac{5}{6}$

29. $4-3k > -2(k+3)-k$
$4-3k > -2k-6-k$
$4 > -6$
$\mathbb{R}; (-\infty, \infty)$

31. a. $A \cup B = \{0, 3, 4, 6, 8, 9, 12\}$

b. $A \cap B = \{0, 12\}$

c. $A \cup C = \{-2, 0, 4, 8, 12\}$

d. $A \cap C = \{4, 8\}$

e. $B \cup C = \{-2, 0, 3, 4, 6, 8, 9, 12\}$

f. $B \cap C = \{\ \}$

33. a. $C \cup D = \mathbb{R}$

b. $C \cap D = \{x \mid -1 \le x < 9\}$

c. $C \cup F = \{x \mid x < 9\}$

d. $C \cap F = \{x \mid x < -8\}$

e. $D \cup F = \{x \mid x < -8 \text{ or } x \ge -1\}$

f. $D \cap F = \{\ \}$

35. $A \cup B = \{-1, 0, 1, 2, 3, 4, 5\}$

37. $M \cap S = \{1\}$

39. a. $x < 4$ and $x \ge -2$

$[-2, 4)$

−2 4

b. $x < 4$ or $x \ge -2$

$(-\infty, \infty)$

41. a. $m + 1 \le 6$ or $\frac{1}{3}m < -2$

$m \le 5$ or $m < -6$

$(-\infty, 5]$

5

b. $m + 1 \le 6$ and $\frac{1}{3}m < -2$

$m \le 5$ and $m < -6$

$(-\infty, -6)$

−6

43. a. $-\frac{2}{3}y > -12$ and $2.08 \ge 0.65y$

$y > 18$ and $y \le 3.2$

$(-\infty, 3.2]$

3.2

b. $-\frac{2}{3}y > -12$ or $2.08 \ge 0.65y$

$y > 18$ or $y \le 3.2$

$(-\infty, 18)$

18

45. a. $3(x-2) + 2 \le x - 8$ or $4(x+1) + 2 > -2x + 4$

$3x - 6 + 2 \le x - 8$ or $4x + 4 + 2 > -2x + 4$

$2x \le -4$ or $6x > -2$

$x \le -2$ or $x > -\frac{1}{3}$

$(-\infty, -2] \cup \left(-\frac{1}{3}, \infty\right)$

−2 $-\frac{1}{3}$

b. $3(x-2) + 2 \le x - 8$ and $4(x+1) + 2 > -2x + 4$

$x \le -2$ and $x > -\frac{1}{3}$

$\{\ \}$

47. $-2.8 < y$ and $y \le 5$

49. $-3 < -2x + 1 \le 9$

$-4 < -2x \le 8$

$2 > x \ge -4$ or $-4 \le x < 2$

$[-4, 2)$

−4 2

51. $1 \le \frac{5x-4}{2} < 3$

$2 \le 5x - 4 < 6$

$6 \le 5x < 10$

$\frac{6}{5} \le x < 2$

$\left[\frac{6}{5}, 2\right)$

$\frac{6}{5}$ 2

53. $-2 \le \frac{-2x+1}{-3} \le 4$

$6 \ge -2x+1 \ge -12$

$5 \ge -2x \ge -13$

$-\frac{5}{2} \le x \le \frac{13}{2}$

$\left[-\frac{5}{2}, \frac{13}{2}\right]$

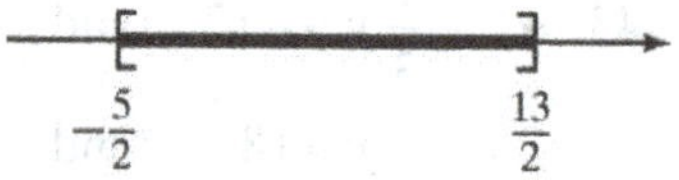

55. $12.0 \le x \le 15.2$ g/dL

57. $90 \le d \le 110$ ft

59. $\frac{88+92+100+80+90+2.5x}{7.5} \ge 92$

$\frac{450+2.5x}{7.5} \ge 92$

$450+2.5x \ge 690$

$2.5x \ge 240$

$x \ge 96$

Marilee needs to score at least 96 on the final exam.

61. Let t represent the time it takes for the car to be more than 16 miles ahead of the truck.

$50t > 40t + 16$

$10t > 16$

$t > 1.6$

It will take more than 1.6 hr or 1 hr 36 min.

63. Let l represent the length of the garden. Then the perimeter of the garden is $(2l + 200)$

$200 + 2l \le 800$

$2l \le 600$

$l \le 300$

The length must be 300 ft or less.

65. Let x represent the average scores that would produce a nonnegative handicap of 72 or less.

$0 \le 0.9(220 - x) \le 72$

$0 \le 220 - x \le 80$

$-220 \le -x \le -140$

$220 \ge x \ge 140$ or $140 \le x \le 220$

An average score in league play between 140 and 220, inclusive, would produce a handicap of 72 or less.

67. $A = P + Prt$

$5000 \le 4000 + 4000(0.05)t$

$1000 \le 200t$

$5 \le t$

At least 5 yr is required.

69. $C = \frac{5}{9}(F - 32)$

$35 \ge \frac{5}{9}(F - 32)$

$63 \ge F - 32$

$95 \ge F$ or $F \le 95$

Hypothermia would set in for a core body temperature below 95°F.

71. **a.** Let s represent the amount of sales.

$25{,}000 + 0.1s > 30{,}000 + 0.08s$

$0.02s > 5000$

$s > 250{,}000$

b. Job A

73. $T = -1.83a + 212$

$-1.83a + 212 < 200$

$-1.83a < -12$

$a > 6.6$

Water will boil at temperatures less than 200°F at altitudes of 6600 ft or more.

75. **a.** $x - 2 \ge 0$

$x \ge 2$

$\{x \mid x \ge 2\}$

b. $2 - x \ge 0$

$2 \ge x$ or $x \le 2$

$\{x \mid x \le 2\}$

77. **a.** $x + 4 \ge 0$

$x \ge -4$

$\{x \mid x \ge -4\}$

b. $\mathbb{R}$

79. **a.** $2x-9\ge 0$

$2x\ge 9$

$x\ge \frac{9}{2}$

$\left\{x \middle| x\ge \frac{9}{2}\right\}$

b. $2x-9\ge 0$

$2x\ge 9$

$x\ge \frac{9}{2}$

$\left\{x \middle| x\ge \frac{9}{2}\right\}$

81. $cd>a$ False

83. If $a>c$, then $ad<cd$. True

85. $(-\infty,2)\cap(-3,4]\cap[1,3]=[1,2)$

87. $[(-\infty,-2)\cup(4,\infty)]\cap[-5,3)=[-5,-2)$

89. $2x+1<-3$ or $2x+3>3$

$2x<-4$ or $2x>0$

$x<-2$ or $x>0$

$(-\infty,-2)\cup(0,\infty)$

91. $-2\le 6d-4\le 2$

$2\le 6d\le 6$

$\frac{2}{6}\le d\le 1$

$\frac{1}{3}\le d\le 1$

$\left[\frac{1}{3},1\right]$

93. $-11<6y+7$ and $6y+7<-5$

$-18<6y$ and $6y<-12$

$-3<y$ and $y<-2$

$(-3,-2)$

95. $6<-\frac{1}{2}p+4$

$2<-\frac{1}{2}p$

$-4>p$ or $p<-4$

$(-\infty,-4)$

97. $-1<\frac{4-x}{-2}\le 3$

$-1<\frac{x-4}{2}\le 3$

$-2<x-4\le 6$

$2<x\le 10$

$(2,10]$

99. The steps are the same with the following exception. If both sides of an inequality are multiplied or divided by a negative real number, then the direction of the inequality sign must be reversed.

101. The statement $-3>w>-1$ is equivalent to $w<-3$ and $w>-1$. No real number is less than -3 and simultaneously greater than -1.

Section 1.8 Absolute Value Equations and Inequalities

1. $|x-4|$ or $|4-x|$

3. $-4<2x-10<4$

$6<2x<14$

$3<x<7$

$\{x\,|\,3<x<7\}$; $(3,7)$

5. $5m\le -20$ or $5m\ge 20$

$m\le -4$ or $m\ge 4$

$\{m\,|\,m\le -4 \text{ or } m\ge 4\}$; $(-\infty,-4]\cup[4,\infty)$

7. absolute; $\{k,-k\}$

9. $-k$; k

11. $\mathbb{R}$

13. a. $|p|=6$

$p=6$ or $p=-6$

$\{6,-6\}$

b. $|p|=0$

$p=0$ or $p=-0$

$\{0\}$

c. $|p|=-6$

$\{\ \}$

15. a. $|x-3|=4$

$x-3=4$ or $x-3=-4$

$x=7$ or $x=-1$

$\{7,-1\}$

b. $|x-3|=0$

$x-3=0$ or $x-3=-0$

$x=3$ or $x=3$

$\{3\}$

c. $|x-3|=-7$

$\{\ \}$

17. $2|3x-4|+7=9$

$2|3x-4|=2$

$|3x-4|=1$

$3x-4=1$ or $3x-4=-1$

$3x=5$ or $3x=3$

$x=\frac{5}{3}$ or $x=1$

$\left\{\frac{5}{3},1\right\}$

19. $-3=-|c-7|+1$

$-4=-|c-7|$

$4=|c-7|$

$c-7=4$ or $c-7=-4$

$c=11$ or $c=3$

$\{11,3\}$

21. $2=8+|11y+4|$

$-6=|11y+4|$

$\{\ \}$

23. $\left|4-\frac{1}{2}w\right|-\frac{1}{3}=\frac{1}{2}$

$6\left(\left|4-\frac{1}{2}w\right|-\frac{1}{3}\right)=6\left(\frac{1}{2}\right)$

$|24-3w|-2=3$

$|24-3w|=5$

$24-3w=5$ or $24-3w=-5$

$-3w=-19$ or $-3w=-29$

$w=\frac{19}{3}$ or $w=\frac{29}{3}$

$\left\{\frac{19}{3},\frac{29}{3}\right\}$

25. $|3y+5|=|y+1|$

$3y+5=y+1 \quad \text{or} \quad 3y+5=-(y+1)$

$2y=-4 \quad \text{or} \quad 4y=-6$

$y=-2 \quad \text{or} \quad y=-\frac{3}{2}$

$\left\{-2,-\frac{3}{2}\right\}$

27. $\left|\frac{1}{4}w\right|=|4w|$

$\frac{1}{4}w=4w \quad \text{or} \quad \frac{1}{4}w=-4w$

$w=0 \quad \text{or} \quad w=0$

$\{0\}$

29. $|x+4|=|x-7|$

$x+4=x-7 \quad \text{or} \quad x+4=-(x-7)$

$4=-7 \quad \text{or} \quad 2x=3$

$x=\frac{3}{2}$

$\left\{\frac{3}{2}\right\}$

31. $|2p-1|=|1-2p|$

$2p-1=1-2p \quad \text{or} \quad 2p-1=-(1-2p)$

$4p=2 \quad \text{or} \quad 0=0$

$p=\frac{1}{2}$

$\mathbb{R}$

33. a. $|x|=7$

$x=7 \text{ or } x=-7$

$\{7,-7\}$

b. $|x|<7$

$-7<x<7$

$(-7,7)$

c. $|x|>7$

$x<-7 \text{ or } x>7$

$(-\infty,-7)\cup(7,\infty)$

35. a. $|a+9|+2=6$

$|a+9|=4$

$a+9=4 \quad \text{or} \quad a+9=-4$

$a=-5 \quad \text{or} \quad a=-13$

$\{-13,-5\}$

b. $|a+9|+2\le 6$

$|a+9|\le 4$

$-4\le a+9\le 4$

$-13\le a\le -5$

$[-13,-5]$

c. $|a+9|+2\ge 6$

$|a+9|\ge 4$

$a+9\le -4 \quad \text{or} \quad a+9\ge 4$

$a\le -13 \quad \text{or} \quad a\ge -5$

$(-\infty,-13]\cup[13,\infty)$

37. $3|4-x|-2<16$

$3|4-x|<18$

$|4-x|<6$

$-6<4-x<6$

$-10<-x<2$

$10>x>-2 \text{ or } -2<x<10$

$(-2,10)$

39. $2|x+3|-4\ge 6$

$2|x+3|\ge 10$

$|x+3|\ge 5$

$x+3\le -5 \quad \text{or} \quad x+3\ge 5$

$x\le -8 \quad \text{or} \quad x\ge 2$

$(-\infty,-8]\cup[2,\infty)$

41. $-11 \le 5 - |2p+4|$
$-16 \le -|2p+4|$
$16 \ge |2p+4|$
$|2p+4| \le 16$
$-16 \le 2p+4 \le 16$
$-20 \le 2p \le 12$
$-10 \le p \le 6$
$[-10, 6]$

43. $10 < |-5c-4| + 2$
$8 < |-5c-4|$
$|-5c-4| > 8$
$-5c-4 < -8$ or $-5c-4 > 8$
$-5c < -4$ or $-5c > 12$
$c > \frac{4}{5}$ or $c < -\frac{12}{5}$
$\left(-\infty, -\frac{12}{5}\right) \cup \left(\frac{4}{5}, \infty\right)$

45. $\left|\frac{y+3}{6}\right| < 2$
$-2 < \frac{y+3}{6} < 2$
$-12 < y+3 < 12$
$-15 < y < 9$
$(-15, 9)$

47. $\left|\frac{1}{2}p - 6\right| \ge 0.01$
$\frac{1}{2}p - 6 \le -0.01$ or $\frac{1}{2}p - 6 \ge 0.01$
$\frac{1}{2}p \le 5.99$ or $\frac{1}{2}p \ge 6.01$
$p \le 11.98$ or $p \ge 12.02$
$(-\infty, 11.98] \cup [12.02, \infty)$

49. a. $|x| = -9$
$\{\ \}$

b. $|x| < -9$
$\{\ \}$

c. $|x| > -9$
$\mathbb{R}$; $(-\infty, \infty)$

51. a. $18 = 4 - |y-7|$
$14 = -|y-7|$
$-14 = |y-7|$
$\{\ \}$

b. $18 \le 4 - |y-7|$
$14 \le -|y-7|$
$-14 \ge |y-7|$
$\{\ \}$

c. $18 \ge 4 - |y-7|$
$14 \ge -|y-7|$
$-14 \le |y-7|$
$\mathbb{R}$; $(-\infty, \infty)$

53. a. $|z| = 0$
$z = 0$ or $z = -0$
$\{0\}$

b. $|z| < 0$
$\{\ \}$

c. $|z| \le 0$
$z = 0$
$\{0\}$

d. $|z| > 0$
$\{z \mid z < 0 \text{ or } z > 0\}$; $(-\infty, 0) \cup (0, \infty)$

e. $|z| \ge 0$
$\mathbb{R}$; $(-\infty, \infty)$

55. a. $|k+4| = 0$
$k+4 = 0$ or $k+4 = -0$
$k = -4$ or $k = -4$
$\{-4\}$

b. $|k+4|<0$

$\{\ \}$

c. $|k+4|\le 0$

$k+4=0$

$k=-4$

$\{-4\}$

d. $|k+4|>0$

$k+4<-0 \quad \text{or} \quad k+4>0$

$k<-4 \quad \text{or} \quad k>-4$

$\{k \mid k<-4 \text{ or } k>-4\}$;

$(-\infty,-4)\cup(-4,\infty)$

e. $|k+4|\ge 0$

$\mathbb{R}$; $(-\infty,\infty)$

57. a. $|x-4|=6$ or $|4-x|=6$

b. $|x-4|=6$

$x-4=-6 \quad \text{or} \quad x-4=6$

$x=-2 \quad \text{or} \quad x=10$

$\{-2,10\}$

59. a. $|v-16|<0.01$ or $|16-v|<0.01$

b. $|v-16|<0.01$

$-0.01<v-16<0.01$

$15.99<v<16.01$

$(15.99,16.01)$

61. a. $|x-4|>1$ or $|4-x|>1$

b. $|x-4|>1$

$x-4<-1 \quad \text{or} \quad x-4>1$

$x<3 \quad \text{or} \quad x>5$

$(-\infty,3)\cup(5,\infty)$

63. $0<|x-c|<\delta$ or $0<|c-x|<\delta$

65. a. $|t-36.5|\le 1.5$ or $|36.5-t|\le 1.5$

b. $|t-36.5|\le 1.5$

$-1.5\le t-36.5\le 1.5$

$35\le t\le 38$

$[35,38]$; If the refrigerator is set to $36.5°\text{F}$, the actual temperature would be between $35°\text{F}$ and $38°\text{F}$, inclusive.

67. a. $|x-0.51|\le 0.03$ or $|0.51-x|\le 0.03$

b. $|x-0.51|\le 0.03$

$-0.03\le x-0.51\le 0.03$

$0.48\le x\le 0.54$

$[0.48,0.54]$; The candidate is expected to receive between 48% of the vote and 54% of the vote, inclusive.

69. a. $\left|\dfrac{x-500}{\sqrt{250}}\right|<1.96$

$-1.96<\dfrac{x-500}{5\sqrt{10}}<1.96$

$-1.96(5\sqrt{10})<x-500<1.96(5\sqrt{10})$

$-30<x-500<30$

$470<x<530$

$[470,530]$; In a group of 1000 jurors selected at random, it would be reasonable to have between 470 and 530 women, inclusive.

b. Yes, because 560 is above the "reasonable" range.

71. $|3x-1|>7$

73. $|2z|\le 4$

75. $-3\le x\le 7$

$-3-2\le x-2\le 7-2$

$-5\le x\le 5$

$|x-2|\le 5$

77.
$$\begin{aligned} x &< 4 &\quad \text{or} \quad x &> 10 \\ x-7 &< 4-7 &\quad \text{or} \quad x-7 &> 10-7 \\ x-7 &< -3 &\quad \text{or} \quad x-7 &> 3 \end{aligned}$$
$$|x-7| > 3$$

79. The absolute value of any nonzero real number is greater than or equal to zero. Therefore, no real number x has an absolute value of -5.

81. The inequality $|x-3| \le 0$ will be true only for values of x for which $x-3=0$ (the absolute value will never be less than 0). The solution set is {3}. The inequality $|x-3| > 0$ is true for all values of x excluding 3. The solution set is $\{x \mid x < 3 \text{ or } x > 3\}$.

83. $|x| + x < 11$
$$\begin{aligned} x + x &< 11 &\quad \text{or} \quad -x + x &< 11 \\ 2x &< 11 &\quad \text{or} \quad 0 &< 11 \\ x &< \frac{11}{2} && \end{aligned}$$
$$\left(-\infty, \frac{11}{2}\right)$$

85. $1 < |x| < 9$
$$\begin{aligned} 1 < x < 9 \quad \text{or} \quad 1 &< -x < 9 \\ -1 &> x > -9 \\ -9 &< x < -1 \end{aligned}$$
$$(-9, -1) \cup (1, 9)$$

87. $5 \le |2x+1| \le 7$
$$\begin{aligned} 5 \le 2x+1 \le 7 &\quad \text{or} \quad 5 \le -2x-1 \le 7 \\ 4 \le 2x \le 6 &\quad \text{or} \quad 6 \le -2x \le 8 \\ 2 \le x \le 3 &\quad \text{or} \quad -3 \ge x \ge -4 \\ &\quad \text{or} \quad -4 \le x \le -3 \end{aligned}$$
$$(-4, -3) \cup (2, 3)$$

89.
$$|p - \hat{p}| < z\sqrt{\frac{\hat{p}\hat{q}}{n}}$$
$$-z\sqrt{\frac{\hat{p}\hat{q}}{n}} < p - \hat{p} < z\sqrt{\frac{\hat{p}\hat{q}}{n}}$$
$$\hat{p} - z\sqrt{\frac{\hat{p}\hat{q}}{n}} < p < \hat{p} + z\sqrt{\frac{\hat{p}\hat{q}}{n}}$$

Problem Recognition Exercises: Recognizing and Solving Equations and Inequalities

1. a. Equation in quadratic form and a polynomial equation

b. Let $u = x^2 - 5$
$$\begin{aligned} (x^2-5)^2 - 5(x^2-5) + 4 &= 0 \\ u^2 - 5u + 4 &= 0 \\ (u-4)(u-1) &= 0 \end{aligned}$$
$$\begin{aligned} u = 4 &\quad \text{or} \quad u = 1 \\ x^2 - 5 = 4 &\quad \text{or} \quad x^2 - 5 = 1 \\ x^2 = 9 &\quad \text{or} \quad x^2 = 6 \\ x = \pm 3 &\quad \text{or} \quad x = \pm\sqrt{6} \end{aligned}$$
$$\{\pm 3, \pm\sqrt{6}\}$$

3. a. Radical equation

b. $\sqrt[3]{2y-5} - 4 = -1$
$$\begin{aligned} \sqrt[3]{2y-5} &= 3 \\ \left(\sqrt[3]{2y-5}\right)^3 &= (3)^3 \\ 2y - 5 &= 27 \\ 2y &= 32 \\ y &= 16 \end{aligned}$$
$$\{16\}$$

5. a. Rational equation

b.

$$\frac{2}{w-3}+\frac{5}{w+1}=1$$

$$(w-3)(w+1)\left(\frac{2}{w-3}+\frac{5}{w+1}\right)=(w-3)(w+1)(1)$$

$$2(w+1)+5(w-3)=(w-3)(w+1)$$

$$2w+2+5w-15=w^2-2w-3$$

$$7w-13=w^2-2w-3$$

$$0=w^2-9w+10$$

$$w=\frac{-(-9)\pm\sqrt{(-9)^2-4(1)(10)}}{2(1)}=\frac{9\pm\sqrt{41}}{2}$$

$$\left\{\frac{9\pm\sqrt{41}}{2}\right\}$$

7. a. Compound inequality

b. $-2(m+2)<-m+5$ and $6\geq m+3$

$-2m-4<-m+5$ and $3\geq m$

$-m<9$ and $m\leq 3$

$m>-9$

$(-9,3]$

9. a. Quadratic equation

b. $(2p+1)(p+5)=2p+40$

$2p^2+11p+5=2p+40$

$2p^2+9p-35=0$

$(2p-5)(p+7)=0$

$p=\frac{5}{2}$ or $p=-7$

$\left\{\frac{5}{2},-7\right\}$

11. a. Linear inequality

b.

$$\frac{a-4}{2}-\frac{3a+1}{4}\leq-\frac{a}{8}$$

$$8\left(\frac{a-4}{2}-\frac{3a+1}{4}\right)\leq 8\left(-\frac{a}{8}\right)$$

$$4(a-4)-2(3a+1)\leq -a$$

$$4a-16-6a-2\leq -a$$

$$-a\leq 18$$

$$a\geq -18$$

$[-18,\infty)$

13. a. Compound inequality

b. $-1\leq\frac{6-x}{-5}\leq 7$

$-1\leq\frac{x-6}{5}\leq 7$

$-5\leq x-6\leq 35$

$1\leq x\leq 41$ $\quad [1,41]$

15. a. Absolute value equation

b. $|4x-5|=|3x-2|$

$4x-5=3x-2$ or $4x-5=-3x+2$

$x=3$ or $7x=7$

$x=1$

$\{1,3\}$

17. a. Absolute value inequality

b. $-|x+4|+8>3$

$-|x+4|>-5$

$|x+4|<5$

$-5<x+4<5$

$-9<x<1$

$(-9,1)$

19. a. Radical equation

b. $c^{2/3}=16$

$\left(c^{2/3}\right)^{3/2}=\pm(16)^{3/2}$

$c=\pm 64$ $\quad \{\pm 64\}$

Chapter 1 Review Exercises

1.
$$\frac{3}{x^2-4}+\frac{4}{2x-7}=\frac{2}{3}$$
$$\frac{3}{(x+2)(x-2)}+\frac{4}{2x-7}=\frac{2}{3}$$
$$x\neq 2, x\neq 2, x\neq \frac{7}{2}$$

3.
$$\frac{4}{5}x-\frac{2}{3}=\frac{7}{10}x-2$$
$$30\left(\frac{4}{5}x-\frac{2}{3}\right)=30\left(\frac{7}{10}x-2\right)$$
$$24x-20=21x-60$$
$$3x=-40$$
$$x=-\frac{40}{3} \qquad \left\{-\frac{40}{3}\right\}$$

5.
$$x-5+2(x-4)=3(x+1)-5$$
$$x-5+2x-8=3x+3-5$$
$$3x-13=3x-2$$
$$-13=-2$$
$$\{\ \}$$

7.
$$(y-4)^2=(y+3)^2$$
$$y^2-8y+16=y^2+6y+9$$
$$7=14y$$
$$\frac{7}{14}=y$$
$$\frac{1}{2}=y$$
$$\left\{\frac{1}{2}\right\}$$

9.
$$\frac{1}{m-1}=\frac{5m}{m^2+3m-4}-\frac{3}{m+4}$$
$$\frac{1}{m-1}=\frac{5m}{(m-1)(m+4)}-\frac{3}{m+4}$$
$$(m-1)(m+4)\left(\frac{1}{m-1}\right)=(m-1)(m+4)\left(\frac{5m}{(m-1)(m+4)}-\frac{3}{m+4}\right)$$
$$1(m+4)=5m-3(m-1)$$
$$m+4=5m-3m+3$$
$$1=m$$
$\{\ \}$; The value 1 does not check.

11.
$$t_a=\frac{t_1+t_2}{2}$$
$$2t_a=t_1+t_2$$
$$2t_a-t_1=t_2 \text{ or } t_2=2t_a-t_1$$

13.
$$A=\frac{1}{41}c+\frac{1}{36}h$$
$$11=\frac{1}{41}c+\frac{1}{36}(288)$$
$$11=\frac{1}{41}c+8$$
$$3=\frac{1}{41}c$$
$$123=c \text{ or } c=123 \text{ mi}$$

15. Let x represent the amount invested in the 10-yr Treasury note. Then, $(x+4000)$ is the amount invested in the 15-yr bond.

	10-yr Note	15-yr Bond	Total
Principal	x	$x+4000$	
Interest ($I=Prt$)	$x(0.035)(10)$	$(x+4000)(0.041)(15)$	10,180

$$x(0.035)(10)+(x+4000)(0.041)(15)=10{,}180$$
$$0.35x+0.615x+2460=10{,}180$$
$$0.965x+2460=10{,}180$$
$$0.965x=7720$$
$$x=8000$$

$x+4000=8000+4000=12{,}000$

Cassandra invested \$8000 in the Treasury note and \$12,000 in the bond.

17. Let x represent the amount of the pure sand (in cubic feet). 250 ft^2 is the amount of the concrete mix that is 50% sand. Therefore, $(x+250)$ is the amount of the resulting 70% sand mixture.

	100% Sand	50% Sand	70% Sand
Amount of Mixture	x	250	$x+250$
Pure Sand	x	$0.5(250)$	$0.7(x+250)$

$$x+0.5(250)=0.7(x+250)$$
$$x+125=0.7x+175$$
$$0.3x=50$$
$$x=166\frac{2}{3}$$

$166\frac{2}{3}$ ft^2 of sand should be mixed with the 50% sand mixture.

19. Let x represent the speed of the boat traveling north. Then, $(x+6)$ is the speed of the boat traveling south.

	Distance	Rate	Time
Northbound	$3x$	x	3
Southbound	$3(x+6)$	$x+6$	3

$$3x+3(x+6)=66$$
$$3x+3x+18=66$$
$$6x+18=66$$
$$6x=48$$
$$x=8$$

$x+6=8+6=14$

The northbound boat travels 8 mph and the southbound boat travels 14 mph.

21. a. $C = 5x$

b. $C = 80$

$5x = 80$

$x = 16$

The dancer will save money on the 17th dance during a 3-month period.

23. Let t represent the time it takes the second pump to drain the pond by itself.

$$\frac{1\text{ job}}{22\text{ hr}} + \frac{1\text{ job}}{t\text{ hr}} = \frac{1\text{ job}}{10\text{ hr}}$$

$$110t\left(\frac{1}{22} + \frac{1}{t}\right) = 110t\left(\frac{1}{10}\right)$$

$$5t + 110 = 11t$$

$$110 = 6t$$

$$t = \frac{55}{3} \approx 18.3\text{ hr}$$

25. Let x represent the number of turtles in the pond.

$$\frac{12}{x} = \frac{3}{36}$$

$$3x = 432$$

$$x = 144$$

There are approximately 144 turtles in the pond.

27. $\sqrt{-12} = i\sqrt{12} = 2i\sqrt{3}$

29. a. Real part: 3; Imaginary part: −7

b. Real part: 0; Imaginary part: 2

31. $$\left(\frac{2}{3} + \frac{3}{5}i\right) - \left(\frac{1}{6} + \frac{2}{5}i\right)$$

$$= \left(\frac{20}{30} + \frac{18}{30}i\right) - \left(\frac{5}{30} + \frac{12}{30}i\right)$$

$$= \left(\frac{20}{30} - \frac{5}{30}\right) + \left(\frac{18}{30} - \frac{12}{30}\right)i$$

$$= \frac{15}{30} + \frac{6}{30}i = \frac{1}{2} + \frac{1}{5}i$$

33. $$\sqrt{-5}\left(\sqrt{11} + \sqrt{-3}\right) = i\sqrt{5}\left(\sqrt{11} + i\sqrt{3}\right)$$

$$= i\sqrt{55} + i^2\sqrt{15}$$

$$= i\sqrt{55} + (-1)\sqrt{15}$$

$$= -\sqrt{15} + i\sqrt{55}$$

35. $$(4 - 6i)^2 = (4)^2 - 2(4)(6i) + (6i)^2$$

$$= 16 - 48i + 36i^2 = 16 - 48i + 36(-1)$$

$$= 16 - 48i - 36 = -20 - 48i$$

37. $(8 - 3i)(8 + 3i) = (8)^2 + (3)^2 = 64 + 9 = 73$

39. $$\left(6 - \sqrt{5}i\right)^{-1} = \frac{1}{6 - \sqrt{5}i} = \frac{1\left(6 + \sqrt{5}i\right)}{\left(6 - \sqrt{5}i\right)\left(6 + \sqrt{5}i\right)}$$

$$= \frac{6 + \sqrt{5}i}{6^2 + 5} = \frac{6 + \sqrt{5}i}{36 + 5} = \frac{6 + \sqrt{5}i}{41}$$

$$= \frac{6}{41} + \frac{\sqrt{5}}{41}i$$

41. $$3y^2 - 4y = 8 - 6y$$

$$3y^2 + 2y - 8 = 0$$

$$(3y - 4)(y + 2) = 0$$

$$3y - 4 = 0 \quad \text{or} \quad y + 2 = 0$$

$$3y = 4 \qquad y = -2$$

$$y = \frac{4}{3}$$

$$\left\{\frac{4}{3}, -2\right\}$$

43. $$10t^2 + 1210 = 0$$

$$\frac{10t^2}{10} + \frac{1210}{10} = \frac{0}{10}$$

$$t^2 + 121 = 0$$

$$t^2 = -121$$

$$t = \pm\sqrt{-121} = \pm 11i$$

$$\{\pm 11i\}$$

45. $x^2-5=(x+2)(x-4)$

$x^2-5=x^2-4x+2x-8$

$x^2-5=x^2-2x-8$

$2x=-3$

$x=-\frac{3}{2}$

$\left\{-\frac{3}{2}\right\}$

47. $x^2+18x+n=x^2+18x+\left[\frac{1}{2}(18)\right]^2$

$=x^2+18x+(9)^2$

$=x^2+18x+81=(x+9)^2$

$n=81; (x+9)^2$

49. a. $x^2-10x=-9$

$x^2-10x+9=0$

$(x-1)(x-9)=0$

$x-1=0$ or $x-9=0$

$x=1$ $\quad x=9$

$\{1, 9\}$

b. $x^2-10x=-9$

$x^2-10x+\left[\frac{1}{2}(-10)\right]^2=-9+\left[\frac{1}{2}(-10)\right]^2$

$x^2-10x+25=-9+25$

$(x-5)^2=16$

$x-5=\pm\sqrt{16}$

$x=5\pm 4$

$x=5-4$ or $x=5+4$

$x=1$ $\quad x=9$

$\{1, 9\}$

c. $x^2-10x=-9$

$x^2-10x+9=0$

$a=1, b=-10, c=9$

$x=\frac{-b\pm\sqrt{b^2-4ac}}{2a}$

$=\frac{-(-10)\pm\sqrt{(-10)^2-4(1)(9)}}{2(1)}$

$=\frac{10\pm\sqrt{100-36}}{2}=\frac{10\pm\sqrt{64}}{2}=\frac{10\pm 8}{2}$

$=5\pm 4=1$ or 9

$\{1, 9\}$

51. False

53. a. $4x^2-20x+25=0$

$b^2-4ac=(-20)^2-4(4)(25)$

$=400-400$

$=0$

b. The discriminant is 0; there is one real solution.

55. a. $5t(t+1)=4t-11$

$5t^2+5t=4t-11$

$5t^2+t+11=0$

$b^2-4ac=(1)^2-4(5)(11)$

$=1-220$

$=-219$

b. $-219<0$; there are two imaginary solutions.

57. $(x-h)^2+(y-k)^2=r^2$

$(y-k)^2=r^2-(x-h)^2$

$y-k=\pm\sqrt{r^2-(x-h)^2}$

$y=k\pm\sqrt{r^2-(x-h)^2}$

59. Let x and $(x+3)$ represent the width and length of the decorative area. Then $(x+1)$ and $(x+4)$ are the width and length of the rug.

$$(x+1)(x+4)=108$$
$$x^2+4x+x+4=108$$
$$x^2+5x-104=0$$
$$(x-8)(x+13)=0$$
$$x-8=0 \quad \text{or} \quad x+13=0$$
$$x=8 \quad \cancel{x=-13}$$
$$x+1=8+1=9$$
$$x+4=8+4=12$$
The rug is 9 ft by 12 ft.

61. Let x and $(1.6x)$ represent the width and length of the screen.
$$a^2+b^2=c^2$$
$$(x)^2+(1.6x)^2=(50)^2$$
$$x^2+2.56x^2=2500$$
$$3.56x^2=2500$$
$$x^2=\frac{2500}{3.56}$$
$$x=\pm\sqrt{\frac{2500}{3.56}}\approx\pm26.5$$
$$1.6x=1.6(26.5)=42.4$$
The width is 26.5 in. and the length is 42.4 in.

63.
$$S=\frac{1}{2}n(n+1)$$
$$4186=\frac{1}{2}n(n+1)$$
$$4186=n^2+n$$
$$n^2+n-4186=0$$
$$(n-91)(n+46)=0$$
$$n-91=0 \quad \text{or} \quad n+46=0$$
$$n=91 \quad \cancel{n=-46}$$
The value of n is 91.

65. a.
$$s=-\frac{1}{2}gt^2+v_0t+s_0$$
$$s=-\frac{1}{2}(32)t^2+(200)t+2$$
$$s=-16t^2+200t+2$$

b.
$$s=-16t^2+200t+2$$
$$80=-16t^2+200t+2$$
$$-40=8t^2-100t-1$$
$$0=8t^2-100t+39$$
$$t=\frac{-b\pm\sqrt{b^2-4ac}}{2a}$$
$$=\frac{-(-100)\pm\sqrt{(-100)^2-4(8)(39)}}{2(8)}$$
$$=\frac{100\pm\sqrt{8752}}{16}\approx 0.4 \text{ or } 12.1$$
The mortar will be at an 80-ft height 0.4 sec after launch.

67.
$$4x^3-6x^2-20x+30=0$$
$$4x^3-20x-6x^2+30=0$$
$$4x(x^2-5)-6(x^2-5)=0$$
$$(x^2-5)(4x-6)=0$$
$$(x^2-5)(2x-3)=0$$
$$x^2=5 \quad \text{or} \quad 2x-3=0$$
$$x=\pm\sqrt{5} \quad \text{or} \quad x=\frac{3}{2}$$
$$\left\{\pm\sqrt{5},\frac{3}{2}\right\}$$

69.
$$\sqrt{k+7}-\sqrt{3-k}=2$$
$$\sqrt{k+7}=\sqrt{3-k}+2$$
$$\left(\sqrt{k+7}\right)^2=\left(\sqrt{3-k}+2\right)^2$$
$$k+7=3-k+4\sqrt{3-k}+4$$
$$2k=4\sqrt{3-k}$$
$$(2k)^2=\left(4\sqrt{3-k}\right)^2$$
$$4k^2=16(3-k)$$
$$4k^2=48-16k$$
$$4k^2+16k-48=0$$
$$k^2+4k-12=0$$
$$(k-2)(k+6)=0$$
$$k=2 \quad \text{or} \quad k=-6$$

Check: $k = 2$

$$\sqrt{k+7} - \sqrt{3-k} = 2$$
$$\sqrt{(2)+7} - \sqrt{3-(2)} \stackrel{?}{=} 2$$
$$\sqrt{9} - \sqrt{1} \stackrel{?}{=} 2$$
$$3 - 1 \stackrel{?}{=} 2$$
$$2 \stackrel{?}{=} 2 \checkmark \text{ true}$$

Check: $k = -6$

$$\sqrt{k+7} - \sqrt{3-k} = 2$$
$$\sqrt{(-6)+7} - \sqrt{3-(-6)} \stackrel{?}{=} 2$$
$$\sqrt{1} - \sqrt{9} \stackrel{?}{=} 2$$
$$1 - 3 \stackrel{?}{=} 2$$
$$-2 \stackrel{?}{=} 2 \text{ false}$$

$\{2\}$; The value -6 does not check.

71. Let $u = v^{-1}$.

$$11v^{-2} + 23v^{-1} + 2 = 0$$
$$11u^2 + 23u + 2 = 0$$
$$(11u+1)(u+2) = 0$$
$$u = -\frac{1}{11} \quad \text{or} \quad u = -2$$
$$v^{-1} = -\frac{1}{11} \quad \text{or} \quad v^{-1} = -2$$
$$v = -11 \quad \text{or} \quad v = -\frac{1}{2}$$

$\left\{-\frac{1}{2}, -11\right\}$

73. $-2\sqrt{3m+4} - 3 = 5$

$$-2\sqrt{3m+4} = 8$$
$$\sqrt{3m+4} = -4$$

$\{\ \}$

75. $p^{2/3} = 7$

$$\left(p^{2/3}\right)^{3/2} = \pm(7)^{3/2}$$
$$p = \pm 7^{3/2}$$

$\left\{\pm 7^{3/2}\right\}$

77. $2(y-5)^{5/4} = 22$

$$(y-5)^{5/4} = 11$$
$$\left[(y-5)^{5/4}\right]^{4/5} = (11)^{4/5}$$
$$y - 5 = 11^{4/5}$$
$$y = 5 + 11^{4/5}$$

$\left\{5 + 11^{4/5}\right\}$

79. Let $u = d^{1/3}$.

$$6d^{2/3} - 7d^{1/3} - 3 = 0$$
$$6u^2 - 7u - 3 = 0$$
$$(3u+1)(2u-3) = 0$$
$$u = -\frac{1}{3} \quad \text{or} \quad u = \frac{3}{2}$$
$$d^{1/3} = -\frac{1}{3} \quad \text{or} \quad d^{1/3} = \frac{3}{2}$$
$$\left(d^{1/3}\right)^3 = \left(-\frac{1}{3}\right)^3 \quad \text{or} \quad \left(d^{1/3}\right)^3 = \left(\frac{3}{2}\right)^3$$
$$d = -\frac{1}{27} \quad \text{or} \quad d = \frac{27}{8}$$

$\left\{-\frac{1}{27}, \frac{27}{8}\right\}$

81. Let $u = \frac{4}{w} + 1$.

$$2\left(\frac{4}{w}+1\right)^2 - 10\left(\frac{4}{w}+1\right) = 0$$
$$2u^2 - 10u = 0$$
$$2u(u-5) = 0$$
$$u = 0 \quad \text{or} \quad u = 5$$
$$\frac{4}{w} + 1 = 0 \quad \text{or} \quad \frac{4}{w} + 1 = 5$$
$$\frac{4}{w} = -1 \quad \text{or} \quad \frac{4}{w} = 4$$
$$w = -4 \quad \text{or} \quad w = 1$$

$\{1, -4\}$

83. $m=\frac{1}{2}\sqrt{2a^2+2b^2-2c^2}$

$2m=\sqrt{2a^2+2b^2-2c^2}$

$(2m)^2=\left(\sqrt{2a^2+2b^2-2c^2}\right)^2$

$4m^2=2a^2+2b^2-2c^2$

$2m^2=a^2+b^2-c^2$

$a^2=2m^2-b^2+c^2$

$a=\pm\sqrt{2m^2-b^2+c^2}$

85. $\frac{a_1t_1}{v_1}=\frac{a_2t_2}{v_2}$

$v_1v_2\left(\frac{a_1t_1}{v_1}\right)=v_1v_2\left(\frac{a_2t_2}{v_2}\right)$

$v_2a_1t_1=v_1a_2t_2$

$v_2=\frac{v_1a_2t_2}{a_1t_1}$

87. $-0.6+0.2x<0.8x-1.8$

$1.2<0.6x$

$2<x$ or $x>2$

$\{x\mid x>2\}$; $(2,\infty)$

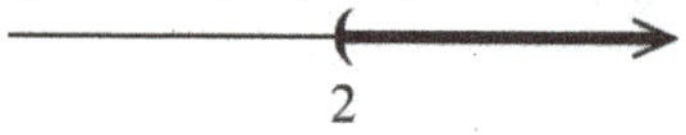

89. $9-[5-4(t-1)]\ge3\{2-[5-(t+2)]\}$

$9-[5-4t+4]\ge3\{2-[5-t-2]\}$

$9-[-4t+9]\ge3\{2-[-t+3]\}$

$9+4t-9\ge3\{2+t-3\}$

$4t\ge3\{t-1\}$

$4t\ge3t-3$

$t\ge-3$

$\{t\mid t\ge-3\}$; $[-3,\infty)$

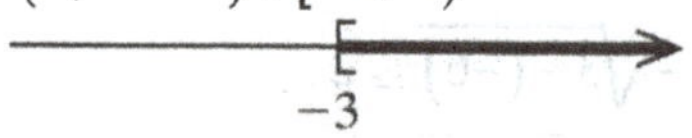

91. **a.** $X\cup Y=\mathbb{R}$

b. $X\cap Y=\{x\mid -2\le x<7\}$

c. $X\cup Z=\{x\mid x<7\}$

d. $X\cap Z=\{x\mid x<-3\}$

e. $Y\cup Z=\{x\mid x<-3 \text{ or } x\ge-2\}$

f. $Y\cap Z=\{\ \}$

93. **a.** $-2(x-1)+4<x+3$ or $5(x+2)-3\le4x+1$

$-2x+2+4<x+3$ or $5x+10-3\le4x+1$

$-3x<-3$ or $x\le-6$

$x>1$ or $x\le-6$

$(-\infty,-6]<(1,\infty)$;

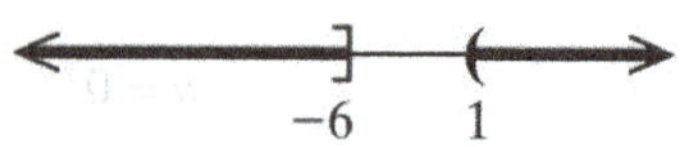

b. $-2(x-1)+4<x+3$ and $5(x+2)-3\le4x+1$

$x>1$ and $x\le-6$ $\{\ \}$

95. $0<\frac{-3x+9}{-4}<6$

$0<\frac{3x-9}{4}<6$

$0<3x-9<24$

$9<3x<33$

$3<x<11$ $(3,11)$

97. Let x represent the September rainfall.

$$\frac{8.54+5.79+8.63+x}{4}>7.83$$
$$\frac{22.96+x}{4}>7.83$$
$$22.96+x>31.32$$
$$x>8.36$$

More than 8.36 in. is needed.

99. a. $x-12\ge 0$

$x\ge 12$

$\{x\mid x\ge 12\}$

b. $12-x\ge 0$

$12\ge x$ or $x\le 12$

$\{x\mid x\le 12\}$

101. a. $|w+2|+1=6$

$|w+2|=5$

$w+2=5$ or $w+2=-5$

$w=3$ or $w=-7$

$\{-7,3\}$

b. $|w+2|+1<6$

$|w+2|<5$

$-5<w+2<5$

$-7<w<3$

$(-7,3)$

c. $|w+2|+1\ge 6$

$|w+2|\ge 5$

$w+2\le -5$ or $w+2\ge 5$

$w\le -7$ or $w\ge 3$

$(-\infty,-7]\cup[3,\infty)$

103. a. $|y+5|-3=-3$

$|y+5|=0$

$y+5=0$

$y=-5$

$\{-5\}$

b. $|y+5|-3<-3$

$|y+5|<0$ $\quad\{\ \}$

c. $|y+5|-3\le -3$

$|y+5|\le 0$

$y+5=0$

$y=-5$

$\{-5\}$

d. $|y+5|-3>-3$

$|y+5|>0$

$y+5<0$ or $y+5>0$

$y<-5$ or $y>-5$

$(-\infty,-5)\cup(-5,\infty)$

e. $|y+5|-3\ge -3$

$|y+5|\ge 0$

$(-\infty,\infty)$

105. $-5=-|5x+1|-4$

$-1=-|5x+1|$

$1=|5x+1|$

$5x+1=1$ or $5x+1=-1$

$5x=0$ or $5x=-2$

$x=0$ or $x=-\frac{2}{5}$

$\left\{0,-\frac{2}{5}\right\}$

107. $|2x|=\left|\frac{1}{2}x\right|$

$2x=\frac{1}{2}x$ or $2x=-\frac{1}{2}x$

$4x=x$ or $4x=-x$

$3x=0$ or $5x=0$

$x=0$ or $x=0$

$\{0\}$

109. $|0.5x-8|<0.01$

$-0.01<0.5x-8<0.01$

$7.99<0.5x<8.01$

$15.98<x<16.02$

$(15.98,16.02)$

Chapter 1 Test

1. $\sqrt{-25}\cdot\sqrt{-4}=5i\cdot 2i=10i^2=10(-1)=-10$

3. $(4-7i)(6+2i)=24+8i-42i-14i^2$
$=24-34i-14(-1)$
$=24-34i+14-38-34i$

5. $\frac{4+3i}{2-5i}=\frac{(4+3i)(2+5i)}{(2-5i)(2+5i)}$
$=\frac{8+20i+6i+15i^2}{2^2+5^2}=\frac{8+26i+15(-1)}{4+25}$
$=\frac{-7+26i}{29}=-\frac{7}{29}+\frac{26}{29}i$

7. a. $x^2+25=10x$
$x^2-10x+25=0$
$b^2-4ac=(-10)^2-4(1)(25)$
$=100-100=0$

b. Because the discriminant is 0, there is one real solution.

9. $3y+2[5(y-4)-2]=5y+6(7+y)-3$
$3y+2(5y-20-2)=5y+42+6y-3$
$3y+2(5y-22)=11y+39$
$3y+10y-44=11y+39$
$13y-44=11y+39$
$2y=83$
$y=\frac{83}{2}$

$\left\{\frac{83}{2}\right\}$

11. $0.4(w+1)+0.8=0.1w+0.3(4+w)$
$0.4w+0.4+0.8=0.1w+1.2+0.3w$
$0.4w+1.2=0.4w+1.2$
$0=0$

$\mathbb{R}$

13. $(3x-4)^2-2=11$
$(3x-4)^2=13$
$3x-4=\pm\sqrt{13}$
$3x=4\pm\sqrt{13}$
$x=\frac{4\pm\sqrt{13}}{3}$

$\left\{\frac{4\pm\sqrt{13}}{3}\right\}$

15. $6t(2t+1)=5-5t$
$12t^2+11t-5=0$
$(3t-1)(4t+5)=0$
$3t-1=0$ or $4t+5=0$
$3t=1$ $\quad$ $4t=-5$
$t=\frac{1}{3}$ $\quad$ $t=-\frac{5}{4}$

$\left\{\frac{1}{3},-\frac{5}{4}\right\}$

17. $12y^3+24y^2=3y+6$
$12y^3+24y^2-3y-6=0$
$12y^3-3y+24y^2-6=0$
$3y(4y^2-1)+6(4y^2-1)=0$
$y(4y^2-1)+2(4y^2-1)=0$
$(4y^2-1)(y+2)=0$
$(2y+1)(2y-1)(y+2)=0$
$2y+1=0$ or $2y-1=0$ or $y+2=0$
$2y=-1$ $\quad$ $2y=1$ $\quad$ $y=-2$
$y=-\frac{1}{2}$ $\quad$ $y=\frac{1}{2}$

$\left\{\pm\frac{1}{2},-2\right\}$

19. $\sqrt{2d} = 1 - \sqrt{d+7}$

$\left(\sqrt{2d}\right)^2 = \left(1 - \sqrt{d+7}\right)^2$

$2d = 1 - 2\sqrt{d+7} + d + 7$

$2\sqrt{d+7} = 8 - d$

$\left(2\sqrt{d+7}\right)^2 = (8-d)^2$

$4(d+7) = 64 - 16d + d^2$

$4d + 28 = d^2 - 16d + 64$

$0 = d^2 - 20d + 36$

$0 = (d-2)(d-18)$

$d = 2$ or $d = 18$

Check: $d = 2$

$\sqrt{2d} = 1 - \sqrt{d+7}$

$\sqrt{2(2)} \stackrel{?}{=} 1 - \sqrt{(2)+7}$

$\sqrt{4} \stackrel{?}{=} 1 - \sqrt{9}$

$2 \stackrel{?}{=} 1 - 3$

$2 \stackrel{?}{=} -2$ false

Check: $d = 18$

$\sqrt{2d} = 1 - \sqrt{d+7}$

$\sqrt{2(18)} \stackrel{?}{=} 1 - \sqrt{(18)+7}$

$\sqrt{36} \stackrel{?}{=} 1 - \sqrt{25}$

$6 \stackrel{?}{=} 1 - 5$

$6 \stackrel{?}{=} -4$ false

$\{\ \}$; The values 2 and 18 do not check.

21. $w^{4/5} - 11 = 0$

$w^{4/5} = 11$

$\left(w^{4/5}\right)^{5/4} = \pm(11)^{5/4}$

$w = \pm 11^{5/4}$

$\left\{\pm 11^{5/4}\right\}$

23. $-2 = |x-3| - 6$

$4 = |x-3|$

$x - 3 = 4$ or $x - 3 = -4$

$x = 7$ or $x = -1$

$\{-1, 7\}$

25. $aP - 4 = Pt + 2$

$aP - Pt = 6$

$P(a-t) = 6$

$P = \dfrac{6}{a-t}$ or $P = -\dfrac{6}{t-a}$

27. $-16t^2 + v_0 t + 2 = 0$

$16t^2 - v_0 t - 2 = 0$

$a = 16, b = -v_0, c = -2$

$t = \dfrac{-b \pm \sqrt{b^2 - 4ac}}{2a}$

$t = \dfrac{-(-v_0) \pm \sqrt{(-v_0)^2 - 4(16)(-2)}}{2(16)}$

$t = \dfrac{v_0 \pm \sqrt{v_0^{\,2} + 128}}{32}$

29. $-2 \le \dfrac{4-x}{3} \le 6$

$-6 \le 4 - x \le 18$

$-10 \le -x \le 14$

$10 \ge x \ge -14$ or $-14 \le x \le 10$

$[-14, 10]$

31. $3(x-5) + 1 \le 4(x+2) + 6$ and $0.3x - 1.6 > 0.2$ $(6, \infty)$

$3x - 15 + 1 \le 4x + 8 + 6$ and $0.3x > 1.8$

$0 \le x$ and $x > 6$

33. $-|8-v| \ge -6$

$|8-v| \le 6$

$-6 \le 8-v \le 6$

$-14 \le -v \le -2$

$14 \ge v \ge 2$

$2 \le v \le 14$

$[2, 14]$

35. a. $|x-13|+4=4$

$|x-13|=0$

$x-13=0$

$x=13 \qquad \{13\}$

b. $|x-13|+4<4$

$|x-13|<0 \qquad \{\ \}$

c. $|x-13|+4 \le 4$

$|x-13| \le 0$

$x-13=0$

$x=13 \qquad \{13\}$

d. $|x-13|+4>4$

$|x-13|>0$

$x-13<-0$ or $x-13>0$

$x<13$ or $x>13$

$(-\infty, 13) \cup (13, \infty)$

e. $|x-13|+4 \ge 4$

$|x-13| \ge 0$

$x-13 \le -0$ or $x-13 \ge 0$

$x \le 13$ or $x \ge 13$

$(-\infty, \infty)$

37. Let x represent the speed of the plane flying to Seattle. Then, $(x+60)$ is the speed of the plane flying to New York City.

	Distance	Rate	Time
Seattle Flight	$2.3x$	x	2.3
New York Flight	$3.3(x+60)$	$x+60$	3.3

$2.3x+3.3(x+60)=2662$

$2.3x+3.3x+198=2662$

$5.6x+198=2662$

$5.6x=2464$

$x=440$

$x+60=440+60=500$

The plane flying to Seattle flies 440 mph, and the plane flying to New York flies 500 mph.

39. Let x represent the patient's LDL cholesterol level. The HDL cholesterol level is 70 mg/dL, and the total cholesterol is $(x+70)$.

$\frac{x+70}{70}=3.8$

$70\left(\frac{x+70}{70}\right)=70(3.8)$

$x+70=266$

$x=196$

$x+70=196+70=266$

The LDL level is 196 mg/dL and the total cholesterol is 266 mg/dL.

41. a. $s=-\frac{1}{2}gt^2+v_0t+s_0$

$=-\frac{1}{2}(32)t^2+(60)t+2$

$=-16t^2+60t+2$

b. $52=-16t^2+60t+2$

$0=-16t^2+60t-50$

$0=8t^2-30t+25$

$t=\frac{-b \pm \sqrt{b^2-4ac}}{2a}$

$=\frac{-(-30) \pm \sqrt{(-30)^2-4(8)(25)}}{2(8)}$

$=\frac{30 \pm \sqrt{100}}{16}=\frac{30 \pm 10}{16}$

$t=\frac{40}{16}=\frac{5}{2}=2.5$ or $t=\frac{20}{16}=\frac{5}{4}=1.25$

The ball will be at a height of 52 ft at times 1.25 sec and 2.5 sec after being kicked.

Chapter 1 Cumulative Review Exercises

1. $\left[(5x+3)^2-(5x-3)^2\right]^2$

$=\left[25x^2+30x+9-(25x^2-30x+9)\right]^2$

$=(25x^2+30x+9-25x^2+30x-9)^2$

$=(60x)^2=3600x^2$

3. $\dfrac{3x^2-x-4}{4x^2-8x-12}\div\dfrac{3x-4}{6x^2-54}$

$=\dfrac{3x^2-x-4}{4(x^2-2x-3)}\cdot\dfrac{6(x^2-9)}{3x-4}$

$=\dfrac{(3x-4)(x+1)}{4(x-3)(x+1)}\cdot\dfrac{6(x+3)(x-3)}{3x-4}$

$=\dfrac{\cancel{(3x-4)}\cancel{(x+1)}}{\cancel{4}_2\cancel{(x-3)}\cancel{(x+1)}}\cdot\dfrac{\cancel{6}^3(x+3)\cancel{(x-3)}}{\cancel{3x-4}}$

$=\dfrac{3(x+3)}{2}$

5. $\dfrac{\frac{1}{5x}-\frac{3}{5}}{\frac{2}{x}+\frac{1}{5}}=\dfrac{5x\cdot\left(\frac{1}{5x}-\frac{3}{5}\right)}{5x\cdot\left(\frac{2}{x}+\frac{1}{5}\right)}=\dfrac{1-3x}{10+x}$

7. $\sqrt[3]{81y^5z^2w^{12}}=\sqrt[3]{27y^3w^{12}\cdot 3y^2z^2}$

$=3yw^4\sqrt[3]{3y^2z^2}$

9. $4x^3-32y^6=4(x^3-8y^6)$

$=4\left[x^3-(2y^2)^3\right]$

$=4\left[(x-2y^2)(x^2+2xy^2+4y^4)\right]$

11. Let x represent the amount borrowed at 4%. Then, $(8000-x)$ is the amount borrowed at 5%.

	4% Interest Loan	5% Interest Loan	Total
Principal	x	$8000-x$	
Interest ($I=Prt$)	$x(0.04)(1)$	$(8000-x)(0.05)(1)$	380

$x(0.04)+(8000-x)(0.05)=380$

$0.04x+400-0.05x=380$

$-0.01x+400=380$

$-0.01x=-20$

$x=2000$

$8000-x=8000-2000=6000$

Stephan borrowed \$6000 at 5% and \$2000 at 4%.

13. $2x(x-4)=2x+5$

$2x^2-8x=2x+5$

$2x^2-10x-5=0$

$a=2, b=-10, c=-5$

$$x=\frac{-b\pm\sqrt{b^2-4ac}}{2a}$$

$$=\frac{-(-10)\pm\sqrt{(-10)^2-4(2)(-5)}}{2(2)}$$

$$=\frac{10\pm\sqrt{100+40}}{4}$$

$$=\frac{10\pm\sqrt{140}}{4}$$

$$=\frac{10\pm2\sqrt{35}}{4}=\frac{5\pm\sqrt{35}}{2}$$

$\left\{\frac{5\pm\sqrt{35}}{2}\right\}$

15. $\sqrt{x+4}-2=x$

$\sqrt{x+4}=x+2$

$\left(\sqrt{x+4}\right)^2=(x+2)^2$

$x+4=x^2+4x+4$

$x^2+3x=0$

$x(x+3)=0$

$x=0$ or $x=-3$

Check: $x=0$

$\sqrt{x+4}-2=x$

$\sqrt{(0)+4}-2\stackrel{?}{=}(0)$

$2-2\stackrel{?}{=}0$

$0\stackrel{?}{=}0$ ✓ true

Check: $x=-3$

$\sqrt{x+4}-2=x$

$\sqrt{(-3)+4}-2\stackrel{?}{=}(-3)$

$1-2\stackrel{?}{=}-3$

$-1\stackrel{?}{=}-3$ false

$\{0\}$

17. $x-9=\frac{72}{x-8}$

$(x-8)(x-9)=(x-8)\left(\frac{72}{x-8}\right)$

$x^2-17x+72=72$

$x^2-17x=0$

$x(x-17)=0$

$x=0$ or $x=17$

$\{0,17\}$

19. $|2x-11|+1\le12$

$|2x-11|\le11$

$-11\le2x-11\le11$

$0\le2x\le22$

$0\le x\le11$

$[0,11]$

Chapter 2 Functions and Graphs

Section 2.1 The Rectangular Coordinate System and Graphing Utilities

1. origin

3. $d=\sqrt{(x_2-x_1)^2+(y_2-y_1)^2}$

5. solution

7. 0

9.

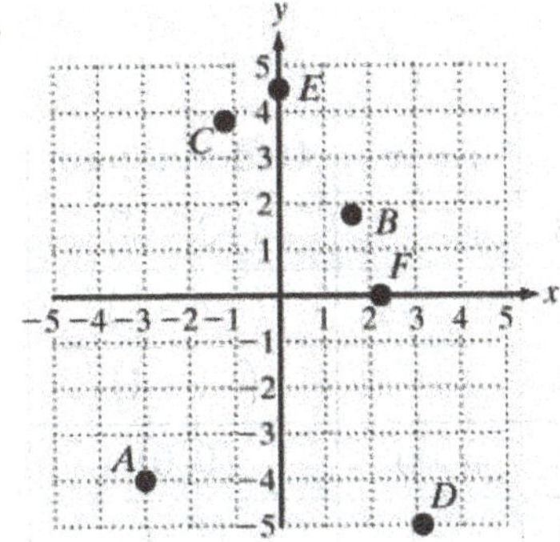

11. a. $d=\sqrt{(x_2-x_1)^2+(y_2-y_1)^2}$

$=\sqrt{[-4-(-2)]^2+(11-7)^2}$

$=\sqrt{(-2)^2+(4)^2}=\sqrt{4+16}$

$=\sqrt{20}=2\sqrt{5}$

b. $M=\left(\frac{x_1+x_2}{2},\frac{y_1+y_2}{2}\right)$

$=\left(\frac{-4+(-2)}{2},\frac{11+7}{2}\right)$

$=\left(\frac{-6}{2},\frac{18}{2}\right)=(-3,9)$

13. a. $d=\sqrt{(x_2-x_1)^2+(y_2-y_1)^2}$

$=\sqrt{[2-(-7)]^2+[5-(-4)]^2}$

$=\sqrt{(9)^2+(9)^2}$

$=\sqrt{81+81}$

$=\sqrt{162}=9\sqrt{2}$

b. $M=\left(\frac{x_1+x_2}{2},\frac{y_1+y_2}{2}\right)$

$=\left(\frac{2+(-7)}{2},\frac{5+(-4)}{2}\right)$

$=\left(\frac{-5}{2},\frac{1}{2}\right)=\left(-\frac{5}{2},\frac{1}{2}\right)$

15. a. $d=\sqrt{(x_2-x_1)^2+(y_2-y_1)^2}$

$=\sqrt{(5.2-2.2)^2+[-6.4-(-2.4)]^2}$

$=\sqrt{(3)^2+(4)^2}$

$=\sqrt{9+16}$

$=\sqrt{25}=5$

b. $M=\left(\frac{x_1+x_2}{2},\frac{y_1+y_2}{2}\right)$

$=\left(\frac{5.2+2.2}{2},\frac{-6.4+(-2.4)}{2}\right)$

$=\left(\frac{7.4}{2},\frac{-8.8}{2}\right)$

$=(3.7,-4.4)$

17. a. $d=\sqrt{(x_2-x_1)^2+(y_2-y_1)^2}$

$=\sqrt{\left(4\sqrt{5}-\sqrt{5}\right)^2+\left[-7\sqrt{2}-\left(-\sqrt{2}\right)\right]^2}$

$=\sqrt{\left(3\sqrt{5}\right)^2+\left(-6\sqrt{2}\right)^2}$

$=\sqrt{45+72}=\sqrt{117}$

b. $M=\left(\frac{x_1+x_2}{2},\frac{y_1+y_2}{2}\right)$

$=\left(\frac{4\sqrt{5}+\sqrt{5}}{2},\frac{-7\sqrt{2}+\left(-\sqrt{2}\right)}{2}\right)$

$=\left(\frac{5\sqrt{5}}{2},\frac{-8\sqrt{2}}{2}\right)=\left(\frac{5\sqrt{5}}{2},-4\sqrt{2}\right)$

19. $d_1 = \sqrt{(3-1)^2 + (1-3)^2} = \sqrt{4+4} = 2\sqrt{2}$

$d_2 = \sqrt{(0-3)^2 + (-2-1)^2} = \sqrt{9+9} = 3\sqrt{2}$

$d_3 = \sqrt{(1-0)^2 + [3-(-2)]^2} = \sqrt{1+25} = \sqrt{26}$

$d_1^{\,2} + d_2^{\,2} \; 0 \; d_3^{\,2}$

$(2\sqrt{2})^2 + (3\sqrt{2})^2 \; 0 \; (\sqrt{26})^2$

$8 + 18 \, 0 \; 26$

$26 \, 0 \; 26$ ✓ True

Yes

21. $d_1 = \sqrt{[5-(-2)]^2 + (0-4)^2} = \sqrt{49+16} = \sqrt{65}$

$d_2 = \sqrt{(-5-5)^2 + (1-0)^2} = \sqrt{100+1} = \sqrt{101}$

$d_3 = \sqrt{[-2-(-5)]^2 + (4-1)^2} = \sqrt{9+9} = 3\sqrt{2}$

$d_1^{\,2} + d_3^{\,2} \; 0 \; d_2^{\,2}$

$(\sqrt{65})^2 + (3\sqrt{2})^2 \; 0 \; (\sqrt{101})^2$

$65 + 18 \, 0 \; 101$

$83 \, 0 \; 101$ False

No

23. a. $x^2 + y = 1$

$(-2)^2 + (-3) \, 0 \; 1$

$4 - 3 \, 0 \; 1$

$1 \, 0 \; 1$ ✓

Yes

b. $x^2 + y = 1$

$(4)^2 + (-17) \, 0 \; 1$

$16 - 17 \, 0 \; 1$

$-1 \, 0 \; 1$ False

No

c. $x^2 + y = 1$

$\left(\frac{1}{2}\right)^2 + \left(\frac{3}{4}\right) 0 \; 1$

$\frac{1}{4} + \frac{3}{4} 0 \; 1$

$1 \, 0 \; 1$ ✓

Yes

25. $x - 3 \neq 0$

$x \neq 3$

$\{x \mid x \neq 3\}$

27. $x - 10 \geq 0$

$x \geq 10$

$\{x \mid x \geq 10\}$

29. $1.5 - x \geq 0$

$-x \geq -1.5$

$x \leq 1.5$

$\{x \mid x \leq 1.5\}$

31. $y = x$

x	y	$y = x$	Ordered pair
-3	-3	$y = -3$	$(-3, -3)$
-2	-2	$y = -2$	$(-2, -2)$
-1	-1	$y = -1$	$(-1, -1)$
0	0	$y = 0$	$(0, 0)$
1	1	$y = 1$	$(1, 1)$
2	2	$y = 2$	$(2, 2)$
3	3	$y = 3$	$(3, 3)$

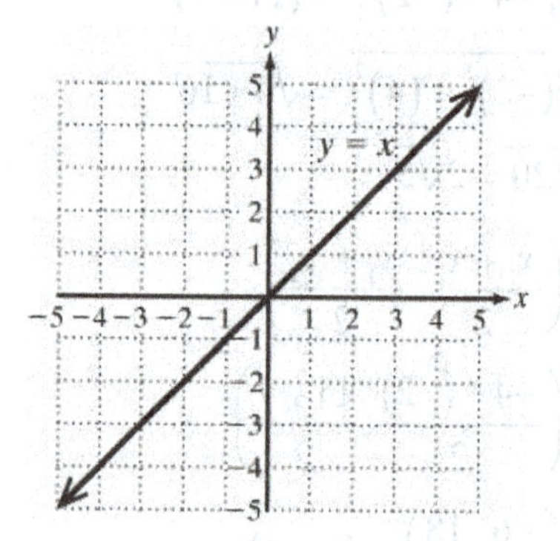

33. $y = \sqrt{x}$

x	y	$y = \sqrt{x}$	Ordered pair
0	0	$y = \sqrt{0} = 0$	$(0, 0)$
1	1	$y = \sqrt{1} = 1$	$(1, 1)$
4	2	$y = \sqrt{4} = 2$	$(4, 2)$
9	3	$y = \sqrt{9} = 3$	$(9, 3)$

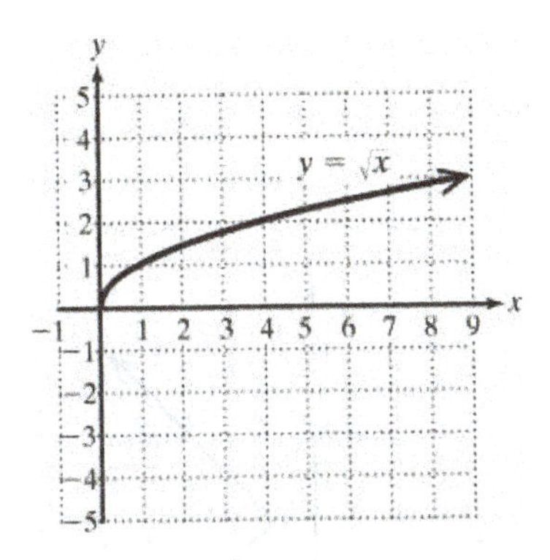

35. $y = x^3$

x	y	$y = x^3$	Ordered pair
−2	−8	$y = (-2)^3 = -8$	$(-2, -8)$
−1	−1	$y = (-1)^3 = -1$	$(-1, -1)$
0	0	$y = (0)^3 = 0$	$(0, 0)$
1	1	$y = (1)^3 = 1$	$(1, 1)$
2	8	$y = (2)^3 = 8$	$(2, 8)$

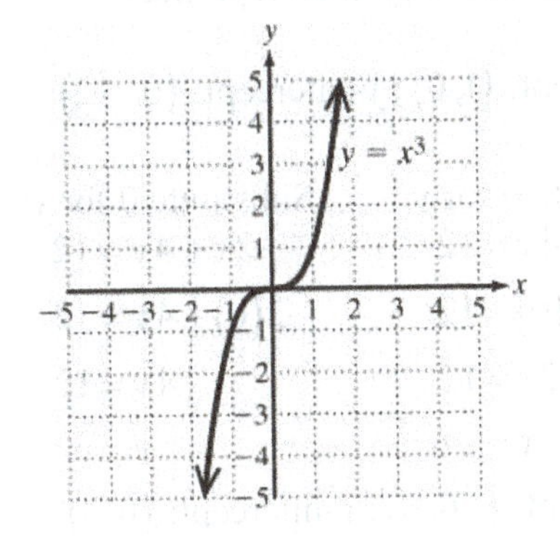

37. $y - |x| = 2$

$y = |x| + 2$

x	y	$y = \lvert x \rvert + 2$	Ordered pair
−3	5	$y = \lvert -3 \rvert + 2 = 5$	$(-3, 5)$
−2	4	$y = \lvert -2 \rvert + 2 = 4$	$(-2, 4)$
−1	3	$y = \lvert -1 \rvert + 2 = 3$	$(-1, 3)$
0	2	$y = \lvert 0 \rvert + 2 = 2$	$(0, 2)$
1	3	$y = \lvert 1 \rvert + 2 = 3$	$(1, 3)$
2	4	$y = \lvert 2 \rvert + 2 = 4$	$(2, 4)$
3	5	$y = \lvert 3 \rvert + 2 = 5$	$(3, 5)$

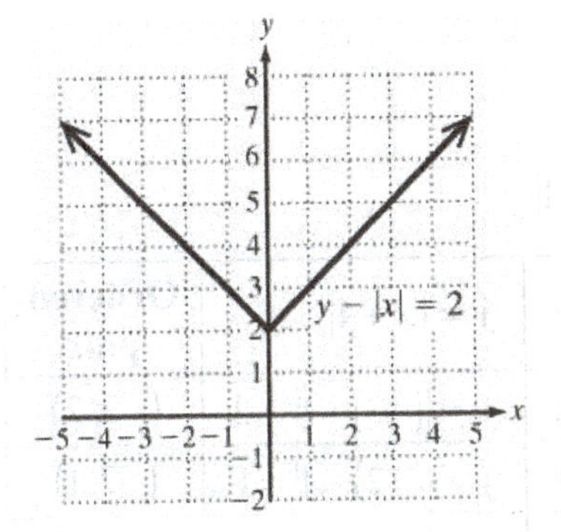

39. $y^2 - x - 2 = 0$

$y^2 = x + 2$

$y = \pm\sqrt{x+2}$

x	y	$y = \pm\sqrt{x+2}$	Ordered pairs
−2	1	$y = \pm\sqrt{(-2)+2} = 0$	$(-2, 0)$
−1	±1	$y = \pm\sqrt{(-1)+2} = \pm 1$	$(-1, 1), (-1, -1)$
2	±2	$y = \pm\sqrt{(2)+2} = \pm 2$	$(2, 2), (2, -2)$
7	±3	$y = \pm\sqrt{(7)+2} = \pm 3$	$(7, 3), (7, -3)$

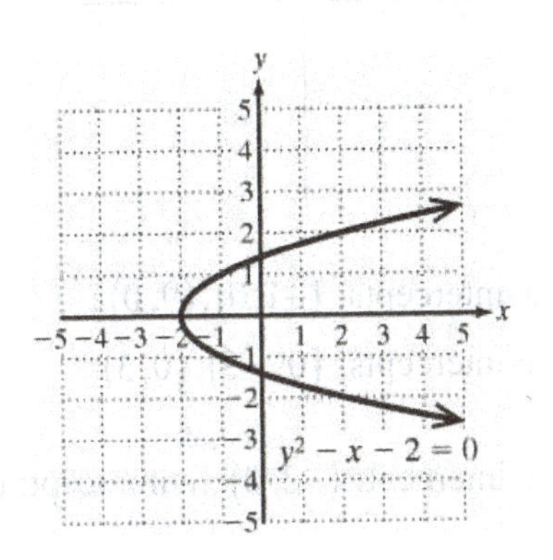

41. $x = |y| + 1$

$|y| = x - 1$

$y = \pm(x - 1)$

x	y	$y = \pm(x-1)$	Ordered pairs
1	0	$y = \pm[(1) - 1] = 0$	$(1, 0)$
2	±1	$y = \pm[(2) - 1] = \pm1$	$(2, 1), (2, -1)$
3	±2	$y = \pm[(3) - 1] = \pm2$	$(3, 2), (3, -2)$
4	±3	$y = \pm[(4) - 1] = \pm3$	$(4, 3), (4, -3)$
5	±4	$y = \pm[(5) - 1] = \pm4$	$(5, 4), (5, -4)$

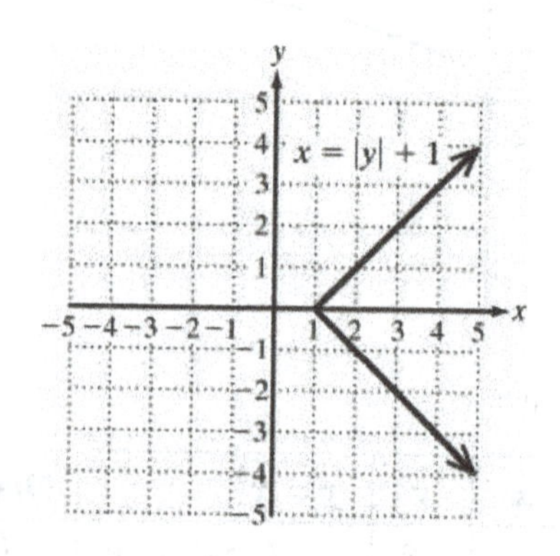

43. $y = |x + 1|$

x	y	$y = \|x+1\|$	Ordered pair
−3	2	$y = \|(-3) + 1\| = 2$	$(-3, 2)$
−2	1	$y = \|(-2) + 1\| = 1$	$(-2, 1)$
−1	0	$y = \|(-1) + 1\| = 0$	$(-1, 0)$
0	1	$y = \|(0) + 1\| = 1$	$(0, 1)$
1	2	$y = \|(1) + 1\| = 2$	$(1, 2)$
2	3	$y = \|(2) + 1\| = 3$	$(2, 3)$
3	4	$y = \|(3) + 1\| = 4$	$(3, 4)$

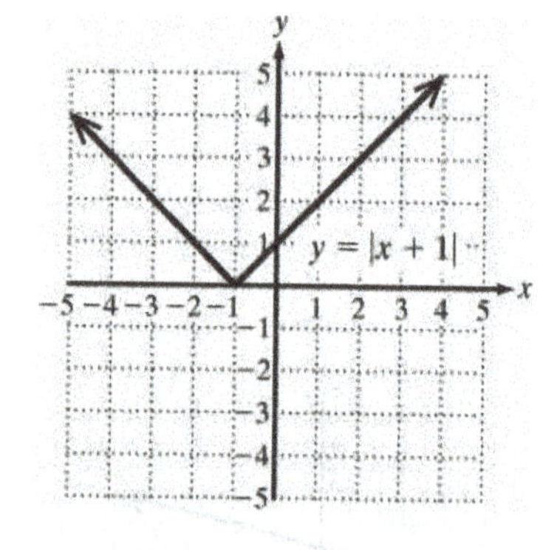

45. x-intercepts: $(-1, 0), (9, 0)$;
y-intercepts: $(0, -3), (0, 3)$

47. x-intercept: $(-2, 0)$; y-intercept: none

49. x-intercept: $(0, 0)$; y-intercept: $(0, 0)$

51. x-intercept: none; y-intercept: none

53. x-intercept: $(1, 0)$; y-intercept: $(0, -2)$

55. Substitute 0 for y:

$$-2x + 4y = 12$$
$$-2x + 4(0) = 12$$
$$-2x = 12$$
$$x = -6$$

Substitute 0 for x:

$$-2x + 4y = 12$$
$$-2(0) + 4y = 12$$
$$4y = 12$$
$$y = 3$$

x-intercept: $(-6, 0)$; y-intercept: $(0, 3)$

57. Substitute 0 for y:

$$x^2 + y = 9$$
$$x^2 + (0) = 9$$
$$x^2 = 9$$
$$x = \pm3$$

Substitute 0 for x:

$$x^2 + y = 9$$
$$(0)^2 + y = 9$$
$$y = 9$$

x-intercepts: $(-3, 0), (3, 0)$; y-intercept: $(0, 9)$

59. Substitute 0 for y:

$$y = |x - 5| - 2$$
$$(0) = |x - 5| - 2$$
$$|x - 5| = 2$$
$$x - 5 = -2 \quad \text{or} \quad x - 5 = 2$$
$$x = 3 \quad \text{or} \quad x = 7$$

Substitute 0 for x:

$y = |x-5| - 2$

$y = |(0)-5| - 2 = 3$

x-intercepts: $(3, 0), (7, 0)$; y-intercept: $(0, 3)$

61. Substitute 0 for y:

$x = y^2 - 1$

$x = (0)^2 - 1 = -1$

Substitute 0 for x:

$x = y^2 - 1$

$(0) = y^2 - 1$

$1 = y^2$

$\pm 1 = y$

x-intercept: $(-1, 0)$; y-intercepts: $(0, -1), (0, 1)$

63. Substitute 0 for y:

$|x| = |y|$

$|x| = |(0)|$

$x = 0$

Substitute 0 for x:

$|x| = |y|$

$|(0)| = |y|$

$0 = y$

x-intercept: $(0, 0)$; y-intercept: $(0, 0)$

65.
$$\frac{(x-3)^2}{4} + \frac{(y-4)^2}{9} = 1$$
$$36\left[\frac{(x-3)^2}{4} + \frac{(y-4)^2}{9}\right] = 36(1)$$
$$9(x-3)^2 + 4(y-4)^2 = 36$$

Substitute 0 for y:
$$9(x-3)^2 + 4(y-4)^2 = 36$$
$$9(x-3)^2 + 4[(0)-4]^2 = 36$$
$$9(x-3)^2 + 64 = 36$$
$$9(x-3)^2 = -28$$
$$(x-3)^2 = -\frac{28}{9}$$
$$x-3 = \pm\sqrt{-\frac{28}{9}}$$
$$x = 3 \pm \sqrt{-\frac{28}{9}}$$

Not a real number.

Substitute 0 for x:
$$9(x-3)^2 + 4(y-4)^2 = 36$$
$$9[(0)-3]^2 + 4(y-4)^2 = 36$$
$$81 + 4(y-4)^2 = 36$$
$$4(y-4)^2 = -45$$
$$(y-4)^2 = -\frac{45}{4}$$
$$y-4 = \pm\sqrt{-\frac{45}{4}}$$
$$y = 4 \pm \sqrt{-\frac{45}{4}}$$

Not a real number.
x-intercept: none; y-intercept: none

67.
$$d_{AC} = \sqrt{[4-(-6)]^2 + (8-10)^2} = \sqrt{100+4} = \sqrt{104} = 2\sqrt{26}$$
$$d_{BC} = \sqrt{(4-6)^2 + (8-0)^2} = \sqrt{4+64} = \sqrt{68} = 2\sqrt{17}$$

Observation tower B is closer.

69. a.
$$l = \sqrt{[1-(-2)]^2 + (-3-0)^2} = \sqrt{9+9} = \sqrt{18} = 3\sqrt{2} \text{ ft}$$
$$w = \sqrt{(3-1)^2 + (1-3)^2} = \sqrt{4+4} = \sqrt{8} = 2\sqrt{2} \text{ ft}$$

b.
$$P = 2l + 2w = 2(3\sqrt{2}) + 2(2\sqrt{2}) = 2(5\sqrt{2}) = 10\sqrt{2} \text{ ft}$$
$$A = lw = (3\sqrt{2})(2\sqrt{2}) = 6(2) = 12 \text{ ft}^2$$

71.
$$C = \left(\frac{-2+4}{2}, \frac{1+3}{2}\right) = \left(\frac{2}{2}, \frac{4}{2}\right) = (1, 2)$$
$$r = \frac{d}{2} = \frac{\sqrt{[4-(-2)]^2 + (3-1)^2}}{2} = \frac{\sqrt{36+4}}{2} = \frac{\sqrt{40}}{2} = \frac{2\sqrt{10}}{2} = \sqrt{10}$$

Center: $(1, 2)$; Radius: $\sqrt{10}$

73. $M=\left(\frac{7+1}{2},\frac{6+(-2)}{2}\right)=\left(\frac{8}{2},\frac{4}{2}\right)=(4,2)$

$h=\sqrt{(4-0)^2+(2-5)^2}=\sqrt{16+9}=\sqrt{25}=5$

$b=\sqrt{(7-1)^2+[6-(-2)]^2}$

$=\sqrt{36+64}=\sqrt{100}=10$

$\text{Area}=\frac{1}{2}bh=\frac{1}{2}(10)(5)=25\text{ m}^2$

75. b

77. c

79. d

81. $M=\left(\frac{1975+1995}{2},\frac{68.8+72.5}{2}\right)$

$=\left(\frac{3970}{2},\frac{141.3}{2}\right)=(1985,70.65)$

70.65 yr

83. $d_{AB}=\sqrt{(4-2)^2+(3-2)^2}=\sqrt{4+1}=\sqrt{5}$

$d_{BC}=\sqrt{(8-4)^2+(5-3)^2}$

$=\sqrt{16+4}=\sqrt{20}=2\sqrt{5}$

$d_{AC}=\sqrt{(2-8)^2+(2-5)^2}$

$=\sqrt{36+9}=\sqrt{45}=3\sqrt{5}$

$d_{AB}+d_{BC}\ 0\ d_{AC}$

$\sqrt{5}+2\sqrt{5}\ 0\ 3\sqrt{5}$

$3\sqrt{5}\ 0\ 3\sqrt{5}$ ✓ True

Collinear

85. $d_{AB}=\sqrt{[1-(-2)]^2+(2-8)^2}$

$=\sqrt{9+36}=\sqrt{45}=3\sqrt{5}$

$d_{BC}=\sqrt{(4-1)^2+(-3-2)^2}=\sqrt{9+25}=\sqrt{34}$

$d_{AC}=\sqrt{(-2-4)^2+[8-(-3)]^2}$

$=\sqrt{36+121}=\sqrt{157}$

$d_{AB}+d_{BC}\ 0\ d_{AC}$

$3\sqrt{5}+\sqrt{34}\ 0\ \sqrt{157}$ False

Not collinear

87. The points (x_1, y_1) and (x_2, y_2) define the endpoints of the hypotenuse d of a right triangle. The lengths of the legs of the triangle are $|x_2-x_1|$ and $|y_2-y_1|$. Applying the Pythagorean theorem produces $d^2=|x_2-x_1|^2+|y_2-y_1|^2$, or equivalently $d=\sqrt{(x_2-x_1)^2+(y_2-y_1)^2}$ for $d\geq 0$.

89. To find the x-intercept(s), substitute 0 for y and solve for x. To find the y-intercept(s), substitute 0 for x and solve for y.

91. $d=\sqrt{(4-5)^2+[6-(-3)]^2+(-1-2)^2}$

$=\sqrt{1+81+9}$

$=\sqrt{91}$

93. $d=\sqrt{(0-3)^2+(-5-7)^2+[1-(-2)]^2}$

$=\sqrt{9+144+9}$

$=\sqrt{162}$

$=9\sqrt{2}$

95.

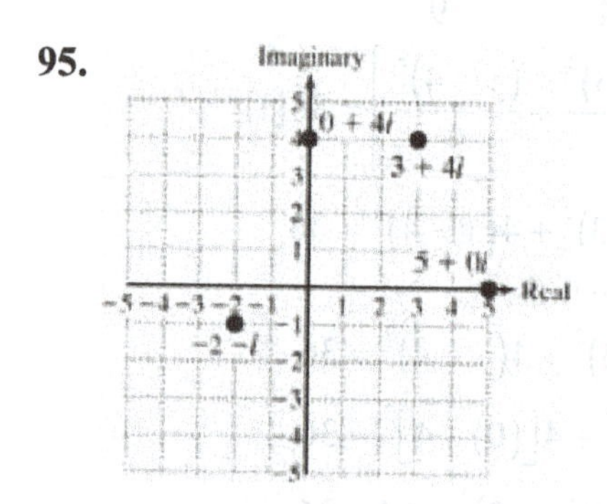

97. $|3-4i|=\sqrt{(3)^2+(-4)^2}=\sqrt{9+16}=\sqrt{25}=5$

99. $|2-7i|=\sqrt{(2)^2+(-7)^2}=\sqrt{4+49}=\sqrt{53}$

101. The viewing window is part of the Cartesian plane shown in the display screen of a calculator. The boundaries of the window are often denoted by [Xmin, Xmax, Xscl] by [Ymin, Ymax, Yscl].

103.

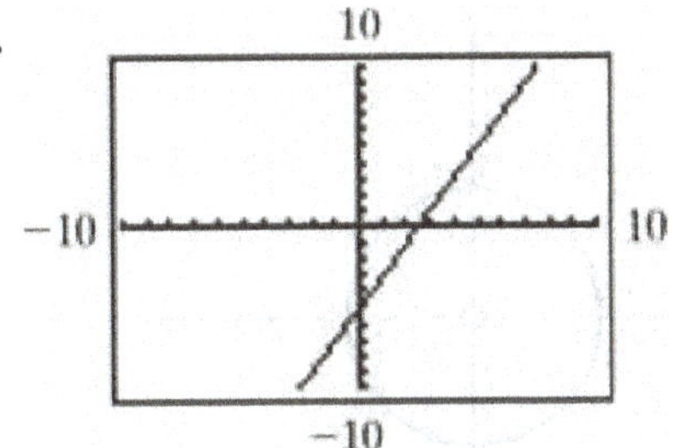

105.

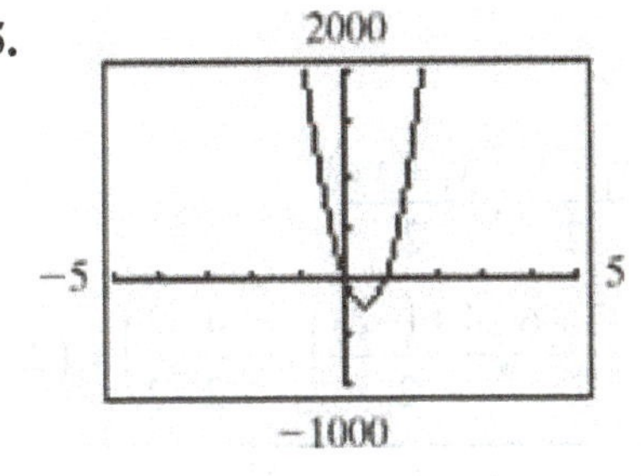

107.

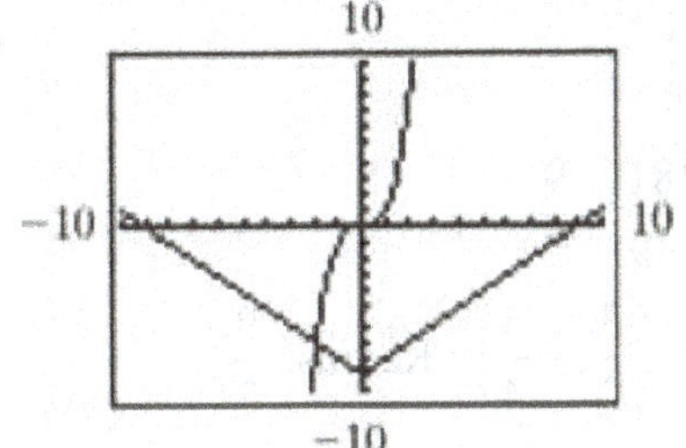

109.

X	Y1	
-2	-15	
-1	0	
0	3	
1	0	
2	-3	
3	0	
4	15	

Y1 ■X^3−3X²−X+3

Section 2.2 Circles

1. a. $d=\sqrt{(x_2-x_1)^2+(y_2-y_1)^2}$

$=\sqrt{[6-(-5)]^2+(7-2)^2}$

$=\sqrt{121+25}=\sqrt{146}$

b. $M=\left(\frac{x_1+x_2}{2},\frac{y_1+y_2}{2}\right)$

$=\left(\frac{-5+6}{2},\frac{2+7}{2}\right)=\left(\frac{1}{2},\frac{9}{2}\right)$

3. Substitute 0 for y:

$x^2+y^2=16$

$x^2+(0)^2=16$

$x^2=16$

$x=\pm 4$

Substitute 0 for x:

$x^2+y^2=16$

$(0)^2+y^2=16$

$y^2=16$

$y=\pm 4$

x-intercepts: $(-4,0),(4,0)$; y-intercepts: $(0,-4),(0,4)$

5. $d=2r$

$10=2r$

$5=r$

5 ft

7. circle; center

9. $(x-h)^2+(y-k)^2=r^2$

11. $(x-2)^2+(y-7)^2=4$

$(2-2)^2+(7-7)^2 \; 0 \; 4$

$0+0 0 \; 4$

$0 0 \; 4$ False

No

13. $(x+1)^2+(y-3)^2=25$

$(-4+1)^2+(7-3)^2 \; 0 \; 25$

$9+16 0 \; 25$

$25 0 \; 25$ ✓ True

Yes

15. $r^2=81$

$r=\sqrt{81}=9$

Center: $(4,-2)$; Radius: 9

17. $r^2=6.25$

$r=\sqrt{6.25}=2.5$

Center: $(0,2.5)$; Radius: 2.5

19. $r^2=20$

$r=\sqrt{20}=2\sqrt{5}$

Center: $(0,0)$; Radius: $2\sqrt{5}$

21. $r^2 = \dfrac{81}{49}$

$r = \sqrt{\dfrac{81}{49}} = \dfrac{9}{7}$

Center: $\left(\dfrac{3}{2}, -\dfrac{3}{4}\right)$; Radius: $\dfrac{9}{7}$

23. a. $(x-h)^2 + (y-k)^2 = r^2$

$[x-(-2)]^2 + (y-5)^2 = (1)^2$

$(x+2)^2 + (y-5)^2 = 1$

b.

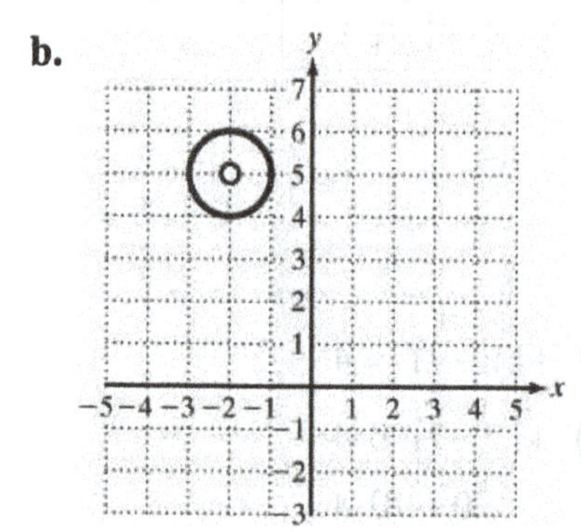

25. a. $(x-h)^2 + (y-k)^2 = r^2$

$[x-(-4)]^2 + [y-(-3)]^2 = \left(\sqrt{11}\right)^2$

$(x+4)^2 + (y+3)^2 = 11$

b.

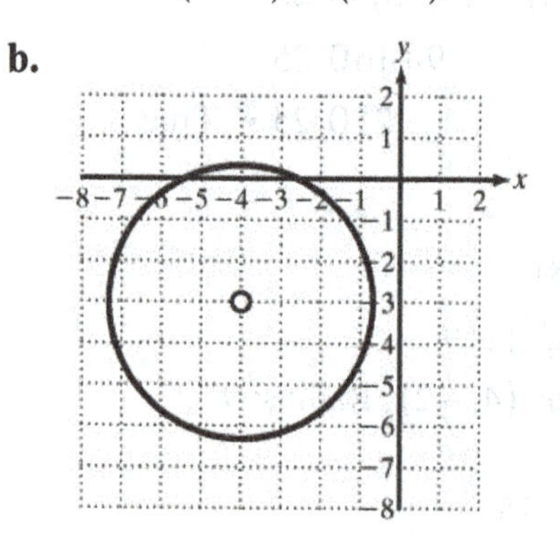

27. a. $(x-h)^2 + (y-k)^2 = r^2$

$(x-0)^2 + (y-0)^2 = (2.6)^2$

$x^2 + y^2 = 6.76$

b.

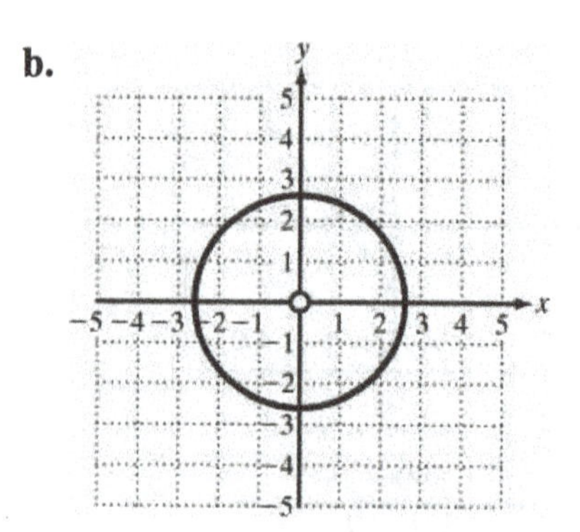

29. a. $C = \left(\dfrac{x_1+x_2}{2}, \dfrac{y_1+y_2}{2}\right)$

$= \left(\dfrac{-2+6}{2}, \dfrac{4+(-2)}{2}\right) = \left(\dfrac{4}{2}, \dfrac{2}{2}\right) = (2,1)$

$r = \sqrt{(x_2-x_1)^2 + (y_2-y_1)^2}$

$= \sqrt{(-2-2)^2 + (4-1)^2}$

$= \sqrt{16+9} = \sqrt{25} = 5$

$(x-h)^2 + (y-k)^2 = r^2$

$(x-2)^2 + (y-1)^2 = (5)^2$

$(x-2)^2 + (y-1)^2 = 25$

b.

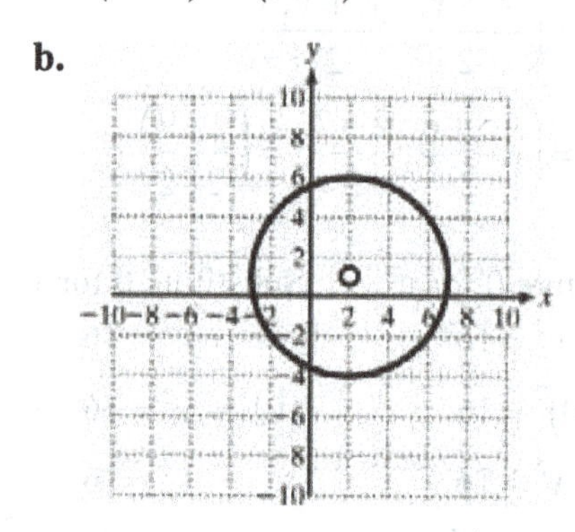

31. a. $r = \sqrt{(x_2-x_1)^2 + (y_2-y_1)^2}$

$= \sqrt{(-2-6)^2 + (-1-5)^2}$

$= \sqrt{64+36} = \sqrt{100} = 10$

$(x-h)^2 + (y-k)^2 = r^2$

$[x-(-2)]^2 + [y-(-1)]^2 = (10)^2$

$(x+2)^2 + (y+1)^2 = 100$

b.

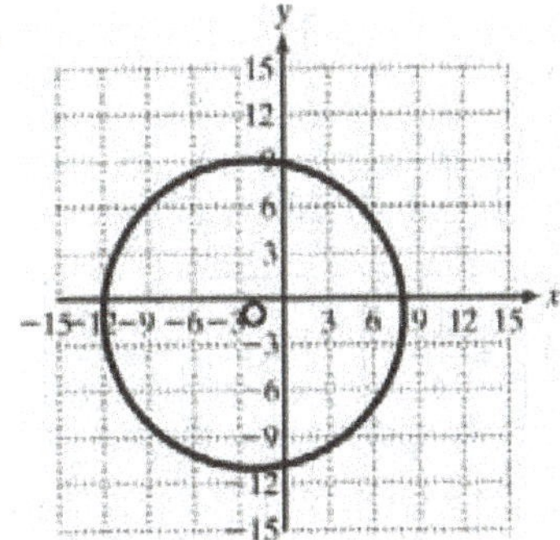

33. a. The circle must touch the y-axis at $(0, 6)$, at one side of a diameter.

$$r = \sqrt{(x_2 - x_1)^2 + (y_2 - y_1)^2}$$
$$= \sqrt{(4-0)^2 + (6-6)^2}$$
$$= \sqrt{16}$$
$$= 4$$
$$(x-h)^2 + (y-k)^2 = r^2$$
$$(x-4)^2 + (y-6)^2 = (4)^2$$
$$(x-4)^2 + (y-6)^2 = 16$$

b.

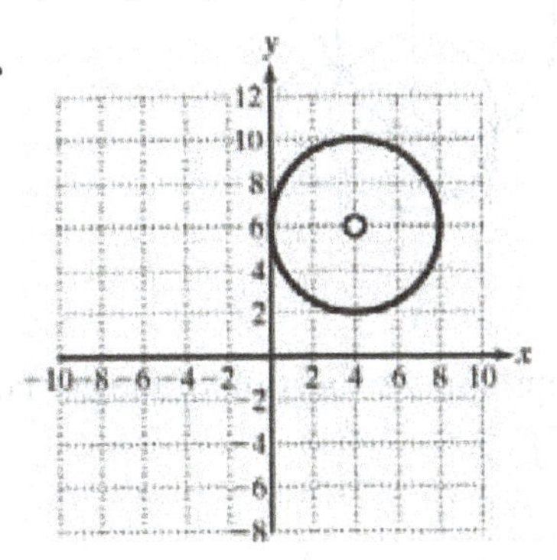

35. a. The circle must touch the x-axis and the y-axis at a distance r from the center. Since the center is in quadrant IV, the center must be $(5, -5)$.

$$(x-h)^2 + (y-k)^2 = r^2$$
$$(x-5)^2 + [y-(-5)]^2 = (5)^2$$
$$(x-5)^2 + (y+5)^2 = 25$$

b.

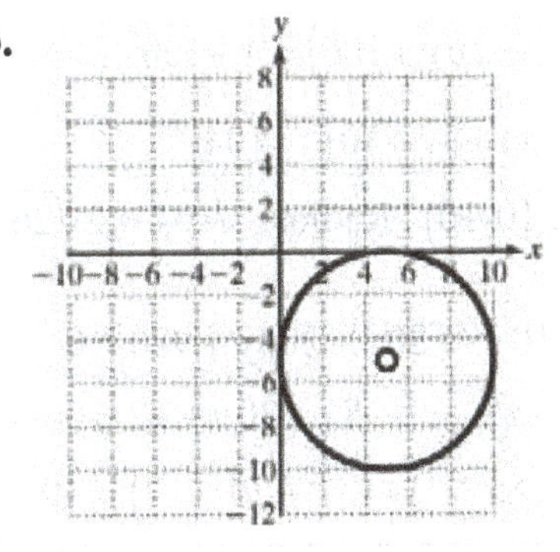

37.
$$(x-h)^2 + (y-k)^2 = r^2$$
$$(x-8)^2 + [y-(-11)]^2 = (5)^2$$
$$(x-8)^2 + (y+11)^2 = 25$$

39. $(x+1)^2 + (y-5)^2 = 0$

The sum of two squares will equal zero only if each individual term is zero. Therefore, $x = -1$ and $y = 5$.

$\{(-1, 5)\}$

41. $(x-17)^2 + (y+1)^2 = -9$

The sum of two squares cannot be negative, so there is no solution.

$\{\ \}$

43.
$$x^2 + y^2 + 6x - 2y + 6 = 0$$
$$(x^2 + 6x \quad) + (y^2 - 2y \quad) = -6$$

$$\boxed{\left[\frac{1}{2}(6)\right]^2 = 9 \quad \left[\frac{1}{2}(-2)\right]^2 = 1}$$

$$(x^2 + 6x + 9) + (y^2 - 2y + 1) = -6 + 9 + 1$$
$$(x+3)^2 + (y-1)^2 = 4$$

Center: $(-3, 1)$; Radius: $\sqrt{4} = 2$

45.
$$x^2 + y^2 - 20y - 4 = 0$$
$$x^2 + (y^2 - 20y \quad) = 4$$

$$\boxed{\left[\frac{1}{2}(-20)\right]^2 = 100}$$

$$x^2+(y^2-20y+100)=4+100$$
$$x^2+(y-10)^2=104$$
Center: $(0,10)$; Radius: $\sqrt{104}=2\sqrt{26}$

47. $10x^2+10y^2-80x+200y+920=0$
$$x^2+y^2-8x+20y+92=0$$
$$(x^2-8x\quad)+(y^2+20y\quad)=-92$$
$$\boxed{\left[\frac{1}{2}(-8)\right]^2=16 \quad \left[\frac{1}{2}(20)\right]^2=100}$$
$$(x^2-8x+16)+(y^2+20y+100)=-92+16+100$$
$$(x-4)^2+(y+10)^2=24$$
Center: $(4,-10)$; Radius: $\sqrt{24}=2\sqrt{6}$

49. $x^2+y^2-4x-18y+89=0$
$$(x^2-4x\quad)+(y^2-18y\quad)=-89$$
$$\boxed{\left[\frac{1}{2}(-4)\right]^2=4 \quad \left[\frac{1}{2}(-18)\right]^2=81}$$
$$(x^2-4x+4)+(y^2-18y+81)=-89+4+81$$
$$(x-2)^2+(y-9)^2=-4$$
Degenerate case: $\{\ \}$

51. $4x^2+4y^2-20y+25=0$
$$x^2+y^2-5y+\frac{25}{4}=0$$
$$x^2+(y^2-5y\quad)=-\frac{25}{4}$$
$$\boxed{\left[\frac{1}{2}(-5)\right]^2=\frac{25}{4}}$$
$$x^2+\left(y^2-5y+\frac{25}{4}\right)=-\frac{25}{4}+\frac{25}{4}$$
$$x^2+\left(y-\frac{5}{2}\right)^2=0$$
Degenerate case (single point): $\left\{\left(0,\frac{5}{2}\right)\right\}$

53. $x^2+y^2-x-\frac{3}{2}y-\frac{3}{4}=0$
$$(x^2-x\quad)+\left(y^2-\frac{3}{2}y\quad\right)=\frac{3}{4}$$
$$\boxed{\left[\frac{1}{2}(-1)\right]^2=\frac{1}{4} \quad \left[\frac{1}{2}\left(-\frac{3}{2}\right)\right]^2=\frac{9}{16}}$$
$$\left(x^2-x+\frac{1}{4}\right)+\left(y^2-\frac{3}{2}y+\frac{9}{16}\right)=\frac{3}{4}+\frac{1}{4}+\frac{9}{16}$$
$$\left(x-\frac{1}{2}\right)^2+\left(y-\frac{3}{4}\right)^2=\frac{25}{16}$$
Center: $\left(\frac{1}{2},\frac{3}{4}\right)$; Radius: $\sqrt{\frac{25}{16}}=\frac{5}{4}$

55. $(x-4)^2+(y-6)^2=(1.5)^2$
$$(x-4)^2+(y-6)^2=2.25$$

57.

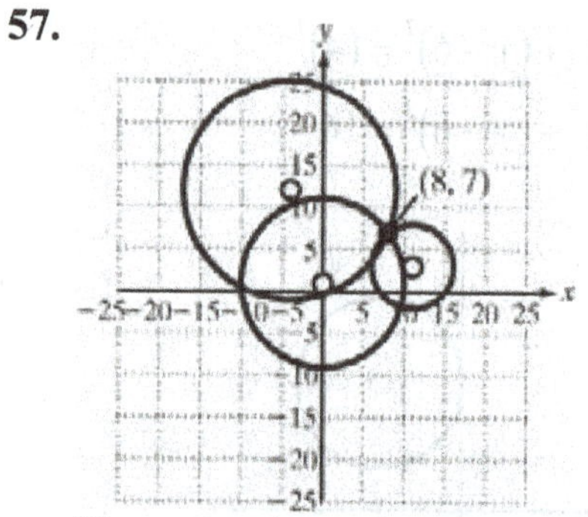

The approximate location of the earthquake is $(8,7)$.

59. $\sqrt{(-2-4)^2+(6-y)^2}=10$
$$\sqrt{36+36-12y+y^2}=10$$
$$\sqrt{72-12y+y^2}=10$$
$$72-12y+y^2=100$$
$$y^2-12y-28=0$$
$$(y+2)(y-14)=0$$
$$y=-2 \quad \text{or} \quad y=14$$
$y=-2$ and $y=14$

61.

$$\sqrt{(2-x)^2+(4-x)^2}=6$$
$$\sqrt{4-4x+x^2+16-8x+x^2}=6$$
$$\sqrt{20-12x+2x^2}=6$$
$$20-12x+2x^2=36$$
$$2x^2-12x-16=0$$
$$x^2-6x-8=0$$
$$x=\frac{-(-6)\pm\sqrt{(-6)^2-4(1)(-8)}}{2(1)}$$
$$=\frac{6\pm\sqrt{68}}{2}$$
$$=\frac{6\pm2\sqrt{17}}{2}=3\pm\sqrt{17}$$

$\left(3+\sqrt{17},3+\sqrt{17}\right)$ and $\left(3-\sqrt{17},3-\sqrt{17}\right)$

63. a.

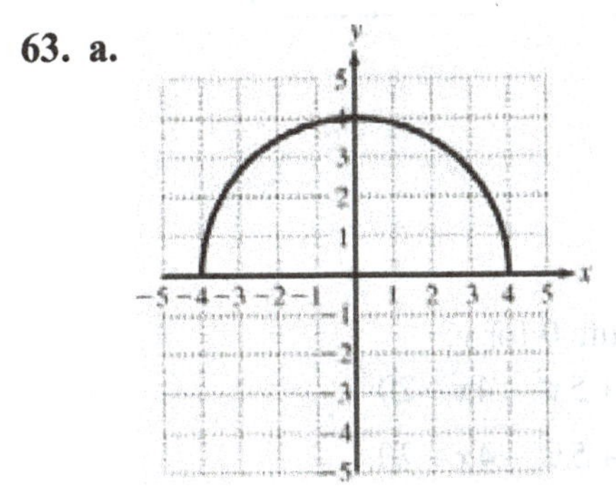

b.

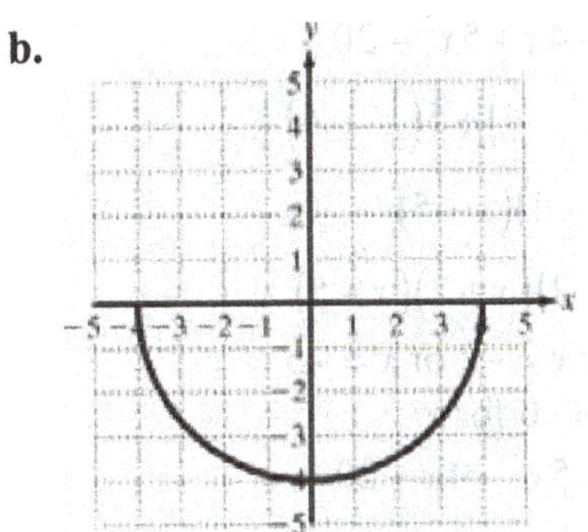

c.

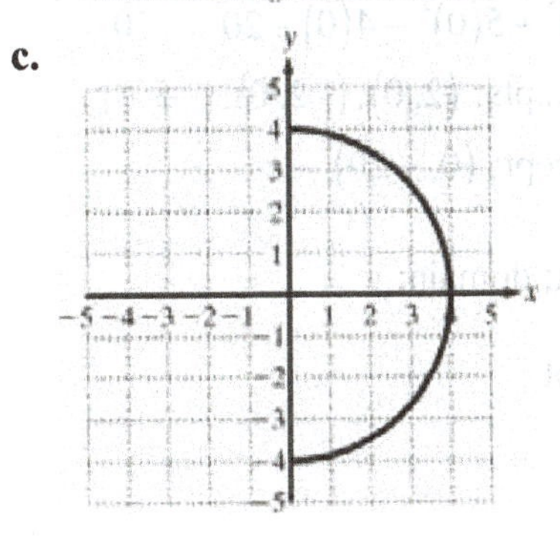

d.

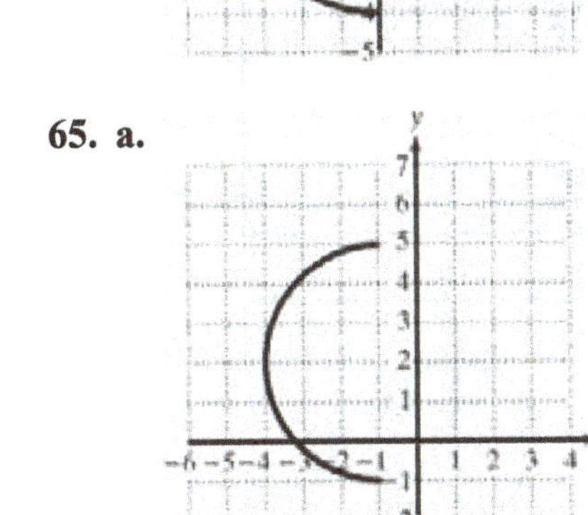

65. a.

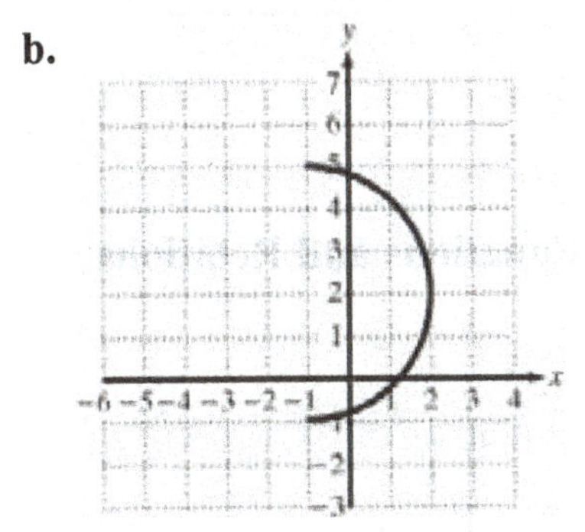

b.

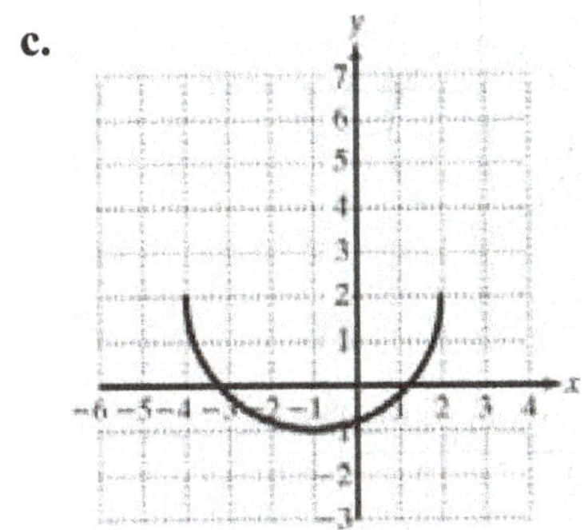

c.

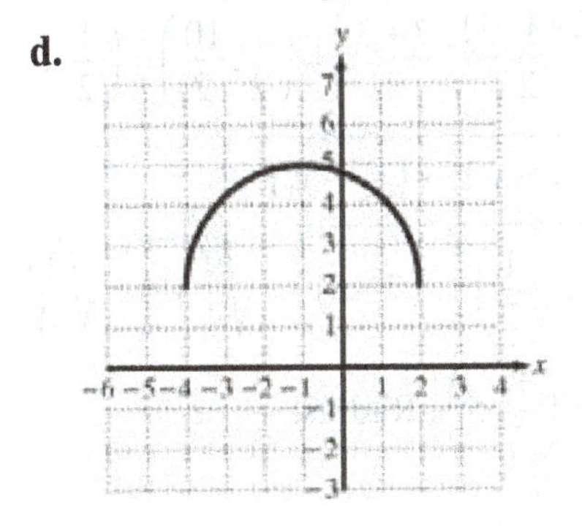

d.

67. A circle is the set of all points in a plane that are equidistant from a fixed point called the center.

69. The calculator does not connect the pixels at the ends of the semicircles because of limited resolution.

71. $[-15.1, 15.1, 1]$ by $[-10, 10, 1]$

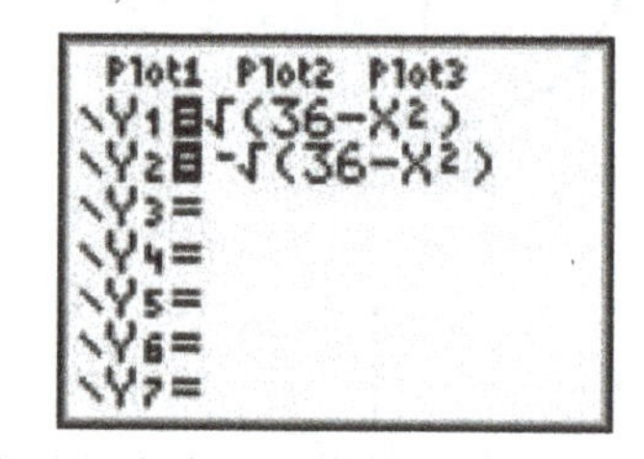

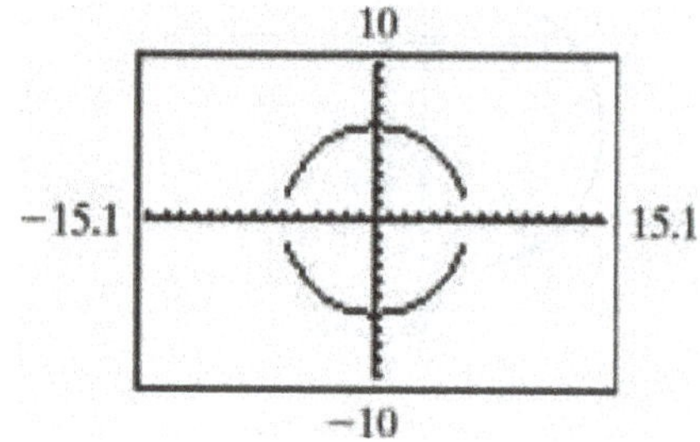

73. $[-14, 39, 5]$ by $[-30, 5, 5]$

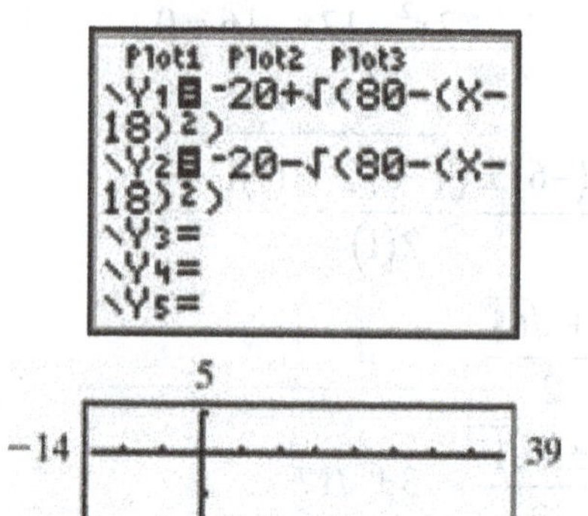

Section 2.3 Functions and Relations

1.

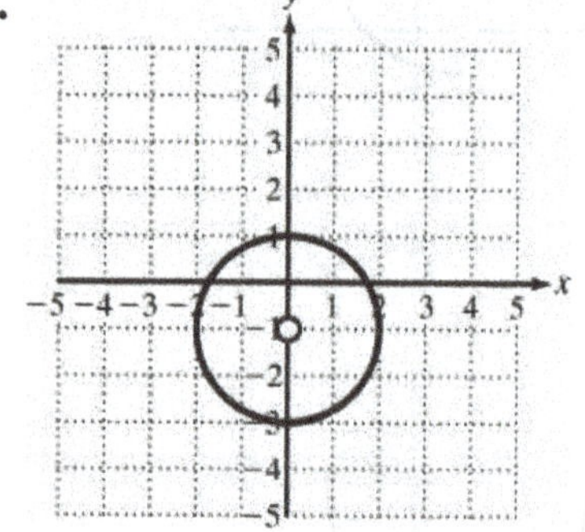

3. $C = \left(\frac{x_1 + x_2}{2}, \frac{y_1 + y_2}{2}\right)$

$= \left(\frac{2+(-1)}{2}, \frac{7+3}{2}\right) = \left(\frac{1}{2}, \frac{10}{2}\right) = \left(\frac{1}{2}, 5\right)$

$r = \sqrt{(x_2 - x_1)^2 + (y_2 - y_1)^2}$

$= \sqrt{\left(2 - \frac{1}{2}\right)^2 + (7-5)^2} = \sqrt{\frac{9}{4} + 4} = \sqrt{\frac{25}{4}} = \frac{5}{2}$

Center: $\left(\frac{1}{2}, 5\right)$; Radius: $\frac{5}{2}$

5. Substitute 0 for y:

$y = x^3 + 5x^2 - 4x - 20$

$0 = x^3 + 5x^2 - 4x - 20$

$0 = x^3 - 4x + 5x^2 - 20$

$0 = x(x^2 - 4) + 5(x^2 - 4)$

$0 = (x^2 - 4)(x + 5)$

$0 = (x - 2)(x + 2)(x + 5)$

$x = 2$ or $x = -2$ or $x = -5$

Substitute 0 for x:

$y = x^3 + 5x^2 - 4x - 20$

$y = (0)^3 + 5(0)^2 - 4(0) - 20 = -20$

x-intercepts: $(2, 0), (-2, 0), (-5, 0)$;

y-intercept: $(0, -20)$

7. relation; domain; y

9. does not

11. y

13. –5

15. a. {(Tom Hanks, 5), (Jack Nicholson, 12), (Sean Penn, 5), (Dustin Hoffman, 7)}

b. {Tom Hanks, Jack Nicholson, Sean Penn, Dustin Hoffman}

c. {5, 12, 7}

d. Yes

17. a. $\{(-4,3),(-2,-3),(1,4),(3,-2),(3,1)\}$

b. $\{-4,-2,1,3\}$

c. $\{3,-3,4,-2,1\}$

d. No

19. False

21. No vertical line intersects the graph in more than one point. This relation is a function.

23. There is at least one vertical line that intersects the graph in more than one point. This relation is not a function.

25. No vertical line intersects the graph in more than one point. This relation is a function.

27. No vertical line intersects the graph in more than one point. This relation is a function.

29. This mapping defines the set of ordered pairs: $\{(-8,3),(-8,4),(-8,5),(-8,6)\}$. All four ordered pairs have the same x value but different y values. This relation is not a function.

31. This mapping defines the set of ordered pairs: $\{(1,4),(2,5),(3,5)\}$. No two ordered pairs have the same x value but different y values. This relation is a function.

33. $(x+1)^2+(y+5)^2=25$
This equation represents the graph of a circle with center $(-1,-5)$ and radius 5. This relation is not a function because it fails the vertical line test.

35. $y=x+3$

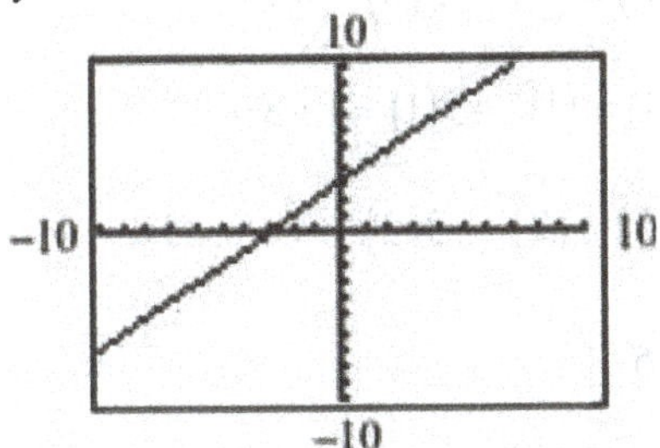

No vertical line intersects the graph in more than one point. This relation is a function.

37. a. $y=x^2$

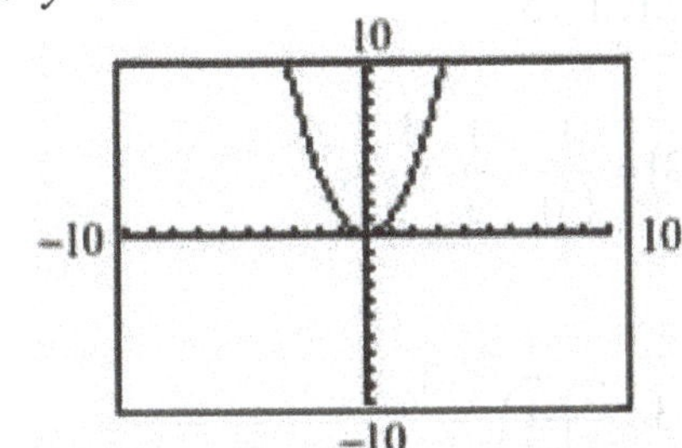

No vertical line intersects the graph in more than one point. This relation is a function.

b. $x=y^2$

$y^2=x$

$y=\pm\sqrt{x}$

x	y	$y=\pm\sqrt{x}$	Ordered pairs
0	0	$y=\pm\sqrt{0}=0$	$(0,0)$
1	±1	$y=\pm\sqrt{1}=\pm1$	$(1,\pm1)$
4	±2	$y=\pm\sqrt{4}=\pm2$	$(4,\pm2)$
9	±3	$y=\pm\sqrt{9}=\pm3$	$(9,\pm3)$

Two or more ordered pairs have the same x value but different y values. This relation is not a function.

39. $(4,1)$

41. $f(x)=x^2+3x$

a. $f(-2)=(-2)^2+3(-2)=4-6=-2$

b. $f(-1)=(-1)^2+3(-1)=1-3=-2$

c. $f(0)=(0)^2+3(0)=0+0=0$

d. $f(1)=(1)^2+3(1)=1+3=4$

e. $f(2)=(2)^2+3(2)=4+6=10$

43. $h(x)=5$

a. $h(-2)=5$ **b.** $h(-1)=5$

c. $h(0)=5$ **d.** $h(1)=5$

e. $h(2)=5$

45. $g(3)=\frac{1}{(3)}=\frac{1}{3}$

47. $g\left(\frac{1}{3}\right)=\frac{1}{\left(\frac{1}{3}\right)}=3$

49. $k(-5)=\sqrt{(-5)+1}=\sqrt{-4}$

Undefined

51. $k(8)=\sqrt{(8)+1}=\sqrt{9}=3$

53. $g(t)=\frac{1}{(t)}=\frac{1}{t}$

55. $k(a+b)=\sqrt{(a+b)+1}=\sqrt{a+b+1}$

57. $f(a+4)=(a+4)^2+3(a+4)$
$=a^2+8a+16+3a+12$
$=a^2+11a+28$

59. $g(0)=\frac{1}{0}$

Undefined

61. $f(9)=7$ **63.** $f(3)=4$

65. $f(x)=6$ when $x=-1$

67. $f(x)=3$ when $x=2$

69. a. $d(t)=18t$
$d(2)=18(2)=36$
Joe rides 36 mi in 2 hr.

b. 40 min is $\frac{40}{60}=\frac{2}{3}$ hr.
$d(t)=18t$
$d\left(\frac{2}{3}\right)=18\left(\frac{2}{3}\right)=12$
12 mi

71. $C(x)=x+0.06x+0.18x$
$C(225)=(225)+0.06(225)+0.18(225)$
$=225+13.5+40.5$
$=279$

If the cost of the food is \$225, then the total bill including tax and tip is \$279.

73. Solve $f(x)=0$:
$f(x)=2x-4$
$0=2x-4$
$2x=4$
$x=2$

Evaluate $f(0)$:
$f(x)=2x-4$
$f(0)=2(0)-4$
$=-4$

x-intercept: $(2,0)$; *y*-intercept: $(0,-4)$

75. Solve $h(x)=0$:
$h(x)=|x|-8$
$0=|x|-8$
$|x|=8$
$x=\pm 8$

Evaluate $h(0)$:
$h(x)=|x|-8$
$h(0)=|(0)|-8$
$=-8$

x-intercepts: $(8,0),(-8,0)$;
y-intercept: $(0,-8)$

77. Solve $p(x)=0$:
$p(x)=-x^2+12$
$0=-x^2+12$
$x^2=12$
$x=\pm\sqrt{12}=\pm 2\sqrt{3}$

Evaluate $p(0)$:
$p(x)=-x^2+12$
$p(0)=-(0)^2+12$
$=12$

x-intercepts: $(2\sqrt{3},0),(-2\sqrt{3},0)$;
y-intercept: $(0,12)$

79. Solve $r(x)=0$:

$r(x)=|x-8|$

$0=|x-8|$

$x-8=0$

$x=8$

Evaluate $r(0)$:

$r(x)=|x-8|$

$r(0)=|(0)-8|$

$=8$

x-intercept: $(8,0)$; y-intercept: $(0,8)$

81. Solve $f(x)=0$:

$f(x)=\sqrt{x}-2$

$0=\sqrt{x}-2$

$\sqrt{x}=2$

$x=4$

Evaluate $f(0)$:

$f(x)=\sqrt{x}-2$

$f(0)=\sqrt{(0)}-2$

$=-2$

x-intercept: $(4,0)$; y-intercept: $(0,-2)$

83. Evaluate $f(0)$:

$f(x)=9.4x+35.7$

$f(0)=9.4(0)+35.7$

$=35.7$

$(0,35.7)$; The y-intercept means that for $x=0$ (the year 2006), the average amount spent on video games per person in the United States was \$35.70.

85. Domain: $\{-3,-2,-1,2,3\}$;

Range: $\{-4,-3,3,4,5\}$

87. Domain: $(-3,\infty)$; Range: $[1,\infty)$

89. Domain: $(-\infty,\infty)$; Range: $(-\infty,\infty)$

91. Domain: $(-\infty,\infty)$; Range: $[-3,\infty)$

93. Domain: $(-5,1]$; Range: $\{-1,1,3\}$

95. $x-4\neq 0$

$x\neq 4$

$(-\infty,4)\cup(4,\infty)$

97. $2t+7\neq 0$

$2t\neq -7$

$t\neq -\frac{7}{2}$

$\left(-\infty,-\frac{7}{2}\right)\cup\left(-\frac{7}{2},\infty\right)$

99. The denominator $|x|+4$ is always positive.

$(-\infty,\infty)$

101. $a+15\geq 0$

$a\geq -15$

$[-15,\infty)$

103. $3-x\geq 0$

$-x\geq -3$

$x\leq 3$

$(-\infty,3]$

105. $3-x>0$

$-x>-3$

$x<3$

$(-\infty,3)$

107. There is no restriction on x.

$(-\infty,\infty)$

109. a. $f(-2)=-4$

b. $f(3)=2$

c. $f(x)=-1$ for $x=-3$, $x=-1$, $x=1$

d. $f(x)=-4$ for $x=-2$, $x=2$

e. x-intercepts: $(0,0)$ and $\left(-\frac{10}{3},0\right)$

f. y-intercept: $(0,0)$

g. Domain: $(-\infty,\infty)$

h. Range: $[-4,\infty)$

111. **a.** $f(-2)=0$

b. $f(3)=5$

c. $f(x)=-1$ for $x=-3$, $x=-1$, $x=1$

d. $f(x)=-4$ for $x=-4$, $x=0$

e. x-intercepts: $(-2, 0)$ and $\left(\frac{4}{3}, 0\right)$

f. y-intercept: $(0, -4)$

g. Domain: $[-4, \infty)$

h. Range: $[-4, 5]$

113. $r(x)+x=400$
$r(x)=400-x$

115. $P(x)=x+x+x$
$P(x)=3x$

117. $C(x)+x=90$
$C(x)=90-x$

119. $f(x)=3x^2-2$

121. If two points in a set of ordered pairs are aligned vertically in a graph, then they have the same x-coordinate but different y-coordinates. This contradicts the definition of a function. Therefore, the points do not define y as a function of x.

123. **a.** $P(s)=4s$

b. $A(s)=s^2$

c. $A(P)=\left(\frac{P}{4}\right)^2$ or $A(P)=\frac{P^2}{16}$

d. $P(A)=4\sqrt{A}$

e. $[d(s)]^2=s^2+s^2=2s^2$
$d(s)=\sqrt{2s^2}$
$d(s)=s\sqrt{2}$

f. $[s(d)]^2+[s(d)]^2=d^2$
$2[s(d)]^2=d^2$
$[s(d)]^2=\frac{d^2}{2}$
$s(d)=\sqrt{\frac{d^2}{2}}$
$s(d)=\frac{d}{\sqrt{2}}$ or $s(d)=\frac{d\sqrt{2}}{2}$

g. $P(d)=4\left(\frac{d\sqrt{2}}{2}\right)$
$P(d)=2\sqrt{2}d$

h. $A(d)=\left(\frac{d\sqrt{2}}{2}\right)^2$
$A(d)=\frac{2d^2}{4}$
$A(d)=\frac{d^2}{2}$

Section 2.4 Linear Equations in Two Variables and Linear Functions

1. **a.** No vertical line intersects the graph in more than one point. This relation is a function.

b. $(-\infty, \infty)$ **c.** $(-\infty, \infty)$

d. $(-2, 0)$ **e.** $(0, 3)$

3. **a.** No, a vertical line is not a function.

b. $\{3\}$ **c.** $(-\infty, \infty)$

d. $(3, 0)$ **e.** None

$$M=\left(\frac{x_1+x_2}{2},\frac{y_1+y_2}{2}\right)$$
$$=\left(\frac{3\sqrt{2}+\sqrt{2}}{2},\frac{\sqrt{5}+\left(-4\sqrt{5}\right)}{2}\right)$$
$$=\left(\frac{4\sqrt{2}}{2},\frac{-3\sqrt{5}}{2}\right)$$

5. a. $$=\left(2\sqrt{2},-\frac{3\sqrt{5}}{3}\right)$$

b. $$d=\sqrt{(x_2-x_1)^2+(y_2-y_1)^2}$$
$$=\sqrt{\left(\sqrt{2}-3\sqrt{2}\right)^2+\left(-4\sqrt{5}-\sqrt{5}\right)^2}$$
$$=\sqrt{\left(-2\sqrt{2}\right)^2+\left(-5\sqrt{5}\right)^2}$$
$$=\sqrt{8+125}=\sqrt{133}$$

7. scatter

9. vertical

11. True

13. zero

15. slope; intercept

17. $m=\dfrac{f(x_2)-f(x_1)}{x_2-x_1}$

19. $-3x+4y=12$

$4y=3x+12$

$y=\dfrac{3}{4}x+3$

x	y
-4	0
-2	$\frac{3}{2}$
0	3
2	$\frac{9}{2}$

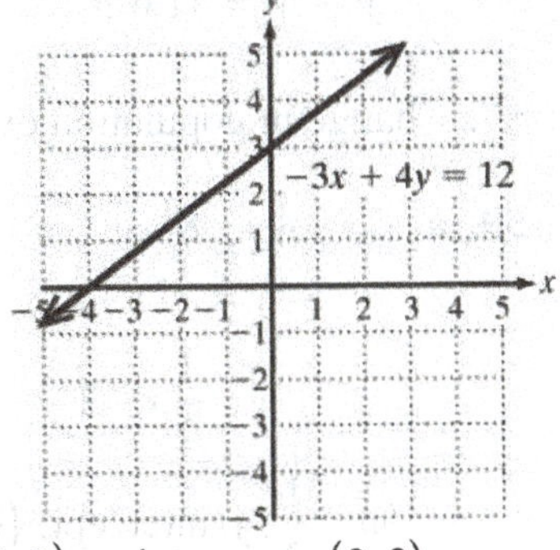

x-intercept: $(-4,0)$; y-intercept: $(0,3)$

21. $2y=-5x+2$

$y=-\dfrac{5}{2}x+1$

x	y
-1	$\frac{7}{2}$
0	1
$\frac{2}{5}$	0
2	-4

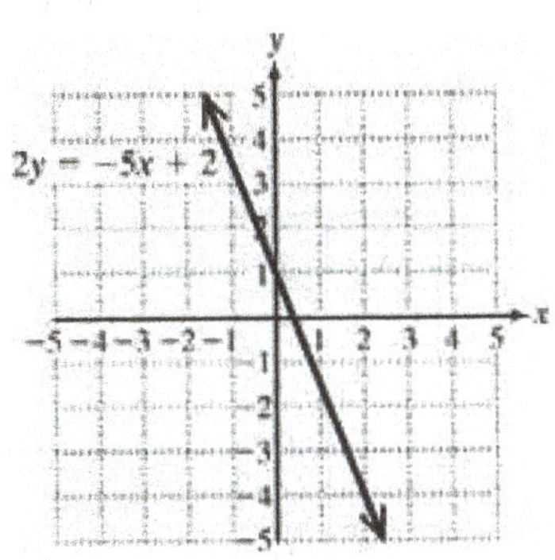

x-intercept: $\left(\frac{2}{5},0\right)$; y-intercept: $(0,1)$

23. $x=-6$

x	y
-6	-2
-6	0
-6	2
-6	4

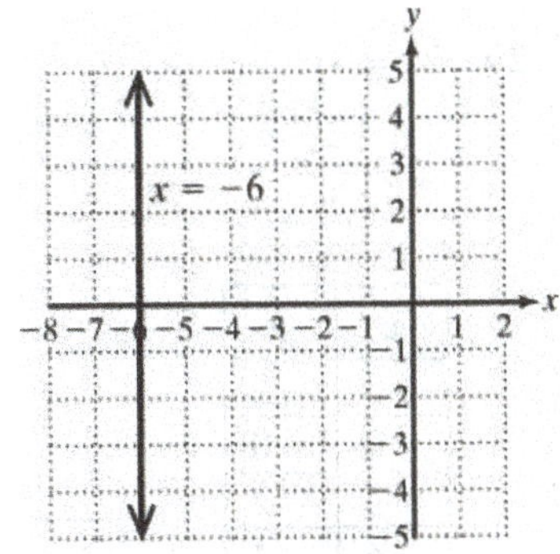

x-intercept: $(-6,0)$; y-intercept: None

25. $5y+1=11$

$5y=10$

$y=2$

x	y
-2	2
0	2
2	2
4	2

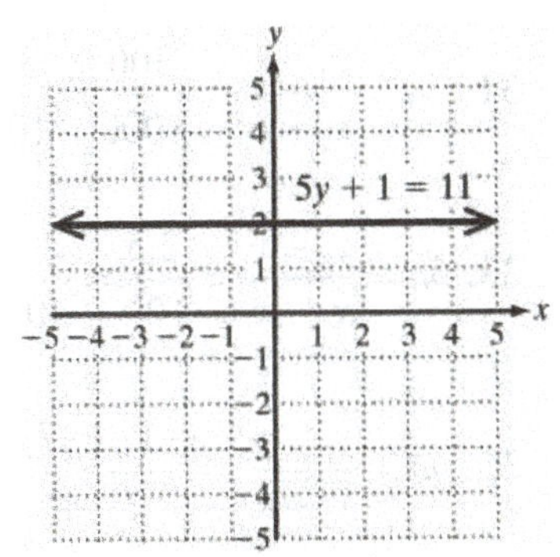

x-intercept: None; y-intercept: $(0,2)$

27. $0.02x + 0.05y = 0.1$

$$2x + 5y = 10$$
$$5y = -2x + 10$$
$$y = -\frac{2}{5}x + 2$$

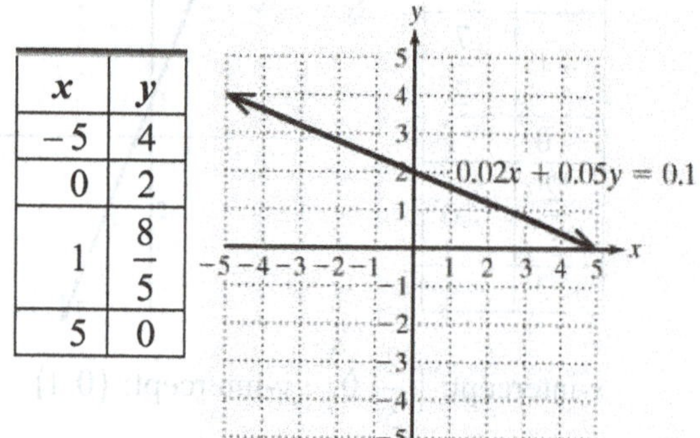

x	y
-5	4
0	2
1	$\frac{8}{5}$
5	0

x-intercept: $(5, 0)$; y-intercept: $(0, 2)$

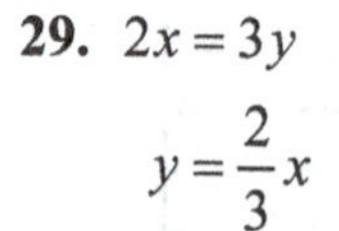

29. $2x = 3y$

$$y = \frac{2}{3}x$$

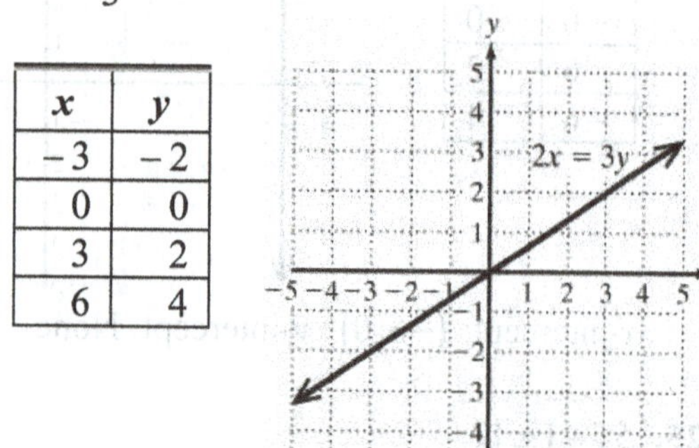

x	y
-3	-2
0	0
3	2
6	4

x-intercept: $(0, 0)$; y-intercept: $(0, 0)$

31. $m = \frac{y_2 - y_1}{x_2 - x_1} = \frac{300}{1000} = \frac{3}{10}$

33. $m = \frac{y_2 - y_1}{x_2 - x_1} = \frac{2.5}{100} = \frac{1}{40}$

35. $m = \frac{y_2 - y_1}{x_2 - x_1} = \frac{-1-(-7)}{2-4} = \frac{6}{-2} = -3$

37. $m = \frac{y_2 - y_1}{x_2 - x_1} = \frac{-39-(-52)}{-22-30} = \frac{13}{-52} = -\frac{1}{4}$

39. $m = \frac{y_2 - y_1}{x_2 - x_1} = \frac{-3.7-4.1}{9.5-2.6} = \frac{-7.8}{6.9} = -\frac{26}{23}$

41. $m = \frac{y_2 - y_1}{x_2 - x_1} = \frac{1-6}{\frac{5}{2} - \frac{3}{4}}$

$$= \frac{-5}{\frac{10}{4} - \frac{3}{4}} = \frac{-5}{\frac{7}{4}} = -\frac{20}{7}$$

43. $m = \frac{y_2 - y_1}{x_2 - x_1}$

$$= \frac{\sqrt{5} - 2\sqrt{5}}{\sqrt{6} - 3\sqrt{6}} = \frac{-\sqrt{5}}{-2\sqrt{6}} = \frac{\sqrt{5}\cdot\sqrt{6}}{2\sqrt{6}\cdot\sqrt{6}} = \frac{\sqrt{30}}{12}$$

45. Use points $(-1, 1)$ and $(0, 4)$:

$$m = \frac{y_2 - y_1}{x_2 - x_1} = \frac{4-1}{0-(-1)} = \frac{3}{1} = 3$$

47. Use points $(0, 2)$ and $(3, 1)$:

$$m = \frac{y_2 - y_1}{x_2 - x_1} = \frac{1-2}{3-0} = \frac{-1}{3} = -\frac{1}{3}$$

49. Use points $(0, -4)$ and $(2, -4)$:

$$m = \frac{y_2 - y_1}{x_2 - x_1} = \frac{-4-(-4)}{2-0} = \frac{0}{2} = 0$$

51. Undefined

53. 0

55.

$$m = \frac{y_2 - y_1}{x_2 - x_1}$$
$$\frac{4}{5} = \frac{y_2 - y_1}{52}$$
$$5(y_2 - y_1) = 208$$
$$y_2 - y_1 = 41.6 \text{ ft}$$

57. Change in population over change in time

59. a. $2x - 4y = 8$

$$-4y = -2x + 8$$
$$y = \frac{1}{2}x - 2$$

$m = \frac{1}{2}$; y-intercept: $(0, -2)$

b.

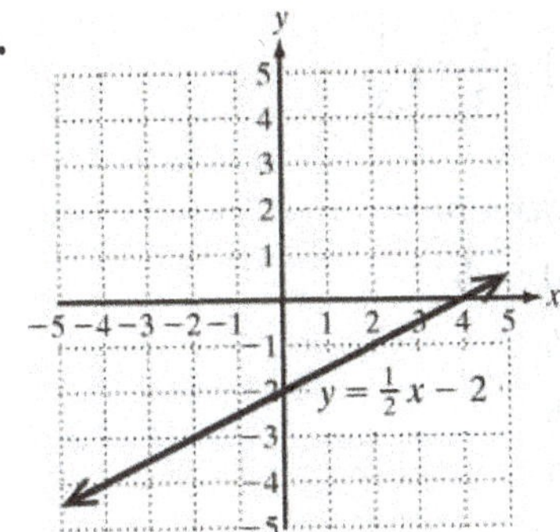

61. a. $3x = 2y - 4$

$2y = 3x + 4$

$y = \frac{3}{2}x + 2$

$m = \frac{3}{2}$; y-intercept: $(0, 2)$

b.

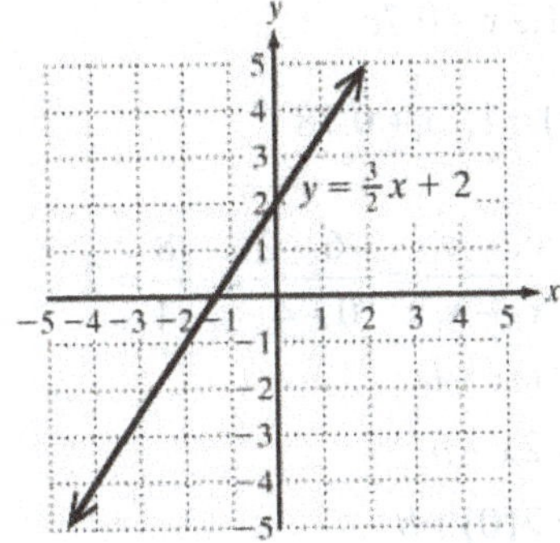

63. a. $3x = 4y$

$y = \frac{3}{4}x$

$m = \frac{3}{4}$; y-intercept: $(0, 0)$

b.

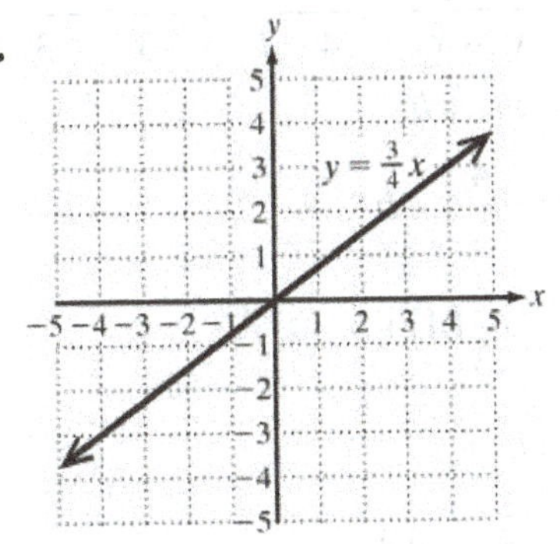

65. a. $2y - 6 = 8$

$2y = 14$

$y = 7$

$m = 0$; y-intercept: $(0, 7)$

b.

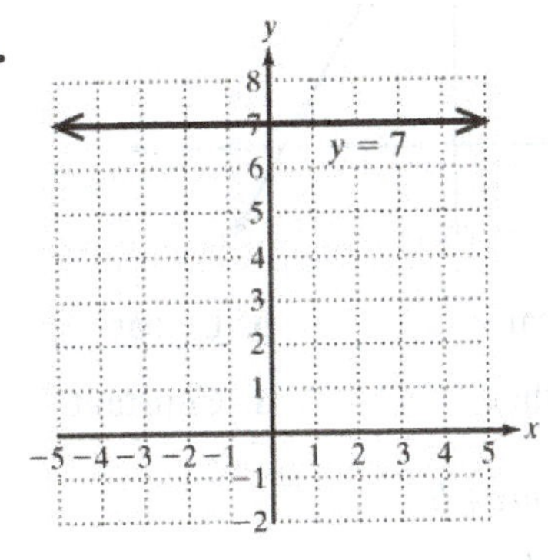

67. a. $0.02x + 0.06y = 0.06$

$2x + 6y = 6$

$6y = -2x + 6$

$y = -\frac{1}{3}x + 1$

$m = -\frac{1}{3}$; y-intercept: $(0, 1)$

b.

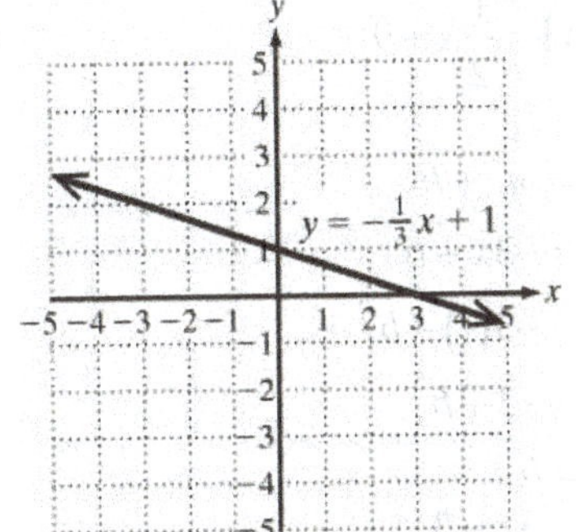

69. a. $\frac{x}{4} + \frac{y}{7} = 1$

$28\left(\frac{x}{4} + \frac{y}{7}\right) = 28(1)$

$7x + 4y = 28$

$4y = -7x + 28$

$y = -\frac{7}{4}x + 7$

$m = -\frac{7}{4}$; y-intercept: $(0, 7)$

b.

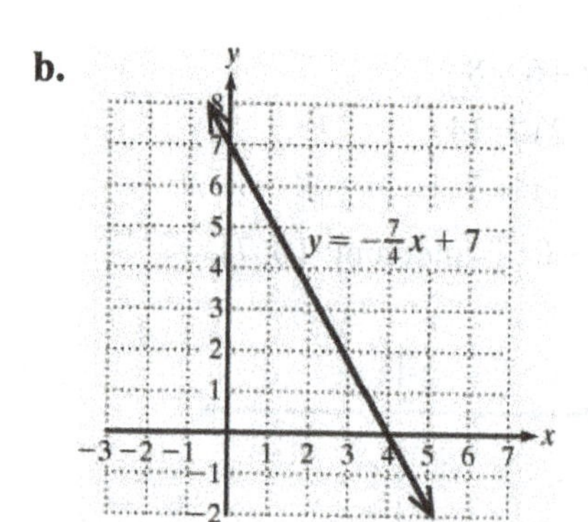

71. a. Linear **b.** Linear

c. Neither **d.** Constant

73. a.
$$y = mx + b$$
$$y = \frac{1}{2}x + b$$
$$9 = \frac{1}{2}(0) + b$$
$$9 = b$$
$$y = \frac{1}{2}x + 9$$

b. $f(x) = \frac{1}{2}x + 9$

75. a.
$$y = mx + b$$
$$y = -3x + b$$
$$-6 = -3(1) + b$$
$$-6 = -3 + b$$
$$-3 = b$$
$$y = -3x - 3$$

b. $f(x) = -3x - 3$

77. a.
$$y = mx + b$$
$$y = \frac{2}{3}x + b$$
$$-3 = \frac{2}{3}(-5) + b$$
$$-3 = -\frac{10}{3} + b$$
$$\frac{1}{3} = b$$
$$y = \frac{2}{3}x + \frac{1}{3}$$

b. $f(x) = \frac{2}{3}x + \frac{1}{3}$

79. a.
$$y = mx + b$$
$$y = 0x + b$$
$$y = b$$
$$5 = b$$
$$y = 5$$

b. $f(x) = 5$

81. a.
$$y = mx + b$$
$$y = 1.2x + b$$
$$5.1 = 1.2(3.6) + b$$
$$5.1 = 4.32 + b$$
$$0.78 = b$$
$$y = 1.2x + 0.78$$

b. $f(x) = 1.2x + 0.78$

83. a.
$$m = \frac{y_2 - y_1}{x_2 - x_1} = \frac{-6 - 2}{0 - 4} = \frac{-8}{-4} = 2$$
$$y = mx + b$$
$$y = 2x + b$$
$$-6 = 2(0) + b$$
$$-6 = b$$
$$y = 2x - 6$$

b. $f(x) = 2x - 6$

85. a.
$$m = \frac{y_2 - y_1}{x_2 - x_1} = \frac{1 - (-3)}{4 - 7} = \frac{4}{-3} = -\frac{4}{3}$$
$$y = mx + b$$
$$y = -\frac{4}{3}x + b$$
$$1 = -\frac{4}{3}(4) + b$$
$$\frac{19}{3} = b$$
$$y = -\frac{4}{3}x + \frac{19}{3}$$

b. $f(x) = -\frac{4}{3}x + \frac{19}{3}$

87. $m = \dfrac{f(x_2) - f(x_1)}{x_2 - x_1} = \dfrac{4-1}{3-1} = \dfrac{3}{2}$

89. **a.** Average rate of change $= \dfrac{f(x_2) - f(x_1)}{x_2 - x_1}$

$= \dfrac{8244 - 6420}{10-5} = \dfrac{1824}{5} = \$364.80\text{/yr}$

b. Average rate of change $= \dfrac{f(x_2) - f(x_1)}{x_2 - x_1}$

$= \dfrac{17{,}452 - 13{,}591}{25-20} = \dfrac{3861}{5} = \$772.20\text{/yr}$

c. Increasing

91. **a.** $f(t) = 0.009t^2 + 2.10t + 182$

$f(0) = 0.009(0)^2 + 2.10(0) + 182 = 182$

$f(10) = 0.009(10)^2 + 2.10(10) + 182$

$= 0.9 + 21 + 182 = 203.9$

Average rate of change $= \dfrac{f(x_2) - f(x_1)}{x_2 - x_1}$

$= \dfrac{203.9 - 182}{10-0} = \dfrac{21.9}{10} \approx 2.2$ million/yr

b. $f(t) = 0.009t^2 + 2.10t + 182$

$f(40) = 0.009(40)^2 + 2.10(40) + 182$

$= 14.4 + 84 + 182 = 280.4$

$f(50) = 0.009(50)^2 + 2.10(50) + 182$

$= 22.5 + 105 + 182 = 309.5$

Average rate of change $= \dfrac{f(x_2) - f(x_1)}{x_2 - x_1}$

$= \dfrac{309.5 - 280.4}{50-40} = \dfrac{29.1}{10} \approx 2.9$ million/yr

c. Increasing

93. $f(x) = x^2 - 3$

$f(0) = (0)^2 - 3 = -3$

$f(1) = (1)^2 - 3 = 1 - 3 = -2$

$f(3) = (3)^2 - 3 = 9 - 3 = 6$

$f(-2) = (-2)^2 - 3 = 4 - 3 = 1$

a. Average rate of change $= \dfrac{f(x_2) - f(x_1)}{x_2 - x_1}$

$= \dfrac{f(1) - f(0)}{1-0} = \dfrac{-2-(-3)}{1} = \dfrac{1}{1} = 1$

b. Average rate of change $= \dfrac{f(x_2) - f(x_1)}{x_2 - x_1}$

$= \dfrac{f(3) - f(1)}{3-1} = \dfrac{6-(-2)}{2} = \dfrac{8}{2} = 4$

c. Average rate of change $= \dfrac{f(x_2) - f(x_1)}{x_2 - x_1}$

$= \dfrac{f(0) - f(-2)}{0-(-2)} = \dfrac{-3-1}{2} = \dfrac{-4}{2} = -2$

95. $h(x) = x^3$

$h(-1) = (-1)^3 = -1$

$h(0) = (0)^3 = 0$

$h(1) = (1)^3 = 1$

$h(2) = (2)^3 = 8$

a. Average rate of change $= \dfrac{h(x_2) - h(x_1)}{x_2 - x_1}$

$= \dfrac{h(0) - h(-1)}{0-(-1)} = \dfrac{0-(-1)}{1} = \dfrac{1}{1} = 1$

b. Average rate of change $= \dfrac{h(x_2) - h(x_1)}{x_2 - x_1}$

$= \dfrac{h(1) - h(0)}{1-0} = \dfrac{1-0}{1} = \dfrac{1}{1} = 1$

c. Average rate of change $= \dfrac{h(x_2) - h(x_1)}{x_2 - x_1}$

$= \dfrac{h(2) - h(1)}{2-1} = \dfrac{8-1}{1} = \dfrac{7}{1} = 7$

97. $m(x)=\sqrt{x}$

$m(0)=\sqrt{0}=0$

$m(1)=\sqrt{1}=1$

$m(4)=\sqrt{4}=2$

$m(9)=\sqrt{9}=3$

a. Average rate of change $=\dfrac{m(x_2)-m(x_1)}{x_2-x_1}$

$=\dfrac{m(1)-m(0)}{1-0}=\dfrac{1-0}{1}=\dfrac{1}{1}=1$

b. Average rate of change $=\dfrac{m(x_2)-m(x_1)}{x_2-x_1}$

$=\dfrac{m(4)-m(1)}{4-1}=\dfrac{2-1}{3}=\dfrac{1}{3}$

c. Average rate of change $=\dfrac{m(x_2)-m(x_1)}{x_2-x_1}$

$=\dfrac{m(9)-m(4)}{9-4}=\dfrac{3-2}{5}=\dfrac{1}{5}$

99. a. $\{-1\}$ **b.** $(-\infty,-1)$ **c.** $[-1,\infty)$

101. a. $\{2\}$ **b.** $(-\infty,2)$ **c.** $[2,\infty)$

103. a. $\{-5\}$ **b.** $[-5,\infty)$ **c.** $(-\infty,-5]$

105. a. $\{14\}$ **b.** $(14,\infty)$ **c.** $(-\infty,14)$

107. a. $\{8\}$; The number of men and women in college was approximately the same in 1978.

b. $[0,8)$; The number of women in college was less than the number of men in college from 1970 to 1978.

c. $(8,40]$; The number of women in college exceeded the number of men in college from 1978 to 2010.

109. The line will be slanted if both A and B are nonzero. If A is zero and B is not zero, then the equation can be written in the form $y=k$ and the graph is a horizontal line. If B is zero and A is not zero, then the equation can be written in the form $x=k$, and the graph is a vertical line.

111. The slope and y-intercept are easily determined by inspection of the equation.

113.

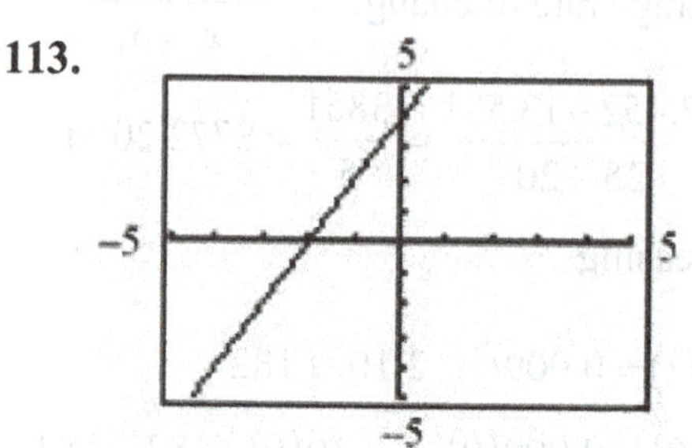

The x-intercept of the line is $(-2,0)$. The base of the triangle is 2 units. The y-intercept of the line is $(0,4)$. The height of the triangle is 4.

$A=\frac{1}{2}bh=\frac{1}{2}(2)(4)=4\text{ units}^2$

115.

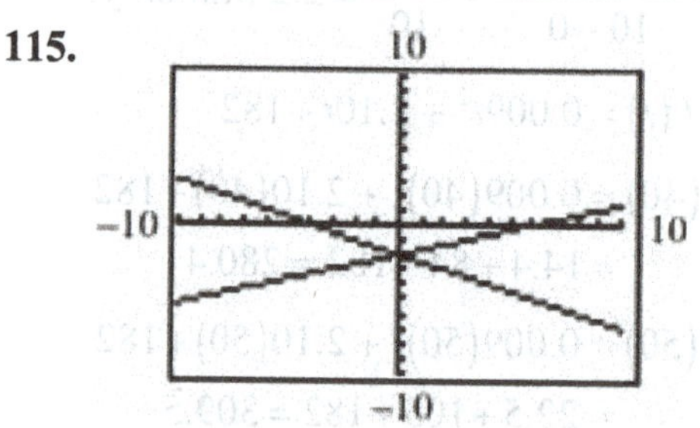

The x-intercepts of the two lines are $(-4,0)$ and $(6,0)$. The base of the triangle is $6-(-4)=10$ units. The y-intercept of the two lines is $(0,-2)$. The height of the triangle is 2.

$A=\frac{1}{2}bh=\frac{1}{2}(10)(2)=10\text{ units}^2$

117. a. $Ax + By = C$

$$By = -Ax + C$$

$$y = -\frac{A}{B}x + \frac{C}{B}$$

b. $m = -\frac{A}{B}$ **c.** $\left(0, \frac{C}{B}\right)$

119. a. $3.1 - 2.2(t+1) = 6.3 + 1.4t$

$$3.1 - 2.2t - 2.2 = 6.3 + 1.4t$$

$$-2.2t + 0.9 = 1.4t + 6.3$$

$$-3.6t = 5.4$$

$$t = -1.5$$

$\{-1.5\}$

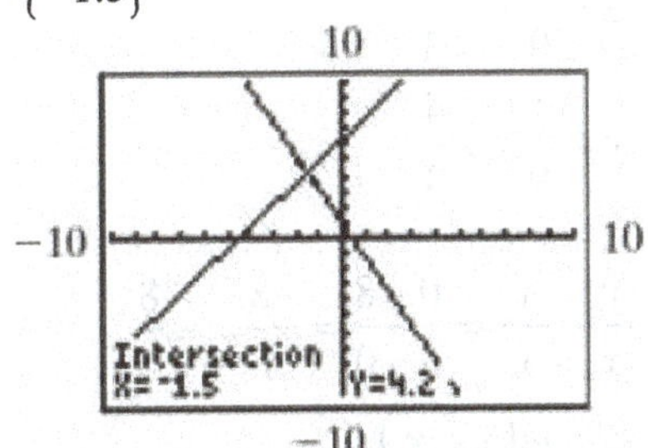

b. $(-\infty, -1.5)$

c. $(-1.5, \infty)$

121. a. $|2x - 3.8| - 4.6 = 7.2$

$$|2x - 3.8| = 11.8$$

$$2x - 3.8 = 11.8 \text{ or } 2x - 3.8 = -11.8$$

$$2x = 15.6 \text{ or } 2x = -8$$

$$x = 7.8 \text{ or } x = -4$$

$\{-4, 7.8\}$

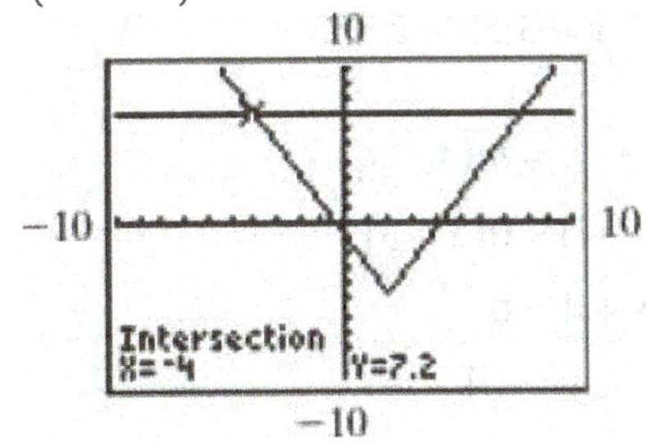

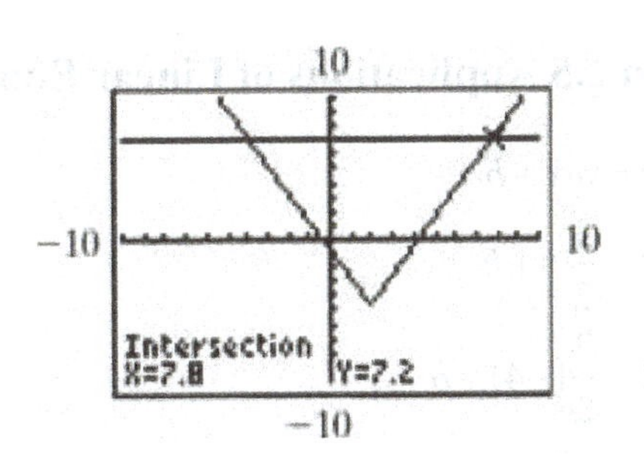

b. $(-\infty, -4] \cup [7.8, \infty)$

c. $[-4, 7.8]$

123. a. $2\sqrt{4z - 3} - 14 = 0$

$$2\sqrt{4z - 3} = 14$$

$$\sqrt{4z - 3} = 7$$

$$4z - 3 = 49$$

$$4z = 52$$

$$z = 13$$

$\{13\}$

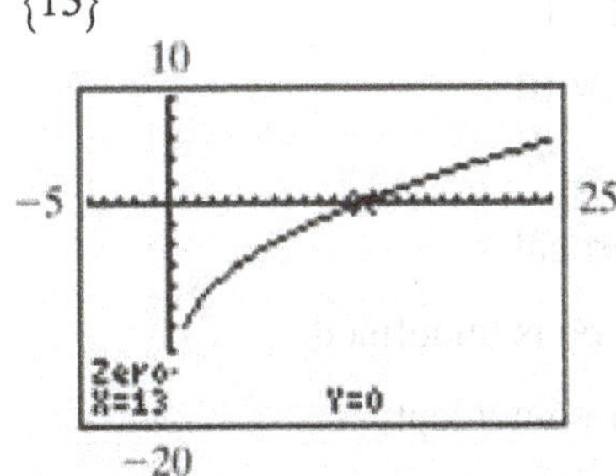

b. $(13, \infty)$ **c.** $\left[\frac{3}{4}, 13\right)$

125.

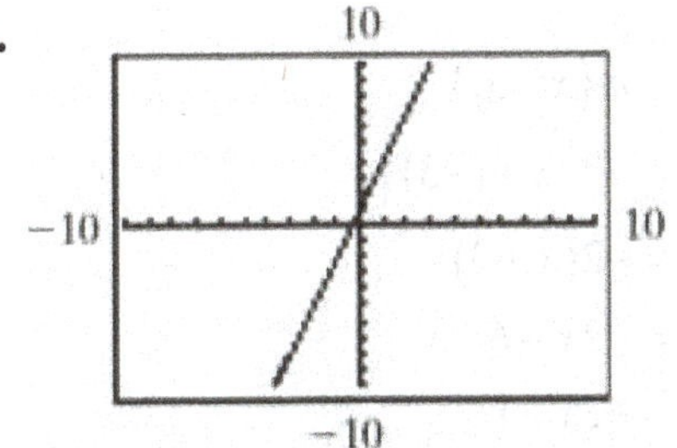

The lines are not exactly the same. The slopes are different.

Section 2.5 Applications of Linear Equations and Modeling

1. $y = mx + b$

$y = \frac{3}{2}x + b$

$2 = \frac{3}{2}(-4) + b$

$2 = -6 + b$

$8 = b$

$y = \frac{3}{2}x + 8$

3. $-6x + 3y = y + 3$

$2y = 6x + 3$

$y = 3x + \frac{3}{2}$

$m = 3$; y-intercept: $\left(0, \frac{3}{2}\right)$

5. $2x + 1 = 11$

$2x = 10$

$x = 5$

a. Vertical

b. Slope is undefined.

c. No y-intercept

7. $y - y_1 = m(x - x_1)$

9. -1

11. $y - y_1 = m(x - x_1)$

$y - 5 = -2[x - (-3)]$

$y - 5 = -2(x + 3)$

$y - 5 = -2x - 6$

$y = -2x - 1$

13. $y - y_1 = m(x - x_1)$

$y - 0 = \frac{2}{3}[x - (-1)]$

$y = \frac{2}{3}(x + 1)$

$y = \frac{2}{3}x + \frac{2}{3}$

15. $y - y_1 = m(x - x_1)$

$y - 2.6 = 1.2(x - 3.4)$

$y - 2.6 = 1.2x - 4.08$

$y = 1.2x - 1.48$

17. $m = \frac{y_2 - y_1}{x_2 - x_1} = \frac{1 - 2}{-3 - 6} = \frac{-1}{-9} = \frac{1}{9}$

$y - y_1 = m(x - x_1)$

$y - 2 = \frac{1}{9}(x - 6)$

$y - 2 = \frac{1}{9}x - \frac{2}{3}$

$y = \frac{1}{9}x + \frac{4}{3}$

19. $m = \frac{y_2 - y_1}{x_2 - x_1} = \frac{0 - 8}{5 - 0} = \frac{-8}{5} = -\frac{8}{5}$

$y - y_1 = m(x - x_1)$

$y - 0 = -\frac{8}{5}(x - 5)$

$y = -\frac{8}{5}x + 8$

21. $m = \frac{y_2 - y_1}{x_2 - x_1} = \frac{3.7 - 5.1}{1.9 - 2.3} = \frac{-1.4}{-0.4} = 3.5$

$y - y_1 = m(x - x_1)$

$y - 5.1 = 3.5(x - 2.3)$

$y - 5.1 = 3.5x - 8.05$

$y = 3.5x - 2.95$

23. $y - y_1 = m(x - x_1)$

$y - (-4) = 0(x - 3)$

$y + 4 = 0$

$y = -4$

25. A line with an undefined slope is a vertical line. Every coordinate has the same x-value, so the equation is $x = \frac{2}{3}$.

27. Undefined

29. **a.** $m = \frac{3}{11}$ **b.** $m = -\frac{1}{\frac{3}{11}} = -\frac{11}{3}$

31. **a.** $m = -6$ **b.** $m = -\frac{1}{-6} = \frac{1}{6}$

33. **a.** $m = 1$ **b.** $m = -\frac{1}{1} = -1$

35. $m_1 \ 0 \ -\frac{1}{m_2}$

$2 0 \ -\frac{1}{-\frac{1}{2}}$

$2 0 \ -2$ ✓ True

Perpendicular

37. $8x - 5y = 3$

$-5y = -8x + 3$

$y = \frac{8}{5}x - \frac{3}{5}$

$2x = \frac{5}{4}y + 1$

$\frac{5}{4}y = 2x - 1$

$y = \frac{8}{5}x - \frac{4}{5}$

$m_1 = m_2$; Parallel

39. $2x = 6$

$x = 3$

$m =$ undefined

$5 = y$

$y = 5$

$m = 0$

Perpendicular

41. $6x = 7y$

$y = \frac{6}{7}x$

$\frac{7}{2}x - 3y = 0$

$-3y = -\frac{7}{2}x$

$y = \frac{7}{6}x$

$m_1 \neq m_2$

$m_1 \ 0 \ -\frac{1}{m_2}$

$\frac{6}{7} 0 \ -\frac{1}{\frac{7}{6}}$

$\frac{6}{7} 0 \ -\frac{6}{7}$ False

Neither

43. $2x + y = 6$

$y = -2x + 6;\ m = -2$

$y - y_1 = m(x - x_1)$

$y - 5 = -2(x - 2)$

$y - 5 = -2x + 4$

$y = -2x + 9$ or $2x + y = 9$

45. $x - 5y = 1$

$-5y = -x + 1$

$y = \frac{1}{5}x - \frac{1}{5};\ m = \frac{1}{5}$

$-\frac{1}{m} = -\frac{1}{\frac{1}{5}} = -5$

$y - y_1 = m(x - x_1)$

$y - (-4) = -5(x - 6)$

$y + 4 = -5x + 30$

$y = -5x + 26$ or $5x + y = 26$

47. $3x = 7y + 5$

$7y = 3x - 5$

$y = \frac{3}{7}x - \frac{5}{7};\ m = \frac{3}{7}$

$y - y_1 = m(x - x_1)$

$y - 8 = \frac{3}{7}(x - 6)$

$y - 8 = \frac{3}{7}x - \frac{18}{7}$

$y = \frac{3}{7}x + \frac{38}{7}$ or $7y = 3x + 38$

$3x - 7y = -38$

49. $2x = 4 - y$

$y = -2x + 4;\ m = -2$

$-\frac{1}{m} = -\frac{1}{-2} = \frac{1}{2} = 0.5$

$y - y_1 = m(x - x_1)$

$y - 6.4 = 0.5(x - 2.2)$

$y - 6.4 = 0.5x - 1.1$

$y = 0.5x + 5.3$ or $10y = 5x + 53$

$5x - 10y = -53$

51. A line that is parallel to the x-axis is a horizontal line with slope 0. Every coordinate has the same y-value, so the equation is $y = 6$.

53. A line that is perpendicular to the y-axis is a horizontal line with slope 0. Every coordinate has the same y-value, so the equation is $y = -\frac{3}{4}$.

55. A line that is parallel to a vertical line is also a vertical line with undefined slope. Every coordinate has the same x-value, so the equation is $x = -61.5$.

57. a. $S(x) = 0.12x + 400$ for $x \geq 0$

b. $S(x) = 0.12(8000) + 400$

$= 960 + 400$

$= 1360$

$S(8000) = 1360$ means that the sales person will make $1360 if $8000 in merchandise is sold for the week.

59. a. $W(t) = 120{,}000 - 2400(2.7)t$ for $0 < t \leq 4.5$

b. $W(2.5) = 120{,}000 - 2400(2.7)(2.5)$

$= 120{,}000 - 16{,}200$

$= 103{,}800$

$W(2.5) = 103{,}800$ means that 2.5 hr into the flight, the mass is 103,800 kg.

61. a. $T(x) = 0.019x + 172$ for $x > 0$

b. $T(80{,}000) = 0.019(80{,}000) + 172$

$= 1520 + 172 = 1692$

$T(80{,}000) = 1692$ means that the property tax is $1692 for a home with a taxable value of 80,000.

63. a. $C(x) = 34.5x + 2275$

b. $R(x) = 80x$

c. $P(x) = R(x) - C(x)$

$= 80x - (34.5x + 2275)$

$= 45.5x - 2275$

d. $P(x) = 0$

$45.5x - 2275 = 0$

$45.5x = 2275$

$x = 50$ items

65. a. $\{730\}$

b. $[0, 730)$

c. $(730, \infty)$

67. a. $C(x) = 0.24(12)x + 790 = 2.88x + 790$

b. $R(x) = 6x$

c. $P(x) = R(x) - C(x)$

$= 6x - (2.88x + 790)$

$= 3.12x - 790$

d. $P(x) > 0$

$3.12x - 790 > 0$

$3.12x > 790$

$x > 253.2$

The business will make a profit if it produces and sells 254 dozen or more cookies.

e. $P(x) = 3.12x - 790$

$P(150) = 3.12(150) - 790$

$= 468 - 790 = -322$

The business will lose $322.

69. a. $m=\frac{y_2-y_1}{x_2-x_1}=\frac{41.8-50.8}{19-1}=\frac{-9}{18}=-\frac{1}{2}$

$y-y_1=m(x-x_1)$

$y-50.8=-0.5(x-1)$

$y-50.8=-0.5x+0.5$

$y=-0.5x+51.3$

b. $m=-0.5$; The slope means that alcohol usage among high school students 1 month prior to the CDC survey has dropped by an average rate of 0.5% per yr.

c. The y-intercept is (0, 51.3) and means that in the year 1990, approximately 51.3% of high school students used alcohol within 1 month prior to the date that the survey was taken.

d. $y=-0.5x+51.3$

$y=-0.5(20)+51.3=-10+51.3=41.3$

Approximately 41.3%

e. No. There is no guarantee that the linear trend will continue well beyond the last observed data point.

71. a. $m=\frac{y_2-y_1}{x_2-x_1}=\frac{13.0-11.2}{14-4}=\frac{1.8}{10}=0.18$

$y-y_1=m(x-x_1)$

$y-11.2=0.18(x-4)$

$y-11.2=0.18x-0.72$

$y=0.18x+10.48$

b. $m=0.18$ means that enrollment in public colleges increased at an average rate of 0.18 million per yr (180,000 per yr).

c. The y-intercept is (0, 10.48) and means that in the year 1990, there were approximately 10,480,000 students enrolled in public colleges.

d. $y=0.18x+10.48$

$y=0.18(25)+10.48$

$=4.5+10.48=14.98$

Approximately 14.98 million

73. a.

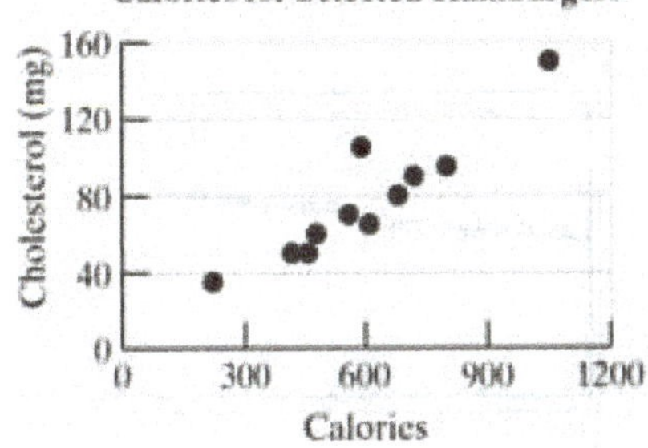

b. $m=\frac{y_2-y_1}{x_2-x_1}=\frac{90-60}{720-480}=\frac{30}{240}=0.125$

$y-y_1=m(x-x_1)$

$y-60=0.125(x-480)$

$y-60=0.125x-60$

$y=0.125x$

$c(x)=0.125x$

c. $m=0.125$ means that the amount of cholesterol increases at an average rate of 0.125 mg per calorie of hamburger.

d. $c(x)=0.125x$

$c(650)=0.125(650)=81.25$ mg

75. Yes; From the graph, the data appear to follow a linear trend.

77. No; From the graph, the data do not appear to follow a linear trend.

79. a. $y=-0.5x+52.3$

b.

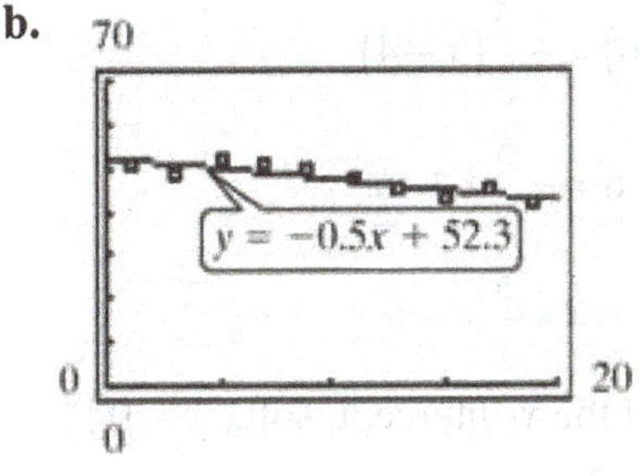

c. $y=-0.5x+52.3$

$y=-0.5(20)+52.3=-10+52.3=42.3$

Approximately 42.3%, which is close to the result of 41.3% obtained in Exercise 69(d).

81. a. $y = 0.175x + 10.46$

b.

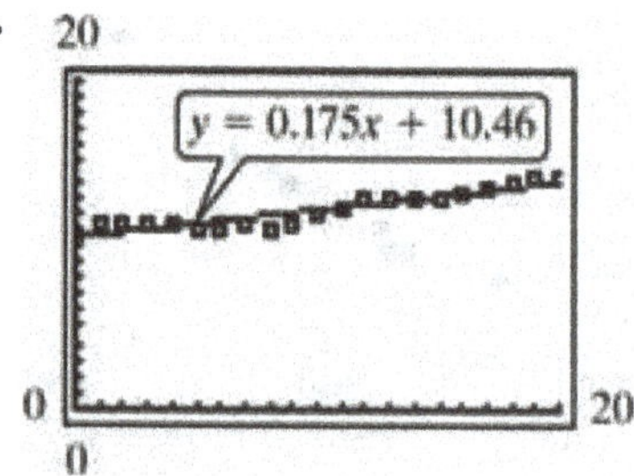

c. $y = 0.175x + 10.46$

$y = 0.175(25) + 10.46$

$= 4.375 + 10.46 = 14.835$

Approximately 14.8 million

83. a. $y = 0.136x - 4.27$

b.

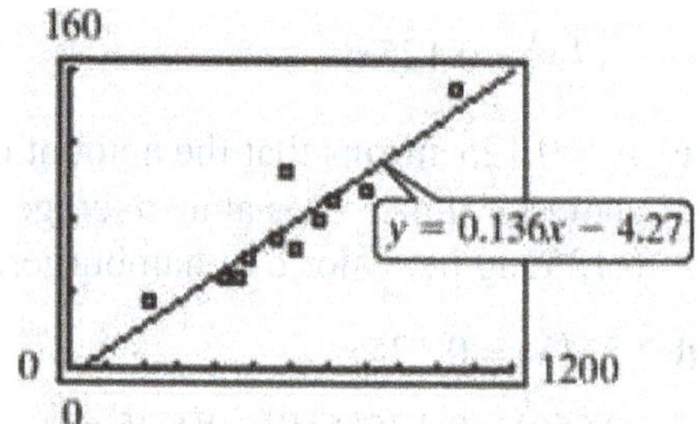

c. $y = 0.136x - 4.27$

$y = 0.136(650) - 4.27 = 88.4 - 4.27 = 84.13$

Approximately 84 mg

85. $m = \dfrac{y_2 - y_1}{x_2 - x_1} = \dfrac{-1-(-6)}{2-4} = \dfrac{5}{-2} = -\dfrac{5}{2}$

$y - y_1 = m(x - x_1)$

$y - (-6) = -\dfrac{5}{2}(x - 4)$

$y + 6 = -\dfrac{5}{2}x + 10$

$y = -\dfrac{5}{2}x + 4$

To find the x-intercept, solve $y = 0$.

$0 = -\dfrac{5}{2}x + 4$

$\dfrac{5}{2}x = 4$

$x = \dfrac{8}{5}$ $\quad\left(\dfrac{8}{5}, 0\right)$

87. Use the points $(0, 4)$ and $(3, 11)$.

$m = \dfrac{y_2 - y_1}{x_2 - x_1} = \dfrac{11-4}{3-0} = \dfrac{7}{3}$

$y - y_1 = m(x - x_1)$

$y - 4 = \dfrac{7}{3}(x - 0)$

$y - 4 = \dfrac{7}{3}x$

$y = \dfrac{7}{3}x + 4$ or $f(x) = \dfrac{7}{3}x + 4$

89. Use the points $(1, 6)$ and $(-3, 2)$.

$m = \dfrac{y_2 - y_1}{x_2 - x_1} = \dfrac{2-6}{-3-1} = \dfrac{-4}{-4} = 1$

$y - y_1 = m(x - x_1)$

$y - 6 = 1(x - 1)$

$y - 6 = x - 1$

$y = x + 5$ or $h(x) = x + 5$

91. If the slopes of the two lines are the same and the y-intercepts are different, then the lines are parallel. If the slope of one line is the opposite of the reciprocal of the slope of the other line, then the lines are perpendicular.

93. Profit is equal to revenue minus cost.

95. $m = \dfrac{y_2 - y_1}{x_2 - x_1} = \dfrac{3-(-1)}{1-(-3)} = \dfrac{4}{4} = 1$

$-\dfrac{1}{m} = -\dfrac{1}{1} = -1$

$y - y_1 = m(x - x_1)$

$y - 3 = -1(x - 1)$

$y - 3 = -x + 1$

$y = -x + 4$

97. Let $c=-2$.

$$\left(c, c^3+1\right)=\left(-2,(-2)^3+1\right)=(-2,-7)$$
$$m=3c^2=3(-2)^2=3(4)=12$$
$$y-y_1=m\left(x-x_1\right)$$
$$y-(-7)=12[x-(-2)]$$
$$y+7=12(x+2)$$
$$y+7=12x+24$$
$$y=12x+17$$

99. $M=\left(\dfrac{x_1+x_2}{2}, \dfrac{y_1+y_2}{2}\right)$

$$=\left(\frac{-2+4}{2}, \frac{9+7}{2}\right)=\left(\frac{2}{2}, \frac{16}{2}\right)=(1,8)$$
$$m=\frac{y_2-y_1}{x_2-x_1}=\frac{-2-8}{5-1}=\frac{-10}{4}=-\frac{5}{2}$$
$$y-y_1=m\left(x-x_1\right)$$
$$y-8=-\frac{5}{2}(x-1)$$
$$y-8=-\frac{5}{2}x+\frac{5}{2}$$
$$y=-\frac{5}{2}x+\frac{21}{2} \text{ for } 1 \leq x \leq 5$$

Problem Recognition Exercises: Comparing Graphs of Equations

1.

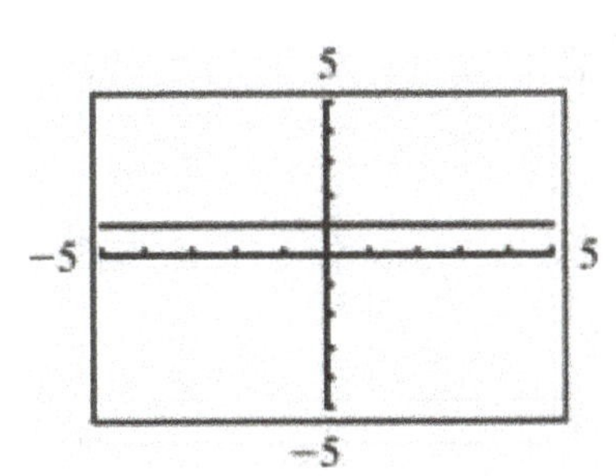

3.

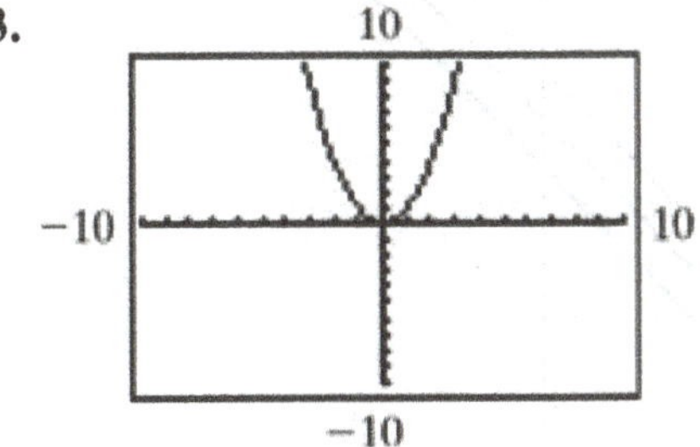

5.

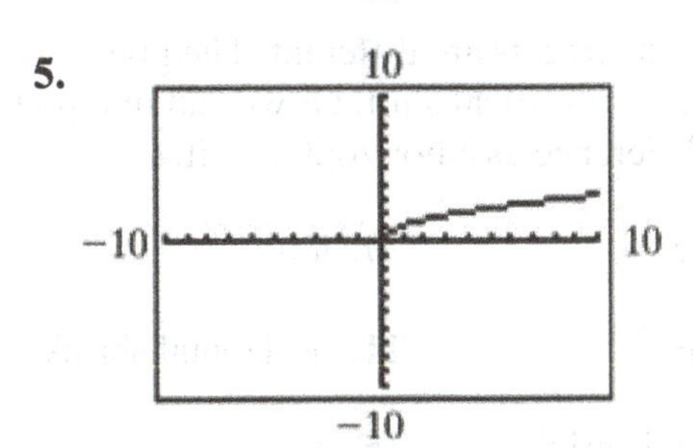

7.

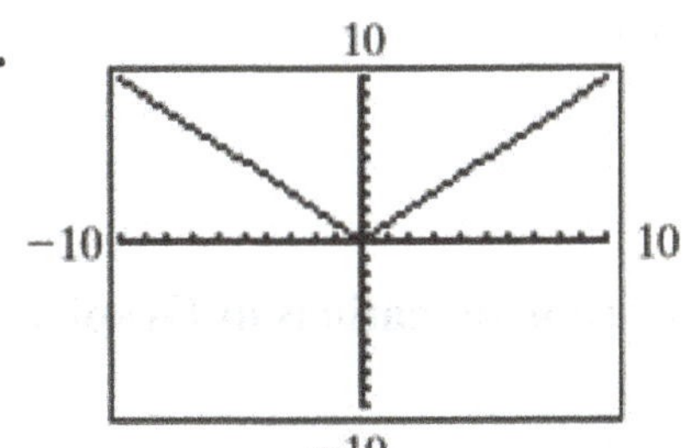

9.

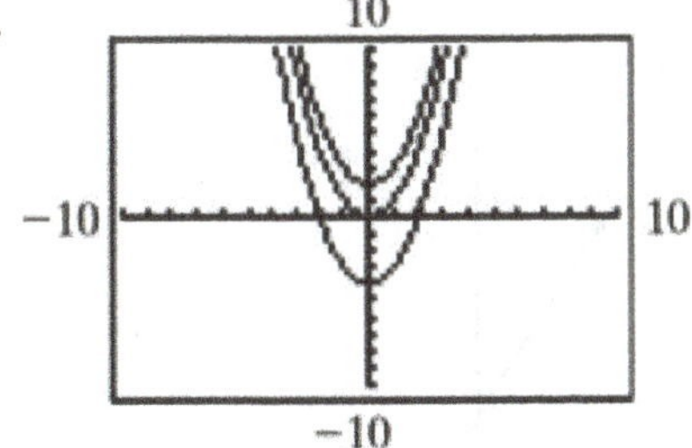

The graphs have the shape of $y=x^2$ with a vertical shift.

11.

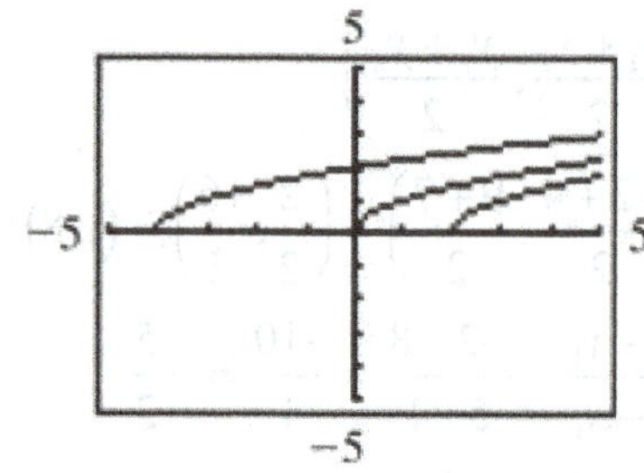

The graphs have the shape of $y=\sqrt{x}$ with a horizontal shift.

13.

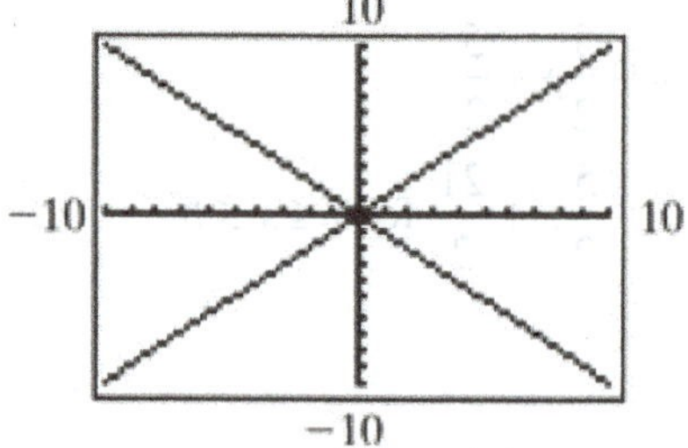

The graph of $g(x)=-|x|$ has the shape of the graph of $y=|x|$ but is reflected across the x-axis.

15.

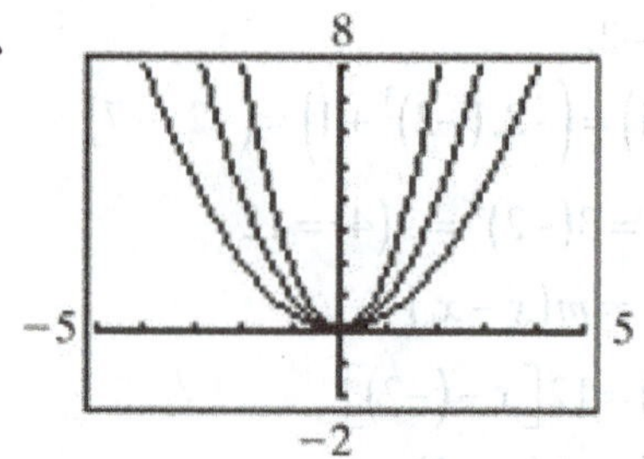

The graphs have the shape of $y=x^2$ but show a vertical shrink or stretch.

17.

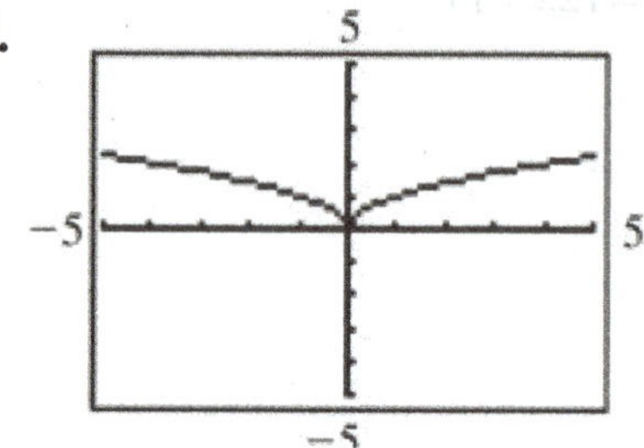

The graph of $g(x)=\sqrt{-x}$ has the shape of the graph of $y=\sqrt{x}$ but is reflected across the y-axis.

Section 2.6 Transformations of Graphs

1. a. $m=-\dfrac{3}{2}$

b. $(0,1)$

c.

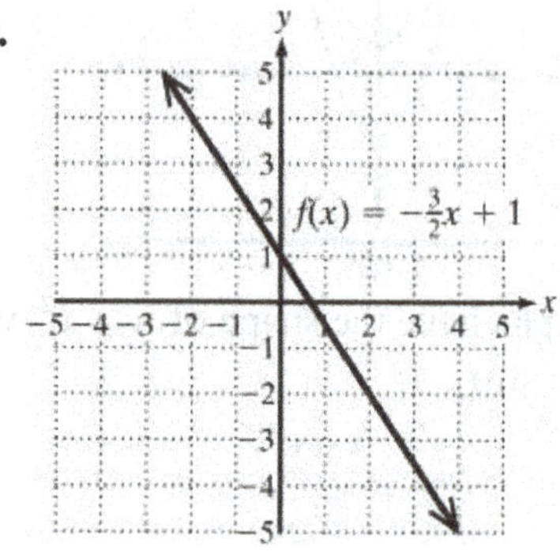

3.

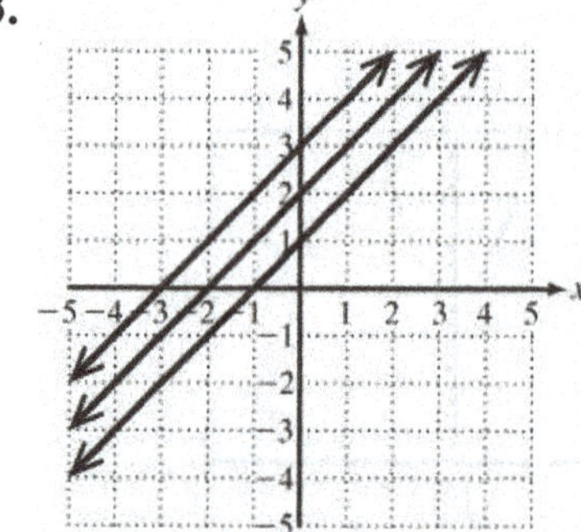

The y-intercepts are different. The graphs differ by a vertical shift. Or we can interpret the difference as a horizontal shift.

5. linear

7. left

9. down

11. horizontal shrink

13. vertical shrink

15. e

17. b

19. a

21. The graph of f is the graph of $f(x)=|x|$ shifted upward 1 unit.

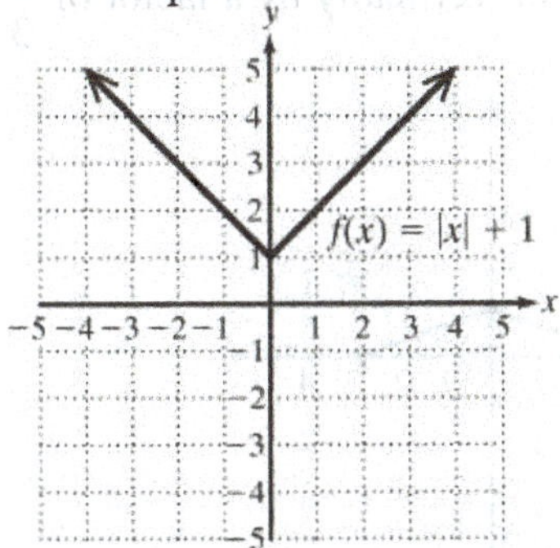

23. The graph of k is the graph of $f(x)=x^3$ shifted downward 2 units.

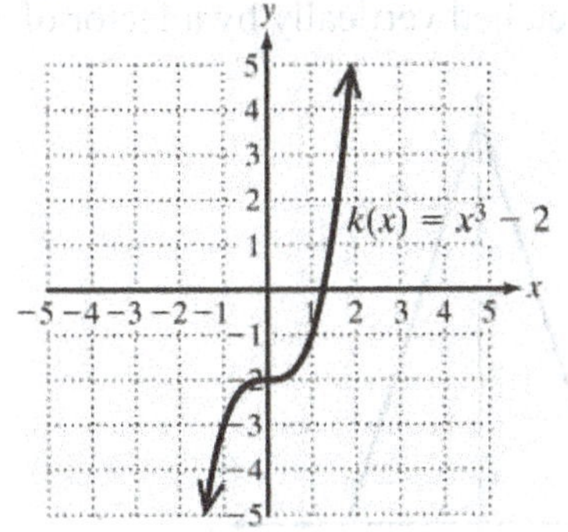

25. The graph of g is the graph of $f(x)=\sqrt{x}$ shifted to the left 5 units.

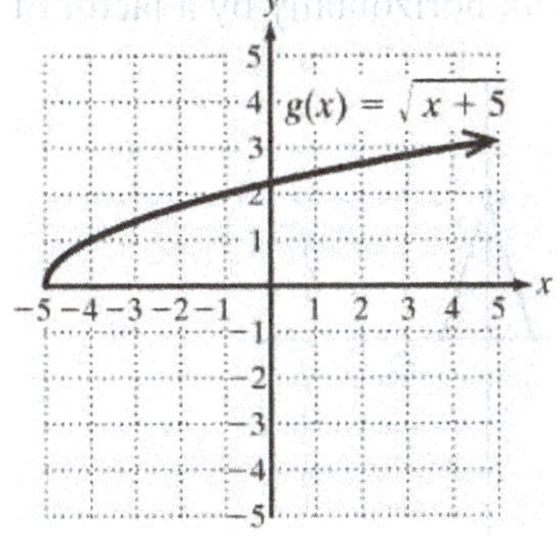

27. The graph of r is the graph of $f(x)=x^2$ shifted to the right 4 units.

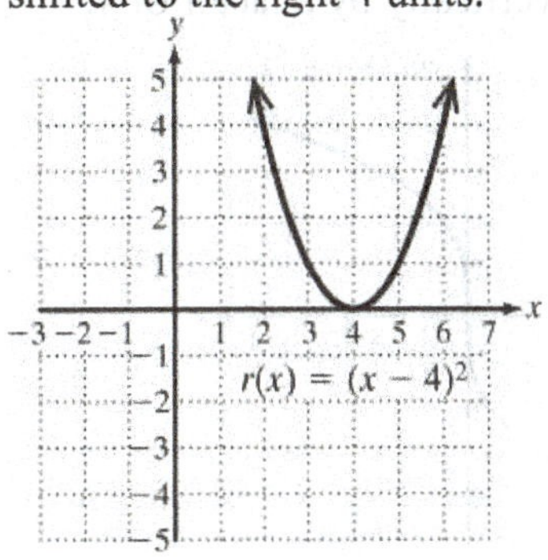

29. The graph of a is the graph of $f(x)=\sqrt{x}$ shifted to the left 1 unit and downward 3 units.

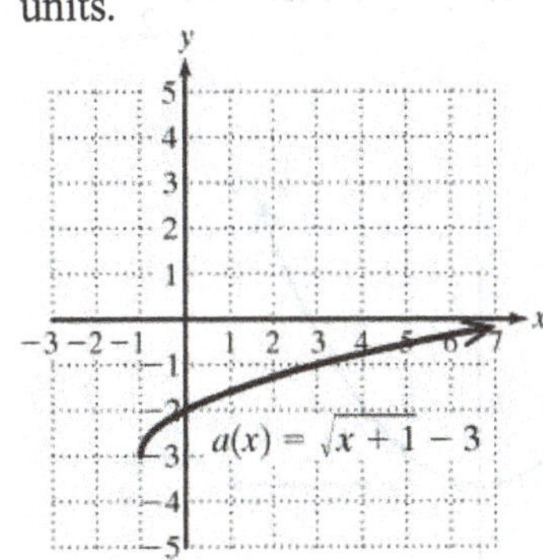

31. The graph of c is the graph of $f(x)=x^2$ shifted to the right 3 units and upward 1 unit.

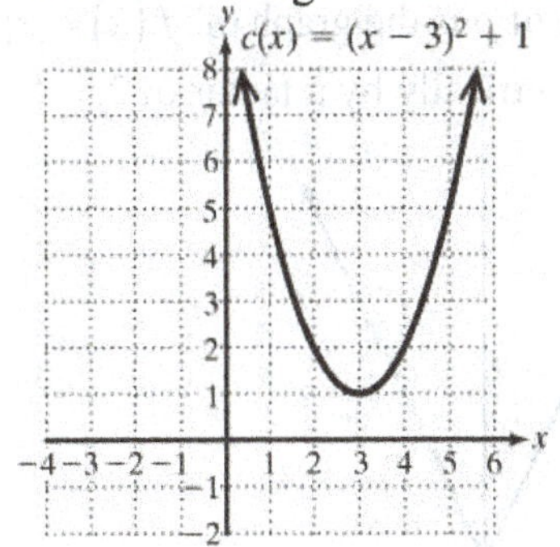

33. The graph of m is the graph of $f(x)=\sqrt[3]{x}$ stretched vertically by a factor of 4.

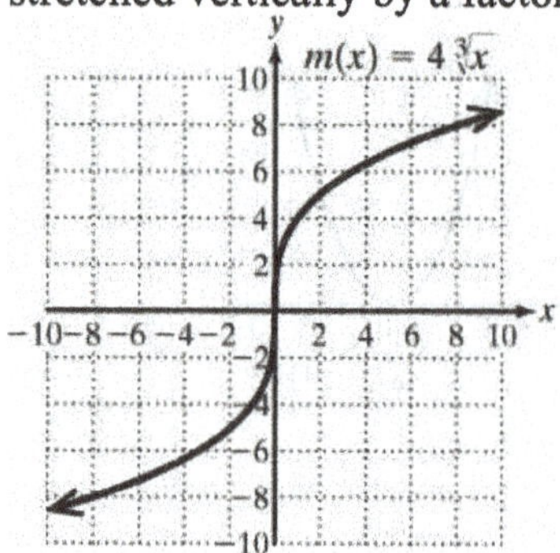

35. The graph of r is the graph of $f(x)=x^2$ shrunk vertically by a factor of $\frac{1}{2}$.

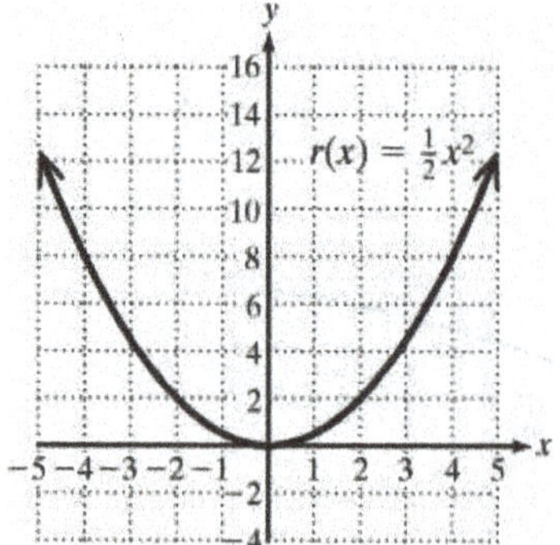

37. $p(x)=|2x|=2|x|$
The graph of p is the graph of $f(x)=|x|$ stretched vertically by a factor of 2.

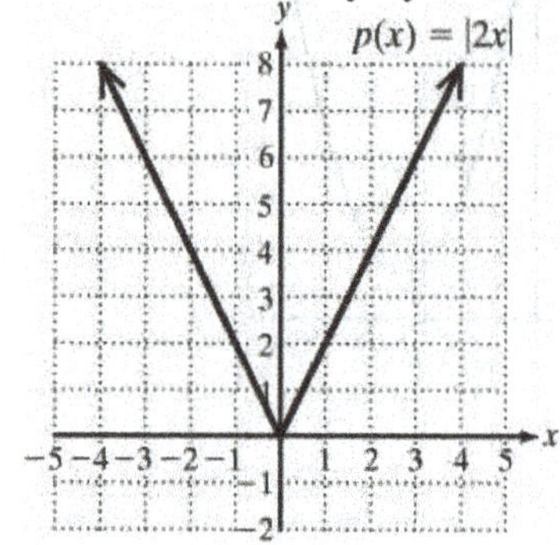

39. The graph of the function is the graph of $f(x)$ shrunk vertically by a factor of $\frac{1}{3}$.

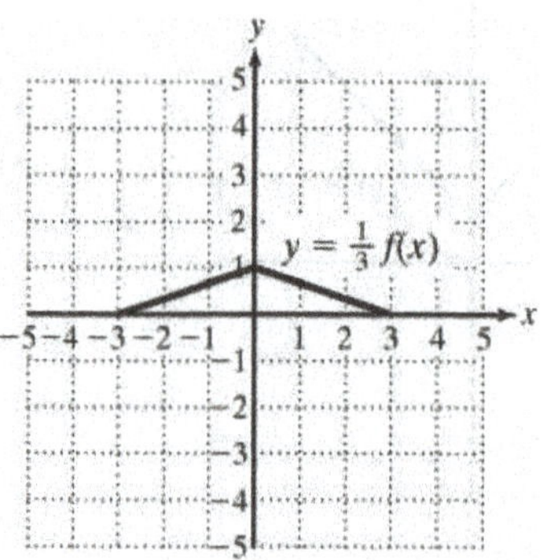

41. The graph of the function is the graph of $f(x)$ stretched vertically by a factor of 3.

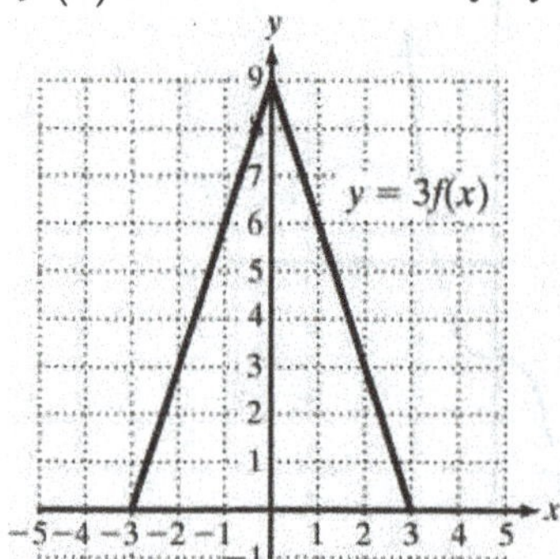

43. The graph of the function is the graph of $f(x)$ shrunk horizontally by a factor of 3.

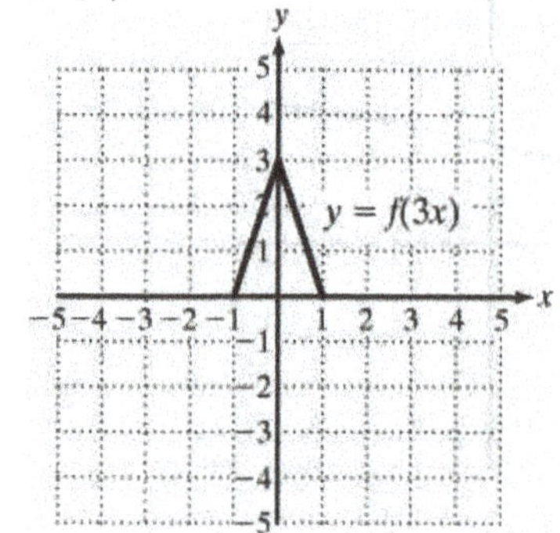

45. The graph of the function is the graph of $f(x)$ stretched horizontally by a factor of $\frac{1}{3}$.

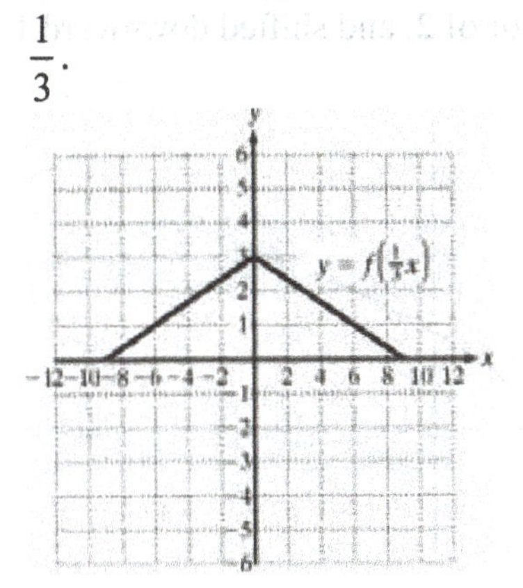

47. The graph of f is the graph of $f(x) = \frac{1}{x}$ reflected across the x-axis.

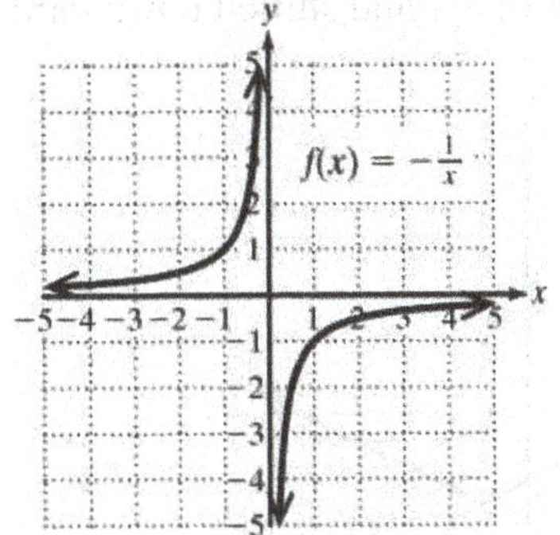

49. The graph of h is the graph of $f(x) = x^3$ reflected across the x-axis.

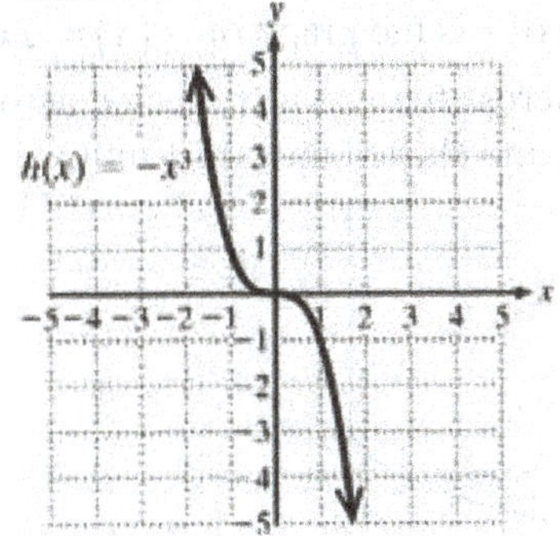

51. The graph of p is the graph of $f(x) = x^3$ reflected across the y-axis.

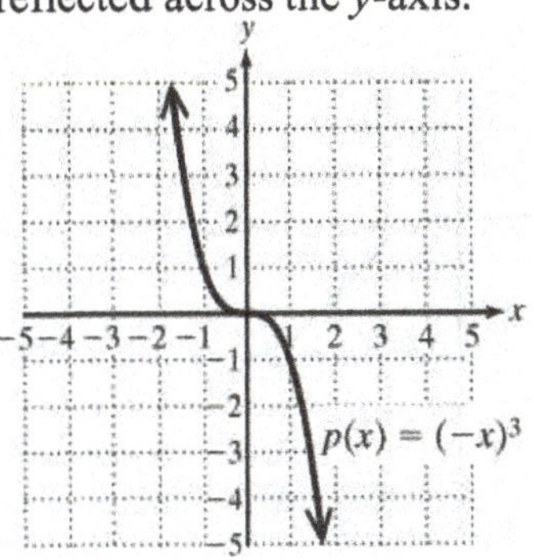

53. The graph of the function is the graph of $f(x)$ reflected across the y-axis.

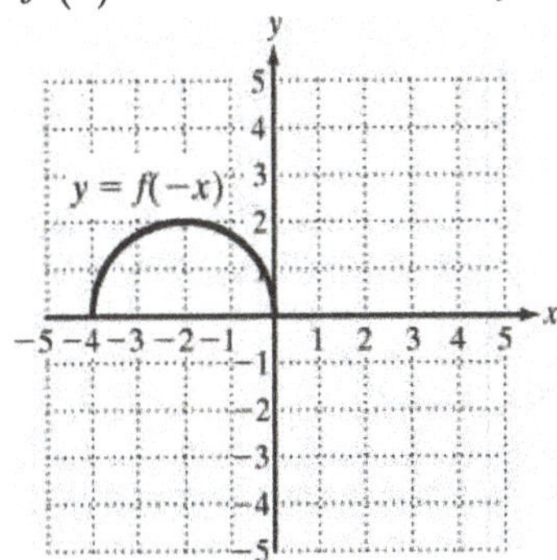

55. The graph of the function is the graph of $f(x)$ reflected across the x-axis.

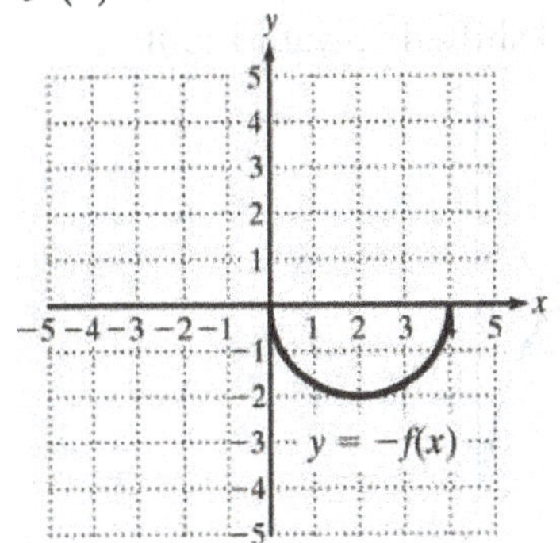

57. The graph of the function is the graph of $f(x)$ reflected across the y-axis.

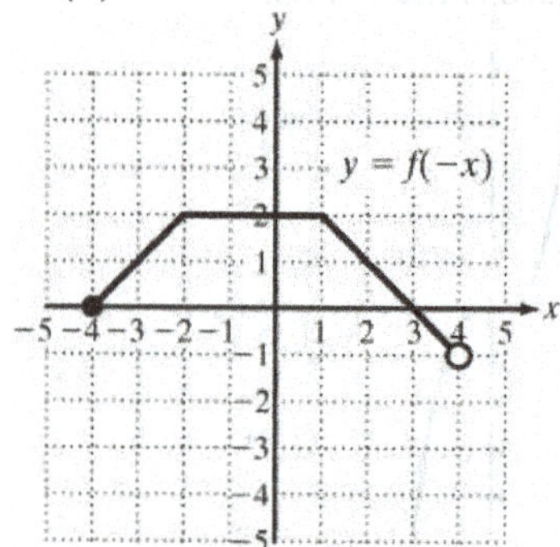

59. The graph of the function is the graph of $f(x)$ reflected across the x-axis.

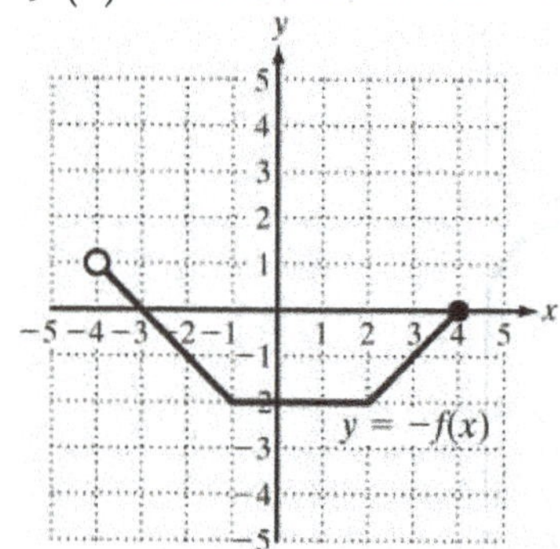

61. The graph of v is the graph of $f(x)=x^2$ shifted to the left 2 units, reflected across the x-axis, and shifted upward 1 unit.

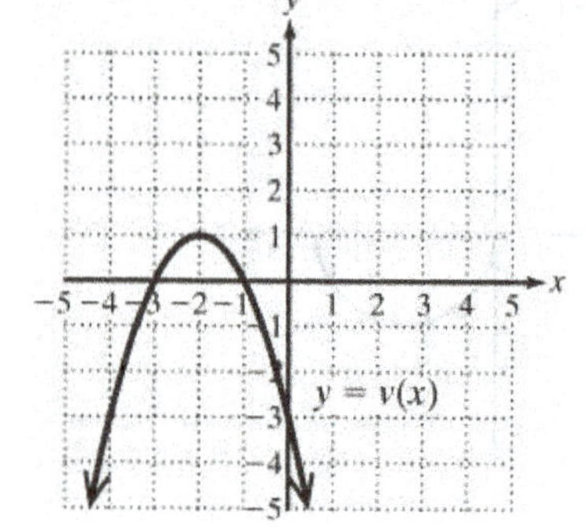

63. The graph of f is the graph of $f(x)=\sqrt{x}$ shifted to the left 3 units, stretched vertically by a factor of 2, and shifted downward 1 unit.

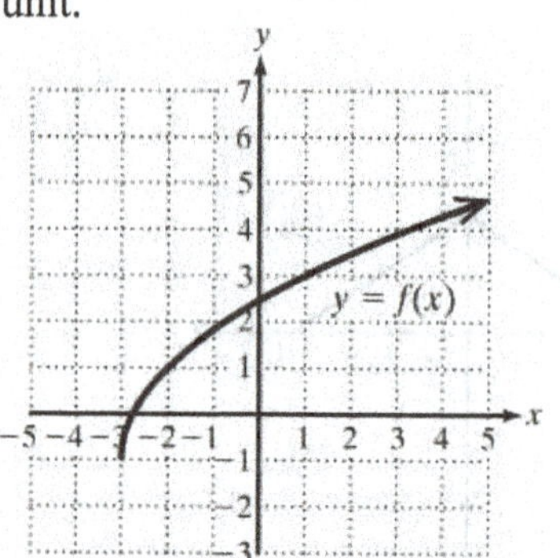

65. The graph of p is the graph of $f(x)=|x|$ shifted to the right 1 unit, shrunk vertically by a factor of $\frac{1}{2}$, and shifted downward 2 units.

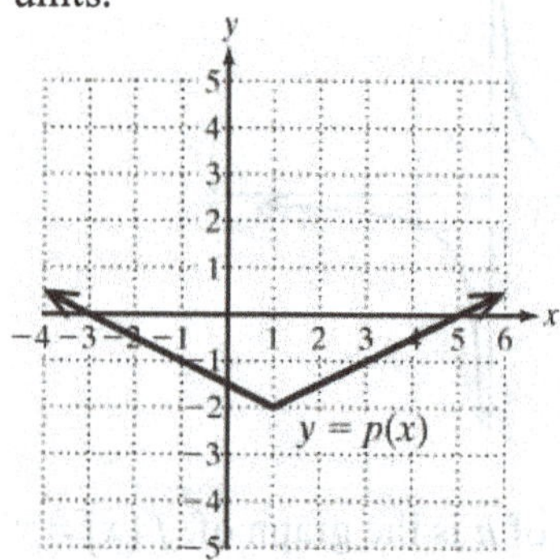

67. The graph of r is the graph of $f(x)=\sqrt{x}$ reflected across the y-axis, reflected across the x-axis, and shifted upward 1 unit.

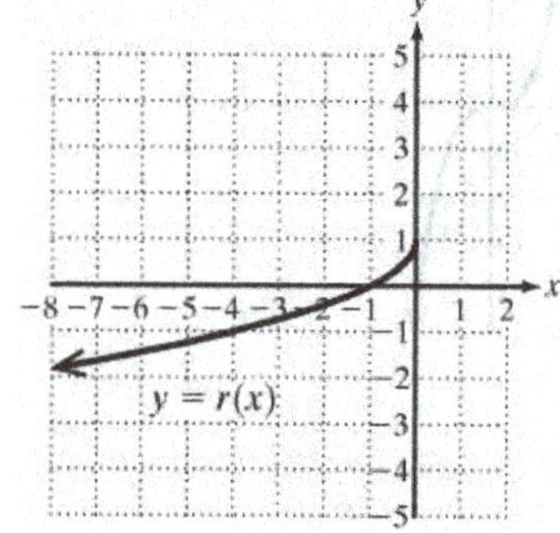

69. $f(x)=\sqrt{-x+3}=\sqrt{-(x-3)}$

The graph of f is the graph of $f(x)=\sqrt{x}$ reflected across the y-axis and shifted right 3 units.

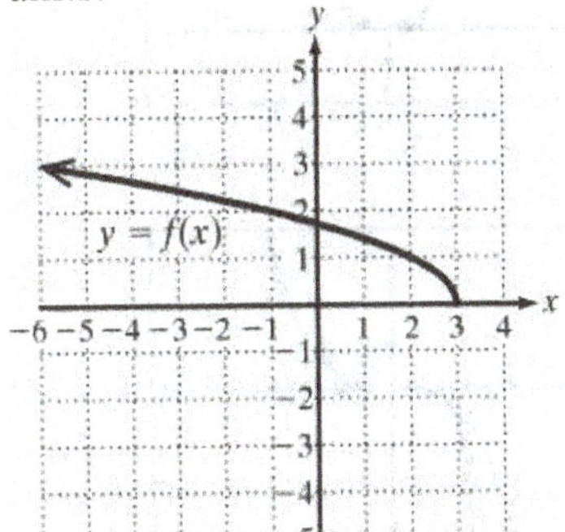

71. $n(x)=(2x+6)^2=\left[2(x+3)\right]^2=4(x+3)^2$

The graph of n is the graph of $f(x)=x^2$ shifted to the left by 3 units and stretched vertically by a factor of 4.

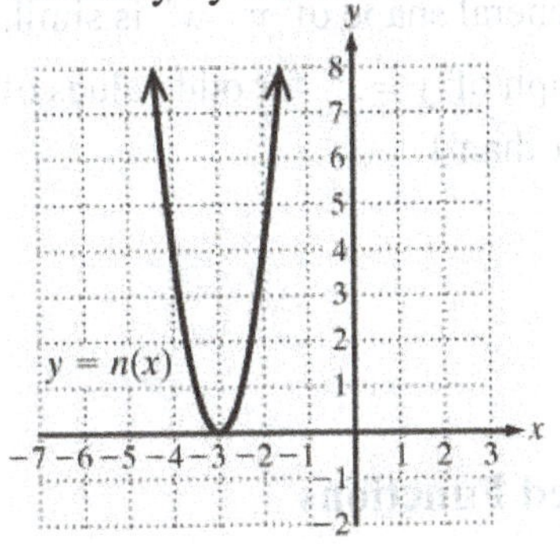

73. The graph of the function is the graph of $f(x)$ shifted to the right 1 unit, reflected across the x-axis, and shifted upward 2 units.

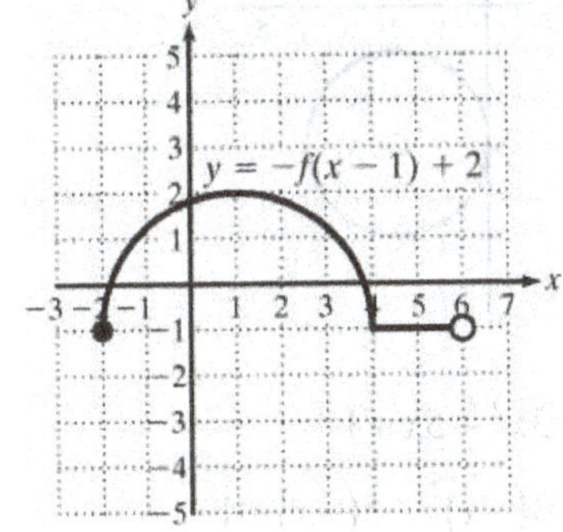

75. The graph of the function is the graph of $f(x)$ shifted to the right 2 units, stretched vertically by a factor of 2, and shifted downward 3 units.

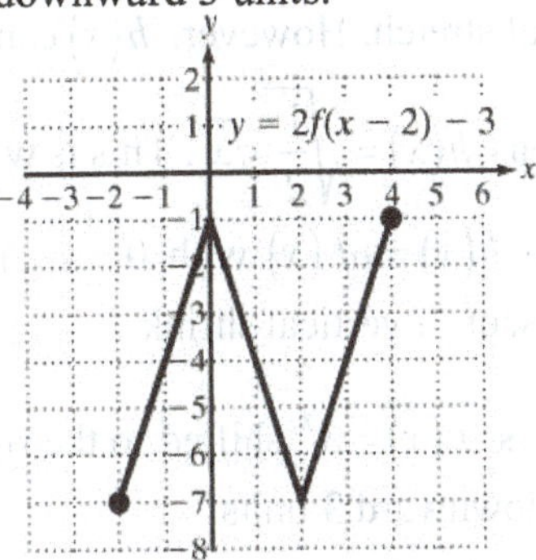

77. The graph of the function is the graph of $f(x)$ shrunk horizontally by a factor of 2, stretched vertically by a factor of 3, and reflected across the x-axis.

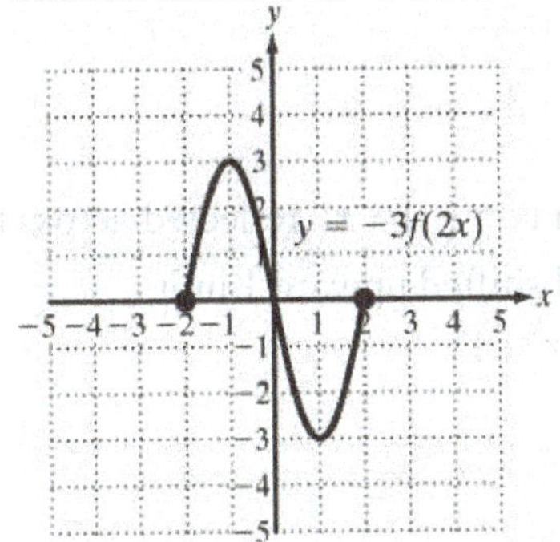

79. The graph of the function is the graph of $f(x)$ reflected across the y-axis and shifted downward 2 units.

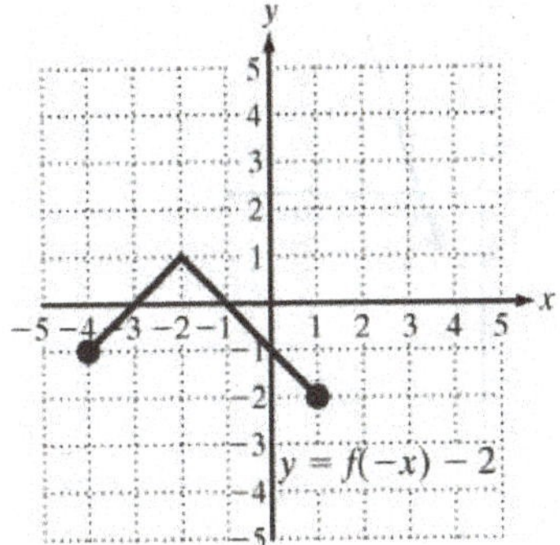

81. c

83. b

85. As written, $h(x)=\sqrt{\frac{1}{2}x}$ is in the form $h(x)=f(ax)$ with $0<a<1$. This indicates a horizontal stretch. However, $h(x)$ can also be written as $h(x)=\sqrt{\frac{1}{2}}\cdot\sqrt{x}$. This is written in the form $h(x)=af(x)$ with $0<a<1$. This represents a vertical shrink.

87. The graph is $f(x)=x^2$ shifted to the right 2 units and downward 3 units.
$f(x)=(x-2)^2-3$

89. The graph is $f(x)=\frac{1}{x}$ shifted to the left 3 units.
$f(x)=\frac{1}{x+3}$

91. The graph is $f(x)=x^3$ reflected across the x-axis and shifted upward 1 unit.
$f(x)=-x^3+1$

93. a.

b. 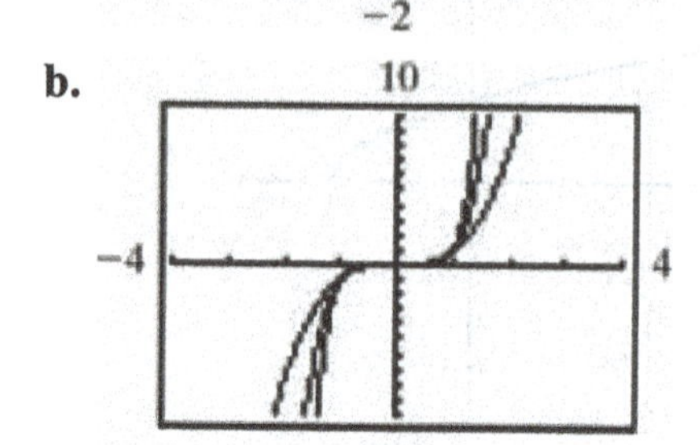

c. The general shape of $y=x^n$ is similar to the graph of $y=x^2$ for even values of n greater than 1.

d. The general shape of $y=x^n$ is similar to the graph of $y=x^3$ for odd values of n greater than 1.

Section 2.7 Analyzing Graphs of Functions and Piecewise-Defined Functions

1. 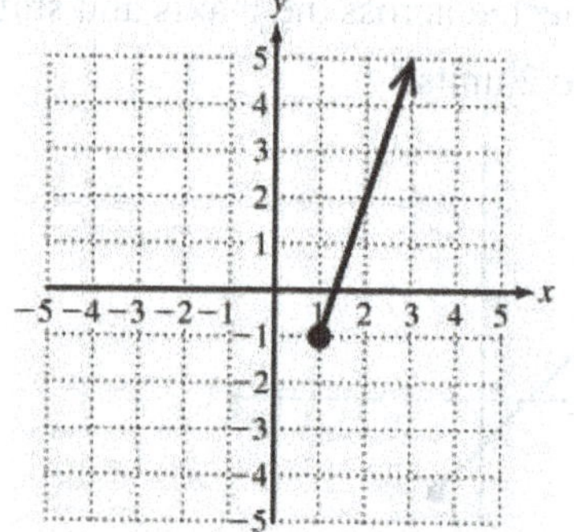

3. 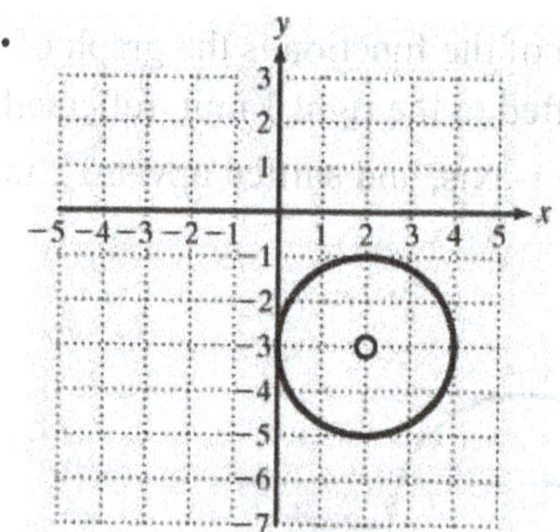

5. $f(x)=-3x^2-5x+1$
$f(-x)=-3(-x)^2-5(-x)+1$
$=-3x^2+5x+1$

7. y

9. origin

11. origin

13. $y = x^2 + 3$

Replace x by $-x$.

$y = (-x)^2 + 3$

$y = x^2 + 3$

This equation *is* equivalent to the original equation, so the graph is symmetric with respect to the y-axis.

Replace y by $-y$.

$-y = x^2 + 3$

$y = -x^2 - 3$

This equation is *not* equivalent to the original equation, so the graph is not symmetric with respect to the x-axis.

Replace x by $-x$ and y by $-y$.

$-y = (-x)^2 + 3$

$-y = x^2 - 3$

$y = -x^2 - 3$

This equation is *not* equivalent to the original equation, so the graph is not symmetric with respect to the origin.

15. $x = -|y| - 4$

Replace x by $-x$.

$-x = -|y| - 4$

$x = |y| + 4$

This equation is *not* equivalent to the original equation, so the graph is not symmetric with respect to the y-axis.

Replace y by $-y$.

$x = -|-y| - 4$

$x = -|y| - 4$

This equation *is* equivalent to the original equation, so the graph is symmetric with respect to the x-axis.

Replace x by $-x$ and y by $-y$.

$x = -|y| - 4$

$-x = -|-y| - 4$

$x = |y| + 4$

This equation is *not* equivalent to the original equation, so the graph is not symmetric with respect to the origin.

17. $x^2 + y^2 = 3$

Replace x by $-x$.

$(-x)^2 + y^2 = 3$

$x^2 + y^2 = 3$

This equation *is* equivalent to the original equation, so the graph is symmetric with respect to the y-axis.

Replace y by $-y$.

$x^2 + (-y)^2 = 3$

$x^2 + y^2 = 3$

This equation *is* equivalent to the original equation, so the graph is symmetric with respect to the x-axis.

Replace x by $-x$ and y by $-y$.

$(-x)^2 + (-y)^2 = 3$

$x^2 + y^2 = 3$

This equation *is* equivalent to the original equation, so the graph is symmetric with respect to the origin.

This equation *is* equivalent to the original equation, so the graph is symmetric with respect to the x-axis.

Replace x by $-x$ and y by $-y$.

$|-x| + |-y| = 4$

$|x| + |y| = 4$

This equation *is* equivalent to the original equation, so the graph is symmetric with respect to the origin.

19. $y = |x| + 2x + 7$

Replace x by $-x$.

$y = |-x| + 2(-x) + 7$

$y = |x| - 2x + 7$

This equation is *not* equivalent to the original equation, so the graph is not symmetric with respect to the y-axis.

Replace y by $-y$.

$-y = |x| + 2x + 7$

$y = -|x| - 2x - 7$

This equation is *not* equivalent to the original equation, so the graph is not symmetric with respect to the x-axis.

Replace x by $-x$ and y by $-y$.

$$-y=|-x|+2(-x)+7$$
$$-y=|x|-2x+7$$
$$y=-|x|+2x-7$$

This equation is *not* equivalent to the original equation, so the graph is not symmetric with respect to the origin.

21. $x^2=5+y^2$

Replace x by $-x$.

$$(-x)^2=5+y^2$$
$$x^2=5+y^2$$

This equation *is* equivalent to the original equation, so the graph is symmetric with respect to the y-axis.

Replace y by $-y$.

$$x^2=5+(-y)^2$$
$$x^2=5+y^2$$

This equation *is* equivalent to the original equation, so the graph is symmetric with respect to the x-axis.

Replace x by $-x$ and y by $-y$.

$$(-x)^2=5+(-y)^2$$
$$x^2=5+y^2$$

This equation *is* equivalent to the original equation, so the graph is symmetric with respect to the origin.

23. $y=\frac{1}{2}x-3$

Replace x by $-x$.

$$y=\frac{1}{2}(-x)-3$$
$$y=-\frac{1}{2}x-3$$

This equation is *not* equivalent to the original equation, so the graph is not symmetric with respect to the y-axis.

Replace y by $-y$.

$$-y=\frac{1}{2}x-3$$
$$y=-\frac{1}{2}x+3$$

This equation is *not* equivalent to the original equation, so the graph is not symmetric with respect to the x-axis.

Replace x by $-x$ and y by $-y$.

$$-y=\frac{1}{2}(-x)-3$$
$$-y=-\frac{1}{2}x-3$$
$$y=\frac{1}{2}x+3$$

This equation is *not* equivalent to the original equation, so the graph is not symmetric with respect to the origin.

25. y-axis symmetry

27. The function is symmetric to the origin. Therefore, the function is an odd function.

29. The function is symmetric to the y-axis. Therefore, the function is an even function.

31. The function is not symmetric with respect to either the y-axis or the origin. Therefore, the function is neither even nor odd.

33. a. $f(x)=4x^2-3|x|$

$$f(-x)=4(-x)^2-3|-x|=4x^2-3|x|$$

b. Yes **c.** Even

35. a. $h(x)=4x^3-2x$

$$h(-x)=4(-x)^3-2(-x)=-4x^3+2x$$

b. $-h(x)=-(4x^3-2x)=-4x^3+2x$

c. Yes **d.** Odd

37. a. $m(x)=4x^2+2x-3$

$$m(-x)=4(-x)^2+2(-x)-3$$
$$=4x^2-2x-3$$

b. $-m(x)=-(4x^2+2x-3)=-4x^2-2x+3$

c. No **d.** No

e. Neither

39. $f(x) = 3x^6 + 2x^2 + |x|$

$f(-x) = 3(-x)^6 + 2(-x) + |-x|$

$f(-x) = 3x^6 + 2x^2 + |x|$

$f(-x) = f(x)$

The function is even.

41. $k(x) = 13x^3 + 12x$

$k(-x) = 13(-x)^3 + 12(-x)$

$k(-x) = -13x^3 - 12x$

$-k(x) = -(13x^3 + 12x) = -13x^3 - 12x$

$k(-x) = -k(x)$

The function is odd.

43. $n(x) = \sqrt{16 - (x-3)^2}$

$n(-x) = \sqrt{16 - (-x-3)^2}$

$-n(x) = -\sqrt{16 - (x-3)^2}$

$n(-x) \neq n(x)$

$n(-x) \neq -n(x)$

The function is neither even nor odd.

45. $q(x) = \sqrt{16 + x^2}$

$q(-x) = \sqrt{16 + (-x)^2} = \sqrt{16 + x^2}$

$q(-x) = q(x)$

The function is even.

47. $h(x) = 5x$

$h(-x) = 5(-x) = -5x$

$-h(x) = -(5x) = -5x$

$h(-x) = -h(x)$

The function is odd.

49. **a.** Use the second rule.

$f(3) = (3)^2 + 3 = 9 + 3 = 12$

b. Use the first rule.

$f(-2) = -3(-2) + 7 = 6 + 7 = 13$

c. Use the second rule.

$f(-1) = (-1)^2 + 3 = 1 + 3 = 4$

d. Use the third rule. $f(4) = 5$

e. Use the third rule. $f(5) = 5$

51. **a.** Use the second rule. $h(-1.7) = 1$

b. Use the first rule. $h(-2.5) = 2$

c. Use the second rule. $h(0.05) = -1$

d. Use the fourth rule. $h(-2) = 1$

e. Use the second rule. $h(0) = -1$

53. c

55. d

57. **a.**

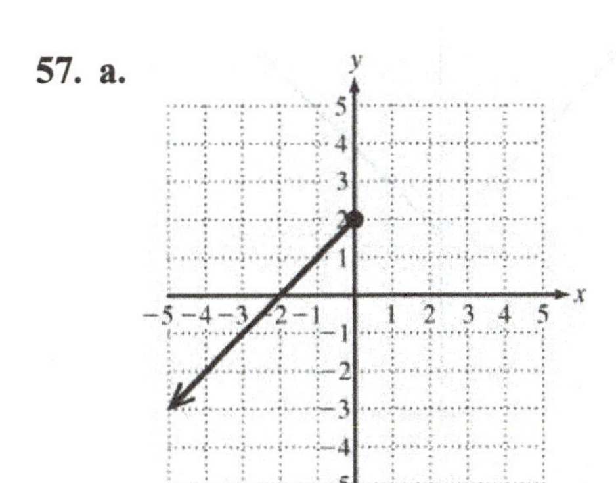

b.

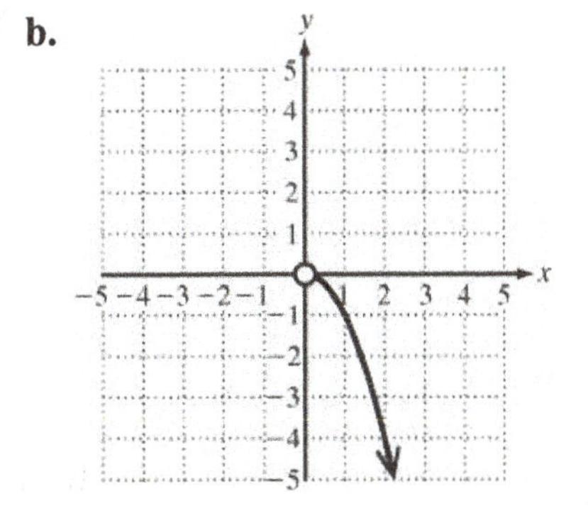

c.

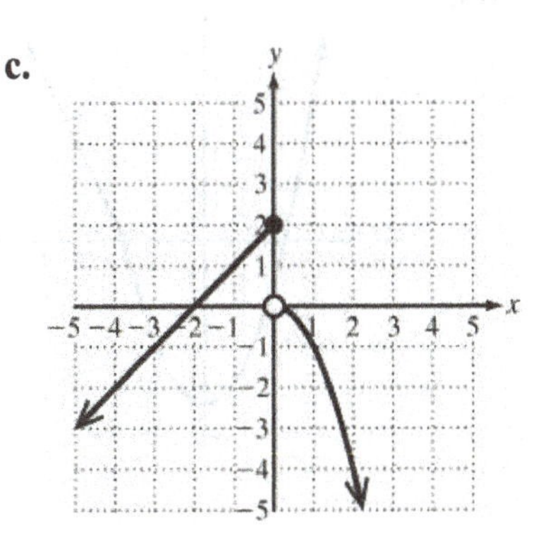

59. a.

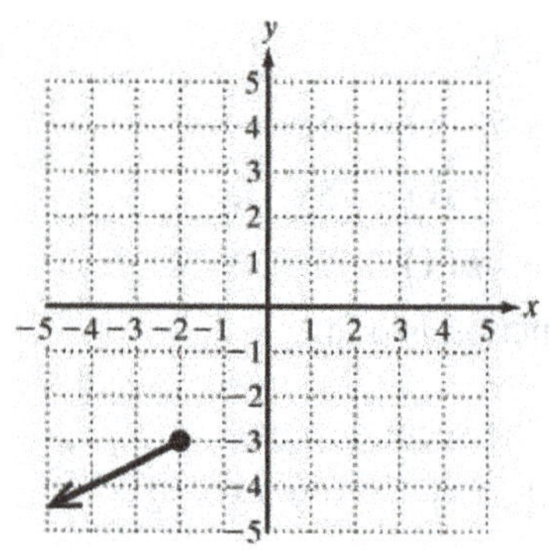

b.

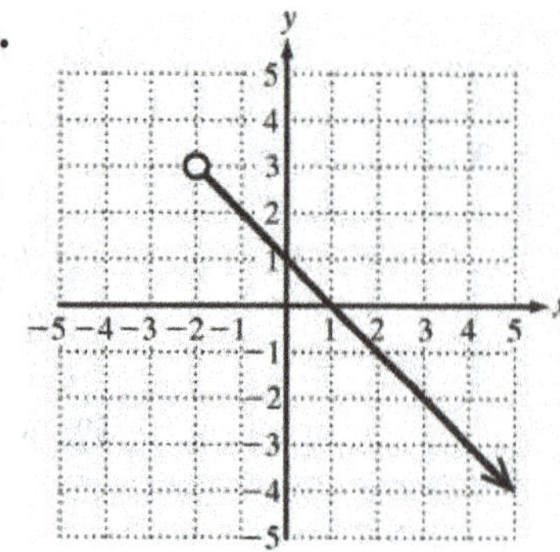

c.

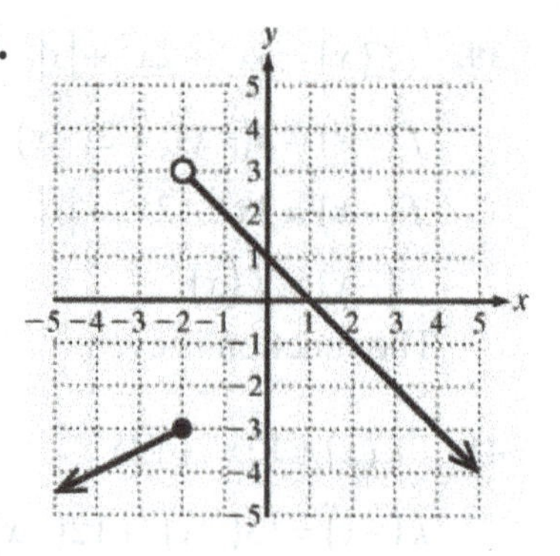

61.

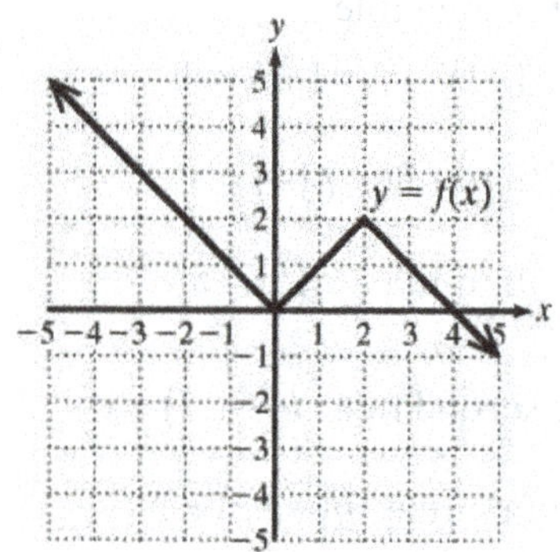

63.

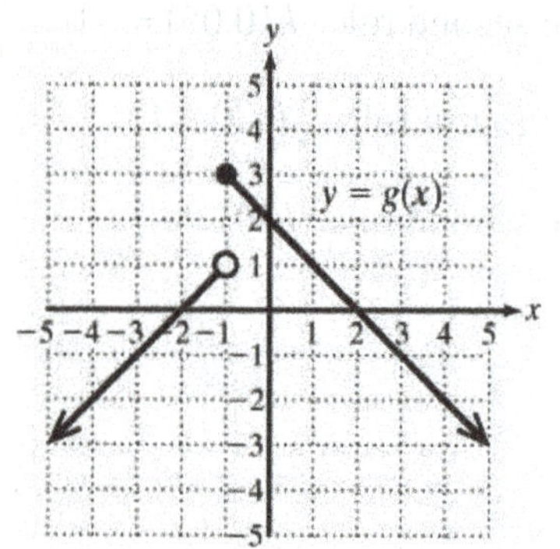

65.

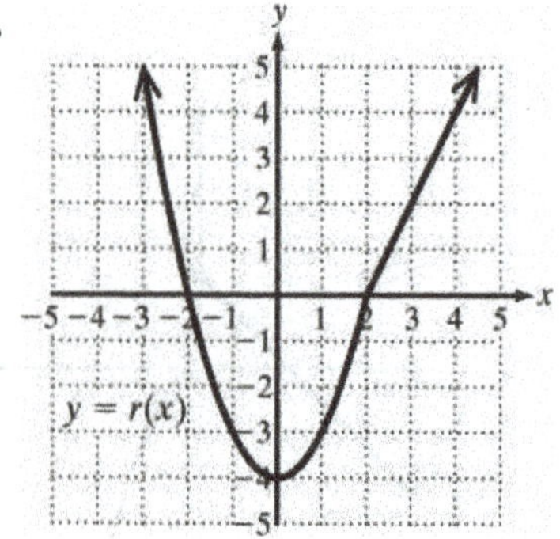

67.

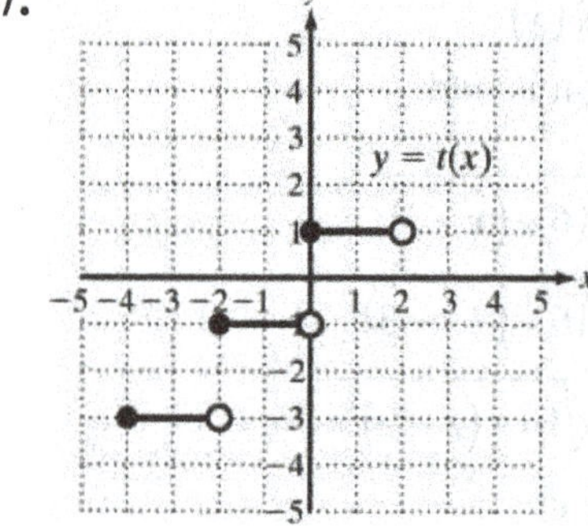

69.

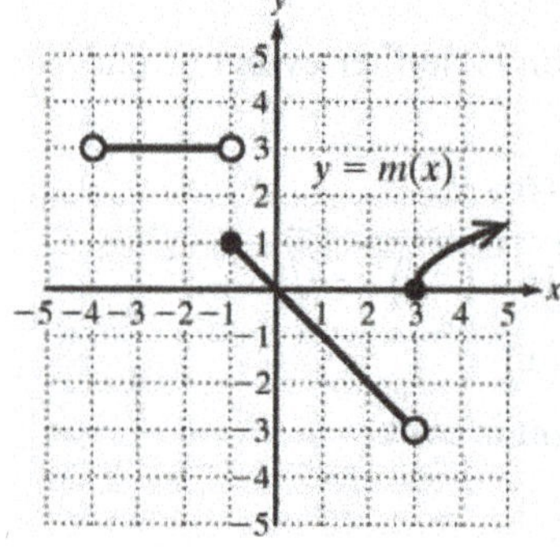

71. a.

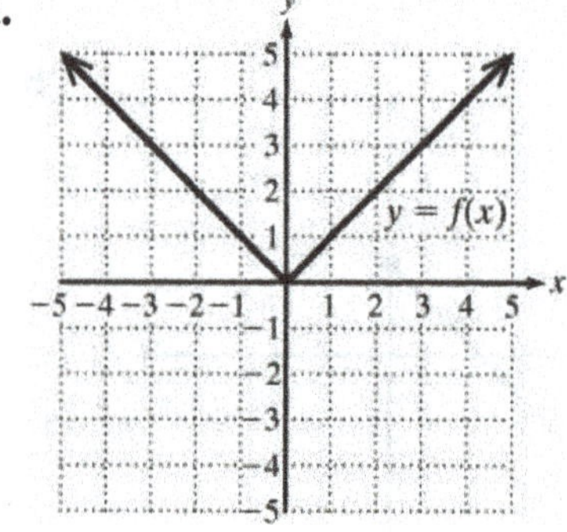

b. $y = |x|$

73. $f(x)=[\![x]\!]$
$f(-4.2)=[\![-4.2]\!]=-5$

75. $f(x)=[\![x]\!]$
$f(-0.09)=[\![-0.09]\!]=-1$

77. $f(x)=[\![x]\!]$
$f(0.09)=[\![0.09]\!]=0$

79. $f(x)=[\![x]\!]$
$f(-9)=[\![-9]\!]=-9$

81.

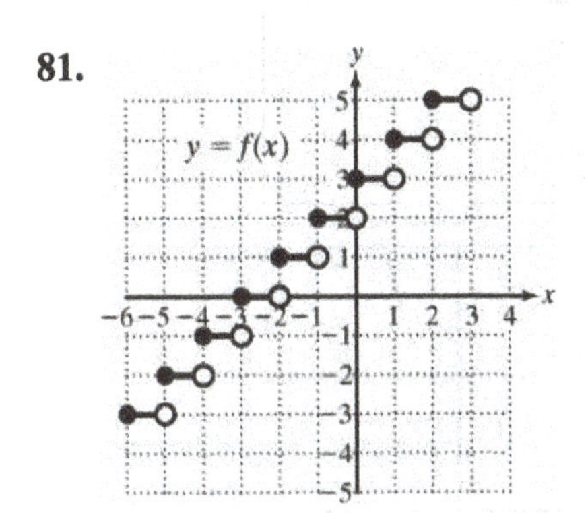

83.

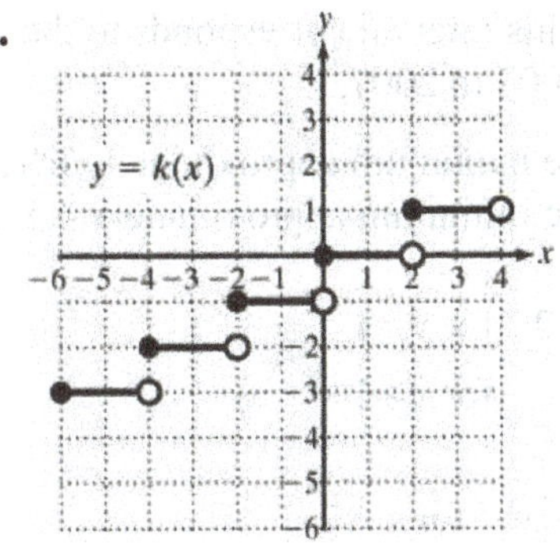

85. a. $C(x)=\begin{cases}0.44 & \text{for } 0<x\le 1\\ 0.61 & \text{for } 1<x\le 2\\ 0.78 & \text{for } 2<x\le 3\\ 0.95 & \text{for } 3<x\le 3.5\end{cases}$

b.

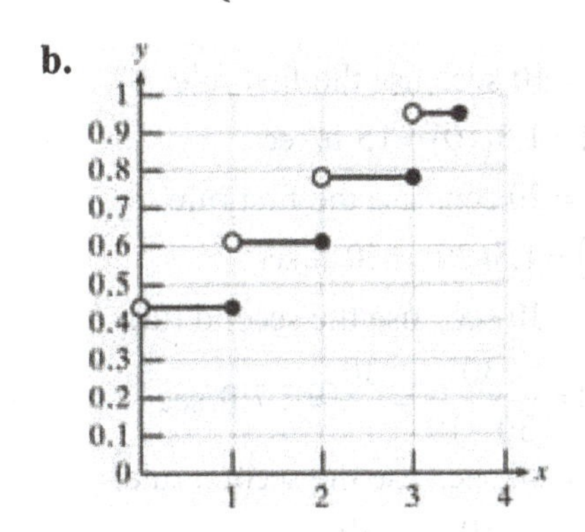

87. $S(x)=\begin{cases}2000 & \text{for } 2<x\le 3\\ 2000+0.05(x-40{,}000) & \text{for } 3<x\le 3.5\end{cases}$

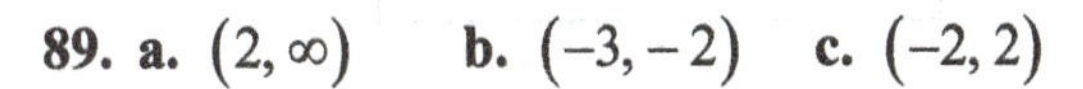

89. a. $(2,\infty)$ **b.** $(-3,-2)$ **c.** $(-2,2)$

91. a. $(-\infty,\infty)$

b. Never decreasing

c. Never constant

93. a. $(1,\infty)$

b. $(-\infty,1)$

c. Never constant

95. a. $(-\infty,-2)\cup(2,\infty)$

b. Never decreasing

c. $(-2,2)$

97. At $x=1$, the function has a relative minimum of -3.

99. At $x=-2$, the function has a relative minimum of 0. At $x=0$, the function has a relative maximum of 2. At $x=2$, the function has a relative minimum of 0.

101. At $x=-2$, the function has a relative minimum of -4. At $x=0$, the function has a relative maximum of 0. At $x=2$, the function has a relative minimum of -4.

103. a. (0, 3) and (7, 9); these intervals correspond to the years 2000 to 2003 and 2007 to 2009.

b. (3, 7); this interval corresponds to the years 2003 to 2007.

c. Relative maximum approximately 6%; Relative minimum approximately 4.5%.

105. $f(x)=\begin{cases} -2 & \text{for } x<1 \\ 3 & \text{for } x\geq 1 \end{cases}$

107. $f(x)=\begin{cases} -|x| & \text{for } x<2 \\ -2 & \text{for } x\geq 2 \end{cases}$

109. $f(x)=\begin{cases} \frac{1}{x} & \text{for } x<0 \\ x & \text{for } x>0 \end{cases}$

111. For $t=10$ sec, use the first rule.
$s(10)=1.5(10)=15$ ft/sec
For $t=20$ sec, use the first rule.
$s(20)=1.5(20)=30$ ft/sec
For $t=30$ sec, use the second rule.
$s(30)=\frac{30}{30-19}=\frac{30}{11}\approx 2.7$ ft/sec
For $t=40$ sec, use the second rule.
$s(40)=\frac{30}{40-19}=\frac{30}{21}\approx 1.4$ ft/sec

113. Graph b

115. **a.** 2 **b.** –4
c. 4 **d.** 3
e. –3 **f.** 4

117. If replacing y by $-y$ in the equation results in an equivalent equation, then the graph is symmetric with respect to the x-axis. If replacing x by $-x$ in the equation results in an equivalent equation, then the graph is symmetric with respect to the y-axis. If replacing both x by $-x$ and y by $-y$ results in an equivalent equation, then the graph is symmetric with respect to the origin.

119. At $x=1$, there are two different y values. The relation contains the ordered pairs (1, 2) and (1, 3).

121. A relative maximum of a function is the greatest function value relative to other points on the function nearby.

123. If the average rate of change is negative and then positive, the function is decreasing and then increasing. Therefore it must have a relative minimum at a.

125. **a.** Concave down **b.** Decreasing

127. **a.** Concave up **b.** Decreasing

129.

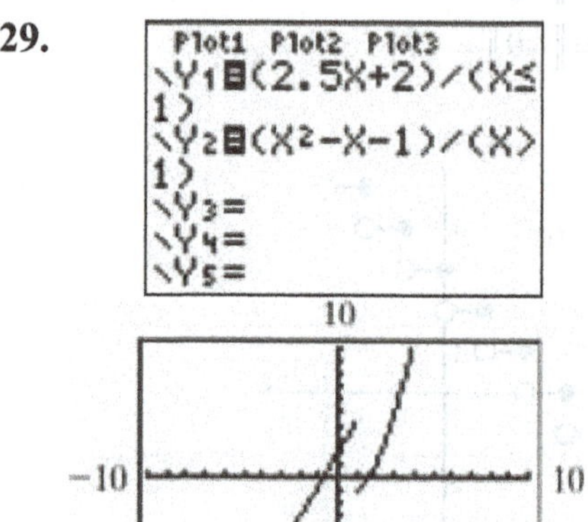

131.

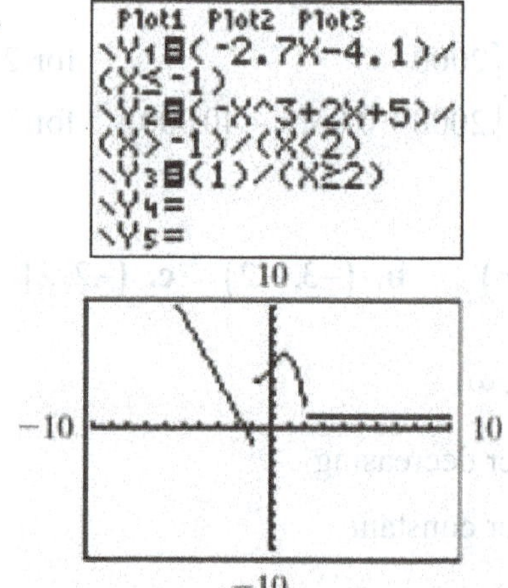

The graph does *not* have a gap at $x=2$, however, the limited resolution on the calculator gives the appearance of a gap.

133.

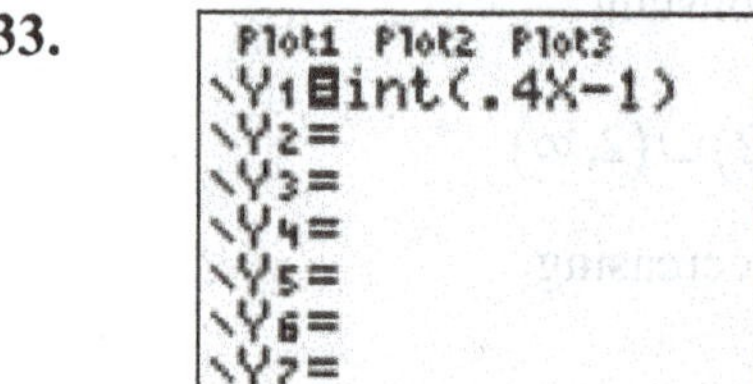

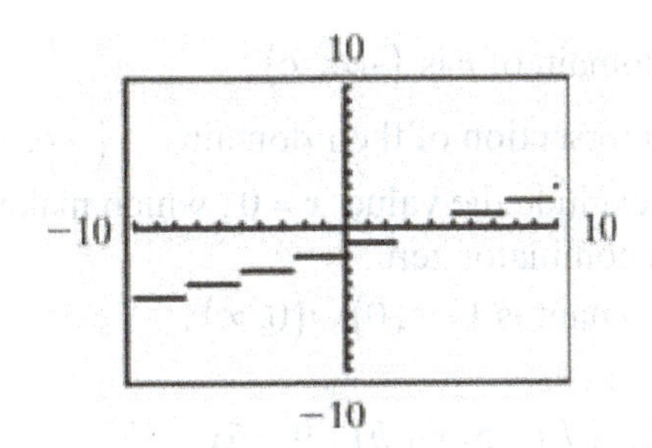

135.

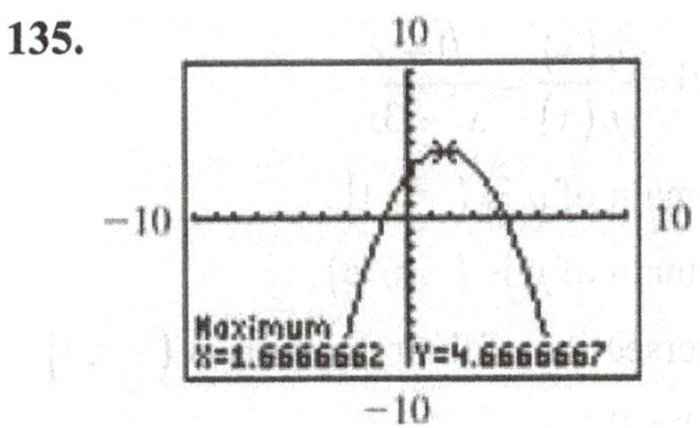

Relative maximum 4.667

Section 2.8 Algebra of Functions and Function Composition

1. $5-2x\ge 0$

$-2x\ge -5$

$x\le \dfrac{5}{2}$

$\left(-\infty, \dfrac{5}{2}\right]$

3. $81-x^2\ne 0$

$-x^2\ne -81$

$x^2\ne 81$

$x\ne \pm 9$

$(-\infty,-9)\cup(-9,9)\cup(9,\infty)$

5. $f(x)=3x^2+5x-1$

$f(t+2)=3(t+2)^2+5(t+2)-1$

$=3(t^2+4t+4)+5t+10-1$

$=3t^2+12t+12+5t+10-1$

$=3t^2+17t+21$

7. $f(x)$; $g(x)$

9. $\dfrac{f(x+h)-f(x)}{h}$

11. $(f+g)(x)=|x|+3$; Graph d

13. $(f+g)(x)=x^2+(-4)=x^2-4$; Graph a

15. $(f-g)(3)=f(3)-g(3)$

$=-2(3)-|3+4|=-6-7=-13$

17. $(f\cdot g)(-1)=f(-1)\cdot g(-1)$

$=-2(-1)\cdot|-1+4|=2\cdot 3=6$

19. $(g+h)(0)=g(0)+h(0)$

$=|0+4|+\dfrac{1}{0-3}$

$=4-\dfrac{1}{3}=\dfrac{11}{3}$

21. $\left(\dfrac{f}{g}\right)(8)=\dfrac{f(8)}{g(8)}=\dfrac{-2(8)}{|8+4|}=\dfrac{-16}{12}=-\dfrac{4}{3}$

23. $\left(\dfrac{g}{f}\right)(0)=\dfrac{g(0)}{f(0)}=\dfrac{|0+4|}{-2(0)}=\dfrac{4}{0}$ Undefined

25. $(r-p)(x)=r(x)-p(x)$

$=-3x-(x^2+3x)$

$=3x-x^2-3x=-x^2-6x$

The domain of r is $(-\infty,\infty)$.

The domain of p is $(-\infty,\infty)$.

The intersection of their domains is $(-\infty,\infty)$.

27. $(p\cdot q)(x)=p(x)\cdot q(x)=(x^2+3x)\cdot\sqrt{1-x}$

$1-x\ge 0$

$-x\ge -1$

$x\le 1$

The domain of p is $(-\infty,\infty)$.

The domain of q is $(-\infty,1]$.

The intersection of their domains is $(-\infty,1]$.

29. $\left(\frac{q}{p}\right)(x)=\frac{q(x)}{p(x)}=\frac{\sqrt{1-x}}{x^2+3x}$

The domain of q is $(-\infty, 1]$.

The domain of p is $(-\infty, \infty)$.

The intersection of their domains is $(-\infty, 1]$.

$$x^2+3x=0$$
$$x(x+3)=0$$
$$x=0 \text{ or } x=-3$$

Also exclude the values $x=0$ and $x=-3$, which make the denominator zero.

The domain is $(-\infty,-3)\cup(-3,0)\cup(0,1]$.

31. $\left(\frac{p}{q}\right)(x)=\frac{p(x)}{q(x)}=\frac{x^2+3x}{\sqrt{1-x}}$

The domain of p is $(-\infty, \infty)$.

The domain of q is $(-\infty, 1]$.

The intersection of their domains is $(-\infty, 1]$.

Also exclude the value $x=1$, which makes the denominator zero.

The domain is $(-\infty, 1)$.

33. $(r+q)(x)=r(x)+q(x)$
$=-3x+\sqrt{1-x}$

The domain of r is $(-\infty, \infty)$.

The domain of q is $(-\infty, 1]$.

The intersection of their domains is $(-\infty, 1]$.

35. $\left(\frac{p}{r}\right)(x)=\frac{p(x)}{r(x)}=\frac{x^2+3x}{-3x}=\frac{x+3}{-3}$

The domain of p is $(-\infty, \infty)$.

The domain of r is $(-\infty, \infty)$.

The intersection of their domains is $(-\infty, \infty)$.

Also exclude the value $x=0$, which makes the denominator zero.

The domain is $(-\infty, 0)\cup(0,\infty)$.

37. a. $f(x+h)=5(x+h)+9=5x+5h+9$

b. $\frac{f(x+h)-f(x)}{h}=\frac{5x+5h+9-(5x+9)}{h}$
$=\frac{5x+5h+9-5x-9}{h}$
$=\frac{5h}{h}=5$

39. a. $f(x+h)=(x+h)^2+4(x+h)$
$=x^2+2xh+h^2+4x+4h$

b. $\frac{f(x+h)-f(x)}{h}$
$=\frac{x^2+2xh+h^2+4x+4h-(x^2+4x)}{h}$
$=\frac{x^2+2xh+h^2+4x+4h-x^2-4x}{h}$
$=\frac{2xh+h^2+4h}{h}=2x+h+4$

41. $\frac{f(x+h)-f(x)}{h}=\frac{-2(x+h)+5-(-2x+5)}{h}$
$=\frac{-2x-2h+5+2x-5}{h}$
$=\frac{-2h}{h}=-2$

43. $\frac{f(x+h)-f(x)}{h}=\frac{-5(x+h)^2-4(x+h)+2-(-5x^2-4x+2)}{h}=\frac{-5(x^2+2xh+h^2)-4x-4h+2+5x^2+4x-2}{h}$
$=\frac{-5x^2-10xh-5h^2-4x-4h+2+5x^2+4x-2}{h}=\frac{-10xh-5h^2-4h}{h}=-10x-5h-4$

45. $\frac{f(x+h)-f(x)}{h}=\frac{(x+h)^3-x^3}{h}=\frac{x^3+3x^2h+3xh^2+h^3-x^3}{h}=3x^2+3xh+h^2$

47. $\dfrac{f(x+h)-f(x)}{h}=\dfrac{\frac{1}{x+h}-\frac{1}{x}}{h}$

$=\dfrac{\frac{x}{x(x+h)}-\frac{x+h}{x(x+h)}}{h}$

$=\dfrac{\frac{x-(x+h)}{x(x+h)}}{h}$

$=\dfrac{-\frac{h}{x(x+h)}}{h}$

$=-\dfrac{1}{x(x+h)}$

49. a. $\dfrac{f(x+h)-f(x)}{h}=\dfrac{4\sqrt{x+h}-4\sqrt{x}}{h}$

b. $\dfrac{4\sqrt{1+1}-4\sqrt{1}}{1}=4\sqrt{2}-4\approx 1.6569$

$\dfrac{4\sqrt{1+0.1}-4\sqrt{1}}{0.1}=\dfrac{4\sqrt{1.1}-4}{0.1}\approx 1.9524$

$\dfrac{4\sqrt{1+0.01}-4\sqrt{1}}{0.01}=\dfrac{4\sqrt{1.01}-4}{0.01}\approx 1.9950$

$\dfrac{4\sqrt{1+0.001}-4\sqrt{1}}{0.001}=\dfrac{4\sqrt{1.001}-4}{0.001}\approx 1.9995$

c. 2

51. $f(g(8))=f\left(\sqrt{2(8)}\right)=f\left(\sqrt{16}\right)$

$=f(4)=(4)^3-4(4)$

$=64-16=48$

53. $h(f(1))=h(1^3-4(1))=h(-3)$

$=2(-3)+3=-6+3=-3$

55. $(f\circ g)(18)=f(g(18))=f\left(\sqrt{2(18)}\right)$

$=f\left(\sqrt{36}\right)=f(6)$

$=(6)^3-4(6)$

$=216-24=192$

57. $(g\circ f)(5)=g(f(5))=g\left((5)^3-4(5)\right)$

$=g(125-20)=g(105)$

$=\sqrt{2(105)}=\sqrt{210}$

59. $(h\circ f)(-3)=h(f(-3))=h\left((-3)^3-4(-3)\right)$

$=h(-27+12)=h(-15)$

$=2(-15)+3=-30+3=-27$

61. $(g\circ f)(1)=g(f(1))=g\left((1)^3-4(1)\right)$

$=g(1-4)=g(-3)$

$=\sqrt{2(-3)}=\sqrt{-6}$

Undefined

63. a. $(f\circ g)(x)=f(g(x))$

$=2(g(x))^2+4=2x^2+4$

b. $(g\circ f)(x)=g(f(x))=(f(x))^2$

$=(2x+4)^2=4x^2+16x+16$

c. No

65. $(n\circ p)(x)=n(p(x))$

$=p(x)-5=x^2-9x-5$

Domain: $(-\infty,\infty)$

67. $(m\circ n)(x)=m(n(x))=\sqrt{n(x)+8}$

$=\sqrt{x-5+8}=\sqrt{x+3}$

$x+3\geq 0$

$x\geq -3$

Domain: $[-3,\infty)$

69. $(q\circ n)(x)=q(n(x))=\dfrac{1}{n(x)-10}$

$=\dfrac{1}{x-5-10}=\dfrac{1}{x-15}$

$x-15\neq 0$

$x\neq 15$

Domain: $(-\infty,15)\cup(15,\infty)$

71. $(q \circ r)(x) = q(r(x))$

$$= \frac{1}{r(x)-10} = \frac{1}{|2x+3|-10}$$

$|2x+3|-10 \neq 0$

$|2x+3| \neq 10$

$2x+3 \neq 10$ or $2x+3 \neq -10$

$2x \neq 7$ or $2x \neq -13$

$x \neq \frac{7}{2}$ or $x \neq -\frac{13}{2}$

Domain: $\left(-\infty, -\frac{13}{2}\right) \cup \left(-\frac{13}{2}, \frac{7}{2}\right) \cup \left(\frac{7}{2}, \infty\right)$

73. $(n \circ r)(x) = n(r(x)) = r(x) - 5 = |2x+3| - 5$

Domain: $(-\infty, \infty)$

75. $(n \circ n)(x) = n(n(x)) = n(x) - 5$

$= x - 5 - 5 = x - 10$

Domain: $(-\infty, \infty)$

77. a. $C(x) = 21.95x$

b. $T(a) = a + 0.06a + 10.99 = 1.06a + 10.99$

c. $(T \circ C)(x) = T(C(x))$

$= 1.06(C(x)) + 10.99$

$= 1.06(21.95x) + 10.99$

$= 23.267x + 10.99$

d. $(T \circ C)(4) = 23.267(4) + 10.99 = 104.058$

The total cost to purchase 4 boxes of stationery is \$104.06.

79. a. $r(t) = 80t$ **b.** $d(r) = 7.2r$

c. $(d \circ r)(t) = d(r(t)) = 7.2(r(t))$

$= 7.2(80t) = 576t$

This function represents the distance traveled (in ft) in t minutes.

d. $(d \circ r)(30) = 576(30) = 17{,}280$ means that the bicycle will travel 17,280 ft (approximately 3.27 mi) in 30 min.

81. $f(x) = x^2$ and $g(x) = x + 7$

83. $f(x) = \sqrt[3]{x}$ and $g(x) = 2x + 1$

85. $f(x) = |x|$ and $g(x) = 2x^2 - 3$

87. $f(x) = \frac{5}{x}$ and $g(x) = x + 4$

89. a. $(f+g)(0) = f(0) + g(0) = 3 + (-2) = 1$

b. $(g-f)(2) = g(2) - f(2) = 0 - 1 = -1$

c. $(g \cdot f)(-1) = g(-1) \cdot f(-1) = -2 \cdot 3 = -6$

d. $\left(\frac{g}{f}\right)(1) = \frac{g(1)}{f(1)} = \frac{-1}{2} = -\frac{1}{2}$

e. $(f \circ g)(4) = f(g(4)) = f(2) = 1$

f. $(g \circ f)(0) = g(f(0)) = g(3) = 1$

g. $g(f(4)) = g(-1) = -2$

91. a. $(h+k)(-1) = h(-1) + k(-1) = 2 + (-3) = -1$

b. $(h \cdot k)(4) = h(4) \cdot k(4) = -1 \cdot 0 = 0$

c. $\left(\frac{k}{h}\right)(-3) = \frac{k(-3)}{h(-3)} = \frac{-3}{0}$ Undefined

d. $(k-h)(1) = k(1) - h(1) = -1 - 2 = -3$

e. $(k \circ h)(4) = k(h(4)) = k(-1) = -3$

f. $(h \circ k)(-2) = h(k(-2)) = h(-3) = 0$

g. $h(k(3)) = h(-1) = 2$

93. $(f+g)(4) = f(4) + g(4) = -2 + 3 = 1$

95. $(g \circ f)(2) = g(f(2)) = g(4) = 3$

97. $(g \circ g)(6) = g(g(6)) = g(0) = 6$

99. $(f \circ g)(5) = f(g(5)) = f(7)$ Undefined

101. $(k+p)(4)=k(4)+p(4)=6+2=8$

103. $(k\circ p)(2)=k(p(2))=k(4)=6$

105. $(k\circ k)(0)=k(k(0))=k(4)=6$

107. $T(x)=(M+W)(x)=M(x)+W(x)$
$=0.08x+5.4+0.03x+1.1=0.11x+6.5$
This function represents the total number of adults (in millions) on probation, on parole, or incarcerated x years after the year 2000.

109. a. $C(x)=2.8x+5000$

b. $R(x)=40x$

c. $(R-C)(x)=R(x)-C(x)$
$=40x-(2.8x+5000)$
$=40x-2.8x-5000$
$=37.2x-5000$
This function represents the profit for selling x CDs.

d. $(R-C)(2400)=37.2(2400)-5000$
$=\$84{,}280$

111. $\left(\frac{H+L}{2}\right)(x)$ represents the average of the high and low temperatures for day x.

113. a. $S_1(x)=x(x+4)=x^2+4x$

b. $S_2(x)=\frac{1}{2}\pi r^2=\frac{1}{2}\pi\left(\frac{x}{2}\right)^2=\frac{1}{2}\pi\frac{x^2}{4}=\frac{1}{8}\pi x^2$

c. $(S_1-S_2)(x)=A_1(x)-A_2(x)$
$=x^2+4x-\frac{1}{8}\pi x^2$
This function represents the area of the region outside the semicircle, but inside the rectangle.

115. The domain of $(f\circ g)(x)$ is the set of real numbers x in the domain of g such that $g(x)$ is in the domain of f.

117. a. $\frac{f(x+h)-f(x)}{h}=\frac{\sqrt{x+h+3}-\sqrt{x+3}}{h}$

b. $\frac{\sqrt{x+h+3}-\sqrt{x+3}}{h}=\frac{\sqrt{x+h+3}-\sqrt{x+3}}{h}\cdot\frac{\sqrt{x+h+3}+\sqrt{x+3}}{\sqrt{x+h+3}+\sqrt{x+3}}$
$=\frac{x+h+3-(x+3)}{h\left(\sqrt{x+h+3}+\sqrt{x+3}\right)}=\frac{h}{h\left(\sqrt{x+h+3}+\sqrt{x+3}\right)}=\frac{1}{\sqrt{x+h+3}+\sqrt{x+3}}$

c. $\frac{1}{\sqrt{x+0+3}+\sqrt{x+3}}=\frac{1}{2\sqrt{x+3}}$

119. a. $\frac{d(t+h)-d(t)}{h}=\frac{-4.84(t+h)^2+88(t+h)-(-4.84t^2+88t)}{h}=\frac{-4.84(t^2+2th+h^2)+88t+88h+4.84t^2-88t}{h}$
$=\frac{-4.84t^2-9.68th-4.84h^2+88h+4.84t^2}{h}=\frac{-9.68th-4.84h^2+88h}{h}=-9.68t-4.84h+88$

b. $-9.68t-4.84h+88=-9.68(0)-4.84(2)+88=-9.68+88=78.32$ ft/sec

c. $-9.68t-4.84h+88=-9.68(2)-4.84(2)+88=-19.36-9.68+88=58.96$ ft/sec

d. $-9.68t-4.84h+88=-9.68(4)-4.84(2)+88=-38.72-9.68+88=39.6$ ft/sec

e. $-9.68t-4.84h+88=-9.68(6)-4.84(2)+88=-58.08-9.68+88=20.24$ ft/sec

121. a.
$$\frac{N(x+h)-N(x)}{h}=\frac{-0.28(x+h)^2+20.8(x+h)+194-\left(-0.28x^2+20.8x+194\right)}{h}$$
$$=\frac{-0.28\left(x^2+2xh+h^2\right)+20.8x+20.8h+194+0.28x^2-20.8x-194}{h}$$
$$=\frac{-0.28x^2-0.56xh-0.28h^2+20.8h+0.28x^2}{h}$$
$$=\frac{-0.56xh-0.28h^2+20.8h}{h}=-0.56x-0.28h+20.8$$

b. $-0.56x-0.28h+20.8=-0.56(10)-0.28(10)+20.8=12.4$

The difference quotient of 12.4 means that cigarette production increased at an average rate of 12.4 billion per yr between the years 1950 and 1960.

c. $-0.56x-0.28h+20.8=-0.56(50)-0.28(10)+20.8=-10$

The difference quotient of -10 means that cigarette production decreased at an average rate of 10 billion per yr between the years 1990 and 2000.

123. $(f\circ f)(x)=f(f(x))$
$$=\frac{1}{f(x)-2}=\frac{1}{\frac{1}{x-2}-2}$$
$$=\frac{1}{\frac{1}{x-2}-\frac{2(x-2)}{x-2}}$$
$$=\frac{1}{\frac{1-2x+4}{x-2}}$$
$$=\frac{1}{\frac{-2x+5}{x-2}}=\frac{x-2}{-2x+5}$$

Domain of f is $(-\infty,2)\cup(2,\infty)$.

$-2x+5\neq 0$

$-2x\neq -5$

$x\neq \frac{5}{2}$

Domain: $(-\infty,2)\cup\left(2,\frac{5}{2}\right)\cup\left(\frac{5}{2},\infty\right)$

125. $(g\circ g)(x)=g(g(x))$
$$=\sqrt{g(x)-3}=\sqrt{\sqrt{x-3}-3}$$

Domain of g is $[3,\infty)$.

$\sqrt{x-3}-3\geq 0$

$\sqrt{x-3}\geq 3$

$x-3\geq 9$

$x\geq 12$ Domain: $[12,\infty)$

127. $(f\circ g\circ h)(x)=f(g(h(x)))=2(g(h(x)))+1$
$$=2\left((h(x))^2\right)+1=2\left(\sqrt[3]{x}\right)^2+1$$

129. $(h\circ g\circ f)(x)=h(g(f(x)))=\sqrt[3]{g(f(x))}$
$$=\sqrt[3]{(f(x))^2}=\sqrt[3]{(2x+1)^2}$$

131. $m(x)=\sqrt[3]{x}$, $n(x)=x+1$, $h(x)=4x$, $k(x)=x^2$

Chapter 2 Review Exercises

1. a. $d=\sqrt{(x_2-x_1)^2+(y_2-y_1)^2}$

$=\sqrt{[4-(-1)]^2+(-2-8)^2}$

$=\sqrt{25+100}$

$=\sqrt{125}$

$=5\sqrt{5}$

b. $M=\left(\frac{x_1+x_2}{2},\frac{y_1+y_2}{2}\right)$

$=\left(\frac{-1+4}{2},\frac{8+(-2)}{2}\right)$

$=\left(\frac{3}{2},\frac{6}{2}\right)$

$=\left(\frac{3}{2},3\right)$

3. a. $4|x-1|+y=18$

$4|(-3)-1|+(2)0\ 18$

$4(4)+2 0\ 18$

$18 0\ 18$ ✓ True

Yes

b. $4|x-1|+y=18$

$4|(5)-1|+(-2)0\ 18$

$4(4)-2 0\ 18$

$14 0\ 18$ False

No

5. x-intercept: $(4,0)$; y-intercepts: $(0,-4),(0,-10)$

7. $y=x^2-2x$

x	y	$y=x^2-2x$	Ordered pair
-2	8	$y=(-2)^2-2(-2)=8$	$(-2,8)$
-1	3	$y=(-1)^2-2(-1)=3$	$(-1,3)$
0	0	$y=(0)^2-2(0)=0$	$(0,0)$
1	-1	$y=(1)^2-2(1)=-1$	$(1,-1)$
2	0	$y=(2)^2-2(2)=0$	$(2,0)$
3	3	$y=(3)^2-2(3)=3$	$(3,3)$
4	8	$y=(4)^2-2(4)=8$	$(4,8)$

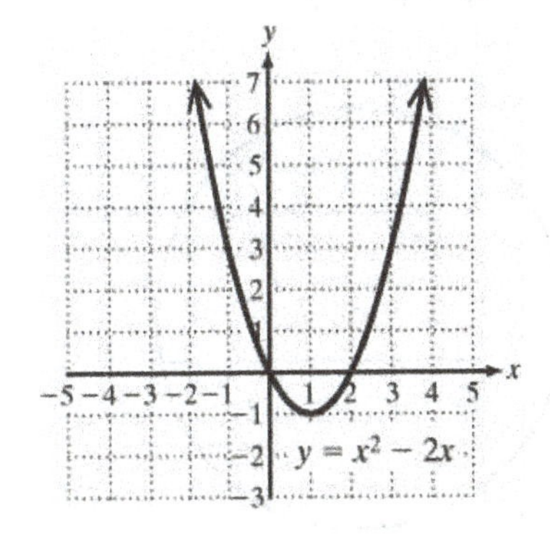

9. $r^2=4$

$r=\sqrt{4}=2$

Center: $(4,-3)$; Radius: 2

11. a. $(x-h)^2+(y-k)^2=r^2$

$[x-(-3)]^2+(y-1)^2=\left(\sqrt{11}\right)^2$

$(x+3)^2+(y-1)^2=11$

b.

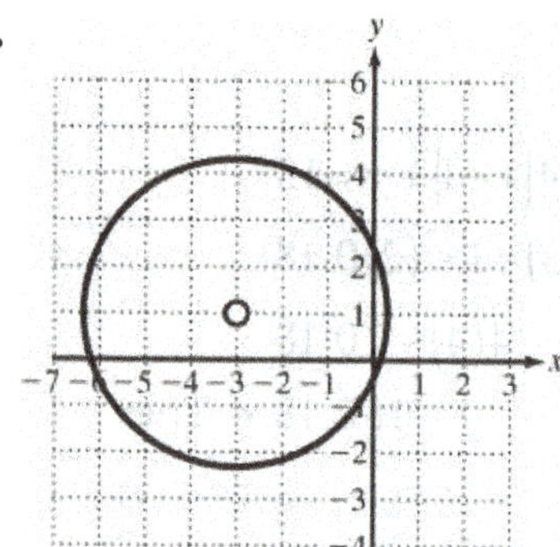

13. a. $C=\left(\dfrac{x_1+x_2}{2},\dfrac{y_1+y_2}{2}\right)$

$=\left(\dfrac{7+1}{2},\dfrac{5+(-3)}{2}\right)=\left(\dfrac{8}{2},\dfrac{2}{2}\right)=(4,1)$

$r=\sqrt{(x_2-x_1)^2+(y_2-y_1)^2}$

$=\sqrt{(7-4)^2+(5-1)^2}=\sqrt{9+16}=\sqrt{25}=5$

$(x-h)^2+(y-k)^2=r^2$

$(x-4)^2+(y-1)^2=(5)^2$

$(x-4)^2+(y-1)^2=25$

b.

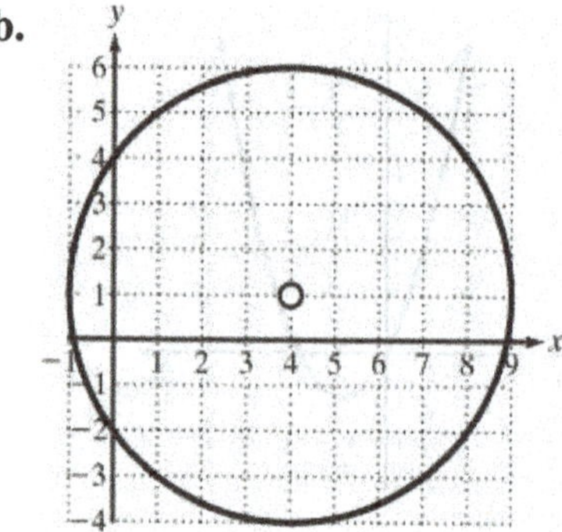

15. a. $x^2+y^2+10x-2y+17=0$

$(x^2+10x\quad)+(y^2-2y\quad)=-17$

$\left[\dfrac{1}{2}(10)\right]^2=25 \quad \left[\dfrac{1}{2}(-2)\right]^2=1$

$(x^2+10x+25)+(y^2-2y+1)=-17+25+1$

$(x+5)^2+(y-1)^2=9$

b. Center: $(-5,1)$; Radius: $\sqrt{9}=3$

17. $(x+3)^2+(y-5)^2=0$

The sum of two squares will equal zero only if each individual term is zero. Therefore, $x=-3$ and $y=5$.

$\{(-3,5)\}$

19. a. {(Dara Torres, 12), (Carl Lewis, 10), (Bonnie Blair, 6), (Michael Phelps, 16)}

b. {Dara Torres, Carl Lewis, Bonnie Blair, Michael Phelps}

c. {12, 10, 6, 16} **d.** Yes

21. There is at least one vertical line that intersects the graph in more than one point. This relation is not a function.

23. $x^2+y-3=4$

$y=-x^2+7$

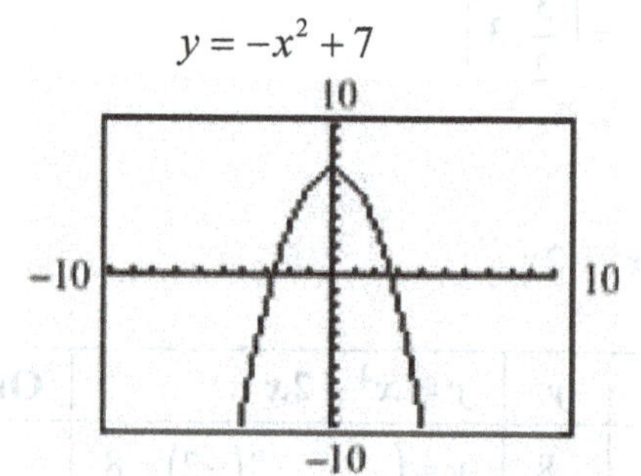

No vertical line intersects the graph in more than one point. This relation is a function.

25. a. $f(1)=5$

b. $f(0)=4$

c. $f(x)=-1$ for $x=3$

27. Solve $p(x)=0$:

$p(x)=|x-3|-1$

$0=|x-3|-1$

$|x-3|=1$

$x-3=1$ or $x-3=-1$

$x=4$ or $x=2$

Evaluate $p(0)$:

$p(x)=|x-3|-1$
$p(0)=|(0)-3|-1=3-1=2$
x-intercepts: $(4,0),(2,0)$; y-intercept: $(0,2)$

29. Domain: $\{-4,-2,0,2,3,5\}$;
Range: $\{-3,0,2,1\}$

31. $x-5\neq 0$
$x\neq 5$
$(-\infty,5)\cup(5,\infty)$

33. There is no restriction on x.
$(-\infty,\infty)$

35. **a.** $f(-2)=-2$ **b.** $f(3)=-1$

c. $f(x)=-1$ for $x=-1,\ x=3$

d. $f(x)=-4$ for $x=-4$

e. x-intercepts: $(0,0)$ and $(2,0)$

f. y-intercept: $(0,0)$

g. Domain: $(-\infty,\infty)$ **h.** Range: $(-\infty,1]$

37. $-2x+4y=8$
$4y=2x+8$
$y=\frac{1}{2}x+2$

x	y
−4	0
−2	1
0	2
2	3

x-intercept: $(-4,0)$; y-intercept: $(0,2)$

39. $y=2$

x	y
−2	2
0	2
2	2
4	2

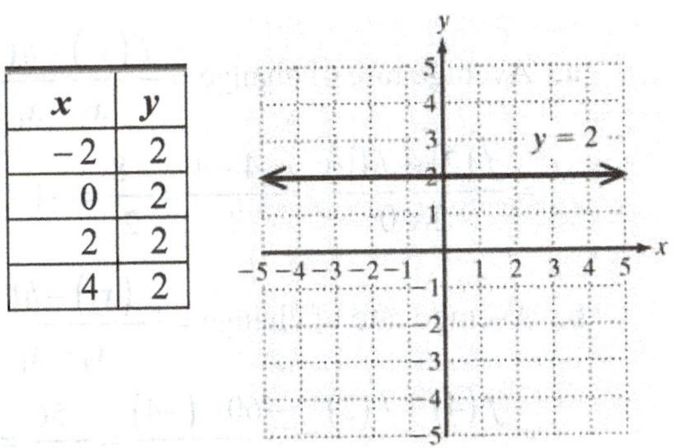

x-intercept: none; y-intercept: $(0,2)$

41. $m=\frac{y_2-y_1}{x_2-x_1}=\frac{-4-(-2)}{-12-4}=\frac{-2}{-16}=\frac{1}{8}$

43. $m=\frac{y_2-y_1}{x_2-x_1}=\frac{f(b)-f(a)}{b-a}$

45. Undefined

47. $\frac{\Delta C}{\Delta t}$ represents the change in cost per change in time.

49. $y=mx+b$
$y=-\frac{2}{3}x+b$
$-5=-\frac{2}{3}(1)+b$
$-5=-\frac{2}{3}+b$
$-\frac{13}{3}=b$
$y=-\frac{2}{3}x-\frac{13}{3}$; $f(x)=-\frac{2}{3}x-\frac{13}{3}$

51. $m=\frac{f(x_2)-f(x_1)}{x_2-x_1}=\frac{3-0}{4-(-3)}=\frac{3}{7}$

53. $f(x)=-x^3+4$
$f(0)=-(0)^3+4=4$
$f(2)=-(2)^3+4=-8+4=-4$
$f(4)=-(4)^3+4=-64+4=-60$

a. Average rate of change $= \dfrac{f(x_2)-h(x_1)}{x_2-x_1}$

$= \dfrac{f(2)-f(0)}{2-0} = \dfrac{-4-4}{2} = \dfrac{-8}{2} = -4$

b. Average rate of change $= \dfrac{f(x_2)-h(x_1)}{x_2-x_1}$

$= \dfrac{f(4)-f(2)}{4-2} = \dfrac{-60-(-4)}{2} = \dfrac{-56}{2} = -28$

55. a. $\dfrac{2}{3}$

b. $-\dfrac{1}{m} = -\dfrac{1}{\frac{2}{3}} = -\dfrac{3}{2}$

57. $y - y_1 = m(x - x_1)$

$y-(-7) = 3[x-(-2)]$

$y+7 = 3(x+2)$

$y+7 = 3x+6$

$y = 3x-1$

59. $m = \dfrac{y_2-y_1}{x_2-x_1} = \dfrac{7.1-5.3}{-0.9-1.1} = \dfrac{1.8}{-2} = -0.9$

$y - y_1 = m(x-x_1)$

$y-5.3 = -0.9(x-1.1)$

$y-5.3 = -0.9x+0.99$

$y = -0.9x+6.29$

61. $2x - y = 4$

$-y = -2x+6$

$y = 2x-6;\ m = 2$

$y - y_1 = m(x-x_1)$

$y-(-6) = 2(x-2)$

$y+6 = 2x-4$

$y = 2x-10$

63. A line that is perpendicular to the y-axis is a horizontal line with slope 0. Every coordinate has the same y-value, so the equation is $y = 7$.

65. a. $C(x) = 1500 + 35x$

b. $R(x) = 60x$

c. $P(x) = R(x) - C(x)$

$= 60x - (1500 + 35x)$

$= 25x - 1500$

d. $P(x) > 0$

$25x - 1500 > 0$

$25x > 1500$

$x > 60$

The studio needs more than 60 private lessons per month to make a profit.

e. $P(x) = 25x - 1500$

$P(82) = 25(82) - 1500$

$= 2050 - 1500$

$= 550$

The studio will make \$550.

67. a. $y = 369.6x + 22{,}111$

b.

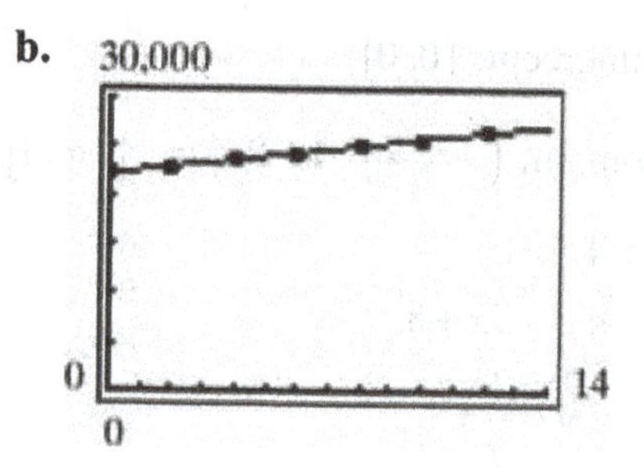

c. $y = 369.6(15) + 22{,}111$

$= 5544 + 22{,}111$

$= 27{,}655$ students

69. The graph of f is the graph of $f(x) = |x|$ shifted downward 4 units.

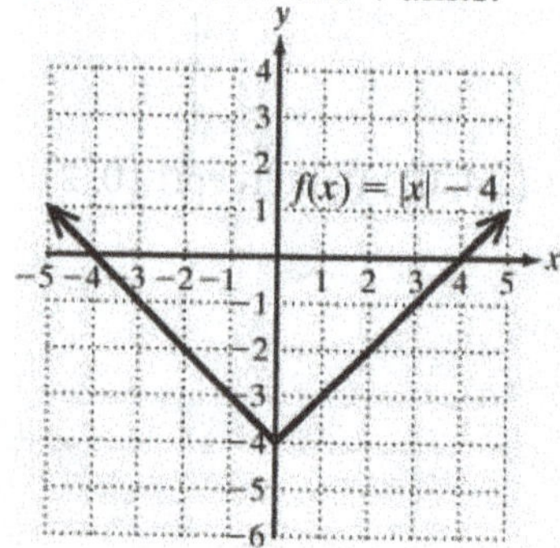

71. The graph of h is the graph of $f(x) = x^2$ shifted to the right 4 units.

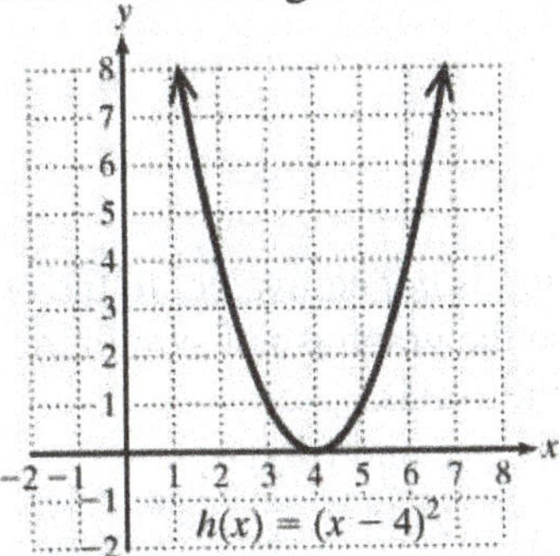

73. The graph of r is the graph of $f(x) = \sqrt{x}$ shifted to the right 3 units and upward 1 unit.

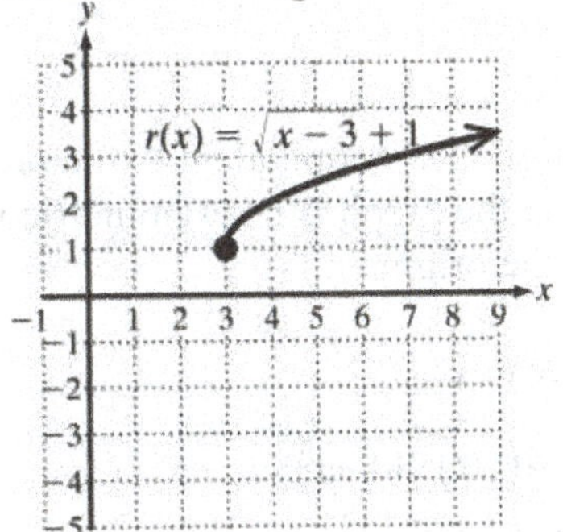

75. The graph of t is the graph of $f(x) = |x|$ reflected across the x-axis and stretched vertically by a factor of 2.

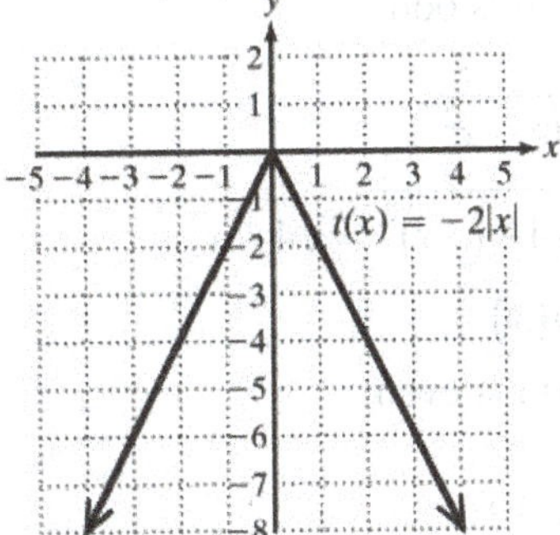

77. $m(x) = \sqrt{-x+5} = \sqrt{-(x-5)}$

The graph of m is the graph of $f(x) = \sqrt{x}$ reflected across the y-axis and shifted right 5 units.

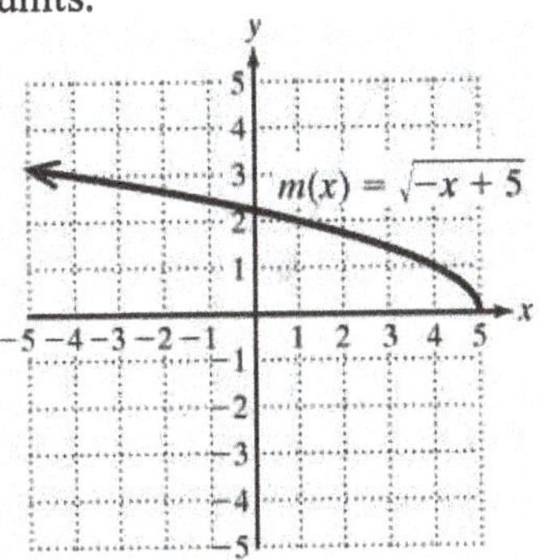

79. The graph of the function is the graph of $f(x)$ shrunk horizontally by a factor of 2.

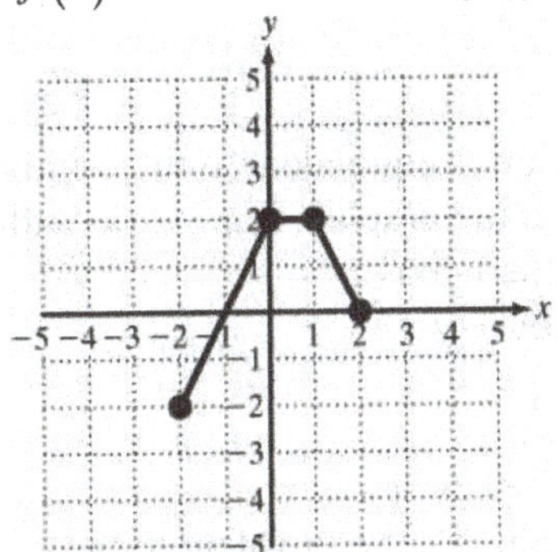

81. The graph of the function is the graph of $f(x)$ shifted to the left 1 unit, reflected across the x-axis, and shifted downward 3 units.

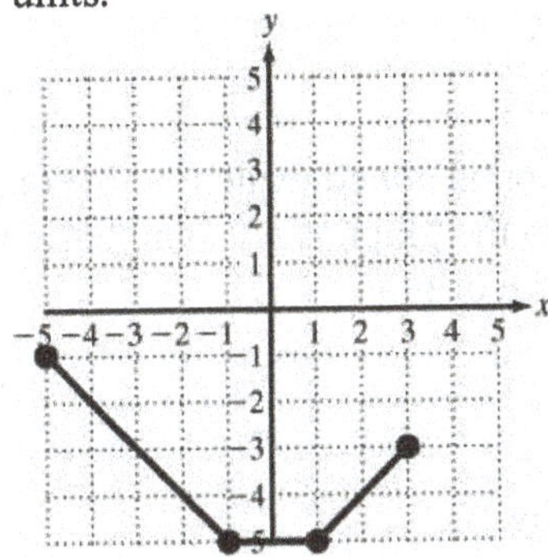

83. The graph of the function is the graph of $f(x)$ shifted to the right 3 units, stretched vertically by a factor of 2, and shifted upward 1 unit.

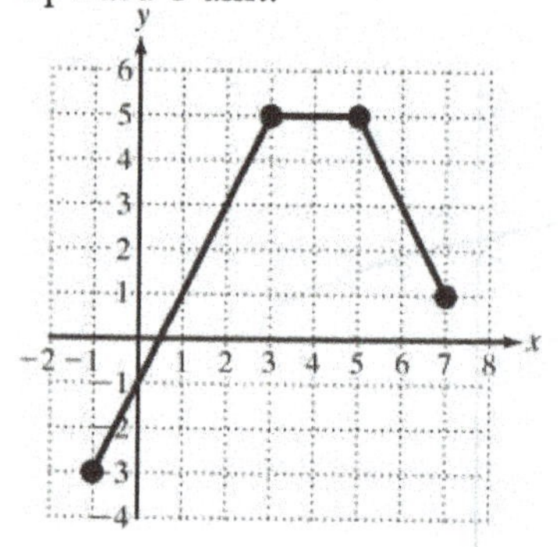

85. $y = x^4 - 3$

Replace x by $-x$.

$y = (-x)^4 - 3$

$y = x^4 - 3$

This equation *is* equivalent to the original equation, so the graph is symmetric with respect to the y-axis.

Replace y by $-y$.

$-y = x^4 - 3$

$y = -x^4 + 3$

This equation is *not* equivalent to the original equation, so the graph is not symmetric with respect to the x-axis.

Replace x by $-x$ and y by $-y$.

$-y = (-x)^4 - 3$

$-y = x^4 - 3$

$y = -x^4 + 3$

This equation is *not* equivalent to the original equation, so the graph is not symmetric with respect to the origin.

87. $y = \frac{1}{3}x - 1$

Replace x by $-x$.

$y = \frac{1}{3}(-x) - 1$

$y = -\frac{1}{3}x - 1$

This equation is *not* equivalent to the original equation, so the graph is not symmetric with respect to the y-axis.

Replace y by $-y$.

$-y = \frac{1}{3}x - 1$

$y = -\frac{1}{3}x + 1$

This equation is *not* equivalent to the original equation, so the graph is not symmetric with respect to the x-axis.

Replace x by $-x$ and y by $-y$.

$-y = \frac{1}{3}(-x) - 1$

$-y = -\frac{1}{3}x - 1$

$y = \frac{1}{3}x + 1$

This equation is *not* equivalent to the original equation, so the graph is not symmetric with respect to the origin.

89. $f(x) = -4x^3 + x$

$f(-x) = -4(-x)^3 + (-x)$

$f(-x) = 4x^3 - x$

$-f(x) = -(-4x^3 + x) = 4x^3 - x$

$f(-x) = -f(x)$

The function is odd.

91. $p(x) = \sqrt{4 - x^2}$

$p(-x) = \sqrt{4 - (-x)^2} = \sqrt{4 - x^2}$

$p(-x) = p(x)$

The function is even.

93. $k(x) = (x - 3)^2$

$k(-x) = (-x - 3)^2$

$-k(x) = -(x - 3)^2$

$k(-x) \neq k(x)$

$k(-x) \neq -k(x)$

The function is neither even nor odd.

95. a. Use the first rule.

$f(-4) = -4(-4) + 2 = 16 + 2 = 18$

b. Use the second rule. $f(-1)=(-1)^2=1$

c. Use the third rule. $f(3)=5$

d. Use the second rule. $f(2)=(2)^2=4$

97.

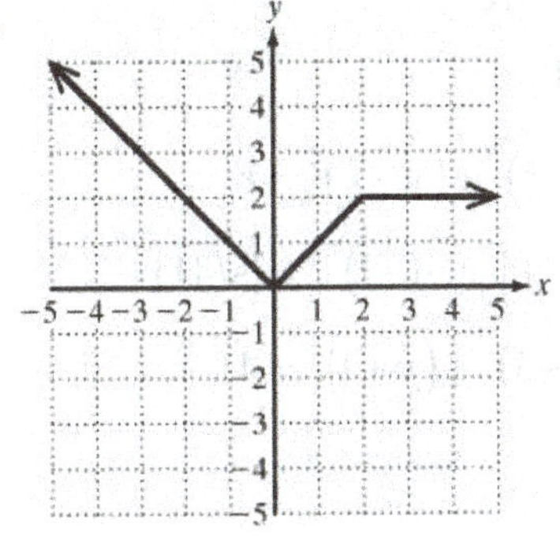

99. a. $f(x)=[\![x-1]\!]$

$f(-1.5)=[\![-1.5-1]\!]=[\![-2.5]\!]=-3$

b. $f(x)=[\![x-1]\!]$

$f(-2)=[\![-2-1]\!]=[\![-3]\!]=-3$

c. $f(x)=[\![x-1]\!]$

$f(0.1)=[\![0.1-1]\!]=[\![-0.9]\!]=-1$

d. $f(x)=[\![x-1]\!]$

$f(6.3)=[\![6.3-1]\!]=[\![5.3]\!]=5$

101. a. $(2,\infty)$ **b.** $(-\infty,2)$

c. Never constant

103. At $x=-2$, the function has a relative maximum of 4.

105. $(f-h)(2)=f(2)-h(2)$

$$=-3(2)-\frac{1}{2+1}$$

$$=-6-\frac{1}{3}$$

$$=-\frac{19}{3}$$

107. $\left(\frac{g}{h}\right)(-5)=\frac{g(-5)}{h(-5)}=\frac{|-5-2|}{\frac{1}{-5+1}}=\frac{7}{\frac{1}{-4}}=-28$

109. $(g\circ f)(5)=g(f(5))=g(-3(5))$

$$=g(-15)=|-15-2|=17$$

111. $(n-m)(x)=n(x)-m(x)$

$$=x^2-4x-(-4x)$$

$$=x^2$$

Domain: $(-\infty,\infty)$

113. $\left(\frac{n}{p}\right)(x)=\frac{n(x)}{p(x)}=\frac{x^2-4x}{\sqrt{x-2}}$

$x-2>0$

$x>2$

Domain: $(2,\infty)$

115. $(q\circ n)(x)=q(n(x))=\frac{1}{n(x)-5}=\frac{1}{x^2-4x-5}$

$x^2-4x-5\neq 0$

$(x-5)(x+1)\neq 0$

$x\neq 5$ and $x\neq -1$

Domain: $(-\infty,-1)\cup(-1,5)\cup(5,\infty)$

117. $\frac{f(x+h)+f(x)}{h}=\frac{-6(x+h)-5-(-6x-5)}{h}$

$$=\frac{-6x-6h-5+6x+5}{h}$$

$$=\frac{-6h}{h}=-6$$

119. $f(x)=x^2$ and $g(x)=x-4$

121. a. $d(t)=60t$

b. $n(d)=\frac{d}{28}$

c. $(n\circ d)(t)=n(d(t))=\frac{d(t)}{28}=\frac{60t}{28}$

This function represents the number of gallons of gasoline used in t hours.

d. $(n\circ d)(7)=\frac{60(7)}{28}=15$ means that 15 gal of gasoline is used in 7 hr.

Chapter 2 Test

1. a. $C=\left(\frac{x_1+x_2}{2},\frac{y_1+y_2}{2}\right)$
$=\left(\frac{8+(-2)}{2},\frac{-5+3}{2}\right)$
$=\left(\frac{6}{2},\frac{-2}{2}\right)$
$=(3,-1)$

b. $r=\sqrt{(x_2-x_1)^2+(y_2-y_1)^2}$
$=\sqrt{(-2-3)^2+[3-(-1)]^2}$
$=\sqrt{25+16}$
$=\sqrt{41}$

c. $(x-h)^2+(y-k)^2=r^2$
$(x-3)^2+[y-(-1)]^2=(\sqrt{41})^2$
$(x-3)^2+(y+1)^2=41$

3. a. $x^2+y^2+14x-10y+70=0$
$(x^2+14x\quad)+(y^2-10y\quad)=-70$

$\left[\frac{1}{2}(14)\right]^2=49$	$\left[\frac{1}{2}(10)\right]^2=25$

$(x^2+14x+49)+(y^2-10y+25)=-70+49-25$
$(x+7)^2+(y-5)^2=4$

b. Center: $(-7,5)$; Radius: $\sqrt{4}=2$

5. There is at least one vertical line that intersects the graph in more than one point. This relation is not a function.

7. a. $f(0)=-2$ **b.** $f(-4)=0$

c. $f(x)=2$ for $x=-2$ and $x=2$

d. $(-\infty,-2)\cup(0,2)$

e. $(-2,0)\cup(2,\infty)$

f. $f(0)=-2$ is a relative minimum.

g. $f(-2)=2$ and $f(2)=2$ are relative maxima.

h. $(-\infty,\infty)$ **i.** $(-\infty,2]$

j. $f(x)=f(-x)$, so the function is even.

9. $4-c\geq 0$
$-c\geq -4$
$c\leq 4$
$(-\infty,4]$

11. $x+3y=4$
$3y=-x+4$
$y=-\frac{1}{3}x+\frac{4}{3}$; $m=-\frac{1}{3}$
$-\frac{1}{m}=-\frac{1}{-\frac{1}{3}}=3$
$y-y_1=m(x-x_1)$
$y-6=3[x-(-2)]$
$y-6=3(x+2)$
$y-6=3x+6$
$y=3x+12$

13. 

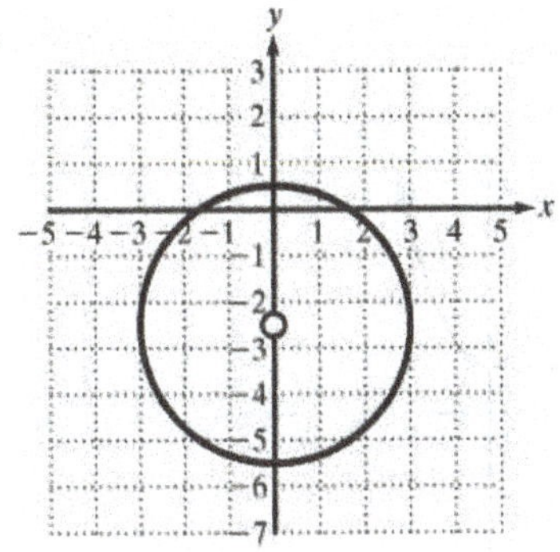

15.

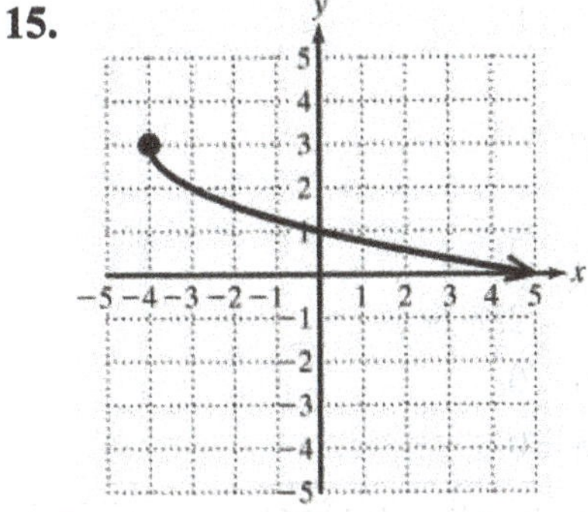

17. $x^2 + |y| = 8$

Replace x by $-x$.

$(-x)^2 + |y| = 8$

$x^2 + |y| = 8$

This equation *is* equivalent to the original equation, so the graph is symmetric with respect to the y-axis.

Replace y by $-y$.

$x^2 + |-y| = 8$

$x^2 + |y| = 8$

This equation *is* equivalent to the original equation, so the graph is symmetric with respect to the x-axis.

Replace x by $-x$ and y by $-y$.

$(-x)^2 + |-y| = 8$

$x^2 + |y| = 8$

This equation *is* equivalent to the original equation, so the graph is symmetric with respect to the origin.

19. $g(x) = x^4 + x^3 + x$

$g(-x) = (-x)^4 + (-x)^3 + (-x) = x^4 - x^3 - x$

$-g(x) = -(x^4 + x^3 + x) = -x^4 - x^3 - x$

$g(-x) \neq g(x);\ g(-x) \neq -g(x)$

The function is neither even nor odd.

21. $(f - h)(6) = f(6) - h(6)$

$= 6 - 4 - \sqrt{6-5}$

$= 2 - 1$

$= 1$

23. $(h \circ f)(1) = h(f(1))$

$= h(1-4)$

$= h(3)$

$= \sqrt{3-5}$

$= \sqrt{-2}$

Undefined

25. $\left(\dfrac{g}{f}\right)(x) = \dfrac{g(x)}{f(x)} = \dfrac{\frac{1}{x-3}}{x-4} = \dfrac{1}{(x-3)(x-4)}$

$(x-3)(x-4) \neq 0$

$x \neq 3$ and $x \neq 4$

Domain: $(-\infty, 3) \cup (3, 4) \cup (4, \infty)$

27. $f(x) = \sqrt[3]{x}$ and $g(x)x - 7$

29. **a.** $m = \dfrac{y_2 - y_1}{x_2 - x_1} = \dfrac{2614 - 1833}{8 - 0} = \dfrac{781}{8} \approx 98$

$y - y_1 = m(x - x_1)$

$y - 1833 = 98(x - 0)$

$y - 1833 = 98x$

$y = 98x + 1833$

b. $y = 98x + 1833$

$y = 98(15) + 1833$

$= 1470 + 1833$

$= \$3303$ million

Chapter 2 Cumulative Review Exercises

1. a. $f(2)=-1$

b. $f(x)=0$ for $x=0$ and $x=3$

c. $(-4,\infty)$ **d.** $[-2,\infty)$

e. $(1,\infty)$ **f.** $(-1,1)$

g. $(-4,-1)$

h. $(f\circ f)(-1)=f(f(-1))=f(2)=-1$

3. $(g\circ f)(x)=g(f(x))=\dfrac{1}{f(x)}=\dfrac{1}{-x^2+3x}$

$-x^2+3x\neq 0$
$x^2-3x\neq 0$
$x(x-3)\neq 0$
$x\neq 0$ and $x\neq 3$

Domain: $(-\infty,0)\cup(0,3)\cup(3,\infty)$

5. $\dfrac{f(x+h)-f(x)}{h}$

$$=\frac{-(x+h)^2+3(x+h)-(-x^2+3x)}{h}$$
$$=\frac{-(x^2+2xh+h^2)+3x+3h+x^2-3x}{h}$$
$$=\frac{-x^2-2xh-h^2+3h+x^2}{h}$$
$$=\frac{-2xh-h^2+3h}{h}=-2x-h+3$$

7. Substitute 0 for $f(x)$:

$f(x)=-x^2+3x$
$0=-x^2+3x$
$x^2-3x=0$
$x(x-3)=0$
$x=0$ or $x=3$

Substitute 0 for x:

$f(x)=-x^2+3x$
$f(0)=-(0)^2+3(0)=0$

x-intercepts: $(0,0)$ and $(3,0)$;
y-intercept: $(0,0)$

9.

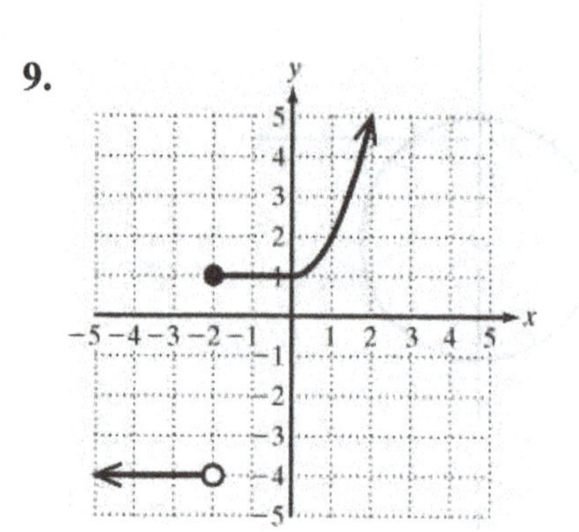

11. $|x-7|$ or $|7-x|$

13. $-3t(t-1)=2t+6$

$-3t^2+3t=2t+6$
$-3t^2+t-6=0$
$3t^2-t+6=0$

$$t=\frac{-(-1)\pm\sqrt{(-1)^2-4(3)(6)}}{2(3)}=\frac{1\pm\sqrt{1-72}}{6}$$
$$=\frac{1\pm\sqrt{-71}}{6}=\frac{1}{6}\pm\frac{\sqrt{71}}{6}i \qquad \left\{\frac{1}{6}\pm\frac{\sqrt{71}}{6}i\right\}$$

15. Let $u=x^{1/5}$.

$x^{2/5}-3x^{1/5}+2=0$
$u^2-3u+2=0$
$(u-1)(u-2)=0$
$u=1$ or $u=2$
$x^{1/5}=1$ $\quad x^{1/5}=2$
$x=1$ $\quad x=32$ $\qquad \{1,32\}$

17. $3\leq -2x+1<7$

$2\leq -2x<6$
$-1\geq x>-3$ or $-3<x\leq -1$ $\qquad (-3,-1]$

19. $3c\sqrt{8c^2d^3}+c^2\sqrt{50d^3}-2d\sqrt{2c^4d}$

$$=3c\sqrt{2^2c^2d^2\cdot 2d}+c^2\sqrt{5^2d^2\cdot 2d}$$
$$-2d\sqrt{(c^2)^2\cdot 2d}$$
$$=6c^2d\sqrt{2d}+5c^2d\sqrt{2d}-2c^2d\sqrt{2d}$$
$$=9c^2d\sqrt{2d}$$

Chapter 3 Polynomial and Rational Functions

Section 3.1 Quadratic Functions and Applications

1. quadratic **3.** vertex

5. downward **7.** k

9. $f(x) = -(x-4)^2 + 1$
$a = -1, h = 4, k = 1$

a. Since $a < 0$, the parabola opens downward.

b. The vertex is $(h, k) = (4, 1)$.

c.
$$\begin{aligned} f(x) &= -(x-4)^2 + 1 \\ 0 &= -(x-4)^2 + 1 \\ -1 &= -(x-4)^2 \\ 1 &= (x-4)^2 \\ \pm\sqrt{1} &= x - 4 \\ 4 \pm 1 &= x \\ x = 3 \quad &\text{or} \quad x = 5 \end{aligned}$$
The x-intercepts are $(3, 0)$ and $(5, 0)$.

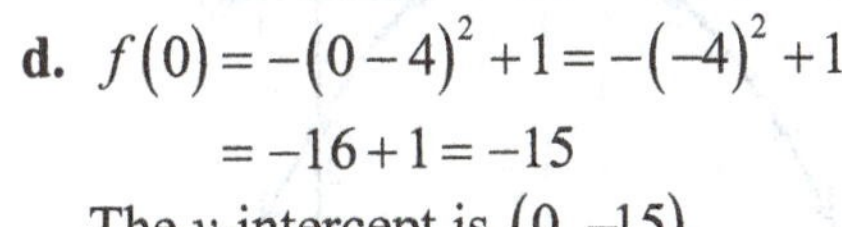

d. $f(0) = -(0-4)^2 + 1 = -(-4)^2 + 1$
$= -16 + 1 = -15$
The y-intercept is $(0, -15)$.

e.

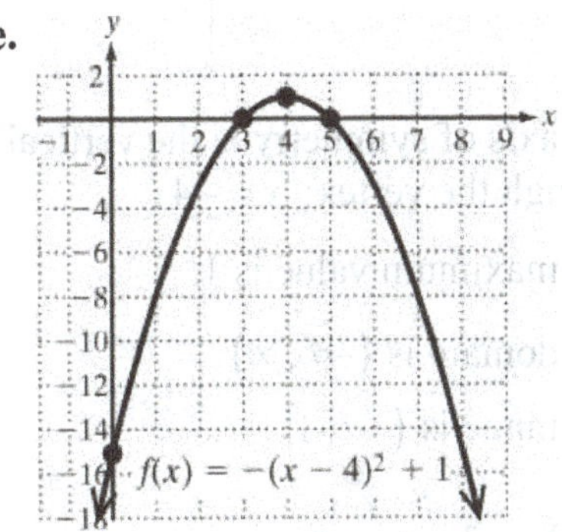

f. The axis of symmetry is the vertical line through the vertex: $x = 4$.

g. The maximum value is 1.

h. The domain is $(-\infty, \infty)$.

The range is $(-\infty, 1]$.

11. $h(x) = 2(x+1)^2 - 8 = 2[x-(-1)]^2 + (-8)$
$a = 2, h = -1, k = -8$

a. Since $a > 0$, the parabola opens upward.

b. The vertex is $(h, k) = (-1, -8)$.

c.
$$\begin{aligned} h(x) &= 2(x+1)^2 - 8 \\ 0 &= 2(x+1)^2 - 8 \\ 8 &= 2(x+1)^2 \\ 4 &= (x+1)^2 \\ \pm\sqrt{4} &= x + 1 \\ -1 \pm 2 &= x \\ x = -3 \quad &\text{or} \quad x = 1 \end{aligned}$$
The x-intercepts are $(-3, 0)$ and $(1, 0)$.

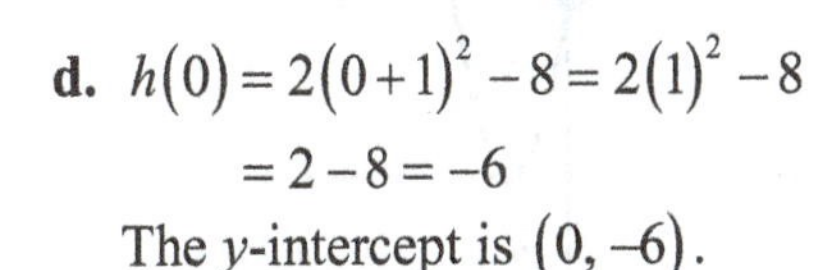

d. $h(0) = 2(0+1)^2 - 8 = 2(1)^2 - 8$
$= 2 - 8 = -6$
The y-intercept is $(0, -6)$.

e.

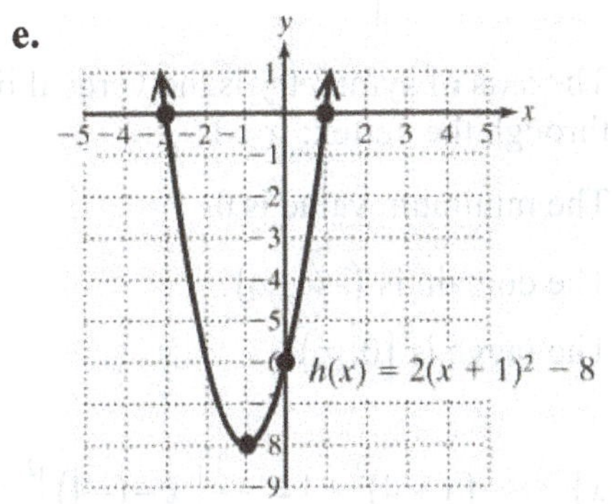

f. The axis of symmetry is the vertical line through the vertex: $x = -1$.

g. The minimum value is -8.

h. The domain is $(-\infty, \infty)$.

The range is $[-8, \infty)$.

13. $m(x)=3(x-1)^2=3(x-1)^2+0$
$a=3, h=1, k=0$

a. Since $a>0$, the parabola opens upward.

b. The vertex is $(h,k)=(1,0)$.

c. $m(x)=3(x-1)^2$

$$0=3(x-1)^2$$
$$0=(x-1)^2$$
$$\pm\sqrt{0}=x-1$$
$$1=x$$

The x-intercept is $(1,0)$.

d. $m(0)=3(0-1)^2=3(-1)^2=3$

The y-intercept is $(0,3)$.

e.

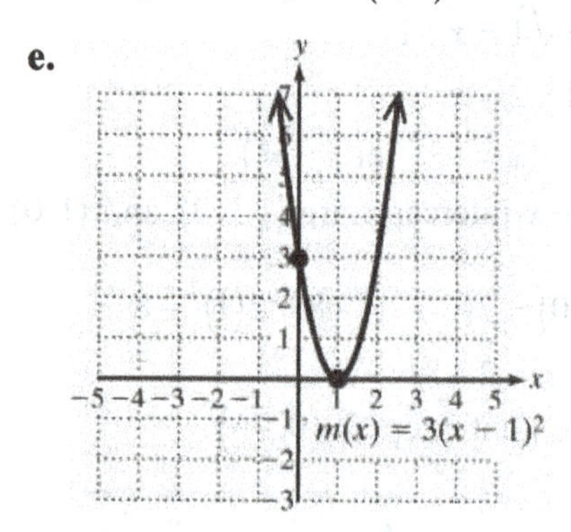

f. The axis of symmetry is the vertical line through the vertex: $x=1$.

g. The minimum value is 0.

h. The domain is $(-\infty,\infty)$.

The range is $[0,\infty)$.

15. $p(x)=-\frac{1}{5}(x+4)^2+1=-\frac{1}{5}[x-(-4)]^2+1$

$a=-\frac{1}{5}, h=-4, k=1$

a. Since $a<0$, the parabola opens downward.

b. The vertex is $(h,k)=(-4,1)$.

c. $p(x)=-\frac{1}{5}(x+4)^2+1$

$$0=-\frac{1}{5}(x+4)^2+1$$
$$-1=-\frac{1}{5}(x+4)^2$$
$$5=(x+4)^2$$
$$\pm\sqrt{5}=x+4$$
$$-4\pm\sqrt{5}=x$$
$$x=-4+\sqrt{5} \quad \text{or} \quad x=-4-\sqrt{5}$$

The x-intercepts are $\left(-4-\sqrt{5},0\right)$ and $\left(-4+\sqrt{5},0\right)$.

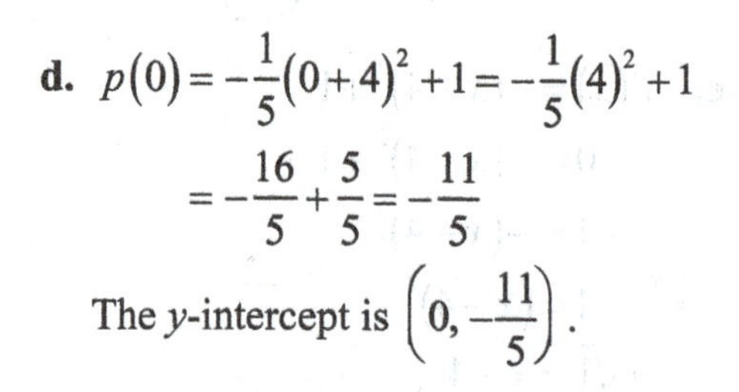

d. $p(0)=-\frac{1}{5}(0+4)^2+1=-\frac{1}{5}(4)^2+1$

$$=-\frac{16}{5}+\frac{5}{5}=-\frac{11}{5}$$

The y-intercept is $\left(0,-\frac{11}{5}\right)$.

e.

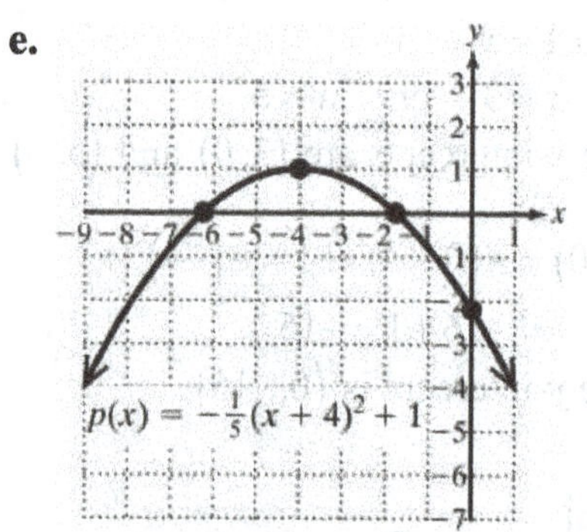

f. The axis of symmetry is the vertical line through the vertex: $x=-4$.

g. The maximum value is 1.

h. The domain is $(-\infty,\infty)$.

The range is $(-\infty,1]$.

17. a. $f(x)=x^2+6x+5$

$$f(x)=\left(x^2+6x\quad\right)+5 \qquad \boxed{\left[\frac{1}{2}(6)\right]^2=9}$$
$$f(x)=\left(x^2+6x+9\right)-9+5$$
$$f(x)=(x+3)^2-4$$

b. The vertex is $(h, k) = (-3, -4)$.

c. $f(x) = x^2 + 6x + 5$

$0 = x^2 + 6x + 5$

$0 = (x+5)(x+1)$

$x = -5 \quad \text{or} \quad x = -1$

The x-intercepts are $(-5, 0)$ and $(-1, 0)$.

d. $f(0) = (0)^2 + 6(0) + 5 = 5$

The y-intercept is $(0, 5)$.

e.

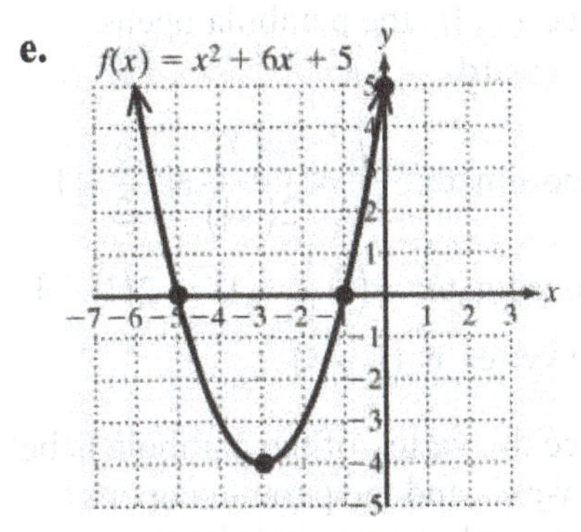

f. The axis of symmetry is the vertical line through the vertex: $x = -3$.

g. The minimum value is -4.

h. The domain is $(-\infty, \infty)$.

The range is $[-4, \infty)$.

19. a. $p(x) = 3x^2 - 12x - 7$

$p(x) = 3(x^2 - 4x \quad\quad) - 7 \quad \boxed{\left[\frac{1}{2}(-4)\right]^2 = 4}$

$p(x) = 3(x^2 - 4x + 4 - 4) - 7$

$p(x) = 3(x^2 - 4x + 4) + 3(-4) - 7$

$p(x) = 3(x-2)^2 - 19$

b. The vertex is $(h, k) = (2, -19)$.

c. $p(x) = 3x^2 - 12x - 7$

$0 = 3x^2 - 12x - 7$

$$x = \frac{-(-12) \pm \sqrt{(-12)^2 - 4(3)(-7)}}{2(3)}$$

$$= \frac{12 \pm \sqrt{228}}{6} = \frac{12 \pm 2\sqrt{57}}{6} = \frac{6 \pm \sqrt{57}}{3}$$

$$x = \frac{6 - \sqrt{57}}{3} \quad \text{or} \quad x = \frac{6 + \sqrt{57}}{3}$$

The x-intercepts are $\left(\frac{6 - \sqrt{57}}{3}, 0\right)$ and $\left(\frac{6 + \sqrt{57}}{3}, 0\right)$.

d. $p(0) = 3(0)^2 - 12(0) - 7 = -7$

The y-intercept is $(0, -7)$.

e.

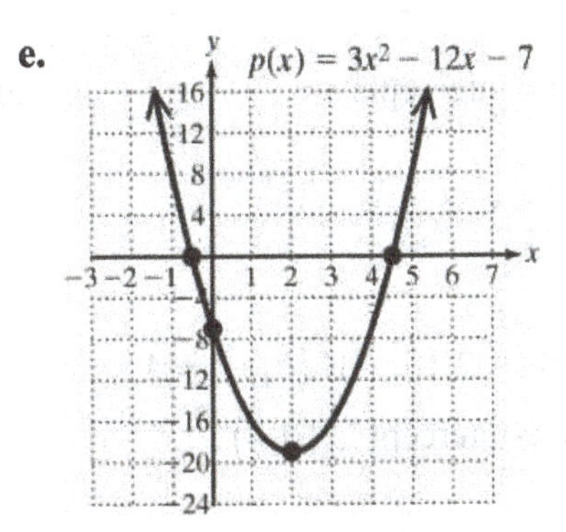

f. The axis of symmetry is the vertical line through the vertex: $x = 2$.

g. The minimum value is -19.

h. The domain is $(-\infty, \infty)$.

The range is $[-19, \infty)$.

21. a. $c(x) = -2x^2 - 10x + 4$

$c(x) = -2(x^2 + 5x \quad\quad) + 4$

$\boxed{\left[\frac{1}{2}(5)\right]^2 = \frac{25}{4}}$

$c(x) = -2\left(x^2 + 5x + \frac{25}{4} - \frac{25}{4}\right) + 4$

$c(x) = -2\left(x^2 + 5x + \frac{25}{4}\right) - 2\left(-\frac{25}{4}\right) + 4$

$c(x) = -2\left(x + \frac{5}{2}\right)^2 + \frac{33}{2}$

b. The vertex is $(h, k) = \left(-\frac{5}{2}, \frac{33}{2}\right)$.

c. $c(x) = -2x^2 - 10x + 4$

$$0 = -2x^2 - 10x + 4$$

$$x = \frac{-(-10) \pm \sqrt{(-10)^2 - 4(-2)(4)}}{2(-2)}$$

$$= \frac{10 \pm \sqrt{132}}{-4}$$

$$= \frac{-10 \pm 2\sqrt{33}}{4} = \frac{-5 \pm \sqrt{33}}{2}$$

$$x = \frac{-5 - \sqrt{33}}{2} \text{ or } x = \frac{-5 + \sqrt{33}}{2}$$

The x-intercepts are $\left(\frac{-5 - \sqrt{33}}{2}, 0\right)$ and $\left(\frac{-5 + \sqrt{33}}{2}, 0\right)$.

d. $c(0) = -2(0)^2 - 10(0) + 4 = 4$
The y-intercept is $(0, 4)$.

e.

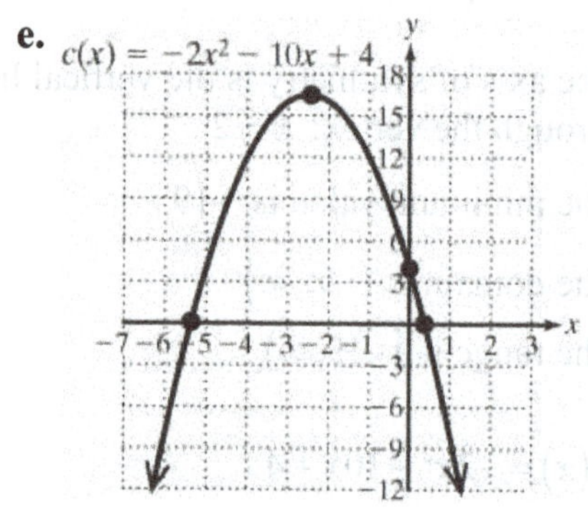

f. The axis of symmetry is the vertical line through the vertex: $x = -\frac{5}{2}$.

g. The maximum value is $\frac{33}{2}$.

h. The domain is $(-\infty, \infty)$.
The range is $\left(-\infty, \frac{33}{2}\right]$.

23. $f(x) = 3x^2 - 42x - 91$

x-coordinate: $\frac{-b}{2a} = \frac{-(-42)}{2(3)} = \frac{42}{6} = 7$

y-coordinate: $f(7) = 3(7)^2 - 42(7) - 91 = -238$
The vertex is $(7, -238)$.

25. $k(a) = -\frac{1}{3}a^2 + 6a + 1$

x-coordinate: $\frac{-b}{2a} = \frac{-(6)}{2\left(-\frac{1}{3}\right)} = \frac{-6}{-\frac{2}{3}} = 9$

y-coordinate: $k(9) = -\frac{1}{3}(9)^2 + 6(9) + 1 = 28$
The vertex is $(9, 28)$.

27. $g(x) = -x^2 + 2x - 4$

a. Since $a < 0$, the parabola opens downward.

b. x-coordinate: $\frac{-b}{2a} = \frac{-(2)}{2(-1)} = \frac{-2}{-2} = 1$

y-coordinate: $g(1) = -(1)^2 + 2(1) - 4 = -3$
The vertex is $(1, -3)$.

c. Since the vertex of the parabola is below the x-axis and the parabola opens downward, the parabola cannot cross or touch the x-axis. Therefore, there are no x-intercepts.

d. $g(0) = -(0)^2 + 2(0) - 4 = -4$
The y-intercept is $(0, -4)$.

e.

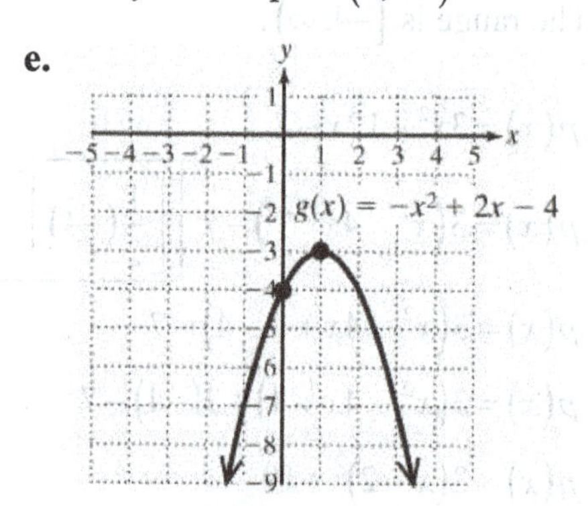

f. The axis of symmetry is the vertical line through the vertex: $x = 1$.

g. The maximum value is -3.

h. The domain is $(-\infty, \infty)$.
The range is $(-\infty, -3]$.

29. $f(x) = 5x^2 - 15x + 3$

a. Since $a > 0$, the parabola opens upward.

b. x-coordinate: $\frac{-b}{2a}=\frac{-(-15)}{2(5)}=\frac{15}{10}=\frac{3}{2}$

y-coordinate:

$f\left(\frac{3}{2}\right)=5\left(\frac{3}{2}\right)^2-15\left(\frac{3}{2}\right)+3=-\frac{33}{4}$

The vertex is $\left(\frac{3}{2},-\frac{33}{4}\right)$.

c. $f(x)=5x^2-15x+3$

$0=5x^2-15x+3$

$$x=\frac{-(-15)\pm\sqrt{(-15)^2-4(5)(3)}}{2(5)}$$

$$=\frac{15\pm\sqrt{165}}{10}$$

$x=\frac{15-\sqrt{165}}{10}$ or $x=\frac{15+\sqrt{165}}{10}$

The x-intercepts are $\left(\frac{15-\sqrt{165}}{10},0\right)$ and $\left(\frac{15+\sqrt{165}}{10},0\right)$.

d. $f(0)=5(0)^2-15(0)+3=3$

The y-intercept is $(0,3)$.

e.

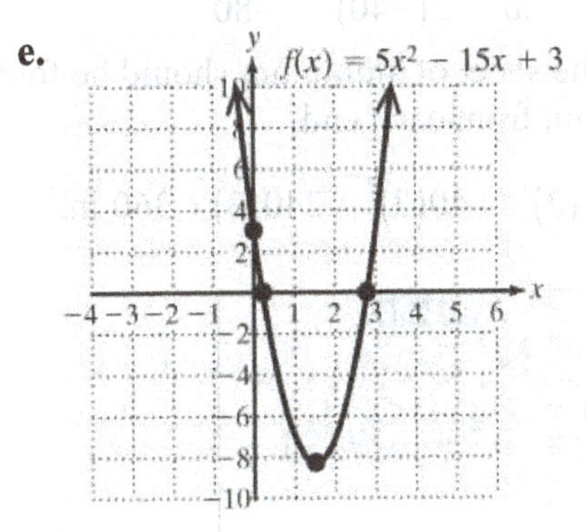

f. The axis of symmetry is the vertical line through the vertex: $x=\frac{3}{2}$.

g. The minimum value is $-\frac{33}{4}$.

h. The domain is $(-\infty,\infty)$.

The range is $\left[-\frac{33}{4},\infty\right)$.

31. a. The time at which the population will be at a maximum is the t-coordinate of the vertex.

$$t=\frac{-b}{2a}=\frac{-(82{,}000)}{2(-1718)}=\frac{82{,}000}{3436}\approx 24 \text{ hr}$$

b. The maximum population is the value of $P(t)$ at the vertex.

$P(24)=-1718(24)^2+82{,}000(24)+10{,}000$

$\approx 988{,}000$

33. a. The time at which the skateboarder will be at his maximum height is the t-coordinate of the vertex.

$$t=\frac{-b}{2a}=\frac{-(5.4)}{2(-4.9)}=\frac{-5.4}{-9.8}\approx 0.55 \text{ sec}$$

b. The maximum height is the value of $h(t)$ at the vertex.

$h(0.55)=-4.9(0.55)^2+5.4(0.55)+3$

≈ 4.5 m

35. a. The horizontal distance at which the water will be at its maximum height is the x-coordinate of the vertex.

$$x=\frac{-b}{2a}=\frac{-(0.576)}{2(-0.026)}=\frac{-0.576}{-0.052}\approx 11.1 \text{ m}$$

b. The maximum height is the value of $h(x)$ at the vertex.

$h(11.1)=-0.026(11.1)^2+0.576(11.1)+3$

≈ 6.2 m

c. The distance of the firefighter from the house is the value of x when $h(x)=6$.

$6=-0.026x^2+0.576x+3$

$0=-0.026x^2+0.576x-3$

$$x=\frac{-0.576\pm\sqrt{(0.576)^2-4(-0.026)(-3)}}{2(-0.026)}$$

$$=\frac{0.576\pm\sqrt{0.019776}}{0.052}$$

$x\approx 8$ or $x\approx 14$

Since we want the downward branch, we use the second value of x. The firefighter is 14 m from the house.

37. Let x represent the first number.
Let y represent the second number.
We know that
$x+y=24$
$y=24-x$
Let P represent the product.
$P=xy$
$P(x)=x(24-x)=24x-x^2=-x^2+24x$
Function P is a quadratic function with a negative leading coefficient. The graph of the parabola opens downward, so the vertex is the maximum point on the function. The x-coordinate of the vertex is the value of x that will maximize the product.
$x=\frac{-b}{2a}=\frac{-(24)}{2(-1)}=\frac{-24}{-2}=12$
$y=24-x=24-12=12$
The numbers are 12 and 12.

39. Let x represent the first number.
Let y represent the second number.
We know that
$x-y=10$
$-y=10-x$
$y=x-10$
Let P represent the product.
$P=xy$
$P(x)=x(x-10)=x^2-10x$
Function P is a quadratic function with a positive leading coefficient. The graph of the parabola opens upward, so the vertex is the minimum point on the function. The x-coordinate of the vertex is the value of x that will minimize the product.
$x=\frac{-b}{2a}=\frac{-(-10)}{2(1)}=\frac{10}{2}=5$
$y=x-10=5-10=-5$
The numbers are 5 and -5.

41. We know that
$2x+y=160$
$y=160-2x$
Let A represent the area.
$A=xy$
$A(x)=x(160-2x)=160x-2x^2=-x^2+160x$
Function A is a quadratic function with a negative leading coefficient. The graph of the parabola opens downward, so the vertex is the maximum point on the function. The x-coordinate of the vertex is the value of x that will maximize the area.
$x=\frac{-b}{2a}=\frac{-(160)}{2(-2)}=\frac{-160}{-4}=40$
$y=160-2x=160-2(40)=160-80=80$
The dimensions are 40 ft by 80 ft.

43. a. Let V represent the volume.
$V=lwh$
$V(x)=20(12-2x)(x)$
$=20(12x-2x^2)$
$=240x-40x^2$
$=-40x^2+240x$

b. Function V is a quadratic function with a negative leading coefficient. The graph of the parabola opens downward, so the vertex is the maximum point on the function. The x-coordinate of the vertex is the value of x that will maximize the volume.
$x=\frac{-b}{2a}=\frac{-(240)}{2(-40)}=\frac{-240}{-80}=3$
The sheet of aluminum should be folded 3 in. from each end.

c. $V(3)=-40(3)^2+240(3)=360$ in.^{3}

45. a.
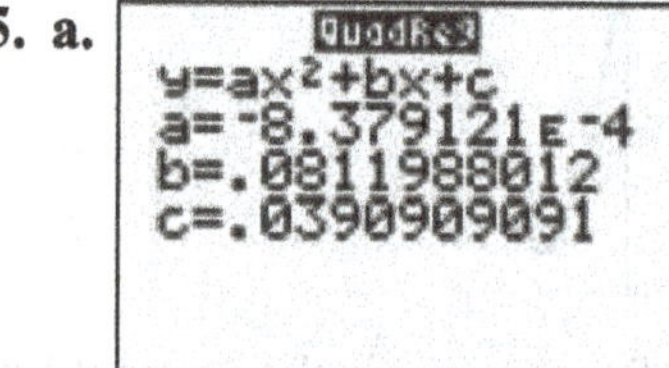

$y(t)=-0.000838t^2+0.0812t+0.040$

b. From the graph, the time when the population is the greatest is the t-coordinate of the vertex.
$t=\frac{-b}{2a}=\frac{-(0.0812)}{2(-0.000838)}=\frac{-0.0812}{-0.001676}$
≈ 48 hr

c. The maximum population of the bacteria is the $y(t)$ value at the vertex.

$$y(48) = -0.000838(48)^2 + 0.0812(48) + 0.040 \approx 2\text{ g}$$

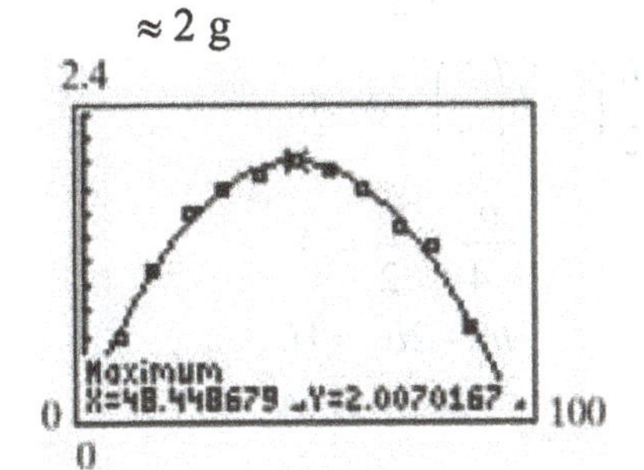

47. a. $x = 0$ cm. From the table, $v(0) = 195.6$ cm/sec.

$$t = \frac{d}{r} = \frac{3000\text{ cm}}{195.6\text{ cm/sec}} \approx 15.3\text{ sec}$$

b. $x = 9$ cm. From the table, $v(9) = 180.0$ cm/sec.

$$t = \frac{d}{r} = \frac{3000\text{ cm}}{180.0\text{ cm/sec}} \approx 16.7\text{ sec}$$

c.

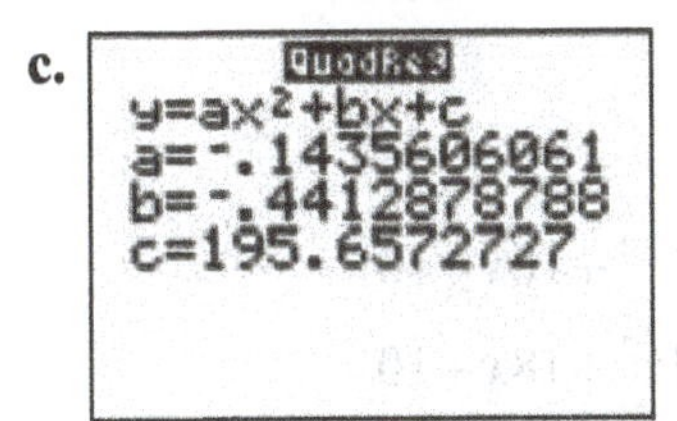

$$v(x) = -0.1436x^2 - 0.4413x + 195.7$$

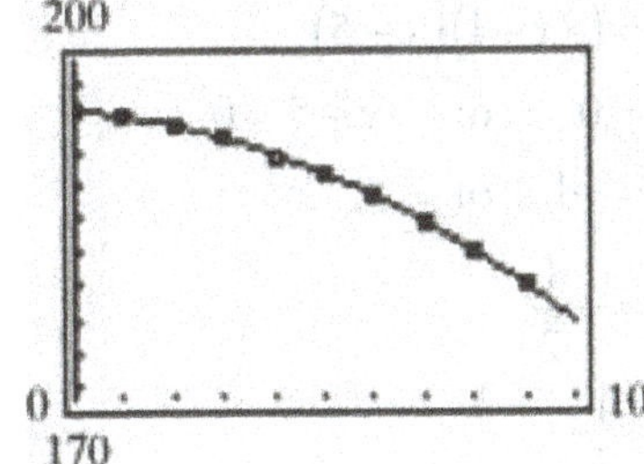

d. $v(5.5) = -0.1436(5.5)^2 - 0.4413(5.5) + 195.7 \approx 188.9$ cm/sec

49. False

51. True

53. $b^2 - 4ac = (12)^2 - 4(4)(9) = 144 - 144 = 0$
Discriminant is 0; one x-intercept

55. $b^2 - 4ac = (-5)^2 - 4(-1)(8) = 25 + 32 = 57$
Discriminant is 57; two x-intercepts

57. $b^2 - 4ac = (6)^2 - 4(-3)(-11) = 36 - 132 = -96$
Discriminant is -96; no x-intercepts

59. Graph g

61. Graph h

63. Graph f

65. Graph c

67. For a parabola opening upward, such as the graph of $f(x) = x^2$, the minimum value is the y-coordinate of the vertex. There is no maximum value because the y values of the function become arbitrarily large for large values of $|x|$.

69. No function defined by $y = f(x)$ can have two y-intercepts because the graph would fail the vertical line test.

71. Because a parabola is symmetric with respect to the vertical line through the vertex, the x-coordinate of the vertex must be equidistant from the x-intercepts. Therefore, given $y = f(x)$, the x-coordinate of the vertex is 4 because 4 is midway between 2 and 6. The y-coordinate of the vertex is $f(4)$.

73. $(h, k) = (2, -3)$

$$f(x) = a(x-2)^2 + (-3) = a(x-2)^2 - 3$$
$$f(0) = 5$$
$$a(0-2)^2 - 3 = 5$$
$$4a = 8$$
$$a = 2$$
$$f(x) = 2(x-2)^2 - 3$$

75. $(h, k) = (4, 6)$

$$f(x) = a(x-4)^2 + 6$$
$$f(1) = 3$$
$$a(1-4)^2 + 6 = 3$$
$$9a = -3$$
$$a = -\frac{1}{3}$$
$$f(x) = -\frac{1}{3}(x-4)^2 + 6$$

77. Find the x-coordinate of the vertex:

$$\frac{-b}{2a} = \frac{-12}{2(2)} = \frac{-12}{4} = -3$$
$$f(-3) = -9$$
$$2(-3)^2 + 12(-3) + c = -9$$
$$18 - 36 + c = -9$$
$$-18 + c = -9$$
$$c = 9$$

79. Find the x-coordinate of the vertex:

$$\frac{-b}{2a} = \frac{-b}{2(-1)} = \frac{-b}{-2} = \frac{b}{2}$$
$$f\left(\frac{b}{2}\right) = 8$$
$$-\left(\frac{b}{2}\right)^2 + b\left(\frac{b}{2}\right) + 4 = 8$$
$$-\frac{b^2}{4} + \frac{b^2}{2} = 4$$
$$-b^2 + 2b^2 = 16$$
$$b^2 = 16$$
$$b = \pm 4$$

$b = -4$ or $b = 4$

Section 3.2 Introduction to Polynomial Functions

1. $f(x) = -6x^2 + 18x - 7$

x-coordinate: $\dfrac{-b}{2a} = \dfrac{-(18)}{2(-6)} = \dfrac{-18}{-12} = \dfrac{3}{2}$

y-coordinate: $f\left(\frac{3}{2}\right) = -6\left(\frac{3}{2}\right)^2 + 18\left(\frac{3}{2}\right) - 7$

$$= -6\left(\frac{9}{4}\right) + 27 - 7$$
$$= -\frac{27}{2} + 20$$
$$= \frac{13}{2}$$

The vertex is $\left(\frac{3}{2}, \frac{13}{2}\right)$.

3. $f(x) = -4x^2 + 18x + 10$

$$0 = -4x^2 + 18x + 10$$
$$0 = 2x^2 - 9x - 5$$
$$0 = (2x+1)(x-5)$$

$2x + 1 = 0$ or $x - 5 = 0$

$2x = -1$ or $x = 5$

$$x = -\frac{1}{2}$$

5 and $-\frac{1}{2}$

5. $-16x^7 + 48x^6 + 9x^5 - 27x^4$

$$= -x^4(16x^3 - 48x^2 - 9x + 27)$$
$$= -x^4\left[16x^2(x-3) - 9(x-3)\right]$$
$$= -x^4(x-3)(16x^2 - 9)$$
$$= -x^4(x-3)(4x-3)(4x+3)$$

7. polynomial

9. is not

11. 1

13. False

15. up; down

17. 3; 4

19. 5

21. cross

23. f has at least one zero on the interval $[a, b]$.

25. origin

27. Multiply the leading terms from each factor.

$$-0.6x^2(10x)^3(x)^4 = -0.6x^2(1000x^3)(x^4) = -600x^9$$

29. The leading coefficient is negative and the degree is even. The end behavior is down to the left and down to the right.

31. The leading coefficient is positive and the degree is odd. The end behavior is down to the left and up to the right.

33. Multiply the leading terms from each factor.

$$-4(x)(2x)^2(x)^4 = -4(x)(4x^2)(x^4) = -16x^7$$

The leading coefficient is negative and the degree is odd. The end behavior is up to the left and down to the right.

35. Multiply the leading terms from each factor.

$$-2x^2(-x)(2x)^3 = -2x^2(-x)(8x^3) = 16x^6$$

The leading coefficient is positive and the degree is even. The end behavior is up to the left and up to the right.

37. $f(x) = x^3 + 2x^2 - 25x - 50$

$$0 = x^3 + 2x^2 - 25x - 50$$
$$0 = x^2(x+2) - 25(x+2)$$
$$0 = (x+2)(x^2 - 25)$$
$$0 = (x+2)(x-5)(x+5)$$
$$x = -2, x = 5, x = -5$$

Each zero is of multiplicity 1.

39. $h(x) = -6x^3 - 9x^2 + 60x$

$$0 = -6x^3 - 9x^2 + 60x$$
$$0 = -3x(2x^2 + 3x - 20)$$
$$0 = -3x(2x-5)(x+4)$$
$$x = 0, x = \frac{5}{2}, x = -4$$

Each zero is of multiplicity 1.

41. $m(x) = x^5 - 10x^4 + 25x^3$

$$0 = x^5 - 10x^4 + 25x^3$$
$$0 = x^3(x^2 - 10x + 25)$$
$$0 = x^3(x-5)^2$$
$$x = 0, x = 5$$

The zero 0 has multiplicity 3. The zero 5 has multiplicity 2.

43. $p(x) = -3x(x+2)^3(x+4)$

$$0 = -3x(x+2)^3(x+4)$$
$$x = 0, x = -2, x = -4$$

The zero 0 has multiplicity 1. The zero -2 has multiplicity 3. The zero -4 has multiplicity 1.

45. $t(x) = 5x(3x-5)(2x+9)(x-\sqrt{3})(x+\sqrt{3})$

$$0 = 5x(3x-5)(2x+9)(x-\sqrt{3})(x+\sqrt{3})$$
$$x = 0, x = \frac{5}{3}, x = -\frac{9}{2}, x = \sqrt{3}, x = -\sqrt{3}$$

Each zero is of multiplicity 1.

47. $c(x) = \left[x - (3-\sqrt{5})\right]\left[x - (3+\sqrt{5})\right]$

$$0 = \left[x - (3-\sqrt{5})\right]\left[x - (3+\sqrt{5})\right]$$
$$x = 3 - \sqrt{5}, x = 3 + \sqrt{5}$$

Each zero is of multiplicity 1.

49. $f(x) = 2x^3 - 7x^2 - 14x + 30$

$$f(1) = 2(1)^3 - 7(1)^2 - 14(1) + 30 = 2 - 7 - 14 + 30 = 11$$
$$f(2) = 2(2)^3 - 7(2)^2 - 14(2) + 30 = 16 - 28 - 28 + 30 = -10$$

$f(3) = 2(3)^3 - 7(3)^2 - 14(3) + 30$
$= 54 - 63 - 42 + 30 = -21$
$f(4) = 2(4)^3 - 7(4)^2 - 14(4) + 30$
$= 128 - 112 - 56 + 30 = -10$
$f(5) = 2(5)^3 - 7(5)^2 - 14(5) + 30$
$= 250 - 175 - 70 + 30 = 35$

a. Since $f(1)$ and $f(2)$ have opposite signs, the intermediate value theorem guarantees that the function has at least one zero on the interval $[1, 2]$.

b. Since $f(2)$ and $f(3)$ have the same sign, the intermediate value theorem does not guarantee that the function has at least one zero on the interval $[2, 3]$.

c. Since $f(3)$ and $f(4)$ have the same sign, the intermediate value theorem does not guarantee that the function has at least one zero on the interval $[3, 4]$.

d. Since $f(4)$ and $f(5)$ have opposite signs, the intermediate value theorem guarantees that the function has at least one zero on the interval $[4, 5]$.

51. a. Since $Y_1(-4)$ and $Y_1(-3)$ have opposite signs, the intermediate value theorem guarantees that the function has at least one zero on the interval $[-4, -3]$.

b. Since $Y_1(-3)$ and $Y_1(-2)$ have opposite signs, the intermediate value theorem guarantees that the function has at least one zero on the interval $[-3, -2]$.

c. Since $Y_1(-2)$ and $Y_1(-1)$ have the same sign, the intermediate value theorem does not guarantee that the function has at least one zero on the interval $[-2, -1]$.

d. Since $Y_1(-1)$ and $Y_1(0)$ have the same sign, the intermediate value theorem does not guarantee that the function has at least one zero on the interval $[-1, 0]$.

53. $f(x) = 4x^3 - 8x^2 - 25x + 50$
$f(-3) = 4(-3)^3 - 8(-3)^2 - 25(-3) + 50$
$= -108 - 72 + 75 + 50$
$= -55$
$f(-2) = 4(-2)^3 - 8(-2)^2 - 25(-2) + 50$
$= -32 - 32 + 50 + 50$
$= 36$

a. Since $f(-3)$ and $f(-2)$ have opposite signs, the intermediate value theorem guarantees that the function has at least one zero on the interval $[-3, -2]$.

b. $f(x) = 4x^3 - 8x^2 - 25x + 50$
$0 = 4x^3 - 8x^2 - 25x + 50$
$0 = 4x^2(x-2) - 25(x-2)$
$0 = (x-2)(4x^2 - 25)$
$0 = (x-2)(2x-5)(2x+5)$
$x = 2, x = \frac{5}{2}, x = -\frac{5}{2}$

The zero on the interval $[-3, -2]$ is $-\frac{5}{2}$.

55. The graph is not smooth. It cannot represent a polynomial function.

57. The graph is smooth and continuous. It can represent a polynomial function.

a. The function has 2 turning points, so the minimum degree is 3.

b. The end behavior is down on the left and up on the right. The leading coefficient is positive and the degree is odd.

c. –4 (odd multiplicity); –1 (odd multiplicity); 3 (odd multiplicity)

59. The graph is smooth and continuous. It can represent a polynomial function.

a. The function has 5 turning points, so the minimum degree is 6.

b. The end behavior is down on the left and down on the right. The leading coefficient is negative and the degree is even.

c. –4 (odd multiplicity); –3 (odd multiplicity); –1 (even multiplicity); 2 (odd multiplicity); $\frac{7}{2}$ (odd multiplicity)

61. The graph is not continuous. It cannot represent a polynomial function.

63. $f(x) = x^3 - 5x^2$

The leading term is x^3. The end behavior is down to the left and up to the right.

$f(0) = (0)^3 - 5(0)^2 = 0$

The y-intercept is $(0, 0)$.

$0 = x^3 - 5x^2$

$0 = x^2(x - 5)$

$x = 0, x = 5$

The zeros of the function are 0 (multiplicity 2) and 5 (multiplicity 1).

Test for symmetry:

$f(-x) = (-x)^3 - 5(-x)^2 = -x^3 - 5x^2$

$f(x)$ is neither even nor odd.

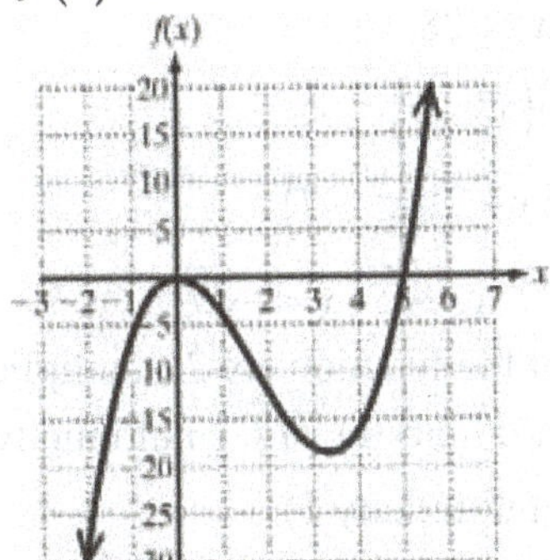

65. $f(x) = \frac{1}{2}(x - 2)(x + 1)(x + 3)$

The leading term is $\frac{1}{2}(x)(x)(x) = \frac{1}{2}x^3$. The end behavior is down to the left and up to the right.

$f(0) = \frac{1}{2}(0 - 2)(0 + 1)(0 + 3) = \frac{1}{2}(-2)(1)(3)$

$= -3$

The y-intercept is $(0, -3)$.

The zeros of the function are 2 (multiplicity 1), –1 (multiplicity 1), and –3 (multiplicity 1).

Test for symmetry:

$f(-x) = \frac{1}{2}(-x - 2)(-x + 1)(-x + 3)$

$f(x)$ is neither even nor odd.

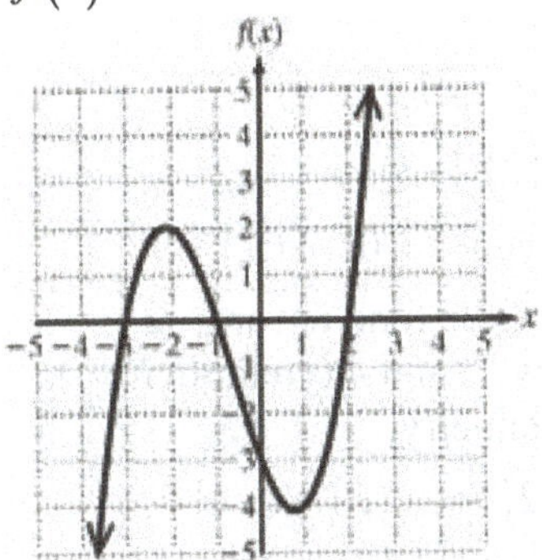

67. $k(x) = x^4 + 2x^3 - 8x^2$

The leading term is x^4. The end behavior is up to the left and up to the right.

$k(0) = (0)^4 + 2(0)^3 - 8(0)^2 = 0$

The y-intercept is $(0, 0)$.

$0 = x^4 + 2x^3 - 8x^2$

$0 = x^2(x^2 + 2x - 8)$

$0 = x^2(x + 4)(x - 2)$

$x = 0, x = -4, x = 2$

The zeros of the function are 0 (multiplicity 2), –4 (multiplicity 1), and 2 (multiplicity 1).

Test for symmetry:

$k(-x) = (-x)^4 + 2(-x)^3 - 8(-x)^2$

$= x^4 - 2x^3 - 8x^2$

$k(x)$ is neither even nor odd.

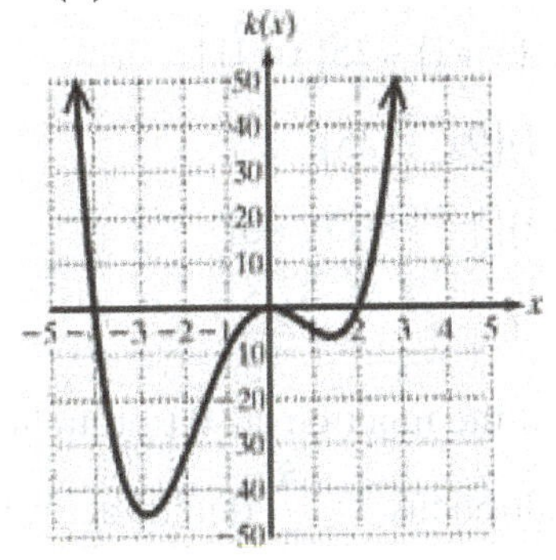

69. $k(x)=0.2(x+2)^2(x-4)^3$

The leading term is $0.2(x)^2(x)^3=0.2x^5$. The end behavior is down to the left and up to the right.

$k(0)=0.2(0+2)^2(0-4)^3$

$=0.2(2)^2(-4)^3=-51.2$

The y-intercept is $(0,-51.2)$.

The zeros of the function are -2 (multiplicity 2) and 4 (multiplicity 3).

Test for symmetry:

$k(-x)=0.2(-x+2)^2(-x-4)^3$

$k(x)$ is neither even nor odd.

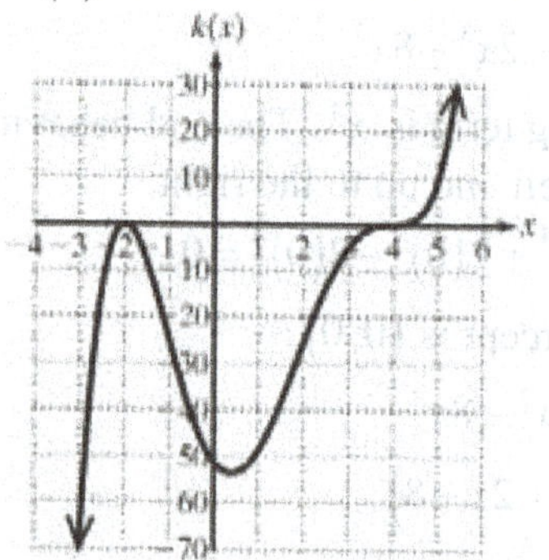

71. $p(x)=9x^5+9x^4-25x^3-25x^2$

The leading term is x^5. The end behavior is down to the left and up to the right.

$p(0)=9(0)^5+9(0)^4-25(0)^3-25(0)^2=0$

The y-intercept is $(0,0)$.

$0=9x^5+9x^4-25x^3-25x^2$

$0=x^2(9x^3+9x^2-25x-25)$

$0=x^2\left[9x^2(x+1)-25(x+1)\right]$

$0=x^2(x+1)(9x^2-25)$

$0=x^2(x+1)(3x-5)(3x+5)$

$x=0,\ x=-1,\ x=\frac{5}{3},\ x=-\frac{5}{3}$

The zeros of the function are 0 (multiplicity 2), -1 (multiplicity 1), $\frac{5}{3}$ (multiplicity 1), and $-\frac{5}{3}$ (multiplicity 1).

Test for symmetry:

$p(-x)$

$=9(-x)^5+9(-x)^4-25(-x)^3-25(-x)^2$

$=-9x^5+9x^4+25x^3-25x^2$

$p(x)$ is neither even nor odd.

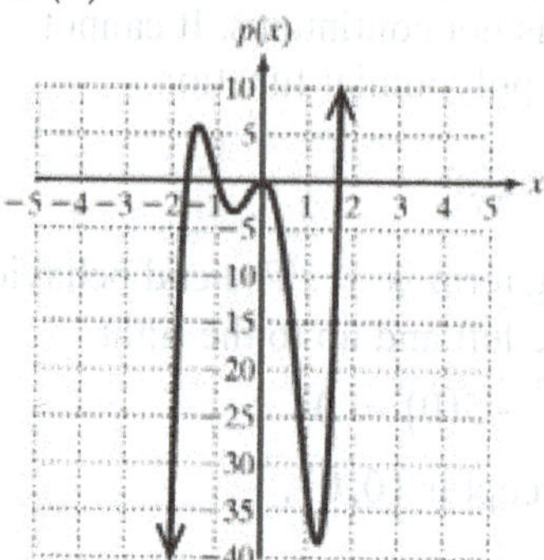

73. $t(x)=-x^4+11x^2-28$

The leading term is $-x^4$. The end behavior is down to the left and down to the right.

$t(0)=-(0)^4+11(0)^2-28=-28$

The y-intercept is $(0,-28)$.

$0=-x^4+11x^2-28$

$0=x^4-11x^2+28$

$0=(x^2-4)(x^2-7)$

$0=(x-2)(x+2)(x^2-7)$

$x=2,\ x=-2,\ x=\sqrt{7},\ x=-\sqrt{7}$

The zeros of the function are 2 (multiplicity 1), -2 (multiplicity 1), $\sqrt{7}$ (multiplicity 1), and $-\sqrt{7}$ (multiplicity 1).

Test for symmetry:

$t(-x)=-(-x)^4+11(-x)^2-28$

$=-x^4+11x^2-28$

$t(x)$ is even, so it is symmetric with respect to the y-axis.

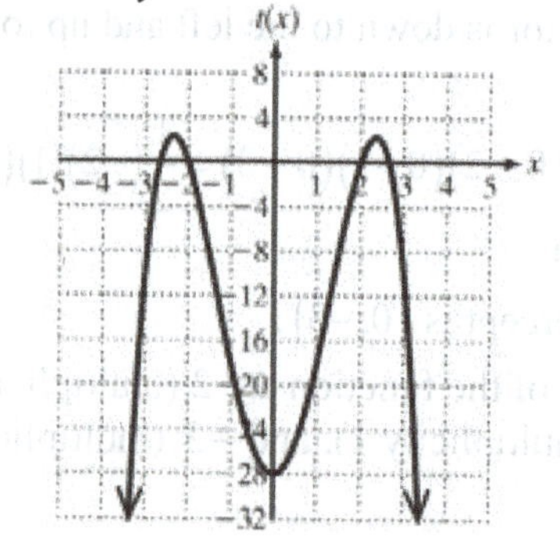

75. False **77.** False

79. True **81.** False

83. False **85.** True

87. **a.** The acceleration is increasing over the interval $(0,12)\cup(68,184)$.

b. The acceleration is decreasing over the interval $(12,68)\cup(184,200)$.

c. The graph shows 3 turning points.

d. The minimum degree of a polynomial function with 3 turning points is 4. The leading coefficient would be negative since the end behavior of the graph is down on the left and down on the right.

e. The acceleration is greatest approximately 184 sec after launch.

f. The maximum acceleration is approximately 2.85 G-forces.

89. The x-intercepts are the real solutions to the equation $f(x)=0$.

91. A function is continuous if its graph can be drawn without lifting the pencil from the paper.

93. b

95. a

97. **a.** $f(3)=(3)^2-3(3)+2=2$

$f(4)=(4)^2-3(4)+2=6$

b. By the intermediate value theorem, because $f(3)=2$ and $f(4)=6$, then f must take on every value between 2 and 6 on the interval $[3,4]$.

c. $f(x)=x^2-3x+2$

$4=x^2-3x+2$

$0=x^2-3x-2$

$$x=\frac{-(-3)\pm\sqrt{(-3)^2-4(1)(-2)}}{2(1)}$$

$$=\frac{3\pm\sqrt{17}}{2}\approx 3.56 \text{ or } -0.56$$

On the interval $[3,4]$,

$$x=\frac{3+\sqrt{17}}{2}\approx 3.56.$$

99.

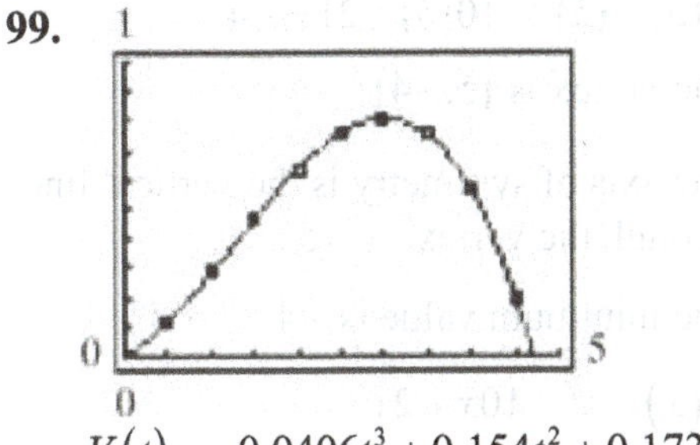

$V(t)=-0.0406t^3+0.154t^2+0.173t-0.0024$

101.

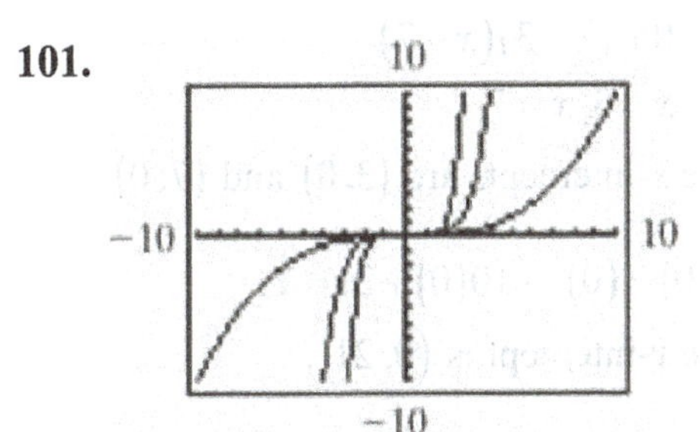

The end behavior is down to the left and up to the right.

103. Window b is better.

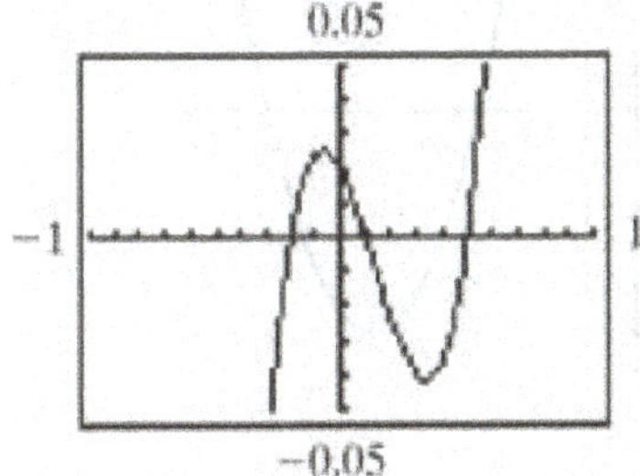

105.

100
−10 30
−200

Section 3.3 Division of Polynomials and the Remainder and Factor Theorems

1. $g(x) = x^2 - 10x + 21$

a. Since $a > 0$, the parabola opens upward.

b. x-coordinate: $\dfrac{-b}{2a} = \dfrac{-(-10)}{2(1)} = \dfrac{10}{2} = 5$

y-coordinate:

$g(5) = (5)^2 - 10(5) + 21 = -4$

The vertex is $(5, -4)$.

c. The axis of symmetry is the vertical line through the vertex: $x = 5$.

d. The minimum value is -4.

e. $g(x) = x^2 - 10x + 21$

$$0 = x^2 - 10x + 21$$
$$0 = (x-3)(x-7)$$
$$x = 3, x = 7$$

The x-intercepts are $(3, 0)$ and $(7, 0)$.

f. $g(0) = (0)^2 - 10(0) + 21 = 21$

The y-intercept is $(0, 21)$.

g.

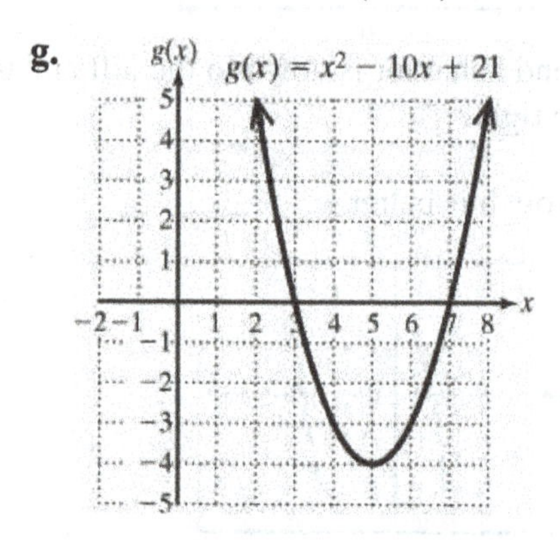

3. $42\overline{)25,019}$ quotient 595

$$\begin{array}{r} 25,019 \\ -210 \\ \hline 401 \\ -378 \\ \hline 239 \\ -210 \\ \hline 29 \end{array}$$

$595\dfrac{29}{42}$

Check: $42(595) + 29 = 24,990 + 29 = 25,019$

5. $\dfrac{-6x^5}{3x} = -2x^4$

7. Quotient: $(x+3)$; Remainder: 0

9. Dividend: $f(x)$; Divisor: $d(x)$; Quotient: $q(x)$; Remainder: $r(x)$

11. $f(c)$

13. True

15. True

17. a. $2x-5\overline{)6x^2 + 9x + 5}$ quotient $3x + 12$

$$\begin{array}{r} 6x^2 + 9x + 5 \\ -(6x^2 - 15x) \\ \hline 24x + 5 \\ -(24x - 60) \\ \hline 65 \end{array}$$

b. Dividend: $6x^2 + 9x + 5$; Divisor: $2x - 5$; Quotient: $3x + 12$; Remainder: 65

c. $(2x-5)(3x+12) + 65$

$$= 6x^2 + 24x - 15x - 60 + 65$$
$$= 6x^2 + 9x + 5$$

19. $x-4\overline{)3x^3 - 11x^2 + 0x - 10}$ quotient $3x^2 + x + 4$

$$\begin{array}{r} 3x^3 - 11x^2 + 0x - 10 \\ -(3x^3 - 12x^2) \\ \hline x^2 + 0x \\ -(x^2 - 4x) \\ \hline 4x - 10 \\ -(4x - 16) \\ \hline 6 \end{array}$$

$3x^2 + x + 4 + \dfrac{6}{x-4}$

21.

$$\begin{array}{r} 4x^3-20x^2+13x+4 \\ x+2\overline{\big)4x^4-12x^3-27x^2+30x+8} \\ -\underline{\left(4x^4+8x^3\right)} \\ -20x^3-27x^2 \\ -\underline{\left(-20x^3-40x^2\right)} \\ 13x^2+30x \\ -\underline{\left(13x^2+26x\right)} \\ 4x+8 \\ -\underline{\left(4x+8\right)} \\ 0 \end{array}$$

$4x^3-20x^2+13x+4$

23.

$$\begin{array}{r} 3x^3-6x^2+2x-4 \\ 2x+4\overline{\big)6x^4+0x^3-20x^2+0x-16} \\ -\underline{\left(6x^4+12x^3\right)} \\ -12x^3-20x^2 \\ -\underline{\left(-12x^3-24x^2\right)} \\ 4x^2+0x \\ -\underline{\left(4x^2+8x\right)} \\ -8x-16 \\ -\underline{\left(-8x-16\right)} \\ 0 \end{array}$$

$3x^3-6x^2+2x-4$

25.

$$\begin{array}{r} x^3+4x^2-5x-2 \\ x^2+5\overline{\big)x^5+4x^4+0x^3+18x^2-20x-10} \\ -\underline{\left(x^5+0x^4+5x^3\right)} \\ 4x^4-5x^3+18x^2 \\ -\underline{\left(4x^4+0x^3+20x^2\right)} \\ -5x^3-2x^2-20x \\ -\underline{\left(-5x^3+0x^2-25x\right)} \\ -2x^2+5x-10 \\ -\underline{\left(-2x^2+0x-10\right)} \\ 5x \end{array}$$

$x^3+4x^2-5x-2+\dfrac{5x}{x^2+5}$

27.

$$\begin{array}{r} 3x^2+1 \\ 2x^2+x-3\overline{\big)6x^4+3x^3-7x^2+6x-5} \\ -\underline{\left(6x^4+3x^3-9x^2\right)} \\ 2x^2+6x-5 \\ -\underline{\left(2x^2+x-3\right)} \\ 5x-2 \end{array}$$

$3x^2+1+\dfrac{5x-2}{2x^2+x-3}$

29.

$$\begin{array}{r} x^2+3x+9 \\ x-3\overline{\big)x^3+0x^2+0x-27} \\ -\underline{\left(x^3-3x^2\right)} \\ 3x^2+0x \\ -\underline{\left(3x^2-9x\right)} \\ 9x-27 \\ -\underline{\left(9x-27\right)} \\ 0 \end{array}$$

x^2+3x+9

31. **a.** Dividend: $2x^4-5x^3-5x^2-4x+29$

b. Divisor: $x-3$

c. Quotient: $2x^3+x^2-2x-10$

d. Remainder: -1

33. **a.** Dividend: $x^3-2x^2-25x-4$

b. Divisor: $x+4$

c. Quotient: x^2-6x-1

d. Remainder: 0

35.

$$\begin{array}{r|rrr} \underline{-6} & 4 & 15 & 1 \\ & & -24 & 54 \\ \hline & 4 & -9 & \underline{|55} \end{array}$$

$4x-9+\dfrac{55}{x+6}$

37.

$$\begin{array}{r|rrr} \underline{4} & 5 & -17 & -12 \\ & & 20 & 12 \\ \hline & 5 & 3 & \underline{|0} \end{array}$$

$5x+3$

39. $\begin{array}{r|rrrrr} -2 & -5 & 0 & -3 & -8 & 4 \\ & & 10 & -20 & 46 & -76 \\ \hline & -5 & 10 & -23 & 38 & -72 \end{array}$

$-5x^3+10x^2-23x+38+\dfrac{-72}{x+2}$

41. $\begin{array}{r|rrrrrr} 3 & 4 & -25 & -58 & 232 & 198 & -63 \\ & & 12 & -39 & -291 & -177 & 63 \\ \hline & 4 & -13 & -97 & -59 & 21 & 0 \end{array}$

$4x^4-13x^3-97x^2-59x+21$

43. $\begin{array}{r|rrrrrr} -2 & 1 & 0 & 0 & 0 & 0 & 32 \\ & & -2 & 4 & -8 & 16 & -32 \\ \hline & 1 & -2 & 4 & -8 & 16 & 0 \end{array}$

$x^4-2x^3+4x^2-8x+16$

45. a. $f(x)=2x^4-5x^3+x^2-7$

$f(4)=2(4)^4-5(4)^3+(4)^2-7$
$=2(256)-5(64)+16-7$
$=512-320+16-7=201$

b. $\begin{array}{r|rrrrr} 4 & 2 & -5 & 1 & 0 & -7 \\ & & 8 & 12 & 52 & 208 \\ \hline & 2 & 3 & 13 & 52 & 201 \end{array}$

The remainder is 201.

47. a. $\begin{array}{r|rrrrr} -1 & 2 & 1 & -49 & 79 & 15 \\ & & -2 & 1 & 48 & -127 \\ \hline & 2 & -1 & -48 & 127 & -112 \end{array}$

By the remainder theorem, $f(-1)=-112$.

b. $\begin{array}{r|rrrrr} 3 & 2 & 1 & -49 & 79 & 15 \\ & & 6 & 21 & -84 & -15 \\ \hline & 2 & 7 & -28 & -5 & 0 \end{array}$

By the remainder theorem, $f(3)=0$.

c. $\begin{array}{r|rrrrr} 4 & 2 & 1 & -49 & 79 & 15 \\ & & 8 & 36 & -52 & 108 \\ \hline & 2 & 9 & -13 & 27 & 123 \end{array}$

By the remainder theorem, $f(4)=123$.

d. $\begin{array}{r|rrrrr} \frac{5}{2} & 2 & 1 & -49 & 79 & 15 \\ & & 5 & 15 & -85 & -15 \\ \hline & 2 & 6 & -34 & -6 & 0 \end{array}$

By the remainder theorem, $f\left(\dfrac{5}{2}\right)=0$.

49. a. $\begin{array}{r|rrrr} 1 & 5 & -4 & -15 & 12 \\ & & 5 & 1 & -14 \\ \hline & 5 & 1 & -14 & -2 \end{array}$

By the remainder theorem, $h(1)=-2$.

b. $\begin{array}{r|rrrr} \frac{4}{5} & 5 & -4 & -15 & 12 \\ & & 4 & 0 & -12 \\ \hline & 5 & 0 & -15 & 0 \end{array}$

By the remainder theorem, $h\left(\dfrac{4}{5}\right)=0$.

c. $\begin{array}{r|rrrr} \sqrt{3} & 5 & -4 & -15 & 12 \\ & & 5\sqrt{3} & 15-4\sqrt{3} & -12 \\ \hline & 5 & -4+5\sqrt{3} & -4\sqrt{3} & 0 \end{array}$

By the remainder theorem, $h(\sqrt{3})=0$.

d. $\begin{array}{r|rrrr} -1 & 5 & -4 & -15 & 12 \\ & & -5 & 9 & 6 \\ \hline & 5 & -9 & -6 & 18 \end{array}$

By the remainder theorem, $h(-1)=18$.

51. a. $\begin{array}{r|rrrrr} 2 & 1 & 3 & -7 & 13 & -10 \\ & & 2 & 10 & 6 & 38 \\ \hline & 1 & 5 & 3 & 19 & 28 \end{array}$

By the remainder theorem, $f(2)=28$.
Since $f(2)\neq 0$, 2 is not a zero of $f(x)$.

b. $\begin{array}{r|rrrrr} -5 & 1 & 3 & -7 & 13 & -10 \\ & & -5 & 10 & -15 & 10 \\ \hline & 1 & -2 & 3 & -2 & 0 \end{array}$

By the remainder theorem, $f(-5)=0$.
Since $f(-5)=0$, -5 is a zero of $f(x)$.

53. a.
$$\begin{array}{r|rrrr} -2 & 2 & 3 & -22 & -33 \\ & & -4 & 2 & 40 \\ \hline & 2 & -1 & -20 & \underline{|7} \end{array}$$

By the remainder theorem, $p(-2)=7$.
Since $p(-2)\neq 0$, -2 is not a zero of $p(x)$.

b.
$$\begin{array}{r|rrrr} -\sqrt{11} & 2 & 3 & -22 & -33 \\ & & -2\sqrt{11} & 22-3\sqrt{11} & 33 \\ \hline & 2 & 3-2\sqrt{11} & -3\sqrt{11} & \underline{|0} \end{array}$$

By the remainder theorem, $p\left(-\sqrt{11}\right)=0$.
Since $p\left(-\sqrt{11}\right)=0$, $-\sqrt{11}$ is a zero of $p(x)$.

55. a.
$$\begin{array}{r|rrrr} 5i & 1 & -2 & 25 & -50 \\ & & 5i & -25-10i & 50 \\ \hline & 1 & -2+5i & -10i & \underline{|0} \end{array}$$

By the remainder theorem, $m(5i)=0$.
Since $m(5i)=0$, $5i$ is a zero of $m(x)$.

b.
$$\begin{array}{r|rrrr} -5i & 1 & -2 & 25 & -50 \\ & & -5i & -25+10i & 50 \\ \hline & 1 & -2-5i & 10i & \underline{|0} \end{array}$$

By the remainder theorem, $m(-5i)=0$.
Since $m(-5i)=0$, $-5i$ is a zero of $m(x)$.

57. a.
$$\begin{array}{r|rrrr} 6+i & 1 & -11 & 25 & 37 \\ & & 6+i & -31+i & -37 \\ \hline & 1 & -5+i & -6+i & \underline{|0} \end{array}$$

By the remainder theorem, $g(6+i)=0$.
Since $g(6+i)=0$, $6+i$ is a zero of $g(x)$.

b.
$$\begin{array}{r|rrrr} 6-i & 1 & -11 & 25 & 37 \\ & & 6-i & -31-i & -37 \\ \hline & 1 & -5-i & -6-i & \underline{|0} \end{array}$$

By the remainder theorem, $g(6-i)=0$.

Since $g(6-i)=0$, $6-i$ is a zero of $g(x)$.

59. a.
$$\begin{array}{r|rrrrr} -5 & 1 & 11 & 41 & 61 & 30 \\ & & -5 & -30 & -55 & -30 \\ \hline & 1 & 6 & 11 & 6 & \underline{|0} \end{array}$$

By the factor theorem, since $f(-5)=0$, $x+5$ is a factor of $f(x)$.

b.
$$\begin{array}{r|rrrrr} 2 & 1 & 11 & 41 & 61 & 30 \\ & & 2 & 26 & 134 & 390 \\ \hline & 1 & 13 & 67 & 195 & \underline{|420} \end{array}$$

By the factor theorem, since $f(2)\neq 0$, $x-2$ is not a factor of $f(x)$.

61. a.
$$\begin{array}{r|rrrr} 1 & 2 & 1 & -16 & -8 \\ & & 2 & 3 & -13 \\ \hline & 2 & 3 & -13 & \underline{|-21} \end{array}$$

By the factor theorem, since $f(1)\neq 0$, $x-1$ is not a factor of $f(x)$.

b.
$$\begin{array}{r|rrrr} 2\sqrt{2} & 2 & 1 & -16 & -8 \\ & & 4\sqrt{2} & 16+2\sqrt{2} & 8 \\ \hline & 2 & 1+4\sqrt{2} & 2\sqrt{2} & \underline{|0} \end{array}$$

By the factor theorem, since $f\left(2\sqrt{2}\right)=0$, $x-2\sqrt{2}$ is a factor of $f(x)$.

63. a.
$$\begin{array}{r|rrr} 2+5i & 1 & -4 & 29 \\ & & 2+5i & -29 \\ \hline & 1 & -2+5i & \underline{|0} \end{array}$$

Yes, $2+5i$ is a factor of $f(x)$.

b.
$$\begin{array}{r|rrr} 2-5i & 1 & -4 & 29 \\ & & 2-5i & -29 \\ \hline & 1 & -2-5i & \underline{|0} \end{array}$$

Yes, $2-5i$ is a factor of $f(x)$.

c. $x = \frac{-(-4) \pm \sqrt{(-4)^2 - 4(1)(29)}}{2(1)}$

$= \frac{4 \pm \sqrt{-100}}{2}$

$= \frac{4 \pm 10i}{2}$

$= 2 \pm 5i$

The solution set is $\{2 \pm 5i\}$.

d. The zeros are $2 \pm 5i$.

65. a.

$\underline{-1}\rfloor$	2	1	−37	−36
		−2	1	36
	2	−1	−36	$\lfloor 0$

$f(x) = (x+1)(2x^2 - x - 36)$

$f(x) = (x+1)(2x-9)(x+4)$

b. $(x+1)(2x-9)(x+4) = 0$

$x = -1, x = \frac{9}{2}, x = -4$

The solution set is $\left\{-1, \frac{9}{2}, -4\right\}$.

67. a.

$\frac{1}{4}\rfloor$	20	39	−3	−2
		5	11	2
	20	44	8	$\lfloor 0$

$f(x) = \left(x - \frac{1}{4}\right)(20x^2 + 44x + 8)$

$f(x) = 4\left(x - \frac{1}{4}\right)(5x^2 + 11x + 2)$

$f(x) = (4x-1)(5x+1)(x+2)$

b. $(4x-1)(5x+1)(x+2) = 0$

$x = \frac{1}{4}, x = -\frac{1}{5}, x = -2$

The solution set is $\left\{\frac{1}{4}, -\frac{1}{5}, -2\right\}$.

69. a.

$\underline{3}\rfloor$	9	−33	19	−3
		27	−18	3
	9	−6	1	$\lfloor 0$

$f(x) = (x-3)(9x^2 - 6x + 1)$

$f(x) = (x-3)(3x-1)^2$

b. $(x-3)(3x-1)^2 = 0$

$x = 3, x = \frac{1}{3}$

The solution set is $\left\{3, \frac{1}{3}\right\}$.

71. $f(x) = (x-2)(x-3)(x+4)$

$= (x-2)(x^2 + x - 12)$

$= x^3 + x^2 - 12x - 2x^2 - 2x + 24$

$= x^3 - x^2 - 14x + 24$

73. $f(x) = (x-1)\left(x - \frac{3}{2}\right)(x)^2$

$= x^2\left(x^2 - \frac{5}{2}x + \frac{3}{2}\right)$

$= x^4 - \frac{5}{2}x^3 + \frac{3}{2}x^2$

or $f(x) = a\left(x^4 - \frac{5}{2}x^3 + \frac{3}{2}x^2\right)$

$= 2\left(x^4 - \frac{5}{2}x^3 + \frac{3}{2}x^2\right)$

$= 2x^4 - 5x^3 + 3x^2$

75. $f(x) = (x - 2\sqrt{11})(x + 2\sqrt{11})$

$= x^2 - (2\sqrt{11})^2$

$= x^2 - 44$

77. $f(x) = (x+2)(x-3i)(x+3i)$

$= (x+2)(x^2 + 9)$

$= x^3 + 2x^2 + 9x + 18$

79. $f(x)=a\left(x+\frac{2}{3}\right)\left(x-\frac{1}{2}\right)(x-4)$
$=6\left(x+\frac{2}{3}\right)\left(x-\frac{1}{2}\right)(x-4)$
$=3\left(x+\frac{2}{3}\right)\cdot 2\left(x-\frac{1}{2}\right)(x-4)$
$=(3x+2)(2x-1)(x-4)$
$=(3x+2)(2x^2-9x+4)$
$=6x^3-27x^2+12x+4x^2-18x+8$
$=6x^3-23x^2-6x+8$

81. $f(x)=[x-(7+8i)][x-(7-8i)]$
$=[(x-7)-8i][(x-7)+8i]$
$=(x-7)^2-(8i)^2$
$=x^2-14x+49+64=x^2-14x+113$

83. Direct substitution is easier.
$p(x)=2x^{452}-4x^{92}$
$p(1)=2(1)^{452}-4(1)^{92}=2-4=-2$

85. a. $f(x)=x^{100}-1$
$f(1)=(1)^{100}-1=1-1=0$
Yes, since $f(1)=0$, $(x-1)$ is a factor of $f(x)$.

b. $f(-1)=(-1)^{100}-1=1-1=0$
Yes, since $f(-1)=0$, $(x+1)$ is a factor of $f(x)$.

c. $g(x)=x^{99}-1$
$g(1)=(1)^{99}-1=1-1=0$
Yes, since $g(1)=0$, $(x-1)$ is a factor of $g(x)$.

d. $g(-1)=(-1)^{99}-1=-1-1=-2$
No, since $g(-1)\neq 0$, $(x+1)$ is not a factor of $g(x)$.

e. Yes **f.** No

87. Since x is not a factor, zero is not a zero of the polynomial. False.

89.
$$\begin{array}{r|rrrr} -4 & 4 & 13 & -5 & m \\ & & -16 & 12 & -28 \\ \hline & 4 & -3 & 7 & \underline{|0} \end{array}$$
$m+(-28)=0$
$m=28$

91.
$$\begin{array}{r|rrrr} -2 & 4 & 5 & m & 2 \\ & & -8 & 6 & -2m-12 \\ \hline & 4 & -3 & m+6 & \underline{|0} \end{array}$$
$2+(-2m-12)=0$
$-2m-10=0$
$-2m=10$
$m=-5$

93. $\frac{x^2-x-12}{x-4}=x+3+\frac{r}{x-4}$
$\frac{(x+3)(x-4)}{x-4}=x+3+\frac{r}{x-4}$
$x+3=x+3+\frac{r}{x-4}$
$0=\frac{r}{x-4}$
$0=r$

95. a. $V(x)=x(x+1)(2x+1)-2(1)(x)$
$=x(2x^2+3x+1)-2x$
$=2x^3+3x^2+x-2x$
$=2x^3+3x^2-x$

b.
$$\begin{array}{r|rrrr} 6 & 2 & 3 & -1 & 0 \\ & & 12 & 90 & 534 \\ \hline & 2 & 15 & 89 & \underline{|534} \end{array}$$
The volume is 534 cm^3.

97. The divisor must be of the form $(x-c)$, where c is a constant.

99. Compute $f(c)$ either by direct substitution or by using the remainder theorem. The remainder theorem states that $f(c)$ is equal to the remainder obtained after dividing $f(x)$ by $(x-c)$.

101. a.
$$\begin{array}{r|rrrr} 5 & 1 & -5 & 1 & -5 \\ & & 5 & 0 & 5 \\ \hline & 1 & 0 & 1 & \underline{|0} \end{array}$$

$f(x)=(x-5)(x^2+1)$

$x^2+1=0$

$x^2=-1$

$x=\pm\sqrt{1}=\pm i$

$f(x)=(x-5)(x-i)(x+i)$

b. $(x-5)(x-i)(x+i)=0$

$x=5, x=i, x=-i$

The solution set is $\{5, i, -i\}$.

103. a.
$$\begin{array}{r|rrrrr} -1 & 1 & 2 & -2 & -6 & -3 \\ & & -1 & -1 & 3 & 3 \\ \hline & 1 & 1 & -3 & -3 & \underline{|0} \end{array}$$

$f(x)=(x+1)(x^3+x^2-3x-3)$

$=(x+1)[x^2(x+1)-3(x+1)]$

$=(x+1)(x+1)(x^2-3)$

$=(x+1)^2(x-\sqrt{3})(x+\sqrt{3})$

b. $(x+1)^2(x-\sqrt{3})(x+\sqrt{3})=0$

$x=-1, x=\sqrt{3}, x=-\sqrt{3}$

The solution set is $\{-1, \sqrt{3}, -\sqrt{3}\}$.

105. a. The solution appears to be 3.

b.
$$\begin{array}{r|rrrr} 3 & 5 & 7 & -58 & -24 \\ & & 15 & 66 & 24 \\ \hline & 5 & 22 & 8 & \underline{|0} \end{array}$$

Since the remainder is zero, 3 is a solution.

c. $f(x)=(x-3)(5x^2+22x+8)$

$=(x-3)(x+4)(5x+2)$

$(x-3)(x+4)(5x+2)=0$

$x=3, x=-4, x=-\frac{2}{5}$

The solution set is $\left\{3, -4, -\frac{2}{5}\right\}$.

Section 3.4 Zeros of Polynomials

1. a. $-4, \frac{2}{5}$

b. $\sqrt{10}, 4-\sqrt{2}$

c. $-4, \frac{2}{5}, \sqrt{10}, 4-\sqrt{2}$

d. $6i, 3+8i$

e. $-4, \frac{2}{5}, \sqrt{10},, 6i, 3+8i\ 4-\sqrt{2}$

3. $f(x)=-3x^4(x-2)^3(4x-7)^2$

$0=-3x^4(x-2)^3(4x-7)^2$

$x=0, x=2, x=\frac{7}{4}$

The zeros of the function are 0 (multiplicity 4), 2 (multiplicity 3), and $\frac{7}{4}$ (multiplicity 2).

5. a. $f(x)=2x^3-7x^2+12x-31$

$f(3)=2(3)^3-7(3)^2+12(3)-31$

$=2(27)-7(9)+36-31$

$=54-63+36-31=-4$

b. $\underline{3}|$ 2 −7 12 −31

6 −3 27

2 −1 9 $\lfloor -4$

By the remainder theorem, $f(3) = -4$.

7. zeros **9.** n

11. Descartes'; $f(-x)$

13. greater than

15. $\dfrac{\text{Factors of } 4}{\text{Factors of } 1} = \dfrac{\pm 1, \pm 2, \pm 4}{\pm 1} = \pm 1, \pm 2, \pm 4$

17. $\dfrac{\text{Factors of } -6}{\text{Factors of } 4}$

$= \dfrac{\pm 1, \pm 2, \pm 3, \pm 6}{\pm 1, \pm 2, \pm 4}$

$= \pm 1, \pm 2, \pm 3, \pm 6, \pm \frac{1}{2}, \pm \frac{3}{2}, \pm \frac{1}{4}, \pm \frac{3}{4}$

19. $\dfrac{\text{Factors of } 8}{\text{Factors of } -12}$

$= \dfrac{\pm 1, \pm 2, \pm 4, \pm 8}{\pm 1, \pm 2, \pm 3, \pm 4, \pm 6, \pm 12}$

$= \pm 1, \pm 2, \pm 4, \pm 8, \pm \frac{1}{2}, \pm \frac{1}{3}, \pm \frac{2}{3}, \pm \frac{4}{3}, \pm \frac{8}{3},$

$\pm \frac{1}{6}, \pm \frac{1}{12}$

21. 7 and $\frac{5}{3}$ are not ratios of any of the factors of 12 over the factors of 2.

23. $\dfrac{\text{Factors of } 2}{\text{Factors of } 2} = \dfrac{\pm 1, \pm 2}{\pm 1, \pm 2} = \pm 1, \pm 2, \pm \frac{1}{2}$

Use synthetic division and the remainder theorem to determine if any of the numbers in the list is a zero of $p(x)$. (Successful tests of zeros are shown.)

$\underline{1}|$ 2 −1 −5 2 2

2 1 −4 −2

2 1 −4 −2 $\lfloor 0$

The remainder is zero. Therefore, 1 is a zero of $p(x)$.

$-\frac{1}{2}|$ 2 −1 −5 2 2

−1 1 2 −2

2 −2 −4 4 $\lfloor 0$

The remainder is zero. Therefore, $-\frac{1}{2}$ is a zero of $p(x)$.

25. $\dfrac{\text{Factors of } 20}{\text{Factors of } 1} = \dfrac{\pm 1, \pm 2, \pm 4, \pm 5, \pm 10, \pm 20}{\pm 1}$

$= \pm 1, \pm 2, \pm 4, \pm 5, \pm 10, \pm 20$

$\underline{5}|$ 1 −7 6 20

5 −10 −20

1 −2 −4 $\lfloor 0$

$f(x) = (x-5)(x^2 - 2x - 4)$

$0 = (x-5)(x^2 - 2x - 4)$

$x = 5$ or $x = \dfrac{-(-2) \pm \sqrt{(-2)^2 - 4(1)(-4)}}{2(1)}$

$x = \dfrac{2 \pm \sqrt{20}}{2} = \dfrac{2 \pm 2\sqrt{5}}{2} = 1 \pm \sqrt{5}$

The zeros are 5, $1 \pm \sqrt{5}$.

27. $\dfrac{\text{Factors of } 7}{\text{Factors of } 5} = \dfrac{\pm 1, \pm 7}{\pm 1, \pm 5} = \pm 1, \pm 7, \pm \frac{1}{5}, \pm \frac{7}{5}$

$\frac{1}{5}|$ 5 −1 −35 7

1 0 −7

5 0 −35 $\lfloor 0$

$h(x) = \left(x - \frac{1}{5}\right)(5x^2 - 35)$

$0 = \left(x - \frac{1}{5}\right)(5x^2 - 35)$

$x = \frac{1}{5}$ or $5x^2 = 35$

$x^2 = 7$

$x = \pm\sqrt{7}$

The zeros are $\frac{1}{5}, \pm\sqrt{7}$.

29. $\dfrac{\text{Factors of } -16}{\text{Factors of } 3}$

$= \dfrac{\pm 1, \pm 2, \pm 4, \pm 8, \pm 16}{\pm 1, \pm 3}$

$= \pm 1, \pm 2, \pm 4, \pm 8, \pm 16,$

$\pm\frac{1}{3}, \pm\frac{2}{3}, \pm\frac{4}{3}, \pm\frac{8}{3}, \pm\frac{16}{3}$

$$\begin{array}{r|rrrrr} \underline{2} & 3 & -1 & -36 & 60 & -16 \\ & & 6 & 10 & -52 & 16 \\ \hline & 3 & 5 & -26 & 8 & \underline{|0} \end{array}$$

$m(x) = (x-2)(3x^3 + 5x^2 - 26x + 8)$

Find the zeros of the quotient.

$$\begin{array}{r|rrrr} \underline{2} & 3 & 5 & -26 & 8 \\ & & 6 & 22 & -8 \\ \hline & 3 & 11 & -4 & \underline{|0} \end{array}$$

$m(x) = (x-2)^2(3x^2 + 11x - 4)$

$m(x) = (x-2)^2(3x-1)(x+4)$

$0 = (x-2)^2(3x-1)(x+4)$

$x = 2, x = \frac{1}{3}, x = -4$

The zeros are 2 (multiplicity 2), $\frac{1}{3}$, -4.

31. $\dfrac{\text{Factors of } 20}{\text{Factors of } 1} = \dfrac{\pm 1, \pm 2, \pm 4, \pm 5, \pm 10, \pm 20}{\pm 1}$

$= \pm 1, \pm 2, \pm 4, \pm 5, \pm 10, \pm 20$

$$\begin{array}{r|rrrr} \underline{-2} & 1 & -4 & -2 & 20 \\ & & -2 & 12 & -20 \\ \hline & 1 & -6 & 10 & \underline{|0} \end{array}$$

$q(x) = (x+2)(x^2 - 6x + 10)$

$0 = (x+2)(x^2 - 6x + 10)$

$x = -2 \quad \text{or} \quad x = \dfrac{-(-6) \pm \sqrt{(-6)^2 - 4(1)(10)}}{2(1)}$

$x = \dfrac{6 \pm \sqrt{-4}}{2} = \dfrac{6 \pm 2i}{2} = 3 \pm i$

The zeros are -2, $3 \pm i$.

33. $t(x) = x^4 - x^2 - 90$

$t(x) = (x^2 - 10)(x^2 + 9)$

$0 = (x^2 - 10)(x^2 + 9)$

$x^2 = 10 \quad \text{or} \quad x^2 = -9$

$x = \pm\sqrt{10} \quad \text{or} \quad x = \pm 3i$

The zeros are $\pm\sqrt{10}, \pm 3i$.

35. one

37. 7

39. a.

$$\begin{array}{r|rrrrr} \underline{2-5i} & 1 & -4 & 22 & 28 & -203 \\ & & 2-5i & -29 & -14+35i & 203 \\ \hline & 1 & -2-5i & -7 & 14+35i & \underline{|0} \end{array}$$

Since $2-5i$ is a zero, $2+5i$ is also a zero.

$$\begin{array}{r|rrrr} \underline{2+5i} & 1 & -2-5i & -7 & 14+35i \\ & & 2+5i & 0 & -14-35i \\ \hline & 1 & 0 & -7 & \underline{|0} \end{array}$$

$f(x) = [x-(2-5i)][x-(2+5i)](x^2 - 7)$

Find the remaining two zeros.

$x^2 - 7 = 0$

$x^2 = 7$

$x = \pm\sqrt{7}$

The zeros are $2 \pm 5i$, $\pm\sqrt{7}$.

b. $f(x) =$

$[x-(2-5i)][x-(2+5i)](x-\sqrt{7})(x+\sqrt{7})$

c. The solution set is $\{2 \pm 5i, \pm\sqrt{7}\}$.

41. a.

$$\begin{array}{r|rrrr} \underline{4+i} & 3 & -28 & 83 & -68 \\ & & 12+3i & -67-4i & 68 \\ \hline & 3 & -16+3i & 16-4i & \underline{|0} \end{array}$$

Since $4+i$ is a zero, $4-i$ is also a zero.

$$\begin{array}{r|rrr} \underline{4-i} & 3 & -16+3i & 16-4i \\ & & 12-3i & -16+4i \\ \hline & 3 & -4 & \underline{|0} \end{array}$$

$f(x) = [x-(4+i)][x-(4-i)](3x-4)$

Find the remaining zero.
$3x-4=0$
$3x=4$
$x=\frac{4}{3}$

The zeros are $4\pm i$, $\frac{4}{3}$.

b. $f(x)=[x-(4+i)][x-(4-i)](3x-4)$

c. The solution set is $\left\{4\pm i, \frac{4}{3}\right\}$.

43. a.

$-\frac{1}{4}$	4	37	117	87	−193	−52
		−1	−9	−27	−15	52
	4	36	108	60	−208	0

Find the zeros of the quotient.

$-3+2i$	4	36	108	60	−208
		$-12+8i$	$-88+24i$	$-108-32i$	208
	4	$24+8i$	$20+24i$	$-48-32i$	0

Since $-3+2i$ is a zero, $-3-2i$ is also a zero.

$-3-2i$	4	$24+8i$	$20+24i$	$-48-32i$
		$-12-8i$	$-36-24i$	$48+32i$
	4	12	−16	0

$$f(x)=[x-(-3+2i)][x-(-3-2i)]\left(x+\frac{1}{4}\right)(4x^2+12x-16)$$
$$f(x)=[x-(-3+2i)][x-(-3-2i)](4x+1)(x^2+3x-4)$$
$$f(x)=[x-(-3+2i)][x-(-3-2i)](4x+1)(x-1)(x+4)$$
$$0=[x-(-3+2i)][x-(-3-2i)](4x+1)(x-1)(x+4)$$
$$x=-3+2i, x=-3-2i, x=-\frac{1}{4}, x=1, x=-4$$

The zeros are $-3\pm 2i, -\frac{1}{4}, 1, -4$

b. $f(x)=[x-(-3+2i)][x-(-3-2i)](4x+1)(x-1)(x+4)$

c. The solution set is $\left\{-3\pm 2i, -\frac{1}{4}, 1, -4\right\}$.

45. $f(x)=a(x-6i)(x+6i)\left(x-\frac{4}{5}\right)=a(x^2+36)\left(x-\frac{4}{5}\right)=a\left(x^3-\frac{4}{5}x^2+36x-\frac{144}{5}\right)$ Let $a=5$.

$$=5\left(x^3-\frac{4}{5}x^2+36x-\frac{144}{5}\right)=5x^3-4x^2+180x-144$$

47. $f(x)=a(x+4)(x-2)^3=a(x+4)(x-2)(x-2)^2=a(x+4)(x-2)(x^2-4x+4)$

$=a(x+4)(x^3-4x^2+4x-2x^2+8x-8)=a(x+4)(x^3-6x^2+12x-8)$

$=a(x^4-6x^3+12x^2-8x+4x^3-24x^2+48x-32)=a(x^4-2x^3-12x^2+40x-32)$

$f(0)=a\left[(0)^4-2(0)^3-12(0)^2+40(0)-32\right]=160$

$-32a=160$

$a=-5$

$f(x)=-5(x^4-2x^3-12x^2+40x-32)$

$f(x)=-5x^4+10x^3+60x^2-200x+160$

49. $f(x)=a\left(x+\frac{4}{3}\right)^2\left(x-\frac{1}{2}\right)$

$=a\left(x^2+\frac{8}{3}x+\frac{16}{9}\right)\left(x-\frac{1}{2}\right)$

$=a\left(x^3-\frac{1}{2}x^2+\frac{8}{3}x^2-\frac{4}{3}x+\frac{16}{9}x-\frac{8}{9}\right)$

$=a\left(x^3+\frac{13}{6}x^2+\frac{4}{9}x-\frac{8}{9}\right)$

$f(0)=a\left[(0)^3+\frac{13}{6}(0)^2+\frac{4}{9}(0)-\frac{8}{9}\right]=-16$

$-\frac{8}{9}a=-16$

$a=18$

$f(x)=18\left(x^3+\frac{13}{6}x^2+\frac{4}{9}x-\frac{8}{9}\right)$

$f(x)=18x^3+39x^2+8x-16$

51. $f(x)=x^4\left[x-(7-4i)\right]\left[x-(7+4i)\right]$

$=x^4(x-7)^2-(4i)^2$

$=x^4(x^2-14x+49+16)$

$=x^4(x^2-14x+65)$

$=x^6-14x^5+65x^4$

53. $f(x)=x^6-2x^4+4x^3-2x^2-5x-6$

3 sign changes in $f(x)$. The number of possible positive real zeros is either 3 or 1.

$f(-x)=(-x)^6-2(-x)^4+4(-x)^3-2(-x)^2-5(-x)-6$

$f(-x)=x^6-2x^4-4x^3-2x^2+5x-6$

3 sign changes in $f(-x)$. The number of possible negative real zeros is either 3 or 1.

55. $k(x)=-8x^7+5x^6-3x^4+2x^3-11x^2+4x-3$

6 sign changes in $k(x)$. The number of possible positive real zeros is either 6, 4, 2, or 0.

$k(-x)=-8(-x)^7+5(-x)^6-3(-x)^4+2(-x)^3-11(-x)^2+4(-x)-3$

$k(-x)=8x^7+5x^6-3x^4-2x^3-11x^2-4x-3$

1 sign change in $k(-x)$. The number of possible negative real zeros is 1.

57. $p(x)=0.11x^4+0.04x^3+0.31x^2+0.27x+1.1$

0 sign changes in $p(x)$. The number of possible positive real zeros is 0.

$p(-x)=0.11(-x)^4+0.04(-x)^3+0.31(-x)^2+0.27(-x)+1.1$

$p(-x)=0.11x^4-0.04x^3+0.31x^2-0.27x+1.1$

4 sign changes in $p(-x)$. The number of possible negative real zeros is either 4, 2, or 0.

59. $v(x)=\frac{1}{8}x^6+\frac{1}{6}x^4+\frac{1}{3}x^2+\frac{1}{10}$

0 sign changes in $v(x)$. The number of possible positive real zeros is 0.

$v(-x)=\frac{1}{8}(-x)^6+\frac{1}{6}(-x)^4+\frac{1}{3}(-x)^2+\frac{1}{10}$

$v(-x)=\frac{1}{8}x^6+\frac{1}{6}x^4+\frac{1}{3}x^2+\frac{1}{10}$

0 sign changes in $v(-x)$. The number of possible negative real zeros is 0.

61. $f(x)=x^8+5x^6+6x^4-x^3$

$f(x)=x^3(x^5+5x^3+6x-1)$

1 sign change. The number of possible positive real zeros is 1.

$(-x)^5+5(-x)^3+6(-x)-1=-x^5-5x^3-6x-1$

0 sign changes. The number of possible negative real zeros is 0.

$f(x)$ has 4 real zeros; 1 positive real zero, no negative real zeros, and the number 0 is a zero of multiplicity 3.

63. a.

2	1	6	0	5	1	−3
		2	16	32	74	150
	1	8	16	37	75	147

The remainder and all coefficients of the quotient are nonnegative. Therefore, 2 is an upper bound for the real zeros of $f(x)$.

b.

−2	1	6	0	5	1	−3
		−2	−8	16	−42	82
	1	4	−8	21	−41	79

The signs of the quotient do not alternate. Therefore, we cannot conclude that −2 is a lower bound for the real zeros of $f(x)$.

65. a.

6	8	−42	33	28
		48	36	414
	8	6	69	442

The remainder and all coefficients of the quotient are nonnegative. Therefore, 6 is an upper bound for the real zeros of $f(x)$.

b.

−1	8	−42	33	28
		−8	50	−83
	8	−50	83	−55

The signs of the quotient alternate. Therefore, −1 is a lower bound for the real zeros of $f(x)$.

67. a.

3	2	11	0	−63	−50	40
		6	51	153	270	660
	2	17	51	90	220	700

The remainder and all coefficients of the quotient are nonnegative. Therefore, 3 is an upper bound for the real zeros of $f(x)$.

b.

−6	2	11	0	−63	−50	40
		−12	6	−36	594	−3264
	2	−1	6	−99	544	−3224

The signs of the quotient alternate. Therefore, −6 is a lower bound for the real zeros of $f(x)$.

69. True

71. False

73. $\frac{\text{Factors of } 28}{\text{Factors of } 8}=\frac{\pm1,\pm2,\pm4,\pm7,\pm14,\pm28}{\pm1,\pm2,\pm4,\pm8}$

$=\pm1,\pm2,\pm4,\pm7,\pm14,\pm28,\pm\frac{1}{2},\pm\frac{1}{4},\pm\frac{1}{8},\pm\frac{7}{2},\pm\frac{7}{4}$

From Exercise 65, we know that 6 and −1 are upper and lower bounds of $f(x)$, respectively. We can restrict the list of possible rational zeros to those on the interval $(-1,6)$:

$1,2,4,\pm\frac{1}{2},\pm\frac{1}{4},\pm\frac{1}{8},\frac{7}{2},\frac{7}{4}$

$$\begin{array}{r|rrrr} 4 & 8 & -42 & 33 & 28 \\ & & 32 & -40 & -28 \\ \hline & 8 & -10 & -7 & \underline{|0} \end{array}$$

Find the zeros of the quotient.

$0 = 8x^2 - 10x - 7$

$0 = (4x-7)(2x+1)$

$x = \frac{7}{4}, x = -\frac{1}{2}$

The zeros are $\frac{7}{4}, -\frac{1}{2}$, and 4 (each with multiplicity 1).

75. $\frac{\text{Factors of } 40}{\text{Factors of } 2}$

$= \frac{\pm 1, \pm 2, \pm 4, \pm 5, \pm 8, \pm 10, \pm 20, \pm 40}{\pm 1, \pm 2}$

$= \pm 1, \pm 2, \pm 4, \pm 5, \pm 8, \pm 10, \pm 20, \pm 40,$

$\pm \frac{1}{2}, \pm \frac{5}{2}$

From Exercise 67, we know that 3 and –6 are upper and lower bounds of $f(x)$, respectively. We can restrict the list of possible rational zeros to those on the interval $(-6, 3)$:

$\pm 1, \pm 2, -4, -5, \pm \frac{1}{2}, \pm \frac{5}{2}$

$$\begin{array}{r|rrrrrr} -2 & 2 & 11 & 0 & -63 & -50 & 40 \\ & & -4 & -14 & 28 & 70 & -40 \\ \hline & 2 & 7 & -14 & -35 & 20 & \underline{|0} \end{array}$$

Find the zeros of the quotient.

$$\begin{array}{r|rrrrr} -4 & 2 & 7 & -14 & -35 & 20 \\ & & -8 & 4 & 40 & -20 \\ \hline & 2 & -1 & -10 & 5 & \underline{|0} \end{array}$$

Find the zeros of the quotient.

$$\begin{array}{r|rrrr} \frac{1}{2} & 2 & -1 & -10 & 5 \\ & & 1 & 0 & -5 \\ \hline & 2 & 0 & -10 & \underline{|0} \end{array}$$

Find the zeros of the quotient.

$2x^2 - 10 = 0$

$2x^2 = 10$

$x^2 = 5$

$x = \pm\sqrt{5}$

The zeros are $\pm\sqrt{5}, \frac{1}{2}, -2$, and -4 (each with multiplicity 1).

77. $\frac{\text{Factors of } 9}{\text{Factors of } 4}$

$= \frac{\pm 1, \pm 3, \pm 9}{\pm 1, \pm 2, \pm 4}$

$= \pm 1, \pm 3, \pm 9, \pm \frac{1}{2}, \pm \frac{1}{4}, \pm \frac{3}{2}, \pm \frac{3}{4}, \pm \frac{9}{2}, \pm \frac{9}{4}$

$$\begin{array}{r|rrrrr} -3 & 4 & 20 & 13 & -30 & 9 \\ & & -12 & -24 & 33 & -9 \\ \hline & 4 & 8 & -11 & 3 & \underline{|0} \end{array}$$

Find the zeros of the quotient.

$$\begin{array}{r|rrrr} -3 & 4 & 8 & -11 & 3 \\ & & -12 & 12 & -3 \\ \hline & 4 & -4 & 1 & \underline{|0} \end{array}$$

Find the zeros of the quotient.

$4x^2 - 4x + 1 = 0$

$(2x-1)^2 = 0$

$x = \frac{1}{2}$

The zeros are –3, and $\frac{1}{2}$ (each with multiplicity 2).

79. $f(x) = x^2(x^4 + 2x^3 + 11x^2 + 20x + 10)$

$\frac{\text{Factors of } 10}{\text{Factors of } 1} = \frac{\pm 1, \pm 2, \pm 5, \pm 10}{\pm 1}$

$= \pm 1, \pm 2, \pm 5, \pm 10$

$$\begin{array}{r|rrrrr} -1 & 1 & 2 & 11 & 20 & 10 \\ & & -1 & -1 & -10 & -10 \\ \hline & 1 & 1 & 10 & 10 & \underline{|0} \end{array}$$

Find the zeros of the quotient.

$$\begin{array}{r|rrrr} -1 & 1 & 1 & 10 & 10 \\ & & -1 & 0 & -10 \\ \hline & 1 & 0 & 10 & \lfloor 0 \end{array}$$

Find the zeros of the quotient.

$x^2+10=0$

$x^2=-10$

$x=\pm i\sqrt{10}$

The zeros are 0 (with multiplicity 2), -1 (with multiplicity 2), and $\pm i\sqrt{10}$ (each with multiplicity 1).

81. $f(x)=x^3(x^2-10x+34)$

Find the zeros of $x^2-10x+34$.

$$x=\frac{-(-10)\pm\sqrt{(-10)^2-4(1)(34)}}{2(1)}$$

$$=\frac{10\pm\sqrt{-36}}{2}=\frac{10\pm 6i}{2}=5\pm 3i$$

The zeros are 0 (with multiplicity 3), and $5\pm 3i$ (each with multiplicity 1).

83. $f(x)=-(x^3-3x^2+9x+13)$

$$\frac{\text{Factors of }13}{\text{Factors of }1}=\frac{\pm1,\pm13}{\pm1}=\pm1,\pm13$$

$$\begin{array}{r|rrrr} -1 & 1 & -3 & 9 & 13 \\ & & -1 & 4 & -13 \\ \hline & 1 & -4 & 13 & \lfloor 0 \end{array}$$

Find the zeros of $x^2-4x+13$.

$$x=\frac{-(-4)\pm\sqrt{(-4)^2-4(1)(13)}}{2(1)}$$

$$=\frac{4\pm\sqrt{-36}}{2}=\frac{4\pm 6i}{2}=2\pm 3i$$

The zeros are -1 and $2\pm 3i$ (each with multiplicity 1).

85. False

87. False

89. True

91. **a.** True **b.** True

c. True **d.** True

93. **a.** $f(2)=2(2)^2-7(2)+4=8-14+4=-2$

$f(3)=2(3)^2-7(3)+4=18-21+4=1$

Since $f(2)$ and $f(3)$ have opposite signs, the intermediate value theorem guarantees that f has at least one real zero between 2 and 3.

b. $$x=\frac{-(-7)\pm\sqrt{(-7)^2-4(2)(4)}}{2(2)}$$

$$=\frac{7\pm\sqrt{17}}{4}$$

Furthermore, $\frac{7+\sqrt{17}}{4}\approx 2.78$ is on the interval [2, 3].

95. If a polynomial has real coefficients, then all imaginary zeros must come in conjugate pairs. This means that if the polynomial has imaginary zeros, there would be an even number of them. A third-degree polynomial has 3 zeros (including multiplicities). Therefore, it would have either 2 or 0 imaginary zeros, leaving room for either 1 or 3 real zeros.

97. $f(x)$ has no variation in sign, nor does $f(-x)$. By Descartes' rule of signs, there are no positive or negative real zeros. Furthermore, 0 itself is not a zero of $f(x)$ because x is not a factor of $f(x)$. Therefore, there are no real zeros of $f(x)$.

99. $f(x)=x^n-1$ has one sign change. The number of possible positive real zeros is 1. Since n is a positive even integer, $f(-x)=(-x)^n-1=x^n-1$ has one sign change. The number of possible negative real zeros is 1. Therefore, there are 2 possible real roots, and $n-2$ possible imaginary roots.

101. $108 = l\left(\frac{1}{2}bh\right)$

$108 = (x+3)\left(\frac{1}{2}\right)(x)\left(\frac{2}{3}x\right)$

$108 = \frac{1}{3}\left(x^3 + 3x^2\right)$

$324 = x^3 + 3x^2$

$0 = x^3 + 3x^2 - 324$

Graph the function $f(x) = x^3 + 3x^2 - 324$ on a graphing utility. Use the Zero feature to find positive real roots of the equation.

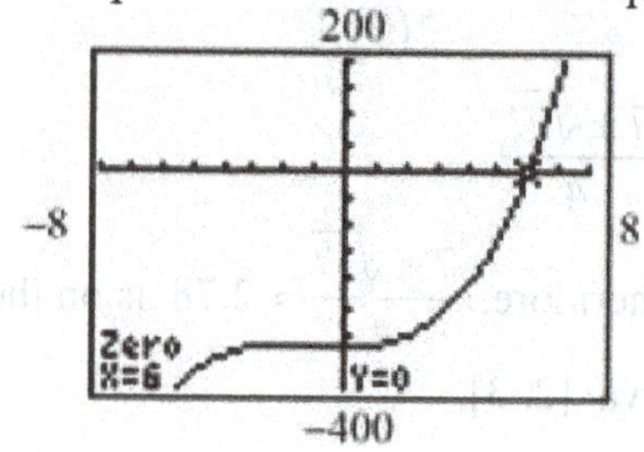

$\frac{2}{3}x = \frac{2}{3}(6) = 4$

$x + 3 = 6 + 3 = 9$

The triangular front has a base of 6 ft and a height of 4 ft. The length is 9 ft.

103. $81 = (10-x)(7-x)\left(\frac{5}{2}-x\right)$

$-162 = (x-10)(x-7)(2x-5)$

$-162 = (x-10)\left(2x^2 - 19x + 35\right)$

$-162 = 2x^3 - 19x^2 + 35x - 20x^2 + 190x - 350$

$0 = 2x^3 - 39x^2 + 225x - 188$

Graph the function $f(x) = 2x^3 - 39x^2 + 225x - 188$ on a graphing utility. Use the Zero feature to find positive real roots of the equation.

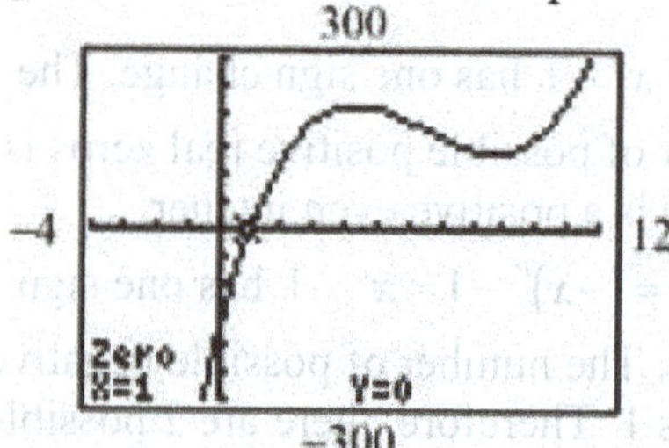

Each dimension was decreased by 1 in.

105. $6 = 2x\left(4 - x^2\right)$

$3 = x\left(4 - x^2\right)$

$3 = 4x - x^3$

$0 = x^3 - 4x + 3$

$\frac{\text{Factors of } 3}{\text{Factors of } 1} = \frac{\pm 1, \pm 3}{\pm 1} = \pm 1, \pm 3$

$$\begin{array}{r|rrrr} 1 & 1 & 0 & -4 & 3 \\ & & 1 & 1 & -3 \\ \hline & 1 & 1 & -3 & \boxed{0} \end{array}$$

Find the zeros of $x^2 + x - 3$.

$x = \frac{-(1) \pm \sqrt{(1)^2 - 4(1)(-3)}}{2(1)}$

$= \frac{-1 \pm \sqrt{13}}{2}$

The width must be positive, so use the positive zero.

The width of the rectangle is 2*x*, which is 2 or $2\left(\frac{-1+\sqrt{13}}{2}\right) = -1 + \sqrt{13}$.

When $x = 1$, $y = 4 - x^2 = 4 - (1)^2 = 3$.

When $x = \frac{-1+\sqrt{13}}{2}$,

$y = 4 - x^2 = 4 - \left(\frac{-1+\sqrt{13}}{2}\right)^2$

$= 4 - \frac{1 - 2\sqrt{13} + 13}{4}$

$= \frac{16}{4} - \frac{14 - 2\sqrt{13}}{4}$

$= \frac{2 - 2\sqrt{13}}{4}$

$= \frac{1 - 1\sqrt{13}}{2}$.

The dimensions are either 2 cm by 3 cm or $-1 + \sqrt{13}$ cm by $\frac{1 - 1\sqrt{13}}{2}$ cm.

107. a. $\dfrac{\text{Factors of } -12}{\text{Factors of } 1} = \dfrac{\pm 1, \pm 2, \pm 3, \pm 4, \pm 6, \pm 12}{\pm 1}$

$= \pm 1, \pm 2, \pm 3, \pm 4, \pm 6, \pm 12$

$$\begin{array}{r|rrrrr} 1 & 1 & 2 & 1 & 8 & -12 \\ & & 1 & 3 & 4 & 12 \\ \hline & 1 & 3 & 4 & 12 & \boxed{0} \end{array}$$

Factor the quotient.

$$\begin{array}{r|rrrr} -3 & 1 & 3 & 4 & 12 \\ & & -3 & 0 & -12 \\ \hline & 1 & 0 & 4 & \boxed{0} \end{array}$$

$f(x) = (x+3)(x-1)(x^2+4)$

b. Factor $x^2 + 4$.

$x^2 + 4 = 0$

$x^2 = -4$

$x = \pm 2i$

$f(x) = (x+3)(x-1)(x+2i)(x-2i)$

109. a. $f(x) = x^4 + 2x^2 - 35$

$f(x) = (x^2-5)(x^2+7)$

$0 = (x^2-5)(x^2+7)$

$x^2 - 5 = 0$ or $x^2 + 7 = 0$

$x^2 = 5$ $\quad$ $x^2 = -7$

$x = \pm\sqrt{5}$ $\quad$ $x = \pm\sqrt{7}i$

$f(x) = (x-\sqrt{5})(x+\sqrt{5})(x^2+7)$

b. $f(x) = (x-\sqrt{5})(x+\sqrt{5})(x+\sqrt{7}i)(x-\sqrt{7}i)$

111. $\dfrac{\text{Factors of } -1}{\text{Factors of } 1} = \dfrac{\pm 1}{\pm 1} = \pm 1$

$$\begin{array}{r|rrrrr} 1 & 1 & 0 & 0 & 0 & -1 \\ & & 1 & 1 & 1 & 1 \\ \hline & 1 & 1 & 1 & 1 & \boxed{0} \end{array}$$

Factor the quotient.

$$\begin{array}{r|rrrr} -1 & 1 & 1 & 1 & 1 \\ & & -1 & 0 & -1 \\ \hline & 1 & 0 & 1 & \boxed{0} \end{array}$$

Factor the quotient.

$x^2 + 1 = 0$

$x^2 = -1$

$x = \pm i$

The fourth roots of 1 are 1, -1, i, and $-i$.

113. The number $\sqrt{5}$ is a real solution to the equation $x^2 - 5 = 0$ and a zero of the polynomial $f(x) = x^2 - 5$. However, by the rational zeros theorem, the only possible rational zeros of $f(x)$ are ± 1 and ± 5. This means that $\sqrt{5}$ is irrational.

115. $x^3 - 3x = -2$

$x^3 + (-3)x = (-2)$

$m = -3, n = -2$

$$x = \sqrt[3]{\sqrt{\left(\frac{n}{2}\right)^2 + \left(\frac{m}{3}\right)^3} + \frac{n}{2}} - \sqrt[3]{\sqrt{\left(\frac{n}{2}\right)^2 + \left(\frac{m}{3}\right)^3} - \frac{n}{2}}$$

$$= \sqrt[3]{\sqrt{\left(\frac{-2}{2}\right)^2 + \left(\frac{-3}{3}\right)^3} + \frac{-2}{2}} - \sqrt[3]{\sqrt{\left(\frac{-2}{2}\right)^2 + \left(\frac{-3}{3}\right)^3} - \frac{-2}{2}}$$

$$= \sqrt[3]{\sqrt{1+(-1)} - 1} - \sqrt[3]{\sqrt{1+(-1)} + 1}$$

$$= \sqrt[3]{-1} - \sqrt[3]{1} = -1 + -1 = -2$$

Section 3.5 Rational Functions

1. $f(x) = 4x^4 - 25x^2 + 36$
$f(x) = (4x^2 - 9)(x^2 - 4)$
$f(x) = (2x-3)(2x+3)(x-2)(x+2)$
$0 = (2x-3)(2x+3)(x-2)(x+2)$
$x = \frac{3}{2}, x = -\frac{3}{2}, x = 2, x = -2$
The zeros are $\frac{3}{2}, -\frac{3}{2}, 2, -2$.

3. $h(x) = \frac{6}{x-2}$
$h(1) = \frac{6}{1-2} = \frac{6}{-1} = -6$
$h(1.9) = \frac{6}{1.9-2} = \frac{6}{-0.1} = -60$
$h(1.99) = \frac{6}{1.99-2} = \frac{6}{-0.01} = -600$

5. rational

7. x approaches infinity

9. vertical; c

11. c

13. left; down

15. denominator

17. $x - 5 \neq 0$
$x \neq 5$
$(-\infty, 5) \cup (5, \infty)$

19. $4x^2 + 3x - 1 \neq 0$
$(x+1)(4x-1) \neq 0$
$x \neq -1, x \neq \frac{1}{4}$
$(-\infty, -1) \cup \left(-1, \frac{1}{4}\right) \cup \left(\frac{1}{4}, \infty\right)$

21. $x^2 + 100 \neq 0$
True for all real x.
$(-\infty, \infty)$

23. **a.** 2 **b.** $-\infty$
c. ∞ **d.** 2
e. Never increasing **f.** $(-\infty, 4) \cup (4, \infty)$
g. $(-\infty, 4) \cup (4, \infty)$ **h.** $(-\infty, 2) \cup (2, \infty)$
i. $x = 4$ **j.** $y = 2$

25. **a.** -1 **b.** ∞
c. ∞ **d.** -1
e. $(-\infty, -3)$ **f.** $(-3, \infty)$
g. $(-\infty, -3) \cup (-3, \infty)$ **h.** $(-1, \infty)$
i. $x = -3$ **j.** $y = -1$

27. $x - 4 = 0$
$x = 4$

29. $2x^2 - 9x - 5 = 0$
$(x-5)(2x+1) = 0$
$x - 5, x = -\frac{1}{2}$

31. $x^2 + 5$ has no real zeros. The function has no vertical asymptotes.

33. $2t^2 + 4t - 3 = 0$
$t = \frac{-(4) \pm \sqrt{(4)^2 - 4(2)(-3)}}{2(2)}$
$= \frac{-4 \pm \sqrt{40}}{4}$
$= \frac{-4 \pm 2\sqrt{10}}{4}$
$= \frac{-2 \pm \sqrt{10}}{2}$
$t = \frac{-2 + \sqrt{10}}{2}, t = \frac{-2 - \sqrt{10}}{2}$

35. **a.** The degree of the numerator is 0.
The degree of the denominator is 2.
Since $n < m$, the line $y = 0$ is a horizontal asymptote of p.

b. $\frac{5}{x^2+2x+1}=0$

$5=0$ No solution

The graph does not cross $y=0$.

37. a. The degree of the numerator is 2.
The degree of the denominator is 2.
Since $n=m$, the line $y=\frac{3}{1}$ or equivalently $y=3$ is a horizontal asymptote of *h*.

b. $\frac{3x^2+8x-5}{x^2+3}=3$

$3x^2+8x-5=3x^2+9$

$8x=14$

$x=\frac{7}{4}$

The graph crosses $y=3$ at $\left(\frac{7}{4},3\right)$.

39. a. The degree of the numerator is 4.
The degree of the denominator is 1.
Since $n>m$, the function has no horizontal asymptotes.

b. Not applicable

41. a. The degree of the numerator is 1.
The degree of the denominator is 2.
Since $n<m$, the line $y=0$ is a horizontal asymptote of *t*.

b. $\frac{2x+4}{x^2+7x-4}=0$

$2x+4=0$

$2x=-4$

$x=-2$

The graph crosses $y=0$ at $(-2,0)$.

43. a. $\frac{x^2+3x+1}{2x^2+5}=\frac{\frac{x^2}{x^2}+\frac{3x}{x^2}+\frac{1}{x^2}}{\frac{2x^2}{x^2}+\frac{5}{x^2}}=\frac{1+\frac{3}{x}+\frac{1}{x^2}}{2+\frac{5}{x^2}}$

b. All of the terms approach 0 as $|x|\to\infty$.

c. Since the terms from part (b) approach 0 as $|x|\to\infty$, the horizontal asymptote is $y=\frac{1+0+0}{2+0}$ or $y=\frac{1}{2}$.

45. The expression $\frac{2x^2+3}{x}$ is in lowest terms and the denominator is 0 at $x=0$. *f* has a vertical asymptote at $x=0$.
The degree of the numerator is exactly one greater than the degree of the denominator. Therefore, *f* has no horizontal asymptote, but does have a slant asymptote.

$\frac{2x^2+3}{x}=\frac{2x^2}{x}+\frac{3}{x}=2x+\frac{3}{x}$

The slant asymptote is $y=2x$.

47. The expression $\frac{-3x^2+4x-5}{x+6}$ is in lowest terms and the denominator is 0 at $x=-6$. *h* has a vertical asymptote at $x=-6$.
The degree of the numerator is exactly one greater than the degree of the denominator. Therefore, *h* has no horizontal asymptote, but does have a slant asymptote.

$$\begin{array}{r|rrr} -6 & -3 & 4 & -5 \\ & & 18 & -132 \\ \hline & -3 & 22 & -137 \end{array}$$

The quotient is $-3x+22$.
The slant asymptote is $y=-3x+22$.

49. The expression $\frac{x^3+5x^2-4x+1}{x^2-5}$ is in lowest terms.

$x^2-5=0$

$x^2=5$

$x=\pm\sqrt{5}$

The denominator is 0 at $x=\pm\sqrt{5}$. *p* has vertical asymptotes at $x=\sqrt{5}$ and $x=-\sqrt{5}$.
The degree of the numerator is exactly one greater than the degree of the denominator. Therefore, *p* has no horizontal asymptote, but does have a slant asymptote.

$$\begin{array}{r} x+5 \\ x^2+0x-5\overline{)x^3+5x^2-4x+1} \\ -\left(x^3+0x^2-5x\right) \\ \hline 5x^2+x+1 \\ -\left(5x^2+0x-25\right) \\ \hline x+26 \end{array}$$

The quotient is $x+5$.
The slant asymptote is $y=x+5$.

51. The expression $\frac{2x+1}{x^3+x^2-4x-4}$ is in lowest terms.

$$\frac{2x+1}{x^3+x^2-4x-4}=\frac{2x+1}{x^2(x+1)-4(x+1)}$$
$$=\frac{2x+1}{(x+1)(x^2-4)}$$

The denominator is 0 at $x=-1$, $x=-2$, and $x=2$. *r* has vertical asymptotes at $x=-1$, $x=-2$, and $x=2$.
The degree of the numerator is 1.
The degree of the denominator is 3.
Since $n<m$, the line $y=0$ is a horizontal asymptote of *r*.
r has no slant asymptote.

53. The expression $\frac{4x^3-2x^2+7x-3}{2x^2+4x+3}$ is in lowest terms.

$2x^2+4x+3=0$

$$x=\frac{-(4)\pm\sqrt{(4)^2-4(2)(3)}}{2(2)}=\frac{-4\pm\sqrt{-8}}{4}$$

The denominator is never 0, so *f* has no vertical asymptotes.
The degree of the numerator is exactly one greater than the degree of the denominator. Therefore, *f* has no horizontal asymptote, but does have a slant asymptote.

$$\begin{array}{r} 2x-5 \\ 2x^2+4x+3\overline{)4x^3-2x^2+7x-3} \\ -\left(4x^3+8x^2+6x\right) \\ \hline -10x^2+x-3 \\ -\left(-10x^2-20x-15\right) \\ \hline 21x+12 \end{array}$$

The quotient is $2x-5$.
The slant asymptote is $y=2x-5$.

55. The graph of *f* is the graph of $y=\frac{1}{x}$ with a shift right 3 units.

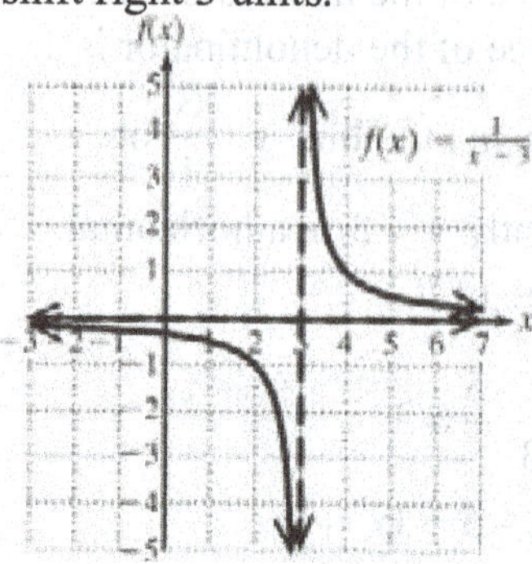

57. The graph of *h* is the graph of $y=\frac{1}{x^2}$ with a shift upward 2 units.

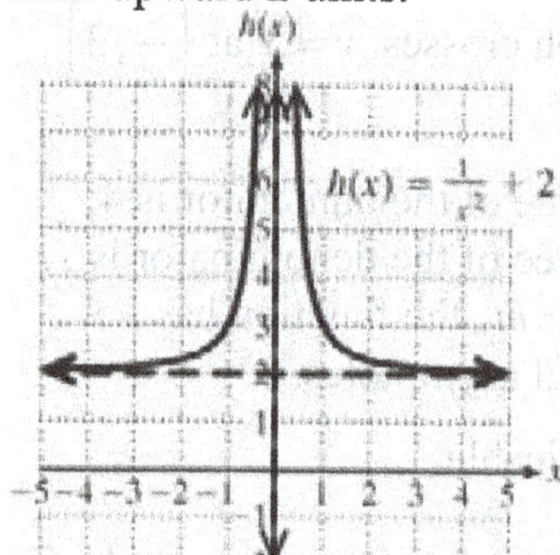

59. The graph of *m* is the graph of $y=\frac{1}{x^2}$ with a shift to the left 4 units and a shift downward 3 units.

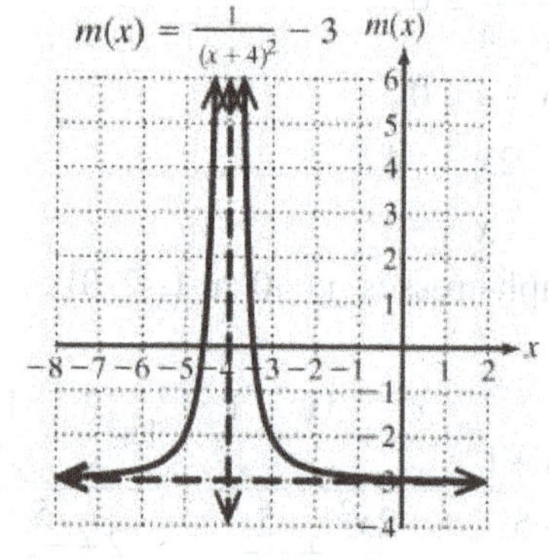

61. The graph of p is the graph of $y=\frac{1}{x}$ reflected across the x-axis.

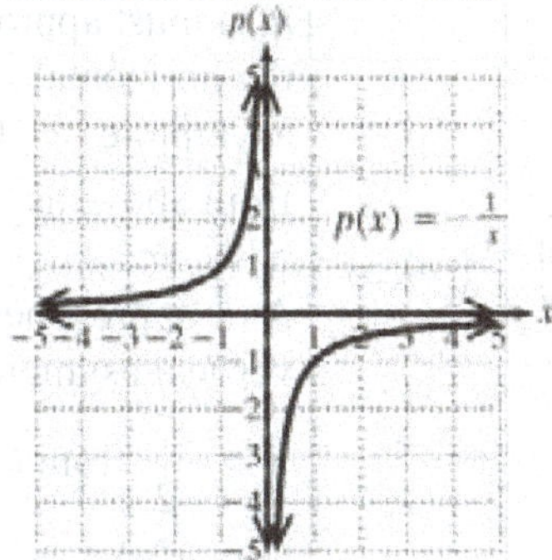

63. a. The numerator is 0 at $x=-3$ and $x=\frac{7}{2}$.

The x-intercepts are $(-3,0)$ and $\left(\frac{7}{2},0\right)$.

b. The denominator is 0 at $x=-2$ and $x=-\frac{1}{4}$. f has vertical asymptotes at $x=-2$ and $x=-\frac{1}{4}$.

c. The degree of the numerator is 2. The degree of the denominator is 2. Since $n=m$, the line $y=\frac{2}{4}$ or equivalently $y=\frac{1}{2}$ is a horizontal asymptote of f.

d. The constant term in the numerator is -21. The constant term in the denominator is 2. All other terms are 0 when evaluated at $x=0$.

The y-intercept is $\left(0,-\frac{21}{2}\right)$.

65. a. The numerator is 0 at $x=\frac{9}{4}$.

The x-intercept is $\left(\frac{9}{4},0\right)$.

b. The denominator is 0 at $x=3$ and $x=-3$. f has vertical asymptotes at $x=3$ and $x=-3$.

c. The degree of the numerator is 1. The degree of the denominator is 2. Since $n<m$, the line $y=0$ is a horizontal asymptote of f.

d. The constant term in the numerator is -9. The constant term in the denominator is -9. All other terms are 0 when evaluated at $x=0$.

The y-intercept is $\left(0,\frac{-9}{-9}\right)=(0,1)$.

67. a. The numerator is 0 at $x=\frac{1}{5}$ and $x=-3$.

The x-intercepts are $\left(\frac{1}{5},0\right)$ and $(-3,0)$.

b. The denominator is 0 at $x=-2$. f has a vertical asymptote at $x=-2$.

c. $f(x)=\frac{(5x-1)(x+3)}{x+2}$

$=\frac{5x^2+14x-3}{x+2}$

The degree of the numerator is exactly one greater than the degree of the denominator. Therefore, f has no horizontal asymptote, but does have a slant asymptote.

$$\begin{array}{r|rrr} -2 & 5 & 14 & -3 \\ & & -10 & -8 \\ \hline & 5 & 4 & \boxed{-11} \end{array}$$

The quotient is $5x+4$.
The slant asymptote is $y=5x+4$.

d. The constant term in the numerator is -3. The constant term in the denominator is 2. All other terms are 0 when evaluated at $x=0$.

The y-intercept is $\left(0,\frac{-3}{2}\right)=\left(0,-\frac{3}{2}\right)$.

69.

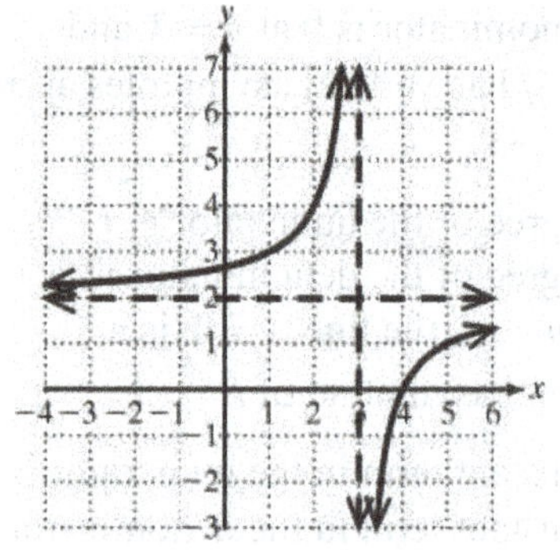

71.

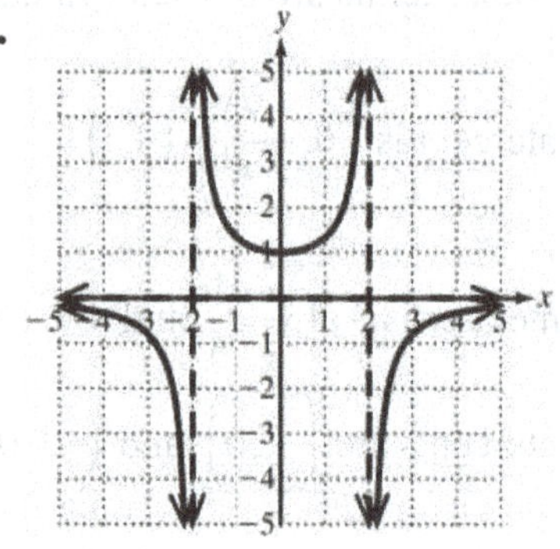

73. $n(0)=\dfrac{-3}{2(0)+7}=\dfrac{-3}{7}=-\dfrac{3}{7}$

The y-intercept is $\left(0,-\dfrac{3}{7}\right)$.

n is never 0. It has no x-intercepts.
n is in lowest terms, and $2x+7$ is 0 for $x=-\dfrac{7}{2}$, which is the vertical asymptote.
The degree of the numerator is 0.
The degree of the denominator is 1.
Since $n<m$, the line $y=0$ is a horizontal asymptote of n.

$0=\dfrac{-3}{2x+7}$

$0=-3$ No solution
n does not cross its horizontal asymptote.

$n(-x)=\dfrac{-3}{2(-x)+7}=\dfrac{-3}{-2x+7}$

$n(-x)\neq n(x)$, $n(-x)\neq -n(x)$
n is neither even nor odd.

Interval	Test Point	Comments
$\left(-\infty,-\dfrac{7}{2}\right)$	$(-4, 3)$	$n(x)$ is positive. $n(x)$ must approach the horizontal asymptote $y=0$ from above as $x\to-\infty$. As x approaches the vertical asymptote $x=-\dfrac{7}{2}$ from the left, $n(x)\to\infty$.
$\left(-\dfrac{7}{2},\infty\right)$	$(-2, -1)$	$n(x)$ is negative. $n(x)$ must approach the horizontal asymptote $y=0$ from below as $x\to\infty$. As x approaches the vertical asymptote $x=-\dfrac{7}{2}$ from the right, $n(x)\to-\infty$.

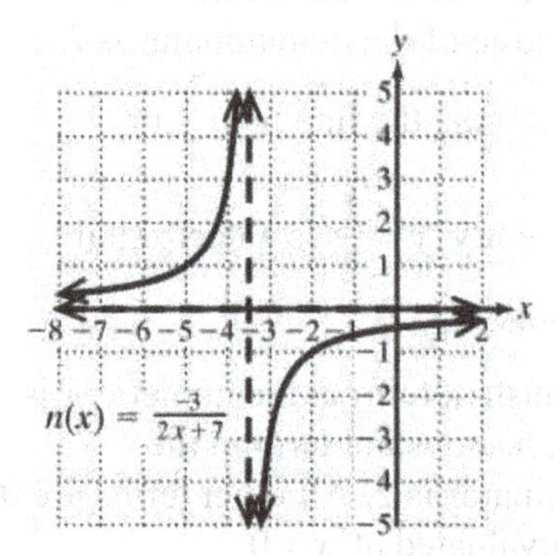

75. $p(0)=\dfrac{6}{(0)^2-9}=\dfrac{6}{-9}=-\dfrac{2}{3}$

The y-intercept is $\left(0,-\dfrac{2}{3}\right)$.

p is never 0. It has no x-intercepts.
p is in lowest terms, and x^2-9 is 0 for $x=3$ and $x=-3$, which are the vertical asymptotes.

The degree of the numerator is 0.
The degree of the denominator is 2.
Since $n < m$, the line $y = 0$ is a horizontal asymptote of p.

$0 = \frac{6}{x^2 - 9}$

$0 = 6$ No solution

p does not cross its horizontal asymptote.

$p(-x) = \frac{6}{(-x)^2 - 9} = \frac{6}{x^2 - 9}$

$p(-x) = p(x)$

p is even.

Interval	Test Point	Comments
$(-\infty, -3)$	$\left(-5, \frac{3}{8}\right)$	$p(x)$ is positive. $p(x)$ must approach the horizontal asymptote $y = 0$ from above as $x \to -\infty$. As x approaches the vertical asymptote $x = -3$ from the left, $p(x) \to \infty$.
$(-3, 0)$	$\left(-2, -\frac{6}{5}\right)$	$p(x)$ is negative. As x approaches the vertical asymptote $x = -3$ from the right, $p(x) \to -\infty$.
$(0, 3)$	$\left(2, -\frac{6}{5}\right)$	$p(x)$ is negative. As x approaches the vertical asymptote $x = 3$ from the left, $p(x) \to -\infty$.
$(3, \infty)$	$\left(5, \frac{3}{8}\right)$	$p(x)$ is positive. $p(x)$ must approach the horizontal asymptote $y = 0$ from above as $x \to \infty$. As x approaches the vertical asymptote $x = 3$ from the right, $p(x) \to \infty$.

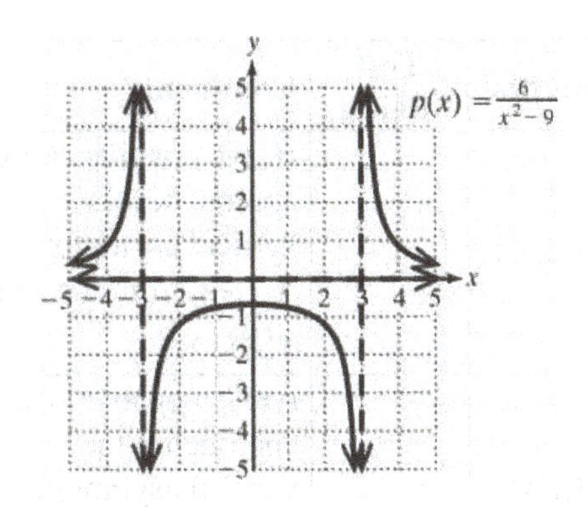

77. $r(0) = \frac{5(0)}{(0)^2 - (0) - 6} = 0$

The y-intercept is $(0, 0)$.

The x-intercept is $(0, 0)$.

r is in lowest terms, and $x^2 - x - 6 = (x-3)(x+2)$ is 0 for $x = 3$ and $x = -2$, which are the vertical asymptotes.
The degree of the numerator is 0.
The degree of the denominator is 2.
Since $n < m$, the line $y = 0$ is a horizontal asymptote of r.

$0 = \frac{5x}{x^2 - x - 6}$

$5x = 0$

$x = 0$

r crosses its horizontal asymptote at $(0, 0)$.

$r(-x) = \frac{5(-x)}{(-x)^2 - (-x) - 6} = \frac{-5x}{x^2 + x - 6}$

$r(-x) \neq r(x)$, $r(-x) \neq -r(x)$

r is neither even nor odd.

Interval	Test Point	Comments
$(-\infty, -2)$	$\left(-3, -\frac{5}{2}\right)$	$r(x)$ is negative. $r(x)$ must approach the horizontal asymptote $y = 0$ from below as $x \to -\infty$. As x approaches the vertical asymptote $x = -2$ from the left, $r(x) \to -\infty$.

Interval	Test Point	Comments
$(-2, 0)$	$\left(-1, \frac{5}{4}\right)$	$r(x)$ is positive. As x approaches the vertical asymptote $x=-2$ from the right, $r(x)\to\infty$.
$(0, 3)$	$\left(1, -\frac{5}{6}\right)$	$r(x)$ is negative. As x approaches the vertical asymptote $x=3$ from the left, $r(x)\to-\infty$.
$(3, \infty)$	$\left(6, \frac{5}{4}\right)$	$r(x)$ is positive. $r(x)$ must approach the horizontal asymptote $y=0$ from above as $x\to\infty$. As x approaches the vertical asymptote $x=3$ from the right, $r(x)\to\infty$.

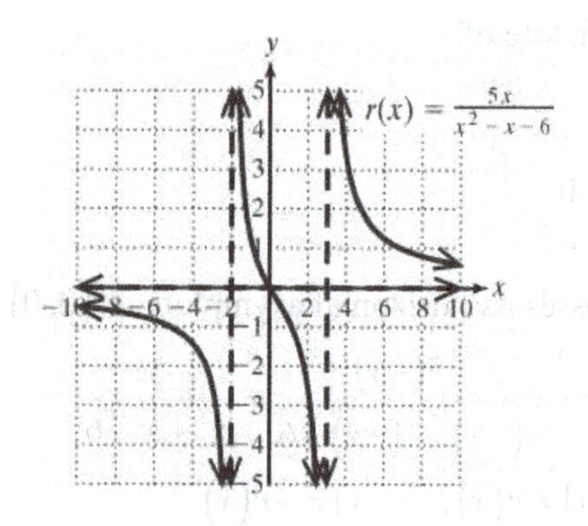

79. $k(0)=\dfrac{5(0)-3}{2(0)-7}=\dfrac{3}{7}$

The y-intercept is $\left(0, \frac{3}{7}\right)$.

$5x-3=0$

$5x=3$

$x=\frac{3}{5}$

The x-intercept is $\left(\frac{3}{5}, 0\right)$.

k is in lowest terms, and $2x-7$ is 0 for $x=\frac{7}{2}$, which is the vertical asymptote.

The degree of the numerator is 1.
The degree of the denominator is 1.
Since $n=m$, the line $y=\frac{5}{2}$ is a horizontal asymptote of k.

$$\frac{5}{2}=\frac{5x-3}{2x-7}$$

$10x-35=10x-6$

$-35=-6$ No solution

k does not cross its horizontal asymptote.

$$k(-x)=\frac{5(-x)-3}{2(-x)-7}=\frac{-5x-3}{-2x-7}$$

$k(-x)\neq k(x)$, $k(-x)\neq -k(x)$

k is neither even nor odd.

Interval	Test Point	Comments
$\left(-\infty, \frac{3}{5}\right)$	$\left(-1, \frac{8}{9}\right)$	$k(x)$ is positive. $k(x)$ must approach the horizontal asymptote $y=\frac{5}{2}$ from above as $x\to-\infty$.
$\left(\frac{3}{5}, \frac{7}{2}\right)$	$(3, -12)$	$k(x)$ is negative. As x approaches the vertical asymptote $x=\frac{7}{2}$ from the left, $k(x)\to-\infty$.
$\left(\frac{7}{2}, \infty\right)$	$(4, 17)$	$k(x)$ is positive. $k(x)$ must approach the horizontal asymptote $y=\frac{5}{2}$ from above as $x\to\infty$. As x approaches the vertical asymptote $x=\frac{7}{2}$ from the right, $k(x)\to\infty$.

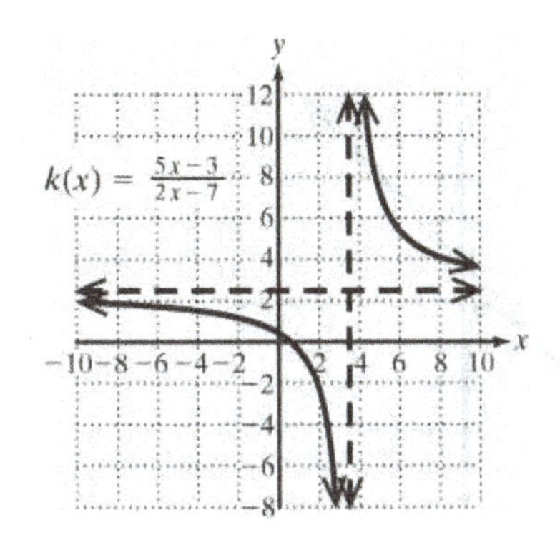

81. $g(0) = \dfrac{3(0)^2 - 5(0) - 2}{(0)^2 + 1} = \dfrac{-2}{1} = -2$

The y-intercept is $(0, -2)$.

$3x^2 - 5x - 2 = 0$

$(3x+1)(x-2) = 0$

$x = -\frac{1}{3}, x = 2$

The x-intercepts are $\left(-\frac{1}{3}, 0\right)$ and $(2, 0)$.

g is in lowest terms, and $x^2 + 1$ is never 0, so there are no vertical asymptotes.
The degree of the numerator is 2.
The degree of the denominator is 2.

Since $n = m$, the line $y = \frac{3}{1}$, or equivalently $y = 3$, is a horizontal asymptote of g.

$$3 = \frac{3x^2 - 5x - 2}{x^2 + 1}$$

$3x^2 + 3 = 3x^2 - 5x - 2$

$5x = -5$

$x = -1$

g crosses its horizontal asymptote at $(-1, 3)$.

$$g(-x) = \frac{3(-x)^2 - 5(-x) - 2}{(-x)^2 + 1} = \frac{3x^2 + 5x - 2}{x^2 + 1}$$

$g(-x) \neq g(x)$, $g(-x) \neq -g(x)$

g is neither even nor odd.

Interval	Test Point	Comments
$(-\infty, -1)$	$(-2, 4)$	Since $g(-2) = 4$ is above the horizontal asymptote $y = 3$, $g(x)$ must approach the horizontal asymptote from above as $x \to -\infty$.
$\left(-1, -\frac{1}{3}\right)$	$\left(-\frac{1}{2}, 1\right)$	Plot the point $\left(-\frac{1}{2}, 1\right)$ between the horizontal asymptote and the x-intercept of $\left(-\frac{1}{3}, 0\right)$.
$\left(-\frac{1}{3}, 2\right)$	$(0, -2)$	The point $(0, -2)$ is the y-intercept.
$(2, \infty)$	$(3, 1)$	Since $g(3) = 1$ is below the horizontal asymptote $y = 3$, $g(x)$ must approach the horizontal asymptote from below as $x \to \infty$.

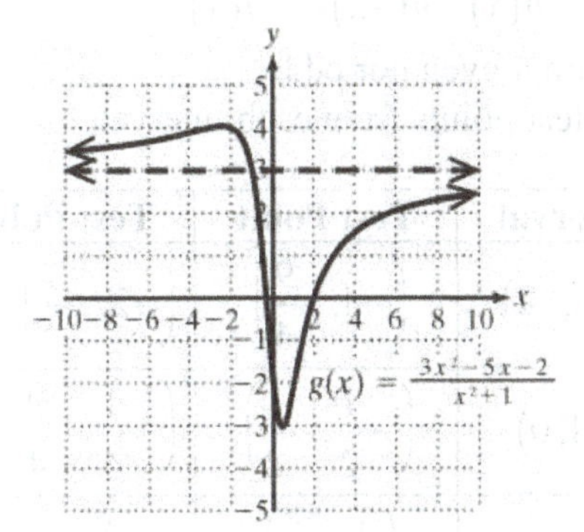

83. $n(0)=\dfrac{(0)^2+2(0)+1}{(0)}$ undefined

There is no y-intercept.

$x^2+2x+1=0$

$(x+1)(x+1)=0$

$x=-1$

The x-intercept is $(-1,0)$.

n is in lowest terms, and the denominator is 0 when $x=0$, which is the vertical asymptote.

The degree of the numerator is exactly one greater than the degree of the denominator. Therefore, n has no horizontal asymptote, but does have a slant asymptote.

$n(x)=\dfrac{x^2+2x+1}{x}=\dfrac{x^2}{x}+\dfrac{2x}{x}+\dfrac{1}{x}=x+2+\dfrac{1}{x}$

The quotient is $x+2$.

The slant asymptote is $y=x+2$.

$x+2=\dfrac{x^2+2x+1}{x}$

$x^2+2x=x^2+2x+1$

$0=1$ No solution

n does not cross its slant asymptote.

$n(-x)=\dfrac{(-x)^2+2(-x)+1}{(-x)}=\dfrac{x^2-2x+1}{-x}$

$n(-x)\neq n(x)$, $n(-x)\neq -n(x)$

n is neither even nor odd.

Select test points from each interval.

Interval	Test Point	Test Point
$(-\infty,-1)$	$\left(-4,-\dfrac{9}{4}\right)$	$\left(-2,-\dfrac{1}{2}\right)$
$(-1,0)$	$\left(-\dfrac{1}{2},-\dfrac{1}{2}\right)$	$\left(-\dfrac{1}{4},-\dfrac{9}{4}\right)$
$(0,\infty)$	$\left(\dfrac{1}{4},\dfrac{25}{4}\right)$	$\left(2,\dfrac{9}{2}\right)$

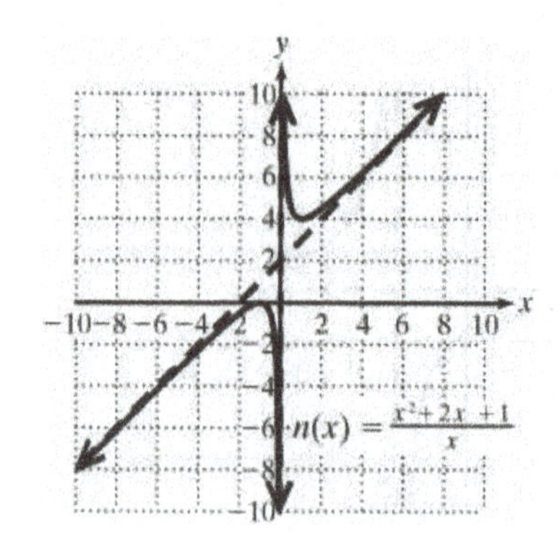

85. $f(0)=\dfrac{(0)^2+7(0)+10}{(0)+3}=\dfrac{10}{3}$

The y-intercept is $\left(0,\dfrac{10}{3}\right)$.

$x^2+7x+10=0$

$(x+5)(x+2)=0$

$x=-5,\ x=-2$

The x-intercepts are $(-5,0)$ and $(-2,0)$.

f is in lowest terms, and $x+3$ is 0 when $x=-3$, which is the vertical asymptote.

The degree of the numerator is exactly one greater than the degree of the denominator. Therefore, f has no horizontal asymptote, but does have a slant asymptote.

$$\begin{array}{r|rrr} -3 & 1 & 7 & 10 \\ & & -3 & -12 \\ \hline & 1 & 4 & -2 \end{array}$$

The quotient is $x+4$.

The slant asymptote is $y=x+4$.

$x+4=\dfrac{x^2+7x+10}{x+3}$

$x^2+7x+12=x^2+7x+10$

$12=10$ No solution

f does not cross its slant asymptote.

$f(-x)=\dfrac{(-x)^2+7(-x)+10}{(-x)+3}=\dfrac{x^2-7x+10}{-x+3}$

$f(-x)\neq f(x)$, $f(-x)\neq -f(x)$

f is neither even nor odd.

Select test points from each interval.

Interval	Test Point	Test Point
$(-\infty, -5)$	$\left(-7, -\frac{5}{2}\right)$	$\left(-6, -\frac{4}{3}\right)$
$(-5, -3)$	$(-4, 2)$	$\left(-\frac{7}{2}, \frac{9}{2}\right)$
$(-3, -2)$	$\left(-\frac{5}{2}, -\frac{5}{2}\right)$	$\left(-\frac{7}{3}, -\frac{4}{3}\right)$
$(-2, \infty)$	$(-1, 2)$	$\left(1, \frac{9}{2}\right)$

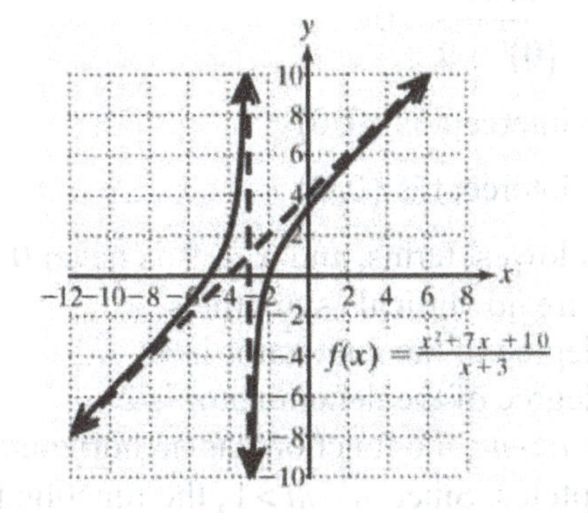

87. $w(0) = \frac{-4(0)^2}{(0)^2+4} = 0$

The y-intercept is $(0, 0)$.

The x-intercept is $(0, 0)$.

w is in lowest terms, and x^2+4 is never 0, so there are no vertical asymptotes.
The degree of the numerator is 2.
The degree of the denominator is 2.

Since $n = m$, the line $y = \frac{-4}{1}$, or equivalently $y = -4$, is a horizontal asymptote of w.

$$-4 = \frac{-4x^2}{x^2+4}$$

$$-4x^2 - 16 = -4x^2$$

$$-16 = 0 \quad \text{No solution}$$

w does not cross its horizontal asymptote.

$$w(-x) = \frac{-4(-x)^2}{(-x)^2+4} = \frac{-4x^2}{x^2+4}$$

$$w(-x) = w(x)$$

w is even.

Interval	Test Point	Comments
$(-\infty, 0)$	$(-2, -2)$	Since $w(-2) = -2$ is above the horizontal asymptote $y = -4$, $w(x)$ must approach the horizontal asymptote from above as $x \to -\infty$.
$(0, \infty)$	$(2, -2)$	Since $w(2) = -2$ is above the horizontal asymptote $y = -4$, $w(x)$ must approach the horizontal asymptote from above as $x \to \infty$.

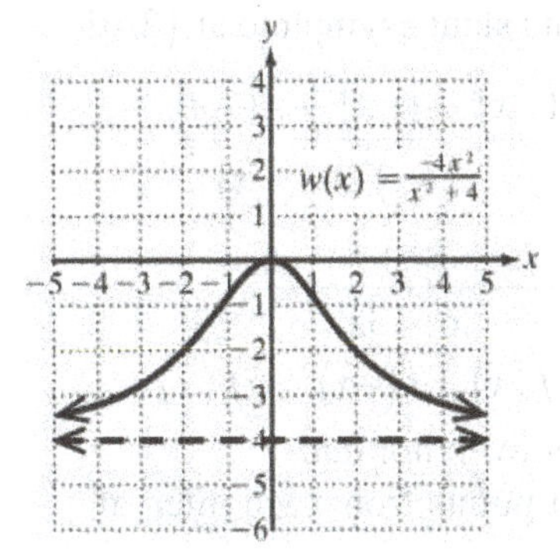

89. $f(0) = \frac{(0)^3+(0)^2-4(0)-4}{(0)^2+3(0)} = \frac{-4}{0}$ undefined

There is no y-intercept.

$$x^3 + x^2 - 4x - 4 = 0$$

$$x^2(x+1) - 4(x+1) = 0$$

$$(x+1)(x^2-4) = 0$$

$$(x+1)(x+2)(x-2) = 0$$

$$x = -1, x = -2, x = 2$$

The x-intercepts are $(-1, 0)$, $(-2, 0)$, and $(2, 0)$.

f is in lowest terms, and $x^2+3x = x(x+3)$ is 0 when $x = 0$ and $x = -3$, which are the vertical asymptotes.
The degree of the numerator is exactly one greater than the degree of the denominator. Therefore, f has no horizontal asymptote, but does have a slant asymptote.

$$\begin{array}{r} x-2 \\ x^2+3x\overline{)x^3+\ x^2-4x-4} \\ -(x^3+3x^2) \\ \hline -2x^2-4x \\ -(-2x^2-6x) \\ \hline 2x-4 \end{array}$$

The quotient is $x-2$.
The slant asymptote is $y=x-2$.

$$x-2=\frac{x^3+x^2-4x-4}{x^2+3x}$$
$$x^3+x^2-6x=x^3+x^2-4x-4$$
$$-2x=-4$$
$$x=2$$
$$y=x-2=2-2=0$$

f crosses its slant asymptote at $(2,0)$.

$$f(-x)=\frac{(-x)^3+(-x)^2-4(-x)-4}{(-x)^2+3(-x)}$$
$$=\frac{-x^3+x^2+4x-4}{x^2-3x}$$
$$f(-x)\neq f(x),\ f(-x)\neq -f(x)$$

f is neither even nor odd.
Select test points from each interval.

Interval	Test Point	Test Point
$(-\infty,-3)$	$\left(-8,-\frac{21}{2}\right)$	$(-4,-9)$
$(-3,-2)$	$\left(-\frac{5}{2},\frac{27}{10}\right)$	
$(-2,-1)$	$\left(-\frac{4}{3},-\frac{1}{3}\right)$	
$(-1,0)$	$\left(-\frac{1}{2},\frac{3}{2}\right)$	
$(0,2)$	$\left(\frac{1}{3},-\frac{14}{3}\right)$	$\left(1,-\frac{3}{2}\right)$
$(2,\infty)$	$\left(3,\frac{10}{9}\right)$	$\left(4,\frac{15}{7}\right)$

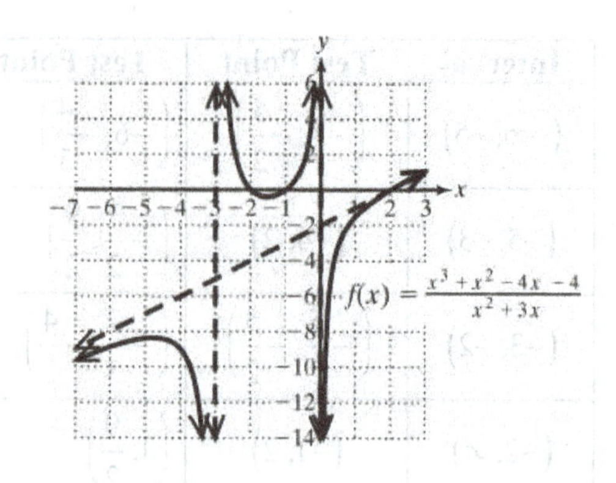

91. $v(0)=\frac{2(0)^4}{(0)^2+9}=0$

The y-intercept is $(0,0)$.

The x-intercept is $(0,0)$.

v is in lowest terms, and x^2+9 is never 0, so there are no vertical asymptotes.
The degree of the numerator is 4.
The degree of the denominator is 2.
Since $n>m$, the function has no horizontal asymptotes. Since $n-m>1$, the function has no slant asymptote.

$$v(-x)=\frac{2(-x)^4}{(-x)^2+9}=\frac{2x^4}{x^2+9}$$
$$v(-x)=v(x)$$

v is even.
Select test points.

The graph passes through $(-3,9)$, $\left(-1,\frac{1}{5}\right)$, $\left(1,\frac{1}{5}\right)$, $(3,9)$.

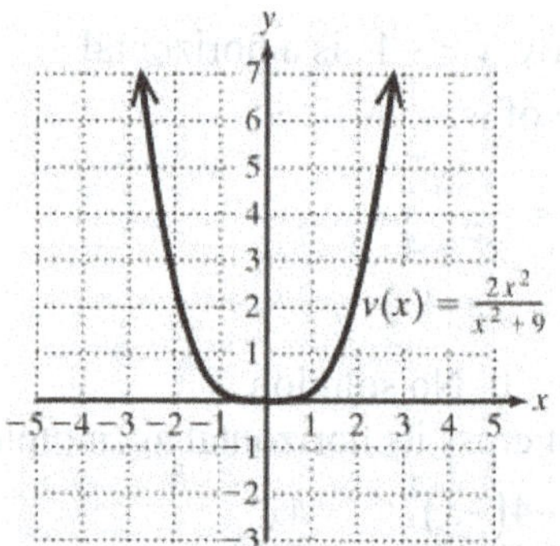

93. a. $C(x)=20x+69.95+39.99$

$C(x)=20x+109.94$

b. $\overline{C}(x)=\dfrac{20x+109.94}{x}$

c. $\overline{C}(5)=\dfrac{20(5)+109.94}{5}\approx 41.99$

$\overline{C}(30)=\dfrac{20(30)+109.94}{30}\approx 23.66$

$\overline{C}(120)=\dfrac{20(120)+109.94}{120}\approx 20.92$

d. The average cost would approach $20 per session. This is the same as the fee paid to the gym in the absence of fixed costs.

95. a. $\overline{C}_1(252)=\dfrac{59.95}{252}\approx 0.24$

$\overline{C}_1(366)=\dfrac{59.95}{366}\approx 0.16$

$\overline{C}_1(400)=\dfrac{59.95}{400}\approx 0.15$

b. $\overline{C}_1(436)=\dfrac{59.95+0.45(436-400)}{252}\approx 0.17$

$\overline{C}_1(582)=\dfrac{59.95+0.45(582-400)}{582}\approx 0.24$

$\overline{C}_1(700)=\dfrac{59.95+0.45(700-400)}{700}\approx 0.28$

c. $\overline{C}_2(x)=\dfrac{79.95}{x}$

d. $\overline{C}_2(252)=\dfrac{79.95}{252}\approx 0.32$

$\overline{C}_2(400)=\dfrac{79.95}{252}\approx 0.20$

$\overline{C}_2(700)=\dfrac{79.95}{252}\approx 0.11$

97. a. $C(25)=\dfrac{600(25)}{100-25}=200$; $200,000

b. $C(50)=\dfrac{600(50)}{100-50}=600$; $600,000

$C(75)=\dfrac{600(75)}{100-75}=1800$; $1,800,000

$C(90)=\dfrac{600(90)}{100-90}=5400$; $5,400,000

c.

$$C(x)=\frac{600x}{100-x}$$

$$1400=\frac{600x}{100-x}$$

$$140{,}000-1400x=600x$$

$$140{,}000=2000x$$

$$70=x$$

70% of the air pollutants can be removed.

99. a. $N(t)=(P+J)(t)$

$N(t)=-0.091t^3+3.48t^2+15.4t+335+23.0t+159$

$N(t)=-0.091t^3+3.48t^2+38.4t+494$

$N(t)$ represents the total number of adults incarcerated in both prisons and jails, t years since 1980.

b. $R(t)=\dfrac{J}{N}(t)$

$R(t)=\dfrac{23.0t+159}{-0.091t^3+3.48t^2+38.4t+494}$

$R(t)$ represents the percentage of incarcerated adults who are in jail, t years since 1980.

c. $R(25)$

$=\dfrac{23.0(25)+159}{-0.091(25)^3+3.48(25)^2+38.4(25)+494}$

≈ 0.333

$R(25)=0.333$ means that in the year 2005, 33.3% of incarcerated adults were in jail.

101. a. $f(x)=\dfrac{1}{[x-(-1)]^2}+3$

$f(x)=\dfrac{1}{(x+1)^2}+3$

b. Domain: $(-\infty,-1)\cup(-1,\infty)$;

Range: $(3,\infty)$

103. $f(x)=\dfrac{2x+7}{x+3}$

Find the quotient.

$$\begin{array}{r|rr} -3 & 2 & 7 \\ & & -6 \\ \hline & 2 & \boxed{1} \end{array}$$

$f(x)=2+\dfrac{1}{x+3}$

The graph of f is the graph of $y=\dfrac{1}{x}$ with a shift to the left 3 units and a shift upward 2 units.

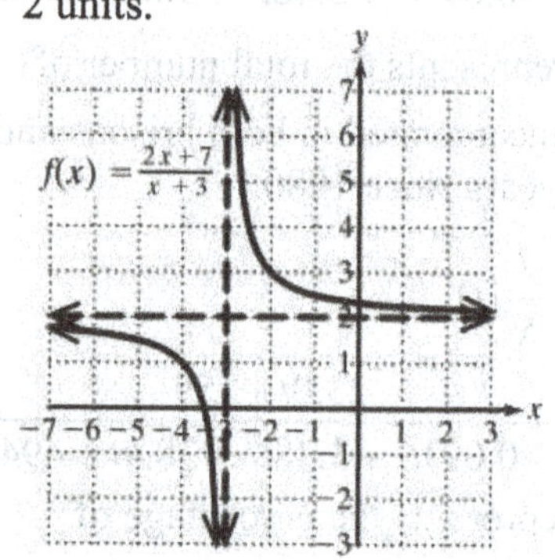

105. The numerator and denominator share a common factor of $x+2$. The value -2 is not in the domain of f. The graph will have a "hole" at $x=-2$ rather than a vertical asymptote.

107. Factors of the numerator: $(x+3)(x+1)$

Factor of the denominator: $(x-2)$

The degree of the numerator equals the degree of the denominator.

$$f(x)=\frac{(x+3)(x+1)}{(x-2)(x-a)}$$

$$f(0)=\frac{3}{4}$$

$$\frac{(0+3)(0+1)}{(0-2)(0-a)}=\frac{3}{4}$$

$$\frac{3}{2a}=\frac{3}{4}$$

$$2a=4$$

$$a=2$$

$$f(x)=\frac{(x+3)(x+1)}{(x-2)(x-2)}=\frac{x^2+4x+3}{x^2-4x+4}$$

109. Factor of the numerator: $\left(x-\dfrac{3}{2}\right)$ or $(2x-3)$

Factors of the denominator: $(x+2)(x-5)$

The degree of the numerator is less than the degree of the denominator.

$$f(x)=a\left[\frac{2x-3}{(x+2)(x-5)}\right]$$

$$f(0)=3$$

$$a\left[\frac{2(0)-3}{(0+2)(0-5)}\right]=3$$

$$a\left(\frac{-3}{-10}\right)=3$$

$$a=10$$

$$f(x)=10\left[\frac{2x-3}{(x+2)(x-5)}\right]=\frac{20x-30}{x^2-3x-10}$$

111. a.

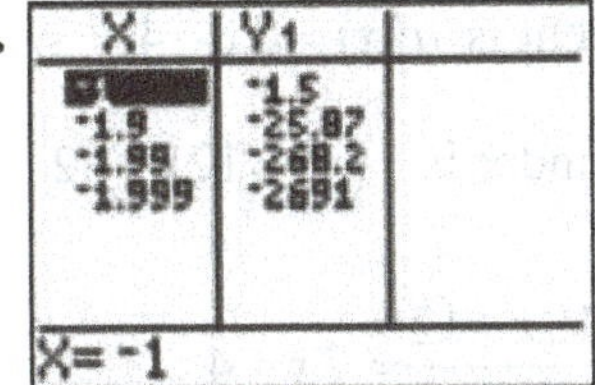

X	Y1
	-1.5
-1.9	-25.87
-1.99	-268.2
-1.999	-2691

X=-1

b.

X	Y1
	-1.667
10	.50887
100	.76729
1000	.79668
10000	.79967

X=1

c. Vertical asymptotes:

$$5x^2+7x-6=0$$
$$(x+2)(5x-3)=0$$
$$x=-2 \text{ and } x=\frac{3}{5};$$

Horizontal asymptote: $y=\frac{4}{5}$

113. a. $f(x)=\dfrac{x^2-5x+4}{x-4}$

$$=\frac{(x-1)\cancel{(x-4)}}{\cancel{x-4}}=x-1$$

No vertical asymptotes

b. On the window $[-9.4, 9.4, 1]$ by $[-6.2, 6.2, 1]$, the calculator shows a discontinuity or "hole" at $(4, 3)$. The graph on the standard viewing window does not show the discontinuity.

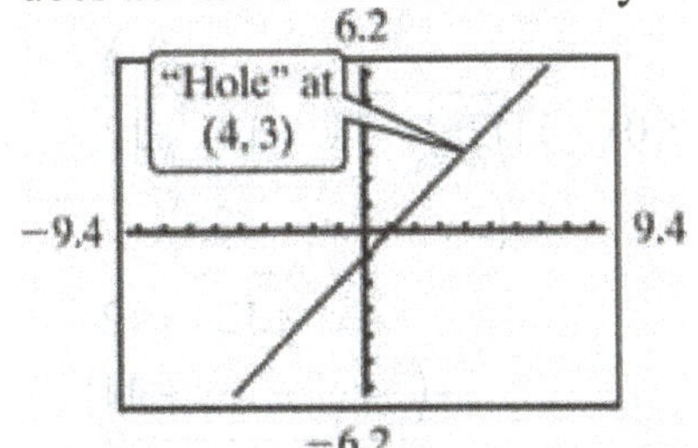

Problem Recognition Exercises: Polynomial and Rational Functions

1. $\dfrac{\text{Factors of } -8}{\text{Factors of } 1}=\dfrac{\pm1, \pm2, \pm4, \pm8}{\pm1}$

$$=\pm1, \pm2, \pm4, \pm8$$

```
2| 1   3  -6  -8
       2  10   8
  ---------------
   1   5   4  |0
```

Factor the quotient.

$$x^2+5x+4=0$$
$$(x+1)(x+4)=0$$
$$x=-1, x=-4$$

The zeros are 2, -1, and -4.

3. The x-intercepts of q are $(-2, 0)$, $(1, 0)$, and $(3, 0)$.

5. The x-intercepts of f are $(2, 0)$, $(-1, 0)$, and $(-4, 0)$.

7. Since the degree of the numerator is the same as the degree of the denominator, the line $y=\frac{1}{1}$, or equivalently $y=1$, is a horizontal asymptote of f.

9. $\dfrac{\text{Factors of } 8}{\text{Factors of } 1}=\dfrac{\pm1, \pm2, \pm4, \pm8}{\pm1}$

$$=\pm1, \pm2, \pm4, \pm8$$

```
4| 1  -4  -2   8
       4   0  -8
  ---------------
   1   0  -2  |0
```

Factor the quotient.

$$x^2-2=0$$
$$x^2=2$$
$$x=\pm\sqrt{2}$$

The zeros are 4, $\sqrt{2}$, and $-\sqrt{2}$.

11 The x-intercepts of d are $(1, 0)$ and $(-2, 0)$.

13. The x-intercepts of g are $(4,0)$, $(\sqrt{2},0)$, and $(-\sqrt{2},0)$.

15. Since the degree of the numerator is the same as the degree of the denominator, the line $y=\frac{1}{1}$, or equivalently $y=1$, is a horizontal asymptote of g.

17. Graph b has the intercepts and asymptotes of function f.

19.
$$\begin{array}{r} 2x-4 \\ x^2+0x-11\overline{)2x^3-4x^2-10x+12} \\ -(2x^3+0x^2-22x) \\ \hline -4x^2+12x+12 \\ -(-4x^2+0x+44) \\ \hline 12x-32 \end{array}$$

a. The quotient is $q(x)=2x-4$.

b. The remainder is $r(x)=12x-32$.

21. $\dfrac{2x^3-4x^2-10x+12}{x^2-11}=2x-4$

$$2x^3-4x^2-10x+12=2x^3-4x^2-22x+44$$
$$-10x+12=-22x+44$$
$$12x=32$$
$$x=\frac{32}{12}=\frac{8}{3}$$

$y=2x-4$

$y=2\left(\frac{8}{3}\right)-4=\frac{16}{3}-\frac{12}{3}=\frac{4}{3}$

The graph crosses the slant asymptote at $\left(\frac{8}{3},\frac{4}{3}\right)$.

Section 3.6 Polynomial and Rational Inequalities

1. $2x^3+5x^2-12x=0$

$x(2x^2+5x-12)=0$

$x(2x-3)(x+4)=0$

$x=0,\ x=\frac{3}{2},\ x=-4$

The solution set is $\left\{0,\frac{3}{2},-4\right\}$.

3. $\dfrac{2x+1}{x-5}=0$

$2x+1=0$

$2x=-1$

$x=-\frac{1}{2}$

The solution set is $\left\{-\frac{1}{2}\right\}$.

5. $x-5=0$

$x=5$

7. polynomial; 2

9. negative

11. **a.** $(-4,-1)$ **b.** $[-4,-1]$
c. $(-\infty,-4)\cup(-1,\infty)$
d. $(-\infty,-4]\cup[-1,\infty)$

13. **a.** $(-\infty,-3)\cup(-3,\infty)$
b. $(-\infty,\infty)$ **c.** $\{\ \}$
d. $\{-3\}$

15. **a.** $(-\infty,-2)\cup(0,3)$ **b.** $(-\infty,-2]\cup[0,3]$
c. $(-2,0)\cup(3,\infty)$ **d.** $[-2,0]\cup[3,\infty)$

17. **a.** $(0,3)\cup(3,\infty)$ **b.** $[0,\infty)$
c. $(-\infty,0)$ **d.** $(-\infty,0]\cup\{3\}$

19. **a.** $(-\infty,\infty)$ **b.** $(-\infty,\infty)$
c. $\{\ \}$ **d.** $\{\ \}$

21. $(5x-3)(x-5)=0$

$x=\frac{3}{5}$ or $x=5$

The boundary points are $\frac{3}{5}$ and 5.

Sign of $(5x-3)$:	$-$	$+$	$+$
Sign of $(x-5)$:	$-$	$-$	$+$
Sign of $(5x-3)(x-5)$:	$+$	$-$	$+$
		$\frac{3}{5}$	5

a. $(5x-3)(x-5)=0$ $\left\{\frac{3}{5}, 5\right\}$

b. $(5x-3)(x-5)<0$ $\left(\frac{3}{5}, 5\right)$

c. $(5x-3)(x-5)\le 0$ $\left[\frac{3}{5}, 5\right]$

d. $(5x-3)(x-5)>0$ $\left(-\infty, \frac{3}{5}\right)\cup(5, \infty)$

e. $(5x-3)(x-5)\ge 0$ $\left(-\infty, \frac{3}{5}\right]\cup[5, \infty)$

23. $-x^2+x+12=0$

$-(x^2-x-12)=0$

$-(x+3)(x-4)=0$

$x=-3$ or $x=4$

The boundary points are -3 and 4.

Sign of $(x+3)$:	$-$	$+$	$+$
Sign of $(x-4)$:	$-$	$-$	$+$
Sign of $(x+3)(x-4)$:	$+$	$-$	$+$
Sign of $-(x+3)(x-4)$:	$-$	$+$	$-$
		-3	4

a. $-x^2+x+12=0$

$\{-3, 4\}$

b. $-x^2+x+12<0$

$(-\infty, -3)\cup(4, \infty)$

c. $-x^2+x+12\le 0$

$(-\infty, -3]\cup[4, \infty)$

d. $-x^2+x+12>0$

$(-3, 4)$

e. $-x^2+x+12\ge 0$

$[-3, 4]$

25. $x^2+12x+36=0$

$(x+6)^2=0$

$x=-6$

The boundary point is -6.

a. $x^2+12x+36=0$

$\{-6\}$

b. $x^2+12x+36<0$

The square of any real number is nonnegative. Therefore, this inequality has no solution.

$\{\ \}$

c. $x^2+12x+36\le 0$

The inequality in part (b) is the same as the inequality in part (a) except that equality is included. The expression $(x+6)^2=0$ for $x=-6$.

$\{-6\}$

d. $x^2+12x+36>0$

The expression $(x+6)^2>0$ for all real numbers except where $(x+6)^2=0$. Therefore, the solution set is all real numbers except -6.

$(-\infty, -6)\cup(-6, \infty)$

e. $x^2+12x+36\ge 0$

The square of any real number is greater than or equal to zero. Therefore, the solution set is all real numbers.

$(-\infty, \infty)$

27. $3w^2 + w < 2(w+2)$

$3w^2 + w < 2w + 4$

$3w^2 - w - 4 < 0$

Find the real zeros of the related equation.

$3w^2 - w - 4 = 0$

$(w+1)(3w-4) = 0$

$w = -1$ or $w = \frac{4}{3}$

The boundary points are -1 and $\frac{4}{3}$.

Sign of $(w+1)$:	$-$	$+$	$+$
Sign of $(3w-4)$:	$-$	$-$	$+$
Sign of $(w+1)(3w-4)$:	$+$	$-$	$+$
		-1	$\frac{4}{3}$

The solution set is $\left(-1, \frac{4}{3}\right)$.

29. $a^2 \geq 3a$

$a^2 - 3a \geq 0$

Find the real zeros of the related equation.

$a^2 - 3a = 0$

$a(a-3) = 0$

$a = 0$ or $a = 3$

The boundary points are 0 and 3.

Sign of (a):	$-$	$+$	$+$
Sign of $(a-3)$:	$-$	$-$	$+$
Sign of $(a)(a-3)$:	$+$	$-$	$+$
		0	3

The solution set is $(-\infty, 0] \cup [3, \infty)$.

31. $10 - 6x > 5x^2$

$-5x^2 - 6x + 10 > 0$

$5x^2 + 6x - 10 < 0$

Find the real zeros of the related equation.

$5x^2 + 6x - 10 = 0$

$$x = \frac{-(6) \pm \sqrt{(6)^2 - 4(5)(-10)}}{2(5)} = \frac{-6 \pm \sqrt{236}}{10} = \frac{-6 \pm 2\sqrt{59}}{10} = \frac{-3 \pm \sqrt{59}}{5}$$

$x = \frac{-3 - \sqrt{59}}{5}$ or $x = \frac{-3 + \sqrt{59}}{5}$

The boundary points are $\frac{-3 - \sqrt{59}}{5}$ and $\frac{-3 + \sqrt{59}}{5}$.

Sign of $\left[x - \left(\frac{-3 - \sqrt{59}}{5}\right)\right]$:	$-$	$+$	$+$
Sign of $\left[x - \left(\frac{-3 + \sqrt{59}}{5}\right)\right]$:	$-$	$-$	$+$
Sign of $\left[x - \left(\frac{-3 - \sqrt{59}}{5}\right)\right]\left[x - \left(\frac{-3 + \sqrt{59}}{5}\right)\right]$:	$+$	$-$	$+$
		$\frac{-3 - \sqrt{59}}{5}$	$\frac{-3 + \sqrt{59}}{5}$

The solution set is $\left(\frac{-3 - \sqrt{59}}{5}, \frac{-3 + \sqrt{59}}{5}\right)$.

33. $(x+4)(x-1)(x-3)\geq 0$

Find the real zeros of the related equation.

$(x+4)(x-1)(x-3)=0$

$x=-4$, $x=1$, or $x=3$

The boundary points are -4, 1, and 3.

Sign of $(x+4)$:	−	+	+	+
Sign of $(x-1)$:	−	−	+	+
Sign of $(x-3)$:	−	−	−	+
Sign of $(x+4)(x-1)(x-3)$:	−	+	−	+
	−4	1	3	

The solution set is $[-4,1]\cup[3,\infty)$.

35. $-5c(c+2)^2(4-c)>0$

$5c(c+2)^2(c-4)>0$

Find the real zeros of the related equation.

$5c(c+2)^2(c-4)=0$

$c(c+2)^2(c-4)=0$

$c=0$, $c=-2$, or $c=4$

The boundary points are -2, 0, and 4.

Sign of $(c+2)^2$:	+	+	+	+
Sign of (c):	−	−	+	+
Sign of $(c-4)$:	−	−	−	+
Sign of $(c)(c+2)^2(c-4)$:	+	+	−	+
	−2	0	4	

The solution set is $(-\infty,-2)\cup(-2,0)\cup(4,\infty)$.

37. $t^4-10t^2+9\leq 0$

Find the real zeros of the related equation.

$t^4-10t^2+9=0$

$(t^2-1)(t^2-9)=0$

$(t+1)(t-1)(t+3)(t-3)=0$

$t=-1$, $t=1$, $t=-3$, or $t=3$

The boundary points are -3, -1, 1, and 3.

Sign of $(t+3)$:	−	+	+	+	+
Sign of $(t+1)$:	−	−	+	+	+
Sign of $(t-1)$:	−	−	−	+	+
Sign of $(t-3)$:	−	−	−	−	+
Sign of $(t+1)(t-1)(t+3)(t-3)$:	+	−	+	−	+
	−3	−1	1	3	

The solution set is $[-3,-1]\cup[1,3]$.

39. $2x^3+5x^2<8x+20$

$2x^3+5x^2-8x-20<0$

Find the real zeros of the related equation.

$2x^3+5x^2-8x-20=0$

$x^2(2x+5)-4(2x+5)=0$

$(2x+5)(x^2-4)=0$

$(2x+5)(x+2)(x-2)=0$

$x=-\frac{5}{2}$, $x=-2$, or $x=2$

The boundary points are $-\frac{5}{2}$, -2, and 2.

Sign of $(2x+5)$:	−	+	+	+
Sign of $(x+2)$:	−	−	+	+
Sign of $(x-2)$:	−	−	−	+
Sign of $(2x+5)(x+2)(x-2)$:	−	+	−	+
	$-\frac{5}{2}$	−2	2	

The solution set is $\left(-\infty,-\frac{5}{2}\right)\cup(-2,2)$.

41. $-2x^4+10x^3-6x^2-18x\geq 0$

$-2x(x^3-5x^2+3x+9)\geq 0$

$x(x^3-5x^2+3x+9)\leq 0$

Find the real zeros of the related equation.

$x(x^3-5x^2+3x+9)=0$

$\frac{\text{Factors of 9}}{\text{Factors of 1}}=\frac{\pm1,\pm3,\pm9}{\pm1}=\pm1,\pm3,\pm9$

$$\begin{array}{r|rrrr} -1 & 1 & -5 & 3 & 9 \\ & & -1 & 6 & -9 \\ \hline & 1 & -6 & 9 & \lfloor 0 \end{array}$$

$x(x+1)(x^2-6x+9)=0$

$x(x+1)(x-3)^2=0$

$x=0$, $x=-1$, or $x=3$

The boundary points are -1, 0, and 3.

Sign of $(x+1)$:	$-$	+	+	+
Sign of (x):	$-$	$-$	+	+
Sign of $(x-3)^2$:	+	+	+	+
Sign of $x(x+1)(x-3)^2$:	+	$-$	+	+
	-1	0	3	

The solution set is $[-1, 0]\cup\{3\}$.

43. $-5u^6+28u^5-15u^4\le 0$

$-u^4(5u^2-28u+15)\le 0$

$u^4(5u^2-28u+15)\ge 0$

Find the real zeros of the related equation.

$u^4(5u^2-28u+15)=0$

$u^4(5u-3)(u-5)=0$

$u=0$, $u=\frac{3}{5}$, or $u=5$

The boundary points are 0, $\frac{3}{5}$, and 5.

Sign of (u^4):	+	+	+	+
Sign of $(5u-3)$:	$-$	$-$	+	+
Sign of $(u-5)$:	$-$	$-$	$-$	+
Sign of $(u^4)(5u-3)(u-5)$:	+	+	$-$	+
	0	$\frac{3}{5}$	5	

The solution set is $\left(-\infty, \frac{3}{5}\right]\cup[5,\infty)$.

45. $6x(2x-5)^4(3x+1)^5(x-4)<0$

$x(2x-5)^4(3x+1)^5(x-4)<0$

Find the real zeros of the related equation.

$x(2x-5)^4(3x+1)^5(x-4)=0$

$x=0$, $x=\frac{5}{2}$, $x=-\frac{1}{3}$, or $x=4$

The boundary points are $-\frac{1}{3}$, 0, $\frac{5}{2}$, and 4.

Sign of $(3x+1)^5$:	$-$	+	+	+	+
Sign of (x):	$-$	$-$	+	+	+
Sign of $(2x-5)^4$:	+	+	+	+	+
Sign of $(x-4)$:	$-$	$-$	$-$	$-$	+
Sign of $x(2x-5)^4(3x+1)^5(x-4)$:	$-$	+	$-$	$-$	+
	$-\frac{1}{3}$	0	$\frac{5}{2}$	4	

The solution set is

$\left(-\infty, -\frac{1}{3}\right)\cup\left(0, \frac{5}{2}\right)\cup\left(\frac{5}{2}, 4\right)$.

47. $(5x-3)^2>-2$

The square of any real number is nonnegative. Therefore, $(5x-3)^2$ is nonnegative, and always greater than 0. The solution set is $(-\infty,\infty)$.

49. $-4\ge(x-7)^2$

The square of any real number is nonnegative. Therefore, $(x-7)^2$ is nonnegative, and always greater than 0. The solution set is $\{\ \}$.

51. $16y^2>24y-9$

$16y^2-24y+9>0$

$(4y-3)^2>0$

The expression $(4y-3)^2>0$ for all real numbers except where $(4y-3)^2=0$. Therefore, the solution set is all real numbers except $\frac{3}{4}$.

The solution set is $\left(-\infty, \frac{3}{4}\right)\cup\left(\frac{3}{4},\infty\right)$.

53. $(x+3)(x+1) \le -1$

$x^2+4x+3+1 \le 0$

$x^2+4x+4 \le 0$

$(x+2)^2 \le 0$

The square of any real number is nonnegative. Therefore, $(x+2)^2$ is nonnegative, and always greater than 0. However, the equality is included.
The solution set is $\{-2\}$.

55. **a.** $(2,3)$ **b.** $[2,3)$

c. $(-\infty,-2)\cup(3,\infty)$ **d.** $(-\infty,-2]\cup(3,\infty)$

57. **a.** $(-\infty,2)\cup(2,\infty)$ **b.** $(-\infty,2)\cup(2,\infty)$

c. $\{\ \}$ **d.** $\{\ \}$

59. $\dfrac{x+2}{x-3}=0$

$x+2=0$

$x=-2$

-2 is a boundary point.
The expression is undefined for $x=3$. This is also a boundary point.

Sign of $(x+2)$:	−	+	+
Sign of $(x-3)$:	−	−	+
Sign of $\dfrac{x+2}{x-3}$:	+	−	+
		−2	3

a. $\dfrac{x+2}{x-3} \le 0$
$[-2,3)$

b $\dfrac{x+2}{x-3} < 0$
$(-2,3)$

c. $\dfrac{x+2}{x-3} \ge 0$
$(-\infty,-2]\cup(3,\infty)$

d. $\dfrac{x+2}{x-3} > 0$
$(-\infty,-2)\cup(3,\infty)$

61. $\dfrac{x^4}{x^2+9}=0$

$x^4=0$

$x=0$

0 is a boundary point.
The denominator is nonzero for all real numbers. Therefore, the only boundary point is 0.

Sign of x^4:	+	+
Sign of x^2+9:	+	+
Sign of $\dfrac{x^4}{x^2+9}$:	+	+
		0

a. $\dfrac{x^4}{x^2+9} \le 0$
$\{0\}$

b $\dfrac{x^4}{x^2+9} < 0$
$\{\ \}$

c. $\dfrac{x^4}{x^2+9} \ge 0$
$(-\infty,\infty)$

d. $\dfrac{x^4}{x^2+9} > 0$
$(-\infty,0)\cup(0,\infty)$

63. $\dfrac{5-x}{x+1} \ge 0$

$\dfrac{x-5}{x+1} \le 0$

The expression is undefined for $x=-1$. This is a boundary point.
Find the real zeros of the related equation.

$\dfrac{x-5}{x+1}=0$

$x-5=0$

$x=5$

The boundary points are -1 and 5.

Sign of $(x+1)$:	−	+	+
Sign of $(x-5)$:	−	−	+
Sign of $\dfrac{x-5}{x+1}$:	+	−	+
		−1	5

The solution set is $(-1,5]$.

65. $\frac{4-2x}{x^2} \le 0$

$\frac{2x-4}{x^2} \ge 0$

$\frac{x-2}{x^2} \ge 0$

The expression is undefined for $x=0$. This is a boundary point.

Find the real zeros of the related equation.

$\frac{x-2}{x^2} = 0$

$x-2=0$

$x=2$

The boundary points are 0 and 2.

Sign of x^2:	+	+	+
Sign of $x-2$:	−	−	+
Sign of $\frac{x-2}{x^2}$:	−	−	+
		0	2

The solution set is $[2, \infty)$.

67. $\frac{x^2-x-2}{x+3} \ge 0$

$\frac{(x+1)(x-2)}{x+3} \ge 0$

The expression is undefined for $x=-3$. This is a boundary point.

Find the real zeros of the related equation.

$\frac{(x+1)(x-2)}{x+3} = 0$

$(x+1)(x-2)=0$

$x=-1$ or $x=2$

The boundary points are -3, -1, and 2.

Sign of $(x+3)$:	−	+	+	+
Sign of $(x+1)$:	−	−	+	+
Sign of $(x-2)$:	−	−	−	+
Sign of $\frac{(x+1)(x-2)}{(x+3)}$:	−	+	−	+
		−3	−1	2

The solution set is $(-3, -1] \cup [2, \infty)$.

69. $\frac{5}{2x-7} > 1$

$\frac{5}{2x-7} - 1 > 0$

$\frac{5}{2x-7} - \frac{2x-7}{2x-7} > 0$

$\frac{12-2x}{2x-7} > 0$

$\frac{-2(x-6)}{2x-7} > 0$

$\frac{x-6}{2x-7} < 0$

The expression is undefined for $x = \frac{7}{2}$. This is a boundary point.

Find the real zeros of the related equation.

$\frac{x-6}{2x-7} = 0$

$x-6=0$

$x=6$

The boundary points are $\frac{7}{2}$ and 6.

Sign of $(2x-7)$:	−	+	+
Sign of $(x-6)$:	−	−	+
Sign of $\frac{(x-6)}{(2x-7)}$:	+	−	+
		$\frac{7}{2}$	6

The solution set is $\left(\frac{7}{2}, 6\right)$.

71. $\frac{2x}{x-2} \le 2$

$\frac{2x}{x-2} - 2 \le 0$

$\frac{2x}{x-2} - 2 \cdot \frac{x-2}{x-2} \le 0$

$\frac{4}{x-2} \le 0$

The expression is undefined for $x=2$. This is a boundary point.

The related equation has no real zeros.

Sign of 4:	+	+
Sign of $(x-2)$:	−	+
Sign of $\frac{4}{(x-2)}$:	−	+
	2	

The solution set is $(-\infty, 2)$.

73. $$\frac{4-x}{x+5} \geq 2$$

$$\frac{4-x}{x+5} - 2 \geq 0$$

$$\frac{4-x}{x+5} - 2 \cdot \frac{x+5}{x+5} \geq 0$$

$$\frac{-3x-6}{x+5} \geq 0$$

$$\frac{x+2}{x+5} \leq 0$$

The expression is undefined for $x=-5$. This is a boundary point.
Find the real zeros of the related equation.

$$\frac{x+2}{x+5} = 0$$

$$x+2=0$$

$$x=-2$$

The boundary points are -5 and -2.

Sign of $(x+5)$:	−	+	+
Sign of $(x+2)$:	−	−	+
Sign of $\frac{(x+2)}{(x+5)}$:	+	−	+
	−5	−2	

The solution set is $(-5, -2]$.

75. $$\frac{x-2}{x^2+4} \leq 0$$

Find the real zeros of the related equation.

$$\frac{x-2}{x^2+4} = 0$$

$$x-2=0$$

$$x=2$$

2 is a boundary point.
The denominator is nonzero for all real numbers. Therefore, the only boundary point is 2.

Sign of $x-2$:	−	+
Sign of x^2+4:	+	+
Sign of $\frac{x-2}{x^2+4}$:	−	+
	2	

The solution set is $(-\infty, 2]$.

77. $$\frac{10}{x+2} \geq \frac{2}{x+2}$$

$$\frac{10}{x+2} - \frac{2}{x+2} \geq 0$$

$$\frac{8}{x+2} \geq 0$$

The expression is undefined for $x=-2$. This is a boundary point.

Sign of 10:	+	+
Sign of $(x+2)$:	−	+
Sign of $\frac{10}{(x+2)}$:	−	+
	−2	

The solution set is $(-2, \infty)$.

79. $$\frac{4}{x+3} > -\frac{2}{x}$$

$$\frac{4}{x+3} + \frac{2}{x} > 0$$

$$\frac{4x}{x(x+3)} + \frac{2(x+3)}{x(x+3)} > 0$$

$$\frac{6(x+1)}{x(x+3)} > 0$$

The expression is undefined for $x=0$ and $x=-3$. These are boundary points.
Find the real zeros of the related equation.

$$\frac{6(x+1)}{x(x+3)} = 0$$

$$x+1=0$$

$$x=-1$$

The boundary points are -3, -1, and 0.

Sign of $(x+3)$:	$-$	$+$	$+$	$+$
Sign of $6(x+1)$:	$-$	$-$	$+$	$+$
Sign of x:	$-$	$-$	$-$	$+$
Sign of $\frac{6(x+1)}{x(x+3)}$:	$-$	$+$	$-$	$+$
		-3	-1	0

The solution set is $(-3, -1)\cup(0, \infty)$.

81.

$$\frac{3}{4-x} \le \frac{6}{1-x}$$
$$\frac{3}{x-4} \ge \frac{6}{x-1}$$
$$\frac{3}{x-4} - \frac{6}{x-1} \ge 0$$
$$\frac{3(x-1)}{(x-4)(x-1)} - \frac{6(x-4)}{(x-4)(x-1)} \ge 0$$
$$\frac{-3x+21}{(x-4)(x-1)} \ge 0$$
$$\frac{-3(x-7)}{(x-4)(x-1)} \ge 0$$
$$\frac{3(x-7)}{(x-4)(x-1)} \le 0$$

The expression is undefined for $x=1$ and $x=4$. These are boundary points.
Find the real zeros of the related equation.

$$\frac{3(x-7)}{(x-4)(x-1)} = 0$$
$$3(x-7)=0$$
$$x=7$$

The boundary points are 1, 4, and 7.

Sign of $(x-1)$:	$-$	$+$	$+$	$+$
Sign of $(x-4)$:	$-$	$-$	$+$	$+$
Sign of $(x-7)$:	$-$	$-$	$-$	$+$
Sign of $\frac{3(x-7)}{(x-4)(x-1)}$:	$-$	$+$	$-$	$+$
		1	4	7

The solution set is $(-\infty, 1)\cup(4, 7]$.

83. $\frac{(2-x)(2x+1)^2}{(x-4)^4} \le 0$

$$\frac{(x-2)(2x+1)^2}{(x-4)^4} \ge 0$$

The expression is undefined for $x=4$. This is a boundary point.
Find the real zeros of the related equation.

$$\frac{(x-2)(2x+1)^2}{(x-4)^4} = 0$$
$$(x-2)(2x+1)^2 = 0$$
$$x=2 \text{ or } x=-\frac{1}{2}$$

The boundary points are $-\frac{1}{2}$, 2, and 4.

Sign of $(2x+1)^2$:	$+$	$+$	$+$	$+$
Sign of $(x-2)$:	$-$	$-$	$+$	$+$
Sign of $(x-4)^4$:	$+$	$+$	$+$	$+$
Sign of $\frac{(x-2)(2x+1)^2}{(x-4)^4}$:	$-$	$-$	$+$	$+$
		$-\frac{1}{2}$	2	4

The solution set is $[2, 4)\cup(4, \infty)$.

85. a. $s(t) = -\frac{1}{2}(32)t^2 + (216)t + (0)$

$s(t) = -16t^2 + 216t$

b. $t = \frac{-b}{2a} = \frac{-(216)}{2(-16)} = \frac{216}{32} = 6.75$

The shell will explode 6.75 sec after launch.

c.

$$-16t^2 + 216t > 200$$
$$-16t^2 + 216t - 200 > 0$$
$$2t^2 - 27t + 25 < 0$$

Find the real zeros of the related equation.

$$2t^2 - 27t + 25 = 0$$
$$(t-1)(2t-25) = 0$$
$$t=1 \text{ or } t=\frac{25}{2}$$

The boundary points are 1 and $\frac{25}{2}$.

Sign of $(t-1)$:	−	+	+
Sign of $(2t-25)$:	−	−	+
Sign of $(t-1)(2t-25)$:	+	−	+
		1	$\frac{25}{2}$

The solution set is $\left(1, \frac{25}{2}\right)$.

The spectators can see the shell between 1 sec and 6.75 sec after launch.

87. $d(v) = 0.06v^2 + 2v$

$$0.06v^2 + 2v < 250$$
$$0.06v^2 + 2v - 250 < 0$$
$$6v^2 + 200v - 25{,}000 < 0$$
$$3v^2 + 100v - 12{,}500 < 0$$

Find the real zeros of the related equation.

$$3v^2 + 100v - 12{,}500 = 0$$
$$(3v + 250)(v - 50) = 0$$
$$v = -\frac{250}{3} \text{ or } v = 50$$

Sign of $(3v+250)$:	−	+	+
Sign of $(v-50)$:	−	−	+
Sign of $(3v+250)(v-50)$:	+	−	+
		$-\frac{250}{3}$	50

The solution set is $\left(-\frac{250}{3}, 50\right)$.

The car will stop within 250 ft if the car is traveling less than 50 mph.

89. a. The degree of the numerator is less than the degree of the denominator. The horizontal asymptote is $y = 0$ and means that the temperature will approach 0°C as time increases without bound.

b.

$$\frac{320}{x^2 + 3x + 10} < 5$$
$$\frac{320}{x^2 + 3x + 10} - 5 < 0$$

Find the real zeros of the related equation.

$$\frac{320}{x^2 + 3x + 10} - 5 = 0$$
$$\frac{320}{x^2 + 3x + 10} = 5$$
$$5x^2 + 15x + 50 = 320$$
$$5x^2 + 15x - 270 = 0$$
$$x^2 + 3x - 54 = 0$$
$$(x+9)(x-6) = 0$$
$$x = -9 \text{ or } x = 6$$

More than 6 hr is required for the temperature to fall below 5°C.

91. Let w represent the width. Then $1.2w$ represents the length.

$$72 \le w(1.2w) \le 96$$
$$72 \le 1.2w^2 \le 96$$
$$60 \le w^2 \le 80$$
$$\sqrt{60} \le w \le \sqrt{80}$$
$$2\sqrt{15} \le w \le 4\sqrt{5}$$

The width should be between $2\sqrt{15}$ ft and $4\sqrt{5}$ ft. This is between approximately 7.7 ft and 8.9 ft.

93. $9 - x^2 \ge 0$

$$x^2 - 9 \le 0$$

Find the real zeros of the related equation.

$$x^2 - 9 = 0$$
$$(x+3)(x-3) = 0$$
$$x = 3 \text{ or } x = -3$$

The boundary points are −3 and 3.

Sign of $(x+3)$:	−	+	+
Sign of $(x-3)$:	−	−	+
Sign of $(x+3)(x-3)$:	+	−	+
		−3	3

$[-3, 3]$

95. $a^2 - 5 \geq 0$

Find the real zeros of the related equation.

$$a^2 - 5 = 0$$

$$\left(a+\sqrt{5}\right)\left(a-\sqrt{5}\right) = 0$$

$$a = -\sqrt{5} \text{ or } a = \sqrt{5}$$

The boundary points are $-\sqrt{5}$ and $\sqrt{5}$.

Sign of $\left(x+\sqrt{5}\right)$:	−	+	+
Sign of $\left(x-\sqrt{5}\right)$:	−	−	+
Sign of $\left(x+\sqrt{5}\right)\left(x-\sqrt{5}\right)$:	+	−	+
	$-\sqrt{5}$	$\sqrt{5}$	

$\left(-\infty, -\sqrt{5}\right] \cup \left[\sqrt{5}, \infty\right)$

97. $2x^2 + 9x - 18 \geq 0$

Find the real zeros of the related equation.

$$2x^2 + 9x - 18 = 0$$

$$(x+6)(2x-3) = 0$$

$$x = -6 \text{ or } x = \frac{3}{2}$$

The boundary points are -6 and $\frac{3}{2}$.

Sign of $(x+6)$:	−	+	+
Sign of $(2x-3)$:	−	−	+
Sign of $(x+6)(2x-3)$:	+	−	+
	-6	$\frac{3}{2}$	

The solution set is $(-\infty, -6] \cup \left[\frac{3}{2}, \infty\right)$.

99. $2x^2 + 9x - 18 > 0$

Find the real zeros of the related equation.

$$2x^2 + 9x - 18 = 0$$

$$(x+6)(2x-3) = 0$$

$$x = -6 \text{ or } x = \frac{3}{2}$$

The boundary points are -6 and $\frac{3}{2}$.

Sign of $(x+6)$:	−	+	+
Sign of $(2x-3)$:	−	−	+
Sign of $(x+6)(2x-3)$:	+	−	+
	-6	$\frac{3}{2}$	

The solution set is $(-\infty, -6) \cup \left(\frac{3}{2}, \infty\right)$.

101. $\dfrac{3x}{x+2} \geq 0$

The expression is undefined for $x = -2$. This is a boundary point.

Find the real zeros of the related equation.

$$\frac{3x}{x+2} = 0$$

$$3x = 0$$

$$x = 0$$

The boundary points are -2 and 0.

Sign of $(x+2)$:	−	+	+
Sign of $(3x)$:	−	−	+
Sign of $\frac{3x}{x+2}$:	+	−	+
	-2	0	

The solution set is $(-\infty, -2) \cup [0, \infty)$.

103. a. $f(x) = (x-a)^2(b-x)(x-c)^3$

Sign of $(x-a)^2$:	+	+	+	+
Sign of $(b-x)$:	+	+	−	−
Sign of $(x-c)^3$:	−	−	−	+
Sign of $(x-a)^2(b-x)(x-c)^3$:	−	−	+	−
	a	b	c	

b. (b, c)

c. $(-\infty, a) \cup (a, b) \cup (c, \infty)$

105. The solution set to the inequality $f(x) < 0$ corresponds to the values of x for which the graph of $y = f(x)$ is below the x-axis.

107. Both the numerator and denominator of the rational expression are positive for all real numbers x. Therefore, the expression cannot be negative for any real number.

109. $\sqrt{2x-6}-2<0$

The radical is real when

$2x-6 \geq 0$

$2x \geq 6$

$x \geq 3$

Find the real zeros of the related equation.

$\sqrt{2x-6}-2=0$

$\sqrt{2x-6}=2$

$2x-6=4$

$2x=10$

$x=5$

The boundary point is 5.

$f\left(\frac{7}{2}\right)=-1 \mid f\left(\frac{15}{2}\right)=1$

$f(x)<0 \mid f(x)>0$

5

The solution set is $[3, 5)$.

111. $\sqrt{4-x}-6 \geq 0$

The radical is real when

$4-x \geq 0$

$-x \geq -4$

$x \leq 4$

Find the real zeros of the related equation.

$\sqrt{4-x}-6=0$

$\sqrt{4-x}=6$

$4-x=36$

$x=-32$

The boundary point is -32.

$f(-45)=1 \mid f(-21)=-1$

$f(x)>0 \mid f(x)<0$

-32

The solution set is $(-\infty, -32]$.

113. $\frac{1}{\sqrt{x-2}-4} \leq 0$

The radical is real when

$x-2 \geq 0$

$x \geq 2$

The denominator is zero when

$\sqrt{x-2}-4=0$

$\sqrt{x-2}=4$

$x-2=16$

$x=18$

The related equation has no zeros.

The boundary point is 18.

$f(11)=-1 \mid f(27)=1$

$f(x)<0 \mid f(x)>0$

18

The solution set is $[2, 18)$.

115. $-3<x^2-6x+5 \leq 5$

Rewrite the left inequality $f(x)$.

$-3<x^2-6x+5$

$0<x^2-6x+8$

$x^2-6x+8>0$

$(x-2)(x-4)>0$

Find the real zeros of the related equation.

$(x-2)(x-4)=0$

$x=2$ or $x=4$

Rewrite the right inequality $g(x)$.

$x^2-6x+5 \leq 5$

$x^2-6x \leq 0$

$x(x-6) \leq 0$

Find the real zeros of the related equation.

$x(x-6)=0$

$x=0$ or $x=6$

The boundary points are 0, 2, 4, and 6.

$f(-1)=15$	$f(1)=3$	$f(3)=-1$	$f(5)=3$	$f(1)=15$
$f(x)>0$	$f(x)>0$	$f(x)<0$	$f(x)>0$	$f(x)>0$
$g(-1)=7$	$g(1)=-5$	$g(3)=-9$	$g(5)=-5$	$g(1)=7$
$g(x)>0$	$g(x)<0$	$g(x)<0$	$g(x)<0$	$g(x)>0$

Boundary points on the number line: 0, 2, 4, 6

The solution set is $[0, 2)\cup(4, 6]$.

117. $|x^2-4|<5$

$|x^2-4|-5<0$

Find the real zeros of the related equation.

$|x^2-4|-5=0$

$|x^2-4|=5$

$x^2-4=5$ or $x^2-4=-5$

$x^2=9$ $\quad x^2=-1$

$x=\pm 3$

The boundary points are -3 and 3.

$f(-4)=7$	$f(0)=-1$	$f(4)=7$
$f(x)>0$	$f(x)<0$	$f(x)>0$

Boundary points on the number line: -3, 3

The solution set is $(-3, 3)$.

119. $|x^2-18|>2$

$|x^2-18|-2>0$

Find the real zeros of the related equation.

$|x^2-18|-2=0$

$|x^2-18|=2$

$x^2-18=2$ or $x^2-18=-2$

$x^2=20$ $\quad x^2=16$

$x=\pm\sqrt{20}$ $\quad x=\pm 4$

$x=\pm 2\sqrt{5}$

The boundary points are $-2\sqrt{5}$, -4, 4, and $2\sqrt{5}$.

$f(-5)=5$	$f(-4.1)=-0.81$	$f(0)=16$	$f(4.1)=-0.81$	$f(5)=5$
$f(x)>0$	$f(x)<0$	$f(x)>0$	$f(x)<0$	$f(x)>0$

Boundary points on the number line: $-2\sqrt{5}$, -4, 4, $2\sqrt{5}$

The solution set is $\left(-\infty, -2\sqrt{5}\right)\cup(-4, 4)\cup\left(2\sqrt{5}, \infty\right)$.

121. a. $0.552x^3 + 4.13x^2 - 1.84x - 3.5 < 6.7$

$0.552x^3 + 4.13x^2 - 1.84x - 10.2 < 0$

b.

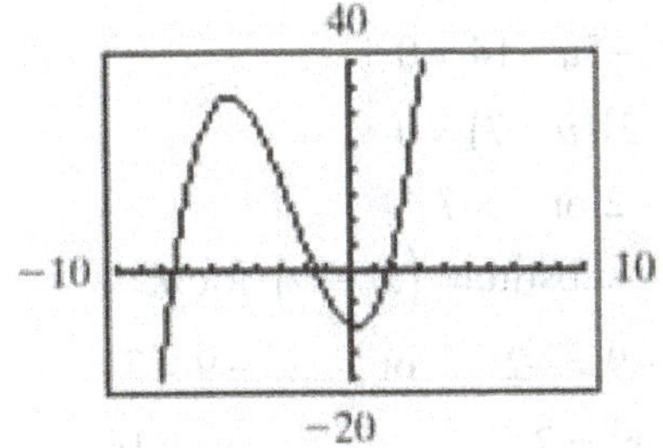

c. The real zeros are approximately -7.6, -1.5, and 1.6.

d. $(-\infty, -7.6) \cup (-1.5, 1.6)$

123. a.

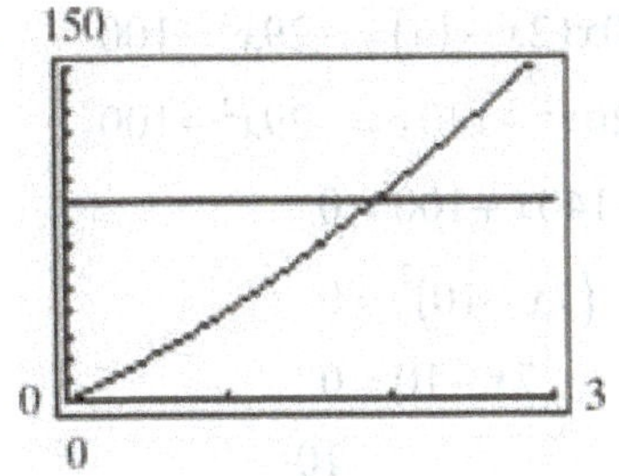

b.

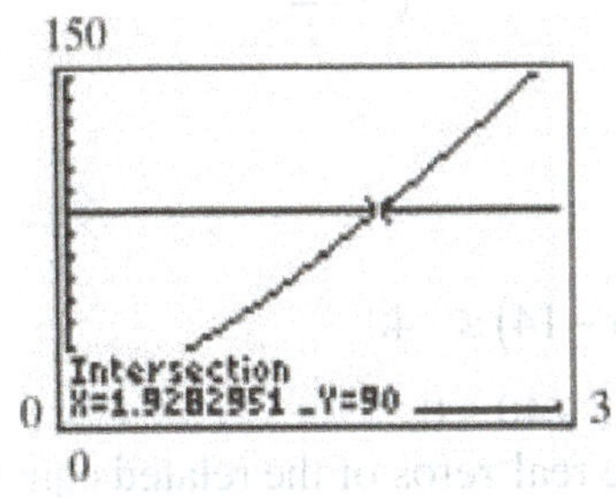

c. The radius should be no more than 1.9 in. to keep the amount of aluminum to at most 90 in.^{2}.

Problem Recognition Exercises: Solving Equations and Inequalities

1. $-\frac{1}{2} \leq -\frac{1}{4}x - 5 < 2$

$\frac{9}{2} \leq -\frac{1}{4}x < 7$

$-18 \geq x > -28$

$-28 < x \leq -18$

$(-28, -18]$

3. $50x^3 - 25x^2 - 2x + 1 = 0$

$25x^2(2x-1) - (2x-1) = 0$

$(2x-1)(25x^2 - 1) = 0$

$2x = 1$ or $25x^2 = 1$

$x = \frac{1}{2}$ $\quad x^2 = \frac{1}{25}$

$x = \pm\frac{1}{5}$

$\left\{\pm\frac{1}{5}, \frac{1}{2}\right\}$

5. $\sqrt[4]{m+4} - 5 = -2$

$\sqrt[4]{m+4} = 3$

$m + 4 = 3^4$

$m + 4 = 81$

$m = 77$

$\{77\}$

7. $|5t - 4| + 2 = 7$

$|5t - 4| = 5$

$5t - 4 = 5$ or $5t - 4 = -5$

$5t = 9$ $\quad 5t = -1$

$t = \frac{9}{5}$ $\quad t = -\frac{1}{5}$

$\left\{-\frac{1}{5}, \frac{9}{5}\right\}$

9. $10x(2x-14)=-29x^2-100$

$20x^2-140x=-29x^2-100$

$49x^2-140x+100=0$

$(7x-10)^2=0$

$7x-10=0$

$x=\frac{10}{7}$

$\left\{\frac{10}{7}\right\}$

11. $x(x-14)\le -40$

$x^2-14x+40\le 0$

Find the real zeros of the related equation.

$x^2-14x+40=0$

$(x-4)(x-10)=0$

$x=4$ or $x=10$

The boundary points are 4 and 10.

Sign of $(x-4)$:	$-$	$+$	$+$
Sign of $(x-10)$:	$-$	$-$	$+$
Sign of $(x-4)(x-10)$:	$+$	$-$	$+$
	4	10	

$[4, 10]$

13. $|x-0.15|=|x+0.05|$

$x-0.15=x+0.05$ or $x-0.15=-(x+0.05)$

$-0.15=0.05$ $\quad$ $x-0.15=-x-0.05$

$2x=0.1$

$x=0.05$

$\{0.05\}$

15. $n^{1/2}+7=10$

$n^{1/2}=3$

$n=9$

$\{9\}$

17. $-2x-5(x+3)=-4(x+2)-3x$

$-2x-5x-15=-4x-8-3x$

$-7x-15=-7x-8$

$-15=8$ No solution

$\{\ \}$

19. $(x^2-9)^2-5(x^2-9)-14=0$

Let $u=(x^2-9)$.

$u^2-5u-14=0$

$(u+2)(u-7)=0$

$u=-2$ or $u=7$

Back substitute (x^2-9) for u.

$x^2-9=-2$ or $x^2-9=7$

$x^2=7$ $\quad$ $x^2=16$

$x=\pm\sqrt{7}$ $\quad$ $x=\pm 4$

$\{\pm 4, \pm\sqrt{7}\}$

21. $|8x-3|+10\le 7$

$|8x-3|\le -3$

Since the absolute value of a number cannot be negative, this inequality has no solution.

$\{\ \}$

23. $x^3-3x^2<6x-8$

$x^3-3x^2-6x+8<0$

Find the real zeros of the related equation.

$x^3-3x^2-6x+8=0$

$\frac{\text{Factors of } -8}{\text{Factors of } 1}=\frac{\pm1, \pm2, \pm4, \pm8}{\pm1}$

$=\pm1, \pm2, \pm4, \pm8$

$$\begin{array}{r|rrrr} 1 & 1 & -3 & -6 & 8 \\ & & 1 & -2 & -8 \\ \hline & 1 & -2 & -8 & \underline{|0} \end{array}$$

$(x-1)(x^2-2x-8)=0$

$(x-1)(x+2)(x-4)=0$

$x=1$, $x=-2$, or $x=4$

The boundary points are -2, 1, and 4.

Sign of $(x+2)$:	$-$	$+$	$+$	$+$
Sign of $(x-1)$:	$-$	$-$	$+$	$+$
Sign of $(x-4)$:	$-$	$-$	$-$	$+$
Sign of $(x-1)(x+2)(x-4)$:	$-$	$+$	$-$	$+$
	-2	1	4	

$(-\infty, -2)\cup(1, 4)$

25. $15-3(x-1)=-2x-(x-18)$
$$15-3x+3=-2x-x+18$$
$$-3x+18=-3x+18$$
$$0=0$$
$(-\infty, \infty)$

27.
$$2<|3-x|-9$$
$$11<|3-x|$$
$$|3-x|>11$$
$$3-x<-11 \quad \text{or} \quad 3-x>11$$
$$-x<-14 \qquad -x>8$$
$$x>14 \qquad x<-8$$
$(-\infty,-8)\cup(14,\infty)$

29. $\frac{1}{3}x+\frac{2}{5}>\frac{5}{6}x-1$
$$10x+12>25x-30$$
$$-15x>-42$$
$$x<\frac{42}{15}$$
$$x<\frac{14}{5}$$
$\left(-\infty,\frac{14}{5}\right)$

Section 3.7 Variation

1.
$$\frac{9}{4}=\frac{3}{32}k$$
$$\frac{9}{4}\left(\frac{32}{3}\right)=k$$
$$24=k$$
$\{24\}$

3.
$$0.24=\frac{1.68}{\sqrt{x}}$$
$$0.24\sqrt{x}=1.68$$
$$\sqrt{x}=7$$
$$x=49$$
$\{49\}$

5. directly

7. constant; variation

9. a. $y=2x$ $\quad y=2(1)=2$
$$y=2(2)=4 \qquad y=2(3)=6$$
$$y=2(4)=8 \qquad y=2(5)=10$$

b. y is also doubled. **c.** y is also tripled.

d. increases **e.** decreases

11. $C=kr$

13. $\overline{C}=\frac{k}{n}$

15. $V=khr^2$

17. $E=\frac{ks}{\sqrt{n}}$

19. $c=\frac{kmn}{t^3}$

21.
$$y=kx$$
$$20=k(8)$$
$$\frac{20}{8}=k$$
$$k=\frac{5}{2}$$

23.
$$p=\frac{k}{q}$$
$$54=\frac{k}{18}$$
$$k=972$$

25. $y = kwv$

$40 = k(40)(0.2)$

$40 = k(8)$

$k = 5$

27. a. $y = kx$

$4 = k(10)$

$\frac{4}{10} = k$

$k = \frac{2}{5}$

$y = \frac{2}{5}x = \frac{2}{5}(5) = 2$

b. $y = \frac{k}{x}$

$4 = \frac{k}{10}$

$k = 40$

$y = \frac{40}{x} = \frac{40}{5} = 8$

29. Let A represent the amount of the medicine. Let w represent the weight of the child.

$A = kw$

$180 = k(40)$

$\frac{180}{40} = k$

$k = 4.5$

$A = 4.5w$

a. $A = 4.5w = 4.5(50) = 225$ mg

b. $A = 4.5w = 4.5(60) = 270$ mg

c. $A = 4.5w = 4.5(70) = 315$ mg

d. $A = 4.5w$

$135 = 4.5w$

$\frac{135}{4.5} = w$

$w = 30$ lb

31. Let C represent the average daily cost to rent a car. Let m represent the number of miles driven.

$C = \frac{k}{m}$

$0.8 = \frac{k}{100}$

$k = 80$

$C = \frac{80}{m}$

a. $C = \frac{80}{m} = \frac{80}{200} = \$0.40$ per mile

b. $C = \frac{80}{m} = \frac{80}{300} \approx \$0.27$ per mile

c. $C = \frac{80}{m} = \frac{80}{400} = \$0.20$ per mile

d. $C = \frac{80}{m}$

$0.16 = \frac{80}{m}$

$0.16m = 80$

$m = \frac{80}{0.16} = 500$ mi

33. Let d represent the distance a bicycle travels in 1 min. Let r represent the rpm of the wheels.

$d = kr$

$440 = k(60)$

$\frac{440}{60} = k$

$k = \frac{22}{3}$

$d = \frac{22}{3}r = \frac{22}{3}(87) = 638$ ft

35. a. Let d represent the stopping distance of the car. Let s represent the speed of the car.

$$d = ks^2$$
$$170 = k(50)^2$$
$$\frac{170}{2500} = k$$
$$k = 0.068$$
$$d = 0.068s^2 = 0.068(70)^2 = 333.2 \text{ ft}$$

b.

$$d = 0.068s^2$$
$$244.8 = 0.068s^2$$
$$\frac{244.8}{0.068} = s^2$$
$$3600 = s^2$$
$$s = 60 \text{ mph}$$

37. Let t represent the number of days to complete the job. Let n represent the number of people working on the job.

$$t = \frac{k}{n}$$
$$12 = \frac{k}{8}$$
$$k = 96$$
$$t = \frac{96}{n}$$

a. $t = \frac{96}{n} = \frac{96}{15} = 6.4$ days

b.

$$t = \frac{96}{n}$$
$$8 = \frac{96}{n}$$
$$8n = 96$$
$$n = 12 \text{ people}$$

39. Let I represent the current. Let V represent the voltage. Let R represent the resistance.

$$I = \frac{kV}{R}$$
$$9 = \frac{k(90)}{10}$$
$$90 = 90k$$
$$k = 1$$
$$I = \frac{V}{R} = \frac{160}{5} = 32 \text{ A}$$

41. Let I represent the amount of interest owed. Let P represent the amount of the principal borrowed. Let t represent the amount of time (in years) that the money is borrowed.

$$I = kPt$$
$$480 = k(4000)(2)$$
$$480 = 8000k$$
$$\frac{480}{8000} = k$$
$$k = 0.06$$
$$I = 0.06(6000)(4) = \$1440$$

43. Let B represent the BMI of an individual. Let w represent the weight of the individual. Let h represent the weight of the individual.

$$B = \frac{kw}{h^2}$$
$$21.52 = \frac{k(150)}{(70)^2}$$
$$105{,}448 = 150k$$
$$\frac{105{,}448}{150} = k$$
$$k = 703$$
$$B = \frac{703(180)}{(68)^2} \approx 27.37$$

45. Let s represent the speed of the canoe. Let ℓ represent the length of the canoe.

$$s = k\sqrt{\ell}$$
$$6.2 = k\sqrt{16}$$
$$\frac{6.2}{\sqrt{16}} = k$$
$$\frac{6.2}{4} = k$$
$$k = 1.55$$
$$s = 1.55\sqrt{\ell} = 1.55\sqrt{25} = 7.75 \text{ mph}$$

47. Let C represent the cost to carpet the room. Let ℓ represent the length of the room. Let w represent the width of the room.

$$C = k\ell w$$
$$3870 = k(10)(15)$$
$$\frac{3870}{150} = k$$
$$k = 25.8$$
$$C = k\ell w = 25.8(18)(24) = \$11{,}145.60$$

49. The data appears to show direct variation. Use the first row to find the constant of variation.

$$y = kx$$
$$16 = k(5)$$
$$\frac{16}{5} = k$$
$$k = 3.2$$
$$y = 3.2x$$

Check the remaining values in the equation. They check.

51. The data appears to show indirect variation. Use the first row to find the constant of variation.

$$y = \frac{k}{x}$$
$$6 = \frac{k}{2}$$
$$k = 12$$
$$y = \frac{12}{x}$$

Check the remaining values in the equation. They check.

53. Make a table of values.

x	1	3
y	3	1

The data appears to show indirect variation. Use the first point to find the constant of variation.

$$y = \frac{k}{x}$$
$$3 = \frac{k}{1}$$
$$k = 3$$
$$y = \frac{3}{x}$$

Check the other point in the equation. It checks.

55. formulas a and c

57. The variable P varies directly as the square of v and inversely as t.

59. **a.** Let S represent the surface area of the sphere. Let r represent the radius of the sphere.

$$S = kr^2$$
$$400\pi = k(10)^2$$
$$\frac{400\pi}{100} = k$$
$$k = 4\pi$$
$$S = 4\pi r^2 = 4\pi(20)^2 = 1600\pi \text{ m}^2$$

b. The surface area is 4 times as great. Doubling the radius results in $(2)^2$ times the surface area of the sphere.

c. The intensity at 20 m should be $\frac{1}{4}$ the intensity at 10 m. This is because the energy from the light is distributed across an area 4 times as great.

d. Let I represent the intensity of light from a source. Let d represent the distance of the light from the source.

$$I = \frac{k}{d^2}$$

$$200 = \frac{k}{(10)^2}$$

$$k = 20{,}000$$

$$I = \frac{20{,}000}{d^2} = \frac{20{,}000}{(20)^2} = 50 \text{ lux}$$

61. $I = \frac{k}{d^2}$

$$I = \frac{k}{(10d)^2} = \frac{k}{100d^2} = \left(\frac{1}{100}\right)\left(\frac{k}{d^2}\right)$$

The intensity is $\frac{1}{100}$ as great.

63. $y = \frac{kx^2}{w^4}$

$$y = \frac{k(2x)^2}{(2w)^4} = \frac{k(4x^2)}{(16w^4)} = \frac{1}{4}\left(\frac{kx^2}{w^4}\right)$$

y will be $\frac{1}{4}$ its original value.

65. $y = kxw^3$

$$y = k\left(\frac{x}{3}\right)(3w)^3 = k\left(\frac{x}{3}\right)(27w^3) = 9(kxw^3)$$

y will be 9 times its original value.

Chapter 3 Review Exercises

1. $f(x) = -(x+5)^2 + 2 = -[x-(-5)]^2 + 2$

Vertex: $(h, k) = (-5, 2)$

3. a. $f(x) = -2x^2 + 4x + 6$

$$f(x) = -2(x^2 - 2x \quad) + 6$$

$$\boxed{\left[\frac{1}{2}(-2)\right]^2 = 1}$$

$$f(x) = -2(x^2 - 2x + 1 - 1) + 6$$

$$f(x) = -2(x^2 - 2x + 1) - 2(-1) + 6$$

$$f(x) = -2(x-1)^2 + 8$$

b. Since $a < 0$, the parabola opens downward.

c. The vertex is $(h, k) = (1, 8)$.

d. $f(x) = -2x^2 + 4x + 6$

$$0 = -2x^2 + 4x + 6$$

$$0 = x^2 - 2x - 3$$

$$0 = (x+1)(x-3)$$

$$x = -1 \quad \text{or} \quad x = 3$$

The x-intercepts are $(-1, 0)$ and $(3, 0)$.

e. $f(0) = -2(0)^2 + 4(0) + 6 = 6$

The y-intercept is $(0, 6)$.

f.

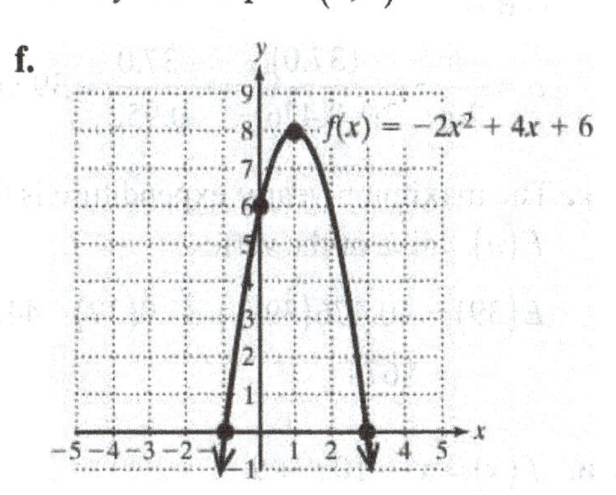

g. The axis of symmetry is the vertical line through the vertex: $x = 1$.

h. The maximum value is 8.

i. The domain is $(-\infty, \infty)$.

The range is $(-\infty, 8]$.

5. We know that

$$2x + y = 180$$
$$y = 180 - 2x$$

Let A represent the area.

$$A = xy$$
$$A(x) = x(180 - 2x)$$
$$= 180x - 2x^2$$
$$= -x^2 + 180x$$

Function A is a quadratic function with a negative leading coefficient. The graph of the parabola opens downward, so the vertex is the maximum point on the function. The x-coordinate of the vertex is the value of x that will maximize the area.

$$x = \frac{-b}{2a} = \frac{-(180)}{2(-2)} = \frac{-180}{-4} = 45$$

$$y = 180 - 2x = 180 - 2(45) = 180 - 90 = 90$$

The dimensions are 45 yd by 90 yd.

7 a.

```
QuadReg
y=ax²+bx+c
a=-.4755357143
b=37.04678571
c=-44.61428571
```

$$E(a) = -0.476a^2 + 37.0a - 44.6$$

b. The age when the yearly expenditure is the greatest is the a-coordinate of the vertex.

$$a = \frac{-b}{2a} = \frac{-(37.0)}{2(-0.476)} = \frac{-37.0}{-0.952} \approx 39 \text{ yr}$$

c. The maximum yearly expenditure is the $E(a)$ value at the vertex.

$$E(39) = -0.476(39)^2 + 37.0(39) - 44.6$$
$$\approx \$674$$

9. a. $f(x) = x^4 - 10x^2 + 9$

The leading term is x^4. The end behavior is up to the left and up to the right.

b. $0 = x^4 - 10x^2 + 9$

$$0 = (x^2 - 9)(x^2 - 1)$$
$$0 = (x - 3)(x + 3)(x - 1)(x + 1)$$
$$x = 3, x = -3, x = 1, x = -1$$

The zeros of the function are 3, -3, 1, and -1 (each with multiplicity 1).

c. The x-intercepts are $(3, 0)$, $(-3, 0)$, $(1, 0)$, and $(-1, 0)$.

d. $f(0) = (0)^4 - 10(0)^2 + 9 = 9$

The y-intercept is $(0, 9)$.

e. $f(-x) = (-x)^4 - 10(-x)^2 + 9$

$$= x^4 - 10x^2 + 9$$

$f(x)$ is even.

f.

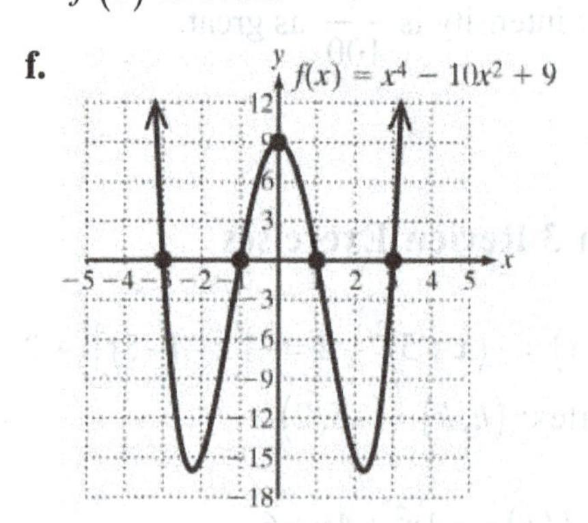

11. a. $f(x) = x^5 - 8x^4 + 13x^3$

The leading term is x^5. The end behavior is down to the left and up to the right.

b. $0 = x^5 - 8x^4 + 13x^3$

$$0 = x^3(x^2 - 8x + 13)$$
$$x = \frac{-(-8) \pm \sqrt{(-8)^2 - 4(1)(13)}}{2(1)}$$
$$x = \frac{8 \pm \sqrt{12}}{2} = \frac{8 \pm 2\sqrt{3}}{2} = 4 \pm \sqrt{3}$$
$$x = 0, x = 4 + \sqrt{3}, x = 4 - \sqrt{3}$$

The zeros of the function are 0 (with multiplicity 3), $4 + \sqrt{3}$ and $4 - \sqrt{3}$ (each with multiplicity 1).

c. The x-intercepts are $(0,0)$, $\left(4+\sqrt{3},0\right)$, and $\left(4-\sqrt{3},0\right)$.

d. $f(0)=(0)^5-8(0)^4+13(0)^3=0$
The y-intercept is $(0,0)$.

e. $f(-x)=(-x)^5-8(-x)^4+13(-x)^3$
$=-x^5-8x^4-13x^3$
$f(x)$ is neither even nor odd.

f.

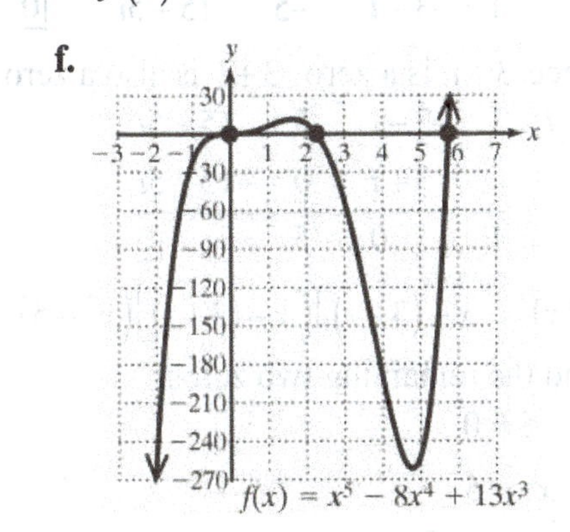

13. False

15. False

17. a.

$$
\begin{array}{r}
-2x^2+3x-9 \\
x^2+x-3\overline{\big)-2x^4+x^3+0x^2+4x-1} \\
-\left(-2x^4-2x^3+6x^2\right) \\
\hline
3x^3-6x^2+4x \\
-\left(3x^3+3x^2-9x\right) \\
\hline
-9x^2+13x-1 \\
-\left(-9x^2-9x+27\right) \\
\hline
22x-28
\end{array}
$$

$$-2x^2+3x-9+\frac{22x-28}{x^2+x-3}$$

b. Dividend: $-2x^4+x^3+4x-1$;
Divisor: x^2+x-3;
Quotient: $-2x^2+3x-9$;
Remainder: $22x-28$

19.

$$
\begin{array}{r|rrrrrr}
-2 & 2 & 0 & 0 & 1 & -5 & 1 \\
 & & -4 & 8 & -16 & 30 & -50 \\
\hline
 & 2 & -4 & 8 & -15 & 25 & \lfloor -49
\end{array}
$$

$$2x^4-4x^3+8x^2-15x+25+\frac{-49}{x+2}$$

21.

$$
\begin{array}{r|rrrrr}
-2 & 3 & 0 & 2 & -4 & 1 \\
 & & -6 & 12 & -28 & 64 \\
\hline
 & 3 & -6 & 14 & -32 & \lfloor 65
\end{array}
$$

By the remainder theorem, $f(-2)=65$.

23. a.

$$
\begin{array}{r|rrrrr}
2 & 3 & 13 & 2 & 52 & -40 \\
 & & 6 & 38 & 80 & 264 \\
\hline
 & 3 & 19 & 40 & 132 & \lfloor 224
\end{array}
$$

By the factor theorem, since $f(2)\neq 0$, $x-2$ is not a factor of $f(x)$.

b.

$$
\begin{array}{r|rrrrr}
\frac{2}{3} & 3 & 13 & 2 & 52 & -40 \\
 & & 2 & 10 & 8 & 40 \\
\hline
 & 3 & 15 & 12 & 60 & \lfloor 0
\end{array}
$$

By the remainder theorem, $f\left(\frac{2}{3}\right)=0$.

Since $f\left(\frac{2}{3}\right)=0$, $\frac{2}{3}$ is a zero of $f(x)$.

25. a.

$$
\begin{array}{r|rrrr}
-4 & 1 & 4 & 9 & 36 \\
 & & -4 & 0 & -36 \\
\hline
 & 1 & 0 & 9 & \lfloor 0
\end{array}
$$

By the factor theorem, since $f(-4)=0$, $x+4$ is a factor of $f(x)$.

b.

$$
\begin{array}{r|rrrr}
3i & 1 & 4 & 9 & 36 \\
 & & 3i & -9+12i & -36 \\
\hline
 & 1 & 4+3i & 12i & \lfloor 0
\end{array}
$$

By the factor theorem, since $f(3i)=0$, $x-3i$ is a factor of $f(x)$.

27. $\underline{\tfrac{2}{3}}|$ 15 −67 26 8

	15	−67	26	8
$\frac{2}{3}$		10	−38	−8
	15	−57	−12	0

$$f(x)=a\left(x-\frac{2}{3}\right)\left(15x^2-57x-12\right)$$
$$=3\left(x-\frac{2}{3}\right)\left(5x^2-19x-4\right)$$
$$=(3x-2)(5x+1)(x-4)$$

29. $f(x)=a\left(x-\frac{1}{4}\right)\left(x+\frac{1}{2}\right)(x-3)$
$$=8\left(x-\frac{1}{4}\right)\left(x+\frac{1}{2}\right)(x-3)$$
$$=4\left(x-\frac{1}{4}\right)\cdot 2\left(x+\frac{1}{2}\right)(x-3)$$
$$=(4x-1)(2x+1)(x-3)$$
$$=(4x-1)\left(2x^2-5x-3\right)$$
$$=8x^3-20x^2-12x-2x^2+5x+3$$
$$=8x^3-22x^2-7x+3$$

31. a. $f(x)$ is a fourth-degree polynomial. It has 4 zeros.

b. $\dfrac{\text{Factors of } -8}{\text{Factors of } 1}=\dfrac{\pm 1, \pm 2, \pm 4, \pm 8}{\pm 1}$
$$=\pm 1, \pm 2, \pm 4, \pm 8$$

c.

	1	4	2	−8	−8
−2		−2	−4	4	8
	1	2	−2	−4	0

Factor the quotient.

	1	2	−2	−4
−2		−2	0	4
	1	0	−2	0

The rational zero is −2 (with multiplicity 2).

d. Find the remaining two zeros.
$$x^2-2=0$$
$$x^2=2$$
$$x=\pm\sqrt{2}$$
The zeros are −2 (with multiplicity 2), $\pm\sqrt{2}$.

33. a.

	1	−6	5	30	−50
$3-i$		$3-i$	−10	$-15+5i$	50
	1	$-3-i$	−5	$15+5i$	0

Since $3-i$ is a zero, $3+i$ is also a zero.

	1	$-3-i$	−5	$15+5i$
$3+i$		$3+i$	0	$-15-5i$
	1	0	−5	0

$$f(x)=[x-(3-i)][x-(3+i)]\left(x^2-5\right)$$
Find the remaining two zeros.
$$x^2-5=0$$
$$x^2=5$$
$$x=\pm\sqrt{5}$$
The zeros are $3\pm i$, $\pm\sqrt{5}$.

b. $f(x)=$
$$[x-(3-i)][x-(3+i)]\left(x-\sqrt{5}\right)\left(x+\sqrt{5}\right)$$

c. The solution set is $\left\{3\pm i, \pm\sqrt{5}\right\}$.

35. $f(x)=a(x-2i)(x+2i)\left(x-\frac{5}{3}\right)$
$$=a\left(x^2+4\right)\left(x-\frac{5}{3}\right)$$
$$=a\left(x^3-\frac{5}{3}x^2+4x-\frac{20}{3}\right) \quad \text{Let } a=3.$$
$$=3\left(x^3-\frac{5}{3}x^2+4x-\frac{20}{3}\right)$$
$$=3x^3-5x^2+12x-20$$

37. $n(x) = x^6 + \frac{1}{3}x^4 + \frac{2}{7}x^3 + 4x^2 + 3$

0 sign changes in $n(x)$. The number of possible positive real zeros is 0.

$n(-x) = (-x)^6 + \frac{1}{3}(-x)^4 + \frac{2}{7}(-x)^3 + 4(-x)^2 + 3$

$n(-x) = x^6 + \frac{1}{3}x^4 - \frac{2}{7}x^3 + 4x^2 + 3$

2 sign changes in $n(-x)$. The number of possible negative real zeros is either 2 or 0.

39. a.

5	1	−4	2	1
		5	5	35
	1	1	7	36

The remainder and all coefficients of the quotient are nonnegative. Therefore, 5 is an upper bound for the real zeros of $f(x)$.

b.

−2	1	−4	2	1
		−2	12	−28
	1	−6	14	−27

The signs of the quotient alternate. Therefore, −2 is a lower bound for the real zeros of $f(x)$.

41. $2x^2 + x - 15 = 0$

$(2x-5)(x+3) = 0$

$x = \frac{5}{2}, x = -3$

43. a. The degree of the numerator is 0.
The degree of the denominator is 2.
Since $n < m$, the line $y = 0$ is a horizontal asymptote of *r*.

b. $\frac{3}{x^2+2x+1} = 0$

$3 = 0$ No solution

The graph does not cross $y = 0$.

45. a. The degree of the numerator is 3.
The degree of the denominator is 1.
Since $n > m$, *k* has no horizontal asymptotes.

b. Not applicable

47. The expression $\frac{-4x^2+5}{3x^2-14x-5}$ is in lowest terms.

$3x^2 - 14x - 5 = 0$

$(3x+1)(x-5) = 0$

$x = -\frac{1}{3}, x = 5$

n has vertical asymptotes at $x = -\frac{1}{3}$ and $x = 5$.
The degree of the numerator is 2.
The degree of the denominator is 2.
Since $n = m$, the line $y = \frac{-4}{3}$ or equivalently $y = -\frac{4}{3}$ is a horizontal asymptote of *n*.

49. $k(0) = \frac{(0)^2}{(0)^2 - (0) - 12} = 0$

The *y*-intercept is $(0, 0)$.
The *x*-intercept is $(0, 0)$.
k is in lowest terms, and $x^2 - x - 12 = (x+3)(x-4)$ is 0 for $x = -3$ and $x = 4$, which are the vertical asymptotes.
The degree of the numerator is 2.
The degree of the denominator is 2.
Since $n = m$, the line $y = \frac{1}{1}$, or equivalently $y = 1$, is a horizontal asymptote of *k*.

$1 = \frac{x^2}{x^2 - x - 12}$

$x^2 - x - 12 = x^2$

$x = -12$

k crosses its horizontal asymptote at $(-12, 1)$.

$k(-x) = \frac{(-x)^2}{(-x)^2 - (-x) - 12} = \frac{x^2}{x^2 + x - 12}$

$k(-x) \neq k(x)$, $k(-x) \neq -k(x)$

k is neither even nor odd.

Interval	Test Point	Comments
$(-\infty, -12)$	$\left(-14, \frac{98}{99}\right)$	$k(x)$ is less than 1. $k(x)$ must approach the horizontal asymptote $y=1$ from below as $x \to -\infty$.
$(-12, -3)$	$(-4, 2)$	$k(x)$ is positive. As x approaches the vertical asymptote $x=-3$ from the left, $k(x) \to \infty$.
$(-3, 0)$	$\left(-2, -\frac{2}{3}\right)$	$k(x)$ is negative. As x approaches the vertical asymptote $x=-3$ from the right, $k(x) \to -\infty$.
$(0, 4)$	$\left(3, -\frac{3}{2}\right)$	$k(x)$ is negative. As x approaches the vertical asymptote $x=4$ from the left, $k(x) \to -\infty$.
$(4, \infty)$	$(6, 2)$	$k(x)$ is positive. As x approaches the vertical asymptote $x=4$ from the right, $k(x) \to \infty$. $k(x)$ is greater than 1. $k(x)$ must approach the horizontal asymptote $y=1$ from above as $x \to \infty$.

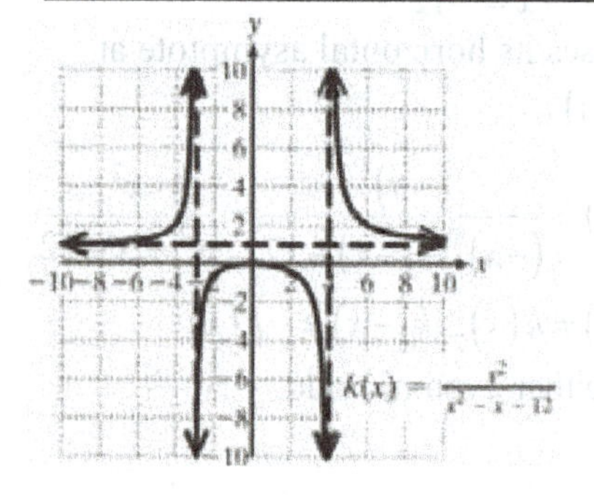

51. $q(0) = \frac{12}{(0)^2+6} = \frac{12}{6} = 2$

The y-intercept is $(0, 2)$.

q is never 0. It has no x-intercepts nor vertical asymptotes.

The degree of the numerator is 0.

The degree of the denominator is 2.

Since $n < m$, the line $y=0$ is a horizontal asymptote of q.

$0 = \frac{12}{x^2+6}$

$0 = 12$ No solution

q does not cross its horizontal asymptote.

$q(-x) = \frac{12}{(-x)^2+6} = \frac{12}{x^2+6}$

$q(-x) = q(x)$

q is even.

Interval	Test Point	Comments
$(-\infty, 0)$	$\left(-3, \frac{4}{5}\right)$	$q(x)$ is positive. $q(x)$ must approach the horizontal asymptote $y=0$ from above as $x \to -\infty$.
$(0, \infty)$	$\left(2, \frac{6}{5}\right)$	$q(x)$ is positive. $q(x)$ must approach the horizontal asymptote $y=0$ from above as $x \to \infty$.

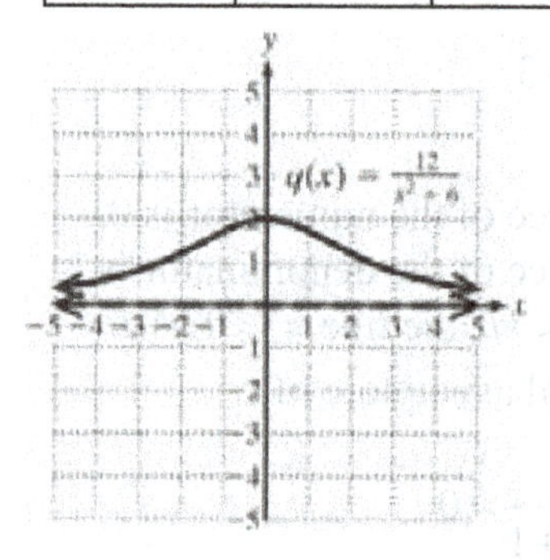

53. a. $(-\infty, -4)$

b. $(-\infty, -4] \cup \{1\}$

c. $(-4, 1) \cup (1, \infty)$

d. $[-4, \infty)$

55. $x^2+7x+10=0$

$(x+5)(x+2)=0$

$x=-5$ or $x=-2$

The boundary points are -5 and -2.

Sign of $(x+5)$:	$-$	$+$	$+$
Sign of $(x+2)$:	$-$	$-$	$+$
Sign of $(x+5)(x+2)$:	$+$	$-$	$+$
		-5	-2

a. $x^2+7x+10=0$

$\{-5,-2\}$

b. $x^2+7x+10<0$

$(-5,-2)$

c. $x^2+7x+10\le 0$

$[-5,-2]$

d. $x^2+7x+10>0$

$(-\infty,-5)\cup(-2,\infty)$

e. $x^2+7x+10\ge 0$

$(-\infty,-5]\cup[-2,\infty)$

57. $t(t-3)\ge 18$

$t^2-3t-18\ge 0$

Find the real zeros of the related equation.

$t^2-3t-18=0$

$(t+3)(t-6)=0$

$t=-3$ or $t=6$

The boundary points are -3 and 6.

Sign of $(t+3)$:	$-$	$+$	$+$
Sign of $(t-6)$:	$-$	$-$	$+$
Sign of $(t+3)(t-6)$:	$+$	$-$	$+$
		-3	6

The solution set is $(-\infty,-3]\cup[6,\infty)$

59. $x^2-2x+4\le 3$

$x^2-2x+1\le 0$

$(x-1)^2\le 0$

The square of any real number is nonnegative. Therefore, $(x-1)^2$ is nonnegative, and always greater than 0. However, the equality is included.

The solution set is $\{1\}$.

61. $z^3-3z^2>10z-24$

$z^3-3z^2-10z+24>0$

Find the real zeros of the related equation.

$z^3-3z^2-10z+24=0$

$$\frac{\text{Factors of 24}}{\text{Factors of 1}}=\frac{\pm1,\pm2,\pm3,\pm4,\pm6,\pm8,\pm12,\pm24}{\pm1}$$

$=\pm1,\pm2,\pm3,\pm4,\pm6,\pm8,\pm12,\pm24$

$$\begin{array}{r|rrrr} 2 & 1 & -3 & -10 & 24 \\ & & 2 & -2 & -24 \\ \hline & 1 & -1 & -12 & \underline{|0} \end{array}$$

$(x-2)(x^2-x-12)=0$

$(x-2)(x+3)(x-4)=0$

$x=2$, $x=-3$, or $x=4$

The boundary points are -3, 2, and 4.

Sign of $(x+3)$:	$-$	$+$	$+$	$+$
Sign of $(x-2)$:	$-$	$-$	$+$	$+$
Sign of $(x-4)$:	$-$	$-$	$-$	$+$
Sign of $(x-2)(x+3)(x-4)$:	$-$	$+$	$-$	$+$
		-3	2	4

The solution set is $(-3,2)\cup(4,\infty)$.

63. $\dfrac{6-2x}{x^2}\ge 0$

$\dfrac{2x-6}{x^2}\le 0$

$\dfrac{x-3}{x^2}\le 0$

The expression is undefined for $x=0$. This is a boundary point.

Find the real zeros of the related equation.

$\dfrac{x-3}{x^2}=0$

$x-3=0$

$x=3$

The boundary points are 0 and 3.

Sign of x^2:	+	+	+
Sign of $x-3$:	−	−	+
Sign of $\frac{x-3}{x^2}$:	−	−	+
		0	3

The solution set is $(-\infty, 0) \cup (0, 3]$.

65. $$\frac{3}{x-2} < -\frac{2}{x}$$

$$\frac{3}{x-2} + \frac{2}{x} < 0$$

$$\frac{3x}{x(x-2)} + \frac{2(x-2)}{x(x-2)} < 0$$

$$\frac{5x-4}{x(x-2)} < 0$$

The expression is undefined for $x = 0$ and $x = 2$. These are boundary points.
Find the real zeros of the related equation.

$$\frac{5x-4}{x(x-2)} = 0$$

$$5x - 4 = 0$$

$$x = \frac{4}{5}$$

The boundary points are 0, $\frac{4}{5}$, and 2.

Sign of (x):	−	+	+	+
Sign of $6(5x-4)$:	−	−	+	+
Sign of $(x-2)$:	−	−	−	+
Sign of $\frac{6(x+1)}{x(x+3)}$:	−	+	−	+
		0	$\frac{4}{5}$	2

The solution set is $(-\infty, 0) \cup \left(\frac{4}{5}, 2\right)$.

67. a. $$\overline{C}(x) = \frac{80 + 40 + 15x}{x}$$

$$\overline{C}(x) = \frac{120 + 15x}{x}$$

b. $$\frac{120 + 15x}{x} < 16$$

$$\frac{120 + 15x}{x} - 16 < 0$$

$$\frac{120 + 15x - 16x}{x} < 0$$

$$\frac{120 - x}{x} < 0$$

$$\frac{x - 120}{x} > 0$$

The expression is undefined for $x = 0$.
This is a boundary point.
Find the real zeros of the related equation

$$\frac{x-120}{x} = 0$$

$$x - 120 = 0$$

$$x = 120$$

The boundary points are 0 and 120.

Sign of $(x-120)$:	−	+	+
Sign of (x):	−	−	+
Sign of $\frac{x-120}{x}$:	+	−	+
		0	120

The solution set is $(-\infty, 0) \cup (120, \infty)$.
The trainer must have more than 120 sessions with his clients for his average cost to drop below $16 per session.

69. $m = kw$

71. $y = \frac{kx\sqrt{z}}{t^3}$

73. $$d = \frac{kc}{x^2}$$

$$1.8 = \frac{k(3)}{(2)^2}$$

$$7.2 = 3k$$

$$k = 2.4$$

75. $F = \frac{k}{L}$

$6.25 = \frac{k}{1.6}$

$k = 10$

$F = \frac{10}{L} = \frac{10}{2} = 5 \text{ lb}$

77. $F = \frac{kq_1q_2}{d^2}$

$F = \frac{k(2q_1)(2q_2)}{\left(\frac{d}{2}\right)^2} = \frac{4kq_1q_2}{\frac{d^2}{4}} = 16\left(\frac{kq_1q_2}{d^2}\right)$

The force will be 16 times as great.

Chapter 3 Test

1. a. $f(x) = 2x^2 - 12x + 16$

$f(x) = 2(x^2 - 6x \quad) + 16$

$\boxed{\left[\frac{1}{2}(-6)\right]^2 = 9}$

$f(x) = 2(x^2 - 6x + 9 - 9) + 16$

$f(x) = 2(x^2 - 6x + 9) + 2(-9) + 16$

$f(x) = 2(x - 3)^2 - 2$

b. Since $a > 0$, the parabola opens upward.

c. The vertex is $(h, k) = (3, -2)$.

d. $f(x) = 2x^2 - 12x + 16$

$0 = 2x^2 - 12x + 16$

$0 = x^2 - 6x + 8$

$0 = (x - 2)(x - 4)$

$x = 2 \quad \text{or} \quad x = 4$

The x-intercepts are $(2, 0)$ and $(4, 0)$.

e. $f(0) = 2(0)^2 - 12(0) + 16 = 16$

The y-intercept is $(0, 16)$.

f.

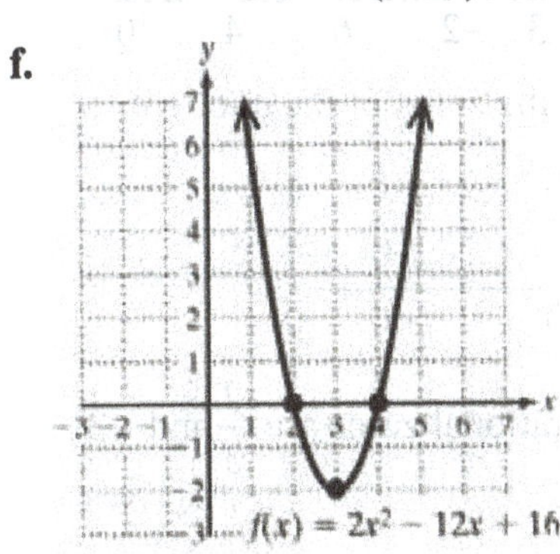

g. The axis of symmetry is the vertical line through the vertex: $x = 3$.

h. The minimum value is -2.

i. The domain is $(-\infty, \infty)$.

The range is $[-2, \infty)$.

3. a. Multiply the leading terms from each factor.

$-0.25x^3(x)^2(x)^4 = -0.25x^9$

b. The leading coefficient is negative and the degree is odd. The end behavior is up to the left and down to the right.

c. $f(x) = -0.25x^3(x - 2)^2(x + 1)^4$

$0 = -0.25x^3(x - 2)^2(x + 1)^4$

$x = 0, x = 2, x = -1$

The zero 0 has multiplicity 3. The zero 2 has multiplicity 2. The zero -1 has multiplicity 4.

5. $f(x) = x^3 - 5x^2 + 2x + 5$

$f(-2) = (-2)^3 - 5(-2)^2 + 2(-2) + 5$
$= -8 - 20 - 4 + 5 = -27$

$f(-1) = (-1)^3 - 5(-1)^2 + 2(-1) + 5$
$= -1 - 5 - 2 + 5 = -3$

$f(0) = (0)^3 - 5(0)^2 + 2(0) + 5 = 5$

$f(1) = (1)^3 - 5(1)^2 + 2(1) + 5$
$= 1 - 5 + 2 + 5 = 3$

$f(2) = (2)^3 - 5(2)^2 + 2(2) + 5$
$= 8 - 20 + 4 + 5 = -3$

a. Since $f(-2)$ and $f(-1)$ have the same sign, the intermediate value theorem does not guarantee that the function has at least one zero on the interval $[-2, -1]$.

b. Since $f(-1)$ and $f(0)$ have opposite signs, the intermediate value theorem guarantees that the function has at least one zero on the interval $[-1, 0]$.

c. Since $f(0)$ and $f(1)$ have the same sign, the intermediate value theorem does not guarantee that the function has at least one zero on the interval $[0, 1]$.

d. Since $f(1)$ and $f(2)$ have opposite signs, the intermediate value theorem guarantees that the function has at least one zero on the interval $[1, 2]$.

7. a.
$$\begin{array}{r|rrrrr} \frac{3}{5} & 5 & 47 & 80 & -51 & -9 \\ & & 3 & 30 & 66 & 9 \\ \hline & 5 & 50 & 110 & 15 & \lfloor 0 \end{array}$$

Since $f\left(\frac{3}{5}\right) = 0$, $\frac{3}{5}$ is a zero of $f(x)$.

b.
$$\begin{array}{r|rrrrr} -1 & 5 & 47 & 80 & -51 & -9 \\ & & -5 & -42 & -38 & 89 \\ \hline & 5 & 42 & 38 & -89 & \lfloor 80 \end{array}$$

Since $f(-1) \neq 0$, -1 is a zero of $f(x)$.

c. Since $f(-1) \neq 0$, $x+1$ is not a factor of $f(x)$.

d.
$$\begin{array}{r|rrrrr} -3 & 5 & 47 & 80 & -51 & -9 \\ & & -15 & -96 & 48 & 9 \\ \hline & 5 & 32 & -16 & -3 & \lfloor 0 \end{array}$$

Since $f(-3) = 0$, $x+3$ is a factor of $f(x)$.

e.
$$\begin{array}{r|rrrrr} -2 & 5 & 47 & 80 & -51 & -9 \\ & & -10 & -74 & -12 & 126 \\ \hline & 5 & 37 & 6 & -63 & \lfloor 117 \end{array}$$

By the remainder theorem, $f(-2) = 117$.

9. a. $f(x)$ is a fourth-degree polynomial. It has 4 zeros.

b. $\dfrac{\text{Factors of } 12}{\text{Factors of } 3} = \dfrac{\pm 1, \pm 2, \pm 3, \pm 4, \pm 6, \pm 12}{\pm 1, \pm 3}$
$= \pm 1, \pm 2, \pm 3, \pm 4, \pm 6, \pm 12, \pm\frac{1}{3}, \pm\frac{2}{3}, \pm\frac{4}{3}$

c.
$$\begin{array}{r|rrrrr} 2 & 3 & 7 & -12 & -14 & 12 \\ & & 6 & 26 & 28 & 28 \\ \hline & 3 & 13 & 14 & 14 & \lfloor 40 \end{array}$$

The remainder and all coefficients of the quotient are nonnegative. Therefore, 2 is an upper bound for the real zeros of $f(x)$.

d.
$$\begin{array}{r|rrrrr} -4 & 3 & 7 & -12 & -14 & 12 \\ & & -12 & 20 & -32 & 184 \\ \hline & 3 & -5 & 8 & -46 & \lfloor 196 \end{array}$$

The signs of the quotient alternate. Therefore, −4 is a lower bound for the real zeros of $f(x)$.

e. We can restrict the list of possible rational zeros to those on the interval $(-4, 2)$:

$\pm 1, -2, -3, \pm\frac{1}{3}, \pm\frac{2}{3}, \pm\frac{4}{3}$

From part (c), the value 2 itself is not a zero of $f(x)$. Likewise, from part (d), the value −4 itself is not a zero. Therefore, 2 and −4 are also eliminated from the list of possible rational zeros.

f.
$$\begin{array}{r|rrrrr} -3 & 3 & 7 & -12 & -14 & 12 \\ & & -9 & 6 & 18 & -12 \\ \hline & 3 & -2 & -6 & 4 & \lfloor 0 \end{array}$$

Find the zeros of the quotient.

$$\begin{array}{r|rrrr} \frac{2}{3} & 3 & -2 & -6 & 4 \\ & & 2 & 0 & -4 \\ \hline & 3 & 0 & -6 & \lfloor 0 \end{array}$$

The rational zeros are $\frac{2}{3}$ and -3.

g. Factor $3x^2 - 6$.

$3x^2 - 6 = 0$

$3x^2 = 6$

$x^2 = 2$

$x = \pm\sqrt{2}$

The zeros are $\frac{2}{3}$, -3, $\pm\sqrt{2}$.

h.

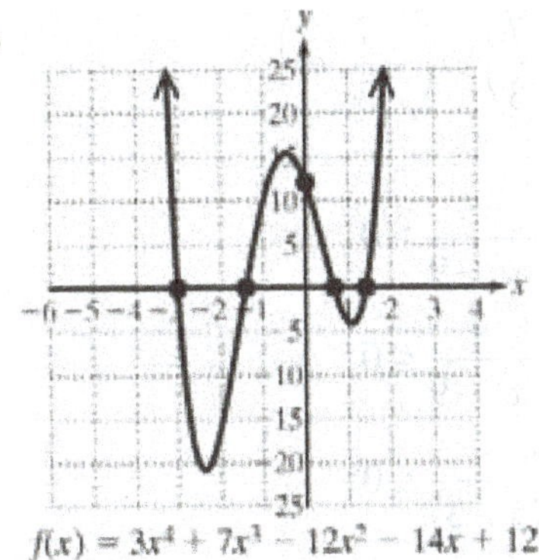

11. $f(x) = -6x^7 - 4x^5 + 2x^4 - 3x^2 + 1$

3 sign changes in $f(x)$. The number of possible positive real zeros is either 3 or 1.

$f(-x) = -6(-x)^7 - 4(-x)^5 + 2(-x)^4 - 3(-x)^2 + 1$

$f(-x) = 6x^7 + 4x^5 + 2x^4 - 3x^2 + 1$

2 sign changes in $f(-x)$. The number of possible negative real zeros is either 2 or 0.

13. The expression $\frac{-3x+1}{4x^2-1}$ is in lowest terms.

$$\frac{-3x+1}{4x^2-1} = \frac{-3x+1}{(2x+1)(2x-1)}$$

The denominator is 0 at $x = -\frac{1}{2}$, and $x = \frac{1}{2}$.

p has vertical asymptotes at $x = -\frac{1}{2}$, and $x = \frac{1}{2}$.

The degree of the numerator is 1.
The degree of the denominator is 2.
Since $n < m$, the line $y = 0$ is a horizontal asymptote of p.
p has no slant asymptote.

15. The graph of m is the graph of $y = \frac{1}{x^2}$ with a shift upward 3 units and a reflection across the x-axis.

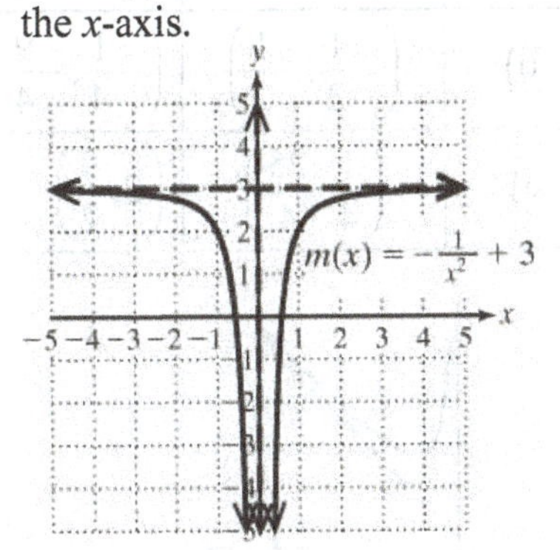

17. $k(0) = \frac{(0)^2 + 2(0) + 1}{(0)}$ undefined

There is no y-intercept.

$x^2 + 2x + 1 = 0$

$(x+1)(x+1) = 0$

$x = -1$

The x-intercept is $(-1, 0)$.

k is in lowest terms, and the denominator is 0 when $x = 0$, which is the vertical asymptote.
The degree of the numerator is exactly one greater than the degree of the denominator.
Therefore, k has no horizontal asymptote, but does have a slant asymptote.

$$k(x) = \frac{x^2 + 2x + 1}{x} = \frac{x^2}{x} + \frac{2x}{x} + \frac{1}{x} = x + 2 + \frac{1}{x}$$

The quotient is $x + 2$.
The slant asymptote is $y = x + 2$.

$x + 2 = \frac{x^2 + 2x + 1}{x}$

$x^2 + 2x = x^2 + 2x + 1$

$0 = 1$ No solution

k does not cross its slant asymptote.

$$k(-x) = \frac{(-x)^2 + 2(-x) + 1}{(-x)} = \frac{x^2 - 2x + 1}{-x}$$

$k(-x) \neq k(x)$, $k(-x) \neq -k(x)$

k is neither even nor odd.
Select test points from each interval.

Interval	Test Point	Test Point
$(-\infty, -1)$	$\left(-4, -\frac{9}{4}\right)$	$\left(-2, -\frac{1}{2}\right)$
$(-1, 0)$	$\left(-\frac{1}{2}, -\frac{1}{2}\right)$	$\left(-\frac{1}{4}, -\frac{9}{4}\right)$
$(0, \infty)$	$\left(\frac{1}{4}, \frac{25}{4}\right)$	$\left(2, \frac{9}{2}\right)$

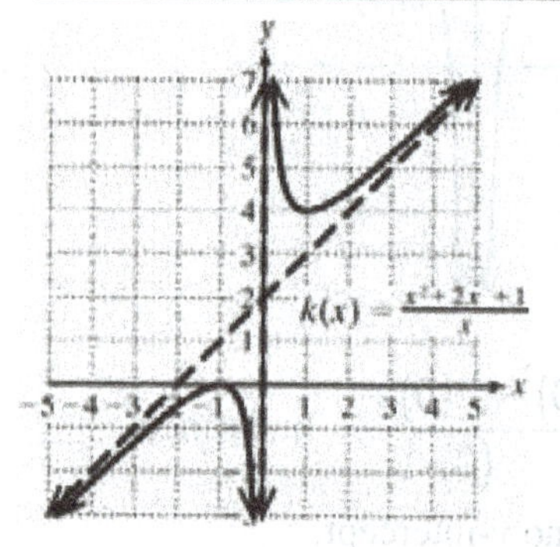

19. $y^3 > 13y - 12$

$y^3 - 13y + 12 > 0$

Find the real zeros of the related equation.

$y^3 - 13y + 12 = 0$

$$\frac{\text{Factors of } 12}{\text{Factors of } 1} = \frac{\pm 1, \pm 2, \pm 3, \pm 4, \pm 6, \pm 12}{\pm 1}$$
$$= \pm 1, \pm 2, \pm 3, \pm 4, \pm 6, \pm 12$$

$$\begin{array}{r|rrrr} 1 & 1 & 0 & -13 & 12 \\ & & 1 & 1 & -12 \\ \hline & 1 & 1 & -12 & \boxed{0} \end{array}$$

$(x-1)(x^2 + x - 12) = 0$

$(x-1)(x+4)(x-3) = 0$

$x = 1$, $x = -4$, or $x = 3$

The boundary points are -4, 1, and 3.

Sign of $(x+4)$:	−	+	+	+
Sign of $(x-1)$:	−	−	+	+
Sign of $(x-3)$:	−	−	−	+
Sign of $(x-1)(x+4)(x-3)$:	−	+	−	+
	−4	1	3	

The solution set is $(-4, 1) \cup (3, \infty)$.

21. $9x^2 + 42x + 49 > 0$

$(3x+7)^2 > 0$

The expression $(3x+7)^2 > 0$ for all real numbers except where $(3x+7)^2 = 0$. Therefore, the solution set is all real numbers except $-\frac{7}{3}$.

$\left(-\infty, -\frac{7}{3}\right) \cup \left(-\frac{7}{3}, \infty\right)$

23. $\frac{-4}{x^2 - 9} \geq 0$

$\frac{4}{x^2 - 9} \leq 0$

$\frac{4}{(x+3)(x-3)} \leq 0$

The expression is undefined for $x = -3$ and $x = 3$. These are boundary points.

The related equation has no real zeros.

Sign of $(x+3)$:	−	+	+
Sign of $(x-3)$:	−	−	+
Sign of $\frac{4}{(x+3)(x-3)}$:	+	−	+
	−3	3	

The solution set is $(-3, 3)$.

25. $E = kv^2$

27. Let A represent the surface area of a cube. Let s represent the length of an edge.

$A = ks^2$

$24 = k(2)^2$

$\frac{24}{4} = k$

$k = 6$

$A = 6s^2 = 6(7)^2 = 294 \text{ ft}^2$

29. Let P represent the pressure on the wall. Let A represent the area of the wall. Let v represent the velocity of the wind.

$P = kAv^2$

$P = kA(3v)^2 = kA(9v^2) = 9(kAv^2)$

The pressure is 9 times as great.

31. a. $y(20) = 140.3$ means that with 20,000 plants per acre, the yield will be 140.3 bushels per acre; $y(30) = 172$ means that with 30,000 plants per acre, the yield will be 172 bushels per acre; $y(60) = 143.5$ means that with 60,000 plants per acre, the yield will be 143.5 bushels per acre.

b. The number of plants per acre that will maximize the yield is the n-coordinate of the vertex.

$n = \frac{-b}{2a} = \frac{-(8.32)}{2(-0.103)} = \frac{-8.32}{-0.206}$

≈ 40.4 thousand plants $\approx 40{,}400$ plants

c. The maximum yield is the value of $y(n)$ at the vertex.

$y(40.4) = -0.103(40.4)^2 + 8.32n + 15.1$

≈ 183 bushels

33 a.

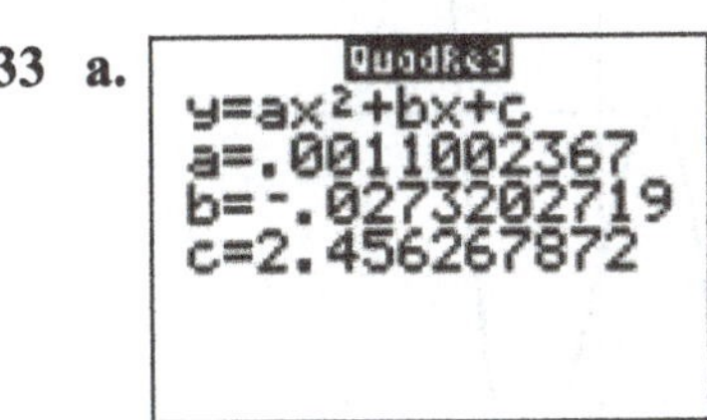

$n(a) = 0.0011a^2 - 0.027a + 2.46$

b. The age when the number of yearly visits is the least is the a-coordinate of the vertex.

$a = \frac{-b}{2a} = \frac{-(-0.027)}{2(0.0011)} = \frac{0.027}{0.0022} \approx 12$ yr

c. The minimum number of yearly visits is the $n(a)$ value at the vertex.

$n(12) = 0.0011(12)^2 - 0.027(12) + 2.46$

≈ 2.3 visits per year

Chapter 3 Cumulative Review Exercises

1. a. $x^2 - 16 = 0$

$x^2 = 16$

$x = 4, x = -4$

b. Since the degree of the numerator is greater than the degree of the denominator by more than 1, the horizontal asymptote is $y = \frac{2}{1} = 2$.

3. $f(x) = 2x^3 - x^2 - 8x - 5$

a. The leading term is $2x^3$. The end behavior is down to the left and up to the right.

b. $\frac{\text{Factors of } -5}{\text{Factors of } 2} = \frac{\pm 1, \pm 5}{\pm 1, \pm 2} = \pm 1, \pm 5, \pm\frac{1}{2}, \pm\frac{5}{2}$

$$\begin{array}{r|rrrr} -1 & 2 & -1 & -8 & -5 \\ & & -2 & 3 & 5 \\ \hline & 2 & -3 & -5 & 0 \end{array}$$

Find the zeros of the quotient.

$2x^2 - 3x - 5 = 0$

$(2x-5)(x+1) = 0$

$x = \frac{5}{2}, x = -1$

The zeros are $\frac{5}{2}$ (multiplicity 1) and -1 (multiplicity 2).

c. The x-intercepts are $\left(\frac{5}{2}, 0\right)$ and $(-1, 0)$.

d. $f(0)=2(0)^3-(0)^2-8(0)-5=-5$

The y-intercept is $(0,-5)$.

e.

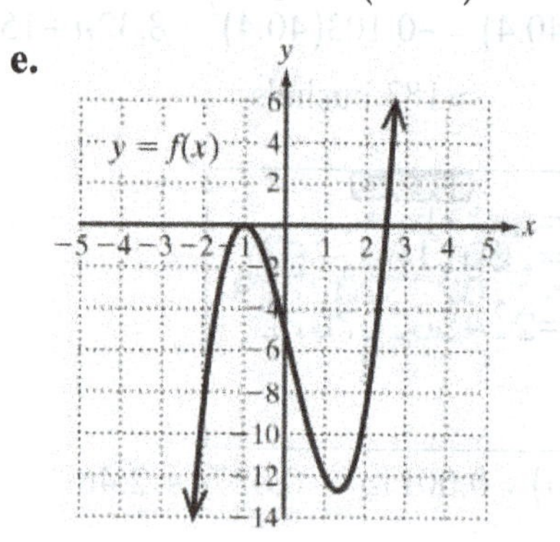

5. $x^2+y^2+8x-14y+56=0$

$(x^2+8x\quad)+(y^2-14y\quad)=-56$

$\left[\frac{1}{2}(8)\right]^2=16 \qquad \left[\frac{1}{2}(-14)\right]^2=49$

$(x^2+8x+16)+(y^2-14y+49)=-56+16+49$

$(x+4)^2+(y-7)^2=9$

Center: $(-4,7)$; Radius: $\sqrt{9}=3$

7. $x=y^2-9$

$x=(0)^2-9$

$x=-9$

The x-intercept is $(-9,0)$.

$x=y^2-9$

$0=y^2-9$

$y^2=9$

$y=\pm 3$

The y-intercepts are $(0,3)$ and $(0,-3)$.

9. $v\,t=\sqrt{m-t}$

$(v_0t)^2=m-t$

$m=(v_0t)^2+t$

11. $125x^6-y^9=(5x^2)^3-(y^3)^3$

$=(5x^2-y^3)(25x^4+5x^2y^3+y^6)$

13. $\sqrt[3]{250z^5xy^{21}}=\sqrt[3]{125z^3y^{21}\cdot 2z^2x}=5zy^7\sqrt[3]{2z^2x}$

15. $-5\le -\frac{1}{4}x+3<\frac{1}{2}$

$20\ge x-12>-2$

$32\ge x>10$

$10<x\le 32$

$(10,32]$

17. $|2x+1|=|x-4|$

$2x+1=x-4$ or $2x+1=-(x-4)$

$x=-5$ $\qquad 2x+1=-x+4$

$3x=3$

$x=1$

$\{-5,1\}$

19. $\frac{49x^2+14x+1}{x}>0$

$\frac{(7x+1)^2}{x}>0$

The expression is undefined for $x=0$. This is a boundary point.

Find the real zeros of the related equation.

$\frac{(7x+1)^2}{x}=0$

$(7x+1)^2=0$

$x=-\frac{1}{7}$

The boundary points are $-\frac{1}{7}$ and 0.

Sign of $(7x+1)^2$:	+	+	+
Sign of (x):	−	−	+
Sign of $\frac{(7x+1)^2}{x}$:	−	−	+
		$-\frac{1}{7}$	0

$(0,\infty)$

Chapter 4 Exponential and Logarithmic Functions

Section 4.1 Inverse Functions

1. $\{(2,1),(3,2),(4,3)\}$

3. $y=x$

5. is

7. x; x

9. (b,a)

11. Yes

13. No

15. Yes

17. No

19. The graph does not define y as a one-to-one function of x because a horizontal line intersects the graph in more than one point.

21. The graph does define y as a one-to-one function of x because no horizontal line intersects the graph in more than one point.

23. The graph does not define y as a one-to-one function of x because a horizontal line intersects the graph in more than one point.

25. The graph does not define y as a one-to-one function of x because a horizontal line intersects the graph in more than one point.

27. The relation does not define y as a function of x, because it fails the vertical line test. If the relation is not a function, it is not a one-to-one function.

29. Assume that $f(a)=f(b)$.

$$5a+4=5b+4$$
$$5a=5b$$
$$a=b$$

Since $f(a)=f(b)$ implies that $a=b$, then f is one-to-one.

31. Assume that $g(a)=g(b)$.

$$a^3+8=b^3+8$$
$$a^3=b^3$$
$$\sqrt[3]{a^3}=\sqrt[3]{b^3}$$
$$a=b$$

Since $g(a)=g(b)$ implies that $a=b$, then f is one-to-one.

33. No; For example the points $(1,-3)$ and $(-1,-3)$ have the same y values but different x values. That is, $m(a)=m(b)=-3$, but $a\neq b$.

35. No; For example the points $(2,3)$ and $(-4,3)$ have the same y values but different x values. That is, $p(a)=p(b)=3$, but $a\neq b$.

37. $(f\circ g)(x)=f(g(x))=f\left(\frac{x-4}{5}\right)$

$$=5\left(\frac{x-4}{5}\right)+4=x-4+4=x \checkmark$$

$$(g\circ f)(x)=g(f(x))=g(5x+4)$$

$$=\frac{(5x+4)-4}{5}=\frac{5x}{5}=x \checkmark$$

Yes, g is the inverse of f.

39. $(m\circ n)(x)=m(n(x))=m\left(\frac{-2+x}{6}\right)$

$$=6\left(\frac{-2+x}{6}\right)-2=-2+x-2$$
$$=x-4$$

No, since $(m\circ n)(x)\neq x$, n is not an inverse of m.

41. $(t \circ v)(x) = t(v(x)) = t\left(\frac{x+4}{x}\right) = \frac{4}{\left(\frac{x+4}{x}\right) - 1}$

$= \frac{4x}{(x+4) - x} = \frac{4x}{4} = x$ ✓

$(v \circ t)(x) = v(t(x)) = v\left(\frac{4}{x-1}\right) = \frac{\left(\frac{4}{x-1}\right) + 4}{\left(\frac{4}{x-1}\right)}$

$= \frac{4 + 4(x-1)}{4} = \frac{4 + 4x - 4}{4} = \frac{4x}{4} = x$ ✓

Yes, v is the inverse of t.

43. a. $(f \circ g)(x) = f(g(x)) = f\left(\frac{x - 2000}{150}\right)$

$= 2000 + 150\left(\frac{x - 2000}{150}\right)$

$= 2000 + x - 2000 = x$ ✓

$(g \circ f)(x) = g(f(x)) = g(2000 + 150x)$

$= \frac{(2000 + 150x) - 2000}{150}$

$= \frac{150x}{150} = x$ ✓

Yes, g is the inverse of f.

b. The value $g(x)$ represents the number of years since the year 2010 based on the number of applicants to the freshman class, x.

45. a. Assume that $f(a) = f(b)$.

$2a - 3 = 2b - 3$

$2a = 2b$

$a = b$

Since $f(a) = f(b)$ implies that $a = b$, then f is one-to-one.

b. $f(x) = 2x - 3$

$y = 2x - 3$

$x = 2y - 3$

$x + 3 = 2y$

$\frac{x+3}{2} = y$

$f^{-1}(x) = \frac{x+3}{2}$

c.

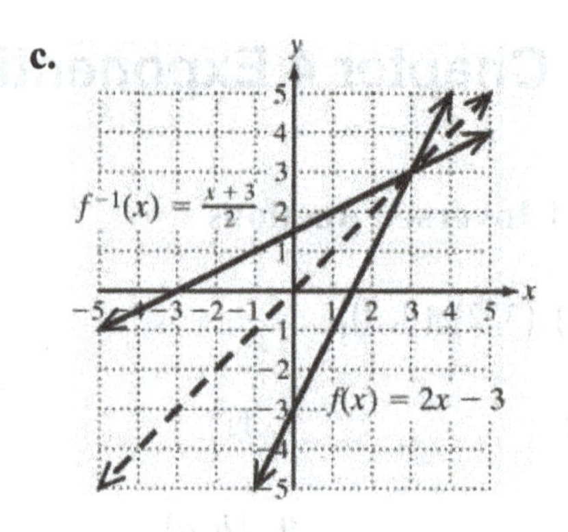

47. $f(x) = \frac{4 - x}{9}$

$y = \frac{4 - x}{9}$

$x = \frac{4 - y}{9}$

$9x = 4 - y$

$y = 4 - 9x$

$f^{-1}(x) = 4 - 9x$

49. $h(x) = \sqrt[3]{x - 5}$

$y = \sqrt[3]{x - 5}$

$x = \sqrt[3]{y - 5}$

$x^3 = y - 5$

$x^3 + 5 = y$

$h^{-1}(x) = x^3 + 5$

51. $m(x) = 4x^3 + 2$

$y = 4x^3 + 2$

$x = 4y^3 + 2$

$x - 2 = 4y^3$

$\frac{x-2}{4} = y^3$

$\sqrt[3]{\frac{x-2}{4}} = y$

$m^{-1}(x) = \sqrt[3]{\frac{x-2}{4}}$

53.
$$c(x)=\frac{5}{x+2}$$
$$y=\frac{5}{x+2}$$
$$x=\frac{5}{y+2}$$
$$x(y+2)=5$$
$$xy+2x=5$$
$$xy=5-2x$$
$$y=\frac{5-2x}{x}$$
$$c^{-1}(x)=\frac{5-2x}{x}$$

55.
$$t(x)=\frac{x-4}{x+2}$$
$$y=\frac{x-4}{x+2}$$
$$x=\frac{y-4}{y+2}$$
$$x(y+2)=y-4$$
$$xy+2x=y-4$$
$$xy-y=-2x-4$$
$$y(x-1)=-(2x+4)$$
$$y=-\frac{2x+4}{x-1}$$
$$t^{-1}(x)=-\frac{2x+4}{x-1}$$

57.
$$f(x)=\frac{(x-a)^3}{b}-c$$
$$y=\frac{(x-a)^3}{b}-c$$
$$x=\frac{(y-a)^3}{b}-c$$
$$x+c=\frac{(y-a)^3}{b}$$
$$b(x+c)=(y-a)^3$$
$$\sqrt[3]{b(x+c)}=y-a$$
$$\sqrt[3]{b(x+c)}+a=y$$
$$f^{-1}(x)=\sqrt[3]{b(x+c)}+a$$

59. a.

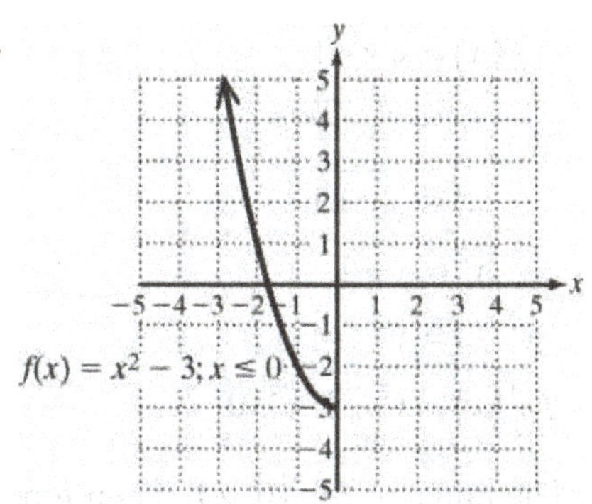

b. Yes **c.** $(-\infty, 0]$

d. $[-3, \infty)$

e.
$$f(x)=x^2-3$$
$$y=x^2-3$$
$$x=y^2-3$$
$$x+3=y^2$$
$$\pm\sqrt{x+3}=y$$
$$f^{-1}(x)=-\sqrt{x+3}$$

f.

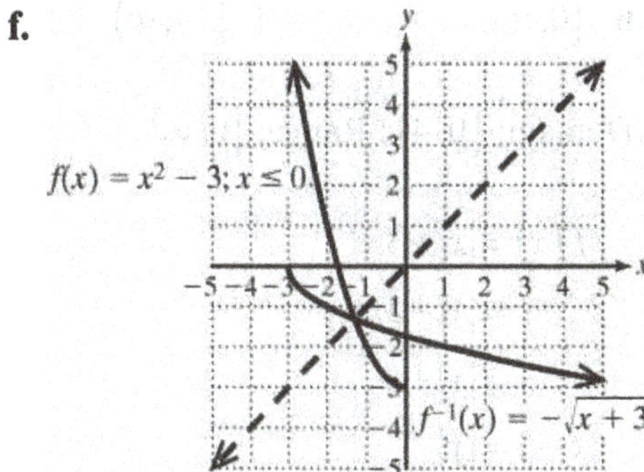

g. $[-3, \infty)$ **h.** $(-\infty, 0]$

61. a.

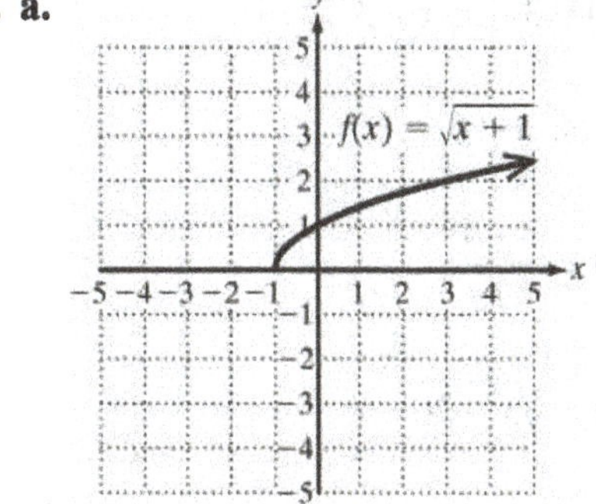

b. Yes **c.** $[-1, \infty)$

d. $[0, \infty)$

e. $f(x)=\sqrt{x+1}$

$$y=\sqrt{x+1}$$
$$x=\sqrt{y+1}$$
$$x^2=y+1$$
$$x^2-1=y$$
$$f^{-1}(x)=x^2-1;\ x\geq 0$$

f. The range of f is $[0,\infty)$. Therefore, the domain of f^{-1} must be $[0,\infty)$.

g.

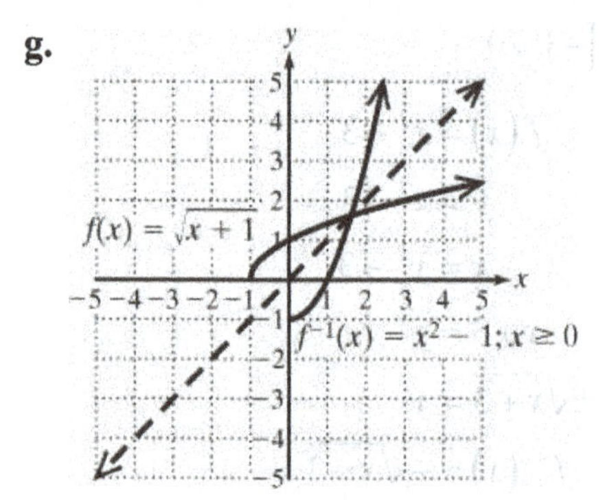

h. $[0,\infty)$ **i.** $[-1,\infty)$

63. Domain: $[0,4)$; Range: $[0,\infty)$

65. $f(x)=|x|+3$

$$y=|x|+3$$
$$x=|y|+3$$
$$x-3=|y|$$
$$y=x-3 \quad \text{or} \quad y=-(x-3)$$
$$y=3-x$$
$$f^{-1}(x)=3-x;\ x\geq 3$$

67. subtracts; $x-6$

69. $f^{-1}(x)=\dfrac{x+4}{7}$

71. $f^{-1}(x)=\sqrt[3]{x-20}$

73. $f^{-1}(x)=\dfrac{x-1}{8}$

75. $q^{-1}(x)=(x-1)^5+4$

77.

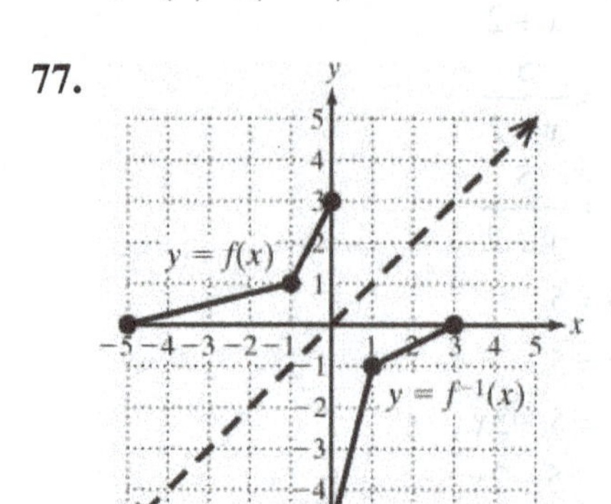

79.

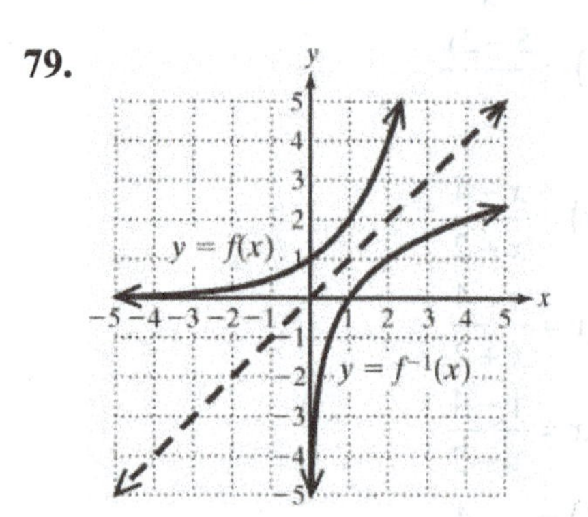

81. a. $f(12)=32$, so $f^{-1}(32)=12$.

b. $f(0.5)=-2.5$, so $f^{-1}(-2.5)=0.5$.

c. $f(10)=26$, so $f^{-1}(26)=10$.

83. True

85. False

87. a. $T(x)=6.33x$

b. $T(x)=6.33x$

$$y=6.33x$$
$$x=6.33y$$
$$\frac{x}{6.33}=y$$
$$T^{-1}(x)=\frac{x}{6.33}$$

c. $T^{-1}(x)$ represents the mass of a mammal based on the amount of air inhaled per breath, x.

d. $T^{-1}(170) = \dfrac{170}{6.33} \approx 27$ means that a mammal that inhales 170 mL of air per breath during normal respiration is approximately 27 kg (this is approximately 60 lb—the size of a Labrador retriever).

89. a. $T(x) = 24x + 108$

b.
$$T(x) = 24x + 108$$
$$y = 24x + 108$$
$$x = 24y + 108$$
$$x - 108 = 24y$$
$$\frac{x-108}{24} = y$$
$$T^{-1}(x) = \frac{x-108}{24}$$

c. $T^{-1}(x)$ represents the taxable value of a home (in \$1000) based on x dollars of property tax paid on the home.

d. $T^{-1}(2988) = \dfrac{2988-108}{24} = 120$ means that if a homeowner is charged \$2998 in property taxes, then the taxable value of the home is \$120,000.

91.
$$V(r) = \frac{4}{3}\pi r^3$$
$$V = \frac{4}{3}\pi r^3$$
$$\frac{3V}{4\pi} = r^3$$
$$\sqrt[3]{\frac{3V}{4\pi}} = r$$
$$r(V) = \sqrt[3]{\frac{3V}{4\pi}}$$

$r(V) = \sqrt[3]{\dfrac{3V}{4\pi}}$ represents the radius of a sphere as a function of its volume.

93. The domain and range of a function and its inverse are reversed.

95. If a horizontal line intersects the graph of a function in more than one point, then the function has at least two ordered pairs with the same y-coordinate but different x-coordinates. This conflicts with the definition of a one-to-one function.

97. a. $2^3 = 8$, so $f(8) = 3$.

b. $2^5 = 32$, so $f(32) = 5$.

c. $2^1 = 2$, so $f(2) = 1$.

d. $2^{-3} = \dfrac{1}{2^3} = \dfrac{1}{8}$, so $f\left(\dfrac{1}{8}\right) = -3$.

99. Let f be an increasing function. Then for every value a and b in the domain of f such that $a < b$ we have $f(a) < f(b)$. Now if $u \neq v$, then either $u < v$ or $v < u$. Then either $f(u) < f(v)$ or $f(v) < f(u)$. In either case, $f(u) \neq f(v)$, and f is one-to-one.

101. a.

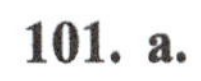

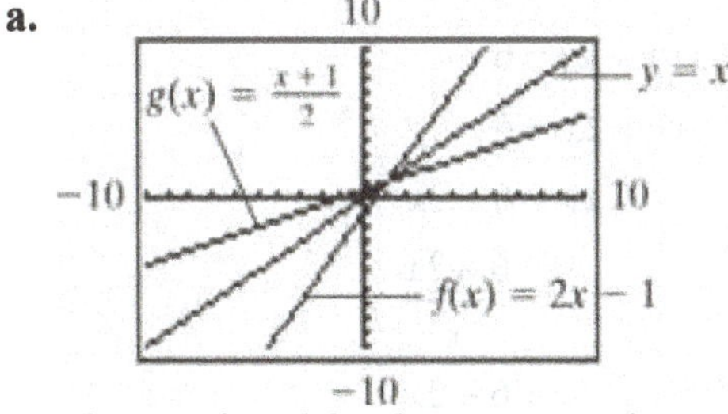

The graphs of f and g appear to be symmetric with respect to the line $y = x$. This suggests that f and g are inverses.

b.

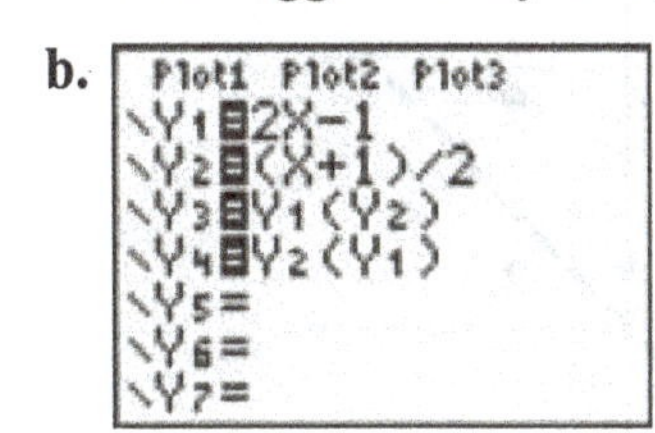

c.

X	Y3	Y4
-3	-3	-3
-2	-2	-2
-1	-1	-1
0	0	0
1	1	1
2	2	2
3	3	3

Y3=-3

The expressions $Y_1(Y_2)$ and $Y_2(Y_1)$ represent the composition of functions $(f \circ g)(x)$ and $(g \circ f)(x)$. In each case, $Y_1(Y_2)$ and $Y_2(Y_1)$ equal the value of x. This suggests that f and g are inverses.

Section 4.2 Exponential Functions

1. y is a linear function of x, so the relation is a function, which we can write as $f(x) = 3x - 5$.

Assume that $f(a) = f(b)$.

$$3a - 5 = 3b - 5$$
$$3a = 3b$$
$$a = b$$

Since $f(a) = f(b)$ implies that $a = b$, then f is one-to-one.

3.
$$f(x) = \frac{6}{x+2}$$
$$y = \frac{6}{x+2}$$
$$x = \frac{6}{y+2}$$
$$x(y+2) = 6$$
$$xy + 2x = 6$$
$$xy = 6 - 2x$$
$$y = \frac{6-2x}{x}$$
$$f^{-1}(x) = \frac{6-2x}{x}$$

5.

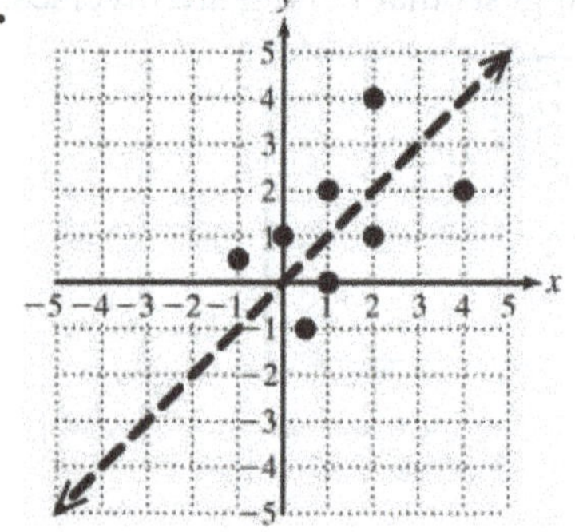

7. b^x

9. increasing

11. $(-\infty, \infty)$

13. $(0, 1)$

15. is not

17. e; natural

19. a. $5^{-1} = 0.2$ **b.** $5^{4.8} \approx 2264.9364$

c. $5^{\sqrt{2}} \approx 9.7385$ **d.** $5^{\pi} \approx 156.9925$

21. a. $\left(\frac{1}{4}\right)^{-3} = 64$ **b.** $\left(\frac{1}{4}\right)^{1.4} = 0.1436$

c. $\left(\frac{1}{4}\right)^{\sqrt{3}} = 0.0906$ **d.** $\left(\frac{1}{4}\right)^{0.5e} = 0.1520$

23. a and d

25.

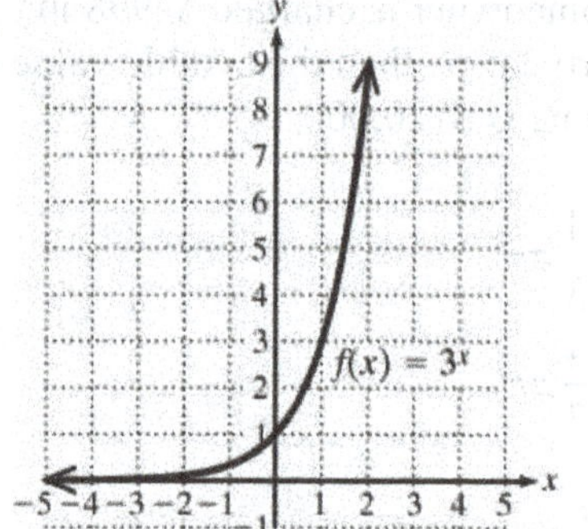

Domain: $(-\infty, \infty)$; Range: $(0, \infty)$

27.

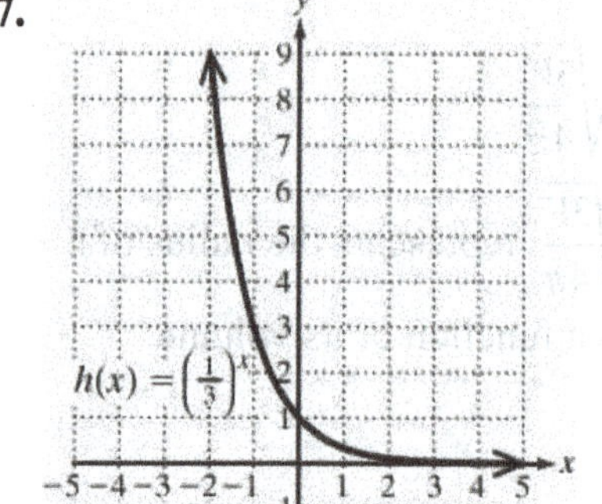

Domain: $(-\infty, \infty)$; Range: $(0, \infty)$

29.

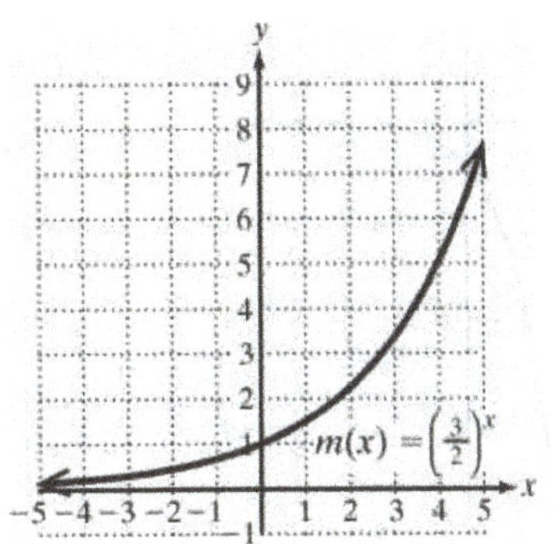

Domain: $(-\infty, \infty)$; Range: $(0, \infty)$

31.

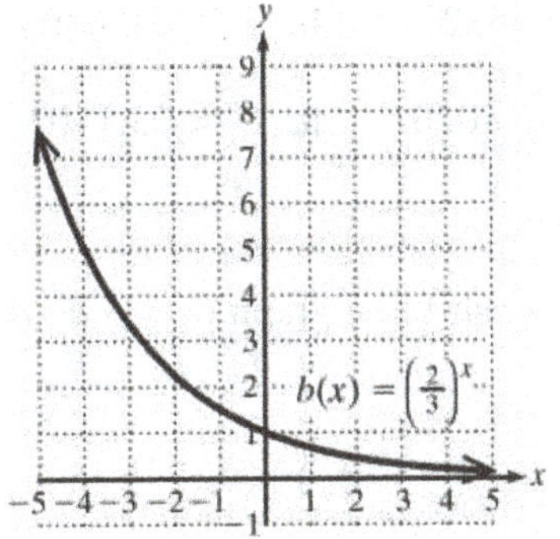

Domain: $(-\infty, \infty)$; Range: $(0, \infty)$

33. $f(x) = 3^x + 2$

Since $k = 2$, shift the graph of $y = 3^x$ up 2 units.

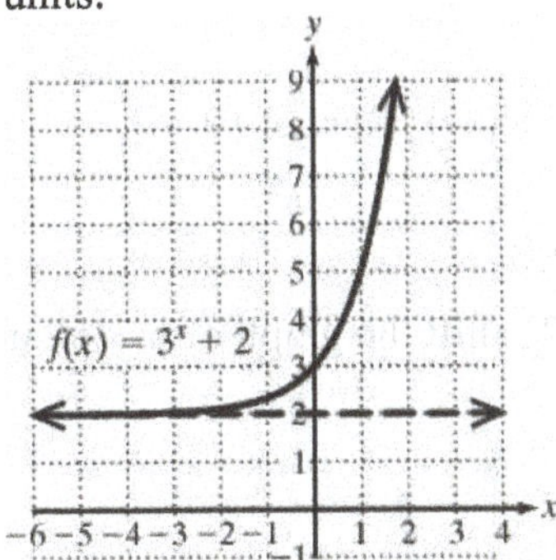

Domain: $(-\infty, \infty)$; Range: $(2, \infty)$

35. $m(x) = 3^{x+2} = 3^{x-(-2)}$

Since $h = -2$, shift the graph of $y = 3^x$ left 2 units.

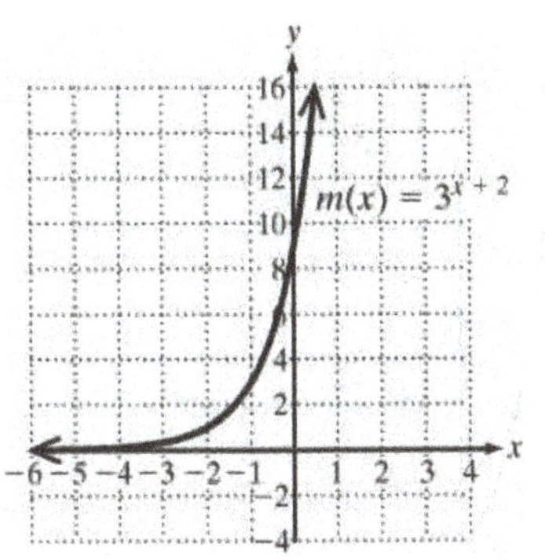

Domain: $(-\infty, \infty)$; Range: $(0, \infty)$

37. $p(x) = 3^{x-4} - 1 = 3^{x-4} + (-1)$

Since $h = 4$ and $k = -1$, shift the graph of $y = 3^x$ right 4 units and down 1 unit.

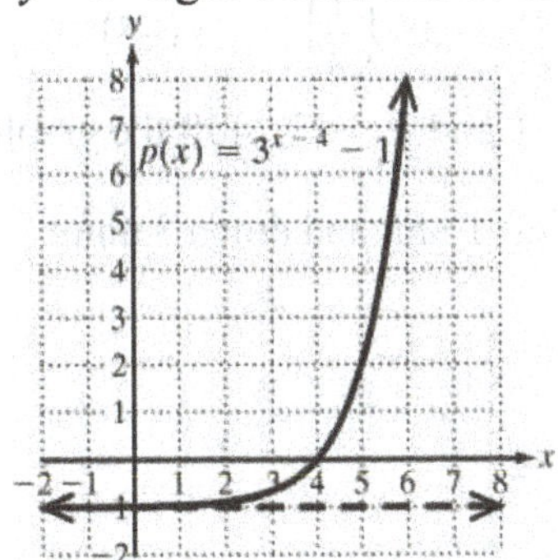

Domain: $(-\infty, \infty)$; Range: $(-1, \infty)$

39. $k(x) = -3^x = (-1)3^x$

Since $a < 0$, reflect the graph of $y = 3^x$ across the x-axis.

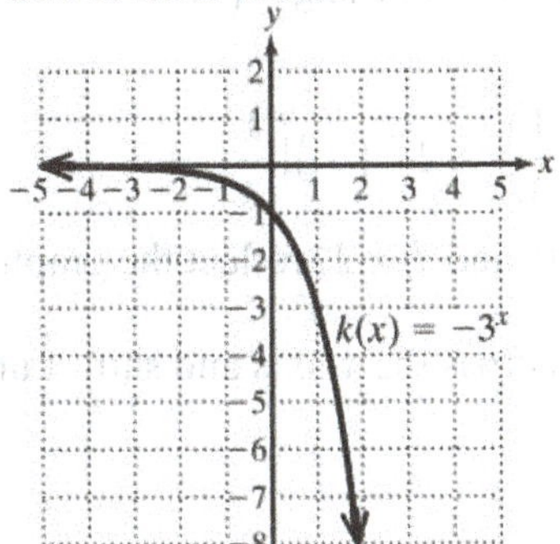

Domain: $(-\infty, \infty)$; Range: $(-\infty, 0)$

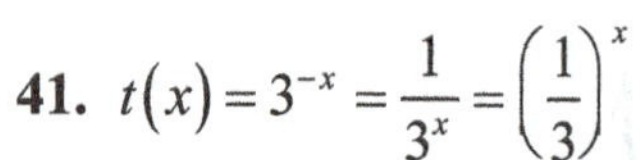

41. $t(x)=3^{-x}=\frac{1}{3^x}=\left(\frac{1}{3}\right)^x$

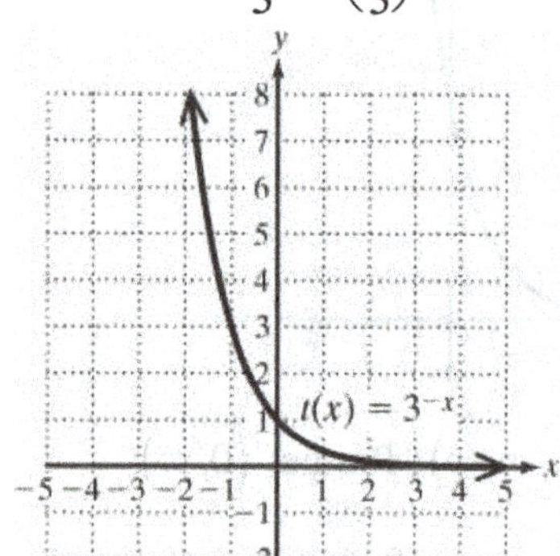

Domain: $(-\infty,\infty)$; Range: $(0,\infty)$

43. $f(x)=\left(\frac{1}{3}\right)^{x+1}-3=\left(\frac{1}{3}\right)^{x-(-1)}+(-3)$

Since $h=-1$ and $k=-3$, shift the graph of $y=\left(\frac{1}{3}\right)^x$ left 1 unit and down 3 units.

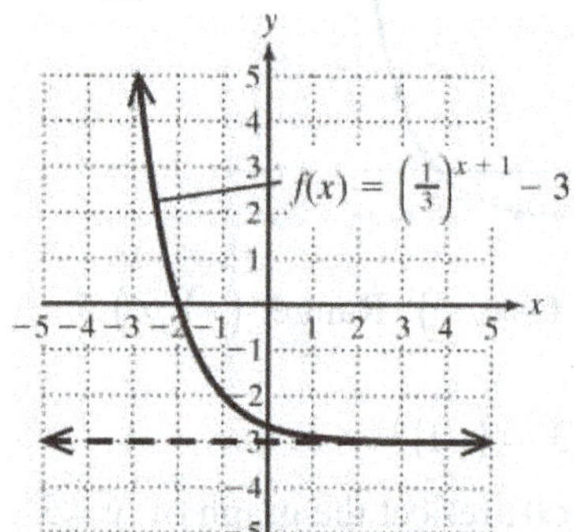

Domain: $(-\infty,\infty)$; Range: $(-3,\infty)$

45. $k(x)=-\left(\frac{1}{3}\right)^x+2=(-1)\left(\frac{1}{3}\right)^x+2$

Since $a<0$ and $k=2$, reflect the graph of $y=\left(\frac{1}{3}\right)^x$ across the x-axis and shift it up 2 units.

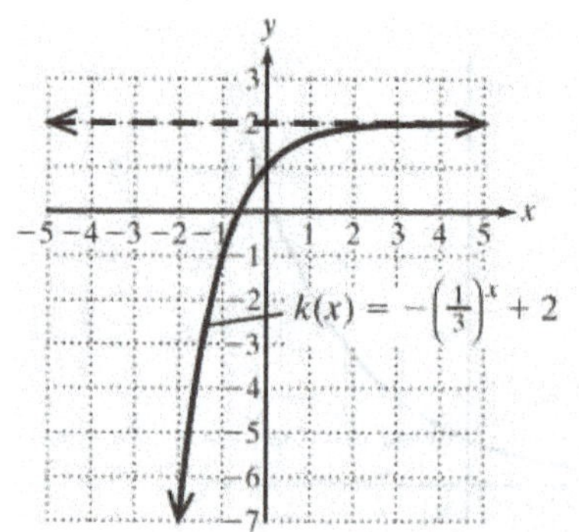

Domain: $(-\infty,\infty)$; Range: $(\infty,2)$

47. a. $e^4\approx 54.5982$ **b.** $e^{-3.2}\approx 0.0408$

c. $e^{\sqrt{13}}\approx 36.8020$ **d.** $e^{\pi}\approx 23.1407$

49. $f(x)=e^{x-4}$

Since $h=4$, shift the graph of $y=e^x$ right 4 units.

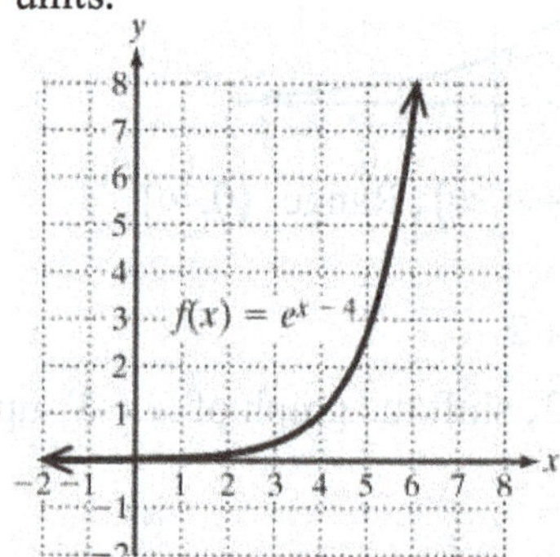

Domain: $(-\infty,\infty)$; Range: $(0,\infty)$

51. $h(x)=e^x+2$

Since $k=2$, shift the graph of $y=e^x$ up 2 units.

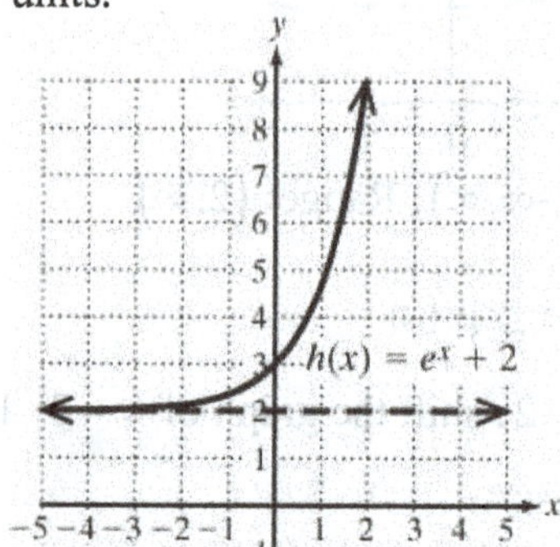

Domain: $(-\infty,\infty)$; Range: $(2,\infty)$

53. $m(x) = -e^x - 3 = (-1)e^x + (-3)$
Since $a < 0$ and $k = -3$, reflect the graph of $y = e^x$ across the x-axis and shift it down 3 units.

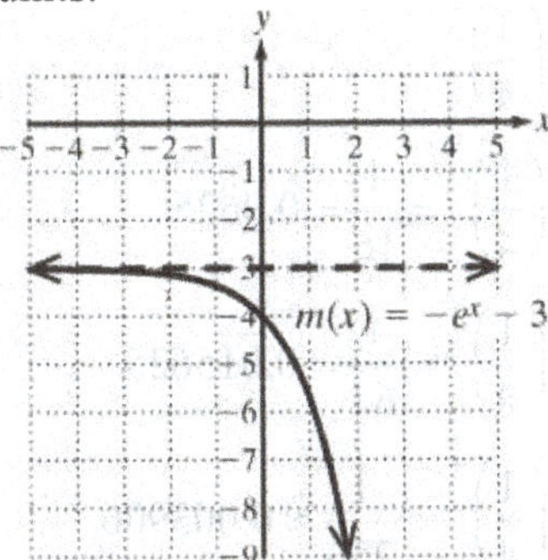

Domain: $(-\infty, \infty)$; Range: $(\infty, -3)$

55. $A = P\left(1 + \frac{r}{n}\right)^{nt}$

$A = 10{,}000\left(1 + \frac{0.04}{n}\right)^{(n \cdot 5)}$

a. For annual compounding, $n = 1$.
$A = 10{,}000\left(1 + \frac{0.04}{1}\right)^{(1 \cdot 5)} \approx 12{,}166.53$

b. For quarterly compounding, $n = 4$.
$A = 10{,}000\left(1 + \frac{0.04}{4}\right)^{(4 \cdot 5)} \approx 12{,}201.90$

c. For monthly compounding, $n = 12$.
$A = 10{,}000\left(1 + \frac{0.04}{12}\right)^{(12 \cdot 5)} \approx 12{,}209.97$

d. For daily compounding, $n = 365$.
$A = 10{,}000\left(1 + \frac{0.04}{365}\right)^{(365 \cdot 5)} \approx 12{,}213.89$

e. For continuous compounding, use $A = Pe^{rt}$.
$A = Pe^{rt} = 10{,}000e^{(0.04 \cdot 5)} \approx 12{,}214.03$

57. $A = Pe^{rt} = 20{,}000e^{10r}$

a. $A = 20{,}000e^{10(0.03)} \approx \$26{,}997.18$

b. $A = 20{,}000e^{10(0.04)} \approx \$29{,}836.49$

c. $A = 20{,}000e^{10(0.055)} \approx \$34{,}665.06$

59. a. $I = Prt = 10{,}000(0.055)(4) = \$2200$

b. $I = Pe^{rt} - P$
$= 10{,}000e^{(0.05)(4)} - 10{,}000$
$\approx 12{,}214.03 - 10{,}000$
$\approx \$2214.03$

c. 5.5% simple interest results in less interest.

61. $I = Prt = 25{,}000(0.052)(30) = \$39{,}000$
$I = Pe^{rt} - P$
$= 25{,}000e^{(0.038)(30)} - 25{,}000$
$\approx 78{,}169.21 - 25{,}000$
$\approx \$53{,}169.21$
3.8% compounded continuously for 30 yr results in more interest.

63. $A(t) = 10\left(\frac{1}{2}\right)^{t/28.9}$

a. $A(28.9) = 10\left(\frac{1}{2}\right)^{28.9/28.9} = 10\left(\frac{1}{2}\right)^1 = 5$

After 28.9 yr, the amount of ^{90}Sr remaining is 5 μg. After one half-life, the amount of substance has been halved.

b. $A(57.8) = 10\left(\frac{1}{2}\right)^{57.8/28.9} = 10\left(\frac{1}{2}\right)^2 = 2.5$

After 57.8 yr, the amount of ^{90}Sr remaining is 2.5 μg. After two half-lives, the amount of substance has been halved, twice.

c. $A(100) = 10\left(\frac{1}{2}\right)^{100/28.9} \approx 0.909$

After 100 yr, the amount of ^{90}Sr remaining is approximately 0.909 μg.

65. $P(t) = 310(1.0097)^t$

a. The base is $1.0097 > 1$, so the graph is an increasing exponential function.

b. $P(0) = 310(1.0097)^0 = 310$
In the year 2010, the U.S. population was approximately 310 million. This is the initial population in 2010.

c. $P(10) = 310(1.0097)^{10} \approx 341$
In the year 2020, the U.S. population will be approximately 341 million if this trend continues.

d. $P(20) = 310(1.0097)^{20} \approx 376$
$P(30) = 310(1.0097)^{30} \approx 414$

e. $P(200) = 310(1.0097)^{200} \approx 2137$
In the year 2210 the U.S population will be approximately 2.137 billion. The model cannot continue indefinitely because the population will become too large to be sustained from the available resources.

67. $P(a) = 760e^{-0.13a}$

a. $P(0) = 760e^{-0.13(0)} = 760e^0 = 760$ mmHg

b. $P(8.848) = 760e^{-0.13(8.848)} \approx 241$ mmHg

69. a. $T(t) = 78 + (350 - 78)e^{-0.046t}$
$T(t) = 78 + 272e^{-0.046t}$

b. $T(10) = 78 + 272e^{-0.046(10)} \approx 250°\text{F}$

c. $T(60) = 78 + 272e^{-0.046(60)} \approx 95.2°\text{F}$
Yes; after 60 min, the cake will be approximately 95.2°F.

71. a. $V(5) = 120(0.8)^5 \approx 39$
The resale value is \$39,000.

b. $V(1) = 120(0.8)^1 = 96$
The depreciation after 1 yr is $120{,}000 - 96{,}000 = 24{,}000$.
$C(h) = (18 + 36 + 22)h = 76h$
$C(800) = 76(800) = 60{,}800$
It costs the farmer $\$60{,}800 + \$24{,}000 = \$84{,}800$ to run the tractor for 800 hr during the first year.

73. a. $P(x) = \left(\frac{1}{4}\right)^x$

$P(2) = \left(\frac{1}{4}\right)^2 = \frac{1}{16} = 0.0625$

$P(3) = \left(\frac{1}{4}\right)^3 = \frac{1}{64} = 0.015625$

$P(4) = \left(\frac{1}{4}\right)^4 = \frac{1}{256} \approx 0.003906$

$P(5) = \left(\frac{1}{4}\right)^5 = \frac{1}{512} \approx 0.000977$

b. The probability decreases.

c. $P(10) = \left(\frac{1}{4}\right)^{10} \approx 9.54 \times 10^{-7}$
It is very unlikely.

75. a. $\{2\}$ **b.** $\{3\}$

c. $\{4\}$

d. x is between 2 and 3.

e. x is between 3 and 4.

77. a. See part (d). **b.** Yes

c. Domain: $(-\infty, \infty)$; Range: $(0, \infty)$

d.

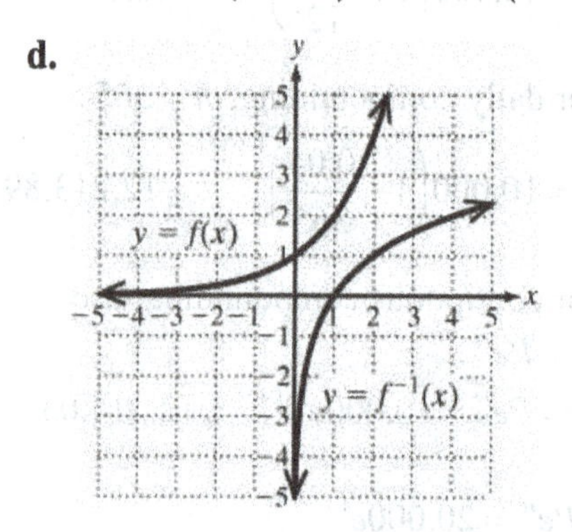

e. Domain: $(0, \infty)$; Range: $(-\infty, \infty)$

f. $f^{-1}(1) = 0$; $f^{-1}(2) = 1$; $f^{-1}(4) = 2$

79. **a.** ∞ **b.** 0

c. ∞ **d.** $-\infty$

81. The range of an exponential function is the set of positive real numbers; that is, 2^x is nonnegative for all values of x in the domain.

83. $f(x) = 2^x$

a. $\dfrac{f(b)-f(a)}{b-a} = \dfrac{1-0.25}{0-(-2)} = \dfrac{0.75}{2} = 0.375$

b. $\dfrac{f(b)-f(a)}{b-a} = \dfrac{4-1}{2-0} = \dfrac{3}{2} = 1.5$

c. $\dfrac{f(b)-f(a)}{b-a} = \dfrac{16-4}{4-2} = \dfrac{12}{2} = 6$

d. $\dfrac{f(b)-f(a)}{b-a} = \dfrac{64-16}{6-4} = \dfrac{48}{2} = 24$

85. $f(x) = \left(\dfrac{1}{2}\right)^x$

a. $\dfrac{f(b)-f(a)}{b-a} = \dfrac{1-4}{0-(-2)} = \dfrac{-3}{2} = -1.5$

b. $\dfrac{f(b)-f(a)}{b-a} = \dfrac{0.25-1}{2-0} = \dfrac{-0.75}{2} = -0.375$

c. $\dfrac{f(b)-f(a)}{b-a} = \dfrac{0.0625-0.25}{4-2}$

$= \dfrac{-0.1875}{2}$

≈ -0.0938

d. $\dfrac{f(b)-f(a)}{b-a} = \dfrac{0.015625-0.0625}{6-4}$

$= \dfrac{-0.046875}{2}$

≈ -0.0234

87. $3x^2e^{-x} - 6xe^{-x} = 0$

$3xe^{-x}(x-2) = 0$

$xe^{-x}(x-2) = 0$

$x = 0$ or $e^{-x} = 0$ or $x = 2$

$e^{-x} = 0$ has no solution because e raised to a power is always greater than 0, so the real solutions are $\{0, 2\}$.

89. **a.** $e^xe^h = e^{x+h}$ **b.** $\left(e^x\right)^2 = e^{2x}$

c. $\dfrac{e^x}{e^h} = e^{x-h}$

d. $e^x \cdot e^{-x} = e^{(x-x)} = e^0 = 1$

e. $e^{-2x} = \left(e^{2x}\right)^{-1} = \dfrac{1}{e^{2x}}$

91. $\left(e^x + e^{-x}\right)^2 = e^{2x} + 2e^0 + e^{-2x}$

$= e^{2x} + 2 + e^{-2x} = \dfrac{e^{4x} + 2e^{2x} + 1}{e^{2x}}$

93. $\left(\dfrac{e^x + e^{-x}}{2}\right)^2 - \left(\dfrac{e^x - e^{-x}}{2}\right)^2$

$= \dfrac{1}{4}\left[\left(e^{2x} + 2 + e^{-2x}\right) - \left(e^{2x} - 2 + e^{-2x}\right)\right]$

$= \dfrac{1}{4}(4) = 1$

95. $\dfrac{f(x+h)-f(x)}{h} = \dfrac{e^{x+h} - e^x}{h} = \dfrac{e^x\left(e^h - 1\right)}{h}$

97.

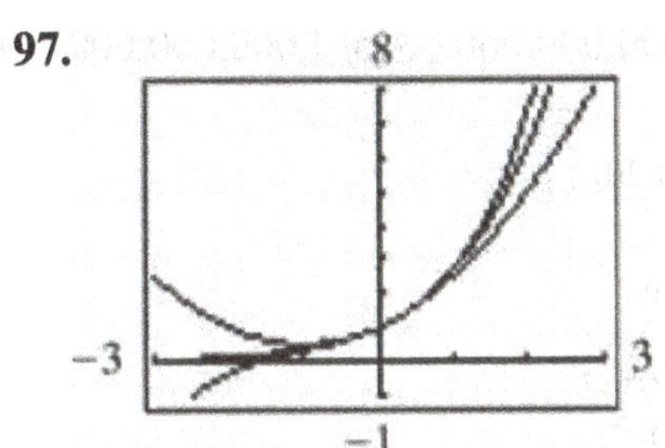

The graphs of Y_2 and Y_3 are close approximations of $Y_1 = e^x$ near $x = 0$.

Section 4.3 Logarithmic Functions

1. $f(x)=\sqrt[3]{x-4}$
$y=\sqrt[3]{x-4}$
$x=\sqrt[3]{y-4}$
$x^3=y-4$
$x^3+4=y$
$f^{-1}(x)=x^3+4$

3. $3^{\boxed{4}}=81$

5. $6^{\boxed{-2}}=\dfrac{1}{36}$

7. logarithmic

9. exponential

11. common; natural

13. 0; 0

15. x; x

17. $x=0$; vertical

19. $\log_8 64=2 \Leftrightarrow 8^2=64$

21. $\log\left(\dfrac{1}{10{,}000}\right)=-4 \Leftrightarrow 10^{-4}=\dfrac{1}{10{,}000}$

23. $\log_4 1=0 \Leftrightarrow 4^0=1$

25. $\log_a b=c \Leftrightarrow a^c=b$

27. $5^3=125 \Leftrightarrow \log_5 125=3$

29. $\left(\dfrac{1}{5}\right)^{-3}=125 \Leftrightarrow \log_{1/5} 125=-3$

31. $10^9=1{,}000{,}000{,}000 \Leftrightarrow \log 1{,}000{,}000{,}000=9$

33. $a^7=b \Leftrightarrow \log_a b=7$

35. $\log_3 9=y$
$3^y=9=3^2$
$y=2$

37. $\log_5 5=y$
$5^y=5=5^1$
$y=1$

39. $\log 100{,}000{,}000=y$
$10^y=100{,}000{,}000=10^8$
$y=8$

41. $\log_2\left(\dfrac{1}{16}\right)=y$
$2^y=\dfrac{1}{16}=\dfrac{1}{2^4}=2^{-4}$
$y=-4$

43. $\log\left(\dfrac{1}{10}\right)=y$
$10^y=\dfrac{1}{10}=\dfrac{1}{10^1}=10^{-1}$
$y=-1$

45. $\ln e^6=y$
$e^y=e^6$
$y=6$

47. $\ln\left(\dfrac{1}{e^3}\right)=y$
$e^y=\dfrac{1}{e^3}=e^{-3}$
$y=-3$

49. $\log_{1/7} 49=y$
$\left(\dfrac{1}{7}\right)^y=49=\left(\dfrac{1}{7}\right)^{-2}$
$y=-2$

51. $\log_{1/2}\left(\dfrac{1}{32}\right)=y$
$\left(\dfrac{1}{2}\right)^y=\dfrac{1}{32}=\left(\dfrac{1}{2}\right)^5$
$y=5$

53. $\log 0.00001=y$
$10^y=0.00001=10^{-5}$
$y=-5$

55. a. $10^4 < 46{,}832 < 10^5$, so $4 < \log 46{,}832 < 5$.
$\log 46{,}832 \approx 4.6705$

b. $10^6 < 1{,}247{,}310 < 10^7$, so
$6 < \log 1{,}247{,}310 < 7$.
$\log 1{,}247{,}310 \approx 6.0960$

c. $10^{-1} < 0.24 < 10^0$, so $-1 < \log 0.24 < 0$.
$\log 0.24 \approx -0.6198$

d. $10^{-6} < 0.0000032 < 10^{-5}$, so
$-6 < \log 0.0000032 < -5$.
$\log 0.0000032 \approx -5.4949$

e. $10^5 < 5.6 \times 10^5 < 10^6$, so
$5 < \log\left(5.6 \times 10^5\right) < 6$.
$\log\left(5.6 \times 10^5\right) \approx 5.7482$

e. $10^{-3} < 5.1 \times 10^{-3} < 10^{-2}$, so
$-3 < \log\left(5.1 \times 10^{-3}\right) < -2$.
$\log\left(5.1 \times 10^{-3}\right) \approx -2.2924$

57. a. $f(94) = \ln 94 \approx 4.5433$

b. $f(0.182) = \ln 0.182 \approx -1.7037$

c. $f\left(\sqrt{155}\right) = \ln\sqrt{155} \approx 2.5217$

d. $f(4\pi) = \ln 4\pi \approx 2.5310$

e. $f\left(3.9 \times 10^9\right) = \ln\left(3.9 \times 10^9\right) \approx 22.0842$

f. $f\left(7.1 \times 10^{-4}\right) = \ln\left(7.1 \times 10^{-4}\right) \approx -7.2502$

59. $\log_4 4^{11} = 11$

61. $\log_c c = 1$

63. $5^{\log_5(x+y)} = x + y$

65. $\ln e^{a+b} = a + b$

67. $\log_{\sqrt{5}} 1 = 0$

69. Interchange the *x*- and *y*-coordinates of the ordered pairs from its inverse function.

x	$y = 3^x$
-2	$\frac{1}{9}$
-1	$\frac{1}{3}$
0	1
1	3
2	9

x	$y = \log_3 x$
$\frac{1}{9}$	-2
$\frac{1}{3}$	-1
1	0
3	1
9	2

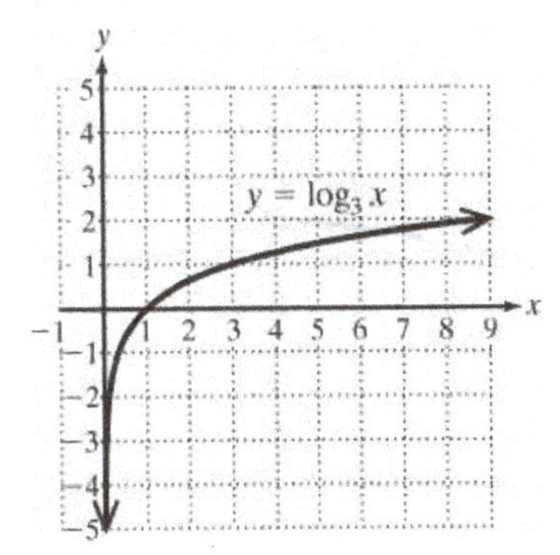

71. Interchange the *x*- and *y*-coordinates of the ordered pairs from its inverse function.

x	$y = \left(\frac{1}{3}\right)^x$
-2	9
-1	3
0	1
1	$\frac{1}{3}$
2	$\frac{1}{9}$

x	$y = \log_{1/3} x$
9	-2
3	-1
1	0
$\frac{1}{3}$	1
$\frac{1}{9}$	2

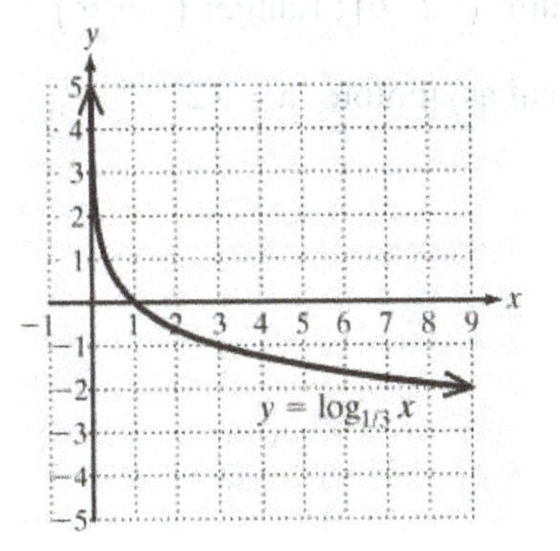

73. Interchange the x- and y-coordinates of the ordered pairs from its inverse function.

x	$y=e^x$
-2	$\frac{1}{e^2}\approx 0.14$
-1	$\frac{1}{e}\approx 0.37$
0	1
1	$e\approx 2.72$
2	$e^2\approx 7.39$

x	$y=\ln x$
0.14	-2
0.37	-1
1	0
2.72	1
7.39	2

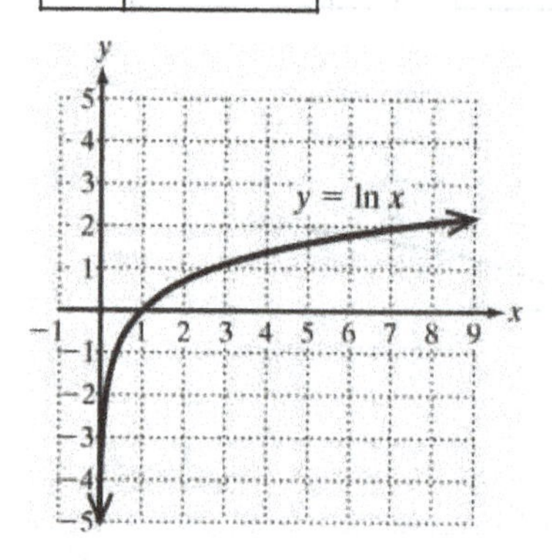

75. a. The graph of $y=\log_3(x+2)$ is the graph of $y=\log_3 x$ shifted to the left 2 units.

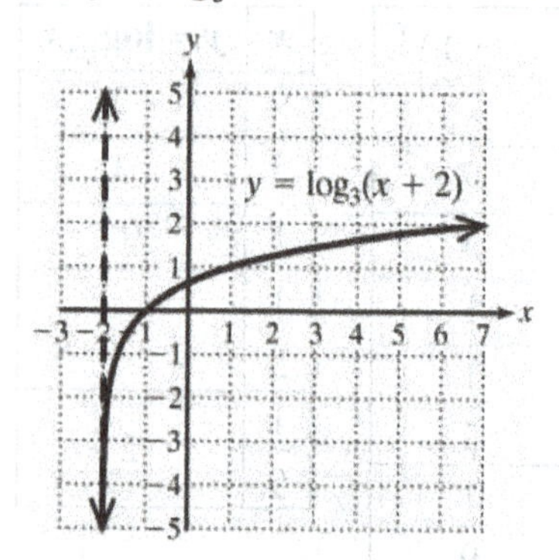

b. Domain: $(-2,\infty)$; Range: $(-\infty,\infty)$

c. Vertical asymptote: $x=-2$

77. a. The graph of $y=2+\log_3 x$ is the graph of $y=\log_3 x$ shifted up 2 units.

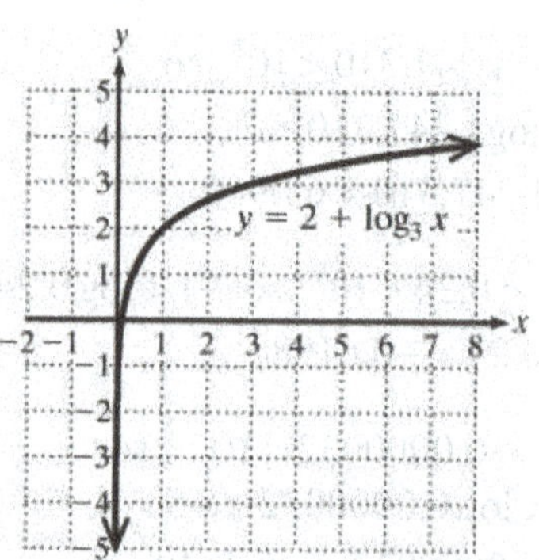

b. Domain: $(0,\infty)$; Range: $(-\infty,\infty)$

c. Vertical asymptote: $x=0$

79. a. The graph of $y=\log_3(x-1)-3$ is the graph of $y=\log_3 x$ shifted right 1 unit and down 3 units.

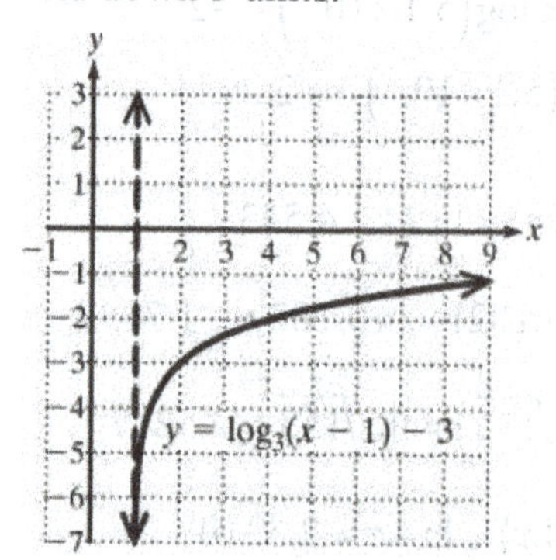

b. Domain: $(1,\infty)$; Range: $(-\infty,\infty)$

c. Vertical asymptote: $x=1$

81. a. The graph of $y=-\log_3 x$ is the graph of $y=\log_3 x$ reflected across the x-axis.

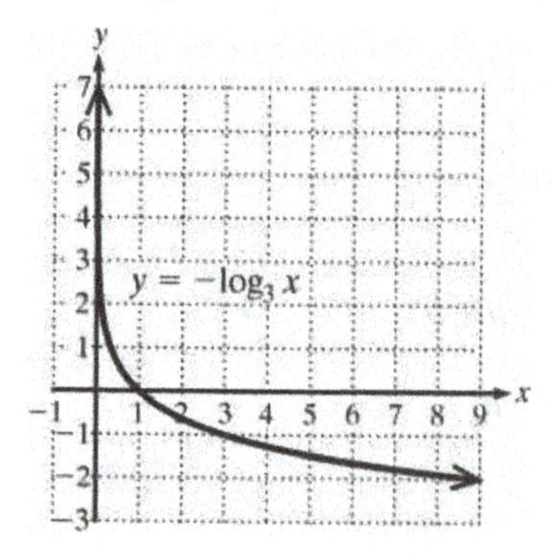

b. Domain: $(0,\infty)$; Range: $(-\infty,\infty)$

c. Vertical asymptote: $x = 0$

83. $f(x) = \log(8 - x)$

$8 - x > 0$

$-x > -8$

$x < 8$ Domain: $(-\infty, 8)$

85. $h(x) = \log_2(6x + 7)$

$6x + 7 > 0$

$6x > -7$

$x > -\frac{7}{6}$ Domain: $\left(-\frac{7}{6}, \infty\right)$

87. $m(x) = \ln(x^2 + 14)$

$x^2 + 14 > 0$

$x^2 > -14$

The square of any real number is positive, so this condition is always true.

Domain: $(-\infty, \infty)$

89. $p(x) = \log(x^2 - x - 12)$

$x^2 - x - 12 > 0$

Find the real zeros of the related equation.

$x^2 - x - 12 = 0$

$(x + 3)(x - 4) = 0$

$x = -3$ or $x = 4$

The boundary points are -3 and 4.

Sign of $(x+3)$:	$-$	$+$	$+$
Sign of $(x-4)$:	$-$	$-$	$+$
Sign of $(x+3)(x-4)$:	$+$	$-$	$+$
	-3	4	

Domain: $(-\infty, -3) \cup (4, \infty)$

91. $r(x) = \log_3(4 - x)^2$

$(4 - x)^2 > 0$

This is true for all real x, except when $4 - x = 0$, or $4 = x$.

Domain: $(-\infty, 4) \cup (4, \infty)$

93. a. $M = \log\left(\frac{I}{I_0}\right) = \log\left(\frac{10^{6.9} I_0}{I_0}\right)$

$= \log 10^{6.9} = 6.9$

b. $M = \log\left(\frac{I}{I_0}\right) = \log\left(\frac{10^{3.2} I_0}{I_0}\right)$

$= \log 10^{3.2} = 3.2$

c. $\frac{10^{6.9} I_0}{10^{3.2} I_0} = 10^{6.9-3.2} = 10^{3.7} \approx 5012$

The Loma Prieta earthquake was approximately 5012 times more intense than an earthquake with magnitude 3.2.

95. a. $L = 10\log\left(\frac{10^{15} I_0}{I_0}\right) = 10\log 10^{15}$

$= 10(15) = 150$ dB

b. $L = 10\log\left(\frac{10^{9} I_0}{I_0}\right) = 10\log 10^{9}$

$= 10(9) = 90$ dB

c. $\frac{10^{15} I_0}{10^{9} I_0} = 10^{15-9} = 10^6 = 1{,}000{,}000$

The sound of a jet plane taking off is 1,000,000 times more intense than the noise from city traffic.

97. a. $\text{pH} = -\log(5.0 \times 10^{-3}) \approx 2.3$

b. $\text{pH} = -\log(1 \times 10^{-2}) = -(-2) = 2$

c. Lemon juice is more acidic.

99. $\varphi(d) = -\log_2 d$

$\varphi(128) = -\log_2 128 = -7$

$\varphi(8) = -\log_2 8 = -3$

$\varphi(4) = -\log_2 4 = -2$

$\varphi(1) = -\log_2 1 = 0$

$\varphi(0.5) = -\log_2 0.5 = -(-1) = 1$

101. a. $\log_3(x+1)=4 \Leftrightarrow 3^4 = x+1$

b. $3^4 = x+1$

$81 = x+1$

$80 = x$

The solution set is $\{80\}$.

c. $\log_3(x+1)=4$

$\log_3(80+1) 0 4$

$\log_3 81 0 4$

$4 0 4$ ✓

103. a. $\log_4(7x-6)=3 \Leftrightarrow 4^3 = 7x-6$

b. $4^3 = 7x-6$

$64 = 7x-6$

$70 = 7x$

$10 = x$

The solution set is $\{10\}$.

c. $\log_4(7x-6)=3$

$\log_4[7(10)-6] 0 3$

$\log_4(70-6) 0 3$

$\log_4 64 0 3$

$3 0 3$ ✓

105. $\log_3(\log_4 64) = \log_3 3 = 1$

107. $\log_{16}(\log_{81} 3) = \log_{16}\left(\frac{1}{4}\right) = -\frac{1}{2}$

109. a. $\log_2 2 + \log_2 4 = 1 + 2 = 3$

b. $\log_2(2\cdot 4) = \log_2 8 = 3$

c. They are the same.

111. a. $\log_4 64 - \log_4 4 = 3 - 1 = 2$

b. $\log_4\left(\frac{64}{4}\right) = \log_4 16 = 2$

c. They are the same.

113. a. $\log_2 2^5 = 5$

b. $5\cdot\log_2 2 = 5\cdot 1 = 5$

c. They are the same.

115. a. $t(r) = \frac{\ln 2}{r}$

$t(0.035) = \frac{\ln 2}{0.035} \approx 19.8$ yr

b. $t(0.04) = \frac{\ln 2}{0.04} \approx 17.3$ yr

$t(0.06) = \frac{\ln 2}{0.06} \approx 11.6$ yr

$t(0.08) = \frac{\ln 2}{0.08} \approx 8.7$ yr

117. a. $\log(6.022\times 10^{23}) \approx 23.7797$

b. $\log(6.626\times 10^{-34}) \approx -33.1787$

c. Given a number $a\times 10^n$, then $\log(a\times 10^n)$ is between n and $n+1$, inclusive.

119. a. $\frac{f(b)-f(a)}{b-a} \approx \frac{0-(-0.3010)}{1-0.5} \approx 0.6020$

b. $\frac{f(b)-f(a)}{b-a} = \frac{1-0}{10-1} = \frac{1}{9} \approx 0.1111$

c. $\frac{f(b)-f(a)}{b-a} \approx \frac{1.3010-1}{20-10} \approx 0.0301$

d. $\frac{f(b)-f(a)}{b-a} \approx \frac{1.4771-1.3010}{30-20} \approx 0.0176$

121. $t(x) = \log_4\left(\frac{x-1}{x-3}\right)$

$\frac{x-1}{x-3} > 0$

The boundary points are 1 and 3.

Sign of $(x-1)$:	$-$	$+$	$+$
Sign of $(x-3)$:	$-$	$-$	$+$
Sign of $\frac{x-1}{x-3}$:	$+$	$-$	$+$
	1	3	

Domain: $(-\infty, 1)\cup(3, \infty)$

123. $s(x) = \ln\left(\sqrt{x+5}-1\right)$

$\sqrt{x+5}-1>0$

$\sqrt{x+5}>1$

$x+5>1$

$x>-4$ Domain: $(-4, \infty)$

125. $c(x) = \log\left(\frac{1}{\sqrt{x-6}}\right)$

$\sqrt{x-6}>0$

$x-6>0$

$x>6$ Domain: $(6, \infty)$

127. a.

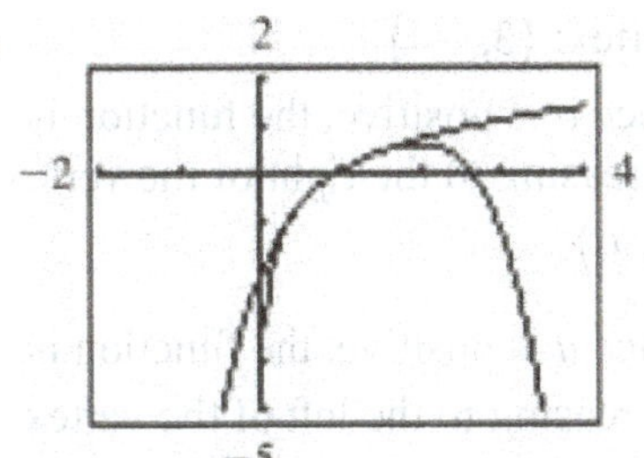

The graphs match closely on the interval $(0, 2)$

b. $\ln 1.5 \approx 0.4010$

129.

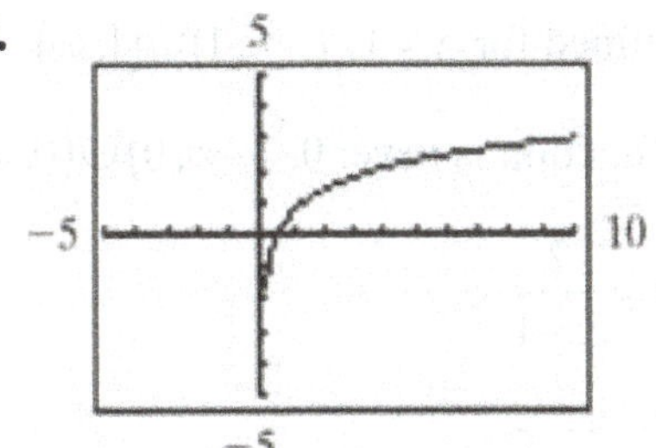

The graphs are the same.

Problem Recognition Exercises: Analyzing Functions

1. $f(x)=3$

a. $(-\infty, \infty)$

b. $\{3\}$

c. $f(x)=3$

$0=3$

No x-intercept

d. $f(x)=3$

$f(0)=3$

y-intercept: $(0, 3)$

e. No asymptotes

f. The slope is 0, so the function is never increasing.

g. The slope is 0, so the function is never deceasing.

h. Graph E

3. $d(x)=(x-3)^2-4=1(x-3)^2+(-4)$

The graph of this function is the graph of $y=x^2$ shifted right 3 units and down 4 units.

a. $(-\infty, \infty)$ **b.** $[-4, \infty)$

c. $d(x)=(x-3)^2-4$

$0=(x-3)^2-4$

$4=(x-3)^2$

$\pm 2=x-3$

$x=1$ or $x=5$

x-intercepts: $(1, 0)$ and $(5, 0)$

d. $d(x)=(x-3)^2-4$

$d(0)=(0-3)^2-4=9-4=5$

y-intercept: $(0, 5)$

e. No asymptotes

f. Vertex: $(3,-4)$
Since a is positive, the function is increasing to the right of the vertex.
$(3,\infty)$

g. Since a is positive, the function is decreasing to the left of the vertex. $(-\infty,3)$

h. Graph N

5. $k(x)=\dfrac{2}{x-1}$

a. Undefined for $x=1$. $(-\infty,1)\cup(1,\infty)$

b. The function is never 0. $(-\infty,0)\cup(0,\infty)$

c. $k(x)=\dfrac{2}{x-1}$

$0=\dfrac{2}{x-1}$

$0=2$

No x-intercept

d. $k(x)=\dfrac{2}{x-1}$

$k(0)=\dfrac{2}{0-1}=-2$

y-intercept: $(0,-2)$

e. Vertical asymptote: $x=1$
The degree of the numerator is 0.
The degree of the denominator is 1.
Horizontal asymptote: $y=0$

f. Never increasing **g.** $(-\infty,1)\cup(1,\infty)$

h. Graph L

7. $p(x)=\left(\dfrac{4}{3}\right)^x$

a. $(-\infty,\infty)$ **b.** $(0,\infty)$

c. $p(x)=\left(\dfrac{4}{3}\right)^x$

$0=\left(\dfrac{4}{3}\right)^x$ Not possible

No x-intercept

d. $p(x)=\left(\dfrac{4}{3}\right)^x$

$p(0)=\left(\dfrac{4}{3}\right)^0=1$

y-intercept: $(0,1)$

e. Horizontal asymptote: $y=0$

f. $(-\infty,\infty)$

g. Never decreasing **h.** Graph M

9. $m(x)=|x-4|-1$
The graph of this function is the graph of $y=|x|$ shifted right 4 units and down 1 unit.

a. $(-\infty,\infty)$ **b.** $[-1,\infty)$

c. $m(x)=|x-4|-1$

$0=|x-4|-1$

$1=|x-4|$

$x-4=1$ or $x-4=-1$

$x=5$ $\quad$ $x=3$

x-intercepts: $(3,0)$ and $(5,0)$

d. $m(x)=|x-4|-1$

$m(0)=|0-4|-1=4-1=3$

y-intercept: $(0,3)$

e. No asymptotes **f.** $(4,\infty)$

g. $(-\infty,4)$ **h.** Graph I

11. $r(x)=\sqrt{3-x}=\sqrt{-(x-3)}$
The graph of this function is the graph of $f(x)=\sqrt{x}$ reflected across the y-axis and shifted right 3 units.

a. $(-\infty,3]$ **b.** $[0,\infty)$

c. $r(x)=\sqrt{3-x}$

$0=\sqrt{3-x}$

$0=3-x$

$x=3$

x-intercept: $(3,0)$

d. $r(x)=\sqrt{3-x}$

$r(0)=\sqrt{3-0}=\sqrt{3}$

y-intercept: $(0, \sqrt{3})$

e. No asymptotes **f.** Never increasing

g. $(-\infty, 3)$ **h.** Graph F

13. $t(x)=e^x+2$

The graph of this function is the graph of $f(x)=e^x$ shifted up 2 units.

a. $(-\infty, \infty)$ **b.** $(2, \infty)$

c. $t(x)=e^x+2$

$0=e^x+2$

$-2=e^x$ No solution

No x-intercept

d. $t(x)=e^x+2$

$t(0)=e^0+2=1+2=3$

y-intercept: $(0, 3)$

e. Horizontal asymptotes: $y=2$

f. $(-\infty, \infty)$

g. Never decreasing **h.** Graph H

Section 4.4 Properties of Logarithms

1. a. $\log_2 4+\log_2 8=2+3=5$

b. $\log_2(4\cdot 8)=\log_2 32=5$

3. a. $\log_4 256-\log_4 16=4-2=2$

b. $\log_4\left(\dfrac{256}{16}\right)=\log_4 16=2$

5. a. $\log_3 81=4$ **b.** $\log_3 3^4=4$

7. 1 **9.** x

11. $\log_b x+\log_b y$

13. $p\log_b x$ **15.** 10; e

17. $\log_5(125z)=\log_5 125+\log_5 z=3+\log_5 z$

19. $\log(8cd)=\log 8+\log c+\log d$

21. $\log_2[(x+y)\cdot z]=\log_2(x+y)+\log_2 z$

23. $\log_{12}\left(\dfrac{p}{q}\right)=\log_{12} p-\log_{12} q$

25. $\ln\left(\dfrac{e}{5}\right)=\ln e-\ln 5=1-\ln 5$

27. $\log\left(\dfrac{m^2+n}{100}\right)=\log(m^2+n)-\log 100$

$=\log(m^2+n)-2$

29. $\log(2x-3)^4=4\log(2x-3)$

31. $\log_6 \sqrt[7]{x^3}=\log_6 x^{3/7}=\dfrac{3}{7}\log_6 x$

33. $\ln 2^{kt}=kt\ln 2$

35. $\log_7\left(\dfrac{1}{7}mn^2\right)=-\log_7 7+\log_7 m+\log_7 n^2$

$=-1+\log_7 m+2\log_7 n$

37. $\log_6\left(\dfrac{p^5}{qt^3}\right)=\log_6 p^5-\log_6 q-\log_6 t^3$

$=5\log_6 p-\log_6 q-3\log_6 t$

39. $\log\left(\dfrac{10}{\sqrt{a^2+b^2}}\right)=\log 10-\log(a^2+b^2)^{1/2}$

$=1-\dfrac{1}{2}\log(a^2+b^2)$

41. $\ln\left(\frac{\sqrt[3]{xy}}{wz^2}\right) = \ln(xy)^{1/3} - \ln w - \ln z^2$

$= \frac{1}{3}(\ln x + \ln y) - \ln w - 2\ln z$

$= \frac{1}{3}\ln x + \frac{1}{3}\ln y - \ln w - 2\ln z$

43. $\ln\sqrt[4]{\frac{a^2+4}{e^3}} = \ln\left(\frac{a^2+4}{e^3}\right)^{1/4} = \frac{1}{4}\ln\left(\frac{a^2+4}{e^3}\right)$

$= \frac{1}{4}\left[\ln(a^2+4) - \ln(e^3)\right]$

$= \frac{1}{4}\ln(a^2+4) - \frac{1}{4}(3)$

$= \frac{1}{4}\ln(a^2+4) - \frac{3}{4}$

45. $\log\left[\frac{2x(x^2+3)^8}{\sqrt{4-3x}}\right]$

$= \log\left[2x(x^2+3)^8\right] - \log\sqrt{4-3x}$

$= \log 2 + \log x + \log\left[(x^2+3)^8\right] - \log(4-3x)^{1/2}$

$= \log 2 + \log x + 8\log(x^2+3) - \frac{1}{2}\log(4-3x)$

47. $\log_5 \sqrt[3]{x\sqrt{5}} = \log_5\left[x(5)^{1/2}\right]^{1/3}$

$= \frac{1}{3}\log_5\left[x(5)^{1/2}\right]$

$= \frac{1}{3}\left[\log_5 x + \log_5(5)^{1/2}\right]$

$= \frac{1}{3}\left(\log_5 x + \frac{1}{2}\log_5 5\right)$

$= \frac{1}{3}\left[\log_5 x + \frac{1}{2}(1)\right]$

$= \frac{1}{3}\log_5 x + \frac{1}{6}$

49. $\log_{15} 3 + \log_{15} 5 = \log_{15}(3\cdot 5) = \log_{15} 15 = 1$

51. $\log_7 98 - \log_7 2 = \log_7\left(\frac{98}{2}\right) = \log_7 49 = 2$

53. $\log 150 - \log 3 - \log 5 = \log 150 - (\log 3 + \log 5)$

$= \log\left(\frac{150}{3\cdot 5}\right) = \log 10 = 1$

55. $2\log_2 x + \log_2 t = \log_2 x^2 + \log_2 t = \log_2(x^2 t)$

57. $4\log_8 m - 3\log_8 n - 2\log_8 p$

$= \log_8 m^4 - \log_8 n^3 - \log_8 p^2$

$= \log_8 m^4 - (\log_8 n^3 + \log_8 p^2)$

$= \log_8\left(\frac{m^4}{n^3 p^2}\right)$

59. $\frac{1}{2}\ln(x+1) - \frac{1}{2}\ln(x-1)$

$= \frac{1}{2}\left[\ln(x+1) - \ln(x-1)\right]$

$= \ln\left[\frac{x+1}{x-1}\right]^{1/2} = \ln\sqrt{\frac{x+1}{x-1}}$

61. $6\log x - \frac{1}{3}\log y - \frac{2}{3}\log z$

$= 6\log x - \frac{1}{3}(\log y + 2\log z)$

$= \log x^6 - \frac{1}{3}\log yz^2 = \log x^6 - \log(yz^2)^{1/3}$

$= \log\left(\frac{x^6}{(yz^2)^{1/3}}\right) = \log\left(\frac{x^6}{\sqrt[3]{yz^2}}\right)$

63. $\frac{1}{3}\log_4 p + \log_4(q^2-16) - \log_4(q-4)$

$= \log_4\left[\frac{p^{1/3}(q^2-16)}{q-4}\right]$

$= \log_4\left[\frac{p^{1/3}(q+4)(q-4)}{q-4}\right]$

$= \log_4\left[p^{1/3}(q+4)\right] = \log_4\left[\sqrt[3]{p}(q+4)\right]$

65. $\frac{1}{2}\left[6\ln(x+2)+\ln x-\ln x^2\right]$

$=\frac{1}{2}\left[\ln(x+2)^6+\ln x-\ln x^2\right]$

$=\frac{1}{2}\left\{\ln\left[\frac{x(x+2)^6}{x^2}\right]\right\}=\frac{1}{2}\left\{\ln\left[\frac{(x+2)^6}{x}\right]\right\}$

$=\ln\left[\frac{(x+2)^6}{x}\right]^{1/2}=\ln\left[\frac{(x+2)^3}{x^{1/2}}\right]$

$=\ln\left[\frac{(x+2)^3}{\sqrt{x}}\right]$

67. $\log(8y^2-7y)+\log y^{-1}$

$=\log\left[(8y^2-7y)(y^{-1})\right]$

$=\log\left[\frac{y(8y-7)}{y}\right]=\log(8y-7)$

69. $\log_b 15=\log_b(3\cdot5)=\log_b 3+\log_b 5$

$\approx 0.565+0.827\approx 1.392$

71. $\log_b 81=\log_b(3^4)=4\log_b 3\approx 4(0.565)\approx 2.26$

73. $\log_b 50=\log_b(2\cdot5^2)=\log_b 2+2\log_b 5$

$\approx 0.356+2(0.827)\approx 2.01$

75. $\log_b\left(\frac{15}{2}\right)=\log_b\left(\frac{3\cdot5}{2}\right)$

$=\log_b 3+\log_b 5-\log_b 2$

$\approx 0.565+0.827-0.356\approx 1.036$

77. $\log_b 100=\log_b(2\cdot5)^2$

$=2\log_b(2\cdot5)$

$=2(\log_b 2+\log_b 5)$

$\approx 2(0.356+0.827)\approx 2.366$

79. a. $8<15<16$

$2^3<15<2^4$

$3<\log_2 15<4$ Between 3 and 4

b. $\log_2 15=\frac{\log 15}{\log 2}\approx 3.9069$

c. $2^{3.9069}\approx 15$

81. a. $1<3<5$

$5^0<3<5^1$

$0<\log_5 3<1$ Between 0 and 1

b. $\log_5 3=\frac{\log 3}{\log 5}\approx 0.6826$

c. $5^{0.6826}\approx 3$

83. a. $0.25<0.3<0.5$

$2^{-2}<0.3<2^{-1}$

$-2<\log_2 0.3<-1$ Between -2 and -1

b. $\log_2 0.3=\frac{\log 0.3}{\log 2}\approx -1.7370$

c. $2^{-1.7370}\approx 0.3$

85. $\log_2(4.68\times10^7)=\frac{\log(4.68\times10^7)}{\log 2}$

≈ 25.4800

$2^{25.4800}\approx 46{,}800{,}000$

87. $\log_4(5.68\times10^{-6})=\frac{\log(5.68\times10^{-6})}{\log 4}\approx -8.7128$

$4^{-8.7128}\approx 5.68\times10^{-6}$

89. $\log e \,0\, \frac{1}{\ln 10}$

$\frac{\ln e}{\ln 10}\,0\,\frac{1}{\ln 10}$

$\frac{1}{\ln 10}\,0\,\frac{1}{\ln 10}$ ✓ True

91. False; $\log_5\left(\frac{1}{125}\right)\neq\frac{1}{\log_5 125}$

(The left side is -3 and the right side is $\frac{1}{3}$.)

93. $\log_4\left(\frac{1}{p}\right) 0 -\log_4 p$

$\log_4(p)^{-1} 0 -\log_4 p$

$-\log_4 p 0 -\log_4 p$ ✓ True

95. False; $\log(10\cdot 10) \neq (\log 10)(\log 10)$
(The left side is 2 and the right side is 1.)

97. $\log_2(7y)+\log_2 10 \log_2(7y)$

$\log_2(7y\cdot 1) 0 \log_2(7y)$

$\log_2(7y) 0 \log_2(7y)$ ✓ True

99. The given statement $\log_5(-5)+\log_5(-25)$ is not defined because the arguments to the logarithmic expressions are not positive real numbers.

101. a. $\frac{\ln(x+h)-\ln x}{h}$

b. $\frac{\ln(x+h)-\ln x}{h}=\frac{1}{h}\left[\ln(x+h)-\ln x\right]$

$=\frac{1}{h}\ln\left(\frac{x+h}{x}\right)$

$=\ln\left(\frac{x+h}{x}\right)^{1/h}$

103. $\log\left(\frac{-b+\sqrt{b^2-4ac}}{2a}\right)+\log\left(\frac{-b-\sqrt{b^2-4ac}}{2a}\right)$

$=\log\left(\frac{-b+\sqrt{b^2-4ac}}{2a}\cdot\frac{-b-\sqrt{b^2-4ac}}{2a}\right)$

$=\log\left[\frac{b^2-(b^2-4ac)}{4a^2}\right]$

$=\log\left(\frac{4ac}{4a^2}\right)$

$=\log\left(\frac{c}{a}\right)$

$=\log c-\log a$

105. $(\log_2 5)(\log_5 9)=\frac{\log 5}{\log 2}\cdot\frac{\log 9}{\log 5}=\frac{\log 9}{\log 2}=\log_2 9$

107. Let $M=\log_b x$ and $N=\log_b y$, which implies that $b^M=x$ and $b^N=y$. Then $\frac{x}{y}=\frac{b^M}{b^N}=b^{M-N}$. Writing the expression $\frac{x}{y}=b^{M-N}$ in logarithmic form, we have

$\log_b\left(\frac{x}{y}\right)=M-N$, or equivalently,

$\log_b\left(\frac{x}{y}\right)=\log_b x-\log_b y$ as desired.

109.

3

−5 10

−3

111.

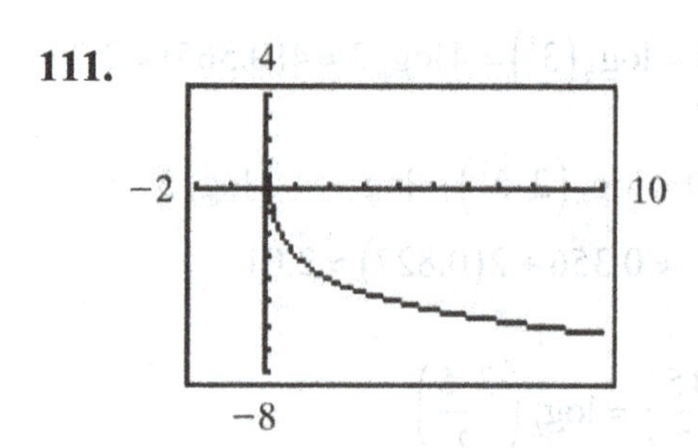

113. a.

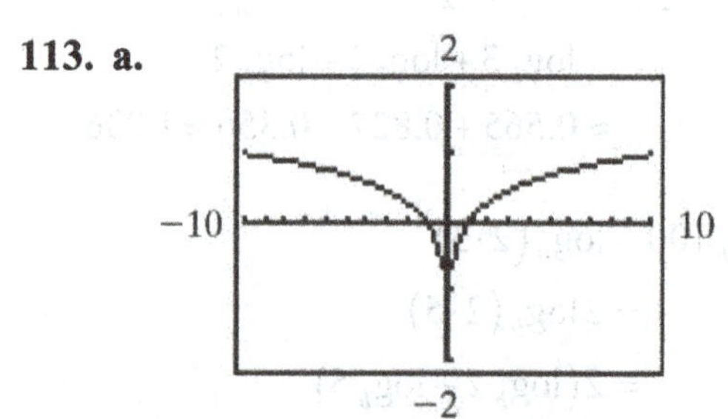

The graphs are the same.

b. $\frac{1}{2}\log x^2=\log\left(x^2\right)^{1/2}=\log\sqrt{x^2}=\log|x|$

Section 4.5 Exponential and Logarithmic Equations

1. $400 = e^{-0.2t} \Leftrightarrow \ln 400 = -0.2t$

3. $\log 11{,}000 = 4w + 6 \Leftrightarrow 10^{4w+6} = 11{,}000$

5. $14x \ln 7 = 2\ln 3 + 7\ln 2$

$$x = \frac{2\ln 3 + 7\ln 2}{14\ln 7} \qquad \left\{\frac{2\ln 3 + 7\ln 2}{14\ln 7}\right\}$$

7. exponential

9. x; y

11. $3^x = 81$

$3^x = 3^4$

$x = 4 \qquad \{4\}$

13. $\sqrt[3]{5} = 5^t$

$5^{1/3} = 5^t$

$\frac{1}{3} = t \qquad \left\{\frac{1}{3}\right\}$

15. $2^{-3y+1} = 16$

$2^{-3y+1} = 2^4$

$-3y + 1 = 4$

$-3y = 3$

$y = -1 \qquad \{-1\}$

17. $11^{3c+1} = \left(\frac{1}{11}\right)^{c-5}$

$11^{3c+1} = \left(11^{-1}\right)^{c-5}$

$11^{3c+1} = 11^{-c+5}$

$3c + 1 = -c + 5$

$4c = 4$

$c = 1 \qquad \{1\}$

19. $8^{2x-5} = 32^{x-6}$

$\left(2^3\right)^{2x-5} = \left(2^5\right)^{x-6}$

$2^{6x-15} = 2^{5x-30}$

$6x - 15 = 5x - 30$

$x = -15 \qquad \{-15\}$

21. $100^{3t-5} = 1000^{3-t}$

$\left(10^2\right)^{3t-5} = \left(10^3\right)^{3-t}$

$10^{6t-10} = 10^{9-3t}$

$6t - 10 = 9 - 3t$

$9t = 19$

$t = \frac{19}{9} \qquad \left\{\frac{19}{9}\right\}$

23. $6^t = 87$

$\log 6^t = \log 87$

$t \log 6 = \log 87$

$t = \frac{\log 87}{\log 6}$

$t \approx 2.4925$

$\left\{\frac{\log 87}{\log 6}\right\}$; $t \approx 2.4925$

25. $1024 = 19^x + 4$

$1020 = 19^x$

$\log 1020 = \log 19^x$

$\log 1020 = x \log 19$

$x = \frac{\log 1020}{\log 19}$

$x \approx 2.3528$

$\left\{\frac{\log 1020}{\log 19}\right\}$; $x \approx 2.3528$

27. $10^{3+4x} - 8100 = 120{,}000$

$10^{3+4x} = 128{,}100$

$\log 10^{3+4x} = \log 128{,}100$

$3 + 4x = \log 128{,}100$

$4x = \log 128{,}100 - 3$

$x = \frac{\log 128{,}100 - 3}{4}$

$x \approx 0.5269$

$\left\{\frac{\log 128{,}100 - 3}{4}\right\}$; $x \approx 0.5269$

29. $21{,}000 = 63{,}000e^{-0.2t}$

$$\frac{1}{3} = e^{-0.2t}$$

$$\ln\left(\frac{1}{3}\right) = \ln e^{-0.2t}$$

$$\ln\left(\frac{1}{3}\right) = -0.2t$$

$$t = \frac{\ln\left(\frac{1}{3}\right)}{-0.2} = \frac{\ln 1 - \ln 3}{-0.2} = \frac{0 - \ln 3}{-0.2}$$

$$t = \frac{\ln 3}{0.2} = \frac{\ln 3}{\frac{1}{5}} = 5\ln 3 \approx 5.4931$$

$\left\{\frac{\ln 3}{0.2}\right\}$ or $\{5\ln 3\}$; $t \approx 5.4931$

31. $3^{6x+5} = 5^{2x}$

$$\ln 3^{6x+5} = \ln 5^{2x}$$

$$(6x+5)\ln 3 = 2x\ln 5$$

$$6x\ln 3 + 5\ln 3 = 2x\ln 5$$

$$6x\ln 3 - 2x\ln 5 = -5\ln 3$$

$$x(6\ln 3 - 2\ln 5) = -5\ln 3$$

$$x = \frac{5\ln 3}{2\ln 5 - 6\ln 3}$$

$$x \approx -1.6286$$

$\left\{\frac{5\ln 3}{2\ln 5 - 6\ln 3}\right\}$; $x \approx -1.6286$

33. $2^{1-6x} = 7^{3x+4}$

$$\ln 2^{1-6x} = \ln 7^{3x+4}$$

$$(1-6x)\ln 2 = (3x+4)\ln 7$$

$$\ln 2 - 6x\ln 2 = 3x\ln 7 + 4\ln 7$$

$$-6x\ln 2 - 3x\ln 7 = 4\ln 7 - \ln 2$$

$$x(-6\ln 2 - 3\ln 7) = 4\ln 7 - \ln 2$$

$$x = \frac{\ln 2 - 4\ln 7}{6\ln 2 + 3\ln 7}$$

$$x \approx -0.7093$$

$\left\{\frac{\ln 2 - 4\ln 7}{6\ln 2 + 3\ln 7}\right\}$; $x \approx -0.7093$

35. $e^{2x} - 9e^{x} - 22 = 0$

$$\left(e^{x}\right)^2 - 9\left(e^{x}\right) - 22 = 0$$

Let $u = e^{x}$.

$$u^2 - 9u - 22 = 0$$

$$(u-11)(u+2) = 0$$

$u = 11$ or $u = -2$

$e^{x} = 11$ or $e^{x} = -2$ No solution

$$\ln e^{x} = \ln 11$$

$$x = \ln 11$$

$$x \approx 2.3979$$

$\{\ln 11\}$; $x \approx 2.3979$

37. $e^{2x} = -9e^{x}$

$$\left(e^{x}\right)^2 + 9\left(e^{x}\right) = 0$$

$$e^{x}\left(e^{x} + 9\right) = 0$$

$e^{x} = 0$ or $e^{x} = -9$

{ }

39. **a.** $\log_2(16-31)\,0\,5 - \log_2 16$

$\log_2(-5)\,0\,5 - 4$

$\log_2(-5)\,0\,1$ ✗

No, the value is not a solution.

b. $\log_2(32-31)\,0\,5 - \log_2 32$

$\log_2 1 0\ 5 - 5$

$0 0\ 0$ ✓

Yes, the value is a solution.

c. $\log_2[16-(-1)]\,0\,5 - \log_2(-1)$

$\log_2(17)\,0\,5 - \log_2(-1)$ ✗

No, the value is not a solution.

41. $\log_4(3w+11) = \log_4(3-w)$

$$3w + 11 = 3 - w$$

$$4w = -8$$

$$w = -2 \qquad \{-2\}$$

43. $\log(x^2+7x)=\log 18$

$x^2+7x=18$

$x^2+7x-18=0$

$(x+9)(x-2)=0$

$x=-9$ or $x=2$ $\{-9, 2\}$

45. $6\log_5(4p-3)=18$

$\log_5(4p-3)=3$

$4p-3=5^3$

$4p-3=125$

$4p=128$

$p=32$ $\{32\}$

47. $\log_8(3y-5)+10=12$

$\log_8(3y-5)=2$

$3y-5=8^2$

$3y-5=64$

$3y=69$

$y=23$ $\{23\}$

49. $\log(p+17)=4.1$

$p+17=10^{4.1}$

$p=10^{4.1}-17\approx 12{,}572.2541$

$\{10^{4.1}-17\}$; $p\approx 12{,}572.2541$

51. $2\ln(4-3t)+1=7$

$2\ln(4-3t)=6$

$\ln(4-3t)=3$

$4-3t=e^3$

$-3t=e^3-4$

$t=\dfrac{4-e^3}{3}\approx -5.3618$

$\left\{\dfrac{4-e^3}{3}\right\}$; $t\approx -5.3618$

53. $\log_2 w-3=-\log_2(w+2)$

$\log_2 w+\log_2(w+2)=3$

$\log_2[w(w+2)]=3$

$w(w+2)=2^3$

$w^2+2w=8$

$w^2+2w-8=0$

$(w-2)(w+4)=0$

$w=2$ or $w=-4$

Check:

$\log_2 w-3=-\log_2(w+2)$

$\log_2(2)-3 0 -\log_2(2+2)$

$\log_2(2)-3 0 -\log_2(4)$

$1-3 0 -2$

$-2 0 -2$ ✓

$\log_2 w-3=-\log_2(w+2)$

$\log_2(-4)-3 0 -\log_2(-4+2)$

$\log_2(-4)-3 0 -\log_2(-2)$

undefined undefined

$\{2\}$; The value -4 does not check.

55. $\log_6(7x-2)=1+\log_6(x+5)$

$\log_6(7x-2)-\log_6(x+5)=1$

$\log_6\left(\dfrac{7x-2}{x+5}\right)=1$

$\dfrac{7x-2}{x+5}=6^1$

$\dfrac{7x-2}{x+5}=6$

$7x-2=6x+30$

$x=32$

$\{32\}$

57. $\log_5 z=3-\log_5(z-20)$

$\log_5 z+\log_5(z-20)=3$

$\log_5[z(z-20)]=3$

$z(z-20)=5^3$

$z^2-20z=125$

$z^2-20z-125=0$

$(z+5)(z-25)=0$

$z=-5$ or $z=25$

Check:

$$\log_5 z = 3 - \log_5(z-20)$$
$$\log_5(-5) 0\ 3 - \log_5(-5-20)$$
$$\log_5(-5) 0\ 3 - \log_5(-25)$$
undefined undefined
$$\log_5 z = 3 - \log_5(z-20)$$
$$\log_5(25) 0\ 3 - \log_5(25-20)$$
$$2\ 0\ 3 - \log_5(5)$$
$$2\ 0\ 3 - 1$$
$$2\ 0\ 2\ ✓$$

$\{25\}$; The value -5 does not check.

59.

$$\ln x + \ln(x-4) = \ln(3x-10)$$
$$\ln x + \ln(x-4) - \ln(3x-10) = 0$$
$$\ln\left[\frac{x(x-4)}{3x-10}\right] = 0$$
$$\frac{x(x-4)}{3x-10} = e^0$$
$$\frac{x^2-4x}{3x-10} = 1$$
$$x^2 - 4x = 3x - 10$$
$$x^2 - 7x + 10 = 0$$
$$(x-2)(x-5) = 0$$

$x = 2$ or $x = 5$

Check:

$$\ln x + \ln(x-4) = \ln(3x-10)$$
$$\ln 2 + \ln(2-4) 0\ \ln[3(2)-10]$$
$$\ln 2 + \ln(-2) 0\ \ln[-4]$$
undefined undefined
$$\ln x + \ln(x-4) = \ln(3x-10)$$
$$\ln 5 + \ln(5-4) 0\ \ln[3(5)-10]$$
$$\ln 5 + \ln 1\ 0\ \ln 5$$
$$\ln 5\ 0\ \ln 5\ ✓$$

$\{5\}$; The value 2 does not check.

61.

$$\log x + \log(x-7) = \log(x-15)$$
$$\log x + \log(x-7) - \log(x-15) = 0$$
$$\log\left[\frac{x(x-7)}{x-15}\right] = 0$$
$$\frac{x(x-7)}{x-15} = 10^0$$
$$\frac{x^2-7x}{x-15} = 1$$
$$x^2 - 7x = x - 15$$
$$x^2 - 8x + 15 = 0$$
$$(x-3)(x-5) = 0$$

$x = 3$ or $x = 5$

Check:

$$\log x + \log(x-7) = \log(x-15)$$
$$\log 3 + \log(3-7) 0\ \log(3-15)$$
$$\log 3 + \log(-4) 0\ \log(-12)$$
undefined undefined
$$\log x + \log(x-7) = \log(x-15)$$
$$\log 5 + \log(5-7) 0\ \log(5-15)$$
$$\log 5 + \log(-2) 0\ \log(-10)$$
undefined undefined

$\{\ \}$; The values 3 and 5 do not check.

63.

$$A = Pe^{rt}$$
$$3(10{,}000) = 10{,}000e^{0.055t}$$
$$3 = e^{0.055t}$$
$$\ln 3 = 0.055t$$
$$\frac{\ln 3}{0.055} = t$$
$$t \approx 20 \text{ yr}$$

65. $A = P\left(1+\frac{r}{n}\right)^{nt}$

Interest is $A - P = \$1000$.

$$4000\left(1+\frac{0.022}{12}\right)^{12t} - 4000 = 1000$$

$$4000\left(1+\frac{0.022}{12}\right)^{12t} = 5000$$

$$\left(1+\frac{0.022}{12}\right)^{12t} = \frac{5}{4}$$

$$\ln\left(1+\frac{0.022}{12}\right)^{12t} = \ln\left(\frac{5}{4}\right)$$

$$12t\ln\left(1+\frac{0.022}{12}\right) = \ln\left(\frac{5}{4}\right)$$

$$t = \frac{\ln\left(\frac{5}{4}\right)}{12\ln\left(1+\frac{0.022}{12}\right)}$$

$$t \approx 10.15$$

10 yr 2 months

67. a. $R = 100(2)^{-t/4.2} = 100(2)^{-2/4.2} \approx 72$ mCi

b.

$$30 = 100(2)^{-t/4.2}$$

$$\frac{3}{10} = (2)^{-t/4.2}$$

$$\ln\left(\frac{3}{10}\right) = \ln(2)^{-t/4.2}$$

$$\ln\left(\frac{3}{10}\right) = -\frac{t\ln 2}{4.2}$$

$$t = -\frac{4.2\ln\left(\frac{3}{10}\right)}{\ln 2} \approx 7.3$$

$7.3 - 2 = 5.3$ days

69.

$$P = e^{-kx}$$

$$0.5 = e^{-kx}$$

$$\ln(0.5) = \ln e^{-kx}$$

$$\ln(0.5) = -kx$$

$$x = \frac{\ln(0.5)}{-k}$$

Ocean: $x = \frac{\ln(0.5)}{-k} = \frac{\ln(0.5)}{-0.0491} \approx 14.1$ m

Tahoe: $x = \frac{\ln(0.5)}{-k} = \frac{\ln(0.5)}{-0.0799} \approx 8.7$ m

Erie: $x = \frac{\ln(0.5)}{-k} = \frac{\ln(0.5)}{-0.1980} \approx 3.5$ m

71.

$$P = e^{-kx}$$

$$0.01 = e^{-kx}$$

$$\ln(0.01) = \ln e^{-kx}$$

$$\ln(0.01) = -kx$$

$$x = \frac{\ln(0.01)}{-k}$$

Ocean: $x = \frac{\ln(0.01)}{-k} = \frac{\ln(0.01)}{-0.0491} \approx 93.8$ m

73.

$$T = 50 + 1550e^{-0.05t}$$

$$100 = 50 + 1550e^{-0.05t}$$

$$50 = 1550e^{-0.05t}$$

$$\frac{1}{31} = e^{-0.05t}$$

$$\ln\left(\frac{1}{31}\right) = \ln e^{-0.05t}$$

$$\ln\left(\frac{1}{31}\right) = -0.05t$$

$$t = \frac{\ln\left(\frac{1}{31}\right)}{-0.05} \approx 69 \text{ or } 1 \text{ hr } 9 \text{ min}$$

75. a. $L = 10\log\left(\frac{I}{I_0}\right)$

$$L = 10\log\left(\frac{3.4\times10^{-8}}{10^{-12}}\right)$$

$$= 10\log\left(3,4\times10^{4}\right) \approx 45.3 \text{ dB}$$

b. $$L = 10\log\left(\frac{I}{10^{-12}}\right)$$
$$\frac{L}{10} = \log\left(\frac{I}{10^{-12}}\right)$$
$$\frac{L}{10} = \log I - \log 10^{-12}$$
$$\frac{L}{10} = \log I + 12\log 10$$
$$\log I = \frac{L}{10} - 12$$
$$I = 10^{\frac{L}{10}-12}$$
$$I = 10^{\frac{30}{10}-12} = 10^{3-12} = 10^{-9}\ \text{W/m}^2$$

77. a. $$L(D) = 8.8 + 5.1\log D$$
$$L(6) = 8.8 + 5.1\log 6 \approx 12.8$$

b. $$L(D) = 8.8 + 5.1\log D$$
$$15.5 = 8.8 + 5.1\log D$$
$$6.7 = 5.1\log D$$
$$\frac{6.7}{5.1} = \log D$$
$$D = 10^{6.7/5.1} \approx 21\ \text{in.}$$

79. a. $$\text{pH} = -\log\left[\text{H}^+\right]$$
$$8.5 = -\log\left[\text{H}^+\right]$$
$$-8.5 = \log\left[\text{H}^+\right]$$
$$10^{-8.5} = \text{H}^+$$
$$\text{H}^+ \approx 3.16\times 10^{-9}\ \text{mol/L}$$

b. $$\text{pH} = -\log\left[\text{H}^+\right]$$
$$2.3 = -\log\left[\text{H}^+\right]$$
$$-2.3 = \log\left[\text{H}^+\right]$$
$$10^{-2.3} = \text{H}^+$$
$$\text{H}^+ \approx 5.01\times 10^{-3}\ \text{mol/L}$$

81. $$S(t) = 94 - 18\ln(t+1)$$
$$65 = 94 - 18\ln(t+1)$$
$$-29 = -18\ln(t+1)$$
$$\frac{29}{18} = \ln(t+1)$$
$$e^{29/18} = t+1$$
$$t = e^{29/18} - 1 \approx 4\ \text{months}$$

83. For $0 < t < 72$:
$$C(t) = -0.088 + 0.89\ln(t+2)$$
$$1.5 = -0.088 + 0.89\ln(t+2)$$
$$1.588 = 0.89\ln(t+2)$$
$$\frac{1.588}{0.89} = \ln(t+2)$$
$$e^{1.588/0.89} = t+2$$
$$t = e^{1.588/0.89} - 2 \approx 4\ \text{min}$$

For $t \geq 72$:
$$C(t) = 4.64e^{-0.003t}$$
$$1.5 = 4.64e^{-0.003t}$$
$$\frac{1.5}{4.64} = e^{-0.003t}$$
$$\ln\left(\frac{1.5}{4.64}\right) = -0.003t$$
$$t = \frac{\ln\left(\frac{1.5}{4.64}\right)}{-0.003} \approx 376\ \text{min}$$

$4\ \text{min} \leq t \leq 376\ \text{min}$

85. $$f(x) = 2^x - 7$$
$$y = 2^x - 7$$
$$x = 2^y - 7$$
$$x + 7 = 2^y$$
$$\log_2(x+7) = y$$
$$f^{-1}(x) = \log_2(x+7)$$

87. $f(x)=\ln(x+5)$
$y=\ln(x+5)$
$x=\ln(y+5)$
$e^x=y+5$
$e^x-5=y$
$f^{-1}(x)=e^x-5$

89. $f(x)=10^{x-3}+1$
$y=10^{x-3}+1$
$x=10^{y-3}+1$
$x-1=10^{y-3}$
$\log(x-1)=y-3$
$\log(x-1)+3=y$
$f^{-1}(x)=\log(x-1)+3$

91. $f(x)=\log(x+7)-9$
$y=\log(x+7)-9$
$x=\log(y+7)-9$
$x+9=\log(y+7)$
$10^{x+9}=y+7$
$10^{x+9}-7=y$
$f^{-1}(x)=10^{x+9}-7$

93. $5^{|x|}-3=122$
$5^{|x|}=125$
$5^{|x|}=5^3$
$|x|=3$
$x=3$ or $x=-3$ $\{-3,3\}$

95. $\log x-2\log 3=2$
$\log\left(\frac{x}{3^2}\right)=2$
$\log\left(\frac{x}{9}\right)=2$
$\frac{x}{9}=10^2$
$x=900$ $\{900\}$

97. $6^{x^2-2}=36$
$6^{x^2-2}=6^2$
$x^2-2=2$
$x^2=4$
$x=\pm4$ $\{-4,4\}$

99. $\log_9|x+4|=\log_9 6$
$|x+4|=6$
$x+4=6$ or $x+4=-6$
$x=2$ $x=-10$ $\{-10,2\}$

101. $x^2e^x=9e^x$
$x^2e^x-9e^x=0$
$e^x(x^2-9)=0$
$e^x(x+3)(x-3)=0$
$e^x=0$ or $x=-3$ or $x=3$
undefined $\{-3,3\}$

103. $\log_3(\log_3 x)=0$
$\log_3 x=3^0$
$\log_3 x=1$
$x=3^1$
$x=3$ $\{3\}$

105. $3|\ln x|-12=0$
$3|\ln x|=12$
$|\ln x|=4$
$\ln x=4$ or $\ln x=-4$
$x=e^4$ $x=e^{-4}$
$x=\frac{1}{e^4}$

$\left\{e^4,\frac{1}{e^4}\right\}$

107. $\log_3 x - \log_3(2x+6) = \frac{1}{2}\log_3 4$

$$\log_3\left(\frac{x}{2x+6}\right) = \log_3 2$$
$$\frac{x}{2x+6} = 2$$
$$x = 4x+12$$
$$-3x = 12$$
$$x = -4$$

Check:

$$\log_3 x - \log_3(2x+6) = \frac{1}{2}\log_3 4$$
$$\log_3(-4) - \log_3[2(-4)+6] \overset{?}{=} \frac{1}{2}\log_3 4$$
$$\log_3(-4) - \log_3[-2] \overset{?}{=} \frac{1}{2}\log_3 4$$

undefined undefined

$\{\ \}$; The value -4 does not check.

109. $2e^x(e^x-3) = 3e^x - 4$

$$2(e^x)^2 - 6e^x - 3e^x + 4 = 0$$
$$2(e^x)^2 - 9e^x + 4 = 0$$

Let $u = e^x$.

$$2u^2 - 9u + 4 = 0$$
$$(2u-1)(u-4) = 0$$

$u = \frac{1}{2}$ or $u = 4$

$e^x = \frac{1}{2}$ $\qquad$ $e^x = 4$

$x = \ln\left(\frac{1}{2}\right)$ $\qquad$ $x = \ln 4$

$\left\{\ln\left(\frac{1}{2}\right), \ln 4\right\}$; $x \approx -0.6931, x \approx 1.3863$

111. If two exponential expressions of the same base are equal, then their exponents must be equal.

113. Take a logarithm of any base b on each side of the equation. Then apply the power property of logarithms to write the product of x and the $\log_b 4$. Finally divide both sides by $\log_b 4$.

115. $\frac{10^x - 13\cdot 10^{-x}}{3} = 4$

$$10^x - 13\cdot 10^{-x} = 12$$
$$10^x(10^x - 13\cdot 10^{-x}) = 12(10^x)$$
$$(10^x)^2 - 12(10^x) - 13 = 0$$

Let $u = 10^x$.

$$u^2 - 12u - 13 = 0$$
$$(u+1)(u-13) = 0$$

$u = -1$ or $u = 13$

$10^x = -1$ $\qquad$ $10^x = 13$

$x = \log(-1)$ $\qquad$ $x = \log 13$

undefined

$\{\log 13\}$; $x \approx 1.1139$

117. $(\ln x)^2 - \ln x^5 = -4$

$$(\ln x)^2 - 5(\ln x) + 4 = 0$$

Let $u = \ln x$.

$$u^2 - 5u + 4 = 0$$
$$(u-1)(u-4) = 0$$

$u = 1$ or $u = 4$

$\ln x = 1$ $\qquad$ $\ln x = 4$

$x = e$ $\qquad$ $x = e^4$

$\{e, e^4\}$; $x \approx 2.7183, x \approx 54.5982$

119. $(\log x)^2 = \log x^2$

$$(\log x)^2 - 2\log x = 0$$
$$\log x(\log x - 2) = 0$$

$\log x = 0$ or $\log x = 2$

$x = 10^0$ $\qquad$ $x = 10^2$

$x = 1$ $\qquad$ $x = 100$

$\{1, 100\}$

121. $\log w + 4\sqrt{\log w} - 12 = 0$

$$\left(\sqrt{\log w}\right)^2 + 4\sqrt{\log w} - 12 = 0$$

Let $u = \sqrt{\log w}$.

$$u^2 + 4u - 12 = 0$$
$$(u+6)(u-2) = 0$$
$$u = -6 \quad \text{or} \quad u = 2$$
$$\sqrt{\log w} = -6 \qquad \sqrt{\log w} = 2$$
$$\text{undefined} \qquad \log w = 4$$
$$w = 10^4 = 10{,}000$$

$\{10{,}000\}$

123. $e^{2x} - 8e^x + 6 = 0$

$$\left(e^x\right)^2 - 8e^x + 6 = 0$$

Let $u = e^x$.

$$u^2 - 8u + 6 = 0$$
$$u = \frac{-(-8) \pm \sqrt{(-8)^2 - 4(1)(6)}}{2(1)}$$
$$= \frac{8 \pm \sqrt{40}}{2} = \frac{8 \pm 2\sqrt{10}}{2} = 4 \pm \sqrt{10}$$
$$u = 4 + \sqrt{10} \quad \text{or} \quad u = 4 - \sqrt{10}$$
$$e^x = 4 + \sqrt{10} \qquad e^x = 4 - \sqrt{10}$$
$$x = \ln\left(4 + \sqrt{10}\right) \qquad x = \ln\left(4 - \sqrt{10}\right)$$

$\left\{\ln\left(4 + \sqrt{10}\right), \ln\left(4 - \sqrt{10}\right)\right\}$; $x \approx 1.9688$, $x \approx -0.1771$

125. $\log_5 \sqrt{6c+5} + \log_5 \sqrt{c} = 1$

$$\log_5\left(\sqrt{6c+5} \cdot \sqrt{c}\right) = 1$$
$$\log_5\left(\sqrt{6c^2 + 5c}\right) = 1$$
$$\frac{1}{2}\log_5\left(6c^2 + 5c\right) = 1$$
$$\log_5\left(6c^2 + 5c\right) = 2$$
$$6c^2 + 5c = 5^2$$
$$6c^2 + 5c - 25 = 0$$
$$(2c+5)(3c-5) = 0$$
$$c = -\frac{5}{2} \quad \text{or} \quad c = \frac{5}{3}$$

Check:

$$\log_5 \sqrt{6c+5} + \log_5 \sqrt{c} = 1$$
$$\log_5 \sqrt{6\left(-\frac{5}{2}\right) + 5} + \log_5 \sqrt{-\frac{5}{2}} \; 0 \; 1$$
$$\log_5 \sqrt{-10} + \log_5 \sqrt{-\frac{5}{2}} \; 0 \; 1$$
$$\text{undefined} \quad \text{undefined}$$
$$\log_5 \sqrt{6c+5} + \log_5 \sqrt{c} = 1$$
$$\log_5 \sqrt{6\left(\frac{5}{3}\right) + 5} + \log_5 \sqrt{\frac{5}{3}} \; 0 \; 1$$
$$\log_5 \sqrt{15} + \log_5 \sqrt{\frac{5}{3}} \; 0 \; 1$$
$$\log_5\left(\sqrt{15} \cdot \sqrt{\frac{5}{3}}\right) 0 \; 1$$
$$\log_5(5) \, 0 \; 1$$
$$1 0 1 \checkmark$$

$\left\{\frac{5}{3}\right\}$; The value $-\frac{5}{2}$ does not check.

127.

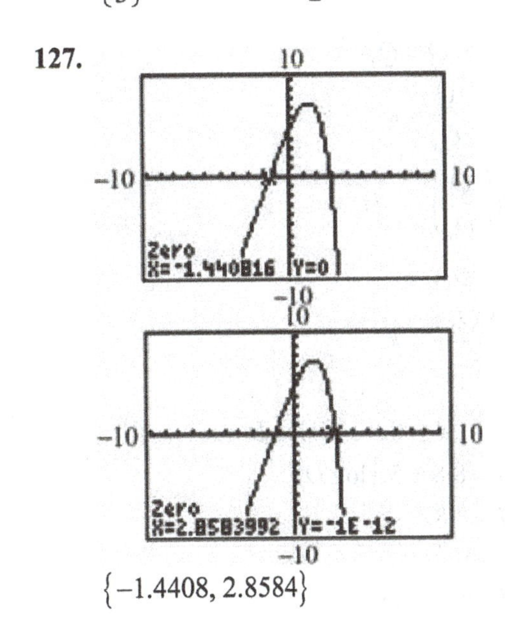

$\{-1.4408, 2.8584\}$

129.

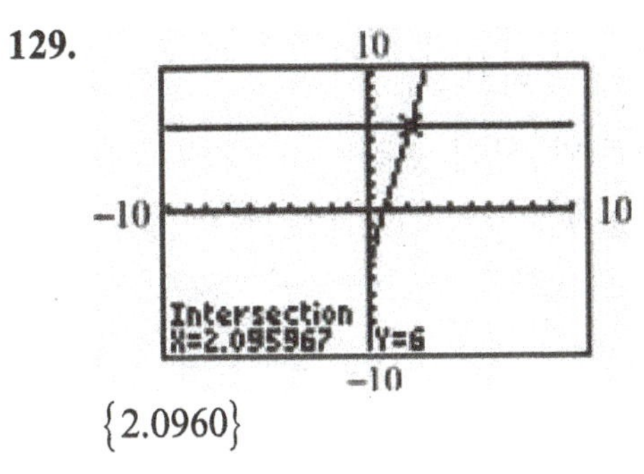

$\{2.0960\}$

Section 4.6 Modeling with Exponential and Logarithmic Functions

1. a and e only

3. $10^{2x-4} = 80{,}600 \Leftrightarrow \log 80{,}600 = 2x - 4$

5.
$$f(x) = e^{x-2}$$
$$y = e^{x-2}$$
$$x = e^{y-2}$$
$$\ln x = y - 2$$
$$2 + \ln x = y \qquad f^{-1}(x) = 2 + \ln x$$

7.
$$5 - 3x > 0$$
$$-3x > -5$$
$$x < \frac{5}{3} \qquad \left(-\infty, \frac{5}{3}\right)$$

9. growth; decay

11. logistic

13.
$$Q = Q_0 e^{-kt}$$
$$\frac{Q}{Q_0} = e^{-kt}$$
$$\ln\left(\frac{Q}{Q_0}\right) = -kt$$
$$-\frac{\ln\left(\frac{Q}{Q_0}\right)}{t} = k$$

15.
$$M = 8.8 + 5.1\log D$$
$$M - 8.8 = 5.1\log D$$
$$\frac{M - 8.8}{5.1} = \log D$$
$$10^{(M-8.8)/5.1} = D$$

17.
$$\text{pH} = -\log\left[\text{H}^+\right]$$
$$-\text{pH} = \log\left[\text{H}^+\right]$$
$$10^{-\text{pH}} = \text{H}^+$$

19.
$$A = P(1+r)^t$$
$$\frac{A}{P} = (1+r)^t$$
$$\ln\left(\frac{A}{P}\right) = \ln(1+r)^t$$
$$\ln\left(\frac{A}{P}\right) = t\ln(1+r)$$
$$\frac{\ln\left(\frac{A}{P}\right)}{\ln(1+r)} = t$$

21.
$$\ln\left(\frac{k}{A}\right) = \frac{-E}{RT}$$
$$\frac{k}{A} = e^{-E/(RT)}$$
$$k = Ae^{-E/(RT)}$$

23. a.
$$A = 12{,}000e^{rt}$$
$$14{,}309.26 = 12{,}000e^{4r}$$
$$\frac{14{,}309.26}{12{,}000} = e^{4r}$$
$$\ln\left(\frac{14{,}309.26}{12{,}000}\right) = 4r$$
$$r = \frac{\ln\left(\frac{14{,}309.26}{12{,}000}\right)}{4} \approx 0.44 \text{ or } 4.4\%$$

b.
$$A = 12{,}000e^{rt}$$
$$20{,}000 = 12{,}000e^{0.044t}$$
$$\frac{20{,}000}{12{,}000} = e^{0.044t}$$
$$\frac{5}{3} = e^{0.044t}$$
$$\ln\left(\frac{5}{3}\right) = 0.044t$$
$$t = \frac{\ln\left(\frac{5}{3}\right)}{0.044} \approx 11.6 \text{ yr}$$

25. a. $A = Pe^{rt}$

$P + 806.07 = Pe^{0.032(3)}$

$P - Pe^{0.096} = -806.07$

$P\left(1 - e^{0.096}\right) = -806.07$

$P = \frac{-806.07}{1 - e^{0.096}} \approx \$8000$

b. $A = Pe^{rt}$

$10{,}000 = 8000e^{0.032t}$

$\frac{10{,}000}{8000} = e^{0.032t}$

$\frac{5}{4} = e^{0.032t}$

$\ln\left(\frac{5}{4}\right) = 0.032t$

$t = \frac{\ln\left(\frac{5}{4}\right)}{0.032} \approx 7 \text{ yr}$

27. a. Australia:

$P(t) = P_0e^{kt}$

$P(t) = 19e^{kt}$

$22.6 = 19e^{k(10)}$

$\frac{22.6}{19} = e^{k(10)}$

$\ln\left(\frac{22.6}{19}\right) = 10k$

$k = \frac{\ln\left(\frac{22.6}{19}\right)}{10} \approx 0.01735$

$P(t) = 19e^{0.01735t}$

Taiwan:

$P(t) = P_0e^{kt}$

$P(t) = 22.9e^{kt}$

$23.7 = 22.9e^{k(10)}$

$\frac{23.7}{22.9} = e^{k(10)}$

$\ln\left(\frac{23.7}{22.9}\right) = 10k$

$k = \frac{\ln\left(\frac{23.7}{22.9}\right)}{10} \approx 0.00343$

$P(t) = 22.9e^{0.00343t}$

b. Australia:

$P(20) = 19e^{0.01735(20)} \approx 26.9 \text{ million}$

Taiwan:

$P(20) = 22.9e^{0.00343(20)} \approx 24.5 \text{ million}$

c. The population growth rate for Australia is greater.

d. Australia:

$30 = 19e^{0.01735t}$

$\frac{30}{19} = e^{0.01735t}$

$\ln\left(\frac{30}{19}\right) = 0.01735t$

$t = \frac{\ln\left(\frac{30}{19}\right)}{0.01735} \approx 26.3 \text{ years}$

In the year 2026

Taiwan:

$30 = 22.9e^{0.00343t}$

$\frac{30}{22.9} = e^{0.00343t}$

$\ln\left(\frac{30}{22.9}\right) = 0.00343t$

$t = \frac{\ln\left(\frac{30}{22.9}\right)}{0.00343} \approx 78.7 \text{ years}$

In the year 2078

29. a. Costa Rica:

$P(t) = 4.3(1.0135)^t$

$P(t) = 4.3e^{(\ln 1.0135)t}$

$P(t) = 4.3e^{0.01341t}$

In 2000, the population is 4.3 million.

Norway:

$P(t) = 4.6(1.0062)^t$

$P(t) = 4.6e^{(\ln 1.0062)t}$

$P(t) = 4.6e^{0.00618t}$

In 2000, the population is 4.6 million.

b. Costa Rica:

$$5 = 4.3e^{0.01341t}$$
$$\frac{5}{4.3} = e^{0.01341t}$$
$$\ln\left(\frac{5}{4.3}\right) = 0.01341t$$
$$t = \frac{\ln\left(\frac{5}{4.3}\right)}{0.01341} \approx 11.2 \text{ years}$$

In the year 2011

Norway:

$$5 = 4.6e^{0.00618t}$$
$$\frac{5}{4.6} = e^{0.00618t}$$
$$\ln\left(\frac{5}{4.6}\right) = 0.00618t$$
$$t = \frac{\ln\left(\frac{5}{4.6}\right)}{0.00618} \approx 13.5 \text{ years}$$

In the year 2013

c. The population growth rate for Costa Rica is greater.

31.
$$Q(t) = Q_0e^{-0.000121t}$$
$$0.78Q_0 = Q_0e^{-0.000121t}$$
$$0.78 = e^{-0.000121t}$$
$$\ln 0.78 = -0.000121t$$
$$t = \frac{\ln 0.78}{-0.000121} \approx 2053 \text{ yr}$$

33. a.
$$Q(t) = Q_0e^{-kt}$$
$$0.5Q_0 = Q_0e^{-k(87.7)}$$
$$0.5 = e^{-k(87.7)}$$
$$\ln 0.5 = -87.7k$$
$$k = \frac{\ln 0.5}{-87.7} \approx 0.0079$$
$$Q(t) = 2e^{-0.0079t}$$

b.
$$Q(t) = 2e^{-0.0079t}$$
$$1.6 = 2e^{-0.0079t}$$
$$0.8 = e^{-0.0079t}$$
$$\ln 0.8 = -0.0079t$$
$$t = \frac{\ln 0.8}{-0.0079} \approx 28 \text{ yr}$$

35. a.
$$Q(t) = Q_0e^{-kt}$$
$$\frac{1}{3}Q_0 = Q_0e^{-k(174)}$$
$$\frac{1}{3} = e^{-k(174)}$$
$$\ln\left(\frac{1}{3}\right) = -174k$$
$$k = \frac{\ln\left(\frac{1}{3}\right)}{-174} \approx 0.0063$$
$$Q(t) = 300e^{-0.0063t}$$

b.
$$Q(t) = Q_0e^{-0.0063t}$$
$$0.5Q_0 = Q_0e^{-0.0063t}$$
$$0.5 = e^{-0.0063t}$$
$$\ln 0.5 = -0.0063t$$
$$t = \frac{\ln 0.5}{-0.0063} \approx 110 \text{ min}$$

37. a.
$$P(t) = 2{,}000{,}000\left(\frac{1}{2}\right)^{t/6}$$
$$= 2{,}000{,}000\left[\left(\frac{1}{2}\right)^t\right]^{1/6}$$
$$= 2{,}000{,}000\left[e^{\ln(1/2)t}\right]^{1/6}$$
$$= 2{,}000{,}000\left[e^{[\ln(1/2)/6]t}\right]$$
$$= 2{,}000{,}000e^{-0.1155t}$$

b. $P(t)=2{,}000{,}000\left(\frac{1}{2}\right)^{t/6}$

$P(0)=2{,}000{,}000\left(\frac{1}{2}\right)^{0/6}=2{,}000{,}000$

$P(6)=2{,}000{,}000\left(\frac{1}{2}\right)^{6/6}=1{,}000{,}000$

$P(12)=2{,}000{,}000\left(\frac{1}{2}\right)^{12/6}=500{,}000$

$P(60)=2{,}000{,}000\left(\frac{1}{2}\right)^{60/6}\approx 1953$

$P(t)=2{,}000{,}000e^{-0.1155t}$

$P(0)=2{,}000{,}000e^{-0.1155t}=2{,}000{,}000$

$P(6)=2{,}000{,}000e^{-0.1155t}\approx 1{,}000{,}000$

$P(12)=2{,}000{,}000e^{-0.1155t}\approx 500{,}000$

$P(60)=2{,}000{,}000e^{-0.1155t}\approx 1956$

39. a. $P(t)=\frac{725}{1+8.295e^{-0.0165t}}$

$P(0)=\frac{725}{1+8.295e^{-0.0165(0)}}=\frac{725}{1+8.295e^{0}}$

$=\frac{725}{1+8.295}\approx 78$

On January 1, 1900, the U.S. population was approximately 78 million.

b. $P(t)=\frac{725}{1+8.295e^{-0.0165t}}$

$P(120)=\frac{725}{1+8.295e^{-0.0165(120)}}$

≈ 338 million

c. $P(t)=\frac{725}{1+8.295e^{-0.0165t}}$

$P(150)=\frac{725}{1+8.295e^{-0.0165(150)}}$

≈ 427 million

d. $P(t)=\frac{725}{1+8.295e^{-0.0165t}}$

$500=\frac{725}{1+8.295e^{-0.0165t}}$

$500\left(1+8.295e^{-0.0165t}\right)=725$

$1+8.295e^{-0.0165t}=1.45$

$8.295e^{-0.0165t}=0.45$

$e^{-0.0165t}=\frac{0.45}{8.295}$

$-0.0165t=\ln\left(\frac{0.45}{8.295}\right)$

$t=\frac{\ln\left(\frac{0.45}{8.295}\right)}{-0.0165}\approx 176.6$

In the year 2076

e. As $t\to\infty$, The denominator of $\frac{8.295}{e^{0.0165t}}$ approaches ∞, so the value of the expression approaches 0.

f. Limiting value: $P(t)=\frac{725}{1+0}=725$ million

41. a. $N(t)=\frac{2.4}{1+15e^{-0.72t}}$

$N(0)=\frac{2.4}{1+15e^{-0.72(0)}}$

$=\frac{2.4}{1+15}\approx 0.15$

150,000 were infected initially.

b. $N(6)=\frac{2.4}{1+15e^{-0.72(6)}}\approx 2$

2,000,000

c. $N(t)=\frac{2.4}{1+15e^{-0.72t}}$

$1=\frac{2.4}{1+15e^{-0.72t}}$

$1+15e^{-0.72t}=2.4$

$15e^{-0.72t}=1.4$

$e^{-0.72t}=\frac{1.4}{15}$

$-0.72t=\ln\left(\frac{1.4}{15}\right)$

$t=\frac{\ln\left(\frac{1.4}{15}\right)}{-0.72}\approx 3.3$ months

d. Limiting value:

$$N(t) = \frac{2.4}{1+0} = 2.4 \text{ million} = 2{,}400{,}000$$

43. exponential

45. logarithmic

47. a. exponential

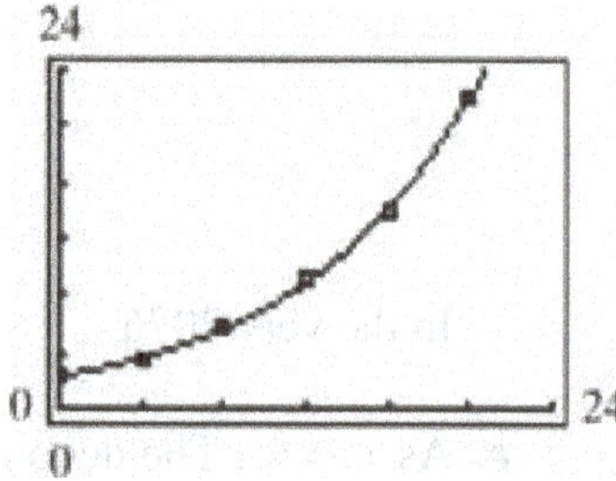

b. $y = 2.3(1.12)^x$

49. a. linear

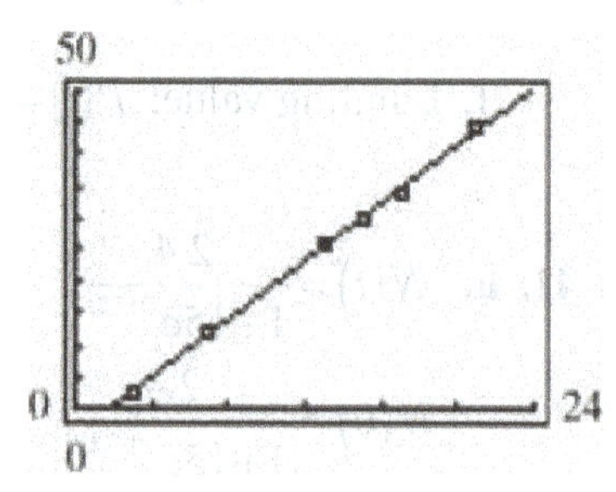

b. $y = 2.28x - 4.08$

51. a. logarithmic

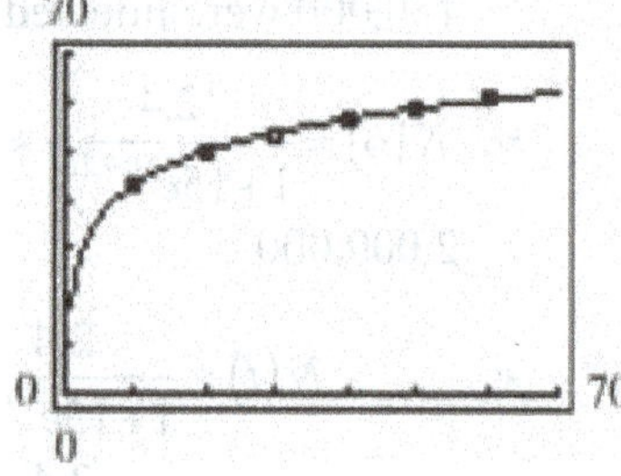

b. $y = 20.7 + 9.72 \ln x$

53. a. logistic

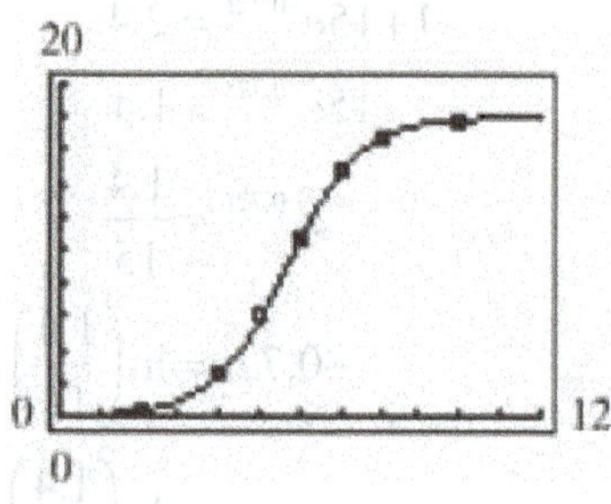

b. $y = \dfrac{18}{1 + 496e^{-1.1x}}$

55. a. $y = 1.4663(1.096)^t$

b. $y = 1.4663(1.096)^t$

$= 1.4663e^{(\ln 1.096)t}$

$= 1.4663e^{0.09167t}$

c. $y = 1.4663e^{0.09167(9 \cdot 12)} \approx 29{,}000$ million

Approximately 29 billion; This is unreasonable because this exceeds the total world population. A logistic model would probably fit the long-term trend better.

d. $y = \dfrac{889}{1 + 576e^{-0.0986t}}$

e. $y = \dfrac{889}{1 + 576e^{-0.0986(9 \cdot 12)}}$

≈ 877 million

57. a. $H(t) = 4.86 + 6.35 \ln t$

b. $25 = 4.86 + 6.35 \ln t$

$20.14 = 6.35 \ln t$

$\dfrac{20.14}{6.35} = \ln t$

$t = e^{20.14/6.35} \approx 24$ yr

c. No, the tree will eventually die.

59. a. $y = mt + b = -2920t + 29{,}200$

b. $y = V_0 b^t$

$y = 29{,}200(0.8)^t$

c. $y = -2920t + 29{,}200$

$y = -2920(5) + 29{,}200 = \$14{,}600$

$y = -2920(10) + 29{,}200 = \$0$

d. $y = 29{,}200(0.8)^t$

$y = 29{,}200(0.8)^5 \approx \$9568$

$y = 29{,}200(0.8)^{10} \approx \$3135$

61. A visual representation of the data can be helpful in determining the type of equation or function that best models the data.

63. An exponential growth model has unbounded growth, whereas a logistic growth model imposes a limiting value on the dependent variable. That is, a logistic growth model has an upper bound restricting the amount of growth.

65. a.
$$P = \frac{\frac{Ar}{12}}{1-\left(1+\frac{r}{12}\right)^{-12t}}$$

$$P\left[1-\left(1+\frac{r}{12}\right)^{-12t}\right] = \frac{Ar}{12}$$

$$P - P\left(1+\frac{r}{12}\right)^{-12t} = \frac{Ar}{12}$$

$$-P\left(1+\frac{r}{12}\right)^{-12t} = \frac{Ar}{12} - P$$

$$\left(1+\frac{r}{12}\right)^{-12t} = 1 - \frac{Ar}{12P}$$

$$\ln\left(1+\frac{r}{12}\right)^{-12t} = \ln\left(1-\frac{Ar}{12P}\right)$$

$$-12t\ln\left(1+\frac{r}{12}\right) = \ln\left(1-\frac{Ar}{12P}\right)$$

$$t = -\frac{\ln\left(1-\frac{Ar}{12P}\right)}{12\ln\left(1+\frac{r}{12}\right)}$$

b. This represents the amount of time (in yr) required to completely pay off a loan of A dollars at interest rate r, by paying P dollars per month.

67. a. $\frac{N(2)-N(1)}{2-1} = \frac{0.52702-0.28911}{1}$
$= 0.23791$

b. $\frac{N(3)-N(2)}{3-2} = \frac{0.87916-0.52702}{1}$
$= 0.35214$

c. $\frac{N(4)-N(3)}{4-3} = \frac{1.3029-0.87916}{1}$
$= 0.42374$

d. $\frac{N(5)-N(4)}{5-4} = \frac{1.7023-1.3029}{1}$
$= 0.39940$

e. $\frac{N(6)-N(5)}{6-5} = \frac{2.0008-1.7023}{1}$
$= 0.29850$

f. $\frac{N(7)-N(6)}{7-6} = \frac{2.1876-2.0008}{1}$
$= 0.18680$

g. The value 0.23791 means that the number of additional computers infected with the virus between month 1 and month 2 increased at a rate of 237,910 per month.

h. The rate of change increases up through approximately month 4 and then begins to decrease. This can be visualized in the graph of the function.

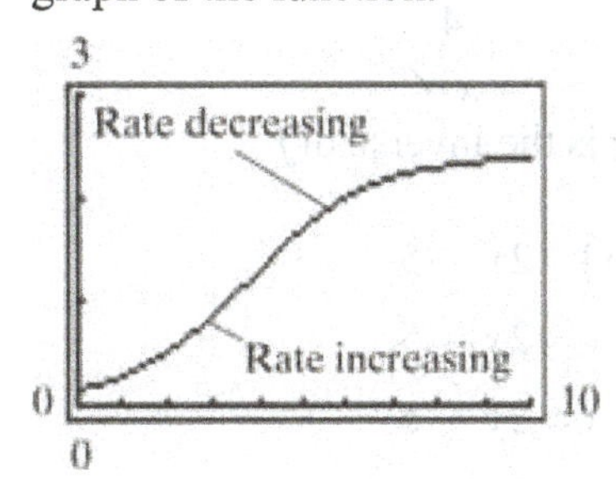

Chapter 4 Review Exercises

1. The relation does not define y as a one-to-one function of x because a horizontal line intersects the graph in more than one point.

3. Assume that $f(a) = f(b)$.

$$a^3 - 1 = b^3 - 1$$
$$a^3 = b^3$$
$$\sqrt[3]{a^3} = \sqrt[3]{b^3}$$
$$a = b$$

Since $f(a) = f(b)$ implies that $a = b$, then f is one-to-one.

5. $(f \circ g)(x) = f(g(x))$

$$= f\left(\frac{x+3}{4}\right)$$
$$= 4\left(\frac{x+3}{4}\right) - 3$$
$$= x + 3 - 3$$
$$= x \checkmark$$

$(g \circ f)(x) = g(f(x))$

$$= g(4x-3)$$
$$= \frac{(4x-3)+3}{4}$$
$$= \frac{4x}{4}$$
$$= x \checkmark$$

Yes, g is the inverse of f.

7.
$$f(x) = 2x^3 - 5$$
$$y = 2x^3 - 5$$
$$x = 2y^3 - 5$$
$$x + 5 = 2y^3$$
$$\frac{x+5}{2} = y^3$$
$$\sqrt[3]{\frac{x+5}{2}} = y$$
$$f^{-1}(x) = \sqrt[3]{\frac{x+5}{2}}$$

9. a.

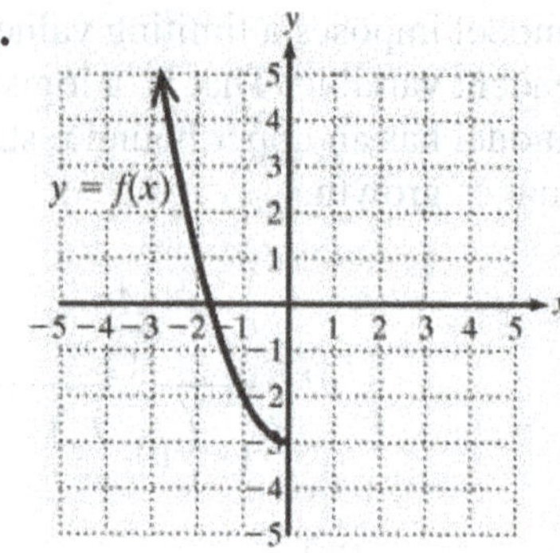

b. Yes

c. $(-\infty, 0]$

d. $[-3, \infty)$

e.
$$f(x) = x^2 - 3$$
$$y = x^2 - 3$$
$$x = y^2 - 3$$
$$x + 3 = y^2$$
$$\pm\sqrt{x+3} = y$$
$$f^{-1}(x) = -\sqrt{x+3}$$

f.

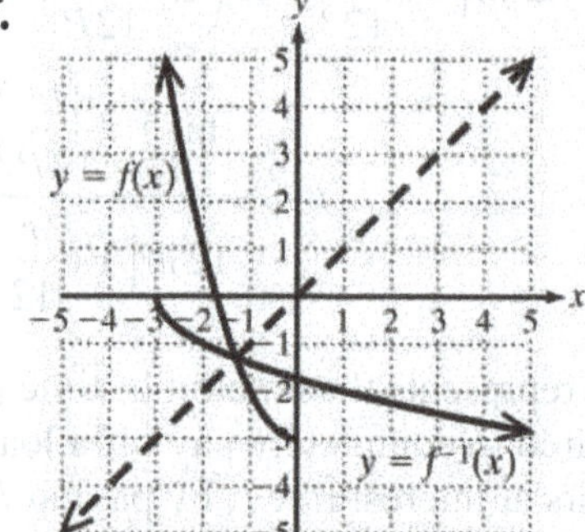

g. $[-3, \infty)$ **h.** $(-\infty, 0]$

11. a.
$$f(x) = 5280x$$
$$y = 5280x$$
$$x = 5280y$$
$$\frac{x}{5280} = y$$
$$f^{-1}(x) = \frac{x}{5280}$$

b. $f^{-1}(x)$ represents the conversion from x feet to $f^{-1}(x)$ miles.

c. $f^{-1}(22{,}176) = \dfrac{22{,}176}{5280} = 4.2 \text{ mi}$

13. a.

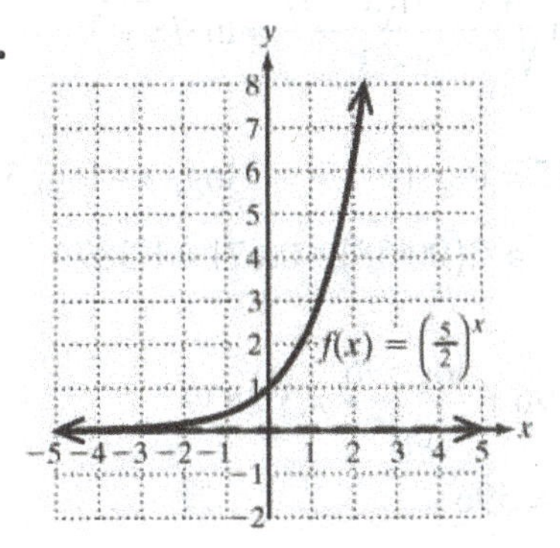

b. Domain: $(-\infty, \infty)$ **c.** Range: $(0, \infty)$

d. $y = 0$

15. a. Since $a < 0$ and $k = 1$, reflect the graph of $y = 3^x$ across the x-axis and shift the graph up 1 unit.

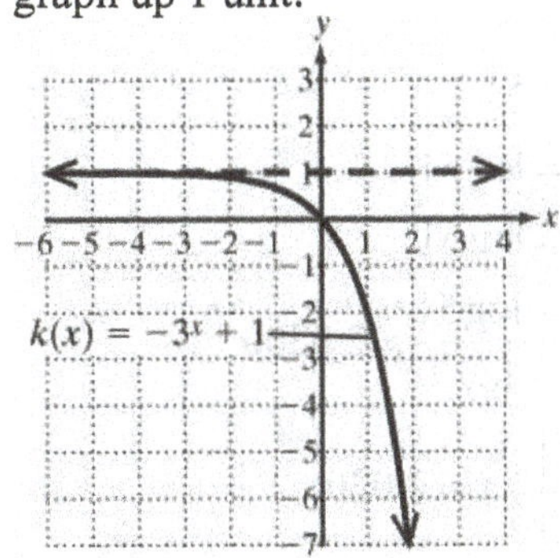

b. Domain: $(-\infty, \infty)$ **c.** Range: $(1, \infty)$

d. $y = 1$

17. The graph of $y = e^x$ is an increasing function.

19. a. $I = Prt = 12{,}000(0.072)(4) = \$3456$

b. $I = Pe^{rt} - P$

$= 12{,}000e^{(0.065)(4)} - 12{,}000$

$\approx 15{,}563.16 - 12{,}000$

$\approx \$3563.16$

c. 7.2% simple interest results in less interest.

21. $\log_b\left(x^2 + y^2\right) = 4 \Leftrightarrow b^4 = x^2 + y^2$

23. $10^6 = 1{,}000{,}000 \Leftrightarrow \log 1{,}000{,}000 = 6$

25. $\log_3 81 = y$

$3^y = 81 = 3^4$

$y = 4$

27. $\log_2\left(\dfrac{1}{64}\right) = y$

$2^y = \dfrac{1}{64} = \dfrac{1}{2^6} = 2^{-6}$

$y = -6$

29. $\log_{11} 1 = y$

$11^y = 1$

$y = 0$

31. $4^{\log_4 7} = 7$

33. $f(x) = \log(x - 4)$

$x - 4 > 0$

$x > 4$

Domain: $(4, \infty)$

35. $h(x) = \log_2\left(x^2 + 4\right)$

$x^2 + 4 > 0$

$x^2 > -4$

The square of any real number is positive, so this condition is always true.

Domain: $(-\infty, \infty)$

37. $m(x) = \log_2(x - 4)^2$

$(x - 4)^2 > 0$

This is true for all real x, except:

$x - 4 = 0$

$x = 4$

Domain: $(-\infty, 4) \cup (4, \infty)$

39. a. The graph of $y=2+\ln x$ is the graph of $y=\ln x$ shifted up 2 units.

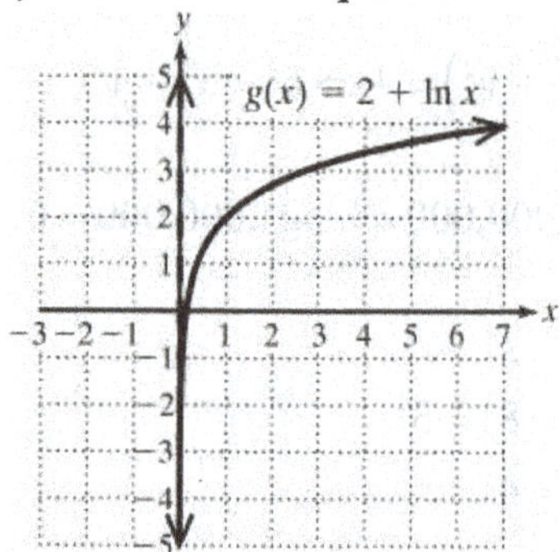

b. Domain: $(0,\infty)$

c. Range: $(-\infty,\infty)$

d. Vertical asymptote: $x=0$

41. a. $\text{pH}=-\log\left(3.16\times10^{-5}\right)\approx4.5$

b. Since the pH is less than 7, tomatoes are acidic.

43. 1

45. x

47. $\log_b x-\log_b y$

49. $\log\left(\frac{100}{\sqrt{c^2+10}}\right)=\log 100-\log\left(c^2+10\right)^{1/2}$

$=2-\frac{1}{2}\log\left(c^2+10\right)$

51. $\ln\left(\frac{\sqrt[3]{ab^2}}{cd^5}\right)=\ln\left(ab^2\right)^{1/3}-\ln c-\ln d^5$

$=\frac{1}{3}(\ln a+2\ln b)-\ln c-5\ln d$

$=\frac{1}{3}\ln a+\frac{2}{3}\ln b-\ln c-5\ln d$

53. $4\log_5 y-3\log_5 x+\frac{1}{2}\log_5 z$

$=\log_5 y^4-\log_5 x^3+\log_5 z^{1/2}$

$=\log_5\left(\frac{y^4\sqrt{z}}{x^3}\right)$

55. $\frac{1}{4}\ln\left(x^2-9\right)-\frac{1}{4}\ln(x-3)$

$=\frac{1}{4}\left[\ln\left(x^2-9\right)-\ln(x-3)\right]=\ln\left[\frac{x^2-9}{x-3}\right]^{1/4}$

$=\ln\sqrt[4]{\frac{(x+3)(x-3)}{x-3}}=\ln\sqrt[4]{x+3}$

57. $\log_b 45=\log_b\left(3^2\cdot5\right)=2\log_b 3+\log_b 5$

$\approx2(0.458)+0.671\approx1.587$

59. $\log_7 596=\frac{\log 596}{\log 7}\approx3.2839$

$7^{3.2839}\approx596$

61.

$$4^{2y-7}=64$$
$$4^{2y-7}=4^3$$
$$2y-7=3$$
$$2y=10$$
$$y=5$$

$\{5\}$

63.

$$7^x=51$$
$$\log 7^x=\log 51$$
$$x\log 7=\log 51$$
$$x=\frac{\log 51}{\log 7}\approx2.0206$$

$\left\{\frac{\log 51}{\log 7}\right\};\ x\approx2.0206$

65.

$$3^{2x+1}=4^{3x}$$
$$\ln 3^{2x+1}=\ln 4^{3x}$$
$$(2x+1)\ln 3=3x\ln 4$$
$$2x\ln 3+\ln 3=3x\ln 4$$
$$2x\ln 3-3x\ln 4=-\ln 3$$
$$x(2\ln 3-3\ln 4)=-\ln 3$$
$$x=\frac{\ln 3}{3\ln 4-2\ln 3}\approx0.5600$$

$\left\{\frac{\ln 3}{3\ln 4-2\ln 3}\right\};\ x\approx0.5600$

67. $400e^{-2t} = 2.989$

$$e^{-2t} = \frac{2.989}{400}$$

$$\ln\left(e^{-2t}\right) = \ln\left(\frac{2.989}{400}\right)$$

$$-2t = \ln\left(\frac{2.989}{400}\right)$$

$$t = \frac{\ln\left(\frac{2.989}{400}\right)}{-2} \approx 2.4483$$

$$\left\{\frac{\ln\left(\frac{2.989}{400}\right)}{-2}\right\}; \ t \approx 2.4483$$

69. $e^{2x} - 3e^x - 40 = 0$

$$\left(e^x\right)^2 - 3\left(e^x\right) - 40 = 0$$

Let $u = e^x$.

$$u^2 - 3u - 40 = 0$$

$$(u-8)(u+5) = 0$$

$u = 8$ or $u = -5$

$e^x = 8$ or $e^x = -5$ No solution

$$\ln e^x = \ln 8$$

$$x = \ln 8 \approx 2.0794$$

$\{\ln 11\}$; $x \approx 2.3979$

71. $\log_5(4p+7) = \log_5(2-p)$

$$4p + 7 = 2 - p$$

$$5p = -5$$

$$p = -1 \qquad \{-1\}$$

73. $2\log_6(4-8y) + 6 = 10$

$$2\log_6(4-8y) = 4$$

$$\log_6(4-8y) = 2$$

$$4 - 8y = 6^2$$

$$4 - 8y = 36$$

$$-8y = 32$$

$$y = -4 \qquad \{-4\}$$

75. $3\ln(n-8) = 6.3$

$$\ln(n-8) = 2.1$$

$$n - 8 = e^{2.1}$$

$$n = e^{2.1} + 8 \approx 16.1662$$

$\left\{e^{2.1} + 8\right\}$; $n \approx 16.1662$

77. $\log_6(3x+2) = \log_6(x+4) + 1$

$$\log_6(3x+2) - \log_6(x+4) = 1$$

$$\log_6\left(\frac{3x+2}{x+4}\right) = 1$$

$$\frac{3x+2}{x+4} = 6^1$$

$$\frac{3x+2}{x+4} = 6$$

$$3x + 2 = 6x + 24$$

$$-3x = 22$$

$$x = -\frac{22}{3}$$

Check:

$$\log_6(3x+2) = \log_6(x+4) + 1$$

$$\log_6\left[3\left(-\frac{22}{3}\right)+2\right] 0 \ \log_6\left(-\frac{22}{3}+4\right)+1$$

$$\log_6(-20) 0 \ \log_6\left(-\frac{10}{3}\right)$$

undefined undefined

$\{\ \}$; The value $-\frac{22}{3}$ does not check.

79. $\log_5(\log_2 x) = 1$

$$\log_2 x = 5^1$$

$$\log_2 x = 5$$

$$x = 2^5$$

$$x = 32 \qquad \{32\}$$

81. $f(x) = 4^x$

$$y = 4^x$$

$$x = 4^y$$

$$\log_4 x = \log_4 4^y$$

$$\log_4 x = y$$

$$f^{-1}(x) = \log_4 x$$

83. a.
$$P = e^{-0.5x}$$
$$0.5 = e^{-0.5x}$$
$$\ln(0.5) = \ln e^{-0.5x}$$
$$\ln(0.5) = -0.5x$$
$$x = \frac{\ln(0.5)}{-0.5} \approx 1.39 \text{ m}$$
Long Island Sound is murky.

b.
$$P = e^{-0.5x}$$
$$0.01 = e^{-0.5x}$$
$$\ln(0.01) = \ln e^{-0.5x}$$
$$\ln(0.01) = -0.5x$$
$$x = \frac{\ln(0.01)}{-0.5} \approx 9.2 \text{ m}$$

85.
$$T = T_f + T_0 e^{-kt}$$
$$-T_0 e^{-kt} = T_f - T$$
$$T_0 e^{-kt} = T - T_f$$
$$e^{-kt} = \frac{T - T_f}{T_0}$$
$$-kt = \ln\left(\frac{T - T_f}{T_0}\right)$$
$$t = -\frac{1}{k}\ln\left(\frac{T - T_f}{T_0}\right)$$

87. a. Since the exponent on e is negative, the population is decreasing.

b.
$$P = 85.5e^{-0.00208t}$$
$$80 = 85.5e^{-0.00208t}$$
$$\frac{80}{85.5} = e^{-0.00208t}$$
$$\ln\left(\frac{80}{85.5}\right) = -0.00208t$$
$$t = \frac{\ln\left(\frac{80}{85.5}\right)}{-0.00208} \approx 32 \text{ yr}$$

89.
$$Q(t) = Q_0 e^{-0.000121t}$$
$$0.712Q_0 = Q_0 e^{-0.000121t}$$
$$0.712 = e^{-0.000121t}$$
$$\ln 0.712 = -0.000121t$$
$$t = \frac{\ln 0.712}{-0.000121} \approx 2800 \text{ yr}$$

b.
$$P(t) = \frac{3000}{1+2e^{-0.37t}}$$
$$P(2) = \frac{3000}{1+2e^{-0.37(2)}} \approx 1535$$

c.
$$P(t) = \frac{3000}{1+2e^{-0.37t}}$$
$$P(4) = \frac{3000}{1+2e^{-0.37(4)}} \approx 2061$$

d.
$$P(t) = \frac{3000}{1+2e^{-0.37t}}$$
$$2800 = \frac{3000}{1+2e^{-0.37t}}$$
$$2800\left(1+2e^{-0.37t}\right) = 3000$$
$$1+2e^{-0.37t} = \frac{15}{14}$$
$$2e^{-0.37t} = \frac{1}{14}$$
$$e^{-0.37t} = \frac{1}{28}$$
$$-0.37t = \ln\left(\frac{1}{28}\right)$$
$$t = \frac{\ln\left(\frac{1}{28}\right)}{-0.37} \approx 9 \text{ yr}$$

e. As $t \to \infty$, The denominator of $\frac{2}{e^{0.37t}}$ approaches ∞, so the value of the expression approaches 0.

f. Limiting value: $P(t) = \frac{3000}{1+0} = 3000$ bass

91. a. $Y_1 = 2.38(1.5)^x$

b.

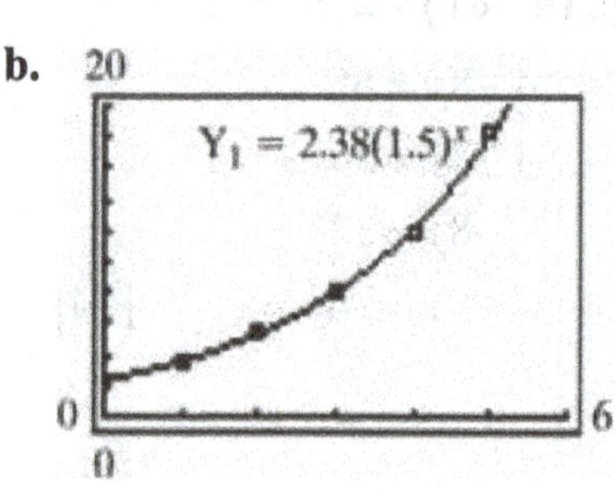

Chapter 4 Test

1. a.
$$f(x)=4x^3-1$$
$$y=4x^3-1$$
$$x=4y^3-1$$
$$x+1=4y^3$$
$$\frac{x+1}{4}=y^3$$
$$\sqrt[3]{\frac{x+1}{4}}=y$$
$$f^{-1}(x)=\sqrt[3]{\frac{x+1}{4}}$$

b.
$$(f\circ f^{-1})(x)=f(f^{-1}(x))$$
$$=f\left(\sqrt[3]{\frac{x+1}{4}}\right)=4\left(\sqrt[3]{\frac{x+1}{4}}\right)^3-1$$
$$=4\left(\frac{x+1}{4}\right)-1=x+1-1=x\ \checkmark$$

$$(f^{-1}\circ f)(x)=f^{-1}(f(x))$$
$$=f^{-1}(4x^3-1)$$
$$=\sqrt[3]{\frac{(4x^3-1)+1}{4}}$$
$$=\sqrt[3]{\frac{4x^3}{4}}=\sqrt[3]{x^3}=x\ \checkmark$$

3.
$$f(x)=\frac{x+3}{x-4}$$
$$y=\frac{x+3}{x-4}$$
$$x=\frac{y+3}{y-4}$$
$$x(y-4)=y+3$$
$$xy-4x=y+3$$
$$xy-y=4x+3$$
$$y(x-1)=4x+3$$
$$y=\frac{4x+3}{x-1}$$
$$f^{-1}(x)=\frac{4x+3}{x-1}$$

5. $f(x)=\log x$

a. Domain: $(0,\infty)$; Range: $(-\infty,\infty)$

b.
$$f(x)=\log x$$
$$y=\log x$$
$$x=\log y$$
$$10^x=10^{\log y}$$
$$10^x=y$$
$$f^{-1}(x)=10^x$$

c. Domain: $(-\infty,\infty)$; Range: $(0,\infty)$

7. a. $f(x)=\left(\frac{1}{3}\right)^x+2$

Since $k=2$, shift the graph of $y=\left(\frac{1}{3}\right)^x$ up 2 units.

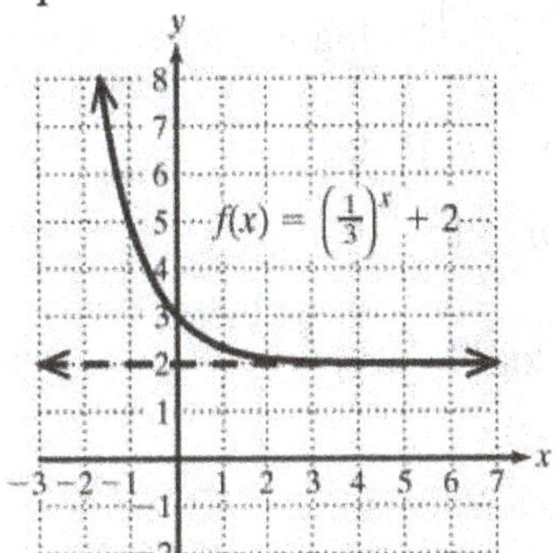

b. Domain: $(-\infty,\infty)$

c. Range: $(2,\infty)$

d. Horizontal asymptote: $y=2$

9. a. The graph of $y=-\ln x$ is the graph of $y=\ln x$ reflected across the x-axis.

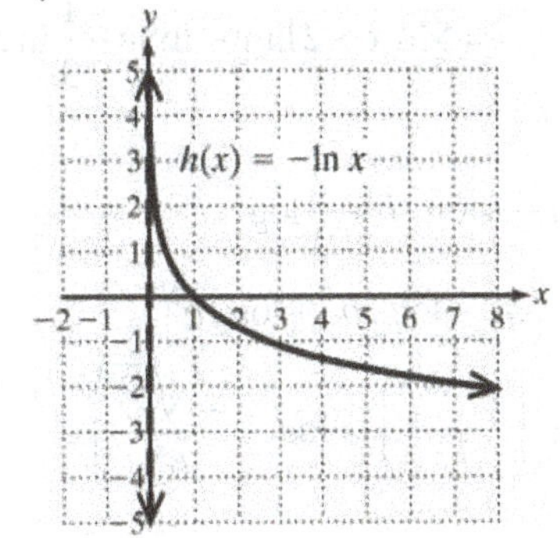

b. Domain: $(0, \infty)$

c. Range: $(-\infty, \infty)$

d. Vertical asymptote: $x = 0$

11. $\ln(x+y) = a \Leftrightarrow e^a = x + y$

13. $\log_9 \frac{1}{81} = y$

$$9^y = \frac{1}{81} = 81^{-1} = (9^2)^{-1} = 9^{-2}$$
$$y = -2$$

15. $\ln e^8 = y$

$$e^y = e^8$$
$$y = 8$$

17. $10^{\log(a^2+b^2)} = y$

$$\log\left(10^{\log(a^2+b^2)}\right) = \log y$$
$$\log(a^2+b^2) = \log y$$
$$a^2 + b^2 = y$$

19. $f(x) = \log(7-2x)$

$$7 - 2x > 0$$
$$-2x > -7$$
$$x < \frac{7}{2}$$

Domain: $\left(-\infty, \frac{7}{2}\right)$

21. $\ln\left(\frac{x^5y^2}{w\sqrt[3]{z}}\right) = \ln(x^5) + \ln(y^2) - \ln w - \ln z^{1/3}$

$$= 5\ln x + 2\ln y - \ln w - \frac{1}{3}\ln z$$

23. $6\log_2 a - 4\log_2 b + \frac{2}{3}\log_2 c$

$$= \log_2 a^6 - \log_2 b^4 + \log_2 c^{2/3}$$
$$= \log_2\left(\frac{a^6c^{2/3}}{b^4}\right) = \log_2\left(\frac{a^6\sqrt[3]{c^2}}{b^4}\right)$$

25. $\log_b 72 = \log_b(2^3 \cdot 3^2) = 3\log_b 2 + 2\log_b 3$

$$\approx 3(0.289) + 2(0.458) \approx 1.783$$

27. $2^{5y+1} = 4^{y-3}$ $\quad\left\{-\frac{7}{3}\right\}$

$$2^{5y+1} = (2^2)^{y-3}$$
$$2^{5y+1} = 2^{2y-6}$$
$$5y + 1 = 2y - 6$$
$$3y = -7$$
$$y = -\frac{7}{3}$$

29. $2^{c+7} = 3^{2c+3}$

$$\ln 2^{c+7} = \ln 3^{2c+3}$$
$$(c+7)\ln 2 = (2c+3)\ln 3$$
$$c\ln 2 + 7\ln 2 = 2c\ln 3 + 3\ln 3$$
$$c\ln 2 - 2c\ln 3 = 3\ln 3 - 7\ln 2$$
$$c(\ln 2 - 2\ln 3) = 3\ln 3 - 7\ln 2$$
$$x = \frac{3\ln 3 - 7\ln 2}{\ln 2 - 2\ln 3} \approx 1.0346$$

$\left\{\frac{3\ln 3 - 7\ln 2}{\ln 2 - 2\ln 3}\right\}$; $x \approx 1.0346$

31. $e^{2x} + 7e^x - 8 = 0$

$$(e^x)^2 + 7(e^x) - 8 = 0$$

Let $u = e^x$.

$$u^2 + 7u - 8 = 0$$
$$(u-1)(u+8) = 0$$

$u = 1$ or $u = -8$

$e^x = 1$ or $e^x = -8$ No solution

$$\ln e^x = \ln 1$$
$$x = \ln 1 = 0 \qquad \{0\}$$

33. $5\ln(x+2) + 1 = 16$

$$\ln(x+2) = 3$$
$$x + 2 = e^3$$
$$x = e^3 - 2 \approx 18.0855$$

$\{e^3 - 2\}$; $x \approx 18.0855$

35. $-3+\log_4 x = -\log_4(x+30)$

$\log_4 x + \log_4(x+30) = 3$

$\log_4[x(x+30)] = 3$

$x(x+30) = 4^3$

$x^2+30x = 64$

$x^2+30x-64 = 0$

$(x+32)(x-2) = 0$

$x = -32$ or $x = 2$

Check:

$-3+\log_4 x = -\log_4(x+30)$

$-3+\log_4(-32) \stackrel{?}{=} -\log_4(-32+30)$

$-3+\log_4(-32) \stackrel{?}{=} -\log_4(-2)$

undefined undefined

$-3+\log_4 x = -\log_4(x+30)$

$-3+\log_4 2 \stackrel{?}{=} -\log_4(2+30)$

$-3 \stackrel{?}{=} -\log_4 32 - \log_4 2$

$-3 \stackrel{?}{=} -(\log_4 32 + \log_4 2)$

$-3 \stackrel{?}{=} -\log_4 64$

$-3 \stackrel{?}{=} -3$ ✓

$\{2\}$; The value -32 does not check.

37. $S = 92 - k\ln(t+1)$

$S-92 = -k\ln(t+1)$

$\dfrac{S-92}{-k} = \ln(t+1)$

$e^{(92-S)/k} = t+1$

$t = e^{(92-S)/k} - 1$

39. a. $A = Pe^{rt}$

$13{,}566.25 = 10{,}000e^{r(5)}$

$\dfrac{13{,}566.25}{10{,}000} = e^{5r}$

$\ln\left(\dfrac{13{,}566.25}{10{,}000}\right) = 5r$

$r = \dfrac{\ln\left(\dfrac{13{,}566.25}{10{,}000}\right)}{5}$

≈ 0.061 or 6.1%

b. $A = Pe^{rt}$

$50{,}000 = 10{,}000e^{0.061t}$

$5 = e^{0.061t}$

$\ln 5 = 0.061t$

$t = \dfrac{\ln 5}{0.061} \approx 26.4$ yr

41. a. $P(t) = \dfrac{1200}{1+2e^{-0.12t}}$

$P(0) = \dfrac{1200}{1+2e^{-0.12)0}} = \dfrac{1200}{1+2e^0} = \dfrac{1200}{1+2} = 400$

400 deer were present when the park service began tracking the herd.

b. $P(t) = \dfrac{1200}{1+2e^{-0.12t}}$

$P(4) = \dfrac{1200}{1+2e^{-0.12(4)}} \approx 536$ deer

c. $P(t) = \dfrac{1200}{1+2e^{-0.12t}}$

$P(8) = \dfrac{1200}{1+2e^{-0.12(8)}} \approx 680$ deer

d. $P(t) = \dfrac{1200}{1+2e^{-0.12t}}$

$900 = \dfrac{1200}{1+2e^{-0.12t}}$

$900(1+2e^{-0.12t}) = 1200$

$1+2e^{-0.12t} = \dfrac{4}{3}$

$2e^{-0.12t} = \dfrac{1}{3}$

$e^{-0.12t} = \dfrac{1}{6}$

$-0.12t = \ln\left(\dfrac{1}{6}\right)$

$t = \dfrac{\ln\left(\dfrac{1}{6}\right)}{-0.12t} \approx 15$ yr

e. As $t \to \infty$, the denominator of $\dfrac{2}{e^{0.12t}}$ approaches ∞, so the value of the expression approaches 0.

f. Limiting value: $P(t) = \dfrac{1200}{1+0} = 1200$ deer

Chapter 4 Cumulative Review Exercises

1. $\dfrac{3x^{-1}-6x^{-2}}{2x^{-2}-x^{-1}}=\dfrac{x^2}{x^2}\cdot\dfrac{3x^{-1}-6x^{-2}}{2x^{-2}-x^{-1}}=\dfrac{3x-6}{2-x}$

$=\dfrac{3(x-2)}{-(x-2)}=-3$

3. $a^3-b^3-a+b=(a-b)(a^2+ab+b^2)-(a-b)$

$=(a-b)(a^2+ab+b^2-1)$

5. $5\le 3+|2x-7|$

$2\le |2x-7|$

$|2x-7|\ge 2$

$2x-7\le -2$ or $2x-7\ge 2$

$2x\le 5$ $\quad$ $2x\ge 9$

$x\le \dfrac{5}{2}$ $\quad$ $x\ge \dfrac{9}{2}$

$\left(-\infty,\dfrac{5}{2}\right]\cup\left[\dfrac{9}{2},\infty\right)$

7. $\sqrt{t+3}+4=t+1$

$\sqrt{t+3}=t-3$

$t+3=(t-3)^2$

$t+3=t^2-6t+9$

$0=t^2-7t+6$

$0=(t-1)(t-6)$

$t=1$ or $t=6$

<u>Check:</u>

$\sqrt{t+3}+4=t+1$

$\sqrt{1+3}+4 0 1+1$

$\sqrt{4}+4 0 2$

$2+4 0 2$

$6 0 2$ ✗

$\sqrt{t+3}+4=t+1$

$\sqrt{6+3}+4 0 6+1$

$\sqrt{9}+4 0 7$

$3+4 0 7$

$7 0 7$ ✓

$\{6\}$; The value 1 does not check.

9. $-x^3-5x^2+4x+20<0$

$-x^2(x+5)+4(x+5)<0$

$(x+5)(-x^2+4)<0$

$(x+5)(x^2-4)>0$

$(x+5)(x+2)(x-2)>0$

Find the real zeros of the related equation.

$(x+5)(x+2)(x-2)=0$

$x=-5$, $x=-2$, or $x=2$

The boundary points are -5, -2, and 2.

Sign of $(x+5)$:	−	+	+	+
Sign of $(x+2)$:	−	−	+	+
Sign of $(x-2)$:	−	−	−	+
Sign of $(x+5)(x+2)(x-2)$:	−	+	−	+
		−5	−2	2

The solution set is $(-5,-2)\cup(2,\infty)$.

11. $(x^2-9)^2-2(x^2-9)-35=0$

Let $u=x^2-9$.

$u^2-2u-35=0$

$(u+5)(u-7)=0$

$u=-5$ or $u=7$

$x^2-9=-5$ $\quad$ $x^2-9=7$

$x^2=4$ $\quad$ $x^2=16$

$x=\pm 2$ $\quad$ $x=\pm 4$

$\{\pm 2,\pm 4\}$

13. $\dfrac{x-4}{x+2}\le 0$

The expression is undefined for $x=-2$. This is a boundary point.

Find the real zeros of the related equation.

$\dfrac{x-4}{x+2}=0$

$x-4=0$

$x=4$

The boundary points are -2 and 4.

Sign of $(x+2)$:	$-$	$+$	$+$
Sign of $(x-4)$:	$-$	$-$	$+$
Sign of $\frac{(x-4)}{(x+2)}$:	$+$	$-$	$+$
		-2	4

The solution set is $(-2, 4]$.

15. a. $f(x) = x^2 - 16x + 55$

Since $a > 0$, the graph of the function opens upward.

b. $x = \frac{-b}{2a} = \frac{-(-16)}{2(1)} = 8$

$y = (8)^2 - 16(8) + 55 = -9$

Vertex: $(8, -9)$

c. The parabola opens upward, so the vertex is the minimum point: $(8, -9)$.

d. The minimum value is -9.

e. $0 = x^2 - 16x + 55$

$0 = (x-5)(x-11)$

$x = 5$ or $x = 11$

x-intercepts: $(5, 0)$ and $(11, 0)$

f. $f(0) = (0)^2 - 16(0) + 55 = 55$

y-intercept: $(0, 55)$

g.

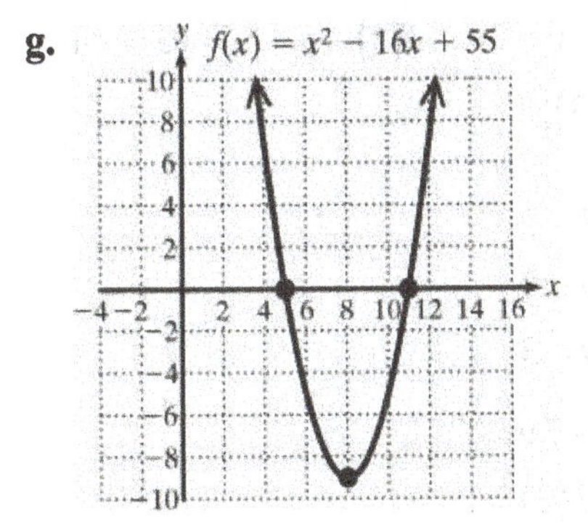

h. The axis of symmetry is $x = 8$.

i. $(-\infty, \infty)$

j. $[-9, \infty)$

17. a. $f(x) = \frac{3x+6}{x-2}$

f is in lowest terms, and $x - 2$ is 0 for $x = 2$, which is the vertical asymptote.

b. The degree of the numerator is 1. The degree of the denominator is 1.

Since $n = m$, the line $y = \frac{3}{1} = 3$ is a horizontal asymptote of f.

c.

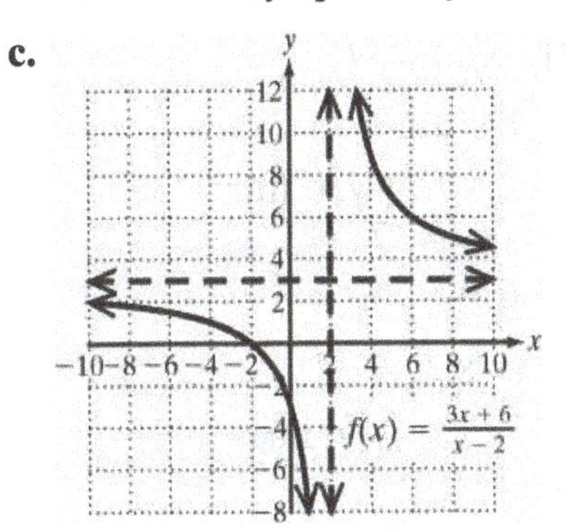

19. $\log 40 + \log 50 - \log 2 = \log\left(\frac{40 \cdot 50}{2}\right)$

$= \log\left(\frac{2000}{2}\right)$

$= \log 1000 = 3$

Chapter 5 Systems of Equations and Inequalities

Section 5.1 Systems of Linear Equations in Two Variables and Applications

1. system

3. substitution; addition **5.** inconsistent

7. a.
$$3x - 5y = -7$$
$$3(1) - 5(2) \stackrel{?}{=} -7$$
$$3 - 10 \stackrel{?}{=} -7$$
$$-7 \stackrel{?}{=} -7 \checkmark \text{ true}$$
$$x - 4y = -7$$
$$(1) - 4(2) \stackrel{?}{=} -7$$
$$1 - 8 \stackrel{?}{=} -7$$
$$-7 \stackrel{?}{=} -7 \checkmark \text{ true}$$
Yes

b.
$$3x - 5y = -7$$
$$3\left(-\frac{2}{3}\right) - 5(1) \stackrel{?}{=} -7$$
$$-2 - 5 \stackrel{?}{=} -7$$
$$-7 \stackrel{?}{=} -7 \checkmark \text{ true}$$
$$x - 4y = -7$$
$$\left(-\frac{2}{3}\right) - 4(1) \stackrel{?}{=} -7$$
$$-\frac{2}{3} - 4 \stackrel{?}{=} -7$$
$$-\frac{14}{3} \stackrel{?}{=} -7 \text{ false}$$
No

9. a.
$$y = \frac{3}{2}x - 5$$
$$(-2) \stackrel{?}{=} \frac{3}{2}(2) - 5$$
$$-2 \stackrel{?}{=} 3 - 5$$
$$-2 \stackrel{?}{=} -2 \checkmark \text{ true}$$
$$6x - 4y = 20$$
$$6(2) - 4(-2) \stackrel{?}{=} 20$$
$$12 + 8 \stackrel{?}{=} 20$$
$$20 \stackrel{?}{=} 20 \checkmark \text{ true}$$
Yes

b.
$$y = \frac{3}{2}x - 5$$
$$(-11) \stackrel{?}{=} \frac{3}{2}(-4) - 5$$
$$-11 \stackrel{?}{=} -6 - 5$$
$$-11 \stackrel{?}{=} -11 \checkmark \text{ true}$$
$$6x - 4y = 20$$
$$6(-4) - 4(-11) \stackrel{?}{=} 20$$
$$-24 + 44 \stackrel{?}{=} 20$$
$$20 \stackrel{?}{=} 20 \checkmark \text{ true}$$
Yes

11. Intersecting lines, one solution.

13. Same line, infinitely many solutions. The equations are dependent.

15.
$$x + 3y = 5$$
$$x = -3y + 5$$
$$3x - 2y = -18$$
$$3(-3y + 5) - 2y = -18$$
$$-9y + 15 - 2y = -18$$
$$-11y = -33$$
$$y = 3$$
$$x = -3y + 5 = -3(3) + 5 = -9 + 5 = -4$$
$$\{(-4, 3)\}$$

17.
$$3y - 7 = 2$$
$$3y = 9$$
$$y = 3$$
$$2x + 7y = 1$$
$$2x + 7(3) = 1$$
$$2x + 21 = 1$$
$$2x = -20$$
$$x = -10$$
$$\{(-10, 3)\}$$

19. $2(x+y)=2-y$

$$2x+2y=2-y$$
$$2x=2-3y$$
$$x=\frac{2-3y}{2}$$

$$4x-1=2-5y$$
$$4\left(\frac{2-3y}{2}\right)-1=2-5y$$
$$2(2-3y)-1=2-5y$$
$$4-6y-1=2-5y$$
$$-y=-1$$
$$y=1$$

$$x=\frac{2-3y}{2}=\frac{2-3(1)}{2}=\frac{2-3}{2}=-\frac{1}{2}$$

$$\left\{\left(-\frac{1}{2},1\right)\right\}$$

$$6y-2=10-7x$$
$$6y-2=10-7\left(\frac{9-3y}{5}\right)$$
$$30y-10=50-7(9-3y)$$
$$30y-10=50-63+21y$$
$$9y=-3$$
$$y=-\frac{1}{3}$$

$$x=\frac{9-3y}{5}=x=\frac{9-3\left(-\frac{1}{3}\right)}{5}=\frac{9+1}{5}=2$$

$$\left\{\left(2,-\frac{1}{3}\right)\right\}$$

21.

$$\begin{aligned} 3x-7y&=1 \xrightarrow{\times -2} & -6x+14y&=-2\\ 6x+5y&=-17 & 6x+5y&=-17\\ & & \hline 19y&=-19\\ & & y&=-1 \end{aligned}$$

$$3x-7y=1$$
$$3x-7(-1)=1$$
$$3x+7=1$$
$$3x=-6$$
$$x=-2 \qquad \{(-2,-1)\}$$

23.

$$11x=-5-4y \qquad 2(x-2y)=22+y$$
$$11x+4y=-5 \qquad 2x-4y=22+y$$
$$2x-5y=22$$

$$\begin{aligned} 11x+4y&=-5 \xrightarrow{\times 5} & 55x+20y&=-25\\ 2x-5y&=22 \xrightarrow{\times 4} & 8x-20y&=88\\ & & \hline 63x&=63\\ & & x&=1 \end{aligned}$$

$$11x=-5-4y$$
$$11(1)=-5-4y$$
$$11=-5-4y$$
$$16=-4y$$
$$-4=y \qquad \{(1,-4)\}$$

25.

$$0.6x+0.1y=0.4 \Rightarrow 6x+y=4$$
$$2x-0.7y=0.3 \Rightarrow 20x-7y=3$$

$$\begin{aligned} 6x+y&=4 \xrightarrow{\times 7} & 42x+7y&=28\\ 20x-7y&=3 & 20x-7y&=3\\ & & \hline 62x&=31\\ & & x&=\frac{1}{2} \end{aligned}$$

$$6x+y=4$$
$$6\left(\frac{1}{2}\right)+y=4$$
$$3+y=4$$
$$y=1 \qquad \left\{\left(\frac{1}{2},1\right)\right\}$$

27.

$$\begin{aligned} 2x+11y&=4 \xrightarrow{\times 3} & 6x+33y&=12\\ 3x-6y&=5 \xrightarrow{\times -2} & -6x+12y&=-10\\ & & \hline 45y&=2\\ & & y&=\frac{2}{45} \end{aligned}$$

$$3x-6y=5$$
$$3x-6\left(\frac{2}{45}\right)=5$$
$$3x-\frac{4}{15}=5$$
$$45x-4=75$$
$$45x=79$$
$$x=\frac{79}{45} \qquad \left\{\left(\frac{79}{45},\frac{2}{45}\right)\right\}$$

29. $3x-4y=6$ $\quad 9x=12y+4$

$9x-12y=4$

$$\begin{array}{l} 3x-4y=6 \xrightarrow{\times -3} -9x+12y=-18 \\ 9x-12y=4 \qquad\quad 9x-12y=4 \\ \hline \qquad\qquad\qquad\qquad 0=-14 \end{array}$$

$\{\ \}$; The system is inconsistent.

31. $3x+y=6$

$y=-3x+6$

$x+\frac{1}{3}y=2$

$x+\frac{1}{3}(-3x+6)=2$

$x-x+2=2$

$2=2$

$\{(x, y)\mid 3x+y=6\}$; The equations are dependent.

33. $2x+4=4-5y$

$2x=5y$

$x=\frac{5}{2}y$

$2+4(x+y)=7y+2$

$4(x+y)=7y$

$4x+4y=7y$

$4x=3y$

$4\left(\frac{5}{2}y\right)=3y$

$10y=3y$

$7y=0$

$y=0$

$x=\frac{5}{2}y=\frac{5}{2}(0)=0$ $\qquad \{(0,0)\}$

35. $5y+800=x$

$x=5y+800$

$3x-10y=1900$

$3(5y+800)-10y=1900$

$15y+2400-10y=1900$

$5y=-500$

$y=-100$

$x=5y+800=5(-100)+800$

$=-500+800=300$ $\qquad \{(300,-100)\}$

37. $x-\frac{3}{2}y=\frac{5}{2}$

$x=\frac{3}{2}y+\frac{5}{2}$

$5(2x+y)=y-x-8$

$10x+5y=y-x-8$

$11x+4y=-8$

$11\left(\frac{3}{2}y+\frac{5}{2}\right)+4y=-8$

$\frac{33}{2}y+\frac{55}{2}+4y=-8$

$33y+55+8y=-16$

$41y=-71$

$y=-\frac{71}{41}$

$x=\frac{3}{2}y+\frac{5}{2}=\frac{3}{2}\left(-\frac{71}{41}\right)+\frac{5}{2}=-\frac{213}{82}+\frac{5}{2}$

$=-\frac{213}{82}+\frac{205}{82}=-\frac{8}{82}=-\frac{4}{41}$

$\left\{\left(-\frac{4}{41},-\frac{71}{41}\right)\right\}$

39. $y=\frac{2}{3}x-1$

$y=\frac{1}{6}x+2$

$\frac{2}{3}x-1=\frac{1}{6}x+2$

$4x-6=x+12$

$3x=18$

$x=6$

$y=\frac{1}{6}x+2=\frac{1}{6}(6)+2=1+2=3$

$\{(6,3)\}$

41. $4(x-2)=6y+3$ $\qquad \frac{1}{4}x-\frac{3}{8}y=-\frac{1}{2}$

$4x-8=6y+3$ $\qquad 2x-3y=-4$

$4x-6y=11$

$$\begin{array}{l} 4x-6y=11 \\ 2x-3y=-4 \xrightarrow{\times -2} \end{array} \quad \begin{array}{r} 4x-6y=11 \\ -4x+6y=8 \\ \hline 0=19 \end{array}$$

$\{\ \}$; The system is inconsistent.

43.
$$2x=\frac{y}{2}+1$$
$$4x=y+2$$
$$4x-2=y$$

$$0.04x-0.01y=0.02$$
$$4x-y=2$$
$$4x-(4x-2)=2$$
$$4x-4x+2=2$$
$$2=2$$

$\{(x, y) \mid 4x-y=2\}$; The equations are dependent.

45. $y=2.4x-1.54$
$y=-3.5x+7.9$

$$2.4x-1.54=-3.5x+7.9$$
$$5.9x=9.44$$
$$x=1.6$$
$$y=2.4x-1.54=2.4(1.6)-1.54$$
$$=3.84-1.54=2.3$$

$\{(1.6, 2.3)\}$

47.
$$\frac{x-2}{8}+\frac{y+1}{2}=-6$$
$$x-2+4y+4=-48$$
$$x=-4y-50$$

$$\frac{x-2}{2}-\frac{y+1}{4}=12$$
$$2(x-2)-(y+1)=48$$
$$2x-4-y-1=48$$
$$-y=-2x+53$$
$$y=2(-4y-50)-53$$
$$y=-8y-100-53$$
$$-9y=153$$
$$y=-17$$
$$x=-4y-50=-4(-17)-50=68-50=18$$

$\{(18, -17)\}$

49. Let x represent the amount of 36% alcohol solution. Let y represent the amount of 20% alcohol solution.

	36% Solution	20% Solution	30% Solution
Amount of mixture	x	y	40
Pure alcohol	$0.36x$	$0.20y$	$0.30(40)=12$

$$\begin{array}{l} x+y=40 \\ 0.36x+0.20y=12 \xrightarrow{\times 100} 36x+20y=1200 \end{array} \quad x+y=40 \xrightarrow{\times -20} \begin{array}{r} -20x-20y=-800 \\ 36x+20y=1200 \\ \hline 16x=400 \\ x=25 \end{array}$$

$$x+y=40$$
$$25+y=40$$
$$y=15$$

25 L of 36% solution and 15 L of 20% solution should be mixed.

51. Let x represent the amount of 100% antifreeze solution. Let y represent the amount of 36% antifreeze solution.

	100% Solution	36% Solution	50% Solution
Amount of mixture	x	y	16
Pure antifreeze	x	$0.36y$	$0.50(16)=8$

$$\begin{aligned} x + \quad y &= 16 \\ x + 0.36y &= 8 \end{aligned} \xrightarrow{\times 100} \begin{aligned} x + y &= 16 \\ 100x + 36y &= 800 \end{aligned}$$

$$x + y = 16 \xrightarrow{\times -100} -100x - 100y = -1600$$

$$\begin{aligned} -100x - 100y &= -1600 \\ 100x + 36y &= 800 \\ \hline -64y &= -800 \\ y &= 12.5 \end{aligned}$$

$$\begin{aligned} x + y &= 16 \\ x + 12.5 &= 16 \\ x &= 3.5 \end{aligned}$$

3.5 L should be replaced.

53. Let x represent the grams of fat in Cherry ice cream. Let y represent the grams of fat in Mint Chocolate Chunk ice cream.

Monique's sundae: $2x + y = 43$

$y = -2x + 43$

Tara's sundae: $x + 2y = 47$

$$\begin{aligned} x + 2(-2x + 43) &= 47 \\ x - 4x + 86 &= 47 \\ -3x &= -39 \\ x &= 13 \end{aligned}$$

$y = -2x + 43 = -2(13) + 43 = -26 + 43 = 17$

Cherry has 13 g of fat and Mint Chocolate Chunk has 17 g of fat.

55. Let x represent the amount borrowed at 4.6%. Let y represent the amount borrowed at 6.2%.

	4.6% interest	**6.2% interest**	**Total**
Principal	x	y	5000
Interest ($I = Prt$)	$x(0.046)(3) = 0.138x$	$y(0.062)(3) = 0.186y$	762

$$\begin{aligned} x + \quad y &= 5000 \\ 0.138x + 0.186y &= 762 \end{aligned} \xrightarrow{\times 1000} \begin{aligned} x + y &= 5000 \\ 138x + 186y &= 762{,}000 \end{aligned}$$

$$x + y = 5000 \xrightarrow{\times -138} -138x - 138y = -690{,}000$$

$$\begin{aligned} -138x - 138y &= -690{,}000 \\ 138x + 186y &= 762{,}000 \\ \hline 48y &= 72{,}000 \\ y &= 1500 \end{aligned}$$

$$\begin{aligned} x + y &= 5000 \\ x + 1500 &= 5000 \\ x &= 3500 \end{aligned}$$

She borrowed \$3500 at 4.6% and \$1500 at 6.2%.

57. Let x represent one employee's weekly salary. Let y represent the other employee's weekly salary.

$$\frac{x + y}{2} = 1350 \qquad x = y + 300$$

$$x + y = 2700$$

$$\begin{aligned} y + 300 + y &= 2700 \\ 2y &= 2400 \\ y &= 1200 \end{aligned}$$

$x = y + 300 = 1200 + 300 = 1500$

One makes \$1200 and the other makes \$1500.

59. Let x represent Josie's walking speed.
Let y represent the speed of the sidewalk.

	Distance (ft)	Rate (ft/sec)	Time (sec)
With sidewalk	200	$x+y$	40
Against sidewalk	90	$x-y$	30

$$d = rt$$
$$200 = (x+y)\cdot 40 = 40x + 40y$$
$$90 = (x-y)\cdot 30 = 30x - 30y$$

$$\begin{aligned} 40x + 40y = 200 &\xrightarrow{\div 4} 10x + 10y = 50 \\ 30x - 30y = 90 &\xrightarrow{\div 3} 10x - 10y = 30 \\ &\qquad\qquad 20x = 80 \\ &\qquad\qquad x = 4 \end{aligned}$$

$$\begin{aligned} 40x + 40y &= 200 \\ 40(4) + 40y &= 200 \\ 160 + 40y &= 200 \\ 40y &= 40 \\ y &= 1 \end{aligned}$$

The sidewalk moves at 1 ft/sec and Josie walks 4 ft/sec on nonmoving ground.

61. Let x represent the speed of one runner Let y represent the speed of the other runner.

	Distance (m)	Rate (m/sec)	Time (sec)
Opposite direction	390	$x+y$	30
Same direction	390	$x-y$	130

$$d = rt$$
$$390 = (x+y)\cdot 30 = 30x + 30y$$
$$390 = (x-y)\cdot 130 = 130x - 130y$$

$$\begin{aligned} 30x + 30y = 390 &\xrightarrow{\div 30} x + y = 13 \\ 130x - 130y = 390 &\xrightarrow{\div 130} x - y = 3 \\ &\qquad\qquad 2x = 16 \\ &\qquad\qquad x = 8 \end{aligned}$$

$$\begin{aligned} 30x + 30y &= 390 \\ 30(8) + 30y &= 390 \\ 240 + 30y &= 390 \\ 30y &= 150 \\ y &= 5 \end{aligned}$$

The speeds are 8 m/sec and 5 m/sec.

63. a. $m = \frac{y_2 - y_1}{x_2 - x_1} = \frac{66.8 - 56.8}{10 - 2} = \frac{10}{8} = 1.25$

$$\begin{aligned} y - y_1 &= m(x - x_1) \\ y - 56.8 &= 1.25(x - 2) \\ y - 56.8 &= 1.25x - 2.5 \\ y &= 1.25x + 54.3 \end{aligned}$$

b. $m = \frac{y_2 - y_1}{x_2 - x_1} = \frac{61.3 - 64.5}{10 - 0} = \frac{-3.2}{10} = -0.32$

$$\begin{aligned} y - y_1 &= m(x - x_1) \\ y - 64.5 &= -0.32(x - 0) \\ y - 64.5 &= -0.32x \\ y &= -0.32x + 64.5 \end{aligned}$$

c. $1.25x + 54.3 = -0.32x + 64.5$

$$\begin{aligned} 1.57x &= 10.2 \\ x &\approx 6.5 \\ y &= -0.32x + 64.5 \\ &\approx -0.32(6.5) + 64.5 \approx 62.4 \end{aligned}$$

{(6.5, 62.4)}; The solution indicates that midyear in 2006, the per capita consumption of beef and chicken was approximately equal at 62.4 lb each.

65. a. $C(x) = 52x + 480$

b. $R(x) = 100x$

c. $C(x) = R(x)$

$$\begin{aligned} 52x + 480 &= 100x \\ -48x &= -480 \\ x &= 10 \text{ offices} \end{aligned}$$

d. $28 > 10$; The company will make money.

67. a.

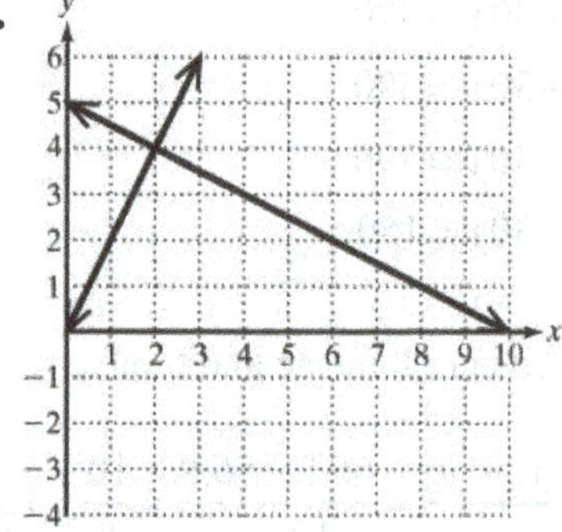

b. Find the intersection of the lines.

$2x = -\frac{1}{2}x + 5$

$4x = -x + 10$

$5x = 10$

$x = 2$

$y = 2x = 2(2) = 4 \qquad (2, 4)$

The slopes are negative reciprocals, so the lines are perpendicular and the triangle is a right triangle.

Length of short leg:

$d = \sqrt{2^2 + 4^2} = \sqrt{4+16} = \sqrt{20} = 2\sqrt{5}$

Length of long leg:

$d = \sqrt{(10-2)^2 + (0-4)^2} = \sqrt{8^2 + (-4)^2}$

$= \sqrt{64+16} = \sqrt{80} = 4\sqrt{5}$

$A = \frac{1}{2}bh = \frac{1}{2}(4\sqrt{5})(2\sqrt{5}) = 20$ square units

69. a. $m = \frac{y_2 - y_1}{x_2 - x_1} = \frac{7-(-1)}{7-(-3)} = \frac{8}{10} = \frac{4}{5}$

$y - y_1 = m(x - x_1)$

$y - (-1) = \frac{4}{5}[x - (-3)]$

$y + 1 = \frac{4}{5}x + \frac{12}{5}$

$y = \frac{4}{5}x + \frac{7}{5}$

b. $m = \frac{y_2 - y_1}{x_2 - x_1} = \frac{1-5}{4-0} = \frac{-4}{4} = -1$

$y - y_1 = m(x - x_1)$

$y - 5 = -1(x - 0)$

$y - 5 = -x$

$y = -x + 5$

c. $\frac{4}{5}x + \frac{7}{5} = -x + 5$

$4x + 7 = -5x + 25$

$9x = 18$

$x = 2$

$y = -x + 5 = -2 + 5 = 3$

The centroid is $(2, 3)$.

71. Let x represent one angle.
Let y represent the other angle.

$x + y = 90$

$x = 2y - 6$

$x + y = 90$

$2y - 6 + y = 90$

$3y = 96$

$y = 32$

$x = 2y - 6 = 2(32) - 6 = 64 - 6 = 58$

The angles are $32°$ and $58°$.

73. $y + 60 + x - 28 = 180$

$y + x = 148$

$x - 28 + x + y = 180$

$y = -2x + 208$

$y + x = 148$

$-2x + 208 + x = 148$

$-x = -60$

$x = 60$

$y = -2x + 208 = -2(60) + 208$

$= -120 + 208 = 88$

$x = 60°,\ y = 88°$

75. For example:

$x + y = a$

$-3 + 5 = a$

$2 = a$

$2x + y = b$

$2(-3) + 5 = b$

$-1 = b$

System: $x + y = 2$

$2x + y = -1$

77. $Cx + 5y = 13$

$C(4) + 5(1) = 13$

$4C + 5 = 13$

$4C = 8$

$C = 2$

$-2x + Dy = -5$

$-2(4) + D(1) = -5$

$-8 + D = -5$

$D = 3$

79.

$$f(x)=mx+b \qquad f(x)=mx+b$$
$$f(3)=-3 \qquad f(-12)=-8$$
$$m(3)+b=-3 \qquad m(-12)+b=-8$$
$$b=-3m-3 \qquad b=12m-8$$

$$-3m-3=12m-8$$
$$-15m=-5$$
$$m=\frac{1}{3}$$

$$b=-3m-3$$
$$b=-3\left(\frac{1}{3}\right)-3$$
$$b=-1-3=-4$$

81. Let $u=\frac{1}{x}$ and $v=\frac{1}{y}$.

$$u+2v=1 \qquad -u+4v=-7$$
$$u=-2v+1 \qquad -u=-4v-7$$
$$u=4v+7$$

$$-2v+1=4v+7$$
$$-6v=6$$
$$v=-1$$
$$u=-2v+1=-2(-1)+1=2+1=3$$

$$u=\frac{1}{x} \qquad v=\frac{1}{y}$$
$$x=\frac{1}{u}=\frac{1}{3} \qquad y=\frac{1}{v}=\frac{1}{-1}=-1$$

$$\left\{\left(\frac{1}{3},-1\right)\right\}$$

83. Let x represent Marta's bicycling speed.
Let y represent Marta's running speed.

Use $d=rt \Leftrightarrow t=\frac{d}{r}$.

$$\frac{12}{x}+\frac{4}{y}=\frac{4}{3}$$
$$\frac{21}{x}+\frac{3}{y}=\frac{5}{3}$$

Let $u=\frac{1}{x}$ and $v=\frac{1}{y}$.

$$12u+4v=\frac{4}{3} \xrightarrow{\div 4} 3u+v=\frac{1}{3}$$
$$21u+3v=\frac{5}{3} \xrightarrow{\div -3} -7u-v=-\frac{5}{9}$$
$$-4u=-\frac{2}{9}$$
$$u=\frac{1}{18}$$

$$12u+4v=\frac{4}{3}$$
$$12\left(\frac{1}{18}\right)+4v=\frac{4}{3}$$
$$\frac{2}{3}+4v=\frac{4}{3}$$
$$4v=\frac{2}{3}$$
$$v=\frac{1}{6}$$

$$u=\frac{1}{x} \qquad v=\frac{1}{y}$$
$$x=\frac{1}{u}=\frac{1}{\frac{1}{18}}=18 \qquad y=\frac{1}{v}=\frac{1}{\frac{1}{6}}=6$$

Marta bicycles 18 mph and runs 6 mph.

85. Let x represent city miles.
Let y represent highway miles.

Use $\text{mpg}=\frac{\text{mi}}{\text{gal}} \Leftrightarrow \text{gal}=\frac{\text{mi}}{\text{mpg}}$.

$$x+y=254$$
$$x=-y+254$$
$$\frac{x}{16}+\frac{y}{22}=14$$
$$176\left(\frac{x}{16}+\frac{y}{22}\right)=176(14)$$
$$11x+8y=2464$$
$$11(-y+254)+8y=2464$$
$$-11y+2794+8y=2464$$
$$-3y=-330$$
$$y=110$$
$$x=-y+254=-110+254=144$$

The truck was driven 144 mi in the city and 110 mi on the highway.

87. $p=0.025x$
$p=-0.04x+104$

a. $0.025x=-0.04x+104$
$$0.065x=104$$
$$x=1600$$
$$p=0.025x=0.025(1600)=40$$
$$\{(1600, 40)\}$$

b. \$40 **c.** 1600 tickets

89. If the system represents two intersecting lines, then the lines intersect in exactly one point. The solution set consists of the ordered pair representing that point. If the lines in the system are parallel, then the lines do not intersect and the system has no solution. If the equations in a system of linear equations represent the same line, then the solution set is the set of points on the line.

91. If the system of equations reduces to a contradiction such as $0=1$, then the system has no solution and is said to be inconsistent.

93.
$$9^2+b^2=c^2$$
$$c^2-b^2=81$$
$$(c-b)(c+b)=81$$

Positive factors of 81: 1, 3, 9, 27, 81

$$\begin{aligned} c-b&=1\\ c+b&=81\\ \hline 2c&=82\\ c&=41\\ c+b&=81\\ 41+b&=81\\ b&=40 \end{aligned} \qquad \begin{aligned} c-b&=81\\ c+b&=1\\ \hline 2c&=82\\ c&=41\\ c+b&=1\\ 41+b&=1\\ b&=-40 \end{aligned} \qquad \begin{aligned} c-b&=3\\ c+b&=27\\ \hline 2c&=30\\ c&=15\\ c+b&=27\\ 15+b&=27\\ b&=12 \end{aligned}$$

$$\begin{aligned} c-b&=27\\ c+b&=3\\ \hline 2c&=30\\ c&=15\\ c+b&=3\\ 15+b&=3\\ b&=-12 \end{aligned} \qquad \begin{aligned} c-b&=9\\ c+b&=9\\ \hline 2c&=18\\ c&=9\\ c+b&=9\\ 9+b&=9\\ b&=0 \end{aligned}$$

$(9,12,15)$ and $(9,40,41)$

95.
$$\frac{\sqrt{3}}{2}|F_1|=\frac{\sqrt{2}}{2}|F_2|$$
$$\frac{\sqrt{3}}{2}|F_1|-\frac{\sqrt{2}}{2}|F_2|=0$$
$$\frac{1}{2}|F_1|+\frac{\sqrt{2}}{2}|F_2|=50$$

$$\begin{aligned} \frac{\sqrt{3}}{2}|F_1| - \frac{\sqrt{2}}{2}|F_2| &= 0\\ \frac{1}{2}|F_1| + \frac{\sqrt{2}}{2}|F_2| &= 50\\ \hline \frac{1+\sqrt{3}}{2}|F_1| &= 50 \end{aligned}$$
$$|F_1|=\frac{100}{1+\sqrt{3}}$$

$$|F_1|=\frac{100}{1+\sqrt{3}}=\frac{100}{1+\sqrt{3}}\cdot\frac{1-\sqrt{3}}{1-\sqrt{3}}$$
$$=\frac{100(1-\sqrt{3})}{1-3}=\frac{100(1-\sqrt{3})}{-2}$$
$$=50(\sqrt{3}-1)\text{ lb}\approx 36.6\text{ lb}$$

$$\frac{\sqrt{2}}{2}|F_2|=\frac{\sqrt{3}}{2}|F_1|$$
$$|F_2|=\frac{2}{\sqrt{2}}\cdot\frac{\sqrt{3}}{2}|F_1|=\frac{\sqrt{3}}{\sqrt{2}}\cdot|F_1|$$
$$=\frac{\sqrt{3}}{\sqrt{2}}\cdot\frac{\sqrt{2}}{\sqrt{2}}\cdot|F_1|=\frac{\sqrt{6}}{2}\cdot|F_1|$$
$$=\frac{\sqrt{6}}{2}\cdot\left[50(\sqrt{3}-1)\right]=\sqrt{6}\cdot\left[25(\sqrt{3}-1)\right]$$
$$=25\sqrt{2}\cdot(3-\sqrt{3})\text{ lb}\approx 44.8\text{ lb}$$

97.

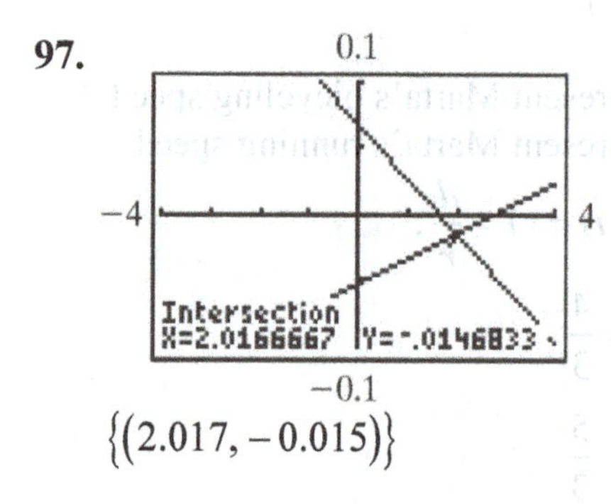

$\{(2.017,-0.015)\}$

99.

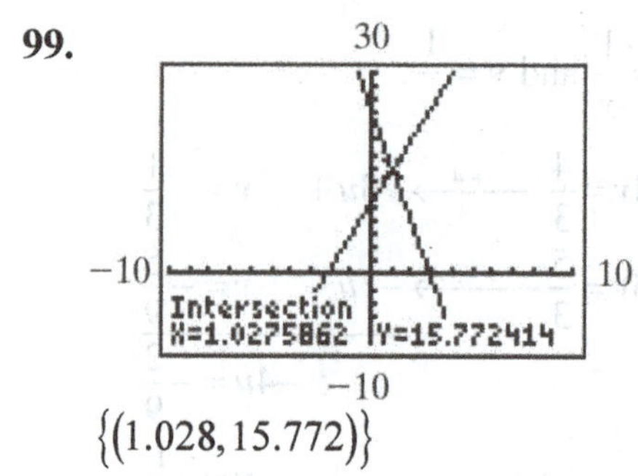

$\{(1.028,15.772)\}$

Section 5.2 Systems of Linear Equations in Three Variables and Applications

1. $x = 5y + 12$

$\frac{1}{2}x = \frac{1}{3} - \frac{1}{3}y$

$3x = 2 - 2y$

$3(5y + 12) = 2 - 2y$

$15y + 36 = 2 - 2y$

$17y = -34$

$y = -2$

$x = 5y + 12 = 5(-2) + 12 = -10 + 12 = 2$

$\{(2, -2)\}$

3. $y = -\frac{3}{4}x + 1$

$3x + 4y = 4$

$3x + 4\left(-\frac{3}{4}x + 1\right) = 4$

$3x - 3x + 4 = 4$

$4 = 4$ $\quad \left\{(x, y) \mid y = -\frac{3}{4}x + 1\right\}$

5. linear

7. plane

9. For example:

Let $x = 0$ and $y = 0$.

$2x + 4y - 6z = 12$

$2(0) + 4(0) - 6z = 12$

$-6z = 12$

$z = -2$

$(0, 0, -2)$

Let $x = 0$ and $z = 0$.

$2x + 4y - 6z = 12$

$2(0) + 4y - 6(0) = 12$

$4y = 12$

$y = 3$

$(0, 3, 0)$

Let $y = 0$ and $z = 0$.

$2x + 4y - 6z = 12$

$2x + 4(0) - 6(0) = 12$

$2x = 12$

$x = 6$ $\quad (6, 0, 0)$

11. a. $-x + 3y - 7z = 7$

$-(2) + 3(3) - 7(0) \stackrel{?}{=} 7$

$-2 + 9 \stackrel{?}{=} 7$

$7 \stackrel{?}{=} 7$ ✓ true

$2x + 4y + z = 16$

$2(2) + 4(3) + (0) \stackrel{?}{=} 16$

$4 + 12 \stackrel{?}{=} 16$

$16 \stackrel{?}{=} 16$ ✓ true

$3x - 5y + 6z = -9$

$3(2) - 5(3) + 6(0) \stackrel{?}{=} -9$

$6 - 15 \stackrel{?}{=} -9$

$-9 \stackrel{?}{=} -9$ ✓ true $\quad$ Yes

b. $-x + 3y - 7z = 7$

$-(-2) + 3(4) - 7(1) \stackrel{?}{=} 7$

$2 + 12 - 7 \stackrel{?}{=} 7$

$7 \stackrel{?}{=} 7$ ✓ true

$2x + 4y + z = 16$

$2(-2) + 4(4) + (1) \stackrel{?}{=} 16$

$-4 + 16 + 1 \stackrel{?}{=} 16$

$13 \stackrel{?}{=} 16$ false $\quad$ No

13. a. $x + y + z = 2$

$(2) + (0) + (0) \stackrel{?}{=} 2$

$2 \stackrel{?}{=} 2$ ✓ true

$x + 2y - z = 2$

$(2) + 2(0) - (0) \stackrel{?}{=} 2$

$2 \stackrel{?}{=} 2$ ✓ true

$3x + 5y - z = 6$

$3(2) + 5(0) - (0) \stackrel{?}{=} 6$

$6 \stackrel{?}{=} 6$ ✓ true $\quad$ Yes

b. $x + y + z = 2$

$(-1) + (2) + (1) \stackrel{?}{=} 2$

$-1 + 2 + 1 \stackrel{?}{=} 2$

$2 \stackrel{?}{=} 2$ ✓ true

$x + 2y - z = 2$

$(-1) + 2(2) - (1) \stackrel{?}{=} 2$

$-1 + 4 - 1 \stackrel{?}{=} 2$

$2 \stackrel{?}{=} 2$ ✓ true

$3x + 5y - z = 6$

$3(-1) + 5(2) - (1) \stackrel{?}{=} 6$

$-3 + 10 - 1 \stackrel{?}{=} 6$

$6 \stackrel{?}{=} 6$ ✓ true $\quad$ Yes

15. $\boxed{A}$ $x-2y+z=-9$

$\boxed{B}$ $3x+4y+5z=9$

$\boxed{C}$ $-2x+3y-z=12$

Eliminate z from equations $\boxed{A}$ and $\boxed{C}$.

$$\begin{array}{ll} \boxed{A} & x-2y+z=-9 \\ \boxed{C} & \underline{-2x+3y-z=12} \\ & -x+y=3 \quad \boxed{D} \end{array}$$

Eliminate z from equations $\boxed{B}$ and $\boxed{C}$.

$$\begin{array}{llcl} \boxed{B} & 3x+4y+5z=9 & & 3x+4y+5z=9 \\ \boxed{C} & -2x+3y-z=12 & \xrightarrow{\text{Multiply by 5}} & \underline{-10x+15y-5z=60} \\ & & & -7x+19y=69 \quad \boxed{E} \end{array}$$

Solve the systems of equations $\boxed{D}$ and $\boxed{E}$.

$$\begin{array}{llcl} \boxed{D} & -x+y=3 & \xrightarrow{\text{Multiply by }-7} & 7x-7y=-21 \\ \boxed{E} & -7x+19y=69 & & \underline{-7x+19y=69} \\ & & & 12y=48 \\ & & & y=4 \end{array}$$

Back substitute.

$$\begin{array}{ll} \boxed{D} & -x+y=3 \\ & -x+4=3 \\ & -x=-1=12 \\ & x=1 \end{array} \qquad \begin{array}{ll} \boxed{A} & x-2y+z=-9 \\ & 1-2(4)+z=-9 \\ & 1-8+z=-9 \\ & z=-2 \end{array} \qquad \{(1,4,-2)\}$$

17. $\boxed{A}$ $4x=3y-2z-5 \rightarrow 4x-3y+2z=-5$

$\boxed{B}$ $2(x+y)=y+z-6 \rightarrow 2x+2y=y+z-6 \rightarrow 2x+y-z=-6$

$\boxed{C}$ $6(x-y)+z=x-5y-8 \rightarrow 6x-6y+z=x-5y-8 \rightarrow 5x-y+z=-8$

Eliminate y and z from equations $\boxed{B}$ and $\boxed{C}$.

$$\begin{array}{ll} \boxed{B} & 2x+y-z=-6 \\ \boxed{C} & \underline{5x-y+z=-8} \\ & 7x=-14 \\ & x=-2 \end{array}$$

Eliminate x from equations $\boxed{A}$ and $\boxed{B}$.

$$\begin{array}{ll} \boxed{A} & 4x-3y+2z=-5 \\ & 4(-2)-3y+2z=-5 \\ & -8-3y+2z=-5 \\ & -3y+2z=3 \quad \boxed{D} \end{array} \qquad \begin{array}{ll} \boxed{B} & 2x+y-z=-6 \\ & 2(-2)+y-z=-6 \\ & -4+y-z=-6 \\ & y-z=-2 \quad \boxed{E} \end{array}$$

Solve the systems of equations [D] and [E].

[D] $-3y+2z=3$

[E] $y-z=-2$ $\xrightarrow{\text{Multiply by 2}}$

$$\begin{array}{rl} -3y+2z= & 3 \\ 2y-2z= & -4 \\ \hline -y \quad = & -1 \\ y= & 1 \end{array}$$

Back substitute.

[E] $y-z=-2$

$1-z=-2$

$z=3$

$\{(-2,1,3)\}$

19. [A] $2x \quad +5z=2$

[B] $3y-7z=9$

[C] $-5x+9y \quad =22$

Eliminate y from equations [B] and [C].

[B] $3y-7z=9$ $\xrightarrow{\text{Multiply by }-3}$ $-9y+21z=-27$

[C] $-5x+9y \quad =22$

$$\begin{array}{rl} -9y+21z= & -27 \\ -5x+9y \quad = & 22 \\ \hline -5x \quad +21z= & -5 \end{array}$$ [D]

Solve the systems of equations [A] and [D].

[A] $2x+5z=2$ $\xrightarrow{\text{Multiply by 5}}$ $10x+25z=10$

[D] $-5x+21z=-5$ $\xrightarrow{\text{Multiply by 2}}$ $-10x+42z=-10$

$$\begin{array}{rl} 10x+25z= & 10 \\ -10x+42z= & -10 \\ \hline 67z= & 0 \\ z= & 0 \end{array}$$

Back substitute.

[A] $2x+5z=2$

$2x+5(0)=2$

$2x=2$

$x=1$

[B] $3y-7z=9$

$3y-7(0)=9$

$3y=9$

$y=3$

$\{(1,3,0)\}$

21. [A] $-4x-3y \quad =0$

[B] $3y+z=-1$

[C] $4x \quad -z=12$

Eliminate x from equations [A] and [C].

[A] $-4x-3y \quad =0$

[C] $4x \quad -z=12$

$$\begin{array}{rl} -4x-3y \quad = & 0 \\ 4x \quad -z= & 12 \\ \hline -3y-z= & 12 \end{array}$$ [D]

Eliminate y and z from equations [B] and [D].

$$\begin{array}{rl} 3y+z= & -1 \\ -3y-z= & 12 \\ \hline 0= & 11 \end{array}$$

The system of equations reduces to a contradiction. There is no solution.

23. $\boxed{A}$ $2x = 3y - 6z - 1 \rightarrow 2x - 3y + 6z = -1$

$\boxed{B}$ $6y = 12z - 10x + 9 \rightarrow 10x + 6y - 12z = 9$

$\boxed{C}$ $3z = 6y - 3x - 1 \rightarrow 3x - 6y + 3z = -1$

Eliminate y from equations $\boxed{B}$ and $\boxed{C}$.

$$\begin{array}{ll} \boxed{B} & 10x + 6y - 12z = 9 \\ \boxed{C} & \underline{3x - 6y + 3z = -1} \\ & 13x \quad\quad - 9z = 8 \quad \boxed{D} \end{array}$$

Eliminate y and z from equations $\boxed{A}$ and $\boxed{B}$.

$$\begin{array}{lll} \boxed{A} & 2x - 3y + 6z = -1 \xrightarrow{\text{Multiply by 2}} & 4x - 6y + 12z = -2 \\ \boxed{B} & 10x + 6y - 12z = 9 & \underline{10x + 6y - 12z = 9} \\ & & 14x \quad\quad = 7 \\ & & x = \frac{1}{2} \end{array}$$

$\boxed{A}$

$$2x - 3y + 6z = -1$$
$$2\left(\frac{1}{2}\right) - 3y + 6z = -1$$
$$1 - 3y + 6z = -1$$
$$-3y + 6z = -2 \quad \boxed{D}$$

$\boxed{B}$

$$3x - 6y + 3z = -1$$
$$3\left(\frac{1}{2}\right) - 6y + 3z = -1$$
$$3 - 12y + 6z = -2$$
$$-12y + 6z = -5 \quad \boxed{E}$$

Solve the systems of equations $\boxed{D}$ and $\boxed{E}$.

$$\begin{array}{lll} \boxed{D} & -3y + 6z = -2 \xrightarrow{\text{Multiply by } -1} & 3y - 6z = 2 \\ \boxed{E} & -12y + 6z = -5 & \underline{-12y + 6z = -5} \\ & & -9y \quad = -3 \\ & & y = \frac{1}{3} \end{array}$$

Back substitute.

$\boxed{A}$

$$2x - 3y + 6z = -1$$
$$2\left(\frac{1}{2}\right) - 3\left(\frac{1}{3}\right) + 6z = -1$$
$$1 - 1 + 6z = -1$$
$$6z = -1$$
$$z = -\frac{1}{6} \qquad \left\{\left(\frac{1}{2}, \frac{1}{3}, -\frac{1}{6}\right)\right\}$$

25. $\boxed{A}$ $x + 2y + 4z = 3$

$\boxed{B}$ $y + 3z = 5$

$\boxed{C}$ $x - 2z = -7$

Eliminate x from equations $\boxed{A}$ and $\boxed{C}$.

$$\begin{array}{lll} \boxed{A} & x + 2y + 4z = 3 \xrightarrow{\text{Multiply by } -1} & -x - 2y - 4z = -3 \\ \boxed{C} & x \quad\quad - 2z = -7 & \underline{x \quad\quad - 2z = -7} \\ & & -2y - 6z = -10 \quad \boxed{D} \end{array}$$

Pair up equations B and D to solve for y and z.

$$\begin{array}{lrcl} \boxed{B} & y+3z = 5 & \xrightarrow{\text{Multiply by 2}} & 2y+6z = 10 \\ \boxed{D} & -2y-6z = -10 & & -2y-6z = -10 \\ & & & \overline{0 = 0} \end{array}$$

The system reduces to the identity $0=0$. It has infinitely many solutions.

27.
$$\begin{array}{llcl} \boxed{A} & 0.2x = 0.1y - 0.6z & \rightarrow & 2x - y + 6z = 0 \\ \boxed{B} & 0.004x + 0.005y - 0.001z = 0 & \rightarrow & 4x + 5y - z = 0 \\ \boxed{C} & 30x = 50z - 20y & \rightarrow & 3x + 2y - 5z = 0 \end{array}$$

Eliminate y from equations A and C.

$$\begin{array}{lrcl} \boxed{A} & 2x - y + 6z = 0 & \xrightarrow{\text{Multiply by 2}} & 4x - 2y + 12z = 0 \\ \boxed{C} & 3x + 2y - 5z = 0 & & 3x + 2y - 5z = 0 \\ & & & \overline{7x + 7z = 0} \quad \boxed{D} \end{array}$$

Eliminate y from equations A and B.

$$\begin{array}{lrcl} \boxed{A} & 2x - y + 6z = 0 & \xrightarrow{\text{Multiply by 5}} & 10x - 5y + 30z = 0 \\ \boxed{B} & 4x + 5y - z = 0 & & 4x + 5y - z = 0 \\ & & & \overline{14x + 30z = 0} \\ & & & 7x + 15z = 0 \quad \boxed{E} \end{array}$$

Solve the systems of equations D and E.

$$\begin{array}{lrcl} \boxed{D} & 7x + 7z = 0 & \xrightarrow{\text{Multiply by }-1} & -7x - 7z = 0 \\ \boxed{E} & 7x + 15z = 0 & & 7x + 15z = 0 \\ & & & \overline{8z = 0} \\ & & & z = 0 \end{array}$$

Back substitute.

$$\begin{array}{lrlr} \boxed{D} & 7x + 7z = 0 & \boxed{A} & 2x - y + 6z = 0 \\ & 7x + 7(0) = 0 & & 2(0) - y + 6(0) = 0 \\ & 7x + 0 = 0 & & 0 - y + 0 = 0 \\ & 7x = 0 & & -y = 0 \\ & x = 0 & & y = 0 \end{array}$$

$\{(0, 0, 0)\}$

29.
$$\begin{array}{llcl} \boxed{A} & \frac{1}{12}x + \frac{1}{4}y + \frac{1}{3}z = \frac{7}{12} & \rightarrow & x + 3y + 4z = 7 \\ \boxed{B} & -\frac{1}{10}x + \frac{1}{2}y - \frac{1}{5}z = -\frac{17}{10} & \rightarrow & -x + 5y - 2z = -17 \\ \boxed{C} & \frac{1}{2}x + \frac{1}{4}y + z = 3 & \rightarrow & 2x + y + 4z = 12 \end{array}$$

Eliminate x from equations A and B.

$$\begin{array}{lr} \boxed{A} & x + 3y + 4z = 7 \\ \boxed{B} & -x + 5y - 2z = -17 \\ & \overline{8y + 2z = -10} \quad \boxed{D} \end{array}$$

Eliminate x and z from equations B and C.

$$\begin{array}{ll} \boxed{B}\ -x+5y-2z=-17 & \xrightarrow{\text{Multiply by 2}} -2x+10y-4z=-34 \\ \boxed{C}\ 2x+y+4z=12 & \phantom{\xrightarrow{\text{Multiply by 2}}} \underline{2x+y+4z=12} \\ & \phantom{\xrightarrow{\text{Multiply by 2}}} 11y=-22 \\ & \phantom{\xrightarrow{\text{Multiply by 2}}} y=-2 \end{array}$$

Back substitute.

$$\begin{array}{ll} \boxed{D}\ 8y+2z=-10 & \boxed{A}\ x+3y+4z=7 \\ 8(-2)+2z=-10 & x+3(-2)+4(3)=7 \\ -16+2z=-10 & x-6+12=7 \\ 2z=6 & x=1 \\ z=3 & \end{array}$$

$\{(1,-2,3)\}$

31. $\boxed{A}\ 3x+2y+5z=12$
$\boxed{B}\ 3y+8z=-8$
$\boxed{C}\ 10z=20$

Solve C for z.

$$\begin{aligned} 10z&=20 \\ z&=2 \end{aligned}$$

Solve B for y.

$$\begin{aligned} 3y+8z&=-8 \\ 3y+8(2)&=-8 \\ y+16&=-8 \\ 3y&=-24 \\ y&=-8 \end{aligned}$$

Solve A for x.

$$\begin{aligned} 3x+2y+5z&=12 \\ 3x+2(-8)+5(2)&=12 \\ 3x-16+10&=12 \\ 3x&=18 \\ x&=6 \end{aligned}$$

$\{(6,-8,2)\}$

33. $\boxed{A}\ \dfrac{x+2}{3}+\dfrac{y-4}{2}+\dfrac{z+1}{6}=8 \rightarrow 2x+4+3y-12+z+1=48 \rightarrow 2x+3y+z=55$

$\boxed{B}\ -\dfrac{x+2}{3}+\dfrac{z+1}{2}=8 \rightarrow -2x-4+3z+3=48 \rightarrow -2x+3z=49$

$\boxed{C}\ \dfrac{y-4}{4}-\dfrac{z+1}{6}=-1 \rightarrow 3y-12-2z-2=-12 \rightarrow 3y-2z=2$

Eliminate x from equations A and B.

$$\begin{array}{l} \boxed{A}\ 2x+3y+z=55 \\ \boxed{B}\ \underline{-2x+3z=49} \\ \quad 3y+4z=104\ \boxed{D} \end{array}$$

Solve the systems of equations C and D.

$$\begin{array}{ll} \boxed{C}\ 3y-2z=2 & \xrightarrow{\text{Multiply by }-1} -3y+2z=-2 \\ \boxed{D}\ 3y+4z=104 & \phantom{\xrightarrow{\text{Multiply by }-1}} \underline{3y+4z=104} \\ & \phantom{\xrightarrow{\text{Multiply by }-1}} 6z=102 \\ & \phantom{\xrightarrow{\text{Multiply by }-1}} z=17 \end{array}$$

Back substitute.

$$\begin{array}{ll} \boxed{D}\ 3y+4z=104 & \boxed{B}\ -2x+3z=49 \\ 3y+4(17)=104 & -2x+3(17)=49 \\ 3y+68=104 & -2x+51=49 \\ 3y=36 & -2x=-2 \\ y=12 & x=1 \end{array}$$

$\{(1,12,17)\}$

35. $\boxed{A}$ $3(x+y)=6-4z+y \rightarrow 3x+3y=6-4z+y \rightarrow 3x+2y+4z=6$

$\boxed{B}$ $4=6y+5z \rightarrow 6y+5z=4 \rightarrow 6y+5z=4$

$\boxed{C}$ $-3x+4y+z=0 \rightarrow -3x+4y+z=0 \rightarrow -3x+4y+z=0$

Eliminate x from equations $\boxed{A}$ and $\boxed{C}$.

$$\begin{array}{rl} \boxed{A} & 3x+2y+4z=6 \\ \boxed{C} & -3x+4y+z=0 \\ \hline & 6y+5z=6 \quad \boxed{D} \end{array}$$

Solve the systems of equations $\boxed{B}$ and $\boxed{D}$.

$$\begin{array}{rlcr} \boxed{B} & 6y+5z=4 & \xrightarrow{\text{Multiply by } -1} & -6y-5z=-4 \\ \boxed{D} & 6y+5z=6 & & 6y+5z=6 \\ \hline & & & 0=2 \end{array}$$

The system of equations reduces to a contradiction. There is no solution.

37. Let x represent the amount invested in the large cap stock fund.
Let y represent the amount invested in the real estate fund.
Let z represent the amount invested in the bond fund.

$\boxed{A}$ $x+y+z=8000 \rightarrow x+y+z=8000$

$\boxed{B}$ $x=2y \rightarrow x-2y=0$

$\boxed{C}$ $0.062x-0.135y+0.044z=66 \rightarrow 62x-135y+44z=66{,}000$

Eliminate x from equations $\boxed{A}$ and $\boxed{B}$ and from equations $\boxed{B}$ and $\boxed{C}$.

$$\begin{array}{rl} \boxed{A} & x+y+z=8000 \\ -1\cdot\boxed{B} & -x+2y=0 \\ \hline & 3y+z=8000 \quad \boxed{D} \end{array}$$

$$\begin{array}{rl} -62\cdot\boxed{B} & -62x+124y=0 \\ \boxed{C} & 62x-135y+44z=66{,}000 \\ \hline & -11y+44z=66{,}000 \\ & -y+4z=6000 \quad \boxed{E} \end{array}$$

Solve the systems of equations $\boxed{D}$ and $\boxed{E}$.

$$\begin{array}{rl} \boxed{D} & 3y+z=8000 \\ 3\cdot\boxed{E} & -3y+12z=18{,}000 \\ \hline & 13z=26{,}000 \\ & z=2000 \end{array}$$

Back substitute.

$$\begin{array}{rl} \boxed{D} & 3y+z=8000 \\ & 3y+2000=8000 \\ & 3y=6000 \\ & y=2000 \end{array}$$

$$\begin{array}{rl} \boxed{B} & x-2y=0 \\ & x-2(2000)=0 \\ & x-4000=0 \\ & x=4000 \end{array}$$

He invested \$4000 in the large cap fund, \$2000 in the real estate fund, and \$2000 in the bond fund.

39. Let x represent the number of free-throws.
Let y represent the number of 2-point shots.
Let z represent the number of 3-point shots.

$\boxed{A}$ $x+2y+3z=26 \rightarrow x+2y+3z=26$
$\boxed{B}$ $y=z+4 \rightarrow y-z=4$
$\boxed{C}$ $x=y+z \rightarrow x-y-z=0$

Eliminate y from equations $\boxed{A}$ and $\boxed{B}$ and from equations $\boxed{B}$ and $\boxed{C}$.

$$\begin{array}{rl} \boxed{A} & x+2y+3z=26 \\ -2\cdot\boxed{B} & \quad -2y+2z=-8 \\ \hline & x \quad\quad +5z=18 \quad \boxed{D} \end{array}$$

$$\begin{array}{rl} \boxed{B} & \quad y-z=4 \\ \boxed{C} & x-y-z=0 \\ \hline & x \quad -2z=4 \quad \boxed{E} \end{array}$$

Solve the systems of equations $\boxed{D}$ and $\boxed{E}$.

$$\begin{array}{rl} -1\cdot\boxed{D} & -x-5z=-18 \\ \boxed{E} & x-2z=4 \\ \hline & -7z=-14 \\ & z=2 \end{array}$$

Back substitute.

$\boxed{D}$ $x+5z=18$
$x+5(2)=18$
$x+10=18$
$x=8$

$\boxed{B}$ $y-z=4$
$y-2=4$
$y=6$

He made eight free-throws, six 2-point shots, and two 3-point shots.

41. Let x represent the percentage of nitrogen.
Let y represent the percentage of phosphorous.
Let z represent the percentage of potassium.

$\boxed{A}$ $z=2y \rightarrow 2y-z=0$
$\boxed{B}$ $x=y+z \rightarrow x-y-z=0$
$\boxed{C}$ $x+y+z=42 \rightarrow x+y+z=42$

Eliminate y and z from equations $\boxed{B}$ and $\boxed{C}$.

$$\begin{array}{rl} \boxed{B} & x-y-z=0 \\ \boxed{C} & x+y+\;\;=42 \\ \hline & 2x \quad =42 \\ & x=21 \end{array}$$

Back substitute.

$\boxed{C}$ $x+y+z=42$
$21+y+z=42$
$y+z=21$ $\boxed{D}$

Eliminate z from equations $\boxed{A}$ and $\boxed{D}$.

$$\begin{array}{rl} \boxed{A} & 2y-z=0 \\ \boxed{D} & \;\;y+z=21 \\ \hline & 3y \quad =21 \\ & y=7 \end{array}$$

Back substitute.

$\boxed{A}$ $2y-z=0$
$2(7)-z=0$
$z=14$

The proper N-P-K label is 21-7-14, which is choice b.

43. Let x represent the length of the shortest side.
Let y represent the length of the middle side.
Let z represent the length of the longest side.

$\boxed{A}$ $x+y+z=55 \rightarrow x+y+z=55$
$\boxed{B}$ $x=z-7 \rightarrow x-z=-7$
$\boxed{C}$ $y=x+z-19 \rightarrow -x+y-z=-19$

Eliminate x and z from equations $\boxed{A}$ and $\boxed{C}$.

$$\begin{array}{rl} \boxed{A} & x+y+z=55 \\ \boxed{C} & -x+y-\;\;=-19 \\ \hline & 2y \quad =36 \\ & y=18 \end{array}$$

Back substitute.

$\boxed{A}$ $x+y+z=55$
$x+18+z=55$
$x+z=37$ $\boxed{D}$

Eliminate z from equations $\boxed{B}$ and $\boxed{D}$.

$$\begin{array}{rl} \boxed{B} & x-z=-7 \\ \boxed{D} & x+z=37 \\ \hline & 2x \quad =30 \\ & x=15 \end{array}$$

Back substitute.

$\boxed{A}$ $x+y+z=55$
$15+18+z=55$
$33+z=55$
$z=22$

The sides are 15 in., 18 in., and 22 in.

45. Let x represent the measure of the largest angle. Let y represent the measure of the middle angle. Let z represent the measure of the smallest angle.

$\boxed{A}\ x+y+z=180 \rightarrow x+y+z=180$

$\boxed{B}\ x=y+z+100 \rightarrow x-y-z=100$

$\boxed{C}\ z=\frac{2}{3}y \rightarrow 2y-3z=0$

Eliminate x from equations $\boxed{A}$ and $\boxed{B}$.

$$\begin{array}{rrcl} \boxed{A} & x+y+z &=& 180 \\ -1\cdot\boxed{B} & -x+y+z &=& -100 \\ \hline & 2y+2z &=& 80 \\ & y+z &=& 40\ \boxed{D} \end{array}$$

Eliminate z from equations $\boxed{C}$ and $\boxed{D}$.

$$\begin{array}{rrcl} \boxed{C} & 2y-3z &=& 0 \\ 3\cdot\boxed{D} & 3y+3z &=& 120 \\ \hline & 5y &=& 120 \\ & y &=& 24 \end{array}$$

Back substitute.

$\boxed{C}$
$$\begin{aligned} 2y-3z&=0 \\ 2(24)-3z&=0 \\ 48-3z&=0 \\ -3z&=-48 \\ z&=16 \end{aligned}$$

$\boxed{A}$
$$\begin{aligned} x+y+z&=180 \\ x+24+16&=180 \\ x+40&=180 \\ x&=140 \end{aligned}$$

The angles are $16°$, $24°$, and $140°$.

47. a. $m=\frac{10-0}{3-1}=\frac{10}{2}=5$ $\qquad m=\frac{15-10}{-2-3}=\frac{5}{-5}=-1$

The slopes are 5 and -1.

b. $y=ax^2+bx+c$

Substitute $(1,0)$: $0=a(1)^2+b(1)+c \rightarrow \boxed{A}\ a+b+c=0$

Substitute $(3,10)$: $10=a(3)^2+b(3)+c \rightarrow \boxed{B}\ 9a+3b+c=10$

Substitute $(-2,15)$: $15=a(-2)^2+b(-2)+c \rightarrow \boxed{C}\ 4a-2b+c=15$

Eliminate b from equations $\boxed{A}$ and $\boxed{B}$ and from equations $\boxed{A}$ and $\boxed{C}$.

$$\begin{array}{rrcl} -3\cdot\boxed{A} & -3a-3b-3c &=& 0 \\ \boxed{B} & 9a+3b+c &=& 10 \\ \hline & 6a-2c &=& 10\ \boxed{D} \end{array}$$

$$\begin{array}{rrcl} 2\cdot\boxed{A} & 2a+2b+2c &=& 0 \\ \boxed{C} & 4a-2b+c &=& 15 \\ \hline & 6a+3c &=& 15\ \boxed{E} \end{array}$$

Solve the system of equations $\boxed{D}$ and $\boxed{E}$.

$$\begin{array}{rrcl} \boxed{D} & 6a-2c &=& 10 \\ -1\cdot\boxed{E} & -6a-3c &=& -15 \\ \hline & -5c &=& -5 \\ & c &=& 1 \end{array}$$

$\boxed{D}$
$$\begin{aligned} 6a-2c&=10 \\ 6a-2(1)&=10 \\ 6a-2&=10 \\ 6a&=12 \\ a&=2 \end{aligned}$$

Back substitute.

$\boxed{A}$
$$\begin{aligned} a+b+c&=0 \\ 2+b+1&=0 \\ b&=-3 \end{aligned}$$

$y=2x^2-3x+1$

c.

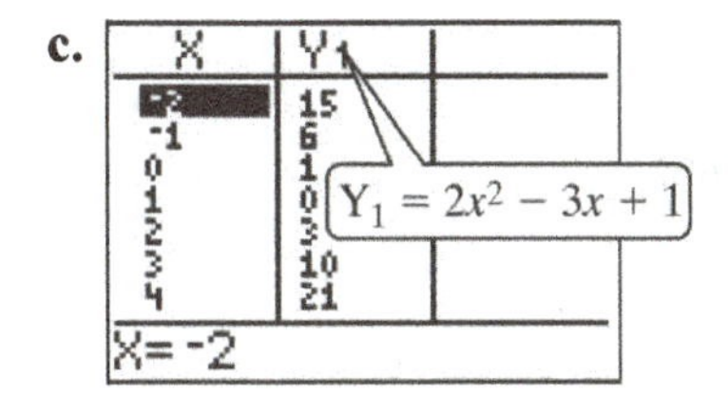

49. $y = ax^2 + bx + c$

Substitute $(0, 6)$: $6 = a(0)^2 + b(0) + c \rightarrow$ [A] $c = 6$

Substitute $(2, -6)$: $-6 = a(2)^2 + b(2) + c \rightarrow$ [B] $4a + 2b + c = -6$

Substitute $(-1, 9)$: $9 = a(-1)^2 + b(-1) + c \rightarrow$ [C] $a - b + c = 9$

Eliminate c from equations [B] and [C].

[B] $4a + 2b + c = -6$

$4a + 2b + 6 = -6$

$4a + 2b = -12$

$2a + b = -6$ [D]

[C] $a - b + c = 9$

$a - b + 6 = 9$

$a - b = 3$ [E]

Solve the system of equations [D] and [E].

[D] $2a + b = -6$

[E] $a - b = 3$

$3a = -3$

$a = -1$

[E] $a - b = 3$

$-1 - b = 3$

$-b = 4$

$b = -4$

$y = -x^2 - 4x + 6$

51. a. $y = ax^2 + bx + c$

Substitute $(0, 6)$: $6 = a(0)^2 + b(0) + c \rightarrow$ [A] $c = 6$

Substitute $(4, 7)$: $7 = a(4)^2 + b(4) + c \rightarrow$ [B] $16a + 4b + c = 7$

Substitute $(9, 9)$: $9 = a(9)^2 + b(9) + c \rightarrow$ [C] $81a + 9b + c = 9$

Eliminate c from equations [B] and [C].

[B] $16a + 4b + c = 7$

$16a + 4b + 6 = 7$

$16a + 4b = 1$ [D]

[C] $81a + 9b + c = 9$

$81a + 9b + 6 = 9$

$81a + 9b = 3$

$27a + 3b = 1$ [E]

Solve the system of equations [D] and [E].

$3 \cdot$ [D] $48a + 12b = 3$

$-4 \cdot$ [E] $-108a - 12b = -4$

$-60a = -1$

$a = \frac{1}{60}$

[D] $16a + 4b = 1$

$16\left(\frac{1}{60}\right) + 4b = 1$

$\frac{4}{15} + 4b = 1$

$4 + 60b = 15$

$b = \frac{11}{60}$

$y = \frac{1}{60}x^2 + \frac{11}{60}x + 6$

b. $y = \frac{1}{60}x^2 + \frac{11}{60}x + 6 = \frac{1}{60}(12)^2 + \frac{11}{60}(12) + 6 = \frac{144}{60} + \frac{132}{60} + 6 = 10.6\%$

53. $s(t)=\frac{1}{2}at^2+v_0t+s_0$

Substitute $(1,30)$: $30=\frac{1}{2}a(1)^2+v_0(1)+s_0 \rightarrow$ $\boxed{A}$ $\frac{1}{2}a+v_0+s_0=30$

Substitute $(2,54)$: $54=\frac{1}{2}a(2)^2+v_0(2)+s_0 \rightarrow$ $\boxed{B}$ $2a+2v_0+s_0=54$

Substitute $(3,82)$: $82=\frac{1}{2}a(3)^2+v_0(3)+s_0 \rightarrow$ $\boxed{C}$ $\frac{9}{2}a+3v_0+s_0=82$

Eliminate v_0 from equations $\boxed{A}$ and $\boxed{B}$ and from equations $\boxed{A}$ and $\boxed{C}$.

$-2\cdot\boxed{A}$ $-a-2v_0-2s_0=-60$

$\boxed{B}$ $2a+2v_0+s_0=54$

$a-s_0=-6$ $\boxed{D}$

$-3\cdot\boxed{A}$ $-\frac{3}{2}a-3v_0-3s_0=-90$

$\boxed{C}$ $\frac{9}{2}a+3v_0+s_0=82$

$3a-2s_0=-8$ $\boxed{D}$

Solve the system of equations $\boxed{D}$ and $\boxed{E}$.

$-2\cdot\boxed{D}$ $-2a+2s_0=12$

$\boxed{E}$ $3a-2s_0=-8$

$a=4$

$\boxed{D}$ $a-s_0=-6$

$4-s_0=-6$

$s_0=10$

Back substitute.

$\boxed{A}$ $\frac{1}{2}a+v_0+s_0=30$

$\frac{1}{2}(4)+v_0+10=30$

$2+v_0+10=30$

$v_0=18$

$a=4, v_0=18, s_0=10$

55. a. $y=ax_1+bx_2+c$

Substitute $(28,0.5,225)$: $225=a(28)+b(0.5)+c \rightarrow$ $\boxed{A}$ $28a+0.5b+c=225$

Substitute $(25,0.8,207)$: $207=a(25)+b(0.8)+c \rightarrow$ $\boxed{B}$ $25a+0.8b+c=207$

Substitute $(18,0.4,154)$: $154=a(18)+b(0.4)+c \rightarrow$ $\boxed{C}$ $18a+0.4b+c=154$

Eliminate c from equations $\boxed{A}$ and $\boxed{B}$ and from equations $\boxed{A}$ and $\boxed{C}$.

$-1\cdot\boxed{A}$ $-28a-0.5b-c=-225$

$\boxed{B}$ $25a+0.8b+c=207$

$-3a+0.3b=-18$

$a-0.1b=6$ $\boxed{D}$

$-1\cdot\boxed{A}$ $-28a-0.5b-c=-225$

$\boxed{B}$ $18a+0.4b+c=154$

$-10a-0.1b=-71$ $\boxed{E}$

Solve the system of equations $\boxed{D}$ and $\boxed{E}$.

$-1\cdot\boxed{D}$ $-a+0.1b=-6$

$\boxed{E}$ $-10a-0.1b=-71$

$-11a=-77$

$a=7$

$\boxed{D}$ $-a+0.1b=-6$

$-7+0.1b=-6$

$0.1b=1$

$b=10$

Back substitute. $\boxed{A}$ $28a+0.5b+c=225$

$28(7)+0.5(10)+c=225$

$196+5+c=225$

$c=24$

$y=7x_1+10x_2+24$

b. $y = 7x_1 + 10x_2 + 24$

$y = 7(20) + 10(0.4) + 24$

$= 140 + 4 + 24 = 168$

$168,000

57. The set of all ordered pairs that are solutions to a linear equation in three variables forms a plane in space.

59. Pair up two equations in the system and eliminate a variable. Choose a different pair of two equations from the system and eliminate the same variable. The result should be a system of two linear equations in two variables. Solve this system using either the substitution or addition method. Then back substitute to find the third variable.

61.
$\boxed{A}\ 2a + b - c + d = 7$

$\boxed{B}\ 3b + 2c - 2d = -11$

$\boxed{C}\ a + 3c + 3d = 14$

$\boxed{D}\ 4a + 2b - 5c = 6$

Eliminate a from equations $\boxed{A}$ and $\boxed{C}$.

$\boxed{A}\ 2a + b - c + d = 7$

$-2 \cdot \boxed{C}\ -2a - 6c - 6d = -28$

$b - 7c - 5d = -21\ \boxed{E}$

Eliminate b from equations $\boxed{B}$ and $\boxed{E}$.

$\boxed{B}\ 3b + 2c - 2d = -11$

$-3 \cdot \boxed{D}\ -3b + 21c + 15d = 63$

$23c + 13d = 52\ \boxed{F}$

Eliminate a and b from equations $\boxed{A}$ and $\boxed{D}$.

$-2 \cdot \boxed{A}\ -4a - 2b + 2c - 2d = -14$

$\boxed{D}\ 4a + 2b - 5c = 6$

$-3c - 2d = -8\ \boxed{G}$

Solve the system of equations $\boxed{F}$ and $\boxed{G}$.

$2 \cdot \boxed{F}\ 46c + 26d = 104$

$13 \cdot \boxed{G}\ -39c - 26d = -104$

$7c = 0$

$c = 0$

Back substitute.

$\boxed{G}\ -3c - 2d = -8$

$-3(0) - 2d = -8$

$-2d = -8$

$d = 4$

$\boxed{E}\ b - 7c - 5d = -21$

$b - 7(0) - 5(4) = -21$

$b - 20 = -21$

$b = -1$

$\boxed{A}\ 2a + b - c + d = 7$

$2a - 1 - 0 + 4 = 7$

$2a = 4$

$a = 2 \qquad \{(2, -1, 0, 4)\}$

63. Let $x = u - 3$, $y = v + 1$, and $z = w - 2$.

$\boxed{A}\ \frac{x}{4} + \frac{y}{3} + \frac{z}{8} = 1 \rightarrow 6x + 8y + 3z = 24$

$\boxed{B}\ \frac{x}{2} + \frac{y}{2} + \frac{z}{4} = 0 \rightarrow 2x + 2y + z = 0$

$\boxed{C}\ \frac{x}{4} - \frac{y}{2} + \frac{z}{2} = -6 \rightarrow x - 2y + 2z = -24$

Eliminate y from equations $\boxed{B}$ and $\boxed{C}$.

$\boxed{B}\ 2x + 2y + z = 0$

$\boxed{C}\ x - 2y + = -24$

$3x + 3z = -24$

$x + z = -8\ \boxed{D}$

Eliminate y from equations $\boxed{A}$ and $\boxed{C}$.

$\boxed{A}\ 6x + 8y + 3z = 24$

$4 \cdot \boxed{C}\ 4x - 8y + = -96$

$10x + 11z = -72\ \boxed{E}$

Solve the systems of equations $\boxed{D}$ and $\boxed{E}$.

$-10 \cdot \boxed{D}\ -10x - 10z = 80$

$\boxed{E}\ 10x + 11z = -72$

$z = 8$

Back substitute.

$\boxed{D}\ x + z = -8$

$x + 8 = -8$

$x = -16$

$\boxed{B}\ 2x + 2y + z = 0$

$2(-16) + 2y + 8 = 0$

$-32 + 2y + 8 = 0$

$2y = 24$

$y = 12$

$x = u - 3 \qquad y = v + 1 \qquad z = w - 2$

$-16 = u - 3 \qquad 12 = v + 1 \qquad 8 = w - 2$

$-13 = u \qquad 11 = v \qquad 10 = w$

$\{(-13, 11, 10)\}$

65. a.

$$x^2+y^2+Ax+By+C=0$$

Substitute $(2,2)$: $(2)^2+(2)^2+A(2)+B(2)+C=0 \rightarrow$ [A] $2A+2B+C=-8$

Substitute $(6,0)$: $(6)^2+(0)^2+A(6)+B(0)+C=0 \rightarrow$ [B] $6A \qquad +C=-36$

Substitute $(7,-3)$: $(7)^2+(-3)^2+A(7)+B(-3)+C=0 \rightarrow$ [C] $7A-3B+C=-58$

Eliminate B from equations [A] and [C].

$$\begin{array}{lrl} 3\cdot\boxed{A} & 6A+6B+3C= & -24 \\ 2\cdot\boxed{C} & 14A-6B+2C= & -116 \\ \hline & 20A- \qquad +5C= & -140 \\ & 4A+ \quad C= & -28 \ \boxed{D} \end{array}$$

Solve the system of equations [B] and [D].

$$\begin{array}{lrl} \boxed{B} & 6A+C= & -36 \\ -1\cdot\boxed{D} & -4A-C= & 28 \\ \hline & 2A \quad = & -8 \\ & A= & -4 \end{array}$$

$$\begin{aligned} \boxed{D} \quad 4A+C&=-28 \\ 4(-4)+C&=-28 \\ -16+C&=-28 \\ C&=-12 \end{aligned}$$

Back substitute.

$$\begin{aligned} \boxed{A} \quad 2A+2B+C&=-8 \\ 2(-4)+2B-12&=-8 \\ -8+2B-12&=-8 \\ 2B&=12 \\ B&=6 \end{aligned}$$

$$x^2+y^2-4x+6y-12=0$$

b.

$$\begin{aligned} x^2+y^2-4x+6y-12&=0 \\ x^2-4x \quad +y^2+6y \quad &=12 \\ x^2-4x+4+y^2+6y+9&=12+4+9 \\ (x-2)^2+(y+3)^2&=25 \\ (x-2)^2+(y+3)^2&=5^2 \end{aligned}$$

Center: $(2,-3)$; Radius: 5

67.

$$c_1x+c_2y+c_3z=1$$

Substitute $(3,1,2)$: $c_1(3)+c_2(1)+c_3(2)=1 \rightarrow$ [A] $3c_1+c_2+2c_3=1$

Substitute $(-1,1,0)$: $c_1(-1)+c_2(1)+c_3(0)=1 \rightarrow$ [B] $-c_1+c_2 \qquad =1$

Substitute $(1,-3,-2)$: $c_1(1)+c_2(-3)+c_3(-2)=1 \rightarrow$ [C] $c_1-3c_2-2c_3=1$

Eliminate c_3 from equations [A] and [C].

$$\begin{array}{lrl} \boxed{A} & 3c_1+c_2+2c_3= & 1 \\ \boxed{C} & c_1-3c_2- \qquad = & 1 \\ \hline & 4c_1-2c_2 \qquad = & 2 \\ & 2c_1-\ c_2= & 1\ \boxed{D} \end{array}$$

Solve the system of equations B and D.

$$\begin{aligned} \boxed{B} \quad -c_1 + c_2 &= 1 \\ \boxed{D} \quad 2c_1 - c_2 &= 1 \\ \hline c_1 \quad &= 2 \end{aligned}$$

$$\begin{aligned} \boxed{B} \quad -c_1 + c_2 &= 1 \\ -2 + c_2 &= 1 \\ c_2 &= 3 \end{aligned}$$

Back substitute.

$$\begin{aligned} \boxed{A} \quad 3c_1 + c_2 + 2c_3 &= 1 \\ 3(2) + 3 + 2c_3 &= 1 \\ 6 + 3 + 2c_3 &= 1 \\ 2c_3 &= -8 \\ c_3 &= -4 \end{aligned}$$

$2x + 3y - 4z = 1$

69.

$$\begin{aligned} \boxed{A} \quad 2A + B + 2C &= 11 \\ \boxed{B} \quad 5A - 2B - 7C &= 26 \\ \boxed{C} \quad 3A - 8B - 4C &= -5 \end{aligned}$$

Eliminate B from equations A and B and equations A and C.

$$\begin{aligned} 2 \cdot \boxed{A} \quad 4A + 2B + 4C &= 22 \\ \boxed{B} \quad 5A - 2B - \quad &= 26 \\ \hline 9A \quad - 3C &= 48 \\ 3A - C &= 16 \quad \boxed{D} \end{aligned}$$

$$\begin{aligned} 8 \cdot \boxed{A} \quad 16A + 8B + 16C &= 88 \\ \boxed{C} \quad 3A - 8B - \quad &= -5 \\ \hline 19A \quad + 12C &= 83 \quad \boxed{E} \end{aligned}$$

Solve the system of equations D and E.

$$\begin{aligned} 12 \cdot \boxed{D} \quad 36A - 12C &= 192 \\ \boxed{E} \quad 19A + 12C &= 83 \\ \hline 55A \quad &= 275 \\ A &= 5 \end{aligned}$$

$$\begin{aligned} \boxed{D} \quad 3A - C &= 16 \\ 3(5) - C &= 16 \\ 15 - C &= 16 \\ C &= -1 \end{aligned}$$

Back substitute.

$$\begin{aligned} \boxed{A} \quad 2A + B + 2C &= 11 \\ 2(5) + B + 2(-1) &= 11 \\ 10 + B - 2 &= 11 \\ B &= 3 \end{aligned}$$

$A = 5, B = 3, C = -1$

Section 5.3 Partial Fraction Decomposition

1. a. $3x-2y=6$

$$-2y=-3x+6$$
$$y=\frac{3}{2}x-3$$

$$4y=6x-12$$
$$y=\frac{3}{2}x-3$$

b. $\frac{3}{2}x-3=\frac{3}{2}x-3$

$-3=-3$ Identity

Infinitely many solutions

c. $\left\{(x,y)\,\middle|\, y=\frac{3}{2}x-3\right\}$

3. $\boxed{A}$ $3A=5B-C-19 \rightarrow 3A-5B+C=-19$

$\boxed{B}$ $B=5C-2A-1 \rightarrow 2A+B-5C=-1$

$\boxed{C}$ $7A+3B+C=13 \rightarrow 7A+3B+C=13$

Eliminate C from equations $\boxed{A}$ and $\boxed{B}$ and equations $\boxed{A}$ and $\boxed{C}$.

$$\begin{array}{rl} 5\cdot\boxed{A} & 15A-25B+5C=-95 \\ \boxed{B} & 2A+B-5C=-1 \\ \hline & 17A-24B=-96 \quad \boxed{D} \end{array}$$

$$\begin{array}{rl} -1\cdot\boxed{A} & -3A+5B-C=19 \\ \boxed{C} & 7A+3B+C=13 \\ \hline & 4A+8B=32 \quad \boxed{E} \end{array}$$

Solve the system of equations $\boxed{D}$ and $\boxed{E}$.

$$\begin{array}{rl} \boxed{D} & 17A-24B=-96 \\ 3\cdot\boxed{E} & 12A+24B=96 \\ \hline & 29A=0 \\ & A=0 \end{array}$$

$\boxed{E}$ $4A+8B=32$

$$4(0)+8B=32$$
$$8B=32$$
$$B=4$$

Back substitute.

$\boxed{A}$ $3A-5B+C=-19$

$$3(0)-5(4)+C=-19$$
$$0-20+C=-19$$
$$C=1$$

$\{0, 4, 1\}$

5.

$$\begin{array}{r} 3x-4 \\ x^2+2x+1\overline{)3x^3+2x^2-x-5} \\ -(3x^3+6x^2+3x) \\ \hline -4x^2-4x-5 \\ -(-4x^2-8x-4) \\ \hline 4x-1 \end{array}$$

$$3x-4+\frac{4x-1}{x^2+2x+1}$$

7. fraction decomposition

9. linear; $Ax+B$

11. $\frac{-x-37}{(x+4)(2x-3)}=\frac{A}{x+4}+\frac{B}{2x-3}$

13. $\frac{8x-10}{x^2-2x}=\frac{8x-10}{x(x-2)}$

$$=\frac{A}{x}+\frac{B}{x-2}$$

15. $\frac{6w-7}{w^2+w-6}=\frac{6w-7}{(w-2)(w+3)}$

$$=\frac{A}{w-2}+\frac{B}{w+3}$$

17. $\frac{x^2+26x+100}{x^3+10x^2+25x}=\frac{x^2+26x+100}{x(x^2+10x+25)}$

$$=\frac{x^2+26x+100}{x(x+5)^2}$$
$$=\frac{A}{x}+\frac{B}{x+5}+\frac{C}{(x+5)^2}$$

19. $$\frac{13x^2+2x+45}{2x^3+18x}=\frac{13x^2+2x+45}{2x\left(x^2+9\right)}$$
$$=\frac{A}{2x}+\frac{Bx+C}{x^2+9}$$

21. $$\frac{2x^3-x^2+13x-5}{x^4+10x^2+25}=\frac{2x^3-x^2+13x-5}{\left(x^2+5\right)^2}$$
$$=\frac{Ax+B}{x^2+5}+\frac{Cx+D}{\left(x^2+5\right)^2}$$

23. $$\frac{5x^2-4x+8}{(x-4)\left(x^2+x+4\right)}=\frac{A}{x-4}+\frac{Bx+C}{x^2+x+4}$$

25. $$\frac{2x^5+3x^3+4x^2+5}{x(x+2)^3\left(x^2+2x+7\right)^2}$$
$$=\frac{A}{x}+\frac{B}{x+2}+\frac{C}{(x+2)^2}+\frac{D}{(x+2)^3}$$
$$+\frac{Ex+F}{x^2+2x+7}+\frac{Gx+H}{\left(x^2+2x+7\right)^2}$$

27. $$\frac{-x-37}{(x+4)(2x-3)}=\frac{A}{x+4}+\frac{B}{2x-3}$$
$$(x+4)(2x-3)\cdot\left[\frac{-x-37}{(x+4)(2x-3)}\right]=(x+4)(2x-3)\cdot\left[\frac{A}{x+4}+\frac{B}{2x-3}\right]$$
$$-x-37=A(2x-3)+B(x+4)$$
$$-x-37=2Ax-3A+Bx+4B$$
$$-x-37=(2A+B)x+(-3A+4B)$$

$$\boxed{1}\quad -1=2A+B \xrightarrow{\text{Multiply by } -4} -8A-4B=4$$
$$\boxed{2}\quad -37=-3A+4B \qquad\qquad -3A+4B=-37$$
$$-11A=-33$$
$$A=3$$

$$\boxed{1}\quad 2A+B=-1$$
$$2(3)+B=-1$$
$$6+B=-1$$
$$B=-7$$

$$\frac{-x-37}{(x+4)(2x-3)}=\frac{3}{x+4}+\frac{-7}{2x-3}$$

29. $$\frac{8x-10}{x^2-2x}=\frac{8x-10}{x(x-2)}=\frac{A}{x}+\frac{B}{x-2}$$
$$x(x-2)\cdot\left[\frac{8x-10}{x(x-2)}\right]=x(x-2)\cdot\left[\frac{A}{x}+\frac{B}{x-2}\right]$$
$$8x-10=A(x-2)+Bx$$
$$8x-10=Ax-2A+Bx$$
$$8x-10=(A+B)x+(-2A)$$

$$-2A=-10 \qquad A=5$$

$$A+B=8 \qquad 5+B=8 \qquad B=3$$

$$\frac{8x-10}{x^2-2x}=\frac{5}{x}+\frac{3}{x-2}$$

31. $\dfrac{6w-7}{w^2+w-6}=\dfrac{6w-7}{(w-2)(w+3)}=\dfrac{A}{w-2}+\dfrac{B}{w+3}$

$$(w-2)(w+3)\cdot\left[\frac{6w-7}{(w-2)(w+3)}\right]=(w-2)(w+3)\cdot\left[\frac{A}{w-2}+\frac{B}{w+3}\right]$$

$$6w-7=A(w+3)+B(w-2)$$
$$6w-7=Aw+3A+Bw-2B$$
$$6w-7=(A+B)w+(3A-2B)$$

$\boxed{1}$ $6=A+B$ $\xrightarrow{\text{Multiply by 2}}$ $2A+2B=12$

$\boxed{2}$ $-7=3A-2B$ $\qquad 3A-2B=-7$

$5A=5$

$A=1$

$\boxed{1}$ $A+B=6$

$1+B=6$

$B=5$

$$\frac{6w-7}{w^2+w-6}=\frac{1}{w-2}+\frac{5}{w+3}$$

33. $\dfrac{x^2+26x+100}{x^3+10x^2+25x}=\dfrac{x^2+26x+100}{x(x^2+10x+25)}=\dfrac{x^2+26x+100}{x(x+5)^2}=\dfrac{A}{x}+\dfrac{B}{x+5}+\dfrac{C}{(x+5)^2}$

$$x(x+5)^2\cdot\left[\frac{x^2+26x+100}{x(x+5)^2}\right]=x(x+5)^2\cdot\left[\frac{A}{x}+\frac{B}{x+5}+\frac{C}{(x+5)^2}\right]$$

$$x^2+26x+100=A(x+5)^2+Bx(x+5)+Cx$$
$$x^2+26x+100=A(x^2+10x+25)+Bx^2+5Bx+Cx$$
$$x^2+26x+100=Ax^2+10Ax+25A+Bx^2+5Bx+Cx$$
$$x^2+26x+100=(A+B)x^2+(10A+5B+C)x+(25A)$$

$25A=100$

$A=4$

$A+B=1$

$4+B=1$

$B=-3$

$10A+5B+C=26$

$10(4)+5(-3)+C=26$

$40-15+C=26$

$C=1$

$$\frac{x^2+26x+100}{x^3+10x^2+25x}=\frac{4}{x}+\frac{-3}{x+5}+\frac{1}{(x+5)^2}$$

35. $\dfrac{13x^2+2x+45}{2x^3+18x}=\dfrac{13x^2+2x+45}{2x(x^2+9)}=\dfrac{A}{2x}+\dfrac{Bx+C}{x^2+9}$

$$2x(x^2+9)\cdot\left[\frac{13x^2+2x+45}{2x(x^2+9)}\right]=2x(x^2+9)\cdot\left[\frac{A}{2x}+\frac{Bx+C}{x^2+9}\right]$$

$$13x^2+2x+45=A(x^2+9)+(Bx+C)\cdot 2x$$
$$13x^2+2x+45=Ax^2+9A+2Bx^2+2Cx$$
$$13x^2+2x+45=(A+2B)x^2+(2C)x+(9A)$$

$9A=45$

$A=5$

$A+2B=13$

$5+2B=13$

$2B=8$

$B=4$

$2C=2$

$C=1$

$$\frac{13x^2+2x+45}{2x^3+18x}=\frac{5}{2x}+\frac{4x+1}{x^2+9}$$

37. $$\begin{array}{r} x-3 \\ x^3+0x^2+7x \overline{)x^4-3x^3+13x^2-28x+28} \\ -\left(x^4+0x^3+7x^2\right) \\ \hline -3x^3+6x^2-28x \\ -\left(-3x^3+0x^2-21x\right) \\ \hline 6x^2-7x+28 \end{array}$$

$$\frac{x^4-3x^3+13x^2-28x+28}{x^3+7x}=x-3+\frac{6x^2-7x+28}{x^3+7x}$$

$$\frac{6x^2-7x+28}{x^3+7x}=\frac{6x^2-7x+28}{x\left(x^2+7\right)}=\frac{A}{x}+\frac{Bx+C}{x^2+7}$$

$7A=28$
$A=4$

$A+B=6$
$4+B=6$
$B=2$

$C=-7$

$$\frac{x^4-3x^3+13x^2-28x+28}{x^3+7x}=x-3+\frac{4}{x}+\frac{2x-7}{x^2+7}$$

39. $$\frac{2x^3-x^2+13x-5}{x^4+10x^2+25}=\frac{2x^3-x^2+13x-5}{\left(x^2+5\right)^2}=\frac{Ax+B}{x^2+5}+\frac{Cx+D}{\left(x^2+5\right)^2}$$

$$\left(x^2+5\right)^2\cdot\left[\frac{2x^3-x^2+13x-5}{\left(x^2+5\right)^2}\right]=\left(x^2+5\right)^2\cdot\left[\frac{Ax+B}{x^2+5}+\frac{Cx+D}{\left(x^2+5\right)^2}\right]$$

$$2x^3-x^2+13x-5=(Ax+B)\left(x^2+5\right)+(Cx+D)$$

$$2x^3-x^2+13x-5=Ax^3+5Ax+Bx^2+5B+Cx+D$$

$$2x^3-x^2+13x-5=Ax^3+Bx^2+(5A+C)x+(5B+D)$$

$A=2$

$B=-1$

$5A+C=13$
$5(2)+C=13$
$10+C=13$
$C=3$

$5B+D=-5$
$5(-1)+D=-5$
$-5+D=-5$
$D=0$

$$\frac{2x^3-x^2+13x-5}{x^4+10x^2+25}=\frac{2x-1}{x^2+5}+\frac{3x}{\left(x^2+5\right)^2}$$

41. $$\frac{5x^2-4x+8}{(x-4)\left(x^2+x+4\right)}=\frac{A}{x-4}+\frac{Bx+C}{x^2+x+4}$$

$$(x-4)\left(x^2+x+4\right)\cdot\left[\frac{5x^2-4x+8}{(x-4)\left(x^2+x+4\right)}\right]=(x-4)\left(x^2+x+4\right)\cdot\left[\frac{A}{x-4}+\frac{Bx+C}{x^2+x+4}\right]$$

$$5x^2-4x+8=A\left(x^2+x+4\right)+(Bx+C)(x-4)$$

$$5x^2-4x+8=Ax^2+Ax+4A+Bx^2-4Bx+Cx-4C$$

$$5x^2-4x+8=(A+B)x^2+(A-4B+C)x+(4A-4C)$$

$$\begin{array}{ll} \boxed{1} & 5 = A + B \\ \boxed{2} & -4 = A - 4B + C \\ \boxed{3} & 8 = 4A - 4C \end{array}$$

$$\begin{array}{ll} \boxed{1} & 5 = A + B \\ \boxed{2} & -4 = A - 4B + C \end{array} \xrightarrow{\text{Multiply by 4}} \begin{array}{r} 4A + 4B = 20 \\ A - 4B + C = -4 \\ \hline 5A + C = 16 \ \boxed{4} \end{array}$$

$$\begin{array}{ll} \boxed{3} & 4A - 4C = 8 \\ \boxed{4} & 5A + C = 16 \end{array} \xrightarrow{\text{Multiply by 4}} \begin{array}{r} 4A - 4C = 8 \\ 20A + 4B = 64 \\ \hline 24A = 72 \\ A = 3 \end{array}$$

$$\begin{array}{r} \boxed{1} \ A + B = 5 \\ 3 + B = 5 \\ B = 2 \end{array} \qquad \begin{array}{r} \boxed{3} \ 4A - 4C = 8 \\ 4(3) - 4C = 8 \\ 12 - 4C = 8 \\ -4C = -4 \\ C = 1 \end{array}$$

$$\frac{5x^2 - 4x + 8}{(x-4)(x^2+x+4)} = \frac{3}{x-4} + \frac{2x+1}{x^2+x+4}$$

43. $\dfrac{4x^3 - 4x^2 + 11x - 7}{x^4 + 5x^2 + 6} = \dfrac{4x^3 - 4x^2 + 11x - 7}{(x^2+2)(x^2+3)} = \dfrac{Ax+B}{x^2+2} + \dfrac{Cx+D}{x^2+3}$

$$(x^2+2)(x^2+3) \cdot \left[\frac{4x^3 - 4x^2 + 11x - 7}{(x^2+2)(x^2+3)}\right] = (x^2+2)(x^2+3) \cdot \left[\frac{Ax+B}{x^2+2} + \frac{Cx+D}{x^2+3}\right]$$

$$4x^3 - 4x^2 + 11x - 7 = (Ax+B)(x^2+3) + (Cx+D)(x^2+2)$$

$$4x^3 - 4x^2 + 11x - 7 = Ax^3 + 3Ax + Bx^2 + 3B + Cx^3 + 2Cx + Dx^2 + 2D$$

$$4x^3 - 4x^2 + 11x - 7 = (A+C)x^3 + (B+D)x^2 + (3A+2C)x + (3B+2D)$$

$$\begin{array}{r} A + C = 4 \\ A = 4 - C \end{array} \qquad \begin{array}{r} 3A + 2C = 11 \\ 3(4-C) + 2C = 11 \\ 12 - 3C + 2C = 11 \\ C = 1 \end{array} \qquad \begin{array}{r} A + C = 4 \\ A + 1 = 4 \\ A = 3 \end{array}$$

$$\begin{array}{r} B + D = -4 \\ B = -4 - D \end{array} \qquad \begin{array}{r} 3B + 2D = -7 \\ 3(-4-D) + 2D = -7 \\ -12 - 3D + 2D = -7 \\ -D = 5 \\ D = -5 \end{array} \qquad \begin{array}{r} B + D = -4 \\ B - 5 = -4 \\ B = 1 \end{array}$$

$$\frac{4x^3 - 4x^2 + 11x - 7}{x^4 + 5x^2 + 6} = \frac{3x+1}{x^2+2} + \frac{x-5}{x^2+3}$$

45.

$$\begin{array}{r} 2x - 5 \\ x^2 - 3x - 10 \overline{)2x^3 - 11x^2 - 4x + 24} \\ -(2x^3 - 6x^2 - 20x) \\ \hline -5x^2 + 16x + 24 \\ -(-5x^2 + 15x + 50) \\ \hline x - 26 \end{array}$$

$$\frac{2x^3-11x^2-4x+24}{x^2-3x-10}=2x-5+\frac{x-26}{x^2-3x-10}$$

$$\frac{x-26}{x^2-3x-10}=\frac{x-26}{(x+2)(x-5)}=\frac{A}{x+2}+\frac{B}{x-5}$$

$$(x+2)(x-5)\cdot\left[\frac{x-26}{(x+2)(x-5)}\right]=(x+2)(x-5)\cdot\left[\frac{A}{x+2}+\frac{B}{x-5}\right]$$

$$x-26=A(x-5)+B(x+2)$$

$$x-26=Ax-5A+Bx+2B$$

$$x-26=(A+B)x+(-5A+2B)$$

$A+B=1$
$A=1-B$

$-5A+2B=-26$
$-5(1-B)+2B=-26$
$-5+5B+2B=-26$
$7B=-21$
$B=-3$

$A+B=1$
$A-3=1$
$A=4$

$$\frac{2x^3-11x^2-4x+24}{x^2-3x-10}=2x-5+\frac{4}{x+2}+\frac{-3}{x-5}$$

47.

$$\begin{array}{r} 3x-4 \\ x^2+2x+1\overline{)3x^3+2x^2-\ x-5} \\ -(3x^3+6x^2+3x) \\ \hline -4x^2-4x-5 \\ -(-4x^2-8x-4) \\ \hline 4x-1 \end{array}$$

$$\frac{3x^3+2x^2-x-5}{x^2+2x+1}=3x-4+\frac{4x-1}{x^2+2x+1}$$

$$\frac{4x-1}{x^2+2x+1}=\frac{4x-1}{(x+1)^2}=\frac{A}{x+1}+\frac{B}{(x+1)^2}$$

$$(x+1)^2\cdot\left[\frac{4x-1}{(x+1)^2}\right]=(x+1)^2\cdot\left[\frac{A}{x+1}+\frac{B}{(x+1)^2}\right]$$

$$4x-1=A(x+1)+B$$

$$4x-1=Ax+A+B$$

$$4x-1=(A)x+(A+B)$$

$A=4$

$A+B=-1$
$4+B=-1$
$B=-5$

$A=4$

$A+B=-1$
$4+B=-1$
$B=-5$

$$\frac{3x^3+2x^2-x-5}{x^2+2x+1}=3x-4+\frac{4}{x+1}+\frac{-5}{(x+1)^2}$$

49. a. $\frac{\text{Factors of } 45}{\text{Factors of } 1} = \frac{\pm 1, \pm 3, \pm 5, \pm 9, \pm 15, \pm 45}{\pm 1} = \pm 1, \pm 3, \pm 5, \pm 9, \pm 15, \pm 45$

$$\begin{array}{r|rrrr} -5 & 1 & -1 & -21 & 45 \\ & & -5 & 30 & -45 \\ \hline & 1 & -6 & 9 & \boxed{0} \end{array}$$

$x^3 - x^2 - 21x + 45 = (x^2 - 6x + 9)(x+5) = (x-3)^2(x+5)$

b. $$\frac{-3x^2 + 35x - 70}{(x-3)^2(x+5)} = \frac{A}{x-3} + \frac{B}{(x-3)^2} + \frac{C}{x+5}$$

$$(x-3)^2(x+5) \cdot \left[\frac{-3x^2 + 35x - 70}{(x-3)^2(x+5)}\right] = (x-3)^2(x+5) \cdot \left[\frac{A}{x-3} + \frac{B}{(x-3)^2} + \frac{C}{x+5}\right]$$

$$-3x^2 + 35x - 70 = A(x-3)(x+5) + B(x+5) + C(x-3)^2$$
$$-3x^2 + 35x - 70 = A(x^2 + 2x - 15) + B(x+5) + C(x^2 - 6x + 9)$$
$$-3x^2 + 35x - 70 = Ax^2 + 2Ax - 15A + Bx + 5B + Cx^2 - 6Cx + 9C$$
$$-3x^2 + 35x - 70 = (A+C)x^2 + (2A + B - 6C)x + (-15A + 5B + 9C)$$

$\boxed{1}$ $A + C = -3$

$\boxed{2}$ $2A + B - 6C = 35$

$\boxed{3}$ $-15A + 5B + 9C = -70$

$-5 \cdot \boxed{2}$ $-10A - 5B + 30C = -175$

$\boxed{3}$ $-15A + 5B + 9C = -70$

$-25A + 39C = -245$ $\boxed{4}$

Solve the system of equations $\boxed{1}$ and $\boxed{4}$.

$25 \cdot \boxed{1}$ $25A + 25C = -75$

$\boxed{4}$ $-25A + 39C = -245$

$64C = -320$

$C = -5$

$\boxed{1}$ $A + C = -3$

$A - 5 = -3$

$A = 2$

$\boxed{2}$ $2A + B - 6C = 35$

$2(2) + B - 6(-5) = 35$

$4 + B + 30 = 35$

$B = 1$

$$\frac{-3x^2 + 35x - 70}{x^3 - x^2 - 21x + 45} = \frac{2}{x-3} + \frac{1}{(x-3)^2} + \frac{-5}{x+5}$$

51. a. $\frac{\text{Factors of } 8}{\text{Factors of } 1} = \frac{\pm 1, \pm 2, \pm 4, \pm 8}{\pm 1} = \pm 1, \pm 2, \pm 4, \pm 8$

$$\begin{array}{r|rrrr} -2 & 1 & 6 & 12 & 8 \\ & & -2 & -8 & -8 \\ \hline & 1 & 4 & 4 & \boxed{0} \end{array}$$

$x^3 + 6x^2 + 12x + 8 = (x^2 + 4x + 2)(x+2) = (x+2)^2(x+2) = (x+2)^3$

b. $$\frac{3x^2 + 8x + 5}{(x+2)^3} = \frac{A}{x+2} + \frac{B}{(x+2)^2} + \frac{C}{(x+2)^3}$$

$$(x+2)^3 \cdot \left[\frac{3x^2 + 8x + 5}{(x+2)^3}\right] = (x+2)^3 \cdot \left[\frac{A}{x+2} + \frac{B}{(x+2)^2} + \frac{C}{(x+2)^3}\right]$$

$$3x^2 + 8x + 5 = A(x+2)^2 + B(x+2) + C$$

$$3x^2+8x+5=A\left(x^2+4x+4\right)+B(x+2)+C$$
$$3x^2+8x+5=Ax^2+4Ax+4A+Bx+2B+C$$
$$3x^2+8x+5=(A)x^2+(4A+B)x+(4A+2B+C)$$

$\boxed{1}$ $A = 3$
$\boxed{2}$ $4A+B = 8$
$\boxed{3}$ $4A+2B+C = 5$

$$4A+B=8$$
$$4(3)+B=8$$
$$12+B=8$$
$$B=-4$$

$$4A+2B+C=5$$
$$4(3)+2(-4)+C=5$$
$$12-8+C=5$$
$$C=1$$

$$\frac{3x^2+8x+5}{x^3+6x^2+12x+8}=\frac{3}{x+2}+\frac{-4}{(x+2)^2}+\frac{1}{(x+2)^3}$$

53. Partial fraction decomposition is a procedure in which a rational expression is written as a sum of two or more simpler rational expressions.

55. A proper rational expression is a rational expression in which the degree of the numerator is less than the degree of the denominator.

57. a.
$$\frac{2}{n(n+2)}=\frac{A}{n}+\frac{B}{n+2}$$
$$n(n+2)\cdot\left[\frac{2}{n(n+2)}\right]=n(n+2)\cdot\left[\frac{A}{n}+\frac{B}{n+2}\right]$$
$$2=A(n+2)+Bn$$
$$2=An+2A+Bn$$
$$2=(A+B)n+(2A)$$

$$2A=2$$
$$A=1$$

$$A+B=0$$
$$1+B=0$$
$$B=-1$$

$$\frac{2}{n(n+2)}=\frac{1}{n}+\frac{-1}{n+2}=\frac{1}{n}-\frac{1}{n+2}$$

b.
$$\frac{2}{1(3)}+\frac{2}{2(4)}+\frac{2}{3(5)}+\frac{2}{4(6)}+\frac{2}{5(7)}\cdots$$
$$=\left(\frac{1}{1}-\frac{1}{3}\right)+\left(\frac{1}{2}-\frac{1}{4}\right)+\left(\frac{1}{3}-\frac{1}{5}\right)$$
$$+\left(\frac{1}{4}-\frac{1}{6}\right)+\left(\frac{1}{5}-\frac{1}{7}\right)\cdots$$

c. As $n\to\infty$, $\frac{1}{n+2}\to 0$.

d.
$$\left(\frac{1}{1}-\frac{1}{3}\right)+\left(\frac{1}{2}-\frac{1}{4}\right)+\left(\frac{1}{3}-\frac{1}{5}\right)+\left(\frac{1}{4}-\frac{1}{6}\right)$$
$$+\left(\frac{1}{5}-\frac{1}{7}\right)\cdots$$
$$=\left(\frac{1}{1}+\frac{1}{2}\right)+\left(-\frac{1}{3}+\frac{1}{3}\right)+\left(-\frac{1}{4}+\frac{1}{4}\right)$$
$$+\left(-\frac{1}{5}+\frac{1}{5}\right)\cdots$$
$$=\frac{3}{2}+0+0+0+\cdots=\frac{3}{2}$$

59.
$$\frac{1}{x(a+bx)}=\frac{A}{x}+\frac{B}{a+bx}$$
$$x(a+bx)\cdot\left[\frac{1}{x(a+bx)}\right]=x(a+bx)\cdot\left[\frac{A}{x}+\frac{B}{a+bx}\right]$$
$$1=A(a+bx)+Bx$$
$$1=aA+bAx+Bx$$
$$1=(bA+B)x+(aA)$$

$$aA=1$$
$$A=\frac{1}{a}$$

$$bA+B=0$$
$$b\frac{1}{a}+B=0$$
$$B=-\frac{b}{a}$$

$$\frac{1}{x(a+bx)}=\frac{\frac{1}{a}}{x}+\frac{-\frac{b}{a}}{a+bx}=\frac{1}{ax}-\frac{b}{a(a+bx)}$$

61. Let $u = e^x$.

$$\frac{5e^x+7}{e^{2x}+3e^x+2} = \frac{5u+7}{u^2+3u+2} = \frac{5u+7}{(u+1)(u+2)} = \frac{A}{u+1} + \frac{B}{u+2}$$

$$(u+1)(u+2)\cdot\left[\frac{5u+7}{(u+1)(u+2)}\right] = (u+1)(u+2)\cdot\left[\frac{A}{u+1}+\frac{B}{u+2}\right]$$

$$5u+7 = A(u+2)+B(u+1)$$
$$5u+7 = Au+2A+Bu+B$$
$$5u+7 = (A+B)u+(2A+B)$$

$A+B=5$
$A=-B+5$

$2A+B=7$
$2(-B+5)+B=7$
$-2B+10+B=7$
$-B=-3$
$B=3$

$A+B=5$
$A+3=5$
$A=2$

$$\frac{5e^x+7}{e^{2x}+3e^x+2} = \frac{A}{u+1}+\frac{B}{u+2} = \frac{2}{e^x+1}+\frac{3}{e^x+2}$$

Section 5.4 Systems of Nonlinear Equations in Two Variables

1. $y=(x-2)^2-1=(x-2)^2+(-1)$
The graph of the equation is the graph of $y=x^2$ shifted to the right 2 units and downward 1 unit.

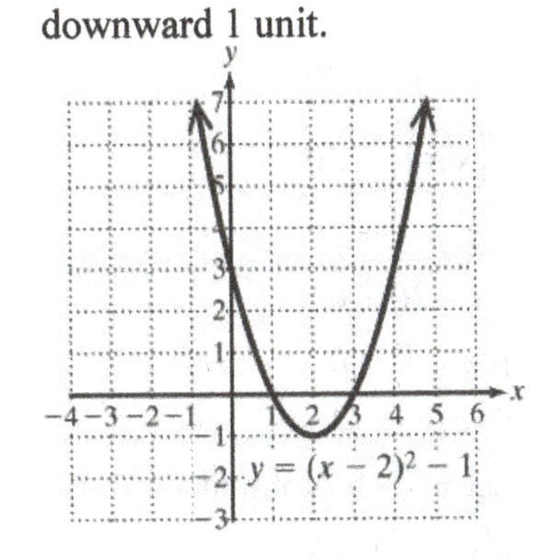

3. $(x-3)^2+(y+2)^2=16$

$(x-3)^2+[y-(-2)]^2=(4)^2$

Center: $(3,-2)$, radius: 4

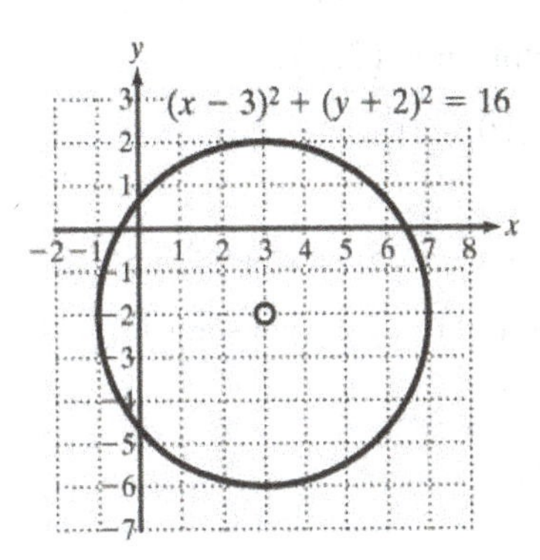

5. $x=4y+7$
$2x+3y=3$

$2x+3y=3$
$2(4y+7)+3y=3$
$8y+14+3y=3$
$11y=-11$
$y=-1$

$x=4y+7$
$x=4(-1)+7$
$x=-4+7$
$x=3$
$\{(3,-1)\}$

7. nonlinear

9. a. A $y = x^2 - 2$

B $2x - y = 2 \rightarrow y = 2x - 2$

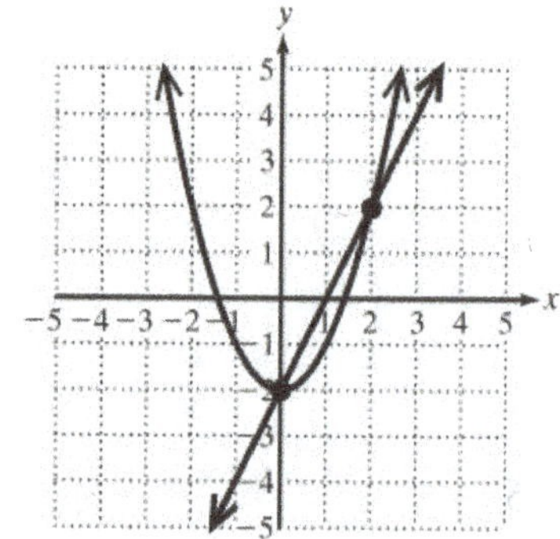

b. B
$$2x - y = 2$$
$$2x - (x^2 - 2) = 2$$
$$2x - x^2 + 2 = 2$$
$$x^2 - 2x = 0$$
$$x(x - 2) = 0$$
$$x = 0 \text{ or } x = 2$$

A $y = x^2 - 2 = (0)^2 - 2 = -2$

The solution is $(0, -2)$.

A $y = x^2 - 2 = (2)^2 - 2 = 4 - 2 = 2$

The solution is $(2, 2)$.

$\{(2, 2), (0, -2)\}$

11. a. A $x^2 + y^2 = 25 \rightarrow x^2 + y^2 = 5^2$

B $x + y = 1 \quad \rightarrow y = -x + 1$

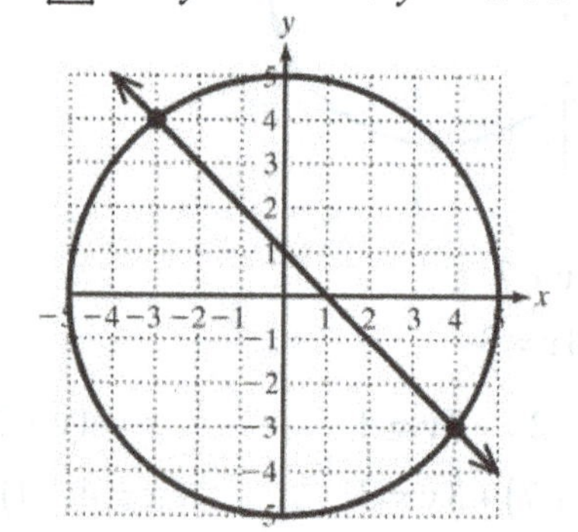

b. A
$$x^2 + y^2 = 25$$
$$x^2 + (-x + 1)^2 = 25$$
$$x^2 + x^2 - 2x + 1 = 25$$
$$2x^2 - 2x - 24 = 0$$
$$x^2 - x - 12 = 0$$
$$(x + 3)(x - 4) = 0$$
$$x = -3 \text{ or } x = 4$$

B $y = -x + 1 = -(-3) + 1 = 3 + 1 = 4$

The solution is $(-3, 4)$.

B $y = -x + 1 = -(4) + 1 = -4 + 1 = -3$

The solution is $(4, -3)$.

$\{(-3, 4), (4, -3)\}$

13. a. A $y = \sqrt{x}$

B $x^2 + y^2 = 20 \rightarrow x^2 + y^2 = (2\sqrt{5})^2$

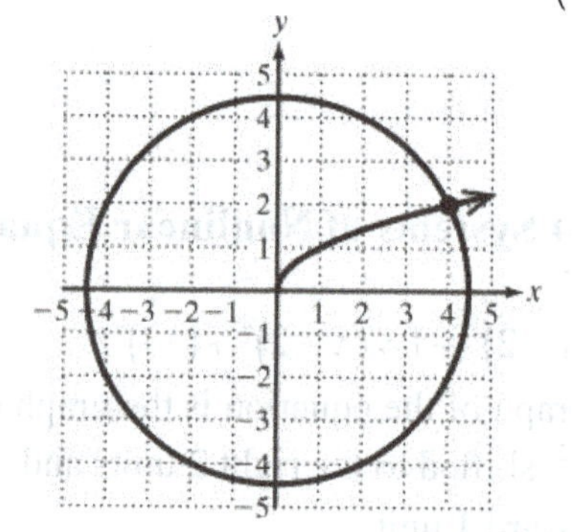

b. B
$$x^2 + y^2 = 20$$
$$x^2 + (\sqrt{x})^2 = 20$$
$$x^2 + x = 20$$
$$x^2 + x - 20 = 0$$
$$(x + 5)(x - 4) = 0$$
$$x = -5 \text{ or } x = 4$$

A $y = \sqrt{x} = \sqrt{-5}$ Not real

A $y = \sqrt{x} = \sqrt{4} = 2$

The solution is $(4, 2)$.

$\{(4, 2)\}$

15. a. $\boxed{A}$ $(x+2)^2+y^2=9 \rightarrow$

$[x-(-2)]^2+(y-0)^2=3^2$

$\boxed{B}$ $y=2x-4$

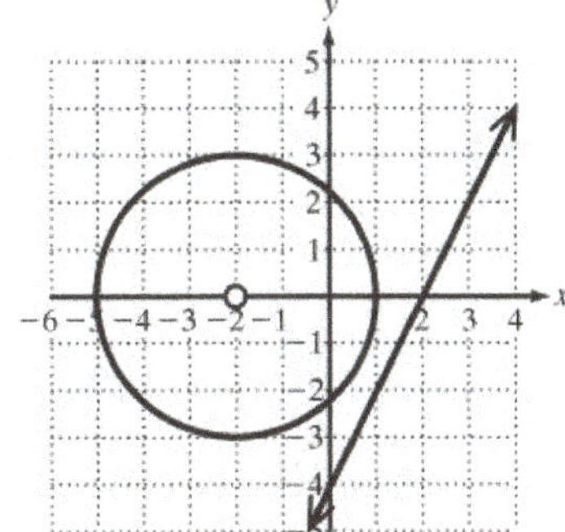

b. $\boxed{A}$

$$(x+2)^2+y^2=9$$
$$(x+2)^2+(2x-4)^2=9$$
$$x^2+4x+4+4x^2-16x+16=9$$
$$5x^2-12x+11=0$$
$$b^2-4ac=(12)^2-4(5)(11)=-76$$

There are no real solutions.

$\{\ \}$

17. a. $\boxed{A}$ $y=x^3$

$\boxed{B}$ $y=x$

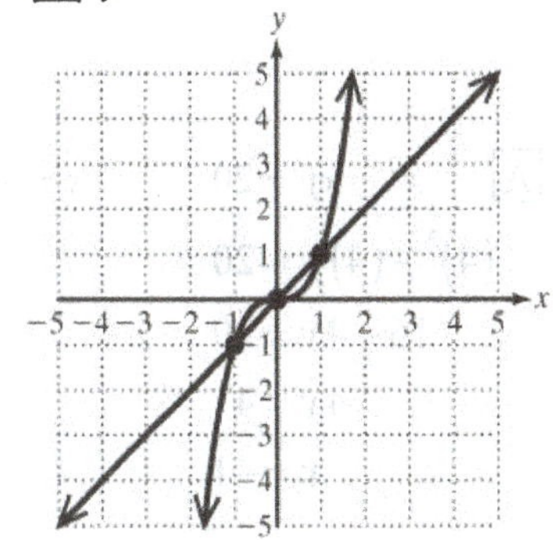

b. $\boxed{A}$ $y=x^3$

$$x=x^3$$
$$0=x^3-x$$
$$0=x(x^2-1)$$
$$0=x(x+1)(x-1)$$
$$x=0 \text{ or } x=-1 \text{ or } x=1$$

$\boxed{B}$ $y=x=0$

The solution is $(0,0)$.

$\boxed{B}$ $y=x=-1$

The solution is $(-1,-1)$.

$\boxed{B}$ $y=x=1$

The solution is $(1,1)$.

$\{(-1,-1),(0,0),(1,1)\}$

19. $\boxed{A}$ $2x^2+3y^2=11$ $\qquad\qquad 2x^2+3y^2=11$

$\boxed{B}$ $x^2+4y^2=8 \xrightarrow{\text{Multiply by }-2.} -2x^2-8y^2=-16$

$$-5y^2=-5$$
$$y^2=1$$
$$y=\pm1$$

$y=1$: $\boxed{B}$ $x^2+4y^2=8$

$$x^2+4(1)^2=8$$
$$x^2+4=8$$
$$x^2=4$$
$$x=\pm2$$

$y=-1$: $\boxed{B}$ $x^2+4y^2=8$

$$x^2+4(-1)^2=8$$
$$x^2+4=8$$
$$x^2=4$$
$$x=\pm2$$

$\{(2,1),(2,-1),(-2,1),(-2,-1)\}$

21.

$$\begin{array}{llcl} \boxed{A} & x^2 - xy = 20 & \xrightarrow{\text{Multiply by 3.}} & 3x^2 - 3xy = 60 \\ \boxed{B} & -2x^2 + 3xy = -44 & & \underline{-2x^2 + 3xy = -44} \\ & & & x^2 = 16 \\ & & & x = \pm 4 \end{array}$$

$x = 4$: $\boxed{A}$

$$\begin{aligned} x^2 - xy &= 20 \\ (4)^2 - (4)y &= 20 \\ 16 - 4y &= 20 \\ -4y &= 4 \\ y &= -1 \end{aligned}$$

$x = -4$: $\boxed{A}$

$$\begin{aligned} x^2 - xy &= 20 \\ (-4)^2 - (-4)y &= 20 \\ 16 + 4y &= 20 \\ 4y &= 4 \\ y &= 1 \end{aligned}$$

$\{(4, -1), (-4, 1)\}$

23.

$$\begin{array}{llcl} \boxed{A} & 5x^2 - 2y^2 = 1 & \xrightarrow{\text{Multiply by 3.}} & 15x^2 - 6y^2 = 3 \\ \boxed{B} & 2x^2 - 3y^2 = -4 & \xrightarrow{\text{Multiply by } -2.} & \underline{-4x^2 + 6y^2 = 8} \\ & & & 11x^2 = 11 \\ & & & x^2 = 1 \\ & & & x = \pm 1 \end{array}$$

$x = 1$: $\boxed{B}$

$$\begin{aligned} 2x^2 - 3y^2 &= -4 \\ 2(1)^2 - 3y^2 &= -4 \\ 2 - 3y^2 &= -4 \\ -3y^2 &= -6 \\ y^2 &= 2 \\ y &= \pm\sqrt{2} \end{aligned}$$

$x = -1$: $\boxed{B}$

$$\begin{aligned} 2x^2 - 3y^2 &= -4 \\ 2(-1)^2 - 3y^2 &= -4 \\ 2 - 3y^2 &= -4 \\ -3y^2 &= -6 \\ y^2 &= 2 \\ y &= \pm\sqrt{2} \end{aligned}$$

$\{(1, \sqrt{2}), (1, -\sqrt{2}), (-1, \sqrt{2}), (-1, -\sqrt{2})\}$

25.

$$\begin{array}{llcl} \boxed{A} & x^2 + y^2 = 1 & \xrightarrow{\text{Multiply by } -9.} & -9x^2 - 9y^2 = -9 \\ \boxed{B} & 9x^2 - 4y^2 = 36 & & \underline{9x^2 - 4y^2 = 36} \\ & & & -13y^2 = 27 \\ & & & y^2 = -\dfrac{27}{13} \\ & & & y = \pm\sqrt{\dfrac{27}{13}}i \end{array}$$

The values for y are all imaginary numbers.

$\{\ \}$

27. $\boxed{A}$ $x^2-4xy+4y^2=1$

$\boxed{B}$ $x+y=4 \quad \to \quad y=-x+4$

$\boxed{A}$
$$\begin{aligned} x^2-4xy+4y^2&=1\\ x^2-4x(-x+4)+4(-x+4)^2&=1\\ x^2+4x^2-16x+4(x^2-8x+16)&=1\\ 5x^2-16x+4x^2-32x+64&=1\\ 9x^2-48x+63&=0\\ 3x^2-16x+21&=0\\ (3x-7)(x-3)&=0\\ x=\frac{7}{3} \quad \text{or} \quad x&=3 \end{aligned}$$

$\boxed{B}$ $y=-x+4=-\left(\frac{7}{3}\right)+4=-\frac{7}{3}+\frac{12}{3}=\frac{5}{3}$

$\boxed{B}$ $y=-x+4=-(3)+4=1$

$\left\{(3,1),\left(\frac{7}{3},\frac{5}{3}\right)\right\}$

29. $\boxed{A}$ $y=x^2+4x+5$

$\boxed{B}$ $y=4x+5$

$\boxed{A}$
$$\begin{aligned} y&=x^2+4x+5\\ 4x+5&=x^2+4x+5\\ 0&=x^2\\ x&=0 \end{aligned}$$

$\boxed{B}$ $y=4x+5=4(0)+5=5$

$\{(0,5)\}$

31. $\boxed{A}$ $y=x^2$

$\boxed{B}$ $y=\frac{1}{x}$

$\boxed{A}$
$$\begin{aligned} y&=x^2\\ \frac{1}{x}&=x^2\\ x^3&=1\\ x&=1 \end{aligned}$$

$\boxed{B}$ $y=\frac{1}{x}=\frac{1}{1}=1$

$\{(1,1)\}$

33. $\boxed{A}$ $x^2+(y-4)^2=25$

$\boxed{B}$ $y=-x^2+9$

$\boxed{A}$
$$\begin{aligned} x^2+(y-4)^2&=25\\ x^2+(-x^2+9-4)^2&=25\\ x^2+(-x^2+5)^2&=25\\ x^2+x^4-10x^2+25&=25\\ x^4-9x^2&=0\\ x^2(x^2-9)&=0\\ x^2(x+3)(x-3)&=0 \end{aligned}$$

$x=0$ or $x=-3$ or $x=3$

$\boxed{B}$ $y=-x^2+9=-(0)^2+9=9$

$\boxed{B}$ $y=-x^2+9=-(-3)^2+9=0$

$\boxed{B}$ $y=-x^2+9=-(3)^2+9=0$

$\{(0,9),(-3,0),(3,0)\}$

35. Let $u=\frac{1}{x^2}$ and $v=\frac{1}{y^2}$.

$\boxed{A}$ $\frac{4}{x^2}-\frac{3}{y^2}=-23 \to 4u-3v=-23$

$\boxed{B}$ $\frac{5}{x^2}+\frac{1}{y^2}=14 \to 5u+v=14 \to v=-5u+14$

$\boxed{A}$
$$\begin{aligned} 4u-3v&=-23\\ 4u-3(-5u+14)&=-23\\ 4u+15u-42&=-23\\ 19u&=19\\ u&=1 \end{aligned}$$

$\boxed{B}$ $v=-5u+14=-5(1)+14=9$

$u=\frac{1}{x^2}$, so $x=\pm\sqrt{\frac{1}{u}}=\pm\sqrt{\frac{1}{1}}=\pm1$

$v=\frac{1}{y^2}$, so $y=\pm\sqrt{\frac{1}{v}}=\pm\sqrt{\frac{1}{9}}=\pm\frac{1}{3}$

$\left\{\left(1,\frac{1}{3}\right),\left(-1,\frac{1}{3}\right),\left(1,-\frac{1}{3}\right),\left(-1,-\frac{1}{3}\right)\right\}$

37. Let x represent one number. Let y represent the other number.

[A] $x+y=12 \rightarrow y=-x+12$

[B] $xy=35$

[B]
$$\begin{aligned} xy &= 35 \\ x(-x+12) &= 35 \\ -x^2+12x-35 &= 0 \\ x^2-12x+35 &= 0 \\ (x-5)(x-7) &= 0 \\ x=5 \quad &\text{or} \quad x=7 \end{aligned}$$

[A] $y=-x+12=-5+12=7$

[A] $y=-x+12=-7+12=5$

The numbers are 5 and 7.

39. Let x represent one number. Let y represent the other number.

[A] $x^2+y^2=29$

[B] $x^2-y^2=21$

$$\begin{aligned} 2x^2 &= 50 \\ x^2 &= 25 \\ x &= \pm 5 \end{aligned}$$

Both numbers must be positive.

$x=5$: [A]
$$\begin{aligned} x^2+y^2 &= 29 \\ (5)^2+y^2 &= 29 \\ 25+y^2 &= 29 \\ y^2 &= 4 \\ y &= \pm 2 \end{aligned}$$

The numbers are 5 and 2.

41. Let x represent one number. Let y represent the other number.

[A] $x-y=2 \rightarrow x=y+2$

[B] $x^2-y^2=44$

[B]
$$\begin{aligned} x^2-y^2 &= 44 \\ (y+2)^2-y^2 &= 44 \\ y^2+4y+4-y^2 &= 44 \\ 4y &= 40 \\ y &= 10 \end{aligned}$$

[A] $x=y+2=10+2=12$

The numbers are 12 and 10.

43. Let x represent one number. Let y represent the other number.

[A] $\frac{x}{y}=\frac{3}{4} \rightarrow 3y=4x \rightarrow y=\frac{4}{3}x$

[B] $x^2+y^2=225$

[B]
$$\begin{aligned} x^2+y^2 &= 225 \\ x^2+\left(\frac{4}{3}x\right)^2 &= 225 \\ x^2+\frac{16}{9}x^2 &= 225 \\ \frac{25}{9}x^2 &= 225 \\ x^2 &= 81 \\ x &= \pm 9 \end{aligned}$$

[A] $y=\frac{4}{3}x=\frac{4}{3}(9)=12$

[A] $y=\frac{4}{3}x=\frac{4}{3}(-9)=-12$

The numbers are 9 and 12 or -9 and -12.

45. Let x represent one dimension. Let y represent the other dimension.

[A] $2x+2y=36 \rightarrow x+y=18 \rightarrow y=-x+18$

[B] $xy=80$

[B]
$$\begin{aligned} xy &= 80 \\ x(-x+18) &= 80 \\ -x^2+18x-80 &= 0 \\ x^2-18x+80 &= 0 \\ (x-8)(x-10) &= 0 \\ x=8 \quad &\text{or} \quad x=10 \end{aligned}$$

[A] $y=-x+18=-8+18=10$

[A] $y=-x+18=-10+18=8$

The rectangle is 10 m by 8 m.

47. Let x represent the length. Let y represent the width.

A $xy=240$

B $2x+2y-6=58$

$$2x+2y=64$$
$$x+y=32$$
$$y=-x+32$$

A
$$xy=240$$
$$x(-x+32)=240$$
$$-x^2+32x-240=0$$
$$x^2-32x+240=0$$
$$(x-12)(x-20)=0$$
$$x=12 \quad \text{or} \quad x=20$$

A $y=-x+32=-12+32=20$

A $y=-x+32=-20+32=12$

The floor is 20 ft by 12 ft.

49. Let x represent the length. Let y represent the width. The floor area is $\frac{288}{6}=48\text{ ft}^2$.

A $xy=48 \quad \rightarrow \quad y=\frac{48}{x}$

B $x^2+y^2=10^2 \quad \rightarrow \quad x^2+y^2=100$

B
$$x^2+y^2=100$$
$$x^2+\left(\frac{48}{x}\right)^2=100$$
$$x^2+\frac{2304}{x^2}=100$$
$$x^4+2304=100x^2$$
$$x^4-100x^2+2304=0$$
$$(x^2-36)(x^2-64)=0$$
$$x=\pm 6 \quad \text{or} \quad x=\pm 8$$

A $y=\frac{48}{x}=\frac{48}{6}=8$

A $y=\frac{48}{x}=\frac{48}{8}=6$

The truck is 6 ft by 6 ft by 8 ft.

51. Let x represent the length. Let y represent the width.

A $16xy=4608 \rightarrow y=\frac{288}{x}$

B $2(16x)+2(16y)+xy=1440$

$$32x+32y+xy=1440$$

B
$$32x+32y+xy=1440$$
$$32x+32\left(\frac{288}{x}\right)+x\left(\frac{288}{x}\right)=1440$$
$$32x+\frac{9216}{x}+288=1440$$
$$32x+\frac{9216}{x}-1152=0$$
$$32x^2-1152x+9216=0$$
$$x^2-36x+288=0$$
$$(x-24)(x-12)=0$$
$$x=24 \quad \text{or} \quad x=12$$

A $y=\frac{288}{x}=\frac{288}{24}=12$

A $y=\frac{288}{x}=\frac{288}{12}=24$

The aquarium is 24 in. by 12 in. by 16 in.

53. Let x represent the length of one side. Let y represent the length of the other side.

A $x+y=11 \quad \rightarrow \quad y=-x+11$

B $x^2+y^2=\left(\sqrt{65}\right)^2 \rightarrow x^2+y^2=65$

B
$$x^2+y^2=65$$
$$x^2+(-x+11)^2=65$$
$$x^2+x^2-22x+121=65$$
$$2x^2-22x+56=0$$
$$x^2-11x+28=0$$
$$(x-4)(x-7)=0$$
$$x=4 \quad \text{or} \quad x=7$$

A $y=-x+11=-4+11=7$

A $y=-x+11=-7+11=4$

The legs are 4 ft and 7 ft.

55. A $y=-\frac{x^2}{192}+\frac{\sqrt{3}}{3}x$

B $y=-\frac{\sqrt{3}}{3}x$

A $y=-\frac{x^2}{192}+\frac{\sqrt{3}}{3}x$

$$-\frac{\sqrt{3}}{3}x=-\frac{x^2}{192}+\frac{\sqrt{3}}{3}x$$

$$\frac{x^2}{192}-\frac{2\sqrt{3}}{3}x=0$$

$$x^2-128\sqrt{3}x=0$$

$$x\left(x-128\sqrt{3}\right)=0$$

$$x=0 \quad \text{or} \quad x=128\sqrt{3}$$

B $y=-\frac{\sqrt{3}}{3}x=-\frac{\sqrt{3}}{3}(0)=0$

B $y=-\frac{\sqrt{3}}{3}x=-\frac{\sqrt{3}}{3}\left(128\sqrt{3}\right)=-128$

The ball will hit the ground at the point $\left(128\sqrt{3},-128\right)$ or approximately $(221.7,-128)$.

57. A system of linear equations contains only linear equations, whereas a nonlinear system has one or more equations that are nonlinear.

59. a. $A(t)=A_0e^{-kt}$

$0.69=A_0e^{-3t}$

b. $A(t)=A_0e^{-kt}$

$0.655=A_0e^{-4t}$

c. A $0.69=A_0e^{-3t} \rightarrow A_0=0.69e^{3k}$

B $0.655=A_0e^{-4t} \rightarrow A_0=0.655e^{4k}$

$$0.69e^{3k}=0.655e^{4k}$$

$$\frac{0.69}{0.655}=\frac{e^{4k}}{e^{3k}}$$

$$\frac{0.69}{0.655}=e^k$$

$$\ln\left(\frac{0.69}{0.655}\right)=k$$

$$0.052\approx k$$

d. $A_0=0.69e^{3k}\approx 0.69e^{3(0.052)}\approx 0.81\ \mu\text{g/dL}$

e. $A(t)\approx 0.81e^{-(0.052)t}$

$A(12)\approx 0.81e^{-(0.052)(12)}\approx 0.43\ \mu\text{g/dL}$

61. a. A $60{,}000=P_0e^{7k} \rightarrow P_0=60{,}000e^{-7k}$

B $80{,}000=P_0e^{12k} \rightarrow P_0=80{,}000e^{-12k}$

$$60{,}000e^{-7k}=80{,}000e^{-12k}$$

$$\frac{e^{-7k}}{e^{-12k}}=\frac{80{,}000}{60{,}000}$$

$$e^{5k}=\frac{4}{3}$$

$$5k=\ln\left(\frac{4}{3}\right)$$

$$k=\frac{1}{5}\ln\left(\frac{4}{3}\right)$$

$$k\approx 0.058$$

b. $P_0=60{,}000e^{-7k}\approx 60{,}000e^{-7(0.058)}\approx 40{,}000$

The original population is 40,000.

c. $P(t)\approx 40{,}000e^{0.058t}$

$$300{,}000\approx 40{,}000e^{0.058t}$$

$$\frac{15}{2}\approx e^{0.058t}$$

$$\ln\left(\frac{15}{2}\right)\approx 0.058t$$

$$\frac{\ln\left(\frac{15}{2}\right)}{0.058}\approx t$$

$$35\approx t$$

The population will reach 300,000 approximately 35 hr after the culture is started.

63. A $y=2^{x+1}\rightarrow \log_2 y=x+1\rightarrow -1+\log_2 y=x$

B $-1+\log_2 y=x$

The equations are the same. The system has infinitely many solutions.

65. $\boxed{A}$ $\log x + 2\log y = 5$

$\boxed{B}$ $2\log x - \log y = 0 \xrightarrow{\text{Multiply by 2.}}$

$$\begin{aligned} \log x + 2\log y &= 5 \\ \underline{4\log x - 2\log y} &\underline{= 0} \\ 5\log x \quad\quad &= 5 \\ \log x &= 1 \\ x &= 10 \end{aligned}$$

$\boxed{A}$

$$\begin{aligned} \log x + 2\log y &= 5 \\ \log 10 + 2\log y &= 5 \\ 1 + 2\log y &= 5 \\ 2\log y &= 4 \\ \log y &= 2 \\ y &= 10^2 = 100 \end{aligned}$$

$\{(10, 100)\}$

67. $\boxed{A}$ $2^x + 2^y = 6 \rightarrow 2^y = -2^x + 6$

$\boxed{B}$ $4^x - 2^y = 14 \rightarrow 2^{2x} - 2^y = 14$

$\boxed{B}$

$$\begin{aligned} 2^{2x} - 2^y &= 14 \\ 2^{2x} - (-2^x + 6) &= 14 \\ 2^{2x} + 2^x - 6 &= 14 \\ 2^{2x} + 2^x - 20 &= 0 \end{aligned}$$

Let $u = 2^x$

$$\begin{aligned} u^2 + u - 20 &= 0 \\ (u+5)(u-4) &= 0 \\ u = -5 \text{ or } u &= 4 \\ 2^x = -5 \text{ or } 2^x &= 4 \\ x &= 2 \end{aligned}$$

$\boxed{A}$

$$\begin{aligned} 2^x + 2^y &= 6 \\ 2^2 + 2^y &= 6 \\ 4 + 2^y &= 6 \\ 2^y &= 2 \\ y &= 1 \end{aligned}$$

$\{(2, 1)\}$

69. $\boxed{A}$ $(x-1)^2+(y+1)^2=5$

$x^2-2x+1+y^2+2y+1=5$

$x^2-2x+y^2+2y-3=0$

$\boxed{B}$ $x^2+(y+4)^2=29$

$x^2=29-(y+4)^2$

$x=\pm\sqrt{29-(y+4)^2}$

$x=\sqrt{29-(y+4)^2}$: $\boxed{A}$

$$x^2-2x+y^2+2y-3=0$$
$$\left(\sqrt{29-(y+4)^2}\right)^2-2\left(\sqrt{29-(y+4)^2}\right)+y^2+2y-3=0$$
$$29-(y+4)^2-2\sqrt{29-(y+4)^2}+y^2+2y-3=0$$
$$29-y^2-8y-16-2\sqrt{29-(y+4)^2}+y^2+2y-3=0$$
$$-2\sqrt{29-(y+4)^2}=6y-10$$
$$\sqrt{29-(y+4)^2}=-3y+5$$
$$29-(y+4)^2=(-3y+5)^2$$
$$29-y^2-8y-16=9y^2-30y+25$$
$$10y^2-22y+12=0$$
$$5y^2-11y+6=0$$
$$(5y-6)(y-1)=0$$
$$y=\frac{6}{5}\text{ or }y=1$$

$x=-\sqrt{29-(y+4)^2}$: $\boxed{A}$

$$x^2-2x+y^2+2y-3=0$$
$$\left(-\sqrt{29-(y+4)^2}\right)^2-2\left(-\sqrt{29-(y+4)^2}\right)+y^2+2y-3=0$$
$$29-(y+4)^2+2\sqrt{29-(y+4)^2}+y^2+2y-3=0$$
$$29-y^2-8y-16+2\sqrt{29-(y+4)^2}+y^2+2y-3=0$$
$$2\sqrt{29-(y+4)^2}=6y-10$$
$$\sqrt{29-(y+4)^2}=3y-5$$
$$29-(y+4)^2=(3y-5)^2$$
$$29-y^2-8y-16=9y^2-30y+25$$
$$10y^2-22y+12=0$$
$$5y^2-11y+6=0$$
$$(5y-6)(y-1)=0$$
$$y=\frac{6}{5}\text{ or }y=1$$

$y=\frac{6}{5}$: B $x^2+(y+4)^2=29$

$$x^2+\left(\frac{6}{5}+4\right)^2=29$$
$$x^2+\left(\frac{26}{5}\right)^2=29$$
$$x^2+\frac{676}{25}=\frac{725}{25}$$
$$x^2=\frac{49}{25}$$
$$x=\pm\frac{7}{5}$$

$y=1$: B $x^2+(y+4)^2=29$

$$x^2+(1+4)^2=29$$
$$x^2+(5)^2=29$$
$$x^2+25=29$$
$$x^2=4$$
$$x=\pm 2$$

Check: $(2,1)$ A $(x-1)^2+(y+1)^2=5$

$$(2-1)^2+(1+1)^2 \stackrel{?}{=} 5$$
$$1+4 \stackrel{?}{=} 5$$
$$5 \stackrel{?}{=} 5 \checkmark \text{ true}$$

B $x^2+(y+4)^2=29$

$$(2)^2+(1+4)^2 \stackrel{?}{=} 29$$
$$4+25 \stackrel{?}{=} 29$$
$$29 \stackrel{?}{=} 29 \checkmark \text{ true}$$

Check: $(-2,1)$ A $(x-1)^2+(y+1)^2=5$

$$(-2-1)^2+(1+1)^2 \stackrel{?}{=} 5$$
$$9+4 \stackrel{?}{=} 5$$
$$13 \stackrel{?}{=} 5 \text{ false}$$

Check: $\left(\frac{7}{5},\frac{6}{5}\right)$ A $(x-1)^2+(y+1)^2=5$

$$\left(\frac{7}{5}-1\right)^2+\left(\frac{6}{5}+1\right)^2 \stackrel{?}{=} 5$$
$$\frac{4}{25}+\frac{121}{25} \stackrel{?}{=} 5$$
$$\frac{125}{25} \stackrel{?}{=} 5$$
$$5 \stackrel{?}{=} 5 \checkmark \text{ true}$$

B $x^2+(y+4)^2=29$

$$\left(\frac{7}{5}\right)^2+\left(\frac{6}{5}+4\right)^2 \stackrel{?}{=} 29$$
$$\frac{49}{25}+\frac{676}{25} \stackrel{?}{=} 29$$
$$\frac{725}{25} \stackrel{?}{=} 29$$
$$29 \stackrel{?}{=} 29 \checkmark \text{ true}$$

Check: $\left(-\frac{7}{5},\frac{6}{5}\right)$ A $(x-1)^2+(y+1)^2=5$

$$\left(-\frac{7}{5}-1\right)^2+\left(\frac{6}{5}+1\right)^2 \stackrel{?}{=} 5$$
$$\frac{144}{25}+\frac{121}{25} \stackrel{?}{=} 5$$
$$\frac{265}{25} \stackrel{?}{=} 5$$
$$\frac{53}{5} \stackrel{?}{=} 5 \text{ false}$$

$$\left\{(2,1),\left(\frac{7}{5},\frac{6}{5}\right)\right\}$$

71. $\boxed{A}$ $2=2\lambda x \;\rightarrow\; x=\dfrac{1}{\lambda}$

$\boxed{B}$ $6=2\lambda y \;\rightarrow\; y=\dfrac{3}{\lambda}$

$\boxed{C}$ $x^2+y^2=10$

$\boxed{C}$ $x^2+y^2=10$

$$\left(\frac{1}{\lambda}\right)^2+\left(\frac{3}{\lambda}\right)^2=10$$

$$\frac{1}{\lambda^2}+\frac{9}{\lambda^2}=10$$

$$\frac{10}{\lambda^2}=10$$

$$10\lambda^2=10$$

$$\lambda^2=1$$

$$\lambda=\pm1$$

$\lambda=1$: $\boxed{A}$ $2=2\lambda x$, $2=2(1)x$, $2=2x$, $1=x$ $\quad$ $\boxed{B}$ $6=2\lambda y$, $6=2(1)y$, $6=2y$, $3=y$

$\lambda=-1$: $\boxed{A}$ $2=2\lambda x$, $2=2(-1)x$, $2=-2x$, $-1=x$ $\quad$ $\boxed{B}$ $6=2\lambda y$, $6=2(-1)y$, $6=-2y$, $-3=y$

$\{(1,3,1),(-1,-3,-1)\}$

73. a. $\boxed{A}$ $x^2+y^2=25$

$\boxed{B}$ $(x-4)^2+(y+2)^2=25$

The circles appear to intersect at $(0,-5)$ and $(4,3)$.

Check: $(0,-5)$

$\boxed{A}$ $x^2+y^2=25$

$$(0)^2+(-5)^2\stackrel{?}{=}25$$

$$0+25\stackrel{?}{=}25$$

$25\stackrel{?}{=}25$ ✓ true

$\boxed{B}$ $(x-4)^2+(y+2)^2=25$

$$(0-4)^2+(-5+2)^2\stackrel{?}{=}25$$

$$16+9\stackrel{?}{=}25$$

$25\stackrel{?}{=}25$ ✓ true

Check: $(4,3)$

$\boxed{A}$ $x^2+y^2=25$

$$(4)^2+(3)^2\stackrel{?}{=}25$$

$$16+9\stackrel{?}{=}25$$

$25\stackrel{?}{=}25$ ✓ true

$\boxed{B}$ $(x-4)^2+(y+2)^2=25$

$$(4-4)^2+(3+2)^2=25$$

$$0+25\stackrel{?}{=}25$$

$25\stackrel{?}{=}25$ ✓ true

$\{(0,-5),(4,3)\}$

b. Find the equation of the line that passes through $(0,-5)$ and $(4,3)$.

$$m=\frac{y_2-y_1}{x_2-x_1}=\frac{3-(-5)}{4-0}=\frac{8}{4}=2$$

$$y-y_1=m(x-x_1)$$

$$y-(-5)_1=2(x-0)$$

$$y+5=2x$$

$$y=2x-5$$

The domain of the line segment is $0\le x\le 4$.

75. $\{(2.359,5.584)\}$

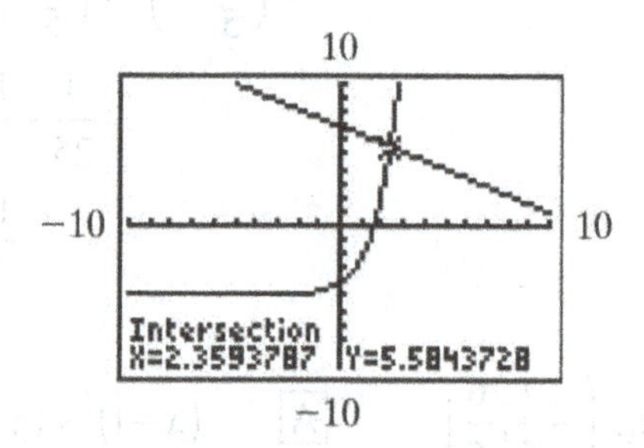

77. $\{(1.538,6.135),(-1.538,6.135),(3.693,-5.135),(-3.693,-5.135)\}$

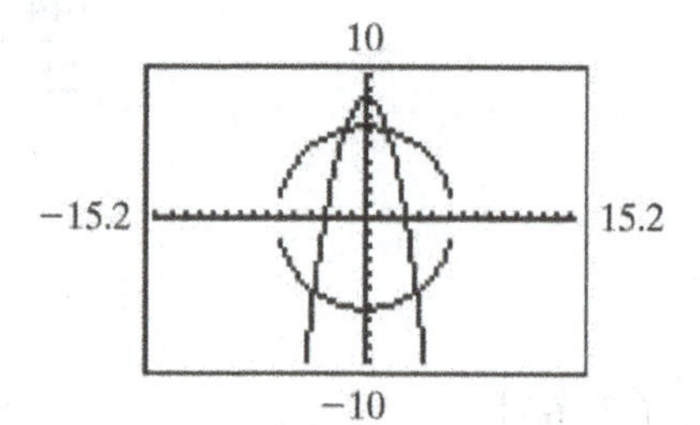

79. { }

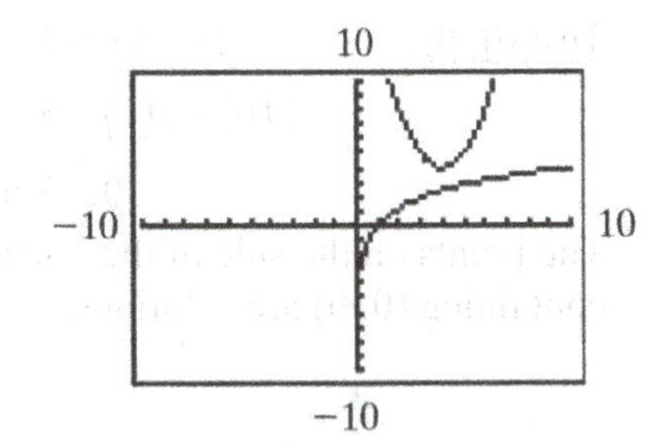

Section 5.5 Inequalities and Systems of Inequalities in Two Variables

1. 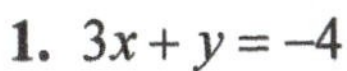

$$3x + y = -4$$
$$y = -3x - 4$$

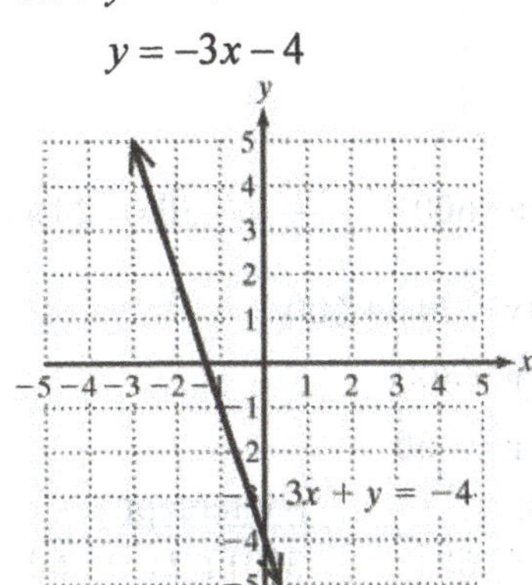

3. $y = x^2 - 4$

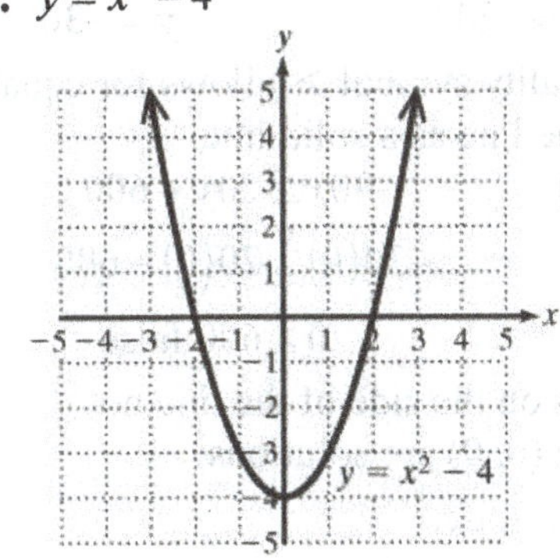

5.
$$3x + y = -4$$
$$y = -3x - 4$$
$$x + 2y = 2$$
$$x + 2(-3x - 4) = 2$$
$$x - 6x - 8 = 2$$
$$-5x = 10$$
$$x = -2$$
$$y = -3x - 4 = -3(-2) - 4 = 6 - 4 = 2$$
$\{(-2, 2)\}$

7. linear

9. above; horizontal

11. II

13. a.
$$3x + 4y < 12$$
$$3(-1) + 4(3) \stackrel{?}{<} 12$$
$$-3 + 12 \stackrel{?}{<} 12$$
$$9 \stackrel{?}{<} 12 \checkmark \text{ Yes}$$

b.
$$3x + 4y < 12$$
$$3(5) + 4(1) \stackrel{?}{<} 12$$
$$15 + 4 \stackrel{?}{<} 12$$
$$19 \stackrel{?}{<} 12 \text{ No}$$

c.
$$3x + 4y < 12$$
$$3(4) + 4(0) \stackrel{?}{<} 12$$
$$12 + 0 \stackrel{?}{<} 12$$
$$12 \stackrel{?}{<} 12 \text{ No}$$

15. a.
$$y \ge (x-3)^2$$
$$30 \stackrel{?}{\ge} (-3-3)^2$$
$$30 \stackrel{?}{\ge} (-6)^2$$
$$30 \stackrel{?}{\ge} 36 \text{ No}$$

b.
$$y \ge (x-3)^2$$
$$4 \stackrel{?}{\ge} (1-3)^2$$
$$4 \stackrel{?}{\ge} (-2)^2$$
$$4 \stackrel{?}{\ge} 4 \checkmark \text{ Yes}$$

c. $y \geq (x-3)^2$

$5 \overset{?}{\geq} (5-3)^2$

$5 \overset{?}{\geq} (2)^2$

$5 \overset{?}{\geq} 4$ ✓ Yes

17. a. $4x - 5y \leq 20 \xrightarrow[\text{equation}]{\text{related}} 4x - 5y = 20$

x-intercept:

$4x - 5(0) = 20$

$4x = 20$

$x = 5$

y-intercept:

$4(0) - 5y = 20$

$-5y = 20$

$y = -4$

The inequality symbol $\leq$ allows for equality, so draw the line as a solid line.

Test (0, 0): $4x - 5y \leq 20$

$4(0) - 5(0) \overset{?}{\leq} 20$

$0 \overset{?}{\leq} 0$ ✓ true

The points on the side of the line containing (0, 0) are solutions.

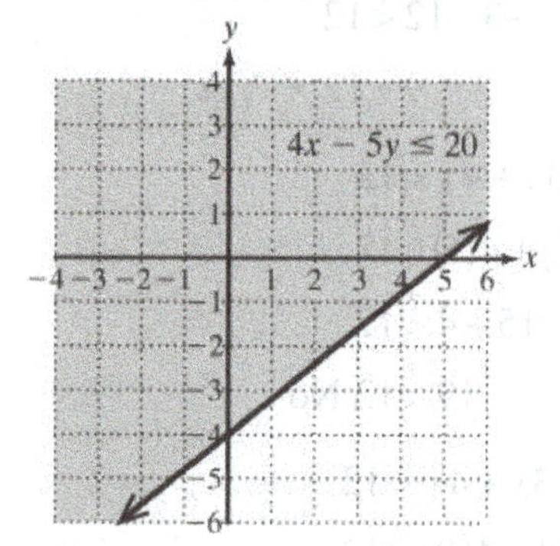

b. The bounding line would be drawn as a dashed line.

c. The bounding line would be dashed and the graph would be shaded strictly below the line.

19. $2x + 5y > 5 \xrightarrow[\text{equation}]{\text{related}} 2x + 5y = 5$

x-intercept:

$2x + 5(0) = 5$

$2x = 5$

$x = \dfrac{5}{2}$

y-intercept:

$2(0) + 5y = 5$

$5y = 5$

$y = 1$

The inequality is strict, so draw the line as a dashed line.

Test (0, 0): $2x + 5y > 5$

$2(0) + 5(0) \overset{?}{>} 5$

$0 \overset{?}{>} 5$ false

The points on the side of the line not containing (0, 0) are solutions.

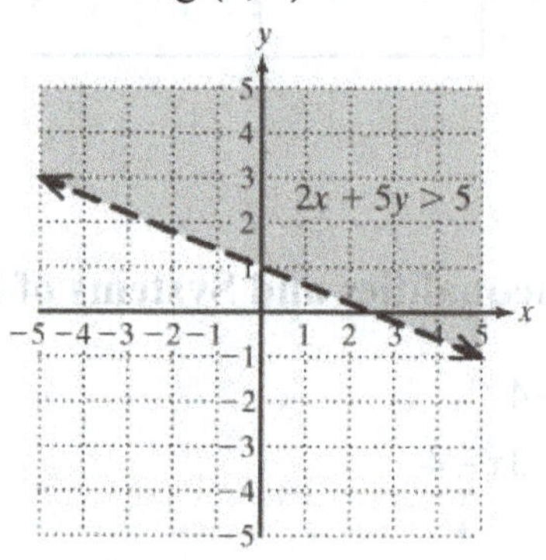

21. $-30x \geq 20y + 600 \xrightarrow[\text{equation}]{\text{related}} -30x = 20y + 600$

$-30x = 20y + 600$

$-30x - 20y = 600$

$3x + 2y = -60$

x-intercept:

$3x + 2(0) = -60$

$3x = -60$

$x = -20$

y-intercept:

$3(0) + 2y = -60$

$2y = -60$

$y = -30$

The inequality symbol $\geq$ allows for equality, so draw the line as a solid line.

Test (0, 0): $-30x \geq 20y + 600$

$-30(0) \overset{?}{\geq} 20(0) + 600$

$0 \overset{?}{\geq} 600$ false

The points on the side of the line not containing (0, 0) are solutions.

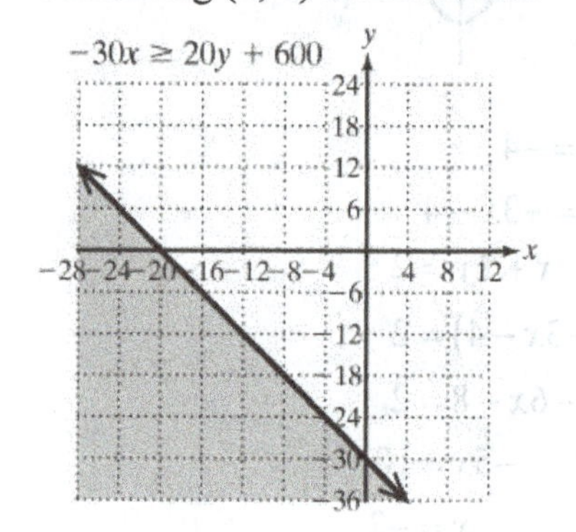

23. $5x \le 6y \xrightarrow[\text{equation}]{\text{related}} 5x = 6y$

$5x = 6y$

$y = \frac{5}{6}x + 0$

The inequality symbol $\le$ allows for equality, so draw the line as a solid line.

Test (1, 1): $5x \le 6y$

$5(1) \overset{?}{\le} 6(1)$

$5 \overset{?}{\le} 6$ ✓ true

The points on the side of the line containing (1, 1) are solutions.

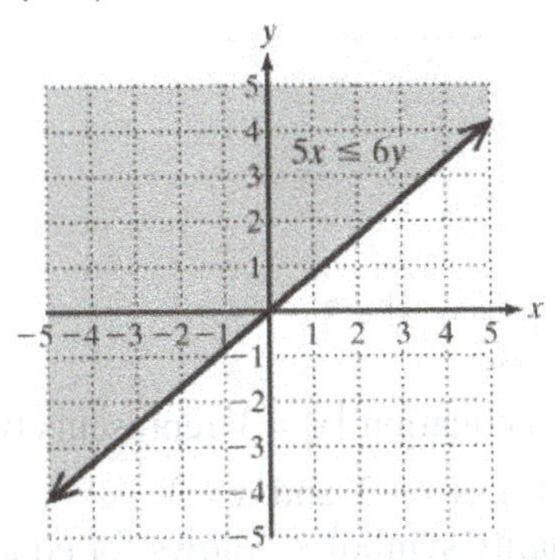

25. $3+2(x+y) > y+3 \xrightarrow[\text{equation}]{\text{related}} 3+2(x+y) = y+3$

$3+2(x+y) = y+3$

$2x+2y = y$

$y = -2x+0$

The inequality is strict, so draw the line as a dashed line.

Test (1, 1): $3+2(x+y) > y+3$

$3+2(1+1) \overset{?}{>} 1+3$

$3+4 \overset{?}{>} 4$

$7 \overset{?}{>} 4$ ✓ true

The points on the side of the line containing (1, 1) are solutions.

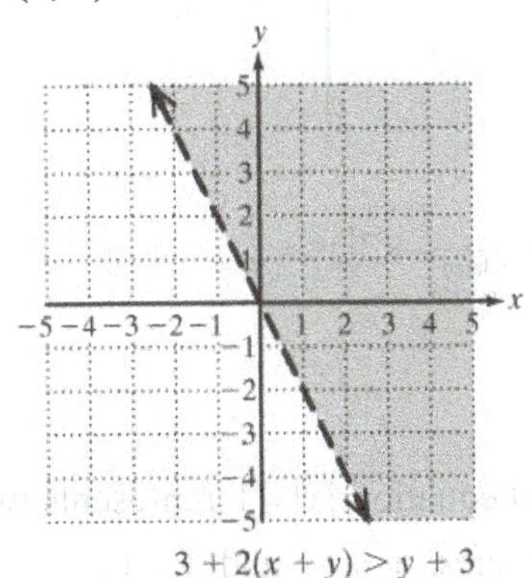

27. $x < 6$

The related equation $x = 6$ is a vertical line. The inequality $x < 6$ represents all points strictly to the left of the line $x = 6$. The inequality is strict, so draw the line as a dashed line.

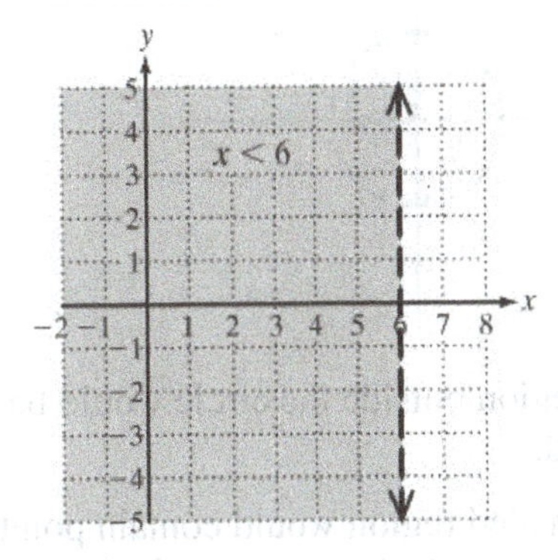

29. $-\frac{1}{2}y + 4 \le 5$

$-\frac{1}{2}y \le 1$

$y \ge -2$

The related equation $y = -2$ is a horizontal line. The inequality $y \ge -2$ represents all points above or on the line $y = -2$. The inequality symbol $\le$ allows for equality, so draw the line as a solid line.

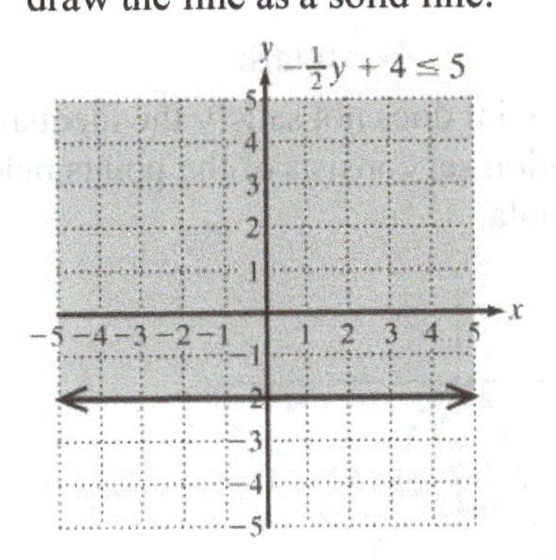

31. a. $x^2 + y^2 < 4 \xrightarrow[\text{equation}]{\text{related}} x^2 + y^2 = 4$

The equation represents a circle centered at $(0, 0)$ with radius 2. The inequality is strict, so draw the circle as a dashed curve.

Test (0, 0): $x^2 + y^2 < 4$

$(0)^2 + (0)^2 \overset{?}{<} 4$

$0 \overset{?}{<} 4$ ✓ true

The test point inside the circle satisfies the inequality. The solution set consists of the points strictly inside the circle.

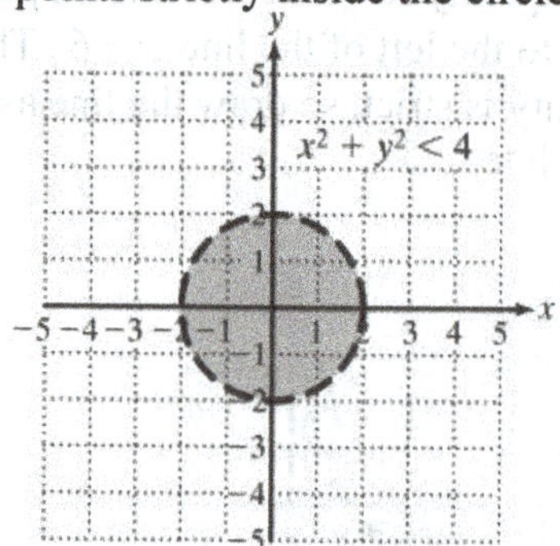

b. The region outside the circle would be shaded.

c. The shaded region would contain points on the circle (solid curve) and points outside the circle.

33. $y < -x^2 \xrightarrow[\text{equation}]{\text{related}} y = -x^2$

The equation represents a parabola opening downward with vertex $(0, 0)$. The inequality is strict, so draw the parabola as a dashed curve.

Test (0, 1): $y < -x^2$

$1 \overset{?}{<} -(0)^2$

$1 \overset{?}{<} 0$ false

The test point does not satisfy the inequality. The solution set consists of the points below the parabola.

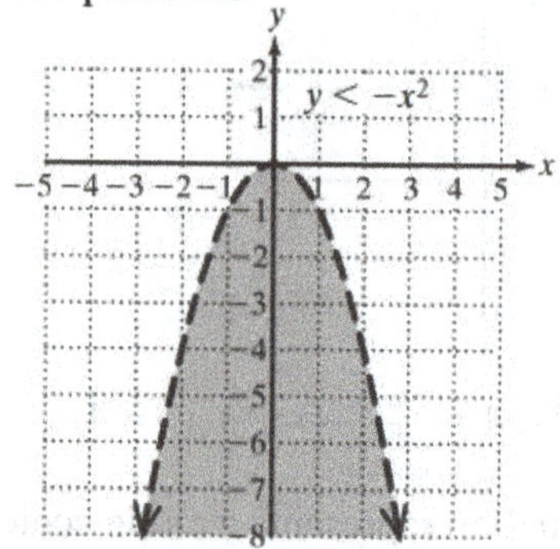

35. $y \le (x-2)^2 + 1 \xrightarrow[\text{equation}]{\text{related}} y = (x-2)^2 + 1$

The equation represents a parabola opening upward with vertex $(2, 1)$. The inequality symbol $\le$ allows for equality, so draw the parabola as a solid curve.

Test (0, 0): $y \le (x-2)^2 + 1$

$0 \overset{?}{\le} (0-2)^2 + 1$

$0 \overset{?}{\le} 5$ ✓ true

The test point satisfies the inequality. The solution set consists of the points below or on the parabola.

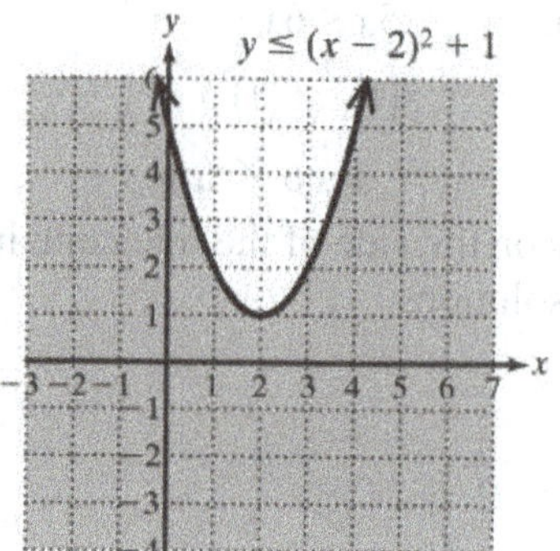

37. $|x| \le 3 \xrightarrow[\text{equation}]{\text{related}} |x| = 3$

The related equation $|x| = 3$ represents two vertical lines, $x = -3$ and $x = 3$.
The inequality symbol $\le$ allows for equality, so draw the lines as solid lines.

Test (0, 0): $|x| \le 3$

$|0| \overset{?}{\le} 3$

$0 \overset{?}{\le} 3$ ✓ true

The test point satisfies the inequality. The solution set consists of all points between or on the lines $x = -3$ and $x = 3$.

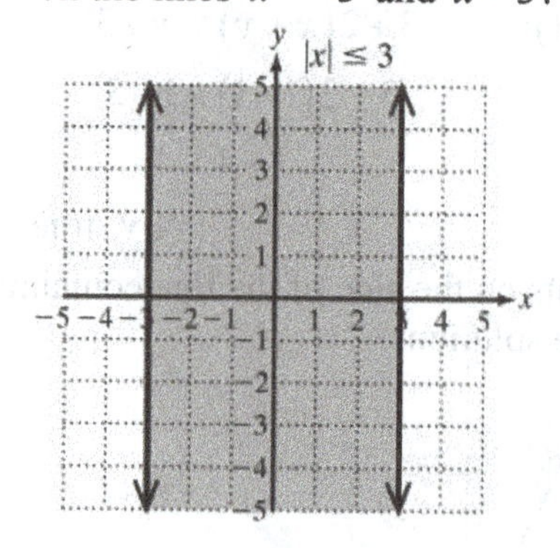

39. $2|y| > 2 \xrightarrow[\text{equation}]{\text{related}} 2|y| = 2$

$2|y| = 2$

$|y| = 1$

The related equation $|y| = 1$ represents two horizontal lines, $y = -1$ and $y = 1$.

The inequality is strict, so draw the lines as dashed lines.

Test (0, 0):
$$2|y| > 2$$
$$2|0| \overset{?}{>} 2$$
$$0 \overset{?}{>} 2 \text{ false}$$

The test point does not satisfy the inequality. The solution set consists of all points strictly outside the lines $y = -1$ and $y = 1$.

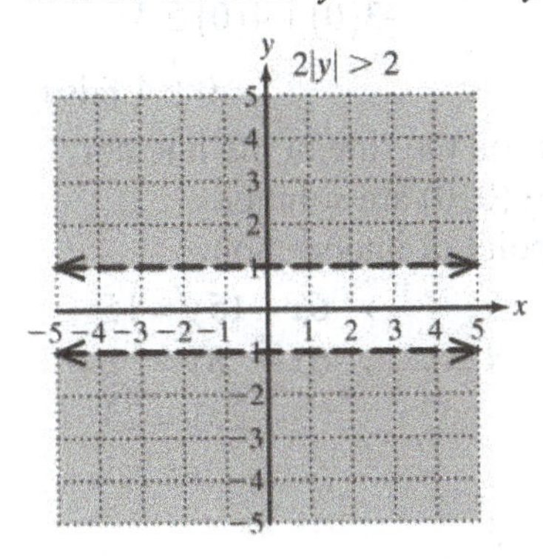

41. $y \geq \sqrt{x} \xrightarrow[\text{equation}]{\text{related}} y = \sqrt{x}$

The equation represents the parent square root function. The inequality symbol $\geq$ allows for equality, so draw the function as a solid curve.

Test (1, 2):
$$y \geq \sqrt{x}$$
$$2 \overset{?}{\geq} \sqrt{1}$$
$$2 \overset{?}{\geq} 1 \checkmark \text{ true}$$

The test point satisfies the inequality. Since the domain of the function is $[0, \infty)$, the y-axis is also a boundary of the solution set. The solution set consists of the points that are above or on the curve, and to the right of or on the y-axis.

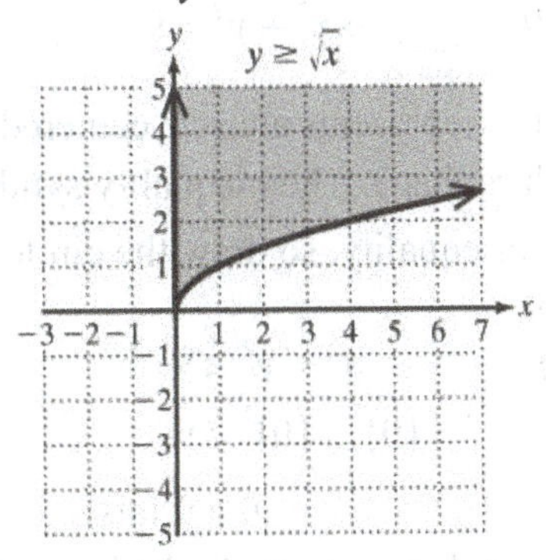

43. a. $y < 2x + 3$
$$1 \overset{?}{<} 2(2) + 3$$
$$1 \overset{?}{<} 7 \checkmark \text{ Yes}$$

b. $x + y \leq 1$
$$2 + 1 \overset{?}{\leq} 1$$
$$3 \overset{?}{\leq} 1 \text{ No}$$

c. No. Since the point $(2, 1)$ is not a solution to the second inequality, it is not a solution to the system of inequalities.

45. a. $x + y < 4$ | $y \leq 2x + 1$
$$0 + 1 \overset{?}{<} 4 \qquad 1 \overset{?}{\leq} 2(0) + 1$$
$$1 \overset{?}{<} 4 \checkmark \text{ true} \qquad 1 \overset{?}{\leq} 1 \checkmark \text{ true}$$

Yes

b. $x + y < 4$
$$3 + 1 \overset{?}{<} 4$$
$$4 \overset{?}{<} 4 \text{ false}$$

No

c. $x + y < 4$ | $y \leq 2x + 1$
$$2 + 0 \overset{?}{<} 4 \qquad 0 \overset{?}{\leq} 2(2) + 1$$
$$2 \overset{?}{<} 4 \checkmark \text{ true} \qquad 0 \overset{?}{\leq} 5 \checkmark \text{ true}$$

Yes

d. $x + y < 4$
$$1 + 4 \overset{?}{<} 4$$
$$5 \overset{?}{<} 4 \text{ false}$$

No

47. $y < \frac{1}{2}x - 4 \xrightarrow[\text{equation}]{\text{related}} y = \frac{1}{2}x - 4$

The inequality is strict, so draw the line as a dashed line.

Test (0, 0):
$$y < \frac{1}{2}x - 4$$
$$0 \overset{?}{<} \frac{1}{2}(0) - 4$$
$$0 \overset{?}{<} -4 \text{ false}$$

The points on the side of the line not containing (0, 0) are solutions.

$y > -2x + 1 \xrightarrow[\text{equation}]{\text{related}} y = -2x + 1$

The inequality is strict, so draw the line as a dashed line.

Test (0, 0): $y > -2x + 1$

$0 \overset{?}{>} -2(0) + 1$

$0 \overset{?}{>} 1$ false

The points on the side of the line not containing (0, 0) are solutions.

Find the point of intersection.

$$\frac{1}{2}x - 4 = -2x + 1$$
$$\frac{1}{2}x = -2x + 5$$
$$x = -4x + 10$$
$$5x = 10$$
$$x = 2$$
$$y = \frac{1}{2}x - 4 = \frac{1}{2}(2) - 4 = -3$$

Graph $(2, -3)$ as an open dot. It is not a solution to either inequality.

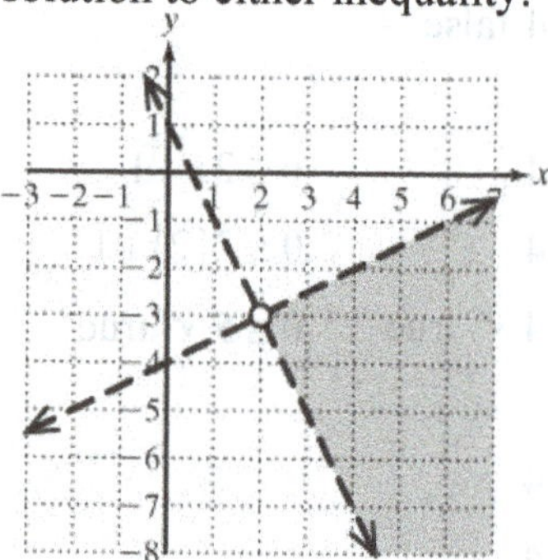

49. $2x + 5y \le 5 \xrightarrow[\text{equation}]{\text{related}} 2x + 5y = 5$

x-intercept: $2x + 5(0) = 5$, $2x = 5$, $x = \frac{5}{2}$

y-intercept: $2(0) + 5y = 5$, $5y = 5$, $y = 1$

The inequality symbol $\le$ allows for equality, so draw the line as a solid line.

Test (0, 0): $2x + 5y \le 5$

$2(0) + 5(0) \overset{?}{\le} 5$

$0 \overset{?}{\le} 5$ ✓ true

The points on the side of the line containing (0, 0) are solutions.

$-3x + 4y \ge 4 \xrightarrow[\text{equation}]{\text{related}} -3x + 4y = 4$

x-intercept: $-3x + 4(0) = 4$, $-3x = 4$, $x = -\frac{4}{3}$

y-intercept: $-3(0) + 4y = 4$, $4y = 4$, $y = 1$

The inequality symbol $\ge$ allows for equality, so draw the line as a solid line.

Test (0, 0): $-3x + 4y \ge 4$

$-3(0) + 4(0) \overset{?}{\ge} 4$

$0 \overset{?}{\ge} 4$ false

The points on the side of the line not containing (0, 0) are solutions.

Find the point of intersection.

$$2x + 5y = 5 \xrightarrow{\times 3} 6x + 15y = 15$$
$$-3x + 4y = 4 \xrightarrow{\times 2} -6x + 8y = 8$$
$$23y = 23$$
$$y = 1$$

$$2x + 5y = 5$$
$$2x + 5(1) = 5$$
$$2x = 0$$
$$x = 0$$

Graph $(0, 1)$ as a closed dot. It is a solution to both inequalities.

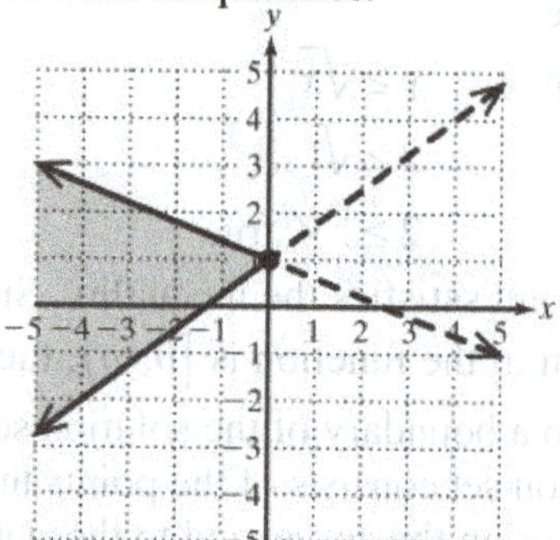

51. $x^2 + y^2 \ge 9 \xrightarrow[\text{equation}]{\text{related}} x^2 + y^2 = 9$

The equation represents a circle centered at $(0, 0)$ with radius 3. The inequality symbol $\ge$ allows for equality, so draw the circle as a solid curve.

Test (0, 0): $x^2 + y^2 \ge 9$

$(0)^2 + (0)^2 \overset{?}{\ge} 9$

$0 \overset{?}{\ge} 9$ false

The test point does not satisfy the inequality. The solution set consists of the points outside or on the circle.

$x^2 + y^2 \le 16 \xrightarrow[\text{equation}]{\text{related}} x^2 + y^2 = 16$

The equation represents a circle centered at $(0, 0)$ with radius 4. The inequality symbol $\le$ allows for equality, so draw the circle as a solid curve.

Test (0, 0): $x^2 + y^2 \le 16$

$(0)^2 + (0)^2 \overset{?}{\le} 16$

$0 \overset{?}{\le} 16$ ✓ true

The test point satisfies the inequality. The solution set consists of the points inside or on the circle.

The graph shows that the circles do not intersect.

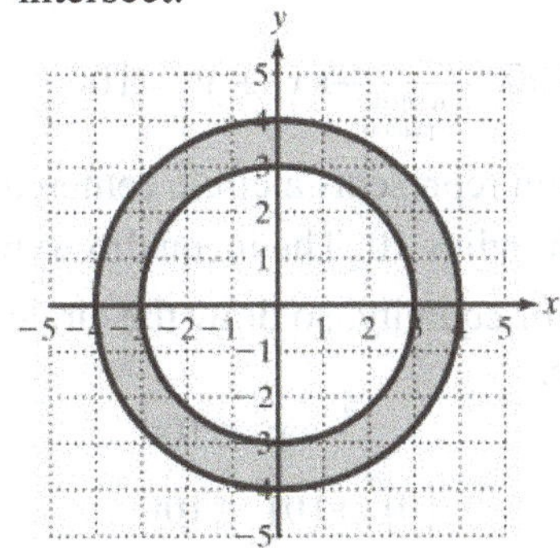

53. $y \ge 3x + 3 \xrightarrow[\text{equation}]{\text{related}} y = 3x + 3$

The inequality symbol $\ge$ allows for equality, so draw the line as a solid line.

Test (0, 0): $y \ge 3x + 3$

$0 \overset{?}{\ge} 3(0) + 3$

$0 \overset{?}{\ge} 3$ false

The points on the side of the line not containing (0, 0) are solutions.

$-3x + y < 1 \xrightarrow[\text{equation}]{\text{related}} y = 3x + 1$

The inequality is strict, so draw the line as a dashed line.

Test (0, 0): $-3x + y < 1$

$-3(0) + 0 \overset{?}{<} 1$

$0 \overset{?}{<} 1$ ✓ true

The points on the side of the line containing (0, 0) are solutions.

Find the point of intersection.

$3x + 3 = 3x + 1$

$3 = 1$

The lines do not intersect; they are parallel. The solutions to the first inequality are above the line, and the solutions to the second inequality are below the line. The solution sets do not overlap. The solution set of the system is the empty set, $\{ \}$.

55. $|x| < 3 \xrightarrow[\text{equation}]{\text{related}} |x| = 3$

The related equation $|x| = 3$ represents two vertical lines, $x = -3$ and $x = 3$.

The inequality is strict, so draw the lines as dashed lines.

Test (0, 0): $|x| < 3$

$|0| \overset{?}{<} 3$

$0 \overset{?}{<} 3$ ✓ true

The test point satisfies the inequality. The solution set consists of all points between the lines $x = -3$ and $x = 3$.

$|y| < 3 \xrightarrow[\text{equation}]{\text{related}} |y| = 3$

The related equation $|y| = 3$ represents two horizontal lines, $y = -3$ and $y = 3$.

The inequality is strict, so draw the lines as dashed lines.

Test (0, 0): $|y| < 3$

$|0| \overset{?}{<} 3$

$0 \overset{?}{<} 3$ ✓ true

The test point satisfies the inequality. The solution set consists of all points between the lines $y = -3$ and $y = 3$.

The lines intersect at $(3, 3)$, $(-3, 3)$, $(3, -3)$, and $(-3, -3)$. These points are not solutions to the strict inequalities, so graph them with open dots.

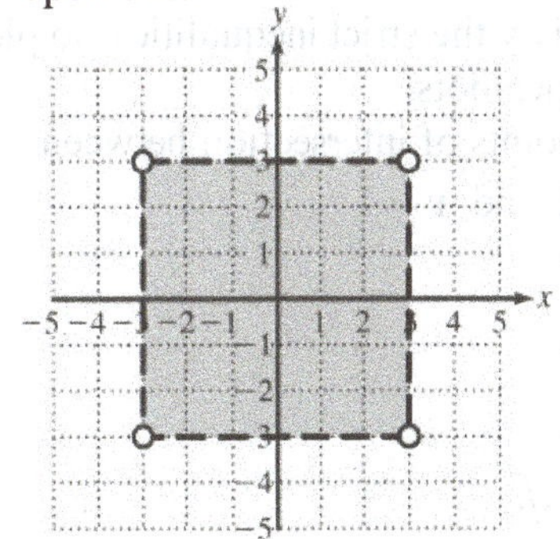

57. $y \ge x^2 - 2 \xrightarrow[\text{equation}]{\text{related}} y = x^2 - 2$

The equation represents a parabola opening upward with vertex $(0, -2)$. The inequality symbol $\ge$ allows for equality, so draw the parabola as a solid curve.

Test (0, 0): $y \ge x^2 - 2$

$0 \overset{?}{\ge} (0)^2 - 2$

$0 \overset{?}{\ge} -2$ ✓ true

The test point satisfies the inequality. The solution set consists of the points above or on the parabola.

$y > x \xrightarrow[\text{equation}]{\text{related}} y = x$

The inequality is strict, so draw the line as a dashed line.

Test (1, 2): $y > x$

$2 \overset{?}{>} 1$ ✓ true

The points on the side of the line containing (1, 2) are solutions.

$y \le 4$

The related equation $y = 4$ is a horizontal line. The inequality $y \le 4$ represents all points below or on the line $y = 4$. The inequality symbol $\le$ allows for equality, so draw the line as a solid line.

Find the points of intersection between $y = x^2 - 2$ and $y = x$:

$x^2 - 2 = x$

$x^2 - x - 2 = 0$

$(x+1)(x-2) = 0$

$x = -1$ or $x = 2$

$y = x = -1$

$y = x = 2$

The solutions are $(-1, -1)$ and $(2, 2)$. They do not satisfy the strict inequalities, so plot them as open dots.

Find the points of intersection between $y = x^2 - 2$ and $y = 4$:

$x^2 - 2 = 4$

$x^2 - 6 = 0$

$x^2 = 6$

$x = \pm\sqrt{6}$

$y = 4$

The solutions are $(\sqrt{6}, 4)$ and $(-\sqrt{6}, 4)$. They satisfy both inequalities, so plot them as closed dots.

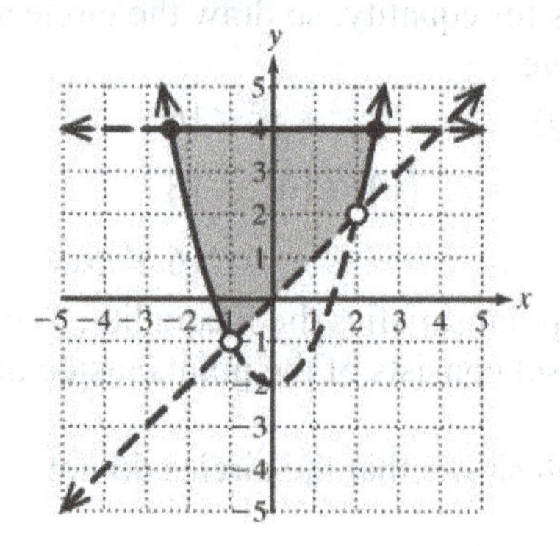

59. $x^2 + y^2 \le 100 \xrightarrow[\text{equation}]{\text{related}} x^2 + y^2 = 100$

The equation represents a circle centered at $(0, 0)$ with radius 10. The inequality symbol $\le$ allows for equality, so draw the circle as a solid curve.

Test (0, 0): $x^2 + y^2 \le 100$

$(0)^2 + (0)^2 \overset{?}{\le} 100$

$0 \overset{?}{\le} 100$ ✓ true

The test point satisfies the inequality. The solution set consists of the points inside or on the circle.

$y < \frac{4}{3}x \xrightarrow[\text{equation}]{\text{related}} y = \frac{4}{3}x$

The inequality is strict, so draw the line as a dashed line.

Test (1, 0): $y < \frac{4}{3}x$

$0 \overset{?}{<} \frac{4}{3}(1)$

$0 \overset{?}{<} \frac{4}{3}$ ✓ true

The points on the side of the line containing (1, 0) are solutions.

$x \le 8$

The related equation $x = 8$ is a vertical line. The inequality $x \le 8$ represents all points above or on the line $x = 8$. The inequality symbol $\le$ allows for equality, so draw the line as a solid line.

Find the points of intersection between $x^2+y^2=100$ and $y=\frac{4}{3}x$:

$$x^2+y^2=100$$
$$x^2+\left(\frac{4}{3}x\right)^2=100$$
$$x^2+\frac{16}{9}x^2=100$$
$$\frac{25}{9}x^2=100$$
$$x^2=36$$
$$x=\pm 6$$
$$y=\frac{4}{3}x=\frac{4}{3}(6)=8$$
$$y=\frac{4}{3}x=\frac{4}{3}(-6)=-8$$

The solutions are $(-6,-8)$ and $(6,8)$. They do not satisfy the inequalities, so plot them as open dots.

Find the points of intersection between $x^2+y^2=100$ and $x=8$:

$$x^2+y^2=100$$
$$(8)^2+y^2=100$$
$$y^2=36$$
$$y=\pm 6$$
$$x=8$$

The solutions are $(8,-6)$ and $(8,6)$. They satisfy the inequalities, so plot them as closed dots.

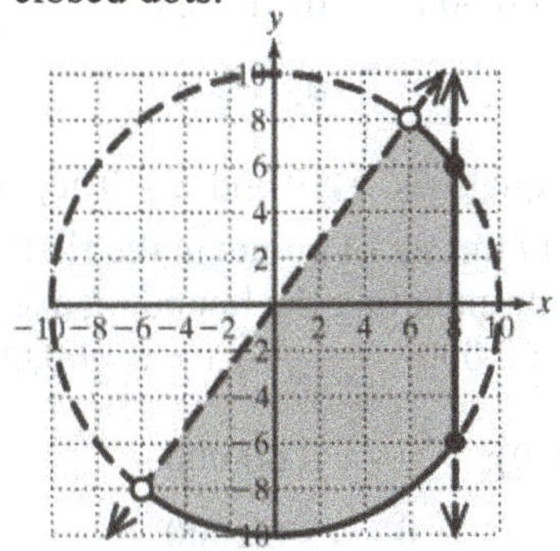

61. $y<e^x \xrightarrow[\text{equation}]{\text{related}} y=e^x$

The equation represents the parent natural exponential function. The inequality is strict, so draw the function as a dashed curve.

Test (0, 0):
$$y<e^x$$
$$0\overset{?}{<}e^0$$
$$0\overset{?}{<}1 \checkmark \text{ true}$$

The test point satisfies the inequality. The solution set consists of the points below the function.

$y>1 \xrightarrow[\text{equation}]{\text{related}} y=1$

The related equation $y=1$ is a horizontal line. The inequality $y>1$ represents all points above the line $y=1$. The inequality is strict, so draw the line as a dashed line.

$x<2$

The related equation $x=2$ is a vertical line. The inequality $x<2$ represents all points to the left of the line $x=2$. The inequality is strict, so draw the line as a dashed line.

Find the point of intersection between $y=e^x$ and $y=1$:

$$y=e^x$$
$$1=e^x$$
$$x=0$$

The solution is $(0,1)$. It does not satisfy the inequalities, so plot it as an open dot.

Find the points of intersection between $y=e^x$ and $x=2$:

$$y=e^x$$
$$y=e^2\approx 7.4$$

The solution is $(2, 7.4)$. It does not satisfy the inequalities, so plot it as an open dot.

The point of intersection between $y=1$ and $x=2$ is $(2,1)$. It does not satisfy the inequalities, so plot it as an open dot.

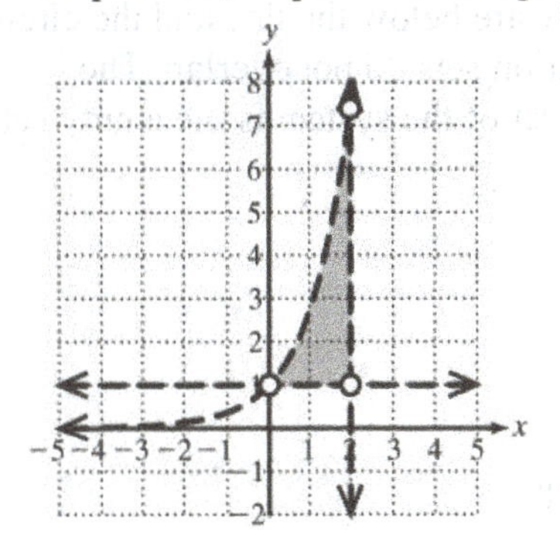

63. $(x+2)^2+(y-3)^2 \le 9$

$\xrightarrow[\text{equation}]{\text{related}} (x+2)^2+(y-3)^2=9$

The equation represents a circle centered at $(-2,3)$ with radius 3. The inequality symbol $\le$ allows for equality, so draw the circle as a solid curve.

Test (0, 0): $(x+2)^2+(y-3)^2 \le 9$

$(0+2)^2+(0-3)^2 \overset{?}{\le} 9$

$4+9 \overset{?}{\le} 9$

$13 \overset{?}{\le} 9$ false

The test point does not satisfy the inequality. The solution set consists of the points inside or on the circle.

$x-y>2 \xrightarrow[\text{equation}]{\text{related}} y=x-2$

The inequality is strict, so draw the line as a dashed line.

Test (0, 0): $x-y>2$

$0-0 \overset{?}{>} 2$

$0 \overset{?}{>} 2$ false

The points on the side of the line not containing (0, 0) are solutions.

Find the points of intersection.

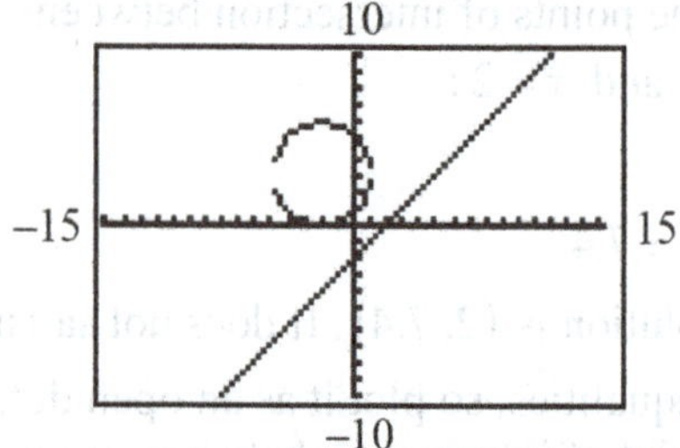

The line does not intersect the circle. The solutions to the first inequality are inside the circle, and the solutions to the second inequality are below the line and the circle. The solution sets do not overlap. The solution set of the system is the empty set, $\{\ \}$.

65. $x \le 6$

67. $y \ge -2$

69. $x+y \le 18$

71. **a.** $x+y \le 9$ **b.** $x \ge 3$

c. $y \le 4$ **d.** $x \ge 0$

e. $y \ge 0$

f. The inequalities $x \ge 0$ and $y \ge 0$ together represent the set of points in the first quadrant, including the bounding axes.

$x+y \le 9 \xrightarrow[\text{equation}]{\text{related}} x+y=9$

Test (0, 0): $x+y \le 9$

$0+0 \overset{?}{\le} 9$

$0 \overset{?}{\le} 9$ ✓ true

The inequality $x+y \le 9$ represents the set of points on and below the line $x+y=9$. In addition, the inequality $x \ge 3$ represents the set of points on or to the right of the line $x=3$, and the inequality $y \le 4$ represents the set of points on or below the line $y=4$.

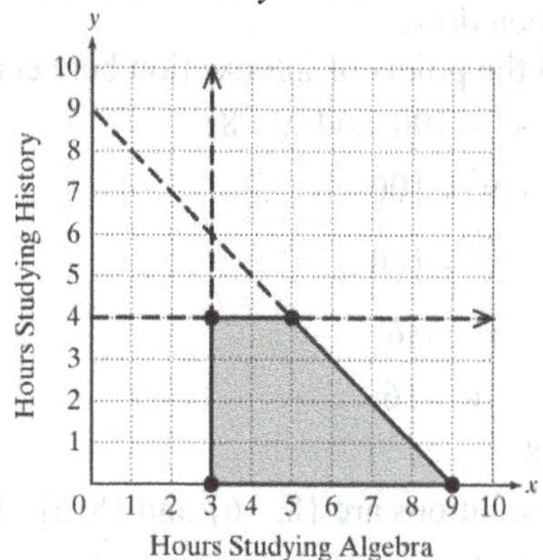

73. **a.** $x+y \le 60{,}000$ **b.** $y \ge 2x$

c. $x \ge 0$ **d.** $y \ge 0$

e. The inequalities $x \ge 0$ and $y \ge 0$ together represent the set of points in the first quadrant, including the bounding axes.

$x+y \le 60{,}000 \xrightarrow[\text{equation}]{\text{related}} x+y=60{,}000$

Test (0, 0): $x+y \le 60{,}000$

$0+0 \overset{?}{\le} 60{,}000$

$0 \overset{?}{\le} 60{,}000$ ✓ true

The inequality $x+y \le 60{,}000$ represents the set of points on and below the line $x+y=60{,}000$.

$y \ge 2x \xrightarrow[\text{equation}]{\text{related}} y = 2x$

Test (5000, 20,000):

$y \ge 2x$

$20{,}000 \overset{?}{\ge} 2(5000)$

$20{,}000 \overset{?}{\ge} 10{,}000$ ✓ true

The inequality $y \ge 2x$ represents the set of points on and above the line $y = 2x$.

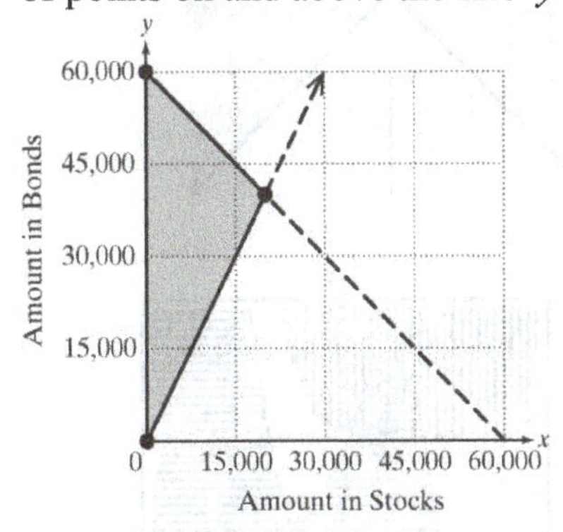

75. These points are the Quadrant I part of a circle with vertex at the origin and radius 3: $x^2 + y^2 < 9$, $x > 0$, $y > 0$

77. Find the equations of the three sides of the triangle.

Line 1: $m = \dfrac{2-(-4)}{3-(-3)} = \dfrac{6}{6} = 1$

$y - 2 = 1(x - 3)$

$y - 2 = x - 3$

$y = x - 1$

Line 2: $m = \dfrac{4-2}{-5-3} = \dfrac{2}{-8} = -\dfrac{1}{4}$

$y - 4 = -\dfrac{1}{4}[x - (-5)]$

$y - 4 = -\dfrac{1}{4}x - \dfrac{5}{4}$

$y = -\dfrac{1}{4}x + \dfrac{11}{4}$

Line 3: $m = \dfrac{-4-4}{-3-(-5)} = \dfrac{-8}{2} = -4$

$y - (-4) = -4[x - (-3)]$

$y + 4 = -4x - 12$

$y = -4x - 16$

Graph the equations to determine the inequalities.

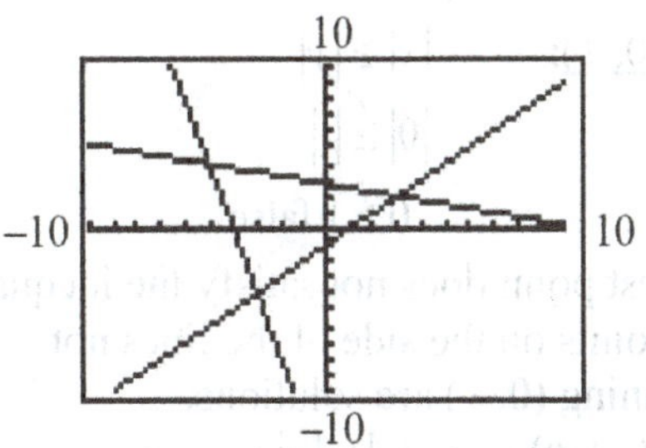

Line 1 is on the right. Line 2 is on the top. Line 3 is on the left.

$y > x - 1$, $y < -\dfrac{1}{4}x + \dfrac{11}{4}$, $y > -4x - 16$

79. a. This is a circle with vertex $(-12, -9)$ and radius 16.

$(x+12)^2 + (y+9)^2 \le 256$

b. The distance from Hawthorne to the point of the earthquake is

$d = \sqrt{9^2 + 12^2} = \sqrt{81 + 144} = \sqrt{225} = 15$ km

Yes, the earthquake could be felt at Hawthorne, which is 15 km from the earthquake.

81. If the inequality is strict—that is, posed with < or >—then the bounding line or curve should be dashed.

83. Find the solution set to each individual inequality in the system. Then to find the solution set for the system of inequalities, take the intersection of the solution sets to the individual inequalities.

85. $|x| \ge |y| \xrightarrow[\text{equation}]{\text{related}} |x| = |y|$

Graph the equivalent form:

$x = y$ and $x = -y$.

The inequality symbol $\ge$ allows for equality, so draw the lines as solid lines.

Test points in each region.

Test (1, 0): $|x| \ge |y|$

$|1| \overset{?}{\ge} |0|$

$1 \overset{?}{\ge} 0$ ✓ true

The test point satisfies the inequality. The points on the side of the lines containing (1, 0) are solutions.

<u>Test (0, 1)</u>: $|x| \geq |y|$

$|0| \overset{?}{\geq} |1|$

$0 \overset{?}{\geq} 1$ false

The test point does not satisfy the inequality. The points on the side of the lines not containing (0, 1) are solutions.

<u>Test $(-1, 0)$</u>: $|x| \geq |y|$

$|-1| \overset{?}{\geq} |0|$

$1 \overset{?}{\geq} 0$ ✓ true

The test point satisfies the inequality. The points on the side of the lines containing $(-1, 0)$ are solutions.

<u>Test $(0, -1)$</u> $|x| \geq |y|$

$|0| \overset{?}{\geq} |-1|$

$0 \overset{?}{\geq} 1$ false

The test point does not satisfy the inequality. The points on the side of the lines not containing $(0, -1)$ are solutions.

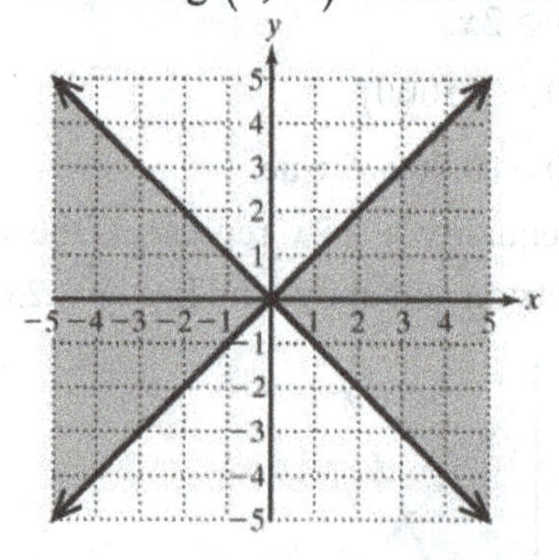

87.

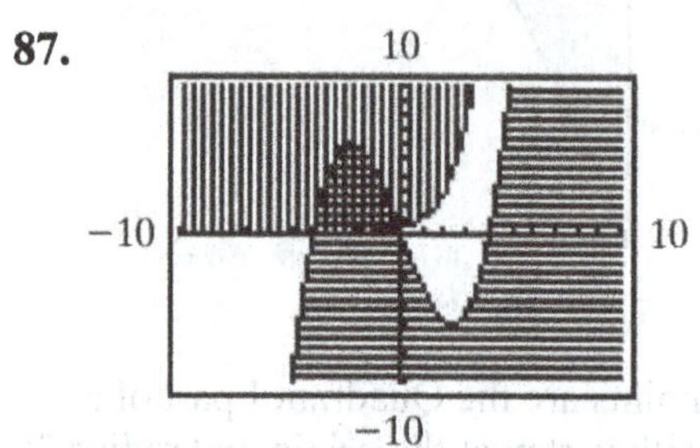

Problem Recognition Exercises: Equations and Inequalities in Two Variables

1. a.

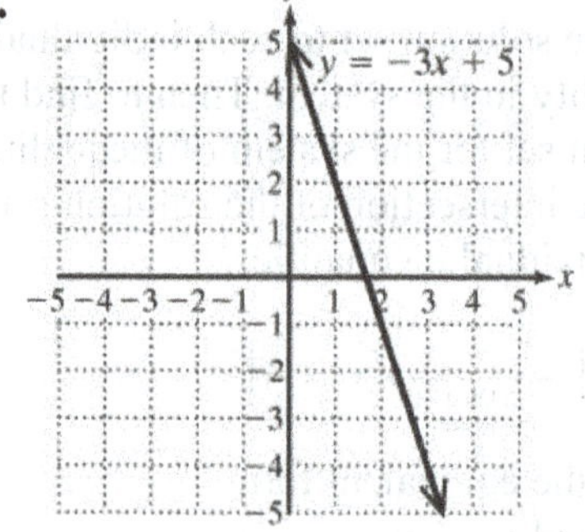

b.

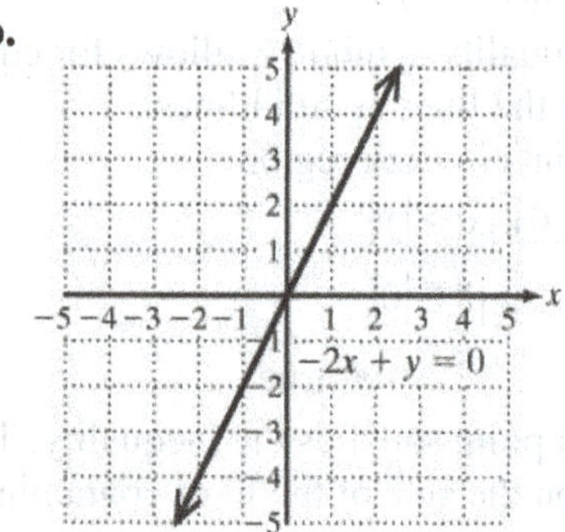

c.

$$-2x + y = 0$$
$$-2x + (-3x + 5) = 0$$
$$-2x - 3x + 5 = 0$$
$$-5x = -5$$
$$x = 1$$
$$y = -3x + 5 = -3(1) + 5 = -3 + 5 = 2$$
$$\{(1, 2)\}$$

d. $y > -3x + 5 \xrightarrow[\text{equation}]{\text{related}} y = -3x + 5$

The inequality is strict, so draw the line as a dashed line.

<u>Test (0, 0)</u>: $y > -3x + 5$

$0 \overset{?}{>} -3(0) + 5$

$0 \overset{?}{>} 2$ false

The points on the side of the line not containing (0, 0) are solutions.

$-2x+y<0 \xrightarrow[\text{equation}]{\text{related}} y=2x$

The inequality is strict, so draw the line as a dashed line.

Test (1, 1): $-2x+y<0$

$-2(1)+1\overset{?}{<}0$

$-1\overset{?}{<}0$ ✓ true

The points on the side of the line containing (1, 1) are solutions.

From part (c), the point of intersection is (1, 2). Graph (1, 2) as an open dot. It is not a solution to either inequality.

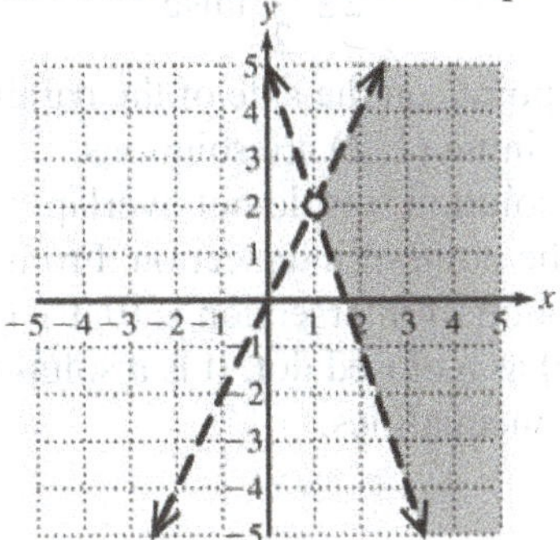

e. $y<-3x+5 \xrightarrow[\text{equation}]{\text{related}} y=-3x+5$

The inequality is strict, so draw the line as a dashed line.

Test (0, 0): $y<-3x+5$

$0\overset{?}{<}-3(0)+5$

$0\overset{?}{<}2$ ✓ true

The points on the side of the line containing (0, 0) are solutions.

$-2x+y>0 \xrightarrow[\text{equation}]{\text{related}} y=2x$

The inequality is strict, so draw the line as a dashed line.

Test (1, 1): $-2x+y>0$

$-2(1)+1\overset{?}{>}0$

$-1\overset{?}{>}0$ false

The points on the side of the line not containing (1, 1) are solutions.

From part (c), the point of intersection is (1, 2). Graph (1, 2) as an open dot. It is not a solution to either inequality.

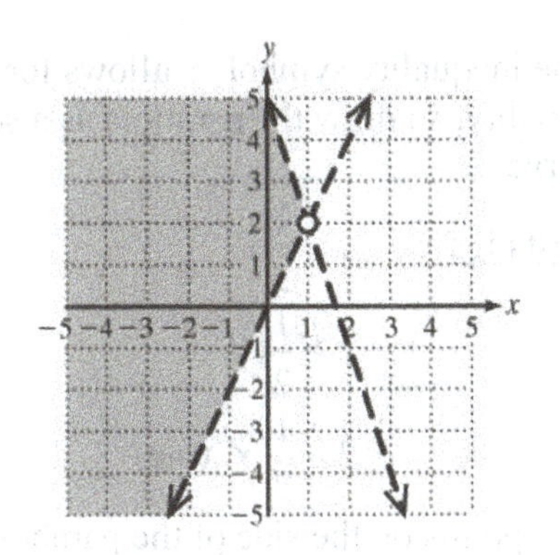

3. a. $y=x^2$

$y=\frac{1}{2}x^2$

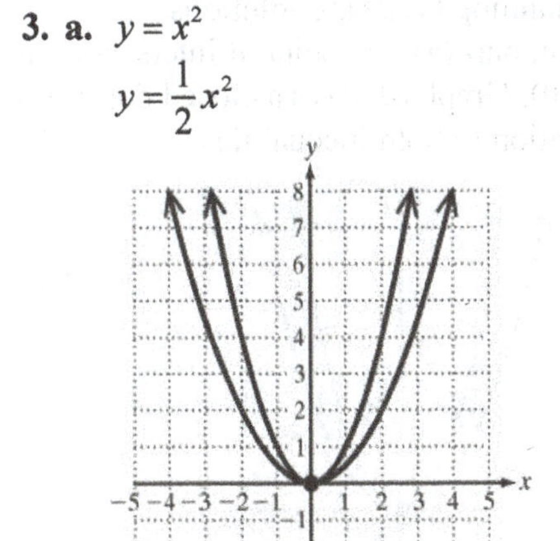

$y=\frac{1}{2}x^2$

$x^2=\frac{1}{2}x^2$

$\frac{1}{2}x^2=0$

$x^2=0$

$x=0$

$y=x^2=(0)^2=0$

$\{(0,0)\}$

b. $y\le x^2 \xrightarrow[\text{equation}]{\text{related}} y=x^2$

The inequality symbol $\le$ allows for equality, so draw the parabola as a solid curve.

Test (1, 2): $y\le x^2$

$2\overset{?}{\le}(1)^2$

$2\overset{?}{\le}1$ false

The points on the side of the parabola not containing (1, 2) are solutions.

$y\ge\frac{1}{2}x^2 \xrightarrow[\text{equation}]{\text{related}} y=\frac{1}{2}x^2$

The inequality symbol $\geq$ allows for equality, so draw the parabola as a solid curve.

Test (1, 2): $y \geq \frac{1}{2}x^2$

$2 \overset{?}{\geq} \frac{1}{2}(1)^2$

$2 \overset{?}{\geq} \frac{1}{2}$ ✓ true

The points on the side of the parabola containing (1, 2) are solutions.
From part (a), the point of intersection is (0, 0). Graph (0, 0) as a closed dot. It is a solution to both inequalities.

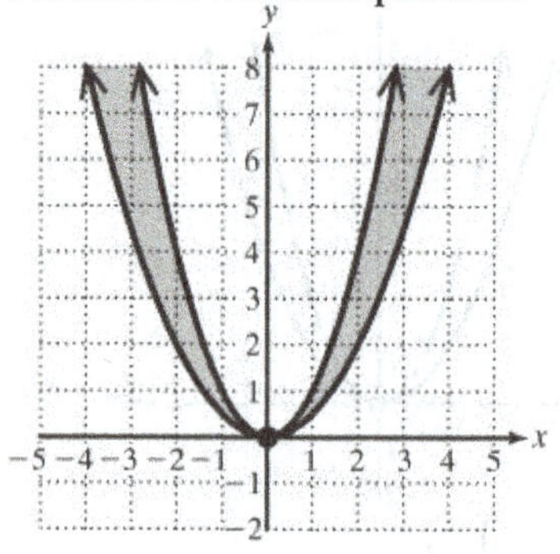

c. $y \geq x^2 \xrightarrow[\text{equation}]{\text{related}} y = x^2$

The inequality symbol $\geq$ allows for equality, so draw parabola as a solid curve.

Test (1, 2): $y \geq x^2$

$2 \overset{?}{\geq} (1)^2$

$2 \overset{?}{\geq} 1$ ✓ true

The points on the side of the parabola containing (1, 2) are solutions.

$y \leq \frac{1}{2}x^2 \xrightarrow[\text{equation}]{\text{related}} y = \frac{1}{2}x^2$

The inequality symbol $\leq$ allows for equality, so draw the parabola as a solid curve.

Test (1, 2): $y \leq \frac{1}{2}x^2$

$2 \overset{?}{\leq} \frac{1}{2}(1)^2$

$2 \overset{?}{\leq} \frac{1}{2}$ false

The points on the side of the parabola not containing (1, 2) are solutions.
The solution sets do not overlap, except for the point of intersection. From part (a), the point of intersection is (0, 0). Graph (0, 0) as a closed dot. It is a solution to both inequalities.

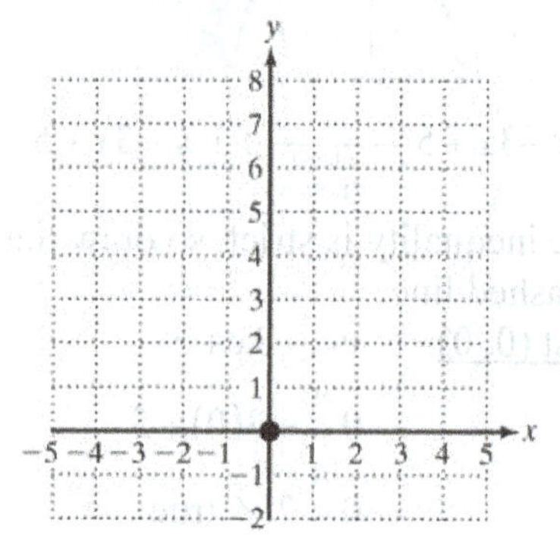

Section 5.6 Linear Programming

1. $x + y \leq 70$

3. $3x + 4y \geq 60$

5. $y \leq 3x$

7. $60 < x \leq 80$

9. linear

11. feasible

13. $z = 0.80x + 1.10y$

15. $z = 0.62x + 0.50y$

17. a. $z = 3x + 2y$

at $(0,0)$ $z = 3(0) + 2(0) = 0$

at $(0,9)$ $z = 3(0) + 2(9) = 18$

at $(4,8)$ $z = 3(4) + 2(8) = 12 + 16 = 28$

at $(7,5)$ $z = 3(7) + 2(5) = 21 + 10 = 31$

at $(9,0)$ $z = 3(9) + 2(0) = 27$

The values $x = 7$ and $y = 5$ produce the maximum value of the objective function.

b. Maximum value: 31

19. a. $z = 1000x + 900y$

at $(5, 45)$ $z = 1000(5) + 900(45)$
$= 5000 + 40{,}500 = 45{,}500$

at $(45, 45)$ $z = 1000(45) + 900(45)$
$= 45{,}000 + 40{,}500 = 85{,}500$

at $(45, 10)$ $z = 1000(45) + 900(10)$
$= 45{,}000 + 9000 = 54{,}000$

at $(25, 15)$ $z = 1000(25) + 900(15)$
$= 25{,}000 + 13{,}500 = 38{,}500$

at $(10, 30)$ $z = 1000(10) + 900(30)$
$= 10{,}000 + 27{,}000 = 37{,}000$

The values $x = 10$ and $y = 30$ produce the minimum value of the objective function.

b. Minimum value: 37,000

21. a. Bounding lines:
$x = 0$, $y = 0$, $x + y = 60$, $y = 2x$

Find the points of intersection between pairs of bounding lines.

$y = 2x$
$y = 2(0)$
$y = 0$ $(0, 0)$

$x + y = 60$
$x + 2x = 60$
$3x = 60$
$x = 20$
$y = 2x = 2(20) = 40$ $(20, 40)$

$x + y = 60$
$x + 0 = 60$
$x = 60$ $(60, 0)$

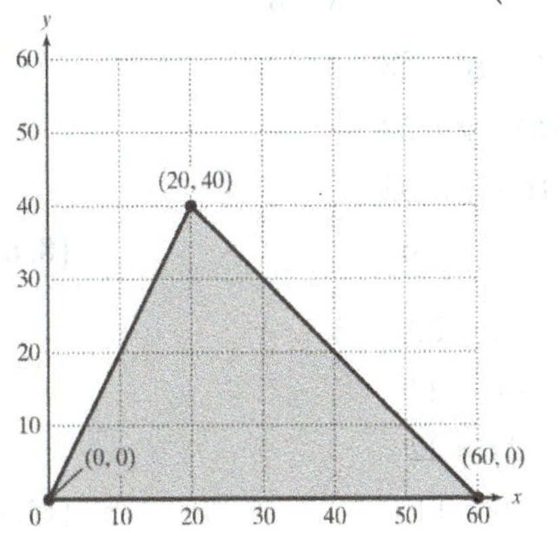

b. $z = 250x + 150y$

at $(0, 0)$ $z = 250(0) + 150(0) = 0$

at $(20, 40)$ $z = 250(20) + 150(40)$
$= 5000 + 6000 = 11{,}000$

at $(60, 0)$ $z = 250(60) + 150(0) = 15{,}000$

The values $x = 60$ and $y = 0$ produce the maximum value of the objective function.

c. Maximum value: 15,000

23. a. Bounding lines:
$x = 0$, $y = 0$,
$3x + y = 50$, $2x + y = 40$

Find the points of intersection between pairs of bounding lines.

$3x + y = 50$
$3(0) + y = 50$
$y = 50$ $(0, 50)$

$3x + y = 50$
$3x + (-2x + 40) = 50$
$x = 10$

$2x + y = 40$
$2(10) + y = 40$
$y = 20$ $(10, 20)$

$2x + y = 40$
$2x + 0 = 40$
$2x = 40$
$x = 20$ $(20, 0)$

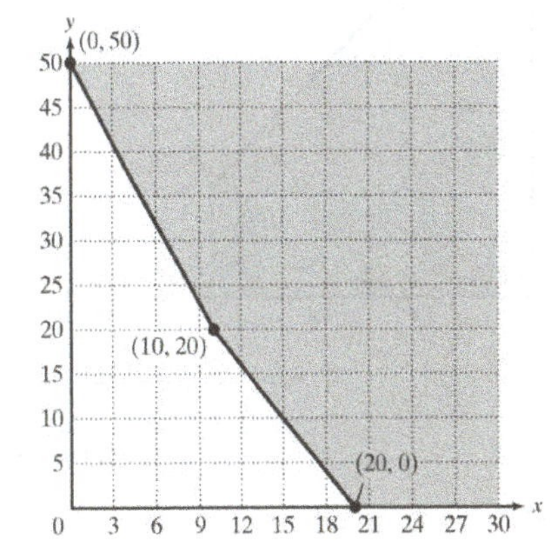

b. $z=3x+2y$

at $(0,50)$ $z=3(0)+2(50)=100$

at $(10,20)$ $z=3(10)+2(20)$

$=30+40=70$

at $(20,0)$ $z=3(20)+2(0)=60$

The values $x=20$ and $y=0$ produce the minimum value of the objective function.

c. Minimum value: 60

25. a. Bounding lines:

$x=0$, $y=0$, $y=40$

$x=36$, $x+y=48$

Find the points of intersection between pairs of bounding lines.

$x=0, y=0$ $(0,0)$

$x=0, y=40$ $(0,40)$

$x+y=48$

$x+40=48$

$x=8$ $(8,40)$

$x+y=48$

$36+y=48$

$y=12$ $(36,12)$

$x=36, y=0$ $(36,0)$

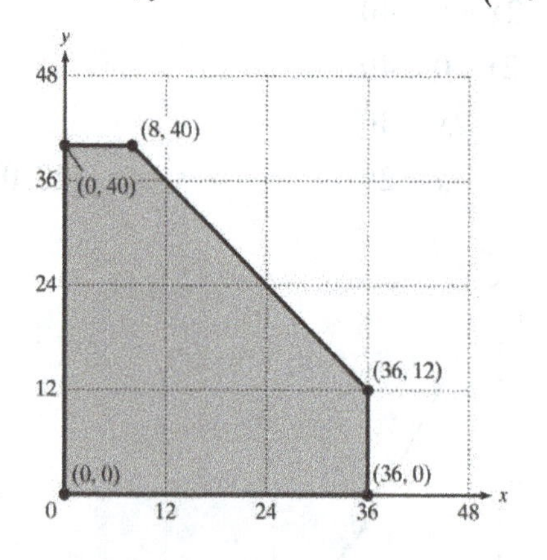

b. $z=150x+90y$

at $(0,0)$ $z=150(0)+90(0)=0$

at $(0,40)$ $z=150(0)+90(40)=3600$

at $(8,40)$ $z=150(8)+90(40)$

$=1200+3600=4800$

at $(36,12)$ $z=150(36)+90(12)$

$=5400+1080=6480$

at $(36,0)$ $z=150(36)+90(0)=5400$

The values $x=36$ and $y=12$ produce the maximum value of the objective function.

c. Maximum value: 6480

27. Bounding lines:

$x=0$, $y=0$, $3x+4y=48$,

$2x+y=22$, $y=9$

Find the points of intersection between pairs of bounding lines.

$x=0, y=0$ $(0,0)$

$x=0, y=9$ $(0,9)$

$3x+4y=48$

$3x+4(9)=48$

$3x+36=48$

$3x=12$

$x=4$ $(4,9)$

$3x+4y=48$

$3x+4(-2x+22)=48$

$3x-8x+88=48$

$-5x=-40$

$x=8$

$2x+y=22$

$2(8)+y=22$

$16+y=22$

$y=6$ $(8,6)$

$2x+y=22$

$2x+0=22$

$2x=22$

$x=11$ $(11,0)$

a. $z = 100x + 120y$

at $(0,0)$ $z = 100(0) + 120(0) = 0$

at $(0,9)$ $z = 100(0) + 120(9) = 1080$

at $(4,9)$ $z = 100(4) + 120(9)$

$= 400 + 1080 = 1480$

at $(8,6)$ $z = 100(8) + 120(6)$

$= 800 + 720 = 1520$

at $(11,0)$ $z = 100(11) + 120(0) = 1100$

The ordered pair $(8,6)$ produces the maximum value of the objective function, 1520.

b. $z = 100x + 140y$

at $(0,0)$ $z = 100(0) + 140(0) = 0$

at $(0,9)$ $z = 100(0) + 140(9) = 1260$

at $(4,9)$ $z = 100(4) + 140(9)$

$= 400 + 1260 = 1660$

at $(8,6)$ $z = 100(8) + 140(6)$

$= 800 + 840 = 1640$

at $(11,0)$ $z = 100(11) + 140(0) = 1100$

The ordered pair $(4,9)$ produces the maximum value of the objective function, 1660.

29. a. Profit: $z = 160x + 240y$

b. $x \ge 0$, $y \ge 0$, $x \le 120$, $y \le 90$, $240x + 320y \le 48{,}000$

c.

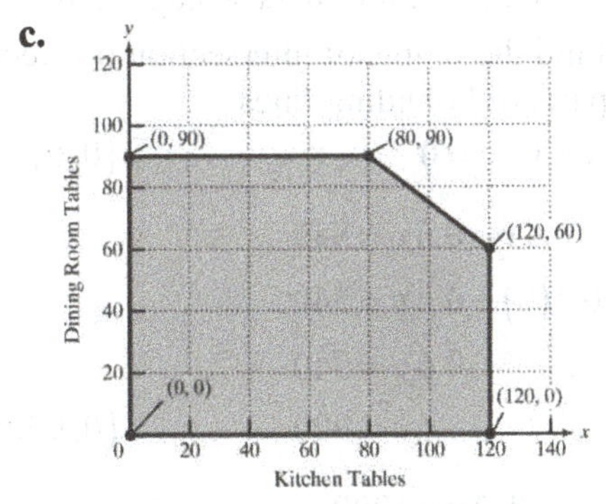

d. Find the points of intersection between pairs of bounding lines.

$x = 0, y = 0$ $\qquad (0,0)$

$x = 0, y = 90$ $\qquad (0,90)$

$240x + 320y = 48{,}000$

$240x + 320(90) = 48{,}000$

$240x + 28{,}800 = 48{,}000$

$240x = 19{,}200$

$x = 80$ $\qquad (80,90)$

$240x + 320y = 48{,}000$

$240(120) + 320y = 48{,}000$

$28{,}800 + 320y = 48{,}000$

$320y = 19{,}200$

$y = 60$ $\qquad (120,60)$

$x = 120, y = 0$ $\qquad (120,0)$

e. $z = 160x + 240y$

at $(0,0)$ $z = 160(0) + 240(0) = 0$

at $(0,90)$ $z = 160(0) + 240(90) = 21{,}600$

at $(80,90)$ $z = 160(80) + 240(90)$

$= 12{,}800 + 21{,}600 = 34{,}400$

at $(120,60)$ $z = 160(120) + 240(60)$

$= 19{,}200 + 14{,}400 = 33{,}600$

at $(120,0)$ $z = 160(120) + 240(0)$

$= 19{,}200$

f. The greatest profit is realized when 80 kitchen tables and 90 dining room tables are produced.

g. The maximum profit is $34,400.

31. a. Let x represent the number of large trees. Let y represent the number of small trees.

Profit: $z = 35x + 30y$

Constraints: $x \ge 0$, $y \ge 0$, $x + y \le 400$, $120x + 80y \le 43{,}200$

Find the points of intersection between pairs of bounding lines.

$x = 0, y = 0$ $\qquad (0,0)$

$x + y = 400$

$0 + y = 400$

$y = 400$ $\qquad (0,400)$

$$120x+80y=43{,}200$$
$$120x+80(-x+400)=43{,}200$$
$$120x-80x+32{,}000=43{,}200$$
$$40x=11{,}200$$
$$x=280$$
$$x+y=400$$
$$280+y=400$$
$$y=120 \qquad (280,120)$$

$$120x+80y=43{,}200$$
$$120x+80(0)=43{,}200$$
$$120x=43{,}200$$
$$x=360 \qquad (360,0)$$

$z=35x+30y$

at $(0,0)$ $z=35(0)+30(0)=0$

at $(0,400)$ $z=35(0)+30(400)=12{,}000$

at $(280,120)$ $z=35(280)+30(120)$
$=9800+3600=13{,}400$

at $(360,0)$ $z=35(360)+30(0)=12{,}600$

280 large trees and 120 small trees would maximize profit.

b. The maximum profit is \$13,400.

c. $z=50x+30y$

at $(0,0)$ $z=50(0)+30(0)=0$

at $(0,400)$ $z=50(0)+30(400)=12{,}000$

at $(280,120)$ $z=50(280)+30(120)$
$=14{,}000+3600=17{,}600$

at $(360,0)$ $z=50(360)+30(0)=18{,}000$

In this case, the nursery should have 360 large trees and no small trees.

33. a. Let x represent the number of trips by the large truck.
Let y represent the number of trips by the small truck.
Cost: $z=150x+120y$
Constraints: $x\ge 0$, $y\ge 0$,
$24x+18y\ge 288$; $x\le \frac{3}{4}y$
Find the points of intersection between pairs of bounding lines.

$$24x+18y=288$$
$$24(0)+18y=288$$
$$18y=288$$
$$y=16 \qquad (0,16)$$

$$24x+18y=288$$
$$24\left(\frac{3}{4}y\right)+18y=288$$
$$18y+18y=288$$
$$36y=288$$
$$y=8$$
$$x=\frac{3}{4}y=\frac{3}{4}(8)=6 \qquad (6,8)$$

$z=150x+120y$

at $(0,16)$ $z=150(0)+120(16)=1920$

at $(6,8)$ $z=150(6)+120(8)$
$=900+960=1860$

The company should make 8 trips with the small truck and 6 trips with the large truck.

b. The minimum cost is \$1860.

35. a. Let x represent the number of grill A units produced.
Let y represent the number of grill B units produced.
Profit: $z=90x+120y$
Constraints: $x\ge 0$, $y\ge 0$,
$x+1.2y\le 1200$; $0.4x+0.6y\le 540$
Find the points of intersection between pairs of bounding lines.
$x=0,\ y=0 \qquad (0,0)$

$$0.4x+0.6y=540$$
$$0.4(0)+0.6y=540$$
$$0.6y=540$$
$$y=900 \qquad (0,900)$$

$$x+1.2y=1200$$
$$x+1.2(0)=1200$$
$$x=1200 \qquad (1200,0)$$

$$0.4x+0.6y=540$$
$$0.4(-1.2y+1200)+0.6y=540$$
$$-0.48y+0.6y=540$$
$$1.08y=540$$
$$y=500$$

$$x+1.2y=1200$$
$$x+1.2(500)=1200$$
$$x+600=1200$$
$$x=600 \qquad (600, 500)$$

$z=90x+120y$

at $(0,0)$ $z=90(0)+120(0)=0$

at $(0,900)$ $z=90(0)+120(900)=108{,}000$

at $(1200,0)$ $z=90(1200)+120(0)=108{,}000$

at $(600,500)$ $z=90(600)+120(500)=54{,}000+60{,}000=114{,}000$

The manufacturer should produce 600 grill A units and 500 grill B units to maximize profit.

b. The maximum profit is \$114,000.

c. $z=110x+120y$

at $(0,0)$ $z=110(0)+120(0)=0$

at $(0,900)$ $z=110(0)+120(900)=108{,}000$

at $(1200,0)$ $z=110(1200)+120(0)=132{,}000$

at $(600,500)$ $z=110(600)+120(500)=66{,}000+60{,}000=126{,}000$

In this case, the manufacturer should produce 1200 grill A units and 0 grill B units.

37. a. Let x represent the number of acres of corn planted.
Let y represent the number of acres of soybeans planted.
Profit: $z=120x+100y$
Constraints: $x\geq 0$, $y\geq 0$, $x+y\leq 1200$, $180x+120y\leq 198{,}000$; $80x+100y\leq 110{,}000$
Find the points of intersection between pairs of bounding lines.

$x=0, y=0 \qquad (0,0)$

$$80x+100y=110{,}000$$
$$80(0)+100y=110{,}000$$
$$100y=110{,}000$$
$$y=1100 \qquad (0,1100)$$

$$180x+120y=198{,}000$$
$$180x+120(0)=198{,}000$$
$$180x=198{,}000$$
$$x=1100 \qquad (1100,0)$$

$$80x+100y=110{,}000$$
$$80x+100(-x+1200)=110{,}000$$
$$80x-100x+120{,}000=110{,}000$$
$$-20x=-10{,}000$$
$$x=500$$

$$x+y=1200$$
$$500+y=1200$$
$$y=700 \qquad (500,700)$$

$$180x+120y=198{,}000$$
$$180x+120(-x+1200)=198{,}000$$
$$180x-120x+144{,}000=198{,}000$$
$$60x=54{,}000$$
$$x=900$$

$$x+y=1200$$
$$900+y=1200$$
$$y=300 \qquad (900,300)$$

$z=120x+100y$

at $(0,0)$ $z=120(0)+100(0)=0$

at $(0,1100)$ $z=120(0)+100(1100)=110{,}000$

at $(1100, 0)$ $z=120(1100)+100(0)$
$=132{,}000$
at $(500, 700)$ $z=120(500)+100(700)$
$=60{,}000+70{,}000$
$=130{,}000$
at $(900, 300)$ $z=120(900)+100(300)$
$=108{,}000+30{,}000$
$=138{,}000$

The farmer should plant 900 acres of corn and 300 acres of soybeans.

b. The maximum profit is $138,000.

c. $z=100x+120y$

at $(0, 0)$ $z=100(0)+120(0)=0$
at $(0, 1100)$ $z=100(0)+120(1100)$
$=132{,}000$
at $(1100, 0)$ $z=100(1100)+120(0)$
$=110{,}000$
at $(500, 700)$ $z=100(500)+120(700)$
$=50{,}000+84{,}000$
$=134{,}000$
at $(900, 300)$ $z=100(900)+120(300)$
$=90{,}000+36{,}000$
$=126{,}000$

In this case, 500 acres of corn and 700 acres of soybeans should be planted.

39. Linear programming is a technique that enables us to maximize or minimize a function under specific constraints.

41. The feasible region for a linear programming application is found by first identifying the constraints on the relevant variables. Then the regions defined by the individual constraints are graphed. The intersection of the constraints define the feasible region.

Chapter 5 Review Exercises

1. a. $2x-3y=0$

$2(1)-3\left(\frac{2}{3}\right)\stackrel{?}{=}0$

$2-2\stackrel{?}{=}0$

$0\stackrel{?}{=}0$ ✓ true

$-5x+6y=-1$

$-5(1)+6\left(\frac{2}{3}\right)\stackrel{?}{=}-1$

$-5+4\stackrel{?}{=}-1$

$-1\stackrel{?}{=}-1$ ✓ true   Yes

b. $2x-3y=0$

$2(6)-3(4)\stackrel{?}{=}0$

$12-12\stackrel{?}{=}0$

$0\stackrel{?}{=}0$ ✓ true

$-5x+6y=-1$

$-5(6)+6(4)\stackrel{?}{=}-1$

$-30+24\stackrel{?}{=}-1$

$-6\stackrel{?}{=}-1$ false   No

3. Intersecting lines, one solution.

5. $5(x-y)=19-2y$ $0.2x+0.7y=-1.7$

$5x-5y=19-2y$ $2x+7y=-17$

$5x-3y=19$

$$\begin{aligned} 5x-3y&=19 \xrightarrow{\times 2} & 10x-6y&=38\\ 2x+7y&=-17 \xrightarrow{\times -5} & -10x-35y&=85\\ && -41y&=123\\ && y&=-3 \end{aligned}$$

$5x-3y=19$

$5x-3(-3)=19$

$5x+9=19$

$5x=10$

$x=2$ $\{(2,-3)\}$

7. $\frac{1}{10}x-\frac{1}{2}y=1$ $2x=10y+6$

$x-5y=10$ $2x-10y=6$

$x=5y+10$

$$2x-10y=6$$
$$2(5y+10)-10y=6$$
$$10y+20-10y=6$$
$$20=6$$

$\{\ \}$; The system is inconsistent.

9. Let x represent the milligrams of calcium in 1 cup of milk. Let y represent the milligrams of calcium in 1 cup of cooked spinach.
Day 1: $3x+y=1140$
$$y=-3x+1140$$
Day 2: $2x+\frac{3}{2}y=960$
$$2x+\frac{3}{2}(-3x+1140)=960$$
$$4x+3(-3x+1140)=1920$$
$$4x-9x+3420=1920$$
$$-5x=-1500$$
$$x=300$$
$$y=-3x+1140=-3(300)+1140$$
$$=-900+1140=240$$
Milk has 300 mg per cup and spinach has 240 mg per cup.

11. Let x represent the speed of the plane in still air. Let y represent the speed of the wind.

	Distance (mi)	Rate (mi/hr)	Time (hr)
With wind	960	$x+y$	2
Against wind	960	$x-y$	$\frac{8}{3}$

$$d=rt$$
$$960=(x+y)\cdot 2=2x+2y$$
$$960=(x-y)\cdot\left(\frac{8}{3}\right)=\frac{8}{3}x-\frac{8}{3}y$$
$$2880=8x-8y$$
$$\begin{array}{ll} 2x+2y=960 \xrightarrow{\times 4} & 8x+8y=3840 \\ 8x-8y=2880 & \underline{8x-8y=2880} \\ & 16x=6720 \\ & x=420 \end{array}$$

$$2x+2y=960$$
$$2(420)+2y=960$$
$$840+2y=960$$
$$2y=120$$
$$y=60$$
The speed of the plane in still air is 420 mph and the speed of the wind is 60 mph.

13. $\boxed{A}$ $3a-4b+2c=-17$
$\boxed{B}$ $2a+3b+c=1$
$\boxed{C}$ $4a+b-3c=7$
Eliminate c from equations $\boxed{A}$ and $\boxed{B}$. and from equations $\boxed{B}$ and $\boxed{C}$.
$$\begin{array}{rl} \boxed{A} & 3a-4b+2c=-17 \\ -2\cdot\boxed{B} & \underline{-4a-6b-2c=-2} \\ & -a-10b=-19 \quad \boxed{D} \end{array}$$
$$\begin{array}{rl} 3\cdot\boxed{B} & 6a+9b+3c=3 \\ \boxed{C} & \underline{4a+b-3c=7} \\ & 10a+10b=10 \\ & a+b=1 \quad \boxed{E} \end{array}$$
Solve the systems of equations $\boxed{D}$ and $\boxed{E}$.
$$\begin{array}{rl} \boxed{D} & -a-10b=-19 \\ \boxed{E} & \underline{a+b=1} \\ & -9b=-18 \\ & b=2 \end{array} \qquad \begin{array}{rl} \boxed{E} & a+b=1 \\ & a+2=1 \\ & a=-1 \end{array}$$
Back substitute.
$$\boxed{A} \quad 3a-4b+2c=-17$$
$$3(-1)-4(2)+2c=-17$$
$$-3-8+2c=-17$$
$$2c=-6$$
$$c=-3 \qquad \{(-1,2,-3)\}$$

15. $\boxed{A}$ $x+2y+z=5$
$\boxed{B}$ $x+y-z=1$
$\boxed{C}$ $4x+7y+2z=16$
Eliminate z from equations $\boxed{A}$ and $\boxed{B}$ and from equations $\boxed{B}$ and $\boxed{C}$.
$$\begin{array}{rl} \boxed{A} & x+2y+z=5 \\ \boxed{B} & \underline{x+y-z=1} \\ & 2x+3y=6 \quad \boxed{D} \end{array}$$

$$\begin{array}{rrl} 2\cdot\boxed{B} & 2x+2y-2z= & 2 \\ \boxed{C} & 4x+7y+2z= & 16 \\ \hline & 6x+9y \quad = & 18 \\ & 2x+3y= & 6 \ \boxed{E} \end{array}$$

Solve the systems of equations $\boxed{D}$ and $\boxed{E}$.

$$\begin{array}{rrl} \boxed{D} & 2x+3y= & 6 \\ -1\cdot\boxed{E} & -2x-3y= & -6 \\ \hline & 0= & 0 \end{array}$$

The system reduces to the identity $0=0$. It has infinitely many solutions.

17. Let x represent the number of seats in Section A. Let y represent the number of seats in Section B Let z represent the number of seats in Section C.

$$\begin{array}{ll} \boxed{A}\ x+y+z=12{,}000 & \rightarrow\ x+\ y+\ z=\ 12{,}000 \\ \boxed{B}\ z=x+y & \rightarrow\ x+\ y-\ z=\ 0 \\ \boxed{C}\ 90x+65y+40z=655{,}000 & \rightarrow\ 18x+13y+8z=131{,}000 \end{array}$$

Eliminate x and y from equations $\boxed{A}$ and $\boxed{B}$.

$$\begin{array}{rrl} \boxed{A} & x+y+\ z= & 12{,}000 \\ -1\cdot\boxed{B} & -x-y+\ z= & 0 \\ \hline & 2z= & 12{,}000 \\ & z= & 6000 \end{array}$$

Back substitute.

$$\begin{array}{rl} \boxed{A} & x+y+z=12{,}000 \\ & x+y+6000=12{,}000 \\ & x+y=6000\ \boxed{D} \end{array}$$

$$\begin{array}{rl} \boxed{B} & 18x+13y+8z=131{,}000 \\ & 18x+13y+8(6000)=131{,}000 \\ & 18x+13y+48{,}000=131{,}000 \\ & 18x+13y=83{,}000\ \boxed{E} \end{array}$$

Solve the system of equations $\boxed{D}$ and $\boxed{E}$.

$$\begin{array}{rrl} -13\cdot\boxed{D} & -13x-13y= & -78{,}000 \\ \boxed{E} & 18x+13y= & 83{,}000 \\ \hline & 5x \quad = & 5000 \\ & x= & 1000 \end{array}$$

Back substitute.

$$\begin{array}{rl} \boxed{A} & x+y+z=12{,}000 \\ & 1000+y+6000=12{,}000 \\ & y=5000 \end{array}$$

There are 1000 seats in Section A, 5000 in Section B, and 6000 in Section C.

19.

$$y=ax^2+bx+c$$

Substitute $(-1,-4)$: $-4=a(-1)^2+b(-1)+c \ \rightarrow\ \boxed{A}\ a-b+c=-4$

Substitute $(1,6)$: $6=a(1)^2+b(1)+c \ \rightarrow\ \boxed{B}\ a+b+c=6$

Substitute $(3,8)$: $8=a(3)^2+b(3)+c \ \rightarrow\ \boxed{C}\ 9a+3b+c=8$

Eliminate b from equations $\boxed{A}$ and $\boxed{B}$ and from equations $\boxed{A}$ and $\boxed{C}$.

$$\begin{array}{rrl} \boxed{A} & a-b+\ c= & -4 \\ \boxed{B} & a+b+\ c= & 6 \\ \hline & 2a \quad +2c= & 2 \\ & a+\ c= & 1\ \boxed{D} \end{array}$$

$$\begin{array}{rrl} 3\cdot\boxed{A} & 3a-3b+3c= & -12 \\ \boxed{C} & 9a+3b+\ c= & 8 \\ \hline & 12a \quad +4c= & -4 \\ & 3a+\ c= & -1\ \boxed{E} \end{array}$$

Solve the system of equations D and E.

$$\begin{array}{lrl} \boxed{D} & a+c= & 1 \\ -1\cdot\boxed{E} & -3a-c= & 1 \\ \hline & -2a = & 2 \\ & a= & -1 \end{array}$$

$$\begin{array}{l} \boxed{D}\quad a+c=1 \\ -1+c=1 \\ c=2 \end{array}$$

Back substitute.

$$\begin{aligned} \boxed{A}\quad a-b+c&=-4 \\ -1-b+2&=-4 \\ -b&=-5 \\ b&=5 \end{aligned}$$

$$y=-x^2+5x+2$$

21. $\dfrac{-x-11}{(x+2)(x-1)}=\dfrac{A}{x+2}+\dfrac{B}{x-1}$

23. $\dfrac{7x^2+19x+15}{2x^3+3x^2}=\dfrac{7x^2+19x+15}{x^2(2x+3)}$

$$=\frac{A}{x}+\frac{B}{x^2}+\frac{C}{2x+3}$$

25. $\dfrac{2x^3-x^2+8x-16}{x^4+5x^2+4}=\dfrac{2x^3-x^2+8x-16}{(x^2+1)(x^2+4)}$

$$=\frac{Ax+B}{x^2+1}+\frac{Cx+D}{x^2+4}$$

27. Use long division to divide the numerator by the denominator. Then perform partial fraction decomposition on the expression (remainder/divisor).

29. $\dfrac{5x+22}{x^2+8x+16}=\dfrac{5x+22}{(x+4)^2}=\dfrac{A}{x+4}+\dfrac{B}{(x+4)^2}$

$$(x+4)^2\cdot\left[\frac{5x+22}{(x+4)^2}\right]=(x+4)^2\cdot\left[\frac{A}{x+4}+\frac{B}{(x+4)^2}\right]$$

$$\begin{aligned} 5x+22&=A(x+4)+B \\ 5x+22&=Ax+4A+B \\ 5x+22&=(A)x+(4A+B) \end{aligned}$$

$$A=5$$

$$\begin{aligned} 4A+B&=22 \\ 4(5)+B&=22 \\ B&=2 \end{aligned}$$

$$\frac{5x+22}{x^2+8x+16}=\frac{5}{x+4}+\frac{2}{(x+4)^2}$$

31. $\dfrac{2x^2+x-10}{x^3+5x}=\dfrac{2x^2+x-10}{x(x^2+5)}=\dfrac{A}{x}+\dfrac{Bx+C}{x^2+5}$

$$x(x^2+5)\cdot\left[\frac{2x^2+x-10}{x(x^2+5)}\right]=x(x^2+5)\cdot\left[\frac{A}{x}+\frac{Bx+C}{x^2+5}\right]$$

$$2x^2+x-10=A(x^2+5)+(Bx+C)x$$

$$2x^2+x-10=Ax^2+5A+Bx^2+Cx$$

$$2x^2+x-10=(A+B)x^2+(C)x+(5A)$$

$5A=-10$ $\quad$ $C=1$ $\quad$ $A+B=2$

$A=-2$ $\quad$ $-2+B=2$

$B=4$

$$\frac{2x^2+x-10}{x^3+5x}=\frac{-2}{x}+\frac{4x+1}{x^2+5}$$

33. a. [A] $y-x^2=1\rightarrow y=x^2+1$

[B] $x-y=-3\rightarrow y=x+3$

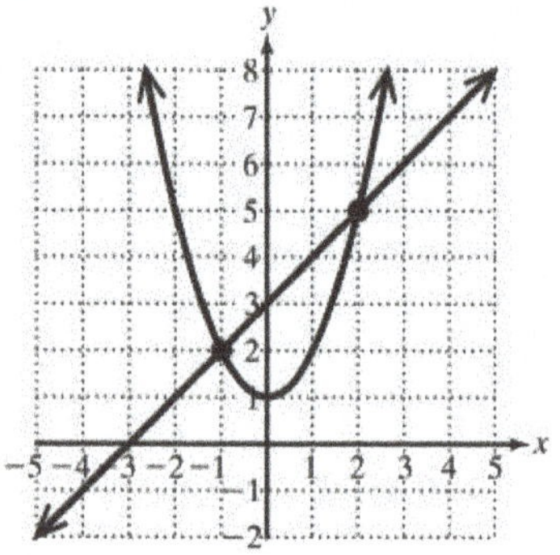

b. [B] $x-y=-3$

$$x-(x^2+1)=-3$$

$$x-x^2-1=-3$$

$$-x^2+x+2=0$$

$$x^2-x-2=0$$

$$(x+1)(x-2)=0$$

$$x=-1 \text{ or } x=2$$

[A] $y=x^2+1=(-1)^2+1=1+1=2$

The solution is $(-1, 2)$.

[A] $y=x^2+1=(2)^2+1=4+1=5$

The solution is $(2, 5)$.

$\{(-1, 2), (2, 5)\}$

35. [A] $3x^2-y^2=-4$ $\xrightarrow{\text{Multiply by 2.}}$ $6x^2-2y^2=-8$

[B] $x^2+2y^2=36$ $\qquad$ $x^2+2y^2=36$

$$7x^2=28$$

$$x^2=4$$

$$x=\pm 2$$

$x=2$: [A] $3x^2-y^2=-4$

$$3(2)^2-y^2=-4$$

$$12-y^2=-4$$

$$-y^2=-16$$

$$y^2=16$$

$$y=\pm 4$$

$x=-2$: [A] $3x^2-y^2=-4$

$$3(-2)^2-y^2=-4$$

$$12-y^2=-4$$

$$-y^2=-16$$

$$y^2=16$$

$$y=\pm 4$$

$\{(2, 4), (2, -4), (-2, 4), (-2, -4)\}$

37. [A] $y=\frac{8}{x}$

[B] $y=\sqrt{x}$

[A] $y=\frac{8}{x}$

$\sqrt{x}=\frac{8}{x}$

$x=\frac{64}{x^2}$

$x^3=64$

$x=4$

[B] $y=\sqrt{x}=\sqrt{4}=\pm 2$

Check: $(4, 2)$

[A] $y=\frac{8}{x}$

$2\overset{?}{=}\frac{8}{4}$

$2\overset{?}{=}2$ ✓ true

[B] $y=\sqrt{x}$

$2\overset{?}{=}\sqrt{4}$

$2\overset{?}{=}2$ ✓ true

Check: $(4, -2)$

[A] $y=\frac{8}{x}$

$-2\overset{?}{=}\frac{8}{4}$

$-2\overset{?}{=}2$ false

$\{(4, 2)\}$

39. Let x represent one number. Let y represent the other number.

[A] $\frac{x}{y}=\frac{4}{3}\rightarrow 4y=3x\rightarrow y=\frac{3}{4}x$

[B] $x^2+y^2=100$

[B] $x^2+y^2=100$

$x^2+\left(\frac{3}{4}\right)^2=100$

$x^2+\frac{9}{16}x^2=100$

$\frac{25}{16}x^2=100$

$x^2=64$

$x=\pm 8$

[A] $y=\frac{3}{4}x=\frac{3}{4}(8)=6$

[A] $y=\frac{3}{4}x=\frac{3}{4}(-8)=-6$

The numbers are 8 and 6 or -8 and -6.

41. Let x represent the length. Let y represent the width.

[A] $xy=288$

[B] $2x+2y=72\rightarrow x+y=36\rightarrow y=-x+36$

[A] $xy=288$

$x(-x+36)=288$

$-x^2+36x-288=0$

$x^2-36x+288=0$

$(x-12)(x-24)=0$

$x=12 \quad \text{or} \quad x=24$

[A] $y=-x+36=-12+36=24$

[A] $y=-x+36=-24+36=12$

The sign is 12 ft by 24 ft.

43. a. $y<(x-4)^2 \xrightarrow[\text{equation}]{\text{related}} y=(x-4)^2$

The equation represents a parabola opening upward with vertex $(4, 0)$. The inequality is strict, so draw the parabola as a dashed curve.

Test (0, 0): $y<(x-4)^2$

$0\overset{?}{<}(0-4)^2$

$0\overset{?}{<}16$ ✓ true

The points on the side of the parabola containing (0, 0) are solutions.

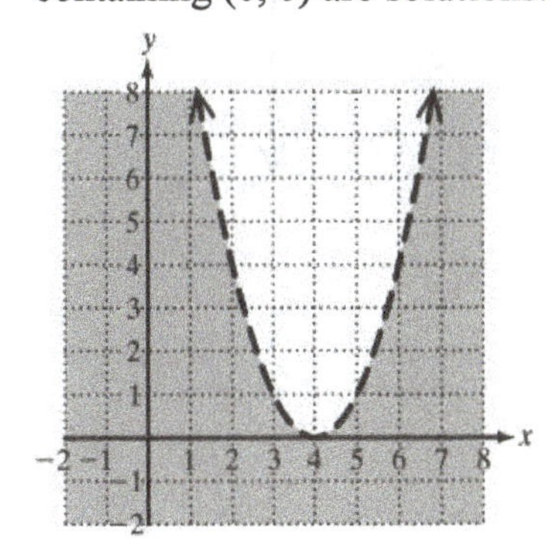

b. $y \geq (x-4)^2 \xrightarrow[\text{equation}]{\text{related}} y = (x-4)^2$

The equation represents a parabola opening upward with vertex $(4, 0)$. The inequality symbol $\geq$ allows for equality, so draw the parabola as a solid curve.

Test (0, 0): $y \leq (x-4)^2$

$0 \overset{?}{\leq} (0-4)^2$

$0 \overset{?}{\leq} 16$ false

The points on the side of the parabola not containing (0, 0) are solutions.

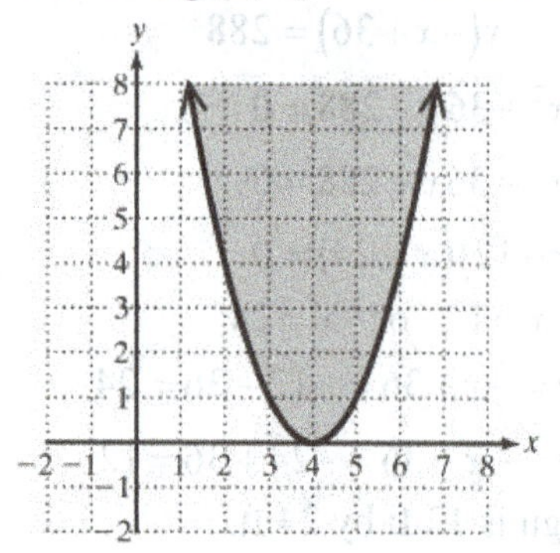

45. $x \leq 3.5$

The related equation $x = 3.5$ is a vertical line. The inequality $x \leq 3.5$ represents all points to the left or on the line $x = 3.5$. The inequality symbol $\leq$ allows for equality, so draw the line as a solid line.

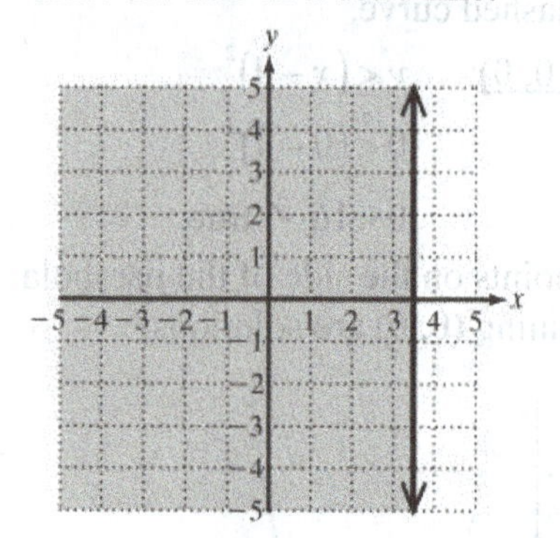

47. $x^2 + (y+2)^2 < 4 \xrightarrow[\text{equation}]{\text{related}} x^2 + (y+2)^2 = 4$

The equation represents a circle centered at $(0, -2)$ with radius 2. The inequality is strict, so draw the circle as a dashed curve.

Test (0, −1): $x^2 + (y+2)^2 < 4$

$(0)^2 + (-1+2)^2 \overset{?}{<} 4$

$1 \overset{?}{<} 4$ ✓ true

The test point inside the circle satisfies the inequality. The solution set consists of the points strictly inside the circle.

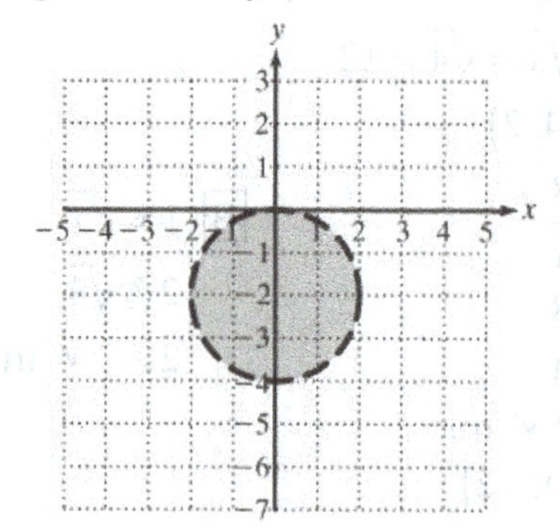

49. a. $x + 2y < 4$

$0 + 2(1) \overset{?}{<} 4$

$2 \overset{?}{<} 4$ ✓ true

$3x - 4y \geq 6$

$3(0) - 4(1) \overset{?}{\geq} 6$

$-4 \overset{?}{\geq} 6$ false

No, the ordered pair is not a solution to the system.

b. $x + 2y < 4$

$1 + 2(-4) \overset{?}{<} 4$

$-7 \overset{?}{<} 4$ ✓ true

$3x - 4y \geq 6$

$3(1) - 4(-4) \overset{?}{\geq} 6$

$19 \overset{?}{\geq} 6$ ✓ true

Yes, the ordered pair is a solution to the system.

51. $x^2+y^2\le 9 \xrightarrow[\text{equation}]{\text{related}} x^2+y^2=9$

The equation represents a circle centered at $(0,0)$ with radius 3. The inequality symbol $\le$ allows for equality, so draw the circle as a solid curve.

Test (0, 0): $x^2+y^2\le 9$

$(0)^2+(0)^2\overset{?}{\le}9$

$0\overset{?}{\le}9$ ✓ true

The test point satisfies the inequality. The solution set consists of the points inside or on the circle.

$(x-1)^2+y^2\ge 4 \xrightarrow[\text{equation}]{\text{related}} (x-1)^2+y^2\ge 4$

The equation represents a circle centered at $(1,0)$ with radius 2. The inequality symbol $\ge$ allows for equality, so draw the circle as a solid curve.

Test (0, 0): $(x-1)^2+y^2\ge 4$

$(0-1)^2+(0)^2\overset{?}{\ge}4$

$1\overset{?}{\ge}4$ false

The test point does not satisfy the inequality. The solution set consists of the points outside or on the circle.

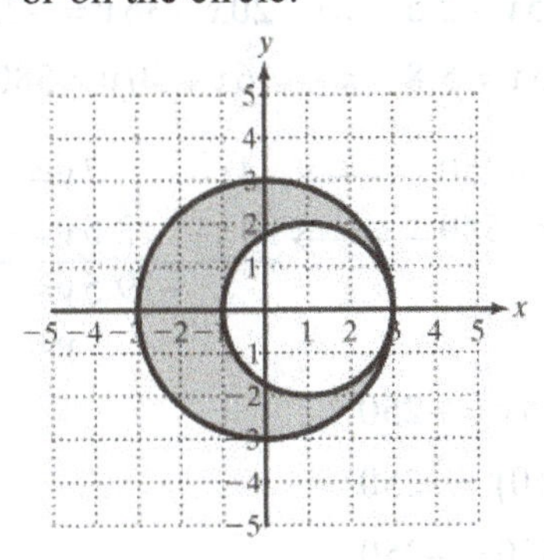

53. $y>e^x \xrightarrow[\text{equation}]{\text{related}} y=e^x$

The equation represents the parent natural exponential function. The inequality is strict, so draw the function as a dashed curve.

Test (0, 0): $y>e^x$

$0\overset{?}{>}e^0$

$0\overset{?}{>}1$ false

The test point does not satisfy the inequality. The solution set consists of the points above the curve.

$y<-x^2-1 \xrightarrow[\text{equation}]{\text{related}} y=-x^2-1$

The equation represents a parabola opening downward with vertex $(0,1)$. The inequality is strict, so draw the parabola as a dashed curve.

Test (0, 0): $y<-x^2-1$

$0\overset{?}{<}-(0)^2-1$

$0\overset{?}{<}-1$ false

The test point does not satisfy the inequality. The solution set consists of the points below the curve.

The solution sets do not overlap. The solution set of the system is the empty set, $\{\ \}$.

55. $z=24x+20y$

57. a. Bounding lines:

$x=0$, $y=0$,

$2x+y=18$, $5x+4y=60$

Find the points of intersection between pairs of bounding lines.

$$5x+4y=60$$
$$5x+4(-2x+18)=60$$
$$5x-8x+72=60$$
$$-3x=-12$$
$$x=4$$

$$2x+y=18$$
$$2(4)+y=18$$
$$8+y=18$$
$$y=10 \qquad (4,10)$$

$$2x+y=18$$
$$2(0)+y=18$$
$$y=18 \qquad (0,18)$$

$$5x+4y=60$$
$$5x+4(0)=60$$
$$5x=60$$
$$x=12 \qquad (12,0)$$

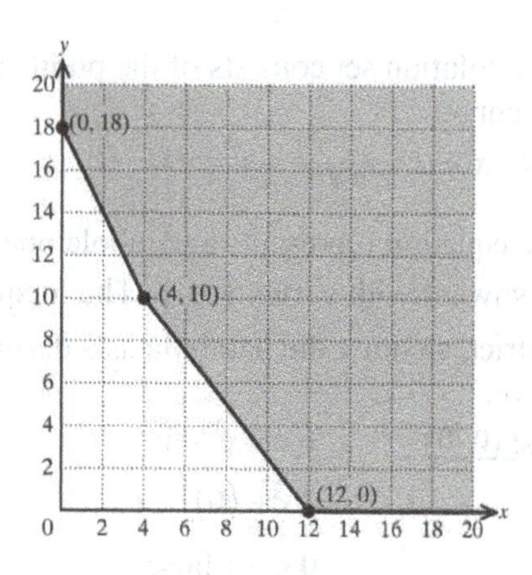

b. $z = 55x + 40y$

at $(4, 10)$ $z = 55(4) + 40(10)$
$= 220 + 400 = 620$

at $(0, 18)$ $z = 55(0) + 40(18) = 720$

at $(12, 0)$ $z = 55(12) + 40(0) = 660$

The values $x = 4$ and $y = 10$ produce the minimum value of the objective function.

c. Minimum value: 620

Chapter 5 Test

1. a.

$$x - 5y = -3$$
$$(7) - 5(2) \stackrel{?}{=} -3$$
$$7 - 10 \stackrel{?}{=} -3$$
$$-3 \stackrel{?}{=} -3 \checkmark \text{ true}$$
$$y = 2x - 12$$
$$(2) \stackrel{?}{=} 2(7) - 12$$
$$2 \stackrel{?}{=} 14 - 12$$
$$2 \stackrel{?}{=} 2 \checkmark \text{ true}$$

Yes

b.

$$x - 5y = -3$$
$$(-3) - 5(0) \stackrel{?}{=} -3$$
$$-3 \stackrel{?}{=} -3 \checkmark \text{ true}$$
$$y = 2x - 12$$
$$(0) \stackrel{?}{=} 2(-3) - 12$$
$$0 \stackrel{?}{=} -6 - 12$$
$$0 \stackrel{?}{=} -18 \text{ false}$$

No

3. a.

$$2x - 4y < 9$$
$$2(-6) - 4(1) \stackrel{?}{<} 9$$
$$-16 \stackrel{?}{<} 9 \checkmark \text{ true}$$
$$-3x + y \ge 4$$
$$-3(-6) + 1 \stackrel{?}{\ge} 4$$
$$19 \stackrel{?}{\ge} 4 \checkmark \text{ true}$$

Yes, the ordered pair is a solution to the system.

b.

$$2x - 4y < 9$$
$$2(1) - 4(4) \stackrel{?}{<} 9$$
$$-14 \stackrel{?}{<} 9 \checkmark \text{ true}$$
$$-3x + y \ge 4$$
$$-3(1) + 4 \stackrel{?}{\ge} 4$$
$$1 \stackrel{?}{\ge} 4 \text{ false}$$

No, the ordered pair is not a solution to the system.

5. $0.2x = 0.35y - 2.5 \Rightarrow 20x - 35y = -250$

$0.16x + 0.5y = 5.8 \Rightarrow 16x + 50y = 580$

$$20x - 35y = -250 \xrightarrow{\div 5} 4x - 7y = -50$$
$$16x + 50y = 580 \xrightarrow{\div -4} -4x - 12.5y = -145$$
$$-19.5y = -195$$
$$y = 10$$

$$20x - 35y = -250$$
$$20x - 35(10) = -250$$
$$20x - 350 = -250$$
$$20x = 100$$
$$x = 5 \qquad \{(5, 10)\}$$

7. $7(x - y) = 3 - 5y$ $\qquad$ $4(3x - y) = -2x$

$7x - 7y = 3 - 5y$ $\qquad$ $12x - 4y = -2x$

$7x - 2y = 3$ $\qquad$ $14x - 4y = 0$

$$7x - 2y = 3 \xrightarrow{\times -2} -14x + 4y = -6$$
$$14x - 4y = 0 \qquad 14x - 4y = 0$$
$$0 = -6$$

{ }; The system is inconsistent.

9. $\boxed{A}\quad x \qquad +4z=10$

$\boxed{B}\quad 3y-2z=9$

$\boxed{C}\quad 2x+5y \qquad =21$

Eliminate x from equations $\boxed{A}$ and $\boxed{C}$.

$$\begin{array}{rrrl} -2\cdot\boxed{A} & -2x & -8z= & -20 \\ \boxed{C} & 2x+5y & = & 21 \\ \hline & 5y-8z= & & 1 \quad \boxed{D} \end{array}$$

Solve the systems of equations $\boxed{B}$ and $\boxed{D}$.

$$\begin{array}{rrl} -4\cdot\boxed{B} & -12y+8z= & -36 \\ \boxed{D} & 5y-8z= & 1 \\ \hline & -7y\quad = & -35 \\ & y= & 5 \end{array}$$

$\boxed{B}\quad 3y-2z=9$

$3(5)-2z=9$

$15-2z=9$

$-2z=-6$

$z=3$

Back substitute.

$\boxed{A}\quad x+4z=10$

$x+4(3)=10$

$x+12=10$

$x=-2$

$\{(-2,5,3)\}$

11. $\boxed{A}\quad (x-4)^2+y^2=25$

$\boxed{B}\quad x-y=3 \rightarrow x=y+3$

$\boxed{A}\quad (x-4)^2+y^2=25$

$(y+3-4)^2+y^2=25$

$(y-1)^2+y^2=25$

$y^2-2y+1+y^2=25$

$2y^2-2y-24=0$

$y^2-y-12=0$

$(y+3)(y-4)=0$

$y=-3$ or $y=4$

$\boxed{B}\quad x=y+3=-3+3=0$

$\boxed{B}\quad x=y+3=4+3=7$

$\{(0,-3),(7,4)\}$

13.

$$\begin{array}{lrcl} \boxed{A} & 2xy-y^2=-24 & \xrightarrow{\text{Multiply by 3.}} & 6xy-3y^2=-72 \\ \boxed{B} & -3xy+2y^2=38 & \xrightarrow{\text{Multiply by 2.}} & -6xy+4y^2=76 \\ \hline & & & y^2=4 \\ & & & y=\pm2 \end{array}$$

$y=2$: $\boxed{A}\quad 2xy-y^2=-24$

$2x(2)-(2)^2=-24$

$4x-4=-24$

$4x=-20$

$x=-5$

$y=-2$: $\boxed{A}\quad 2xy-y^2=-24$

$2x(-2)-(-2)^2=-24$

$-4x-4=-24$

$-4x=-20$

$x=5$

$\{(5,-2),(-5,2)\}$

15. Let x represent the number of pounds of 100% peanuts.
Let y represent the number of pounds of 45% peanuts.

	100% Peanuts	45% Peanuts	56% Peanuts
Amount of mixture	x	y	20
Peanuts	x	$0.45y$	$0.56(20)=11.2$

$$\begin{array}{l} x+y=20 \\ x+0.45y=11.2 \xrightarrow{\times 100} 100x+45y=1120 \end{array} \qquad x+y=20 \xrightarrow{\times -100} \begin{array}{r} -100x-100y=-2000 \\ 100x+45y=1120 \\ \hline -55y=-880 \\ y=16 \end{array}$$

$$x+y=20$$
$$x+16=20$$
$$x=4$$

The manager should mix 4 lb of peanuts with 16 lb of the 45% mixture.

17. Let x represent the amount invested in the first stock.
Let y represent the amount invested in the second stock.
Let z represent the amount invested in the third stock.

$$\boxed{A}\ x+y+z=15{,}000 \rightarrow x+y+z=15{,}000$$
$$\boxed{B}\ -0.08x+0.032y+0.058z=274 \rightarrow -80x+32y+58z=274{,}000$$
$$\boxed{C}\ y=z+2000 \rightarrow y-z=2000$$

Eliminate x from equations $\boxed{A}$ and $\boxed{B}$.

$$\begin{array}{lr} 80\cdot\boxed{A} & 80x+80y+80z=1{,}200{,}000 \\ \boxed{B} & -80x+32y+58z=274{,}000 \\ \hline & 112y+138z=1{,}474{,}000\ \boxed{D} \end{array}$$

Solve the systems of equations $\boxed{C}$ and $\boxed{D}$.

$$\begin{array}{lr} 138\cdot\boxed{C} & 138y-138z=276{,}000 \\ \boxed{D} & 112y+138z=1{,}474{,}000 \\ \hline & 250y=1{,}750{,}000 \\ & y=7000 \end{array}$$

$$\boxed{D}\quad y-z=2000$$
$$7000-z=2000$$
$$-z=-5000$$
$$z=5000$$

Back substitute.

$$\boxed{A}\quad x+y+z=15{,}000$$
$$x+7000+5000=15{,}000$$
$$x=3000$$

Dylan invested \$3000 in the risky stock, \$7000 in the second stock, and \$5000 in the third stock.

19. Let x represent the length. Let y represent the width.

[A] $xy = 1452$

[B] $2x + 2y = 154 \rightarrow x + y = 77 \rightarrow y = -x + 77$

[A]
$$\begin{aligned} xy &= 1452 \\ x(-x+77) &= 1452 \\ -x^2 + 77x - 1452 &= 0 \\ x^2 - 77x + 1452 &= 0 \\ (x-44)(x-33) &= 0 \\ x = 44 \quad &\text{or} \quad x = 33 \end{aligned}$$

[A] $y = -x + 77 = -44 + 77 = 33$

[A] $y = -x + 77 = -33 + 77 = 44$

The screen is 44 in. by 33 in.

21. $\dfrac{-15x+15}{3x^2+x-2} = \dfrac{-15x+15}{(x+1)(3x-2)} = \dfrac{A}{x+1} + \dfrac{B}{3x-2}$

23. $\dfrac{-12x-29}{2x^2+11x+15} = \dfrac{-12x-29}{(x+3)(2x+5)} = \dfrac{A}{x+3} + \dfrac{B}{2x+5}$

$$(x+3)(2x+5)\cdot\left[\frac{-12x-29}{(x+3)(2x+5)}\right] = (x+3)(2x+5)\cdot\left[\frac{A}{x+3} + \frac{B}{2x+5}\right]$$

$$\begin{aligned} -12x-29 &= A(2x+5) + B(x+3) \\ -12x-29 &= 2Ax + 5A + Bx + 3B \\ -12x-29 &= (2A+B)x + (5A+3B) \end{aligned}$$

$$\begin{aligned} 2A + B &= -12 \\ B &= -2A - 12 \end{aligned}$$

$$\begin{aligned} 5A + 3B &= -29 \\ 5A + 3(-2A-12) &= -29 \\ 5A - 6A - 36 &= -29 \\ -A &= 7 \\ A &= -7 \end{aligned}$$

$$\begin{aligned} 2A + B &= -12 \\ 2(-7) + B &= -12 \\ -14 + B &= -12 \\ B &= 2 \end{aligned}$$

$$\frac{-12x-29}{2x^2+11x+15} = \frac{-7}{x+3} + \frac{2}{2x+5}$$

25.
$$\begin{array}{r} x - 2 \\ x^3 - 4x^2 \overline{\smash{)}\, x^4 - 6x^3 + 4x^2 + 20x - 32} \\ \underline{-(x^4 - 4x^3)} \\ -2x^3 + 4x^2 \\ \underline{-(-2x^3 + 8x^2)} \\ -4x^2 + 20x - 32 \end{array}$$

$$\frac{x^4-6x^3+4x^2+20x-32}{x^3-4x^2}=x-2+\frac{-4x^2+20x-32}{x^3-4x^2}$$

$$\frac{-4x^2+20x-32}{x^3-4x^2}=\frac{-4x^2+20x-32}{x^2(x-4)}=\frac{A}{x}+\frac{B}{x^2}+\frac{C}{x-4}$$

$$x^2(x-4)\cdot\left[\frac{-4x^2+20x-32}{x^2(x-4)}\right]=x^2(x-4)\cdot\left[\frac{A}{x}+\frac{B}{x^2}+\frac{C}{x-4}\right]$$

$$-4x^2+20x-32=Ax(x-4)+B(x-4)+Cx^2$$

$$-4x^2+20x-32=Ax^2-4Ax+Bx-4B+Cx^2$$

$$-4x^2+20x-32=(A+C)x^2+(-4A+B)x+(-4B)$$

$$-4B=-32 \qquad B=8$$

$$-4A+B=20 \qquad -4A+8=20 \qquad -4A=12 \qquad A=-3$$

$$A+C=-4 \qquad -3+C=-4 \qquad C=-1$$

$$\frac{x^4-6x^3+4x^2+20x-32}{x^3-4x^2}=x-2+\frac{-3}{x}+\frac{8}{x^2}+\frac{-1}{x-4}$$

27. $$\frac{7x^3+4x^2+63x+15}{x^4+11x^2+18}=\frac{7x^3+4x^2+63x+15}{(x^2+2)(x^2+9)}=\frac{Ax+B}{x^2+2}+\frac{Cx+D}{x^2+9}$$

$$(x^2+2)(x^2+9)\cdot\left[\frac{7x^3+4x^2+63x+15}{(x^2+2)(x^2+9)}\right]=(x^2+2)(x^2+9)\cdot\left[\frac{Ax+B}{x^2+2}+\frac{Cx+D}{x^2+9}\right]$$

$$7x^3+4x^2+63x+15=(Ax+B)(x^2+9)+(Cx+D)(x^2+2)$$

$$7x^3+4x^2+63x+15=Ax^3+9Ax+Bx^2+9B+Cx^3+2Cx+Dx^2+2D$$

$$7x^3+4x^2+63x+15=(A+C)x^3+(B+D)x^2+(9A+2C)x+(9B+2D)$$

$$A+C=7 \qquad A=7-C$$

$$9A+2C=63 \qquad 9(7-C)+2C=63 \qquad 63-9C+2C=63 \qquad -7C=0 \qquad C=0$$

$$A+C=7 \qquad A+0=7 \qquad A=7$$

$$B+D=4 \qquad B=4-D$$

$$9B+2D=15 \qquad 9(4-D)+2D=15 \qquad 36-9D+2D=15 \qquad -7D=-21 \qquad D=3$$

$$B+D=4 \qquad B+3=4 \qquad B=1$$

$$\frac{7x^3+4x^2+63x+15}{x^4+11x^2+18}=\frac{7x+1}{x^2+2}+\frac{3}{x^2+9}$$

29. $(x+3)^2 + y^2 \geq 9 \xrightarrow[\text{equation}]{\text{related}} (x+3)^2 + y^2 = 9$

The equation represents a circle centered at $(-3, 0)$ with radius 3. The inequality symbol $\geq$ allows for equality, so draw the circle as a solid curve.

Test (1, 1):

$$(x+3)^2 + y^2 \geq 9$$
$$(1+3)^2 + (1)^2 \stackrel{?}{\geq} 9$$
$$17 \stackrel{?}{\geq} 9 \checkmark \text{ true}$$

The test point satisfies the inequality. The solution set consists of the points outside or on the circle.

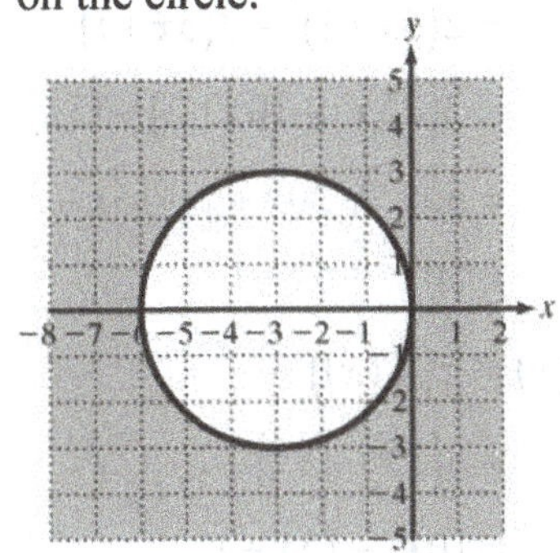

31. $x + y \leq 4 \xrightarrow[\text{equation}]{\text{related}} x + y = 4$ or $y = -x + 4$

The inequality symbol $\leq$ allows for equality, so draw the line as a solid line.

Test (0, 0):

$$x + y \leq 4$$
$$0 + 0 \stackrel{?}{\leq} 4$$
$$0 \stackrel{?}{\leq} 4 \checkmark \text{ true}$$

The points on the side of the line containing (0, 0) are solutions.

$2x - y > -2 \xrightarrow[\text{equation}]{\text{related}} 2x - y = -2$ or $y = 2x + 2$

The inequality is strict, so draw the line as a dashed line.

Test (0, 0):

$$2x - y > -2$$
$$2(0) - 0 \stackrel{?}{>} -2$$
$$0 \stackrel{?}{>} -2 \checkmark \text{ true}$$

The points on the side of the line containing (0, 0) are solutions.

Find the point of intersection.

$$-x + 4 = 2x + 2$$
$$-3x = -2$$
$$x = \frac{2}{3}$$

$$y = -x + 4 = -\frac{2}{3} + 4 = \frac{10}{3}$$

Graph $\left(\frac{2}{3}, \frac{10}{3}\right)$ as an open dot. It is not a solution to first inequality.

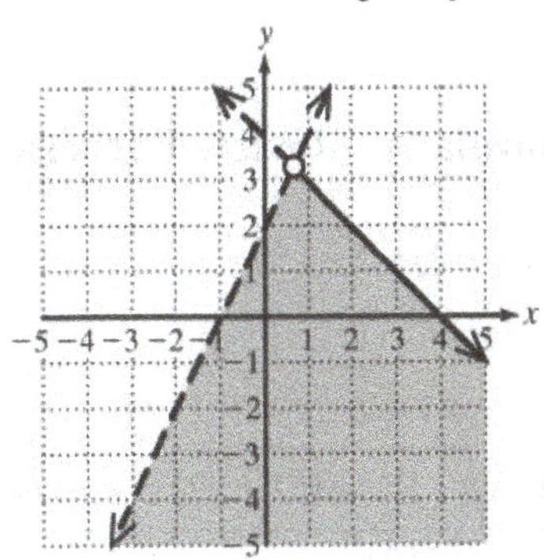

33. $z = 2.4x + 0.55y$

35. a. Bounding lines:

$x = 0$, $y = 0$, $x + y = 48$, $y = 3x$

Find the points of intersection between pairs of bounding lines.

$$y = 3x$$
$$y = 3(0)$$
$$y = 0 \qquad (0, 0)$$

$$x + y = 48$$
$$x + 3x = 48$$
$$4x = 48$$
$$x = 12$$
$$y = 3x = 3(12) = 36 \qquad (12, 36)$$

$$x + y = 48$$
$$x + 0 = 48$$
$$x = 48 \qquad (48, 0)$$

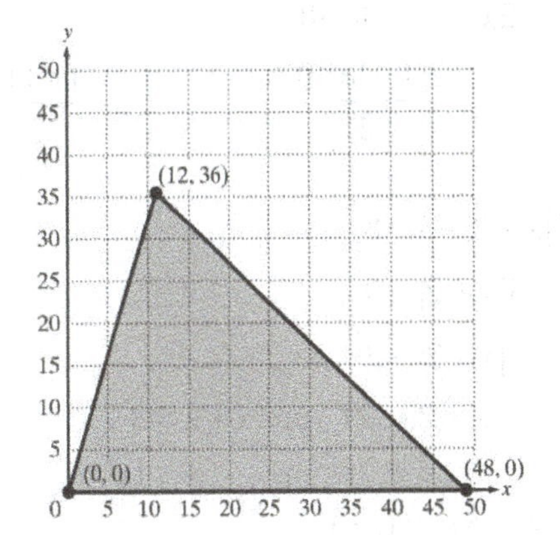

b. $z = 600x + 850y$

at $(0, 0)$ $z = 600(0) + 850(0) = 0$

at $(12, 36)$ $z = 600(12) + 850(36)$

$= 7200 + 30{,}600 = 37{,}800$

at $(48, 0)$ $z = 600(48) + 850(0) = 28{,}800$

The values $x = 12$ and $y = 36$ produce the maximum value of the objective function.

c. Maximum value: 37,800

Chapter 5 Cumulative Review Exercises

1.
$$2x(x+2) = 5x + 7$$
$$2x^2 + 4x - 5x - 7 = 0$$
$$2x^2 - x - 7 = 0$$
$$x = \frac{-(-1) \pm \sqrt{(-1)^2 - 4(2)(-7)}}{2(2)} = \frac{1 \pm \sqrt{57}}{4}$$
$$\left\{\frac{1 \pm \sqrt{57}}{4}\right\}$$

3. $(x^2 - 4)^2 - 7(x^2 - 4) - 60 = 0$

Let $u = x^2 - 4$.
$$u^2 - 7u - 60 = 0$$
$$(u+5)(u-12) = 0$$
$$u = -5 \quad \text{or} \quad u = 12$$
$$x^2 - 4 = -5 \qquad x^2 - 4 = 12$$
$$x^2 = -1 \qquad x^2 = 16$$
$$x = \pm i \qquad x = \pm 4$$
$$\{\pm 4, \pm i\}$$

5. $50e^{2x+1} = 2000$
$$e^{2x+1} = 40$$
$$2x + 1 = \ln 40$$
$$2x = -1 + \ln 40$$
$$x = \frac{-1 + \ln 40}{2} \qquad \left\{\frac{-1 + \ln 40}{2}\right\}$$

7. $3|x+2| - 1 > 8$
$$3|x+2| > 9$$
$$|x+2| > 3$$
$$x + 2 < -3 \quad \text{or} \quad x + 2 > 3$$
$$x < -5 \quad \text{or} \quad x > 1$$
$$(-\infty, -5) \cup (1, \infty)$$

9. a. $(f \circ g)(x) = f(g(x))$
$$= 2(g(x))^2 - 3(g(x))$$
$$= 2(5x+1)^2 - 3(5x+1)$$
$$= 2(25x^2 + 10x + 1) - 15x - 3$$
$$= 50x^2 + 20x + 2 - 15x - 3$$
$$= 50x^2 + 5x - 1$$

b. $(g \circ f)(x) = g(f(x))$
$$= 5(2x^2 - 3x) + 1$$
$$= 10x^2 - 15x + 1$$

11. $\log_5 256 = \dfrac{\log 256}{\log 5} \approx 3.4454$

13. $x + 3y = 6$
$$3y = -x + 6$$
$$y = -\frac{1}{3}x + 2$$

The slope is $-\frac{1}{3}$. The slope of a perpendicular line is $-\left(-\frac{1}{3}\right)^{-1} = 3$.
$$y - y_1 = m(x - x_1)$$
$$y - (-1) = 3(x - 2)$$
$$y + 1 = 3x - 6$$
$$y = 3x - 7$$

15. a. The graph of $f(x)=2|x-1|-3$ is the graph of $f(x)=|x|$ shifted to the right 1 unit, stretched vertically by a factor of 2, and shifted downward 3 units.

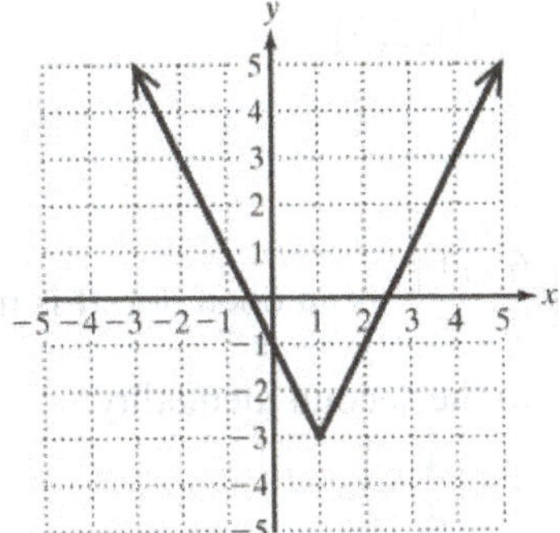

b. Domain: $(-\infty, \infty)$

c. Range: $[-3, \infty)$

17. a. $f(x)=-x^2(x-1)(x+2)^2$

The leading term is $-x^2(x)(x)^2=-x^5$.

The end behavior is up to the left and down to the right.

$f(0)=-(0)^2(0-1)(0+2)^2=0$

The y-intercept is $(0, 0)$.

The zeros of the function are 0 (multiplicity 2), 1 multiplicity 1, and -2 (multiplicity 2).

Test for symmetry:

$$f(-x)=-(-x)^2(-x-1)(-x+2)^2$$
$$=-x^2(-x-1)(-x+2)^2$$

$f(x)$ is neither even nor odd.

b. Domain: $(-\infty, \infty)$

c. Range: $(-\infty, \infty)$

19.
$\boxed{A}$ $5a-2b+3c=10$
$\boxed{B}$ $-3a+b-2c=-7$
$\boxed{C}$ $a+4b-4c=-3$

Eliminate b from equations $\boxed{A}$ and $\boxed{B}$.

$\boxed{A}$ $5a-2b+3c=10$ $\qquad\qquad\qquad 5a-2b+3c=10$
$\boxed{B}$ $-3a+b-2c=-7$ $\xrightarrow{\text{Multiply by 2}}$ $-6a+2b-4c=-14$
$\qquad\qquad\qquad -a \qquad -c=-4$ $\boxed{D}$

Eliminate b from equations $\boxed{B}$ and $\boxed{C}$.

$\boxed{B}$ $-3a+b-2c=-7$ $\xrightarrow{\text{Multiply by }-4}$ $12a-4b+8c=28$
$\boxed{C}$ $a+4b-4c=-3$ $\qquad\qquad\qquad a+4b-4c=-3$
$\qquad\qquad\qquad 13a \qquad +4c=25$ $\boxed{E}$

Solve the systems of equations $\boxed{D}$ and $\boxed{E}$.

$\boxed{D}$ $-a-c=-4$ $\xrightarrow{\text{Multiply by 4}}$ $-4a-4c=-16$
$\boxed{E}$ $13a+4c=25$ $\qquad\qquad\qquad 13a+4c=25$
$\qquad\qquad\qquad 9a=9$
$\qquad\qquad\qquad a=1$

Back substitute.

D $-a-c=-4$

$-1-c=-4$

$-c=-3$

$c=3$

C $a+4b-4c=-3$

$1+4b-4(3)=-3$

$4b=8$

$b=2$

$\{(1,2,3)\}$

21. $3x-y\leq 1 \xrightarrow[\text{equation}]{\text{related}} 3x-y=1$

$3x-y=1$

$-y=-3x+1$

$y=3x-1$

The inequality symbol $\leq$ allows for equality, so draw the line as a solid line.

Test (0, 0): $3x-y\leq 1$

$3(0)-0\overset{?}{\leq}1$

$0\overset{?}{\leq}1$ ✓ true

The points on the side of the line containing (0, 0) are solutions.

$x+2y<4 \xrightarrow[\text{equation}]{\text{related}} x+2y=4$

x-intercept:

$x+2(0)=4$

$x=4$

y-intercept:

$0+2y=4$

$2y=4$

$y=2$

The inequality is strict, so draw the line as a dashed line.

Test (0, 0): $x+2y<4$

$0+2(0)\overset{?}{<}4$

$0\overset{?}{<}4$ ✓ true

The points on the side of the line containing (0, 0) are solutions.

Find the point of intersection.

$x+2y=4$

$x+2(3x-1)=4$

$x+6x-2=4$

$7x=6$

$x=\frac{6}{7}$

$y=3x-1=3\left(\frac{6}{7}\right)-1=\frac{18}{7}-\frac{7}{7}=\frac{11}{7}$

Graph $\left(\frac{6}{7},\frac{11}{7}\right)$ as an open dot. It is not a solution to the second inequality.

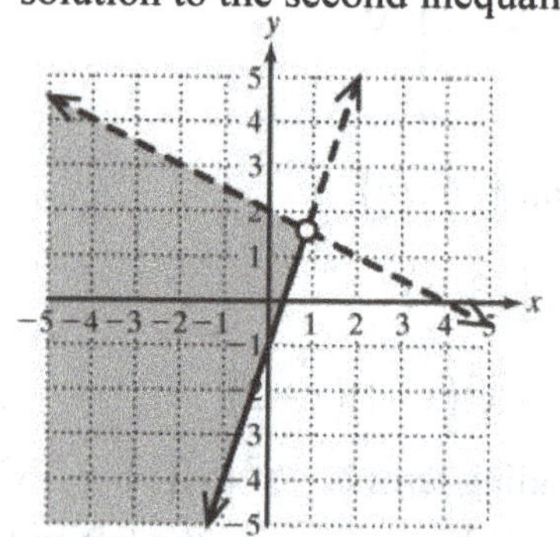

23. $$\frac{10}{\sqrt{5x}}+\frac{4\sqrt{5x}}{x}=\frac{10}{\sqrt{5x}}\cdot\frac{\sqrt{5x}}{\sqrt{5x}}+\frac{4\sqrt{5x}}{x}\cdot\frac{5}{5}$$

$$=\frac{10\sqrt{5x}}{5x}+\frac{20\sqrt{5x}}{5x}$$

$$=\frac{30\sqrt{5x}}{5x}$$

$$=\frac{6\sqrt{5x}}{x}$$

25. $y=kxz^2$

$36=k(10)(3)^2$

$36=90k$

$k=\frac{36}{90}=\frac{2}{5}$

$y=\frac{2}{5}xz^2$

$=\frac{2}{5}(12)(4)^2$

$=\frac{384}{5}$

$=76.8$

Chapter 6 Matrices and Determinants and Applications

Section 6.1 Solving Systems of Linear Equations Using Matrices

1. $x-5y=21$
$x-5(-4)=21$
$x+20=21$
$x=1$ $\{(1,-4)\}$

3. $y+4z=11$
$y+4(2)=11$
$y+8=11$
$y=3$
$x+2y-3z=4$
$x+2(3)-3(2)=4$
$x+6-6=4$
$x=4$ $\{(4,3,2)\}$

5. matrix

7. Elementary

9. Interchange rows 2 and 3.

11. Multiply row 1 by 3 and replace the original row 1 with the result.

13. Add 3 times row 1 to row 2 and replace the original row 2 with the result.

15. $\left[\begin{array}{ccc|c} -3 & 2 & -1 & 4 \\ 8 & 0 & 4 & 12 \\ 0 & 2 & -5 & 1 \end{array}\right]$

17. $4(x-2y)=6y+2 \rightarrow 4x-14y=2$
$3x=5y+7 \rightarrow 3x-5y=7$
$\left[\begin{array}{cc|c} 4 & -14 & 2 \\ 3 & -5 & 7 \end{array}\right]$

19. $\left[\begin{array}{ccc|c} 1 & 0 & 0 & 2 \\ 0 & 1 & 0 & \frac{6}{7} \\ 0 & 0 & 1 & 12 \end{array}\right]$

21. $-4x+6y=11$
$-3x+9y=1$

23. $x+4y+3z=8$
$y+2z=12$
$z=6$

25. $x=8$
$y=-9$
$z=\frac{3}{2}$

27. $\left[\begin{array}{cc|c} 1 & 4 & 2 \\ -3 & 6 & 6 \end{array}\right] \xrightarrow[\text{Interchange rows 1 and 2.}]{R_1 \Leftrightarrow R_2} \left[\begin{array}{cc|c} -3 & 6 & 6 \\ 1 & 4 & 2 \end{array}\right]$

29. $\left[\begin{array}{cc|c} 1 & 4 & 2 \\ -3 & 6 & 6 \end{array}\right] \xrightarrow[\text{Multiply row 1 by 3.}]{3R_1 \Leftrightarrow R_1} \left[\begin{array}{cc|c} 3 & 12 & 6 \\ -3 & 6 & 6 \end{array}\right]$

31. $\left[\begin{array}{cc|c} 1 & 4 & 2 \\ -3 & 6 & 6 \end{array}\right] \xrightarrow[\text{Add } \frac{1}{3} \text{ times row 2 to row 1.}]{\frac{1}{3}R_2 + R_1 \Leftrightarrow R_1} \left[\begin{array}{cc|c} 0 & 6 & 4 \\ -3 & 6 & 6 \end{array}\right]$

33. $\left[\begin{array}{ccc|c} 1 & 5 & 6 & 2 \\ 2 & 1 & 5 & 1 \\ 4 & -2 & -3 & 10 \end{array}\right] \xrightarrow[\text{Interchange rows 2 and 3.}]{R_2 \Leftrightarrow R_3} \left[\begin{array}{ccc|c} 1 & 5 & 6 & 2 \\ 4 & -2 & -3 & 10 \\ 2 & 1 & 5 & 1 \end{array}\right]$

35. $\left[\begin{array}{ccc|c} 1 & 5 & 6 & 2 \\ 2 & 1 & 5 & 1 \\ 4 & -2 & -3 & 10 \end{array}\right] \xrightarrow[\text{Multiply row 3 by } \frac{1}{4}.]{\frac{1}{4}R_3 \Leftrightarrow R_3} \left[\begin{array}{ccc|c} 1 & 5 & 6 & 2 \\ 2 & 1 & 5 & 1 \\ 1 & -\frac{1}{2} & -\frac{3}{4} & \frac{5}{2} \end{array}\right]$

37. $\left[\begin{array}{ccc|c} 1 & 5 & 6 & 2 \\ 2 & 1 & 5 & 1 \\ 4 & -2 & -3 & 10 \end{array}\right]$

$\xrightarrow[\text{Add } -2 \text{ times row 1 to row 2.}]{-2R_1 + R_2 \Leftrightarrow R_2} \left[\begin{array}{ccc|c} 1 & 5 & 6 & 2 \\ 0 & -9 & -7 & -3 \\ 4 & -2 & -3 & 10 \end{array}\right]$

39. No; The element on the main diagonal in the second row is not 1.

41. Yes

43. No; The elements on the main diagonal are not 1 with zeros above and below.

45. Yes

47. Augmented matrix: $\left[\begin{array}{cc|c} 2 & 3 & -13 \\ 1 & 4 & -14 \end{array}\right]$

$R_1 \Leftrightarrow R_2 \longrightarrow \left[\begin{array}{cc|c} 1 & 4 & -14 \\ 2 & 3 & -13 \end{array}\right]$

$-2R_1 + R_2 \to R_2 \longrightarrow \left[\begin{array}{cc|c} 1 & 4 & -14 \\ 0 & -5 & 15 \end{array}\right]$

$-\frac{1}{5}R_2 \to R_2 \longrightarrow \left[\begin{array}{cc|c} 1 & 4 & -14 \\ 0 & 1 & -3 \end{array}\right]$

$-4R_2 + R_1 \to R_1 \longrightarrow \left[\begin{array}{cc|c} 1 & 0 & -2 \\ 0 & 1 & -3 \end{array}\right]$

$\{(-2, -3)\}$

49. Augmented matrix: $\left[\begin{array}{cc|c} 2 & -7 & -41 \\ 3 & -9 & -51 \end{array}\right]$

$-R_1 + R_2 \to R_1 \longrightarrow \left[\begin{array}{cc|c} 1 & -2 & -10 \\ 3 & -9 & -51 \end{array}\right]$

$-3R_1 + R_2 \to R_2 \longrightarrow \left[\begin{array}{cc|c} 1 & -2 & -10 \\ 0 & -3 & -21 \end{array}\right]$

$-\frac{1}{3}R_2 \to R_2 \longrightarrow \left[\begin{array}{cc|c} 1 & -2 & -10 \\ 0 & 1 & 7 \end{array}\right]$

$2R_2 + R_1 \to R_1 \longrightarrow \left[\begin{array}{cc|c} 1 & 0 & 4 \\ 0 & 1 & 7 \end{array}\right]$ $\{(4, 7)\}$

51. $-3(x-6y) = -167 - y \to -3x + 19y = -167$

$14y = 2x - 122 \to -2x + 14y = -122$

Augmented matrix: $\left[\begin{array}{cc|c} -3 & 19 & -167 \\ -2 & 14 & -122 \end{array}\right]$

$-R_1 + R_2 \to R_1 \longrightarrow \left[\begin{array}{cc|c} 1 & -5 & 45 \\ -2 & 14 & -122 \end{array}\right]$

$2R_1 + R_2 \to R_2 \longrightarrow \left[\begin{array}{cc|c} 1 & -5 & 45 \\ 0 & 4 & -32 \end{array}\right]$

$\frac{1}{4}R_2 \to R_2 \longrightarrow \left[\begin{array}{cc|c} 1 & -5 & 45 \\ 0 & 1 & -8 \end{array}\right]$

$5R_2 + R_1 \to R_1 \longrightarrow \left[\begin{array}{cc|c} 1 & 0 & 5 \\ 0 & 1 & -8 \end{array}\right]$

$\{(5, -8)\}$

53. Augmented matrix: $\left[\begin{array}{ccc|c} 3 & 7 & 22 & 83 \\ 1 & 3 & 10 & 37 \\ -2 & -5 & -18 & -66 \end{array}\right]$

$R_1 \Leftrightarrow R_2 \longrightarrow \left[\begin{array}{ccc|c} 1 & 3 & 10 & 37 \\ 3 & 7 & 22 & 83 \\ -2 & -5 & -18 & -66 \end{array}\right]$

$\begin{array}{l} -3R_1 + R_2 \to R_2 \longrightarrow \\ 2R_1 + R_3 \to R_3 \longrightarrow \end{array} \left[\begin{array}{ccc|c} 1 & 3 & 10 & 37 \\ 0 & -2 & -8 & -28 \\ 0 & 1 & 2 & 8 \end{array}\right]$

$R_2 \Leftrightarrow R_3 \longrightarrow \left[\begin{array}{ccc|c} 1 & 3 & 10 & 37 \\ 0 & 1 & 2 & 8 \\ 0 & -2 & -8 & -28 \end{array}\right]$

$\begin{array}{l} -3R_2 + R_1 \to R_1 \longrightarrow \\ 2R_2 + R_3 \to R_3 \longrightarrow \end{array} \left[\begin{array}{ccc|c} 1 & 0 & 4 & 13 \\ 0 & 1 & 2 & 8 \\ 0 & 0 & -4 & -12 \end{array}\right]$

$-\frac{1}{4}R_3 \to R_3 \longrightarrow \left[\begin{array}{ccc|c} 1 & 0 & 4 & 13 \\ 0 & 1 & 2 & 8 \\ 0 & 0 & 1 & 3 \end{array}\right]$

$\begin{array}{l} -4R_3 + R_1 \to R_1 \longrightarrow \\ -2R_3 + R_2 \to R_2 \longrightarrow \end{array} \left[\begin{array}{ccc|c} 1 & 0 & 0 & 1 \\ 0 & 1 & 0 & 2 \\ 0 & 0 & 1 & 3 \end{array}\right]$

$\{(1, 2, 3)\}$

55. Augmented matrix: $\left[\begin{array}{ccc|c} -2 & 4 & 1 & 7 \\ 4 & -13 & 10 & 17 \\ 3 & -9 & 6 & 9 \end{array}\right]$

$R_1 + R_3 \to R_1 \longrightarrow \left[\begin{array}{ccc|c} 1 & -5 & 7 & 16 \\ 4 & -13 & 10 & 17 \\ 3 & -9 & 6 & 9 \end{array}\right]$

$\begin{array}{l} -4R_1 + R_2 \to R_2 \longrightarrow \\ -3R_1 + R_3 \to R_3 \longrightarrow \end{array} \left[\begin{array}{ccc|c} 1 & -5 & 7 & 16 \\ 0 & 7 & -18 & -47 \\ 0 & 6 & -15 & -39 \end{array}\right]$

$-R_3 + R_2 \to R_2 \longrightarrow \left[\begin{array}{ccc|c} 1 & -5 & 7 & 16 \\ 0 & 1 & -3 & -8 \\ 0 & 6 & -15 & -39 \end{array}\right]$

$$\begin{array}{r} 5R_2+R_1 \to R_1 \longrightarrow \\ \\ -6R_2+R_3 \to R_3 \longrightarrow \end{array} \left[\begin{array}{ccc|c} 1 & 0 & -8 & -24 \\ 0 & 1 & -3 & -8 \\ 0 & 0 & 3 & 9 \end{array}\right]$$

$$\tfrac{1}{3}R_3 \to R_3 \longrightarrow \left[\begin{array}{ccc|c} 1 & 0 & -8 & -24 \\ 0 & 1 & -3 & -8 \\ 0 & 0 & 1 & 3 \end{array}\right]$$

$$\begin{array}{r} 8R_3+R_1 \to R_1 \longrightarrow \\ 3R_3+R_2 \to R_2 \longrightarrow \\ \\ \end{array} \left[\begin{array}{ccc|c} 1 & 0 & 0 & 0 \\ 0 & 1 & 0 & 1 \\ 0 & 0 & 1 & 3 \end{array}\right]$$

$\{(0, 1, 3)\}$

57. Augmented matrix: $\left[\begin{array}{ccc|c} -3 & 4 & -15 & -44 \\ 1 & -1 & 4 & 13 \\ 1 & -3 & 14 & 27 \end{array}\right]$

$$R_1 \Leftrightarrow R_2 \longrightarrow \left[\begin{array}{ccc|c} 1 & -1 & 4 & 13 \\ -3 & 4 & -15 & -44 \\ 1 & -3 & 14 & 27 \end{array}\right]$$

$$\begin{array}{r} \\ 3R_1+R_2 \to R_2 \longrightarrow \\ -R_1+R_3 \to R_3 \longrightarrow \end{array} \left[\begin{array}{ccc|c} 1 & -1 & 4 & 13 \\ 0 & 1 & -3 & -5 \\ 0 & -2 & 10 & 14 \end{array}\right]$$

$$\begin{array}{r} R_2+R_1 \to R_1 \longrightarrow \\ \\ 2R_2+R_3 \to R_3 \longrightarrow \end{array} \left[\begin{array}{ccc|c} 1 & 0 & 1 & 8 \\ 0 & 1 & -3 & -5 \\ 0 & 0 & 4 & 4 \end{array}\right]$$

$$\tfrac{1}{4}R_3 \to R_3 \longrightarrow \left[\begin{array}{ccc|c} 1 & 0 & 1 & 8 \\ 0 & 1 & -3 & -5 \\ 0 & 0 & 1 & 1 \end{array}\right]$$

$$\begin{array}{r} -R_3+R_1 \to R_1 \longrightarrow \\ 3R_3+R_2 \to R_2 \longrightarrow \\ \\ \end{array} \left[\begin{array}{ccc|c} 1 & 0 & 0 & 7 \\ 0 & 1 & 0 & -2 \\ 0 & 0 & 1 & 1 \end{array}\right]$$

$\{(7, -2, 1)\}$

59. $2x+8z = 7y-46 \to 2x-7y+8z = -46$

$x = 3y-3z-18 \quad \to x-3y+3z = -18$

$6z = 5y-x-34 \quad \to x-5y+6z = -34$

Augmented matrix: $\left[\begin{array}{ccc|c} 2 & -7 & 8 & -46 \\ 1 & -3 & 3 & -18 \\ 1 & -5 & 6 & -34 \end{array}\right]$

$$R_1 \Leftrightarrow R_2 \longrightarrow \left[\begin{array}{ccc|c} 1 & -3 & 3 & -18 \\ 2 & -7 & 8 & -46 \\ 1 & -5 & 6 & -34 \end{array}\right]$$

$$\begin{array}{r} \\ -2R_1+R_2 \to R_2 \longrightarrow \\ -R_1+R_3 \to R_3 \longrightarrow \end{array} \left[\begin{array}{ccc|c} 1 & -3 & 3 & -18 \\ 0 & -1 & 2 & -10 \\ 0 & -2 & 3 & -16 \end{array}\right]$$

$$-R_2 \to R_2 \longrightarrow \left[\begin{array}{ccc|c} 1 & -3 & 3 & -18 \\ 0 & 1 & -2 & 10 \\ 0 & -2 & 3 & -16 \end{array}\right]$$

$$\begin{array}{r} 3R_2+R_1 \to R_1 \longrightarrow \\ \\ 2R_2+R_3 \to R_3 \longrightarrow \end{array} \left[\begin{array}{ccc|c} 1 & 0 & -3 & 12 \\ 0 & 1 & -2 & 10 \\ 0 & 0 & -1 & 4 \end{array}\right]$$

$$-R_3 \to R_3 \longrightarrow \left[\begin{array}{ccc|c} 1 & 0 & -3 & 12 \\ 0 & 1 & -2 & 10 \\ 0 & 0 & 1 & -4 \end{array}\right]$$

$$\begin{array}{r} 3R_3+R_1 \to R_1 \longrightarrow \\ 2R_3+R_2 \to R_2 \longrightarrow \\ \\ \end{array} \left[\begin{array}{ccc|c} 1 & 0 & 0 & 0 \\ 0 & 1 & 0 & 2 \\ 0 & 0 & 1 & -4 \end{array}\right]$$

$\{(0, 2, -4)\}$

61. $11y+65 = 3x+13z \quad \to -3x+11y-13z = -65$

$x+3z = 3y+15 \quad \to x-3y+3z = 15$

$-2x+4y-7z = -25 \to -2x+4y-7z = -25$

Augmented matrix: $\left[\begin{array}{ccc|c} -3 & 11 & -13 & -65 \\ 1 & -3 & 3 & 15 \\ -2 & 4 & -7 & -25 \end{array}\right]$

$$R_1 \Leftrightarrow R_2 \longrightarrow \left[\begin{array}{ccc|c} 1 & -3 & 3 & 15 \\ -3 & 11 & -13 & -65 \\ -2 & 4 & -7 & -25 \end{array}\right]$$

$$\begin{array}{r} \\ 3R_1+R_2 \to R_2 \longrightarrow \\ 2R_1+R_3 \to R_3 \longrightarrow \end{array} \left[\begin{array}{ccc|c} 1 & -3 & 3 & 15 \\ 0 & 2 & -4 & -20 \\ 0 & -2 & -1 & 5 \end{array}\right]$$

$$\tfrac{1}{2}R_2 \to R_2 \longrightarrow \left[\begin{array}{ccc|c} 1 & -3 & 3 & 15 \\ 0 & 1 & -2 & -10 \\ 0 & -2 & -1 & 5 \end{array}\right]$$

$$\begin{array}{l} 3R_2+R_1 \to R_1 \longrightarrow \\ \\ 2R_2+R_3 \to R_3 \longrightarrow \end{array} \left[\begin{array}{ccc|c} 1 & 0 & -3 & -15 \\ 0 & 1 & -2 & -10 \\ 0 & 0 & -5 & -15 \end{array}\right]$$

$$-\tfrac{1}{5}R_3 \to R_3 \longrightarrow \left[\begin{array}{ccc|c} 1 & 0 & -3 & -15 \\ 0 & 1 & -2 & -10 \\ 0 & 0 & 1 & 3 \end{array}\right]$$

$$\begin{array}{l} 3R_3+R_1 \to R_1 \longrightarrow \\ 2R_3+R_2 \to R_2 \longrightarrow \\ \\ \end{array} \left[\begin{array}{ccc|c} 1 & 0 & 0 & -6 \\ 0 & 1 & 0 & -4 \\ 0 & 0 & 1 & 3 \end{array}\right]$$

$\{(-6, -4, 3)\}$

63. Augmented matrix: $\left[\begin{array}{cccc|c} 1 & 3 & 0 & -3 & -5 \\ 0 & 1 & 0 & -2 & -6 \\ -2 & -4 & 1 & 2 & 1 \\ 0 & 1 & 1 & 0 & 5 \end{array}\right]$

$$2R_1+R_3 \to R_3 \longrightarrow \left[\begin{array}{cccc|c} 1 & 3 & 0 & -3 & -5 \\ 0 & 1 & 0 & -2 & -6 \\ 0 & 2 & 1 & -4 & -9 \\ 0 & 1 & 1 & 0 & 5 \end{array}\right]$$

$$\begin{array}{l} -3R_2+R_1 \to R_1 \longrightarrow \\ \\ -2R_2+R_3 \to R_3 \longrightarrow \\ -R_2+R_4 \to R_4 \longrightarrow \end{array} \left[\begin{array}{cccc|c} 1 & 0 & 0 & 3 & 13 \\ 0 & 1 & 0 & -2 & -6 \\ 0 & 0 & 1 & 0 & 3 \\ 0 & 0 & 1 & 2 & 11 \end{array}\right]$$

$$-R_3+R_4 \to R_4 \longrightarrow \left[\begin{array}{cccc|c} 1 & 0 & 0 & 3 & 13 \\ 0 & 1 & 0 & -2 & -6 \\ 0 & 0 & 1 & 0 & 3 \\ 0 & 0 & 0 & 2 & 8 \end{array}\right]$$

$$\tfrac{1}{2}R_4 \to R_4 \longrightarrow \left[\begin{array}{cccc|c} 1 & 0 & 0 & 3 & 13 \\ 0 & 1 & 0 & -2 & -6 \\ 0 & 0 & 1 & 0 & 3 \\ 0 & 0 & 0 & 1 & 4 \end{array}\right]$$

$$\begin{array}{l} -3R_4+R_1 \to R_1 \longrightarrow \\ 2R_4+R_2 \to R_2 \longrightarrow \\ \\ \\ \end{array} \left[\begin{array}{cccc|c} 1 & 0 & 0 & 0 & 1 \\ 0 & 1 & 0 & 0 & 2 \\ 0 & 0 & 1 & 0 & 3 \\ 0 & 0 & 0 & 1 & 4 \end{array}\right]$$

$\{(1, 2, 3, 4)\}$

65. Augmented matrix: $\left[\begin{array}{cccc|c} 1 & 1 & 0 & 5 & -4 \\ 0 & 1 & 0 & 2 & 3 \\ 0 & -2 & 1 & -3 & -5 \\ 3 & 3 & 0 & 17 & -10 \end{array}\right]$

$$-3R_1+R_4 \to R_4 \longrightarrow \left[\begin{array}{cccc|c} 1 & 1 & 0 & 5 & -4 \\ 0 & 1 & 0 & 2 & 3 \\ 0 & -2 & 1 & -3 & -5 \\ 0 & 0 & 0 & 2 & 2 \end{array}\right]$$

$$\begin{array}{l} -R_2+R_1 \to R_1 \longrightarrow \\ \\ 2R_2+R_3 \to R_3 \longrightarrow \\ \\ \end{array} \left[\begin{array}{cccc|c} 1 & 0 & 0 & 3 & -7 \\ 0 & 1 & 0 & 2 & 3 \\ 0 & 0 & 1 & 1 & 1 \\ 0 & 0 & 0 & 2 & 2 \end{array}\right]$$

$$\tfrac{1}{2}R_4 \to R_4 \longrightarrow \left[\begin{array}{cccc|c} 1 & 0 & 0 & 3 & -7 \\ 0 & 1 & 0 & 2 & 3 \\ 0 & 0 & 1 & 1 & 1 \\ 0 & 0 & 0 & 1 & 1 \end{array}\right]$$

$$\begin{array}{l} -3R_4+R_1 \to R_1 \longrightarrow \\ -2R_4+R_2 \to R_2 \longrightarrow \\ -R_4+R_3 \to R_3 \longrightarrow \\ \\ \end{array} \left[\begin{array}{cccc|c} 1 & 0 & 0 & 0 & -10 \\ 0 & 1 & 0 & 0 & 1 \\ 0 & 0 & 1 & 0 & 0 \\ 0 & 0 & 0 & 1 & 1 \end{array}\right]$$

$\{(-10, 1, 0, 1)\}$

67. Let x represent the amount Andre borrowed from his parents. Let y represent the amount he borrowed from the credit union. Let z represent the amount he borrowed from the bank.

$x + y + z = 20{,}000$

$0.02x + 0.04y + 0.05z = 620 \to$

$2x + 4y + 5z = 62{,}000$

$x = 5z \to x - 5z = 0$

Augmented matrix: $\left[\begin{array}{ccc|c} 1 & 1 & 1 & 20{,}000 \\ 2 & 4 & 5 & 62{,}000 \\ 1 & 0 & -5 & 0 \end{array}\right]$

$$\begin{array}{l} \\ -2R_1+R_2 \to R_2 \longrightarrow \\ -R_1+R_3 \to R_3 \longrightarrow \end{array} \left[\begin{array}{ccc|c} 1 & 1 & 1 & 20{,}000 \\ 0 & 2 & 3 & 22{,}000 \\ 0 & -1 & -6 & -20{,}000 \end{array}\right]$$

$$R_3+R_2 \to R_2 \longrightarrow \left[\begin{array}{ccc|c} 1 & 1 & 1 & 20{,}000 \\ 0 & 1 & -3 & 2000 \\ 0 & -1 & -6 & -20{,}000 \end{array}\right]$$

$$\begin{array}{l} -R_2+R_1 \to R_1 \longrightarrow \\ \\ R_2+R_3 \to R_3 \longrightarrow \end{array} \left[\begin{array}{ccc|c} 1 & 0 & 4 & 18{,}000 \\ 0 & 1 & -3 & 2000 \\ 0 & 0 & -9 & -18{,}000 \end{array}\right]$$

$$-\tfrac{1}{9}R_3 \to R_3 \longrightarrow \left[\begin{array}{ccc|c} 1 & 0 & 4 & 18{,}000 \\ 0 & 1 & -3 & 2000 \\ 0 & 0 & 1 & 2000 \end{array}\right]$$

$$\begin{array}{l} -4R_3+R_1 \to R_1 \longrightarrow \\ 3R_3+R_2 \to R_2 \longrightarrow \end{array} \left[\begin{array}{ccc|c} 1 & 0 & 0 & 10{,}000 \\ 0 & 1 & 0 & 8000 \\ 0 & 0 & 1 & 2000 \end{array}\right]$$

He borrowed \$10,000 from his parents, \$8000 from the credit union, and \$2000 from the bank.

69. Let x represent the number of days Danielle stayed in Washington, D.C. Let y represent number of days she stayed in Atlanta. Let z represent the number of days she stayed in Dallas.

$x+y+z=14$

$200x+100y+150z=2200 \to$

$20x+10y+15z=220$

$z=2x \to -2x+z=0$

Augmented matrix: $\left[\begin{array}{ccc|c} 1 & 1 & 1 & 14 \\ 20 & 10 & 15 & 220 \\ -2 & 0 & 1 & 0 \end{array}\right]$

$$\begin{array}{l} \\ -20R_1+R_2 \to R_2 \longrightarrow \\ 2R_1+R_3 \to R_3 \longrightarrow \end{array} \left[\begin{array}{ccc|c} 1 & 1 & 1 & 14 \\ 0 & -10 & -5 & -60 \\ 0 & 2 & 3 & 28 \end{array}\right]$$

$$-\tfrac{1}{10}R_2 \to R_2 \longrightarrow \left[\begin{array}{ccc|c} 1 & 1 & 1 & 14 \\ 0 & 1 & \frac{1}{2} & 6 \\ 0 & 2 & 3 & 28 \end{array}\right]$$

$$\begin{array}{l} -R_2+R_1 \to R_1 \longrightarrow \\ \\ -2R_2+R_3 \to R_3 \longrightarrow \end{array} \left[\begin{array}{ccc|c} 1 & 0 & \frac{1}{2} & 8 \\ 0 & 1 & \frac{1}{2} & 6 \\ 0 & 0 & 2 & 16 \end{array}\right]$$

$$\tfrac{1}{2}R_3 \to R_3 \longrightarrow \left[\begin{array}{ccc|c} 1 & 0 & \frac{1}{2} & 8 \\ 0 & 1 & \frac{1}{2} & 6 \\ 0 & 0 & 1 & 8 \end{array}\right]$$

$$\begin{array}{l} -\tfrac{1}{2}R_3+R_1 \to R_1 \longrightarrow \\ -\tfrac{1}{2}R_3+R_2 \to R_2 \longrightarrow \end{array} \left[\begin{array}{ccc|c} 1 & 0 & 0 & 4 \\ 0 & 1 & 0 & 2 \\ 0 & 0 & 1 & 8 \end{array}\right]$$

She spent 4 nights in Washington, 2 nights in Atlanta, and 8 nights in Dallas.

71.
$$\frac{5x^2-6x-13}{(x+3)(x-2)^2}=\frac{A}{x+3}+\frac{B}{x-2}+\frac{C}{(x-2)^2}$$

$$(x+3)(x-2)^2\cdot\left[\frac{5x^2-6x-13}{(x+3)(x-2)^2}\right]=(x+3)(x-2)^2\cdot\left[\frac{A}{x+3}+\frac{B}{x-2}+\frac{C}{(x-2)^2}\right]$$

$$5x^2-6x-13=A(x-2)^2+B(x+3)(x-2)+C(x+3)$$

$$5x^2-6x-13=A(x^2-4x+4)+B(x^2+x-6)+Cx+3C$$

$$5x^2-6x-13=Ax^2-4Ax+4A+Bx^2+Bx-6B+Cx+3C$$

$$5x^2-6x-13=(A+B)x^2+(-4A+B+C)x+(4A-6B+3C)$$

Augmented matrix: $\left[\begin{array}{ccc|c} 1 & 1 & 0 & 5 \\ -4 & 1 & 1 & -6 \\ 4 & -6 & 3 & -13 \end{array}\right] \Rightarrow \begin{array}{l} \\ 4R_1+R_2 \to R_2 \longrightarrow \\ -4R_1+R_3 \to R_3 \longrightarrow \end{array} \left[\begin{array}{ccc|c} 1 & 1 & 0 & 5 \\ 0 & 5 & 1 & 14 \\ 0 & -10 & 3 & -33 \end{array}\right]$

$$\frac{1}{5}R_3 \to R_3 \longrightarrow \left[\begin{array}{ccc|c} 1 & 1 & 0 & 5 \\ 0 & 1 & \frac{1}{5} & \frac{14}{5} \\ 0 & -10 & 3 & -33 \end{array}\right] \Rightarrow \begin{array}{l} -R_2 + R_1 \to R_1 \longrightarrow \\ \\ 10R_2 + R_3 \to R_3 \longrightarrow \end{array} \left[\begin{array}{ccc|c} 1 & 0 & -\frac{1}{5} & \frac{11}{5} \\ 0 & 1 & \frac{1}{5} & \frac{14}{5} \\ 0 & 0 & 5 & -5 \end{array}\right]$$

$$\frac{1}{5}R_3 \to R_3 \longrightarrow \left[\begin{array}{ccc|c} 1 & 0 & -\frac{1}{5} & \frac{11}{5} \\ 0 & 1 & \frac{1}{5} & \frac{14}{5} \\ 0 & 0 & 1 & -1 \end{array}\right] \Rightarrow \begin{array}{l} \frac{1}{5}R_3 + R_1 \to R_1 \longrightarrow \\ -\frac{1}{5}R_3 + R_2 \to R_2 \longrightarrow \\ \end{array} \left[\begin{array}{ccc|c} 1 & 0 & -\frac{1}{5} & 2 \\ 0 & 1 & 0 & 3 \\ 0 & 0 & 1 & -1 \end{array}\right]$$

$$\frac{5x^2 - 6x - 13}{(x+3)(x-2)^2} = \frac{2}{x+3} + \frac{3}{x-2} + \frac{-1}{(x-2)^2} = \frac{2}{x+3} + \frac{3}{x-2} - \frac{1}{(x-2)^2}$$

73. Interchanging two rows in an augmented matrix represents interchanging two equations in a system of equations. This operation does not affect the solution set of the system.

75. Reduced row-echelon form is the same format as row-echelon form with the added condition that all elements above the leading 1's must be 0's.

77. $\{(9.32, -17.48, 12.93)\}$

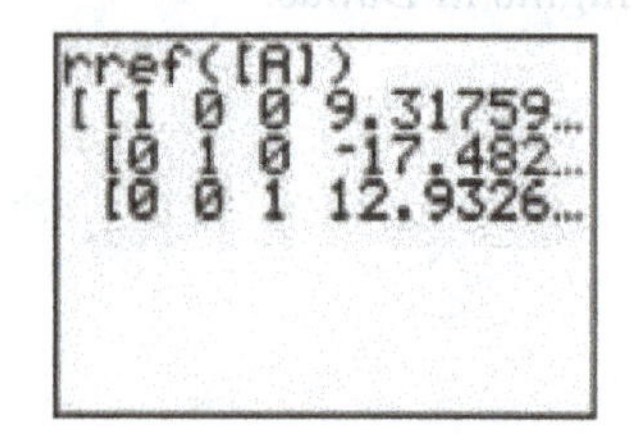

79. a. System of equations:

$2400a + 800b + c \quad 36{,}000$

$2000a + 500b + c \quad 30{,}000$

$3000a + 1000b + c = 44{,}000$

b.

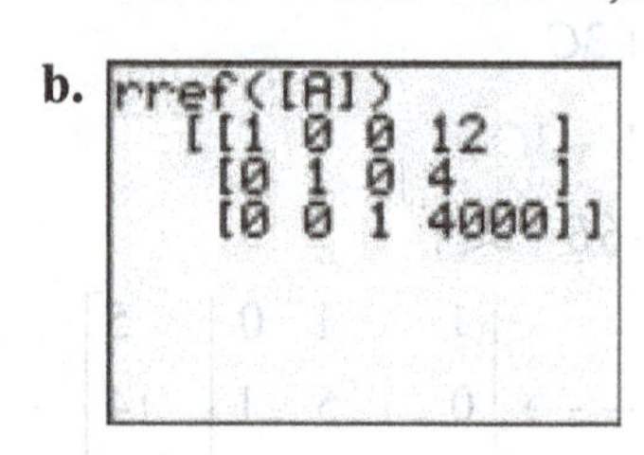

c. $y = 12x_1 + 4x_2 + 4000$

d. $y = 12x_1 + 4x_2 + 4000$

$= 12(2500) + 4(500) + 4000$

$= 30{,}000 + 2000 + 4000$

$= 36{,}000$

$36,000

81. a. $f(x) = ax^2 + bx + c$

$f(-3) = a(-3)^2 + b(-3) + c$

$= 9a - 3b + c = -7.28$

$f(-1) = a(-1)^2 + b(-1) + c$

$= a - b + c = 3.68$

$f(10) = a(10)^2 + b(10) + c$

$= 100a + 10b + c = 18.2$

System of equations:

$9a - 3b + c \quad -7.28$

$a - b + c \quad 3.68$

$100a + 10b + c = 18.2$

b.

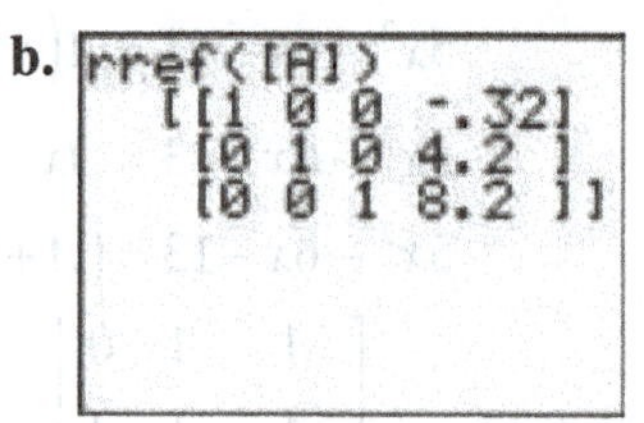

c. $f(x) = -0.32x^2 + 4.2x + 8.2$

Section 6.2 Inconsistent Systems and Dependent Equations

1. Yes

3. $3(x-2y)=10-y \rightarrow 3x-5y=10$

$\frac{3}{10}x=\frac{1}{2}y+1 \quad \rightarrow \frac{3}{10}x-\frac{1}{2}y=1$

$$\begin{aligned} 3x-5y=10 &\longrightarrow 3x-5y=10 \\ \frac{3}{10}x-\frac{1}{2}y=1 &\xrightarrow[\text{by } -10.]{\text{Multiply}} -3x+5y=-10 \\ & \qquad 0=0 \end{aligned}$$

$\{(x, y) \mid 3x-5y=10\}$

5. $y=-1.5x+10$

$$\begin{aligned} 0.03x+0.02y&=0.6 \\ 3x+2y&=60 \\ 3x+2(-1.5x+10)&=60 \\ 3x-3x+20&=60 \\ 20&=60 \end{aligned}$$

$\{\ \}$

7. True

9. False

11. inconsistent

13. The system represents a contradiction: $0=5$. It has no solution.

15. The system represents a system of two equations and three unknowns It has infinitely many solutions.

17. By inspection, the system is in reduced row-echelon form and has one solution.

19. The system represents a system of one equation and three unknowns. It has infinitely many solutions.

21. a. Augmented matrix: $\left[\begin{array}{cc|c} 1 & 2 & 5 \\ 0 & 1 & 0 \end{array}\right]$

$-2R_2+R_1 \rightarrow R_1 \longrightarrow \left[\begin{array}{cc|c} 1 & 0 & 5 \\ 0 & 1 & 0 \end{array}\right]$

$\{(5, 0)\}$

b. The system represents a system of one equation and two unknowns. It has infinitely many solutions.

$\{(x, y) \mid x+2y=5\}$

c. The system represents a contradiction: $0=1$. It has no solution.

$\{\ \}$

23. a. Augmented matrix: $\left[\begin{array}{ccc|c} 1 & 0 & 6 & 3 \\ 0 & 1 & 4 & 5 \\ 0 & 0 & 1 & 0 \end{array}\right]$

$\begin{array}{l} -6R_3+R_1 \rightarrow R_1 \longrightarrow \\ -4R_3+R_2 \rightarrow R_2 \longrightarrow \end{array} \left[\begin{array}{ccc|c} 1 & 0 & 0 & 3 \\ 0 & 1 & 0 & 5 \\ 0 & 0 & 1 & 0 \end{array}\right]$

$\{(3, 5, 0)\}$

b. The system represents a contradiction: $0=1$. It has no solution.

$\{\ \}$

c. The system represents a system of two equation and three unknowns. It has infinitely many solutions.

$x+6z=3 \rightarrow x=-6z+3$

$y+4z=5 \rightarrow y=-4z+5$

$\{(-6z+3, -4z+5, z) \mid z \text{ is any real number}\}$

25. Augmented matrix: $\left[\begin{array}{cc|c} 2 & 4 & 5 \\ 1 & 2 & 4 \end{array}\right]$

$R_1 \Leftrightarrow R_2 \longrightarrow \left[\begin{array}{cc|c} 1 & 2 & 4 \\ 2 & 4 & 5 \end{array}\right]$

$-2R_1+R_2 \rightarrow R_2 \longrightarrow \left[\begin{array}{cc|c} 1 & 2 & 4 \\ 0 & 0 & -3 \end{array}\right]$

$\{\ \}$

27. $2x+7y=10 \quad \rightarrow 2x+7y=10$

$\frac{1}{5}x=1-\frac{7}{10}y \rightarrow \frac{1}{5}x+\frac{7}{10}y=1$

Augmented matrix: $\left[\begin{array}{cc|c} 2 & 7 & 10 \\ \frac{1}{5} & \frac{7}{10} & 1 \end{array}\right]$

$$\tfrac{1}{2}R_1 \to R_1 \longrightarrow \left[\begin{array}{cc|c} 1 & \frac{7}{2} & 5 \\ \frac{1}{5} & \frac{7}{10} & 1 \end{array}\right]$$

$$-\tfrac{1}{5}R_1 + R_2 \to R_2 \longrightarrow \left[\begin{array}{cc|c} 1 & \frac{7}{2} & 5 \\ 0 & 0 & 0 \end{array}\right]$$

$\{(x, y) \mid 2x + 7y = 10\}$

29. Augmented matrix: $\left[\begin{array}{ccc|c} 1 & -3 & 14 & -9 \\ -2 & 7 & -31 & 21 \\ 1 & -5 & 20 & -14 \end{array}\right]$

$$\begin{array}{l} 2R_1 + R_2 \to R_2 \longrightarrow \\ -R_1 + R_3 \to R_3 \longrightarrow \end{array} \left[\begin{array}{ccc|c} 1 & -3 & 14 & -9 \\ 0 & 1 & -3 & 3 \\ 0 & -2 & 6 & -5 \end{array}\right]$$

$$\begin{array}{l} 3R_2 + R_1 \to R_1 \longrightarrow \\ 2R_2 + R_3 \to R_3 \longrightarrow \end{array} \left[\begin{array}{ccc|c} 1 & 0 & 5 & 0 \\ 0 & 1 & -3 & 3 \\ 0 & 0 & 0 & 1 \end{array}\right]$$

$\{\ \}$

31. Augmented matrix: $\left[\begin{array}{ccc|c} 5 & 7 & -11 & 45 \\ 3 & 5 & -9 & 23 \\ 1 & 1 & -1 & 11 \end{array}\right]$

$$R_1 \Leftrightarrow R_3 \quad \left[\begin{array}{ccc|c} 1 & 1 & -1 & 11 \\ 3 & 5 & -9 & 23 \\ 5 & 7 & -11 & 45 \end{array}\right]$$

$$\begin{array}{l} -3R_1 + R_2 \to R_2 \longrightarrow \\ -5R_1 + R_3 \to R_3 \longrightarrow \end{array} \left[\begin{array}{ccc|c} 1 & 1 & -1 & 11 \\ 0 & 2 & -6 & -10 \\ 0 & 2 & -6 & -10 \end{array}\right]$$

$$\tfrac{1}{2}R_2 \to R_2 \longrightarrow \left[\begin{array}{ccc|c} 1 & 1 & -1 & 11 \\ 0 & 1 & -3 & -5 \\ 0 & 2 & -6 & -10 \end{array}\right]$$

$$\begin{array}{l} -R_2 + R_1 \to R_1 \longrightarrow \\ -2R_2 + R_3 \to R_3 \longrightarrow \end{array} \left[\begin{array}{ccc|c} 1 & 0 & 2 & 16 \\ 0 & 1 & -3 & -5 \\ 0 & 0 & 0 & 0 \end{array}\right]$$

$x + 2z = 16 \to x = -2z + 16$

$y - 3z = -5 \to y = 3z - 5$

$\{(-2z + 16, 3z - 5, z) \mid z \text{ is any real number}\}$

33. $2x = 5y - 16z + 40 \to 2x - 5y + 16z = 40$

$2(x + y) = 4z \to 2x + 2y - 4z = 0$

$x - 2y + 7z = 18 \to x - 2y + 7z = 18$

Augmented matrix: $\left[\begin{array}{ccc|c} 2 & -5 & 16 & 40 \\ 2 & 2 & -4 & 0 \\ 1 & -2 & 7 & 18 \end{array}\right]$

$$R_1 \Leftrightarrow R_3 \quad \left[\begin{array}{ccc|c} 1 & -2 & 7 & 18 \\ 2 & 2 & -4 & 0 \\ 2 & -5 & 16 & 40 \end{array}\right]$$

$$\begin{array}{l} -2R_1 + R_2 \to R_2 \longrightarrow \\ -2R_1 + R_3 \to R_3 \longrightarrow \end{array} \left[\begin{array}{ccc|c} 1 & -2 & 7 & 18 \\ 0 & 6 & -18 & -36 \\ 0 & -1 & 2 & 4 \end{array}\right]$$

$$\tfrac{1}{6}R_2 \to R_2 \longrightarrow \left[\begin{array}{ccc|c} 1 & -2 & 7 & 18 \\ 0 & 1 & -3 & -6 \\ 0 & -1 & 2 & 4 \end{array}\right]$$

$$\begin{array}{l} 2R_2 + R_1 \to R_1 \longrightarrow \\ R_2 + R_3 \to R_3 \longrightarrow \end{array} \left[\begin{array}{ccc|c} 1 & 0 & 1 & 6 \\ 0 & 1 & -3 & -6 \\ 0 & 0 & -1 & -2 \end{array}\right]$$

$$-R_3 \to R_3 \longrightarrow \left[\begin{array}{ccc|c} 1 & 0 & 1 & 6 \\ 0 & 1 & -3 & -6 \\ 0 & 0 & 1 & 2 \end{array}\right]$$

$$\begin{array}{l} -R_3 + R_1 \to R_1 \longrightarrow \\ 3R_3 + R_2 \to R_2 \longrightarrow \end{array} \left[\begin{array}{ccc|c} 1 & 0 & 0 & 4 \\ 0 & 1 & 0 & 0 \\ 0 & 0 & 1 & 2 \end{array}\right]$$

$\{(4, 0, 2)\}$

35. Augmented matrix: $\left[\begin{array}{ccc|c} 2 & -5 & -20 & -24 \\ 1 & -3 & -11 & -15 \end{array}\right]$

$$R_1 \Leftrightarrow R_2 \longrightarrow \left[\begin{array}{ccc|c} 1 & -3 & -11 & -15 \\ 2 & -5 & -20 & -24 \end{array}\right]$$

$$-2R_1 + R_2 \to R_2 \longrightarrow \left[\begin{array}{ccc|c} 1 & -3 & -11 & -15 \\ 0 & 1 & 2 & 6 \end{array}\right]$$

$$3R_2 + R_1 \to R_1 \longrightarrow \left[\begin{array}{ccc|c} 1 & 0 & -5 & 3 \\ 0 & 1 & 2 & 6 \end{array}\right]$$

$x - 5z = 3 \to x = 5z + 3$

$y + 2z = 6 \to y = -2z + 6$

$\{(5z + 3, -2z + 6, z) \mid z \text{ is any real number}\}$

37. Augmented matrix: $\left[\begin{array}{ccc|c} 2 & 3 & 4 & 12 \\ -4 & -6 & -8 & -24 \\ 1 & 1.5 & 2 & 6 \end{array}\right]$

$R_1 \Leftrightarrow R_3 \left[\begin{array}{ccc|c} 1 & 1.5 & 2 & 6 \\ -4 & -6 & -8 & -24 \\ 2 & 3 & 4 & 12 \end{array}\right]$

$\begin{array}{l} 4R_1 + R_2 \to R_2 \longrightarrow \\ -2R_1 + R_3 \to R_3 \longrightarrow \end{array} \left[\begin{array}{ccc|c} 1 & 1.5 & 2 & 6 \\ 0 & 0 & 0 & 0 \\ 0 & 0 & 0 & 0 \end{array}\right]$

$x + 1.5y + 2z = 6 \to 2x + 3y + 4z = 12$

$\{(x, y, z) \mid 2x + 3y + 4z = 12\}$

39. $2x - 3y + 9z = -2 \quad \to 2x - 3y + 9z = -2$

$x = 5y - 8z - 15 \quad \to x - 5y + 8z = -15$

$3(x - y) + 6z = 2x - 7 \to x - 3y + 6z = -7$

Augmented matrix: $\left[\begin{array}{ccc|c} 2 & -3 & 9 & -2 \\ 1 & -5 & 8 & -15 \\ 1 & -3 & 6 & -7 \end{array}\right]$

$R_1 \Leftrightarrow R_2 \longrightarrow \left[\begin{array}{ccc|c} 1 & -5 & 8 & -15 \\ 2 & -3 & 9 & -2 \\ 1 & -3 & 6 & -7 \end{array}\right]$

$\begin{array}{l} -2R_1 + R_2 \to R_2 \longrightarrow \\ -R_1 + R_3 \to R_3 \longrightarrow \end{array} \left[\begin{array}{ccc|c} 1 & -5 & 8 & -15 \\ 0 & 7 & -7 & 28 \\ 0 & 2 & -2 & 8 \end{array}\right]$

$\frac{1}{7}R_2 \to R_2 \longrightarrow \left[\begin{array}{ccc|c} 1 & -5 & 8 & -15 \\ 0 & 1 & -1 & 4 \\ 0 & 2 & -2 & 8 \end{array}\right]$

$\begin{array}{l} 5R_2 + R_1 \to R_1 \longrightarrow \\ -2R_2 + R_3 \to R_3 \longrightarrow \end{array} \left[\begin{array}{ccc|c} 1 & 0 & 3 & 5 \\ 0 & 1 & -1 & 4 \\ 0 & 0 & 0 & 0 \end{array}\right]$

$x + 3z = 5 \to x = -3z + 5$

$y - z = 4 \to y = z + 4$

$\{(-3z + 5, z + 4, z) \mid z \text{ is any real number}\}$

41. $x + 4z = 3y + 1 \to x - 3y + 4z = 1$

Augmented matrix: $\left[\begin{array}{ccc|c} -5 & 12 & -20 & -11 \\ 1 & -3 & 4 & 1 \end{array}\right]$

$R_1 \Leftrightarrow R_2 \longrightarrow \left[\begin{array}{ccc|c} 1 & -3 & 4 & 1 \\ -5 & 12 & -20 & -11 \end{array}\right]$

$5R_1 + R_2 \to R_2 \longrightarrow \left[\begin{array}{ccc|c} 1 & -3 & 4 & 1 \\ 0 & -3 & 0 & -6 \end{array}\right]$

$-\frac{1}{3}R_2 \to R_2 \longrightarrow \left[\begin{array}{ccc|c} 1 & -3 & 4 & 1 \\ 0 & 1 & 0 & 2 \end{array}\right]$

$3R_2 + R_1 \to R_1 \longrightarrow \left[\begin{array}{ccc|c} 1 & 0 & 4 & 7 \\ 0 & 1 & 0 & 2 \end{array}\right]$

$x + 4z = 7 \to x = -4z + 7$

$y = 2 \quad \to y = 2$

$\{(-4z + 7, 2, z) \mid z \text{ is any real number}\}$

43. Augmented matrix: $\left[\begin{array}{cccc|c} 1 & -3 & 9 & -14 & 32 \\ 0 & 1 & -3 & 6 & -10 \\ 0 & 1 & -1 & 2 & -4 \\ 1 & -2 & 8 & -12 & 24 \end{array}\right]$

$-R_1 + R_4 \to R_4 \longrightarrow \left[\begin{array}{cccc|c} 1 & -3 & 9 & -14 & 32 \\ 0 & 1 & -3 & 6 & -10 \\ 0 & 1 & -1 & 2 & -4 \\ 0 & 1 & -1 & 2 & -8 \end{array}\right]$

$\begin{array}{l} 3R_2 + R_1 \to R_1 \longrightarrow \\ \\ -R_2 + R_3 \to R_3 \longrightarrow \\ -R_2 + R_4 \to R_4 \longrightarrow \end{array} \left[\begin{array}{cccc|c} 1 & 0 & 0 & 4 & 2 \\ 0 & 1 & -3 & 6 & -10 \\ 0 & 0 & 2 & -4 & 6 \\ 0 & 0 & 2 & -4 & 2 \end{array}\right]$

$\frac{1}{2}R_3 \to R_3 \longrightarrow \left[\begin{array}{cccc|c} 1 & 0 & 0 & 4 & 2 \\ 0 & 1 & -3 & 6 & -10 \\ 0 & 0 & 1 & -2 & 3 \\ 0 & 0 & 2 & -4 & 2 \end{array}\right]$

$\begin{array}{l} \\ 3R_3 + R_2 \to R_2 \longrightarrow \\ \\ -2R_3 + R_4 \to R_4 \longrightarrow \end{array} \left[\begin{array}{cccc|c} 1 & 0 & 0 & 4 & 2 \\ 0 & 1 & 0 & 0 & -1 \\ 0 & 0 & 1 & -2 & 3 \\ 0 & 0 & 0 & 0 & -4 \end{array}\right]$

$\{ \}$

45. a. $x = 2z + 1 = 2(1) + 1 = 2 + 1 = 3$

$y = z - 4 = (1) - 4 = 1 - 4 = -3$

$z = 1 \qquad (3, -3, 1)$

b. $x = 2z + 1 = 2(4) + 1 = 8 + 1 = 9$

$y = z - 4 = (4) - 4 = 4 - 4 = 0$

$z = 4$ $\quad (9, 0, 4)$

c. $x = 2z + 1 = 2(-2) + 1 = -4 + 1 = -3$

$y = z - 4 = (-2) - 4 = -2 - 4 = -6$

$z = -2$ $\quad (-3, -6, -2)$

47. For example:

Let $z = 0$.

$x = 4z = 4(0) = 0$

$y = 6 - z = 6 - 0 = 6$

$z = 0$ $\quad (0, 6, 0)$

Let $z = 1$.

$x = 4z = 4(1) = 4$

$y = 6 - z = 6 - 1 = 5$

$z = 1$ $\quad (4, 5, 1)$

Let $z = 2$.

$x = 4z = 4(2) = 8$

$y = 6 - z = 6 - 2 = 4$

$z = 2$ $\quad (8, 4, 2)$

49. For example:

Let $x = 0$ and $y = 0$.

$2x + 3y + 6z = 6$

$2(0) + 3(0) + 6z = 6$

$6z = 6$

$z = 1$ $\quad (0, 0, 1)$

Let $x = 0$ and $z = 0$.

$2x + 3y + 6z = 6$

$2(0) + 3y + 6(0) = 6$

$3y = 6$

$y = 2$ $\quad (0, 2, 0)$

Let $y = 0$ and $z = 0$.

$2x + 3y + 6z = 6$

$2x + 3(0) + 6(0) = 6$

$2x = 6$

$x = 3$ $\quad (3, 0, 0)$

51. a. $180 + 190 = x_1 + x_3$

b. $x_1 + x_2 = 180 + 220$

c. $x_3 + 120 = x_2 + 90$

d. $180 + 190 = x_1 + x_3 \rightarrow x_1 + x_3 = 370$

$x_1 + x_2 = 180 + 220 \rightarrow x_1 + x_2 = 400$

$x_3 + 120 = x_2 + 90 \rightarrow x_2 - x_3 = 30$

e. Augmented matrix: $\left[\begin{array}{ccc|c} 1 & 0 & 1 & 370 \\ 1 & 1 & 0 & 400 \\ 0 & 1 & -1 & 30 \end{array}\right]$

$-R_1 + R_2 \rightarrow R_2 \longrightarrow \left[\begin{array}{ccc|c} 1 & 0 & 1 & 370 \\ 0 & 1 & -1 & 30 \\ 0 & 1 & -1 & 30 \end{array}\right]$

$-R_2 + R_3 \rightarrow R_3 \longrightarrow \left[\begin{array}{ccc|c} 1 & 0 & 1 & 370 \\ 0 & 1 & -1 & 30 \\ 0 & 0 & 0 & 0 \end{array}\right]$

f. $x_3 = 120$:

$x_1 + x_3 = 370$

$x_1 + 120 = 370$

$x_1 = 250$ vehicles per hour

$x_2 - x_3 = 30$

$x_2 - 120 = 30$

$x_2 = 150$ vehicles per hour

g. $x_3 = 100$:

$x_1 + x_3 = 370$

$x_1 + 100 = 370$

$x_1 = 270$ vehicles per hour

$x_2 - x_3 = 30$

$x_2 - 100 = 30$

$x_2 = 130$ vehicles per hour

$x_3 = 150$:

$x_1 + x_3 = 370$

$x_1 + 150 = 370$

$x_1 = 220$ vehicles per hour

$x_2 - x_3 = 30$

$x_2 - 150 = 30$

$x_2 = 180$ vehicles per hour

$220 \le x_1 \le 270$ vehicles per hour

$130 \le x_2 \le 180$ vehicles per hour

53. $x_4 + 242 = x_1 + 402 \rightarrow x_1 - x_4 = -160$

$x_1 + 425 = x_2 + 275 \rightarrow x_1 - x_2 = -150$

$x_2 + 357 = x_3 + 397 \rightarrow x_2 - x_3 = 40$

$x_3 + 331 = x_4 + 281 \rightarrow x_3 - x_4 = -50$

Augmented matrix: $\left[\begin{array}{cccc|c} 1 & 0 & 0 & -1 & -160 \\ 1 & -1 & 0 & 0 & -150 \\ 0 & 1 & -1 & 0 & 40 \\ 0 & 0 & 1 & -1 & -50 \end{array}\right]$

$-R_1 + R_2 \rightarrow R_2 \longrightarrow \left[\begin{array}{cccc|c} 1 & 0 & 0 & -1 & -160 \\ 0 & -1 & 0 & 1 & 10 \\ 0 & 1 & -1 & 0 & 40 \\ 0 & 0 & 1 & -1 & -50 \end{array}\right]$

$-R_2 \rightarrow R_2 \longrightarrow \left[\begin{array}{cccc|c} 1 & 0 & 0 & -1 & -160 \\ 0 & 1 & 0 & -1 & -10 \\ 0 & 1 & -1 & 0 & 40 \\ 0 & 0 & 1 & -1 & -50 \end{array}\right]$

$-R_2 + R_3 \rightarrow R_3 \longrightarrow \left[\begin{array}{cccc|c} 1 & 0 & 0 & -1 & -160 \\ 0 & 1 & 0 & -1 & -10 \\ 0 & 0 & -1 & 1 & 50 \\ 0 & 0 & 1 & -1 & -50 \end{array}\right]$

$-R_3 \rightarrow R_3 \longrightarrow \left[\begin{array}{cccc|c} 1 & 0 & 0 & -1 & -160 \\ 0 & 1 & 0 & -1 & -10 \\ 0 & 0 & 1 & -1 & -50 \\ 0 & 0 & 1 & -1 & -50 \end{array}\right]$

$-R_3 + R_4 \rightarrow R_4 \longrightarrow \left[\begin{array}{cccc|c} 1 & 0 & 0 & -1 & -160 \\ 0 & 1 & 0 & -1 & -10 \\ 0 & 0 & 1 & -1 & -50 \\ 0 & 0 & 0 & 0 & 0 \end{array}\right]$

a. $x_4 = 220$:

$x_1 - x_4 = -160$

$x_1 - 220 = -160$

$x_1 = 60$ vehicles per hour

$x_2 - x_4 = -10$

$x_2 - 220 = -10$

$x_2 = 210$ vehicles per hour

$x_3 - x_4 = -50$

$x_3 - 220 = -50$

$x_3 = 170$ vehicles per hour

b. $x_4 = 200$:

$x_1 - x_4 = -160$

$x_1 - 200 = -160$

$x_1 = 40$ vehicles per hour

$x_2 - x_4 = -10$

$x_2 - 200 = -10$

$x_2 = 190$ vehicles per hour

$x_3 - x_4 = -50$

$x_3 - 200 = -50$

$x_3 = 150$ vehicles per hour

$x_4 = 250$:

$x_1 - x_4 = -160$

$x_1 - 250 = -160$

$x_1 = 90$ vehicles per hour

$x_2 - x_4 = -10$

$x_2 - 250 = -10$

$x_2 = 240$ vehicles per hour

$x_3 - x_4 = -50$

$x_3 - 250 = -50$

$x_3 = 200$ vehicles per hour

$40 \le x_1 \le 90$ vehicles per hour

$190 \le x_2 \le 240$ vehicles per hour

$150 \le x_3 \le 200$ vehicles per hour

55. a. $80x + 400y + 480z = 9280$

$50x + 350y + 400z = 7800$

$75x + 525y + 600z = 10{,}500$

b. $\frac{1}{80}(80x + 400y + 480z) = \frac{1}{80}(9280)$

$x + 5y + 6z = 116$

$\frac{1}{50}(50x + 350y + 400z) = \frac{1}{50}(7800)$

$x + 7y + 8z = 156$

$\frac{1}{75}(75x + 525y + 600z) = \frac{1}{75}(10{,}500)$

$x + 7y + 8z = 140$

Augmented matrix: $\left[\begin{array}{ccc|c} 1 & 5 & 6 & 116 \\ 1 & 7 & 8 & 156 \\ 1 & 7 & 8 & 140 \end{array}\right]$

$$\begin{array}{r} \\ -R_1+R_2 \to R_2 \longrightarrow \\ -R_1+R_3 \to R_3 \longrightarrow \end{array}\left[\begin{array}{ccc|c} 1 & 5 & 6 & 116 \\ 0 & 2 & 2 & 40 \\ 0 & 2 & 2 & 24 \end{array}\right]$$

$$\begin{array}{r} \\ \frac{1}{2}R_2 \to R_2 \longrightarrow \\ \\ \end{array}\left[\begin{array}{ccc|c} 1 & 5 & 6 & 116 \\ 0 & 1 & 1 & 20 \\ 0 & 2 & 2 & 24 \end{array}\right]$$

$$\begin{array}{r} -5R_2+R_1 \to R_1 \longrightarrow \\ \\ -2R_2+R_3 \to R_3 \longrightarrow \end{array}\left[\begin{array}{ccc|c} 1 & 0 & 1 & 16 \\ 0 & 1 & 1 & 20 \\ 0 & 0 & 0 & -16 \end{array}\right]$$

$\{\ \}$

c. The system of equations reduces to a contradiction. There are no values for x, y, and z that can simultaneously meet the conditions of this problem.

57.

$$5x+7y-11z=45$$
$$5(-2z+16)+7(3z-5)-11z \;0\; 45$$
$$-10z+80+21z-35-11z \;0\; 45$$
$$45 \;0\; 45 \checkmark \text{ true}$$
$$3x+5y-9z=23$$
$$3(-2z+16)+5(3z-5)-9z \;0\; 23$$
$$-6z+48+15z-25-9z \;0\; 23$$
$$23 \;0\; 23 \checkmark \text{ true}$$
$$x+y-z=11$$
$$-2z+16+3z-5-z \;0\; 11$$
$$11 \;0\; 11 \checkmark \text{ true}$$

59. Augmented matrix: $\left[\begin{array}{ccc|c} 1 & 2 & -8 & 0 \\ -5 & -11 & 43 & 0 \\ 1 & 5 & -12 & 0 \end{array}\right]$

$$\begin{array}{r} \\ 5R_1+R_2 \to R_2 \longrightarrow \\ -R_1+R_3 \to R_3 \longrightarrow \end{array}\left[\begin{array}{ccc|c} 1 & 2 & -8 & 0 \\ 0 & -1 & 3 & 0 \\ 0 & 3 & -4 & 0 \end{array}\right]$$

$$\begin{array}{r} \\ -R_2 \to R_2 \longrightarrow \\ \\ \end{array}\left[\begin{array}{ccc|c} 1 & 2 & -8 & 0 \\ 0 & 1 & -3 & 0 \\ 0 & 3 & -4 & 0 \end{array}\right]$$

$$\begin{array}{r} -2R_2+R_1 \to R_1 \longrightarrow \\ \\ -3R_2+R_3 \to R_3 \longrightarrow \end{array}\left[\begin{array}{ccc|c} 1 & 0 & -2 & 0 \\ 0 & 1 & -3 & 0 \\ 0 & 0 & 5 & 0 \end{array}\right]$$

$$\begin{array}{r} \\ \\ \frac{1}{5}R_3 \to R_3 \longrightarrow \end{array}\left[\begin{array}{ccc|c} 1 & 0 & -2 & 0 \\ 0 & 1 & -3 & 0 \\ 0 & 0 & 1 & 0 \end{array}\right]$$

The system consists of independent equations. (0, 0, 0) is the only solution.

61. Augmented matrix: $\left[\begin{array}{ccc|c} 1 & -5 & 13 & 0 \\ -3 & 17 & -45 & 0 \\ 1 & -4 & 10 & 0 \end{array}\right]$

$$\begin{array}{r} \\ 3R_1+R_2 \to R_2 \longrightarrow \\ -R_1+R_3 \to R_3 \longrightarrow \end{array}\left[\begin{array}{ccc|c} 1 & -5 & 13 & 0 \\ 0 & 2 & -6 & 0 \\ 0 & 1 & -3 & 0 \end{array}\right]$$

$$\begin{array}{r} \\ \frac{1}{2}R_2 \to R_2 \longrightarrow \\ \\ \end{array}\left[\begin{array}{ccc|c} 1 & -5 & 13 & 0 \\ 0 & 1 & -3 & 0 \\ 0 & 1 & -3 & 0 \end{array}\right]$$

$$\begin{array}{r} 5R_2+R_1 \to R_1 \longrightarrow \\ \\ -R_2+R_3 \to R_3 \longrightarrow \end{array}\left[\begin{array}{ccc|c} 1 & 0 & -2 & 0 \\ 0 & 1 & -3 & 0 \\ 0 & 0 & 0 & 0 \end{array}\right]$$

$x-2z=0 \qquad y-3z=0$

$x=2z \qquad y=3z$

Infinitely many solutions;

$\{(2z, 3z, z) \mid z \text{ is any real number}\}$

63. If a row of the reduced row-echelon form results in a contradiction (that is, zeros to the left of the vertical bar and a nonzero element to the right), then the system is inconsistent.

65. The equations are equivalent, meaning that they all have the same solution set. The points in the solution set represent a common plane in space.

67.

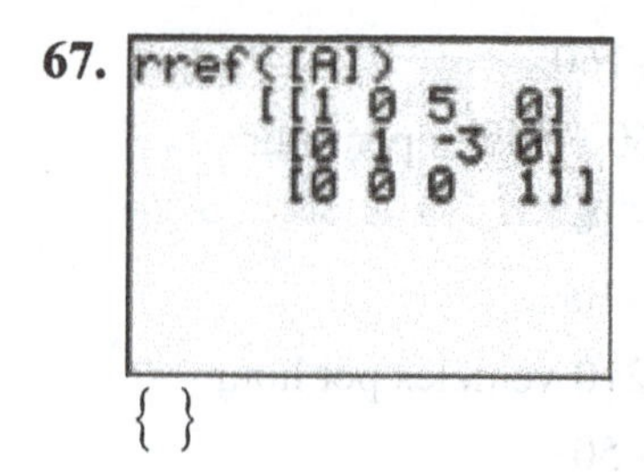

$\{\ \}$

69.

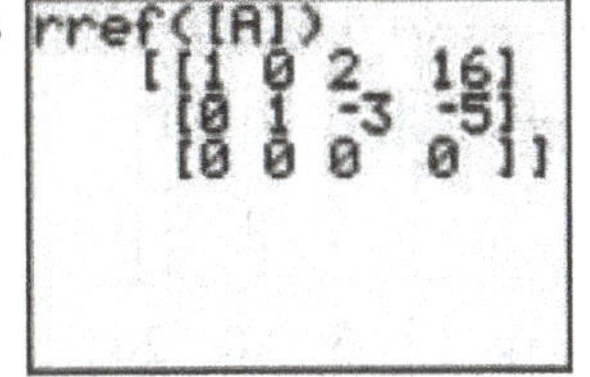

$\{(-2z+16, 3z-5, z) \mid z \text{ is any real number}\}$

71.

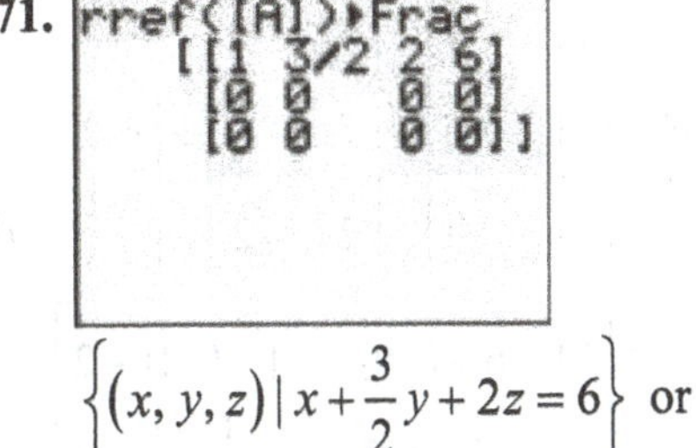

$\left\{(x, y, z) \mid x+\frac{3}{2}y+2z=6\right\}$ or

$\{(x, y, z) \mid 2x+3y+4z=12\}$

Section 6.3 Operations on Matrices

1. $-a$

3. $\frac{2}{3}$

5. $(5, 2)$

7. order; rows; columns

9. square

11. The order of the matrices must be the same, and the corresponding elements must be equal.

13. $\left[-a_{ij}\right]$

15. columns; rows

17. False

19. a. The matrix has 2 rows and 3 columns. 2×3

b. None of these

21. a. The matrix has 3 rows and 1 column. 3×1

b. Column matrix

23. a. The matrix has 2 rows and 2 columns. 2×2

b. Square matrix

25. a_{31} represents the element in the 3rd row, 1st column: $\sqrt{5}$

27. a_{13} represents the element in the 1st row, 3rd column: $\frac{1}{3}$

29. a_{43} represents the element in the 4th row, 3rd column: 2

31. $x=4, y=2, z=10$

33. $-B=\begin{bmatrix} 4 & -6 & -9 \\ -\frac{3}{5} & -1 & -7 \end{bmatrix}$

35. $A+B=\begin{bmatrix} 6 & -1 \\ 7 & \frac{1}{2} \\ 2 & \sqrt{2} \end{bmatrix}+\begin{bmatrix} -9 & 2 \\ 6.2 & 2 \\ \frac{1}{3} & \sqrt{8} \end{bmatrix}$

$=\begin{bmatrix} 6+(-9) & -1+2 \\ 7+6.2 & \frac{1}{2}+2 \\ 2+\frac{1}{3} & \sqrt{2}+\sqrt{8} \end{bmatrix}$

$=\begin{bmatrix} -3 & 1 \\ 13.2 & \frac{5}{2} \\ \frac{7}{3} & 3\sqrt{2} \end{bmatrix}$

37. $C - A + B = \begin{bmatrix} 11 & 4 \\ 1 & -\frac{1}{3} \\ 1 & 6 \end{bmatrix} - \begin{bmatrix} 6 & -1 \\ 7 & \frac{1}{2} \\ 2 & \sqrt{2} \end{bmatrix} + \begin{bmatrix} -9 & 2 \\ 6.2 & 2 \\ \frac{1}{3} & \sqrt{8} \end{bmatrix}$

$= \begin{bmatrix} 11-6+(-9) & 4-(-1)+2 \\ 1-7+6.2 & -\frac{1}{3}-\frac{1}{2}+2 \\ 1-2+\frac{1}{3} & 6-\sqrt{2}+\sqrt{8} \end{bmatrix}$

$= \begin{bmatrix} -4 & 7 \\ 0.2 & \frac{7}{6} \\ -\frac{2}{3} & 6+2\sqrt{2} \end{bmatrix}$

39. Matrix B is a 3×2 matrix, whereas matrix D is a 2×3 matrix. The matrices are of different orders and therefore it is not possible to add them. The sum is undefined.

41. $3A = 3\begin{bmatrix} 2 & 4 & -9 \\ 1 & \sqrt{3} & \frac{1}{2} \end{bmatrix} = \begin{bmatrix} 6 & 12 & -27 \\ 3 & 3\sqrt{3} & \frac{3}{2} \end{bmatrix}$

43. $-2A - 7B = -2\begin{bmatrix} 2 & 4 & -9 \\ 1 & \sqrt{3} & \frac{1}{2} \end{bmatrix} - 7\begin{bmatrix} -1 & 0 & 4 \\ 2 & 9 & \frac{2}{3} \end{bmatrix}$

$= \begin{bmatrix} -4 & -8 & 18 \\ -2 & -2\sqrt{3} & -1 \end{bmatrix} - \begin{bmatrix} -7 & 0 & 28 \\ 14 & 63 & \frac{14}{3} \end{bmatrix}$

$= \begin{bmatrix} 3 & -8 & -10 \\ -16 & -2\sqrt{3}-63 & -\frac{17}{3} \end{bmatrix}$

45. $-4(A+B) = -4A - 4B$

$= -4\begin{bmatrix} 2 & 4 & -9 \\ 1 & \sqrt{3} & \frac{1}{2} \end{bmatrix} - 4\begin{bmatrix} -1 & 0 & 4 \\ 2 & 9 & \frac{2}{3} \end{bmatrix}$

$= \begin{bmatrix} -8 & -16 & 36 \\ -4 & -4\sqrt{3} & -2 \end{bmatrix} - \begin{bmatrix} -4 & 0 & 16 \\ 8 & 36 & \frac{8}{3} \end{bmatrix}$

$= \begin{bmatrix} -4 & -16 & 20 \\ -12 & -4\sqrt{3}-36 & -\frac{14}{3} \end{bmatrix}$

47. $-3A + 5(A - B) = -3A + 5A - 5B = 2A - 5B$

$= 2\begin{bmatrix} 2 & 4 & -9 \\ 1 & \sqrt{3} & \frac{1}{2} \end{bmatrix} - 5\begin{bmatrix} -1 & 0 & 4 \\ 2 & 9 & \frac{2}{3} \end{bmatrix}$

$= \begin{bmatrix} 4 & 8 & -18 \\ 2 & 2\sqrt{3} & 1 \end{bmatrix} - \begin{bmatrix} -5 & 0 & 20 \\ 10 & 45 & \frac{10}{3} \end{bmatrix}$

$= \begin{bmatrix} 9 & 8 & -38 \\ -8 & 2\sqrt{3}-45 & -\frac{7}{3} \end{bmatrix}$

49. $2X - B = A$

$2X = A + B$

$X = \frac{1}{2}(A+B)$

$X = \frac{1}{2}\left(\begin{bmatrix} 1 & 6 \\ 4 & -2 \end{bmatrix} + \begin{bmatrix} 2 & -4 \\ 6 & 9 \end{bmatrix}\right)$

$X = \frac{1}{2}\begin{bmatrix} 3 & 2 \\ 10 & 7 \end{bmatrix} = \begin{bmatrix} \frac{3}{2} & 1 \\ 5 & \frac{7}{2} \end{bmatrix}$

51. $A + 5X = B$

$5X = B - A$

$X = \frac{1}{5}(B - A)$

$X = \frac{1}{5}\left(\begin{bmatrix} 2 & -4 \\ 6 & 9 \end{bmatrix} - \begin{bmatrix} 1 & 6 \\ 4 & -2 \end{bmatrix}\right)$

$X = \frac{1}{5}\begin{bmatrix} 1 & -10 \\ 2 & 11 \end{bmatrix} = \begin{bmatrix} \frac{1}{5} & -2 \\ \frac{2}{5} & \frac{11}{5} \end{bmatrix}$

53. $2A - B = 10X$

$X = \frac{1}{10}(2A - B)$

$X = \frac{1}{10}\left(2\begin{bmatrix} 1 & 6 \\ 4 & -2 \end{bmatrix} - \begin{bmatrix} 2 & -4 \\ 6 & 9 \end{bmatrix}\right)$

$X = \frac{1}{10}\left(\begin{bmatrix} 2 & 12 \\ 8 & -4 \end{bmatrix} - \begin{bmatrix} 2 & -4 \\ 6 & 9 \end{bmatrix}\right)$

$X = \frac{1}{10}\begin{bmatrix} 0 & 16 \\ 2 & -13 \end{bmatrix} = \begin{bmatrix} 0 & \frac{8}{5} \\ \frac{1}{5} & -\frac{13}{10} \end{bmatrix}$

55. **a.** Yes; 4×1 **b.** No

57. **a.** Yes; 5×5 **b.** Yes; 1×1

59. a. $AB=\begin{bmatrix}2&3\\5&7\end{bmatrix}\cdot\begin{bmatrix}1&4\\-1&3\end{bmatrix}=\begin{bmatrix}2(1)+3(-1)&2(4)+3(3)\\5(1)+7(-1)&5(4)+7(3)\end{bmatrix}=\begin{bmatrix}-1&17\\-2&41\end{bmatrix}$

b. $BA=\begin{bmatrix}1&4\\-1&3\end{bmatrix}\cdot\begin{bmatrix}2&3\\5&7\end{bmatrix}=\begin{bmatrix}1(2)+4(5)&1(3)+4(7)\\-1(2)+3(5)&-1(3)+3(7)\end{bmatrix}=\begin{bmatrix}22&31\\13&18\end{bmatrix}$

c. $A^2=\begin{bmatrix}2&3\\5&7\end{bmatrix}\cdot\begin{bmatrix}2&3\\5&7\end{bmatrix}=\begin{bmatrix}2(2)+3(5)&2(3)+3(7)\\5(2)+7(5)&5(3)+7(7)\end{bmatrix}=\begin{bmatrix}19&27\\45&64\end{bmatrix}$

61. a. $AB=\begin{bmatrix}2&4\\-6&3\\1&7\end{bmatrix}\cdot\begin{bmatrix}1&4&-1\\-2&0&10\end{bmatrix}=\begin{bmatrix}2(1)+4(-2)&2(4)+4(0)&2(-1)+4(10)\\-6(1)+3(-2)&-6(4)+3(0)&-6(-1)+3(10)\\1(1)+7(-2)&1(4)+7(0)&1(-1)+7(10)\end{bmatrix}=\begin{bmatrix}-6&8&38\\-12&-24&36\\-13&4&69\end{bmatrix}$

b. $BA=\begin{bmatrix}1&4&-1\\-2&0&10\end{bmatrix}\cdot\begin{bmatrix}2&4\\-6&3\\1&7\end{bmatrix}=\begin{bmatrix}1(2)+4(-6)+(-1)(1)&1(4)+4(3)+(-1)(7)\\(-2)(2)+(0)(-6)+(10)(1)&(-2)(4)+(0)(3)+(10)(7)\end{bmatrix}=\begin{bmatrix}-23&9\\6&62\end{bmatrix}$

c. Not possible

63. a. $AB=\begin{bmatrix}-9&2&3\\-1&5&4\\0&1&7\end{bmatrix}\cdot\begin{bmatrix}-1\\5\\0\end{bmatrix}=\begin{bmatrix}-9(-1)+2(5)+3(0)\\-1(-1)+5(5)+4(0)\\(0)(-1)+1(5)+7(0)\end{bmatrix}=\begin{bmatrix}19\\26\\5\end{bmatrix}$

b. Not possible

c. $A^2=\begin{bmatrix}-9&2&3\\-1&5&4\\0&1&7\end{bmatrix}\cdot\begin{bmatrix}-9&2&3\\-1&5&4\\0&1&7\end{bmatrix}$

$=\begin{bmatrix}-9(-9)+2(-1)+3(0)&-9(2)+2(5)+3(1)&-9(3)+2(4)+3(7)\\-1(-9)+5(-1)+4(0)&-1(2)+5(5)+4(1)&-1(3)+5(4)+4(7)\\(0)(-9)+1(-1)+7(0)&(0)(2)+1(5)+7(1)&(0)(3)+1(4)+7(7)\end{bmatrix}=\begin{bmatrix}79&-5&2\\4&27&45\\-1&12&53\end{bmatrix}$

65. a. $AB=\begin{bmatrix}1&\frac{1}{2}\end{bmatrix}\cdot\begin{bmatrix}-\frac{1}{3}\\2\end{bmatrix}=\begin{bmatrix}1\left(-\frac{1}{3}\right)+\frac{1}{2}(2)\end{bmatrix}=\begin{bmatrix}\frac{2}{3}\end{bmatrix}$

b. $BA=\begin{bmatrix}-\frac{1}{3}\\2\end{bmatrix}\cdot\begin{bmatrix}1&\frac{1}{2}\end{bmatrix}=\begin{bmatrix}-\frac{1}{3}(1)&-\frac{1}{3}\left(\frac{1}{2}\right)\\2(1)&2\left(\frac{1}{2}\right)\end{bmatrix}=\begin{bmatrix}-\frac{1}{3}&-\frac{1}{6}\\2&1\end{bmatrix}$

c. Not possible

67. a. $AB=\begin{bmatrix}4\\-6\end{bmatrix}\cdot\begin{bmatrix}1&2&5&6\end{bmatrix}=\begin{bmatrix}4(1)&4(2)&4(5)&4(6)\\-6(1)&-6(2)&-6(5)&-6(6)\end{bmatrix}=\begin{bmatrix}4&8&20&24\\-6&-12&-30&-36\end{bmatrix}$

b. Not possible

c. Not possible

69. a. $AB = [-5]\cdot[5] = [-5(5)] = [-25]$

b. $BA = [5]\cdot[-5] = [5(-5)] = [-25]$

c. $A^2 = [-5]\cdot[-5] = [-5(-5)] = [25]$

71. $$AB = \begin{bmatrix} 3.1 & -2.3 \\ 1.1 & 6.5 \end{bmatrix}\cdot\begin{bmatrix} 1 & 0 \\ 0 & 1 \end{bmatrix} = \begin{bmatrix} 3.1(1)+(-2.3)(0) & 3.1(0)+(-2.3)(1) \\ 1.1(1)+6.5(0) & 1.1(0)+6.5(1) \end{bmatrix} = \begin{bmatrix} 3.1 & -2.3 \\ 1.1 & 6.5 \end{bmatrix}$$

$$BA = \begin{bmatrix} 1 & 0 \\ 0 & 1 \end{bmatrix}\cdot\begin{bmatrix} 3.1 & -2.3 \\ 1.1 & 6.5 \end{bmatrix} = \begin{bmatrix} 1(3.1)+(0)(1.1) & 1(-2.3)+(0)(6.5) \\ (0)(3.1)+1(1.1) & (0)(-2.3)+1(6.5) \end{bmatrix} = \begin{bmatrix} 3.1 & -2.3 \\ 1.1 & 6.5 \end{bmatrix}$$

73. $$AB = \begin{bmatrix} 1 & 0 & 0 \\ 0 & 1 & 0 \\ 0 & 0 & 1 \end{bmatrix}\cdot\begin{bmatrix} \frac{9}{5} & -3 & \sqrt{6} \\ 5 & \frac{1}{2} & 2 \\ 3 & 0 & 1 \end{bmatrix}$$

$$= \begin{bmatrix} 1\left(\frac{9}{5}\right)+(0)(5)+(0)(3) & 1(-3)+(0)\left(\frac{1}{2}\right)+(0)(0) & 1\left(\sqrt{6}\right)+(0)(2)+(0)(1) \\ (0)\left(\frac{9}{5}\right)+1(5)+(0)(3) & (0)(-3)+1\left(\frac{1}{2}\right)+(0)(0) & (0)\left(\sqrt{6}\right)+1(2)+(0)(1) \\ (0)\left(\frac{9}{5}\right)+(0)(5)+1(3) & (0)(-3)+(0)\left(\frac{1}{2}\right)+1(0) & (0)\left(\sqrt{6}\right)+(0)(2)+1(1) \end{bmatrix} = \begin{bmatrix} \frac{9}{5} & -3 & \sqrt{6} \\ 5 & \frac{1}{2} & 2 \\ 3 & 0 & 1 \end{bmatrix}$$

$$BA = \begin{bmatrix} \frac{9}{5} & -3 & \sqrt{6} \\ 5 & \frac{1}{2} & 2 \\ 3 & 0 & 1 \end{bmatrix}\cdot\begin{bmatrix} 1 & 0 & 0 \\ 0 & 1 & 0 \\ 0 & 0 & 1 \end{bmatrix}$$

$$= \begin{bmatrix} \frac{9}{5}(1)+(-3)(0)+\sqrt{6}(0) & \frac{9}{5}(0)+(-3)(1)+\sqrt{6}(0) & \frac{9}{5}(0)+(-3)(0)+\sqrt{6}(1) \\ 5(1)+\frac{1}{2}(0)+(2)(0) & 5(0)+\frac{1}{2}(1)+(2)(0) & 5(0)+\frac{1}{2}(0)+2(1) \\ 3(1)+(0)(0)+(1)(0) & 3(0)+(0)(1)+(1)(0) & 3(0)+(0)(0)+1(1) \end{bmatrix} = \begin{bmatrix} \frac{9}{5} & -3 & \sqrt{6} \\ 5 & \frac{1}{2} & 2 \\ 3 & 0 & 1 \end{bmatrix}$$

75. a. $$M - D = \begin{bmatrix} \$21{,}930 & \$18{,}505 \\ \$22{,}730 & \$19{,}305 \end{bmatrix} - \begin{bmatrix} \$19{,}963 & \$17{,}109 \\ \$20{,}686 & \$17{,}843 \end{bmatrix} = \begin{bmatrix} \$1967 & \$1396 \\ \$2044 & \$1462 \end{bmatrix}$$

This represents the profit that the dealer clears for each model of car.

b. $$F = 1.06D = 1.06\cdot\begin{bmatrix} \$19{,}963 & \$17{,}109 \\ \$20{,}686 & \$17{,}843 \end{bmatrix} = \begin{bmatrix} \$21{,}161 & \$18{,}136 \\ \$21{,}927 & \$18{,}914 \end{bmatrix}$$

77. a. $CN_1 = \begin{bmatrix} \$0.25 & \$0.40 \\ \$0 & \$0.40 \\ \$0.10 & \$0 \end{bmatrix} \cdot \begin{bmatrix} 24 \\ 100 \end{bmatrix}$

$= \begin{bmatrix} \$46 \\ \$40 \\ \$2.40 \end{bmatrix}$

The matrix CN_1 represents the additional cost for 24 text messages and 100 extra minutes for each of the cell phone plans.

b. $CN_3 = \begin{bmatrix} \$0.25 & \$0.40 \\ \$0 & \$0.40 \\ \$0.10 & \$0 \end{bmatrix} \cdot \begin{bmatrix} 24 & 56 & 30 \\ 100 & 24 & 0 \end{bmatrix}$

$= \begin{bmatrix} \$46 & \$23.60 & \$7.50 \\ \$40 & \$9.60 & \$0 \\ \$2.40 & \$5.60 & \$3 \end{bmatrix}$

The matrix CN_3 represents the additional cost per month for each plan. For example, row 1 represents the cost for plan A for months 1, 2, and 3, respectively.

79. a. $B + CN_1 + T$

$= \begin{bmatrix} \$39.99 \\ \$49.99 \\ \$59.99 \end{bmatrix} + \begin{bmatrix} \$46 \\ \$40 \\ \$2.40 \end{bmatrix} + \begin{bmatrix} \$11.96 \\ \$13.04 \\ \$14.91 \end{bmatrix}$

$= \begin{bmatrix} \$97.95 \\ \$103.03 \\ \$77.30 \end{bmatrix}$

This matrix represents the total cost (including the base cost and tax) for each plan for 24 text messages and 100 additional minutes.

b. $N_1 = \begin{bmatrix} 60 \\ 20 \end{bmatrix}$

$CN_1 = \begin{bmatrix} \$0.25 & \$0.40 \\ \$0 & \$0.40 \\ \$0.10 & \$0 \end{bmatrix} \cdot \begin{bmatrix} 60 \\ 20 \end{bmatrix} = \begin{bmatrix} \$23 \\ \$8 \\ \$6 \end{bmatrix}$

$B + CN_1 + T = \begin{bmatrix} \$39.99 \\ \$49.99 \\ \$59.99 \end{bmatrix} + \begin{bmatrix} \$23 \\ \$8 \\ \$6 \end{bmatrix} + \begin{bmatrix} \$11.96 \\ \$13.04 \\ \$14.91 \end{bmatrix}$

$= \begin{bmatrix} \$74.95 \\ \$71.03 \\ \$80.90 \end{bmatrix}$

Plan B is least expensive.

81. a. $A = \begin{bmatrix} -1 & 0 & 4 \\ 1 & 3 & 2 \end{bmatrix}$

b. $\begin{bmatrix} -1 & 0 & 4 \\ 1 & 3 & 2 \end{bmatrix} + \begin{bmatrix} 2 & 2 & 2 \\ -4 & -4 & -4 \end{bmatrix} = \begin{bmatrix} 1 & 2 & 6 \\ -3 & -1 & -2 \end{bmatrix}$

c. $\begin{bmatrix} -1 & 0 \\ 0 & 1 \end{bmatrix} \cdot \begin{bmatrix} -1 & 0 & 4 \\ 1 & 3 & 2 \end{bmatrix} = \begin{bmatrix} -1(-1)+(0)(1) & -1(0)+(0)(3) & -1(4)+(0)(2) \\ (0)(-1)+1(1) & (0)(0)+1(3) & (0)(4)+1(2) \end{bmatrix} = \begin{bmatrix} 1 & 0 & -4 \\ 1 & 3 & 2 \end{bmatrix}$

This matrix represents the reflection of the triangle across the y-axis.

d. $\begin{bmatrix} 1 & 0 \\ 0 & -1 \end{bmatrix} \cdot \begin{bmatrix} -1 & 0 & 4 \\ 1 & 3 & 2 \end{bmatrix} = \begin{bmatrix} 1(-1)+(0)(1) & 1(0)+(0)(3) & 1(4)+(0)(2) \\ (0)(-1)+(-1)(1) & (0)(0)+(-1)(3) & (0)(4)+(-1)(2) \end{bmatrix} = \begin{bmatrix} -1 & 0 & 4 \\ -1 & -3 & -2 \end{bmatrix}$

This matrix represents the reflection of the triangle across the x-axis.

e. $\begin{bmatrix} 1 & 0 \\ 0 & -1 \end{bmatrix} \cdot A + \begin{bmatrix} -1 & -1 & -1 \\ 2 & 2 & 2 \end{bmatrix} = \begin{bmatrix} -1 & 0 & 4 \\ -1 & -3 & -2 \end{bmatrix} + \begin{bmatrix} -1 & -1 & -1 \\ 2 & 2 & 2 \end{bmatrix} = \begin{bmatrix} -2 & -1 & 3 \\ 1 & -1 & 0 \end{bmatrix}$

This matrix represents the reflection of the triangle across the x-axis, followed by a shift to the left 1 unit and a shift upward 2 units.

83. a. $A = \begin{bmatrix} -2 & 3 & 3 & 0 \\ 1 & 3 & 0 & -1 \end{bmatrix}$

b. $A + \begin{bmatrix} 0 & 0 & 0 & 0 \\ -3 & -3 & -3 & -3 \end{bmatrix}$

c. $A + \begin{bmatrix} -4 & -4 & -4 & -4 \\ 0 & 0 & 0 & 0 \end{bmatrix}$

d. $$\begin{bmatrix} 1 & 0 \\ 0 & -1 \end{bmatrix} \cdot \begin{bmatrix} -2 & 3 & 3 & 0 \\ 1 & 3 & 0 & -1 \end{bmatrix} = \begin{bmatrix} 1(-2)+(0)(1) & 1(3)+(0)(3) & 1(3)+(0)(0) & 1(0)+(0)(-1) \\ (0)(-2)+(-1)(1) & (0)(3)+(-1)(3) & (0)(3)+(-1)(0) & (0)(0)+(-1)(-1) \end{bmatrix}$$
$$= \begin{bmatrix} -2 & 3 & 3 & 0 \\ -1 & -3 & 0 & 1 \end{bmatrix}$$

e. $$\begin{bmatrix} -1 & 0 \\ 0 & 1 \end{bmatrix} \cdot \begin{bmatrix} -2 & 3 & 3 & 0 \\ 1 & 3 & 0 & -1 \end{bmatrix} = \begin{bmatrix} -1(-2)+(0)(1) & -1(3)+(0)(3) & -1(3)+(0)(0) & -1(0)+(0)(-1) \\ (0)(-2)+1(1) & (0)(3)+1(3) & (0)(3)+1(0) & (0)(0)+1(-1) \end{bmatrix}$$
$$= \begin{bmatrix} 2 & -3 & -3 & 0 \\ 1 & 3 & 0 & -1 \end{bmatrix}$$

85. a. $A = \begin{bmatrix} 0 & 6 & 6 \\ 0 & 3 & 0 \end{bmatrix}$

b. $$\begin{bmatrix} \frac{\sqrt{3}}{2} & -\frac{1}{2} \\ \frac{1}{2} & \frac{\sqrt{3}}{2} \end{bmatrix} \cdot \begin{bmatrix} 0 & 6 & 6 \\ 0 & 3 & 0 \end{bmatrix} = \begin{bmatrix} \frac{\sqrt{3}}{2}(0)+\left(-\frac{1}{2}\right)(0) & \frac{\sqrt{3}}{2}(6)+\left(-\frac{1}{2}\right)(3) & \frac{\sqrt{3}}{2}(6)+\left(-\frac{1}{2}\right)(0) \\ \frac{1}{2}(0)+\frac{\sqrt{3}}{2}(0) & \frac{1}{2}(6)+\frac{\sqrt{3}}{2}(3) & \frac{1}{2}(6)+\frac{\sqrt{3}}{2}(0) \end{bmatrix}$$
$$= \begin{bmatrix} 0 & 3\sqrt{3}-\frac{3}{2} & 3\sqrt{3} \\ 0 & 3+\frac{3\sqrt{3}}{2} & 3 \end{bmatrix}$$
$$\approx \begin{bmatrix} 0 & 3.7 & 5.2 \\ 0 & 5.6 & 3 \end{bmatrix}$$

c.

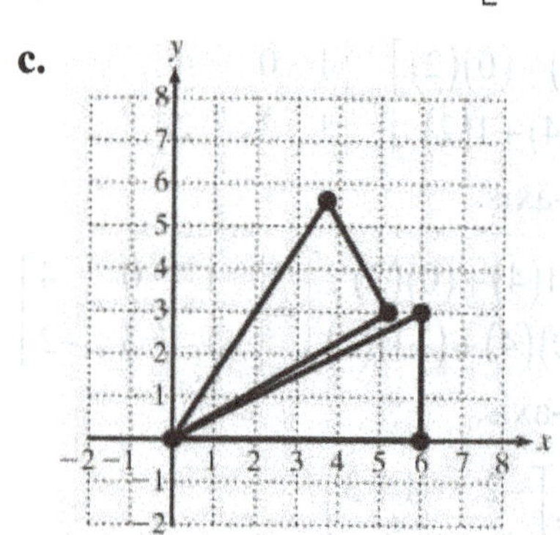

It appears that the triangle was rotated approximately 30° counterclockwise.

87. a. $\begin{bmatrix} 6 & 6 & 6 \\ 6 & 0 & 0 \\ 6 & 6 & 6 \\ 6 & 0 & 0 \\ 6 & 6 & 6 \end{bmatrix}$

b. Dark to medium dark is a change of -1.
White to light gray is a change of $+1$.

$$\begin{bmatrix} 6 & 6 & 6 \\ 6 & 0 & 0 \\ 6 & 6 & 6 \\ 6 & 0 & 0 \\ 6 & 6 & 6 \end{bmatrix} + \begin{bmatrix} -1 & -1 & -1 \\ -1 & 1 & 1 \\ -1 & -1 & -1 \\ -1 & 1 & 1 \\ -1 & -1 & -1 \end{bmatrix} = \begin{bmatrix} 5 & 5 & 5 \\ 5 & 1 & 1 \\ 5 & 5 & 5 \\ 5 & 1 & 1 \\ 5 & 5 & 5 \end{bmatrix}$$

89. $A+B = \begin{bmatrix} a_1 & a_2 \\ a_3 & a_4 \end{bmatrix} + \begin{bmatrix} b_1 & b_2 \\ b_3 & b_4 \end{bmatrix}$

$= \begin{bmatrix} a_1+b_1 & a_2+b_2 \\ a_3+b_3 & a_4+b_4 \end{bmatrix}$

$= \begin{bmatrix} b_1+a_1 & b_2+a_2 \\ b_3+a_3 & b_4+a_4 \end{bmatrix}$

$= \begin{bmatrix} b_1 & b_2 \\ b_3 & b_4 \end{bmatrix} + \begin{bmatrix} a_1 & a_2 \\ a_3 & a_4 \end{bmatrix}$

$= B + A$

91. $A+(-A) = \begin{bmatrix} a_1 & a_2 \\ a_3 & a_4 \end{bmatrix} + \begin{bmatrix} -a_1 & -a_2 \\ -a_3 & -a_4 \end{bmatrix}$

$= \begin{bmatrix} a_1+(-a_1) & a_2+(-a_2) \\ a_3+(-a_3) & a_4+(-a_4) \end{bmatrix}$

$= \begin{bmatrix} 0 & 0 \\ 0 & 0 \end{bmatrix} = 0$

93. $s(tA) = s \cdot \left(t \begin{bmatrix} a_1 & a_2 \\ a_3 & a_4 \end{bmatrix} \right)$

$= s \cdot \begin{bmatrix} ta_1 & ta_2 \\ ta_3 & ta_4 \end{bmatrix}$

$= \begin{bmatrix} sta_1 & sta_2 \\ sta_3 & sta_4 \end{bmatrix}$

$= (st) \begin{bmatrix} a_1 & a_2 \\ a_3 & a_4 \end{bmatrix} = (st)A$

95. $A^2 = \begin{bmatrix} i & 0 \\ 0 & i \end{bmatrix}^2$

$= \begin{bmatrix} i & 0 \\ 0 & i \end{bmatrix} \begin{bmatrix} i & 0 \\ 0 & i \end{bmatrix}$

$= \begin{bmatrix} i(i)+(0)(0) & i(0)+(0)(i) \\ (0)(i)+i(0) & (0)(0)+i(i) \end{bmatrix}$

$= \begin{bmatrix} -1 & 0 \\ 0 & -1 \end{bmatrix}$

$A^3 = A \cdot A^2$

$= \begin{bmatrix} i & 0 \\ 0 & i \end{bmatrix} \begin{bmatrix} -1 & 0 \\ 0 & -1 \end{bmatrix}$

$= \begin{bmatrix} i(-1)+(0)(0) & i(0)+(0)(-1) \\ (0)(-1)+i(0) & (0)(0)+i(-1) \end{bmatrix}$

$= \begin{bmatrix} -i & 0 \\ 0 & -i \end{bmatrix}$

$A^4 = A \cdot A^3$

$= \begin{bmatrix} i & 0 \\ 0 & i \end{bmatrix} \begin{bmatrix} -i & 0 \\ 0 & -i \end{bmatrix}$

$= \begin{bmatrix} i(-i)+(0)(0) & i(0)+(0)(-i) \\ (0)(-i)+i(0) & (0)(0)+i(-i) \end{bmatrix}$

$= \begin{bmatrix} 1 & 0 \\ 0 & 1 \end{bmatrix}$

The entries along the main diagonal in matrix A^n are the same as the value of i^n.

97. a. $\begin{bmatrix} a & 0 \\ 0 & b \end{bmatrix} \begin{bmatrix} c & 0 \\ 0 & d \end{bmatrix}$

$= \begin{bmatrix} a(c)+(0)(0) & a(0)+(0)(d) \\ (0)(c)+(b)(0) & (0)(0)+(b)(d) \end{bmatrix}$

$= \begin{bmatrix} ac & 0 \\ 0 & bd \end{bmatrix}$

b. $\begin{bmatrix} 3 & 0 \\ 0 & 7 \end{bmatrix} \begin{bmatrix} 1 & 0 \\ 0 & 2 \end{bmatrix} = \begin{bmatrix} 3 & 0 \\ 0 & 14 \end{bmatrix}$

99. The number of columns in the first matrix is not equal to the number of rows in the second matrix.

101. To find $-A$, take the additive inverse of each individual element of A. That is, $-A = \left[-a_{ij}\right]$.

103.

```
5[A]+2[B]
       [[17   19]
        [2    27]
        [-15  66]]
```

105.

	A	B	C	D	E	F	G	H	I	J	K
1		1	3			6	2			39	5
2	A=	2	5		B=	-4	1		-3A + 7B =	-34	-8
3		-3	10			0	8			9	26

107.

L1 f_x =MMULT(B1:D3,G1:I3)

	A	B	C	D	E	F	G	H	I	J	K	L	M	N
1		-3	4	8			4	2	0			-32	10	-16
2	A=	0	1	6		C=	1	0	6		AC=	-17	12	-24
3		7	-2	5			-3	2	-5			11	24	-37

Section 6.4 Inverse Matrices and Matrix Equations

1. a. $\begin{bmatrix} 2 & -3 \\ 1 & 5 \end{bmatrix}\begin{bmatrix} x \\ y \end{bmatrix} = \begin{bmatrix} 2x-3y \\ x+5y \end{bmatrix}$

b. $\begin{bmatrix} 2x-3y \\ x+5y \end{bmatrix} = \begin{bmatrix} 14 \\ -19 \end{bmatrix}$

$2x - 3y = 14$

$x + 5y = -19$

3. a. $\begin{bmatrix} 3 & -4 & 6 \\ 2 & 0 & 4 \\ 9 & 2 & 0 \end{bmatrix}\begin{bmatrix} x \\ y \\ z \end{bmatrix} = \begin{bmatrix} 3x-4y+6z \\ 2x+4z \\ 9x+2y \end{bmatrix}$

b. $\begin{bmatrix} 3x-4y+6z \\ 2x+4z \\ 9x+2y \end{bmatrix} = \begin{bmatrix} 17 \\ 14 \\ 11 \end{bmatrix}$

$3x - 4y + 6z = 17$

$2x \qquad + 4z = 14$

$9x + 2y \qquad = 11$

5. identity

7. inverse

9. $\dfrac{1}{ad-bc}\begin{bmatrix} d & -b \\ -c & a \end{bmatrix}$

11. a. $AI_n = \begin{bmatrix} -\frac{7}{8} & \sqrt{5} \\ 5.1 & 8 \end{bmatrix}\begin{bmatrix} 1 & 0 \\ 0 & 1 \end{bmatrix}$

$= \begin{bmatrix} -\frac{7}{8}(1)+\sqrt{5}(0) & -\frac{7}{8}(0)+\sqrt{5}(1) \\ 5.1(1)+8(0) & 5.1(0)+8(1) \end{bmatrix}$

$= \begin{bmatrix} -\frac{7}{8} & \sqrt{5} \\ 5.1 & 8 \end{bmatrix}$ ✓

b. $I_nA = \begin{bmatrix} 1 & 0 \\ 0 & 1 \end{bmatrix}\begin{bmatrix} -\frac{7}{8} & \sqrt{5} \\ 5.1 & 8 \end{bmatrix}$

$= \begin{bmatrix} 1\left(-\frac{7}{8}\right)+(0)(5.1) & 1\left(\sqrt{5}\right)+(0)(8) \\ (0)\left(-\frac{7}{8}\right)+1(5.1) & (0)\left(\sqrt{5}\right)+1(8) \end{bmatrix}$

$= \begin{bmatrix} -\frac{7}{8} & \sqrt{5} \\ 5.1 & 8 \end{bmatrix}$ ✓

13. a. $AI_n = \begin{bmatrix} 1 & -3 & 4 \\ 9 & 5 & 3 \\ 11 & -6 & -4 \end{bmatrix}\begin{bmatrix} 1 & 0 & 0 \\ 0 & 1 & 0 \\ 0 & 0 & 1 \end{bmatrix}$

$$= \begin{bmatrix} 1(1)+(-3)(0)+4(0) & 1(0)+(-3)(1)+4(0) & 1(0)+(-3)(0)+4(1) \\ 9(1)+5(0)+3(0) & 9(0)+5(1)+3(0) & 9(0)+5(0)+3(1) \\ 11(1)+(-6)(0)+(-4)(0) & 11(0)+(-6)(1)+(-4)(0) & 11(0)+(-6)(0)+(-4)(1) \end{bmatrix} = \begin{bmatrix} 1 & -3 & 4 \\ 9 & 5 & 3 \\ 11 & -6 & -4 \end{bmatrix} \checkmark$$

b. $I_nA = \begin{bmatrix} 1 & 0 & 0 \\ 0 & 1 & 0 \\ 0 & 0 & 1 \end{bmatrix}\begin{bmatrix} 1 & -3 & 4 \\ 9 & 5 & 3 \\ 11 & -6 & -4 \end{bmatrix}$

$$= \begin{bmatrix} 1(1)+(0)(9)+(0)(11) & 1(-3)+(0)(5)+(0)(-6) & 1(4)+(0)(3)+(0)(-4) \\ (0)(1)+1(9)+(0)(11) & (0)(-3)+1(5)+(0)(-6) & (0)(4)+1(3)+(0)(-4) \\ (0)(1)+(0)(9)+1(11) & (0)(-3)+(0)(5)+1(-6) & (0)(4)+(0)(3)+1(-4) \end{bmatrix} = \begin{bmatrix} 1 & -3 & 4 \\ 9 & 5 & 3 \\ 11 & -6 & -4 \end{bmatrix} \checkmark$$

15. $AB = \begin{bmatrix} 10 & -3 \\ 4 & -2 \end{bmatrix}\begin{bmatrix} \frac{1}{4} & -\frac{3}{8} \\ \frac{1}{2} & -\frac{5}{4} \end{bmatrix} = \begin{bmatrix} 10\left(\frac{1}{4}\right)+(-3)\left(\frac{1}{2}\right) & 10\left(-\frac{3}{8}\right)+(-3)\left(-\frac{5}{4}\right) \\ 4\left(\frac{1}{4}\right)+(-2)\left(\frac{1}{2}\right) & 4\left(-\frac{3}{8}\right)+(-2)\left(-\frac{5}{4}\right) \end{bmatrix} = \begin{bmatrix} 1 & 0 \\ 0 & 1 \end{bmatrix}$ ✓ Yes

$BA = \begin{bmatrix} \frac{1}{4} & -\frac{3}{8} \\ \frac{1}{2} & -\frac{5}{4} \end{bmatrix}\begin{bmatrix} 10 & -3 \\ 4 & -2 \end{bmatrix} = \begin{bmatrix} \frac{1}{4}(10)+\left(-\frac{3}{8}\right)(4) & \frac{1}{4}(-3)+\left(-\frac{3}{8}\right)(-2) \\ \frac{1}{2}(10)+\left(-\frac{5}{4}\right)(4) & \frac{1}{2}(-3)+\left(-\frac{5}{4}\right)(-2) \end{bmatrix} = \begin{bmatrix} 1 & 0 \\ 0 & 1 \end{bmatrix}$ ✓ Yes

17. $AB = \begin{bmatrix} -2 & -3 & 1 \\ -3 & -3 & 1 \\ -2 & -4 & 1 \end{bmatrix}\begin{bmatrix} 1 & -1 & 0 \\ 1 & 0 & -1 \\ 6 & -2 & -3 \end{bmatrix}$

$$= \begin{bmatrix} -2(1)+(-3)(1)+1(6) & -2(-1)+(-3)(0)+1(-2) & -2(0)+(-3)(-1)+1(-3) \\ -3(1)+(-3)(1)+1(6) & -3(-1)+(-3)(0)+1(-2) & -3(0)+(-3)(-1)+1(-3) \\ -2(1)+(-4)(1)+1(6) & -2(-1)+(-4)(0)+1(-2) & -2(0)+(-4)(-1)+1(-3) \end{bmatrix} = \begin{bmatrix} 1 & 0 & 0 \\ 0 & 1 & 0 \\ 0 & 0 & 1 \end{bmatrix} \checkmark \text{ Yes}$$

$BA = \begin{bmatrix} 1 & -1 & 0 \\ 1 & 0 & -1 \\ 6 & -2 & -3 \end{bmatrix}\begin{bmatrix} -2 & -3 & 1 \\ -3 & -3 & 1 \\ -2 & -4 & 1 \end{bmatrix}$

$$= \begin{bmatrix} 1(-2)+(-1)(-3)+(0)(-2) & 1(-3)+(-1)(-3)+(0)(-4) & 1(1)+(-1)(1)+(0)(1) \\ 1(-2)+(0)(-3)+(-1)(-2) & 1(-3)+(0)(-3)+(-1)(-4) & 1(1)+(0)(1)+(-1)(1) \\ 6(-2)+(-2)(-3)+(-3)(-2) & 6(-3)+(-2)(-3)+(-3)(-4) & 6(1)+(-2)(1)+(-3)(1) \end{bmatrix}$$

$$= \begin{bmatrix} 1 & 0 & 0 \\ 0 & 1 & 0 \\ 0 & 0 & 1 \end{bmatrix} \checkmark \text{ Yes}$$

19. $AB = \begin{bmatrix} 2 & 1 \\ 3 & 4 \end{bmatrix}\begin{bmatrix} 4 & -3 \\ -7 & 6 \end{bmatrix} = \begin{bmatrix} 2(4)+1(-7) & 2(-3)+1(6) \\ 3(4)+4(-7) & 3(-3)+4(6) \end{bmatrix} = \begin{bmatrix} 1 & 0 \\ -16 & 15 \end{bmatrix}$ No

21. Set up matrix $[A\,|\,I_2]$: $\left[\begin{array}{rr|rr} -4 & -3 & 1 & 0 \\ 6 & 5 & 0 & 1 \end{array}\right]$

$-\frac{1}{4}R_1 \to R_1 \longrightarrow \left[\begin{array}{rr|rr} 1 & \frac{3}{4} & -\frac{1}{4} & 0 \\ 6 & 5 & 0 & 1 \end{array}\right]$

$-6R_1 + R_2 \to R_2 \longrightarrow \left[\begin{array}{rr|rr} 1 & \frac{3}{4} & -\frac{1}{4} & 0 \\ 0 & \frac{1}{2} & \frac{3}{2} & 1 \end{array}\right]$

$2R_2 \to R_2 \longrightarrow \left[\begin{array}{rr|rr} 1 & \frac{3}{4} & -\frac{1}{4} & 0 \\ 0 & 1 & 3 & 2 \end{array}\right]$

$-\frac{3}{4}R_2 + R_1 \to R_1 \longrightarrow \left[\begin{array}{rr|rr} 1 & 0 & -\frac{5}{2} & -\frac{3}{2} \\ 0 & 1 & 3 & 2 \end{array}\right]$

$A^{-1} = \begin{bmatrix} -\frac{5}{2} & -\frac{3}{2} \\ 3 & 2 \end{bmatrix}$

23. Set up matrix $[A\,|\,I_2]$: $\left[\begin{array}{rr|rr} -8 & -2 & 1 & 0 \\ 10 & 5 & 0 & 1 \end{array}\right]$

$-\frac{1}{8}R_1 \to R_1 \longrightarrow \left[\begin{array}{rr|rr} 1 & \frac{1}{4} & -\frac{1}{8} & 0 \\ 10 & 5 & 0 & 1 \end{array}\right]$

$-10R_1 + R_2 \to R_2 \longrightarrow \left[\begin{array}{rr|rr} 1 & \frac{1}{4} & -\frac{1}{8} & 0 \\ 0 & \frac{5}{2} & \frac{5}{4} & 1 \end{array}\right]$

$\frac{2}{5}R_2 \to R_2 \longrightarrow \left[\begin{array}{rr|rr} 1 & \frac{1}{4} & -\frac{1}{8} & 0 \\ 0 & 1 & \frac{1}{2} & \frac{2}{5} \end{array}\right]$

$-\frac{1}{4}R_2 + R_1 \to R_1 \longrightarrow \left[\begin{array}{rr|rr} 1 & 0 & -\frac{1}{4} & -\frac{1}{10} \\ 0 & 1 & \frac{1}{2} & \frac{2}{5} \end{array}\right]$

$A^{-1} = \begin{bmatrix} -\frac{1}{4} & -\frac{1}{10} \\ \frac{1}{2} & \frac{2}{5} \end{bmatrix}$

25. Set up matrix $[A\,|\,I_2]$: $\left[\begin{array}{rr|rr} 3 & 7 & 1 & 0 \\ 6 & 14 & 0 & 1 \end{array}\right]$

$\frac{1}{3}R_1 \to R_1 \longrightarrow \left[\begin{array}{rr|rr} 1 & \frac{7}{3} & \frac{1}{3} & 0 \\ 6 & 14 & 0 & 1 \end{array}\right]$

$-6R_1 + R_2 \to R_2 \longrightarrow \left[\begin{array}{rr|rr} 1 & \frac{7}{3} & \frac{1}{3} & 0 \\ 0 & 0 & -2 & 1 \end{array}\right]$

A is a singular matrix.

27. Set up matrix $[A\,|\,I_3]$: $\left[\begin{array}{rrr|rrr} 2 & -7 & 8 & 1 & 0 & 0 \\ 1 & -3 & 3 & 0 & 1 & 0 \\ 1 & -5 & 6 & 0 & 0 & 1 \end{array}\right]$

$R_1 \Leftrightarrow R_2 \longrightarrow \left[\begin{array}{rrr|rrr} 1 & -3 & 3 & 0 & 1 & 0 \\ 2 & -7 & 8 & 1 & 0 & 0 \\ 1 & -5 & 6 & 0 & 0 & 1 \end{array}\right]$

$-2R_1 + R_2 \to R_2 \longrightarrow$
$-R_1 + R_3 \to R_3 \longrightarrow \left[\begin{array}{rrr|rrr} 1 & -3 & 3 & 0 & 1 & 0 \\ 0 & -1 & 2 & 1 & -2 & 0 \\ 0 & -2 & 3 & 0 & -1 & 1 \end{array}\right]$

$-R_2 \to R_2 \longrightarrow \left[\begin{array}{rrr|rrr} 1 & -3 & 3 & 0 & 1 & 0 \\ 0 & 1 & -2 & -1 & 2 & 0 \\ 0 & -2 & 3 & 0 & -1 & 1 \end{array}\right]$

$3R_2 + R_1 \to R_1 \longrightarrow$
$2R_2 + R_3 \to R_3 \longrightarrow \left[\begin{array}{rrr|rrr} 1 & 0 & -3 & -3 & 7 & 0 \\ 0 & 1 & -2 & -1 & 2 & 0 \\ 0 & 0 & -1 & -2 & 3 & 1 \end{array}\right]$

$-R_3 \to R_3 \longrightarrow \left[\begin{array}{rrr|rrr} 1 & 0 & -3 & -3 & 7 & 0 \\ 0 & 1 & -2 & -1 & 2 & 0 \\ 0 & 0 & 1 & 2 & -3 & -1 \end{array}\right]$

$3R_3 + R_1 \to R_1 \longrightarrow$
$2R_3 + R_2 \to R_2 \longrightarrow \left[\begin{array}{rrr|rrr} 1 & 0 & 0 & 3 & -2 & -3 \\ 0 & 1 & 0 & 3 & -4 & -2 \\ 0 & 0 & 1 & 2 & -3 & -1 \end{array}\right]$

$A^{-1} = \begin{bmatrix} 3 & -2 & -3 \\ 3 & -4 & -2 \\ 2 & -3 & -1 \end{bmatrix}$

29. Set up matrix $[A|I_3]$: $\left[\begin{array}{ccc|ccc} 1 & -2 & 2 & 1 & 0 & 0 \\ 2 & -3 & 1 & 0 & 1 & 0 \\ 0 & 1 & -1 & 0 & 0 & 1 \end{array}\right]$

$$-2R_1+R_2 \to R_2 \longrightarrow \left[\begin{array}{ccc|ccc} 1 & -2 & 2 & 1 & 0 & 0 \\ 0 & 1 & -3 & -2 & 1 & 0 \\ 0 & 1 & -1 & 0 & 0 & 1 \end{array}\right]$$

$$\begin{array}{r} 2R_2+R_1 \to R_1 \longrightarrow \\ \\ -R_2+R_3 \to R_3 \longrightarrow \end{array} \left[\begin{array}{ccc|ccc} 1 & 0 & -4 & -3 & 2 & 0 \\ 0 & 1 & -3 & -2 & 1 & 0 \\ 0 & 0 & 2 & 2 & -1 & 1 \end{array}\right]$$

$$\tfrac{1}{2}R_3 \to R_3 \longrightarrow \left[\begin{array}{ccc|ccc} 1 & 0 & -4 & -3 & 2 & 0 \\ 0 & 1 & -3 & -2 & 1 & 0 \\ 0 & 0 & 1 & 1 & -\frac{1}{2} & \frac{1}{2} \end{array}\right]$$

$$\begin{array}{r} 4R_3+R_1 \to R_1 \longrightarrow \\ 3R_3+R_2 \to R_2 \longrightarrow \\ \\ \end{array} \left[\begin{array}{ccc|ccc} 1 & 0 & 0 & 1 & 0 & 2 \\ 0 & 1 & 0 & 1 & -\frac{1}{2} & \frac{3}{2} \\ 0 & 0 & 1 & 1 & -\frac{1}{2} & \frac{1}{2} \end{array}\right]$$

$$A^{-1} = \begin{bmatrix} 1 & 0 & 2 \\ 1 & -\frac{1}{2} & \frac{3}{2} \\ 1 & -\frac{1}{2} & \frac{1}{2} \end{bmatrix}$$

31. Set up matrix $[A|I_3]$: $\left[\begin{array}{ccc|ccc} 5 & 7 & -11 & 1 & 0 & 0 \\ 3 & 5 & -9 & 0 & 1 & 0 \\ 1 & 1 & -1 & 0 & 0 & 1 \end{array}\right]$

$$R_1 \Leftrightarrow R_3 \left[\begin{array}{ccc|ccc} 1 & 1 & -1 & 0 & 0 & 1 \\ 3 & 5 & -9 & 0 & 1 & 0 \\ 5 & 7 & -11 & 1 & 0 & 0 \end{array}\right]$$

$$\begin{array}{r} \\ -3R_1+R_2 \to R_2 \longrightarrow \\ -5R_1+R_3 \to R_3 \longrightarrow \end{array} \left[\begin{array}{ccc|ccc} 1 & 1 & -1 & 0 & 0 & 1 \\ 0 & 2 & -6 & 0 & 1 & -3 \\ 0 & 2 & -6 & 1 & 0 & -5 \end{array}\right]$$

$$-R_2+R_3 \to R_3 \longrightarrow \left[\begin{array}{ccc|ccc} 1 & 1 & -1 & 0 & 0 & 1 \\ 0 & 2 & -6 & 0 & 1 & -3 \\ 0 & 0 & 0 & 1 & -1 & -2 \end{array}\right]$$

A is a singular matrix.

33. Set up matrix $[A|I_3]$: $\left[\begin{array}{ccc|ccc} 0 & 1 & 1 & 1 & 0 & 0 \\ \frac{5}{3} & -1 & -3 & 0 & 1 & 0 \\ -1 & 1 & 2 & 0 & 0 & 1 \end{array}\right]$

$$R_1 \Leftrightarrow -R_3 \left[\begin{array}{ccc|ccc} 1 & -1 & -2 & 0 & 0 & -1 \\ \frac{5}{3} & -1 & -3 & 0 & 1 & 0 \\ 0 & 1 & 1 & 1 & 0 & 0 \end{array}\right]$$

$$-\tfrac{5}{3}R_1+R_2 \to R_2 \longrightarrow \left[\begin{array}{ccc|ccc} 1 & -1 & -2 & 0 & 0 & -1 \\ 0 & \frac{2}{3} & \frac{1}{3} & 0 & 1 & \frac{5}{3} \\ 0 & 1 & 1 & 1 & 0 & 0 \end{array}\right]$$

$$R_2 \Leftrightarrow -R_3 \longrightarrow \left[\begin{array}{ccc|ccc} 1 & -1 & -2 & 0 & 0 & -1 \\ 0 & 1 & 1 & 1 & 0 & 0 \\ 0 & \frac{2}{3} & \frac{1}{3} & 0 & 1 & \frac{5}{3} \end{array}\right]$$

$$\begin{array}{r} R_2+R_1 \to R_1 \longrightarrow \\ \\ -\tfrac{2}{3}R_2+R_3 \to R_3 \longrightarrow \end{array} \left[\begin{array}{ccc|ccc} 1 & 0 & -1 & 1 & 0 & -1 \\ 0 & 1 & 1 & 1 & 0 & 0 \\ 0 & 0 & -\frac{1}{3} & -\frac{2}{3} & 1 & \frac{5}{3} \end{array}\right]$$

$$-3R_3 \to R_3 \longrightarrow \left[\begin{array}{ccc|ccc} 1 & 0 & -1 & 1 & 0 & -1 \\ 0 & 1 & 1 & 1 & 0 & 0 \\ 0 & 0 & 1 & 2 & -3 & -5 \end{array}\right]$$

$$\begin{array}{r} R_3+R_1 \to R_1 \longrightarrow \\ -R_3+R_2 \to R_2 \longrightarrow \\ \\ \end{array} \left[\begin{array}{ccc|ccc} 1 & 0 & 0 & 3 & -3 & -6 \\ 0 & 1 & 0 & -1 & 3 & 5 \\ 0 & 0 & 1 & 2 & -3 & -5 \end{array}\right]$$

$$A^{-1} = \begin{bmatrix} 3 & -3 & -6 \\ -1 & 3 & 5 \\ 2 & -3 & -5 \end{bmatrix}$$

35. Set up matrix $[A|I_4]$: $\left[\begin{array}{cccc|cccc} 1 & -2 & 5 & 0 & 1 & 0 & 0 & 0 \\ 0 & -1 & 2 & 0 & 0 & 1 & 0 & 0 \\ 0 & 1 & -1 & 0 & 0 & 0 & 1 & 0 \\ 2 & -1 & 7 & 1 & 0 & 0 & 0 & 1 \end{array}\right]$

$$\begin{array}{r} \\ \\ \\ -2R_1+R_4 \to R_4 \longrightarrow \end{array} \left[\begin{array}{cccc|cccc} 1 & -2 & 5 & 0 & 1 & 0 & 0 & 0 \\ 0 & -1 & 2 & 0 & 0 & 1 & 0 & 0 \\ 0 & 1 & -1 & 0 & 0 & 0 & 1 & 0 \\ 0 & 3 & -3 & 1 & -2 & 0 & 0 & 1 \end{array}\right]$$

$$\begin{array}{r} \\ -R_2 \to R_2 \longrightarrow \\ \\ \\ \end{array} \left[\begin{array}{cccc|cccc} 1 & -2 & 5 & 0 & 1 & 0 & 0 & 0 \\ 0 & 1 & -2 & 0 & 0 & -1 & 0 & 0 \\ 0 & 1 & -1 & 0 & 0 & 0 & 1 & 0 \\ 0 & 3 & -3 & 1 & -2 & 0 & 0 & 1 \end{array}\right]$$

$$\begin{array}{r} 2R_2+R_1 \to R_1 \longrightarrow \\ \\ -R_2+R_3 \to R_3 \longrightarrow \\ -3R_2+R_4 \to R_4 \longrightarrow \end{array} \left[\begin{array}{cccc|cccc} 1 & 0 & 1 & 0 & 1 & -2 & 0 & 0 \\ 0 & 1 & -2 & 0 & 0 & -1 & 0 & 0 \\ 0 & 0 & 1 & 0 & 0 & 1 & 1 & 0 \\ 0 & 0 & 3 & 1 & -2 & 3 & 0 & 1 \end{array}\right]$$

$$\begin{array}{r} -R_3+R_1 \to R_1 \longrightarrow \\ 2R_3+R_2 \to R_2 \longrightarrow \\ \\ -3R_3+R_4 \to R_4 \longrightarrow \end{array} \left[\begin{array}{cccc|cccc} 1 & 0 & 0 & 0 & 1 & -3 & -1 & 0 \\ 0 & 1 & 0 & 0 & 0 & 1 & 2 & 0 \\ 0 & 0 & 1 & 0 & 0 & 1 & 1 & 0 \\ 0 & 0 & 0 & 1 & -2 & 0 & -3 & 1 \end{array}\right] \qquad A^{-1} = \begin{bmatrix} 1 & -3 & -1 & 0 \\ 0 & 1 & 2 & 0 \\ 0 & 1 & 1 & 0 \\ -2 & 0 & -3 & 1 \end{bmatrix}$$

37. $\begin{bmatrix} 3 & -4 \\ 2 & 1 \end{bmatrix}\begin{bmatrix} x \\ y \end{bmatrix} = \begin{bmatrix} -1 \\ 14 \end{bmatrix}$

39. $\begin{bmatrix} 9 & -6 & 4 \\ 4 & 0 & -1 \\ 0 & 3 & 1 \end{bmatrix}\begin{bmatrix} x \\ y \\ z \end{bmatrix} = \begin{bmatrix} 27 \\ 1 \\ 0 \end{bmatrix}$

41. $\begin{bmatrix} x \\ y \end{bmatrix} = \begin{bmatrix} -\frac{5}{2} & -\frac{3}{2} \\ 3 & 2 \end{bmatrix}\begin{bmatrix} -4 \\ 8 \end{bmatrix}$

$\begin{bmatrix} x \\ y \end{bmatrix} = \begin{bmatrix} -\frac{5}{2}(-4)+\left(-\frac{3}{2}\right)(8) \\ 3(-4)+2(8) \end{bmatrix} = \begin{bmatrix} -2 \\ 4 \end{bmatrix}$

$\{(-2, 4)\}$

43. Set up matrix $[A|I_2]$: $\left[\begin{array}{cc|cc} 3 & 7 & 1 & 0 \\ 4 & 9 & 0 & 1 \end{array}\right]$

$R_1 \Leftrightarrow R_2 \longrightarrow \left[\begin{array}{cc|cc} 4 & 9 & 0 & 1 \\ 3 & 7 & 1 & 0 \end{array}\right]$

$R_1+(-1)R_2 \to R_1 \longrightarrow \left[\begin{array}{cc|cc} 1 & 2 & -1 & 1 \\ 3 & 7 & 1 & 0 \end{array}\right]$

$-3R_1+R_2 \to R_2 \longrightarrow \left[\begin{array}{cc|cc} 1 & 2 & -1 & 1 \\ 0 & 1 & 4 & -3 \end{array}\right]$

$-2R_2+R_1 \to R_1 \longrightarrow \left[\begin{array}{cc|cc} 1 & 0 & -9 & 7 \\ 0 & 1 & 4 & -3 \end{array}\right]$

$\begin{bmatrix} x \\ y \end{bmatrix} = \begin{bmatrix} -9 & 7 \\ 4 & -3 \end{bmatrix}\begin{bmatrix} -5 \\ -7 \end{bmatrix}$

$\begin{bmatrix} x \\ y \end{bmatrix} = \begin{bmatrix} -9(-5)+7(-7) \\ 4(-5)+(-3)(-7) \end{bmatrix} = \begin{bmatrix} -4 \\ 1 \end{bmatrix}$

$\{(-4, 1)\}$

45. $\begin{bmatrix} x \\ y \\ z \end{bmatrix} = \begin{bmatrix} 3 & -2 & -3 \\ 3 & -4 & -2 \\ 2 & -3 & -1 \end{bmatrix}\begin{bmatrix} 1 \\ 0 \\ 2 \end{bmatrix} = \begin{bmatrix} 3(1)+(-2)(0)+(-3)(2) \\ 3(1)+(-4)(0)+(-2)(2) \\ 2(1)+(-3)(0)+(-1)(2) \end{bmatrix} = \begin{bmatrix} -3 \\ -1 \\ 0 \end{bmatrix}$ $\{(-3,-1,0)\}$

47. $\begin{bmatrix} x \\ y \\ z \end{bmatrix} = \begin{bmatrix} 1 & 0 & 2 \\ 1 & -\frac{1}{2} & \frac{3}{2} \\ 1 & -\frac{1}{2} & \frac{1}{2} \end{bmatrix}\begin{bmatrix} -12 \\ -10 \\ 6 \end{bmatrix} = \begin{bmatrix} 1(-12)+(0)(-10)+2(6) \\ 1(-12)+\left(-\frac{1}{2}\right)(-10)+\frac{3}{2}(6) \\ 1(-12)+\left(-\frac{1}{2}\right)(-10)+\frac{1}{2}(6) \end{bmatrix} = \begin{bmatrix} 0 \\ 2 \\ -4 \end{bmatrix}$ $\{(0,2,-4)\}$

49. $\begin{bmatrix} w \\ x \\ y \\ z \end{bmatrix} = \begin{bmatrix} 1 & -3 & -1 & 0 \\ 0 & 1 & 2 & 0 \\ 0 & 1 & 1 & 0 \\ -2 & 0 & -3 & 1 \end{bmatrix}\begin{bmatrix} 3 \\ 1 \\ -1 \\ 5 \end{bmatrix} = \begin{bmatrix} 1(3)+(-3)(1)+(-1)(-1)+(0)(5) \\ (0)(3)+1(1)+2(-1)+(0)(5) \\ (0)(3)+1(1)+1(-1)+(0)(5) \\ (-2)(3)+(0)(1)+(-3)(-1)+1(5) \end{bmatrix} = \begin{bmatrix} 1 \\ -1 \\ 0 \\ 2 \end{bmatrix}$ $\{(1,-1,0,2)\}$

51. False

53. False

55. True

57. False

59. Given $AA^{-1} = I_2$ and $A = A^{-1}$:

$$\begin{bmatrix} a & b \\ c & d \end{bmatrix}\begin{bmatrix} a & b \\ c & d \end{bmatrix} = \begin{bmatrix} 1 & 0 \\ 0 & 1 \end{bmatrix}$$

$$\begin{bmatrix} a^2+bc & ab+bd \\ ac+cd & bc+d^2 \end{bmatrix} = \begin{bmatrix} 1 & 0 \\ 0 & 1 \end{bmatrix}$$

$a^2+bc=1$

$ab+bd=0 \rightarrow b(a+d)=0$

$ac+cd=0 \rightarrow c(a+d)=0$

$bc+d^2=1$

Let $a=0$.

$a^2+bc=1$

$(0)^2+bc=1$

$bc=1$

$ab+bd=0$

$(0)b+bd=0$

$bd=0$

$ac+cd=0$

$(0)c+cd=0$

$cd=0$

Since $bc=1$, neither b nor c can be equal to 0. Therefore, since $bd=0$ and $cd=0$, $d=0$.

Pick b and c such that $bc=1$.

One example: $\begin{bmatrix} 0 & 1 \\ 1 & 0 \end{bmatrix}$

61. a. $A^{-1} = \dfrac{1}{ad-bc}\begin{bmatrix} d & -b \\ -c & a \end{bmatrix}$

$= \dfrac{1}{3(6)-2(5)}\begin{bmatrix} 6 & -2 \\ -5 & 3 \end{bmatrix}$

$= \dfrac{1}{8}\begin{bmatrix} 6 & -2 \\ -5 & 3 \end{bmatrix} = \begin{bmatrix} \frac{3}{4} & -\frac{1}{4} \\ -\frac{5}{8} & \frac{3}{8} \end{bmatrix}$

b. $\left(A^{-1}\right)^{-1} = \dfrac{1}{ad-bc}\begin{bmatrix} d & -b \\ -c & a \end{bmatrix}$

$= \dfrac{1}{\frac{3}{4}\left(\frac{3}{8}\right)-\left(-\frac{1}{4}\right)\left(-\frac{5}{8}\right)}\begin{bmatrix} \frac{3}{8} & \frac{1}{4} \\ \frac{5}{8} & \frac{3}{4} \end{bmatrix}$

$= 8\begin{bmatrix} \frac{3}{8} & \frac{1}{4} \\ \frac{5}{8} & \frac{3}{4} \end{bmatrix} = \begin{bmatrix} 3 & 2 \\ 5 & \frac{6}{8} \end{bmatrix}$

63. $A^{-1}=\dfrac{1}{ad-bc}\begin{bmatrix} d & -b \\ -c & a \end{bmatrix}=\dfrac{1}{ab-(0)(0)}\begin{bmatrix} b & 0 \\ 0 & a \end{bmatrix}=\dfrac{1}{ab}\begin{bmatrix} b & 0 \\ 0 & a \end{bmatrix}=\begin{bmatrix} \frac{1}{a} & 0 \\ 0 & \frac{1}{b} \end{bmatrix}$

65. If $AB=I_n$ and $BA=I_n$, then the matrices are inverses.

67. Set up matrix $[A|I_2]$: $\left[\begin{array}{cc|cc} a & b & 1 & 0 \\ c & d & 0 & 1 \end{array}\right]$

$\frac{1}{a}R_1 \to R_1 \longrightarrow \left[\begin{array}{cc|cc} 1 & \frac{b}{a} & \frac{1}{a} & 0 \\ c & d & 0 & 1 \end{array}\right]$

$-cR_1+R_2 \to R_2 \longrightarrow \left[\begin{array}{cc|cc} 1 & \frac{b}{a} & \frac{1}{a} & 0 \\ 0 & \frac{ad-bc}{a} & -\frac{c}{a} & 1 \end{array}\right]$

$\left(\frac{a}{ad-bc}\right)R_2 \to R_2 \longrightarrow \left[\begin{array}{cc|cc} 1 & \frac{b}{a} & \frac{1}{a} & 0 \\ 0 & 1 & -\frac{c}{ad-bc} & \frac{a}{ad+bc} \end{array}\right]$

$\left(-\frac{b}{a}\right)R_2+R_1 \to R_1 \longrightarrow \left[\begin{array}{cc|cc} 1 & 0 & \frac{d}{ad-bc} & -\frac{b}{ad-bc} \\ 0 & 1 & -\frac{c}{ad-bc} & \frac{a}{ad-bc} \end{array}\right]$

Therefore, $A^{-1}=\begin{bmatrix} \frac{d}{ad-bc} & -\frac{b}{ad-bc} \\ -\frac{c}{ad-bc} & \frac{a}{ad-bc} \end{bmatrix}=\dfrac{1}{ad-bc}\begin{bmatrix} d & -b \\ -c & a \end{bmatrix}$, provided $ad-bc \neq 0$.

69. $x_1=\frac{1}{4}(36+32+x_2+x_3)$

$x_1=\frac{1}{4}(68+x_2+x_3) \to 4x_1=68+x_2+x_3 \to -4x_1+x_2+x_3=-68$

$x_2=\frac{1}{4}(32+40+x_1+x_4)$

$x_2=\frac{1}{4}(72+x_1+x_4) \to 4x_2=72+x_1+x_4 \to x_1-4x_2+x_4=-72$

$x_3=\frac{1}{4}(36+48+x_1+x_4)$

$x_3=\frac{1}{4}(84+x_1+x_4) \to 4x_3=84+x_1+x_4 \to x_1-4x_3+x_4=-84$

$x_4=\frac{1}{4}(40+48+x_2+x_3)$

$x_4=\frac{1}{4}(88+x_2+x_3) \to 4x_4=88+x_2+x_3 \to x_2+x_3-4x_4=-88$

Set up matrix $[A \mid I_4]$:
$$\left[\begin{array}{cccc|cccc} -4 & 1 & 1 & 0 & 1 & 0 & 0 & 0 \\ 1 & -4 & 0 & 1 & 0 & 1 & 0 & 0 \\ 1 & 0 & -4 & 1 & 0 & 0 & 1 & 0 \\ 0 & 1 & 1 & -4 & 0 & 0 & 0 & 1 \end{array}\right]$$

$R_1 \Leftrightarrow R_2 \longrightarrow$
$$\left[\begin{array}{cccc|cccc} 1 & -4 & 0 & 1 & 0 & 1 & 0 & 0 \\ -4 & 1 & 1 & 0 & 1 & 0 & 0 & 0 \\ 1 & 0 & -4 & 1 & 0 & 0 & 1 & 0 \\ 0 & 1 & 1 & -4 & 0 & 0 & 0 & 1 \end{array}\right]$$

$4R_1 + R_2 \rightarrow R_2 \longrightarrow$
$-R_1 + R_3 \rightarrow R_3 \longrightarrow$
$$\left[\begin{array}{cccc|cccc} 1 & -4 & 0 & 1 & 0 & 1 & 0 & 0 \\ 0 & -15 & 1 & 4 & 1 & 4 & 0 & 0 \\ 0 & 4 & -4 & 0 & 0 & -1 & 1 & 0 \\ 0 & 1 & 1 & -4 & 0 & 0 & 0 & 1 \end{array}\right]$$

$R_2 \Leftrightarrow -R_4$
$$\left[\begin{array}{cccc|cccc} 1 & -4 & 0 & 1 & 0 & 1 & 0 & 0 \\ 0 & 1 & 1 & -4 & 0 & 0 & 0 & 1 \\ 0 & 4 & -4 & 0 & 0 & -1 & 1 & 0 \\ 0 & -15 & 1 & 4 & 1 & 4 & 0 & 0 \end{array}\right]$$

$R_3 + R_1 \rightarrow R_1 \longrightarrow$
$-4R_2 + R_3 \rightarrow R_3 \longrightarrow$
$15R_2 + R_4 \rightarrow R_4 \longrightarrow$
$$\left[\begin{array}{cccc|cccc} 1 & 0 & -4 & 1 & 0 & 0 & 1 & 0 \\ 0 & 1 & 1 & -4 & 0 & 0 & 0 & 1 \\ 0 & 0 & -8 & 16 & 0 & -1 & 1 & -4 \\ 0 & 0 & 16 & -56 & 1 & 4 & 0 & 15 \end{array}\right]$$

$-\frac{1}{8}R_3 \rightarrow R_3 \longrightarrow$
$$\left[\begin{array}{cccc|cccc} 1 & 0 & -4 & 1 & 0 & 0 & 1 & 0 \\ 0 & 1 & 1 & -4 & 0 & 0 & 0 & 1 \\ 0 & 0 & 1 & -2 & 0 & \frac{1}{8} & -\frac{1}{8} & \frac{1}{2} \\ 0 & 0 & 16 & -56 & 1 & 4 & 0 & 15 \end{array}\right]$$

$4R_3 + R_1 \rightarrow R_1 \longrightarrow$
$-R_3 + R_2 \rightarrow R_2 \longrightarrow$
$-16R_3 + R_4 \rightarrow R_4 \longrightarrow$
$$\left[\begin{array}{cccc|cccc} 1 & 0 & 0 & -7 & 0 & \frac{1}{2} & \frac{1}{2} & 2 \\ 0 & 1 & 0 & -2 & 0 & -\frac{1}{8} & \frac{1}{8} & \frac{1}{2} \\ 0 & 0 & 1 & -2 & 0 & \frac{1}{8} & -\frac{1}{8} & \frac{1}{2} \\ 0 & 0 & 0 & -24 & 1 & 2 & 2 & 7 \end{array}\right]$$

$-\frac{1}{24}R_4 \rightarrow R_4 \longrightarrow$
$$\left[\begin{array}{cccc|cccc} 1 & 0 & 0 & -7 & 0 & \frac{1}{2} & \frac{1}{2} & 2 \\ 0 & 1 & 0 & -2 & 0 & -\frac{1}{8} & \frac{1}{8} & \frac{1}{2} \\ 0 & 0 & 1 & -2 & 0 & \frac{1}{8} & -\frac{1}{8} & \frac{1}{2} \\ 0 & 0 & 0 & 1 & -\frac{1}{24} & -\frac{1}{12} & -\frac{1}{12} & -\frac{7}{24} \end{array}\right]$$

$$\begin{array}{l} 7R_4+R_1\to R_1 \longrightarrow \\ 2R_4+R_2\to R_2 \longrightarrow \\ 2R_4+R_3\to R_3 \longrightarrow \end{array} \left[\begin{array}{cccc|cccc} 1 & 0 & 0 & 0 & -\frac{7}{24} & -\frac{1}{12} & -\frac{1}{12} & -\frac{1}{24} \\ 0 & 1 & 0 & 0 & -\frac{1}{12} & -\frac{7}{24} & -\frac{1}{24} & -\frac{1}{12} \\ 0 & 0 & 1 & 0 & -\frac{1}{12} & -\frac{1}{24} & -\frac{7}{24} & -\frac{1}{12} \\ 0 & 0 & 0 & 1 & -\frac{1}{24} & -\frac{1}{12} & -\frac{1}{12} & -\frac{7}{24} \end{array}\right]$$

$$\begin{bmatrix} x_1 \\ x_2 \\ x_3 \\ x_4 \end{bmatrix} = \begin{bmatrix} -\frac{7}{24} & -\frac{1}{12} & -\frac{1}{12} & -\frac{1}{24} \\ -\frac{1}{12} & -\frac{7}{24} & -\frac{1}{24} & -\frac{1}{12} \\ -\frac{1}{12} & -\frac{1}{24} & -\frac{7}{24} & -\frac{1}{12} \\ -\frac{1}{24} & -\frac{1}{12} & -\frac{1}{12} & -\frac{7}{24} \end{bmatrix} \begin{bmatrix} -68 \\ -72 \\ -84 \\ -88 \end{bmatrix} = \begin{bmatrix} -\frac{7}{24}(68)+\left(-\frac{1}{12}\right)(72)+\left(-\frac{1}{12}\right)(84)+\left(-\frac{1}{24}\right)(88) \\ -\frac{1}{12}(68)+\left(-\frac{7}{24}\right)(72)+\left(-\frac{1}{24}\right)(84)+\left(-\frac{1}{12}\right)(88) \\ -\frac{1}{12}(68)+\left(-\frac{1}{24}\right)(72)+\left(-\frac{7}{24}\right)(84)+\left(-\frac{1}{12}\right)(88) \\ -\frac{1}{24}(68)+\left(-\frac{1}{12}\right)(72)+\left(-\frac{1}{12}\right)(84)+\left(-\frac{7}{24}\right)(88) \end{bmatrix} = \begin{bmatrix} 36.5 \\ 37.5 \\ 40.5 \\ 41.5 \end{bmatrix}$$

$x_1 = 36.5°$; $x_2 = 37.5°$; $x_3 = 40.5°$; $x_4 = 41.5°$

71.

```
[A]⁻¹
[.29  -.10  .08
 -.09  .12  -.00
 -.12  .00  -.07
 .15  .03  -.06
```

$$A^{-1} = \begin{bmatrix} 0.29 & -0.10 & 0.08 & -0.27 \\ -0.09 & 0.12 & 0.00 & 0.27 \\ -0.12 & 0.00 & -0.07 & 0.25 \\ 0.15 & 0.03 & -0.06 & -0.01 \end{bmatrix}$$

73. $\{(5.35, 41.71, 4.45)\}$

Section 6.5 Determinants and Cramer's Rule

1. $-\frac{1}{3}x = -\frac{1}{6} - \frac{1}{2}y$

$x = \frac{1}{2} + \frac{3}{2}y$

$2x - 3y = 1$

$2\left(\frac{1}{2} + \frac{3}{2}y\right) - 3y = 1$

$1 + 3y - 3y = 1$

$1 = 1$ identity

$\{(x, y) \mid 2x - 3y = 1\}$

3. $\boxed{A}$ $3x + z = 4$

$\boxed{B}$ $2y + 5z = 1$

$\boxed{C}$ $9x + 5y = -1$

$-3 \cdot \boxed{A}$ $-9x - 3z = -12$

$\boxed{C}$ $9x + 5y = -1$

$5y - 3z = -13$ $\boxed{D}$

$5 \cdot \boxed{B}$ $10y + 25z = 5$

$-2 \cdot \boxed{D}$ $-10y + 6z = 26$

$31z = 31$

$z = 1$

$\boxed{A}$ $3x + z = 4$

$3x + 1 = 4$

$3x = 3$

$x = 1$

$\boxed{B}$ $2y + 5z = 1$

$2y + 5(1) = 1$

$2y + 5 = 1$

$2y = -4$

$y = -2$

$\{(1, -2, 1)\}$

5. $\boxed{A}$ $x+y+4z=5$

$\boxed{B}$ $-2x-3y-14z=-14$

$\boxed{C}$ $-3x-y=7$

$$\begin{array}{rrcl} 7\cdot\boxed{A} & 7x+7y+28z &=& 35 \\ 2\cdot\boxed{B} & -4x-6y-28z &=& -28 \\ \hline & 3x+y &=& 7 \quad \boxed{D} \end{array}$$

$$\begin{array}{rrcl} \boxed{C} & -3x-y &=& 7 \\ \boxed{D} & 3x+y &=& 7 \\ \hline & 0 &=& 14 \end{array}$$

$\{\ \}$

7. determinant

9. minor

11. $\begin{vmatrix} b_2 & c_2 \\ b_3 & c_3 \end{vmatrix}$; $\begin{vmatrix} b_1 & c_1 \\ b_3 & c_3 \end{vmatrix}$; $\begin{vmatrix} b_1 & c_1 \\ b_2 & c_2 \end{vmatrix}$

13. does not

15. $|A| = ad-bc$
$= (3)(5)-(-2)(6) = 15+12 = 27$

17. $|C| = ad-bc$
$= \left(\frac{2}{3}\right)(12)-\left(\frac{1}{5}\right)(10) = 8-2 = 6$

19. $|E| = ad-bc$
$= (-3)(0)-(0)(4) = 0-0 = 0$

21. $|G| = ad-bc$
$= (x)(x)-(4)(9)$
$= x^2-36$

23. $|T| = ad-bc$
$= (e^x)(-e^x)-(e^{2x})(4)$
$= -e^{2x}-4e^{2x}$
$= -5e^{2x}$

25. a. $M_{12} = \begin{vmatrix} 4 & -5 \\ -3 & 10 \end{vmatrix}$
$= (4)(10)-(-5)(-3)$
$= 40-15$
$= 25$

b. $(-1)^{i+j} M_{ij} = (-1)^{1+2} M_{12}$
$= (-1)^3(25)$
$= -25$

27. a. $M_{31} = \begin{vmatrix} 11 & 8 \\ -2 & -5 \end{vmatrix}$
$= (11)(-5)-(8)(-2)$
$= -55+16$
$= -39$

b. $(-1)^{i+j} M_{ij} = (-1)^{3+1} M_{31}$
$= (-1)^4(-39)$
$= -39$

29. a. $M_{22} = \begin{vmatrix} -6 & 8 \\ -3 & 10 \end{vmatrix}$
$= (-6)(10)-(8)(-3)$
$= -60+24$
$= -36$

b. $(-1)^{i+j} M_{ij} = (-1)^{2+2} M_{22}$
$= (-1)^4(-36)$
$= -36$

31. Expand cofactors about the elements in the first column.

$|A| = 4(-1)^{1+1}\begin{vmatrix} -1 & 2 \\ 8 & 0 \end{vmatrix} + 0(-1)^{2+1}\begin{vmatrix} 1 & 3 \\ 8 & 0 \end{vmatrix}$
$+5(-1)^{3+1}\begin{vmatrix} 1 & 3 \\ -1 & 2 \end{vmatrix}$
$= 4(1)[(-1)(0)-(2)(8)]$
$+5(1)[(1)(2)-(3)(-1)]$
$= 4(-16)+5(5) = -64+25 = -39$

$|A| \neq 0$, so the matrix is invertible.

33. Expand cofactors about the elements in the first row.

$$|C| = 5(-1)^{1+1}\begin{vmatrix} 3 & 4 \\ -1 & 7 \end{vmatrix} + 1(-1)^{1+2}\begin{vmatrix} 2 & 4 \\ 8 & 7 \end{vmatrix} + 6(-1)^{1+3}\begin{vmatrix} 2 & 3 \\ 8 & -1 \end{vmatrix}$$

$$= 5(1)[(3)(7)-(4)(-1)] + 1(-1)[(2)(7)-(4)(8)] + 6(1)[(2)(-1)-(3)(8)]$$

$$= 5(25) - 1(-18) + 6(-26)$$

$$= 125 + 18 - 156 = -13$$

$|C| \neq 0$, so the matrix is invertible.

35. Expand cofactors about the elements in the first row.

$$|E| = 2(-1)^{1+1}\begin{vmatrix} -1 & 2 \\ 1 & 0 \end{vmatrix} + 0(-1)^{1+2}\begin{vmatrix} 1 & 2 \\ 3 & 0 \end{vmatrix} + 1(-1)^{1+3}\begin{vmatrix} 1 & -1 \\ 3 & 1 \end{vmatrix}$$

$$= 2(1)[(-1)(0)-(2)(1)] + 0(-1)[(1)(0)-(2)(3)] + 1(1)[(1)(1)-(-1)(3)]$$

$$= 2(-2) + 1(4) = -4 + 4 = 0$$

$|E| = 0$, so the matrix is not invertible.

37. Expand cofactors about the elements in the fourth column. Only one cofactor is nonzero. Expand cofactors of the resulting determinant about the elements in the first row.

$$|G| = 1(-1)^{1+4}\begin{vmatrix} 2 & 0 & 3 \\ 0 & 1 & 4 \\ -1 & 2 & 0 \end{vmatrix}$$

$$= 1(-1)\left\{2(-1)^{1+1}\begin{vmatrix} 1 & 4 \\ 2 & 0 \end{vmatrix} + 0(-1)^{1+2}\begin{vmatrix} 0 & 4 \\ -1 & 0 \end{vmatrix} + 3(-1)^{1+3}\begin{vmatrix} 0 & 1 \\ -1 & 2 \end{vmatrix}\right\}$$

$$= -1\{2(1)[(1)(0)-(4)(2)] + 3[(0)(2)-(1)(-1)]\} = -1[2(-8)+3(1)] = -1(-16+3) = 13$$

$|G| \neq 0$, so the matrix is invertible.

39. Expand cofactors about the elements in the third column. Only two cofactors are nonzero. Expand cofactors of the resulting determinants about the elements in the third row.

$$|T| = 1(-1)^{1+3}\begin{vmatrix} -2 & 4 & 5 \\ -1 & 1 & -1 \\ 0 & 5 & 3 \end{vmatrix} + 2(-1)^{4+3}\begin{vmatrix} 3 & 8 & 4 \\ -2 & 4 & 5 \\ -1 & 1 & -1 \end{vmatrix}$$

$$= 1(1)\left\{0(-1)^{3+1}\begin{vmatrix} 4 & 5 \\ 1 & -1 \end{vmatrix} + 5(-1)^{3+2}\begin{vmatrix} -2 & 5 \\ -1 & -1 \end{vmatrix} + 3(-1)^{3+3}\begin{vmatrix} -2 & 4 \\ -1 & 1 \end{vmatrix}\right\}$$

$$+ 2(-1)\left\{-1(-1)^{3+1}\begin{vmatrix} 8 & 4 \\ 4 & 5 \end{vmatrix} + 1(-1)^{3+2}\begin{vmatrix} 3 & 4 \\ -2 & 5 \end{vmatrix} - 1(-1)^{3+3}\begin{vmatrix} 3 & 8 \\ -2 & 4 \end{vmatrix}\right\}$$

$$= 1\{5(-1)[(-2)(-1)-(5)(-1)] + 3(1)[(-2)(1)-(4)(-1)]\}$$

$$-2\{-1(1)[(8)(5)-(4)(4)] + 1(-1)[(3)(5)-(4)(-2)] - 1(1)[(3)(4)-(8)(-2)]\}$$

$$= 1[-5(7)+3(2)] - 2[-1(24)-1(23)-1(28)]$$

$$= (-35+6) - 2(-24-23-28) = -29 - 2(-75) = -29 + 150 = 121$$

$|T| \neq 0$, so the matrix is invertible.

41. $a_1=2, b_1=10, c_1=11, a_2=3, b_2=-5, c_2=6$

$$D=\begin{vmatrix} a_1 & b_1 \\ a_2 & b_2 \end{vmatrix}=\begin{vmatrix} 2 & 10 \\ 3 & -5 \end{vmatrix}=(2)(-5)-(10)(3)$$
$$=-10-30=-40$$
$$D_x=\begin{vmatrix} c_1 & b_1 \\ c_2 & b_2 \end{vmatrix}=\begin{vmatrix} 11 & 10 \\ 6 & -5 \end{vmatrix}=(11)(-5)-(10)(6)$$
$$=-55-60=-115$$
$$D_y=\begin{vmatrix} a_1 & c_1 \\ a_2 & c_2 \end{vmatrix}=\begin{vmatrix} 2 & 11 \\ 3 & 6 \end{vmatrix}=(2)(6)-(11)(3)$$
$$=12-33=-21$$
$$x=\frac{D_x}{D}=\frac{-115}{-40}=\frac{23}{8},\ y=\frac{D_y}{D}=\frac{-21}{-40}=\frac{21}{40}$$
$$\left\{\left(\frac{23}{8},\frac{21}{40}\right)\right\}$$

43. $-10x+4y=7 \rightarrow -10x+4y=7$

$6x=7y+2 \quad \rightarrow 6x-7y=2$

$a_1=-10, b_1=4, c_1=7, a_2=6, b_2=-7, c_2=2$

$$D=\begin{vmatrix} a_1 & b_1 \\ a_2 & b_2 \end{vmatrix}=\begin{vmatrix} -10 & 4 \\ 6 & -7 \end{vmatrix}$$
$$=(-10)(-7)-(4)(6)=70-24=46$$
$$D_x=\begin{vmatrix} c_1 & b_1 \\ c_2 & b_2 \end{vmatrix}=\begin{vmatrix} 7 & 4 \\ 2 & -7 \end{vmatrix}=(7)(-7)-(4)(2)$$
$$=-49-8=-57$$
$$D_y=\begin{vmatrix} a_1 & c_1 \\ a_2 & c_2 \end{vmatrix}=\begin{vmatrix} -10 & 7 \\ 6 & 2 \end{vmatrix}=(-10)(2)-(7)(6)$$
$$=-20-42=-62$$
$$x=\frac{D_x}{D}=\frac{-57}{46}=-\frac{57}{46},\ y=\frac{D_y}{D}=\frac{-62}{46}=-\frac{31}{23}$$
$$\left\{\left(-\frac{57}{46},-\frac{31}{23}\right)\right\}$$

45. $3(x-y)=y+8 \rightarrow 3x-4y=8$

$y=\frac{3}{4}x-2 \quad \rightarrow -\frac{3}{4}x+y=-2$

$a_1=3, b_1=-4, c_1=8, a_2=-\frac{3}{4}, b_2=1, c_2=-2$

$$D=\begin{vmatrix} a_1 & b_1 \\ a_2 & b_2 \end{vmatrix}=\begin{vmatrix} 3 & -4 \\ -\frac{3}{4} & 1 \end{vmatrix}=(3)(1)-(-4)\left(-\frac{3}{4}\right)$$
$$=3-3=0$$
$$D_x=\begin{vmatrix} c_1 & b_1 \\ c_2 & b_2 \end{vmatrix}=\begin{vmatrix} 8 & -4 \\ -2 & 1 \end{vmatrix}=(8)(1)-(-4)(-2)$$
$$=8-8=0$$
$$D_y=\begin{vmatrix} a_1 & c_1 \\ a_2 & c_2 \end{vmatrix}=\begin{vmatrix} 3 & 8 \\ -\frac{3}{4} & -2 \end{vmatrix}$$
$$=(3)(-2)-(8)\left(-\frac{3}{4}\right)=-6+6=0$$

The determinant D is zero as are both D_x and D_y. Therefore, the equations are dependent and the system has infinitely many solutions.

47. $y=-3x+7 \rightarrow 3x+y=7$

$\frac{1}{2}x+\frac{1}{6}y=1 \rightarrow \frac{1}{2}x+\frac{1}{6}y=1$

$a_1=3, b_1=1, c_1=7, a_2=\frac{1}{2}, b_2=\frac{1}{6}, c_2=1$

$$D=\begin{vmatrix} a_1 & b_1 \\ a_2 & b_2 \end{vmatrix}=\begin{vmatrix} 3 & 1 \\ \frac{1}{2} & \frac{1}{6} \end{vmatrix}$$
$$=(3)\left(\frac{1}{6}\right)-(1)\left(\frac{1}{2}\right)=\frac{1}{2}-\frac{1}{2}=0$$
$$D_x=\begin{vmatrix} c_1 & b_1 \\ c_2 & b_2 \end{vmatrix}=\begin{vmatrix} 7 & 1 \\ 1 & \frac{1}{6} \end{vmatrix}$$
$$=(7)\left(\frac{1}{6}\right)-(1)(1)=\frac{7}{6}-1=\frac{1}{6}$$

The determinant D is zero and at least one of the determinants D_x in the numerator is nonzero, so the system has no solution.

49. Use a graphing utility to find the determinants of the matrices.

$$D=\begin{vmatrix} 11 & 0 & -3 \\ 0 & 2 & 9 \\ 4 & 5 & 0 \end{vmatrix}=-471$$
$$D_x=\begin{vmatrix} 1 & 0 & -3 \\ 6 & 2 & 9 \\ -9 & 5 & 0 \end{vmatrix}=-189$$
$$D_y=\begin{vmatrix} 11 & 1 & -3 \\ 0 & 6 & 9 \\ 4 & -9 & 0 \end{vmatrix}=999$$

$$D_z = \begin{vmatrix} 11 & 0 & 1 \\ 0 & 2 & 6 \\ 4 & 5 & -9 \end{vmatrix} = -536$$

$$x = \frac{D_x}{D} = \frac{-189}{-471} = \frac{63}{157}$$

$$y = \frac{D_y}{D} = \frac{999}{-471} = -\frac{333}{157}$$

$$z = \frac{D_z}{D} = \frac{-536}{-471} = \frac{536}{471}$$

$$\left\{\left(\frac{63}{157}, -\frac{333}{157}, \frac{536}{471}\right)\right\}$$

51. Use a graphing utility to find the determinants of the matrices.

$$D = \begin{vmatrix} 2 & -5 & 1 \\ 3 & 7 & -4 \\ 1 & -9 & 2 \end{vmatrix} = -28$$

$$D_x = \begin{vmatrix} 11 & -5 & 1 \\ 8 & 7 & -4 \\ 4 & -9 & 2 \end{vmatrix} = -182$$

$$D_y = \begin{vmatrix} 2 & 11 & 1 \\ 3 & 8 & -4 \\ 1 & 4 & 2 \end{vmatrix} = -42$$

$$D_z = \begin{vmatrix} 2 & -5 & 11 \\ 3 & 7 & 8 \\ 1 & -9 & 4 \end{vmatrix} = -154$$

$$x = \frac{D_x}{D} = \frac{-182}{-28} = \frac{13}{2}$$

$$y = \frac{D_y}{D} = \frac{-42}{-28} = \frac{3}{2}$$

$$z = \frac{D_z}{D} = \frac{-154}{-28} = \frac{11}{2}$$

$$\left\{\left(\frac{13}{2}, \frac{3}{2}, \frac{11}{2}\right)\right\}$$

53. Use a graphing utility to find the determinants of the matrices.

$$D = \begin{vmatrix} 2 & -3 & 1 \\ -4 & 6 & -2 \\ 6 & -9 & 3 \end{vmatrix} = 0$$

$$D_x = \begin{vmatrix} 6 & -3 & 1 \\ -12 & 6 & -2 \\ 18 & -9 & 3 \end{vmatrix} = 0$$

$$D_y = \begin{vmatrix} 2 & 6 & 1 \\ -4 & -12 & -2 \\ 6 & 18 & 3 \end{vmatrix} = 0$$

$$D_z = \begin{vmatrix} 2 & -3 & 6 \\ -4 & 6 & -12 \\ 6 & -9 & -18 \end{vmatrix} = 0$$

The determinant D is zero as are D_x, D_y, and D_z. Therefore, the equations are dependent and the system has infinitely many solutions.

55. Use a graphing utility to find the determinants of the matrices.

$$D = \begin{vmatrix} 1 & -2 & 3 \\ 5 & -7 & 3 \\ 1 & 0 & -5 \end{vmatrix} = 0$$

$$D_x = \begin{vmatrix} -1 & -2 & 3 \\ 1 & -7 & 3 \\ 2 & 0 & -5 \end{vmatrix} = -15$$

The determinant D is zero and at least one of the determinants D_x in the numerator is nonzero, so the system has no solution.

57. Use a graphing utility to find the determinants of the matrices.

$$D = \begin{vmatrix} 1 & 2 & 3 & -4 \\ 0 & 5 & 0 & 1 \\ 1 & 0 & 4 & 0 \\ 0 & 0 & 5 & -2 \end{vmatrix} = -120$$

$$D_{x_2} = \begin{vmatrix} 1 & 3 & 3 & -4 \\ 0 & 9 & 0 & 1 \\ 1 & -1 & 4 & 0 \\ 0 & 8 & 5 & -2 \end{vmatrix} = -226$$

$$x_2 = \frac{D_{x_2}}{D} = \frac{-226}{-120} = \frac{113}{60}$$

59. Expand cofactors about the elements in the third column.

$$\begin{vmatrix} 3 & 6 & 1 \\ 6 & 10 & 1 \\ -3 & -2 & 1 \end{vmatrix} = 1(-1)^{1+3}\left[(6)(-2)-(10)(-3)\right]$$
$$+1(-1)^{2+3}\left[(3)(-2)-(6)(-3)\right]$$
$$+1(-1)^{3+3}\left[(3)(10)-(6)(6)\right]$$
$$=1(1)(18)+1(-1)(12)+1(1)(-6)$$
$$=18-12-6=0$$

Yes, the points are collinear.

61. Expand cofactors about the elements in the third column.

$$\begin{vmatrix} 4 & -3 & 1 \\ 5 & -7 & 1 \\ 8 & -14 & 1 \end{vmatrix} = 1(-1)^{1+3}\left[(5)(-14)-(-7)(8)\right]$$
$$+1(-1)^{2+3}\left[(4)(-14)-(-3)(8)\right]$$
$$+1(-1)^{3+3}\left[(4)(-7)-(-3)(5)\right]$$
$$=1(1)(-14)+1(-1)(-32)$$
$$+1(1)(-13)$$
$$=-14+32-13=5$$

No, the points are not collinear.

63. a. $\begin{vmatrix} x & y & 1 \\ -3 & 2 & 1 \\ -4 & 6 & 1 \end{vmatrix} = 0$

b. $(x)(-1)^{1+1}\left[(2)(1)-(1)(6)\right]+(y)(-1)^{1+2}\left[(-3)(1)-(1)(-4)\right]+(1)(-1)^{1+3}\left[(-3)(6)-(2)(-4)\right]=0$

$$x(1)(-4)+y(-1)(1)+1(1)(-10)=0$$
$$-4x-y-10=0$$
$$y=-4x-10$$

65. $\text{Area} = \pm\frac{1}{2}\begin{vmatrix} 1 & 0 & 1 \\ 7 & -2 & 1 \\ 4 & -5 & 1 \end{vmatrix}$

$$=\pm\frac{1}{2}\left\{(1)(-1)^{1+1}\left[(-2)(1)-(1)(-5)\right]+(0)(-1)^{1+2}\left[(7)(1)-(1)(4)\right]+(1)(-1)^{1+3}\left[(7)(-5)-(-2)(4)\right]\right\}$$
$$=\pm\frac{1}{2}\left[(1)(1)(3)+(1)(1)(-27)\right]=\pm\frac{1}{2}(-24)=12$$

67. a. $|A|=(5)(6)-(2)(-3)=36$

b. $|B|=(-3)(2)-(6)(5)=-36$

c. Rows 1 and 2 are interchanged between matrix A and matrix B. The determinants are opposite in sign.

69. a. $|A|=(1)(1)-(-3)(4)=13$

b. $|B|=(2)(1)-(-6)(4)=26$

c. Row 1 of matrix B is 2 times row 1 of matrix A. The value $|B|=2|A|$.

71. a. $|A|=(1)(4)-(2)(3)=-2$

b. $|B|=(1)(10)-(2)(6)=-2$

c. Row 2 of matrix B is the same as the sum of 3 times row 1 of A and row 2 of A. The value of $|A|$ equals $|B|$.

73. $|A|=(3)(10)-(5)(6)=0$

75. Expand cofactors about the elements in the third column.

$$|A|=(0)(-1)^{1+3}\left[(3)(1)-(-1)(0)\right]$$
$$+(0)(-1)^{2+3}\left[(4)(1)-(-5)(0)\right]$$
$$+(0)(-1)^{3+3}\left[(4)(-1)-(-5)(3)\right]$$
$$=0+0+0=0$$

77. $|I_2| = \begin{vmatrix} 1 & 0 \\ 0 & 1 \end{vmatrix} = (1)(1) - (0)(0) = 1$

79. $\begin{vmatrix} a & 0 & 0 \\ 0 & b & 0 \\ 0 & 0 & c \end{vmatrix} = (a)(-1)^{1+1}[(b)(c) - (0)(0)]$

$+(0)(-1)^{1+2}[(0)(c) - (0)(0)]$

$+(0)(-1)^{1+3}[(0)(0) - (b)(0)]$

$= (a)(1)(bc) = abc$

81. $AB = \begin{bmatrix} 4 & -2 \\ 3 & 1 \end{bmatrix} \cdot \begin{bmatrix} -5 & 1 \\ 3 & 2 \end{bmatrix}$

$= \begin{bmatrix} 4(-5) + (-2)(3) & 4(1) + (-2)(2) \\ 3(-5) + 1(3) & 3(1) + 1(2) \end{bmatrix}$

$= \begin{bmatrix} -26 & 0 \\ -12 & 5 \end{bmatrix}$

$|AB| = \begin{vmatrix} -26 & 0 \\ -12 & 5 \end{vmatrix} = (-26)(5) - (0)(-12) = -130$

$|A| = (4)(1) - (-2)(3) = 10$

$|B| = (-5)(2) - (1)(3) = -13$

So, $|A| \cdot |B| = (10)(-13) = -130$ and, therefore, $|A| \cdot |B| = |AB|$.

83. The minor is the determinant of the matrix obtained by deleting the *i*th row and *j*th column of the original matrix. The cofactor is the product of the minor and the factor $(-1)^{i+j}$.

85. Choose the row or column with the greatest number of zero elements.

87. 10,112

Problem Recognition Exercises: Using Multiple Methods to Solve Systems of Equations

1. a.

$$-3x - 7y = 22$$
$$-3(-3y - 10) - 7y = 22$$
$$9y + 30 - 7y = 22$$
$$2y = -8$$
$$y = -4$$

$$x = -3y - 10$$
$$= -3(-4) - 10$$
$$= 12 - 10$$
$$= 2 \qquad \{(2, -4)\}$$

b. $x = -3y - 10 \rightarrow x + 3y = -10$

$-3x - 7y = 22 \rightarrow -3x - 7y = 22$

Augmented matrix: $\left[\begin{array}{cc|c} 1 & 3 & -10 \\ -3 & -7 & 22 \end{array}\right]$

$3R_1 + R_2 \rightarrow R_2 \longrightarrow \left[\begin{array}{cc|c} 1 & 3 & -10 \\ 0 & 2 & -8 \end{array}\right]$

$\frac{1}{2}R_2 \rightarrow R_2 \longrightarrow \left[\begin{array}{cc|c} 1 & 3 & -10 \\ 0 & 1 & -4 \end{array}\right]$

$y = -4$

$$x + 3y = -10$$
$$x + 3(-4) = -10$$
$$x - 12 = -10$$
$$x = 2$$

$\{(2, -4)\}$

c. Augmented matrix: $\left[\begin{array}{cc|c} 1 & 3 & -10 \\ -3 & -7 & 22 \end{array}\right]$

$3R_1 + R_2 \rightarrow R_2 \longrightarrow \left[\begin{array}{cc|c} 1 & 3 & -10 \\ 0 & 2 & -8 \end{array}\right]$

$\frac{1}{2}R_2 \rightarrow R_2 \longrightarrow \left[\begin{array}{cc|c} 1 & 3 & -10 \\ 0 & 1 & -4 \end{array}\right]$

$-3R_2 + R_1 \rightarrow R_1 \longrightarrow \left[\begin{array}{cc|c} 1 & 0 & 2 \\ 0 & 1 & -4 \end{array}\right]$

$\{(2, -4)\}$

d. Set up matrix $[A\,|\,I_2]$: $\left[\begin{array}{cc|cc} 1 & 3 & 1 & 0 \\ -3 & -7 & 0 & 1 \end{array}\right]$

$$3R_1 + R_2 \to R_2 \longrightarrow \left[\begin{array}{cc|cc} 1 & 3 & 1 & 0 \\ 0 & 2 & 3 & 1 \end{array}\right]$$

$$\tfrac{1}{2}R_2 \to R_2 \longrightarrow \left[\begin{array}{cc|cc} 1 & 3 & 1 & 0 \\ 0 & 1 & \frac{3}{2} & \frac{1}{2} \end{array}\right]$$

$$-3R_2 + R_1 \to R_1 \longrightarrow \left[\begin{array}{cc|cc} 1 & 0 & -\frac{7}{2} & -\frac{3}{2} \\ 0 & 1 & \frac{3}{2} & \frac{1}{2} \end{array}\right]$$

$$\begin{bmatrix} x \\ y \end{bmatrix} = \begin{bmatrix} -\frac{7}{2} & -\frac{3}{2} \\ \frac{3}{2} & \frac{1}{2} \end{bmatrix} \begin{bmatrix} -10 \\ 22 \end{bmatrix}$$

$$\begin{bmatrix} x \\ y \end{bmatrix} = \begin{bmatrix} -\frac{7}{2}(-10) + \left(-\frac{3}{2}\right)(22) \\ \frac{3}{2}(-10) + \frac{1}{2}(22) \end{bmatrix} = \begin{bmatrix} 2 \\ -4 \end{bmatrix}$$

$\{(2, -4)\}$

e. $D = \begin{vmatrix} 1 & 3 \\ -3 & -7 \end{vmatrix} = (1)(-7) - (3)(-3)$

$= -7 + 9 = 2$

$D_x = \begin{vmatrix} -10 & 3 \\ 22 & -7 \end{vmatrix} = (-10)(-7) - (3)(22)$

$= 70 - 66 = 4$

$D_y = \begin{vmatrix} 1 & -10 \\ -3 & 22 \end{vmatrix} = (1)(22) - (-10)(-3)$

$= 22 - 30 = -8$

$x = \dfrac{D_x}{D} = \dfrac{4}{2} = 2$

$y = \dfrac{D_y}{D} = \dfrac{-8}{2} = -4$

$\{(2, -4)\}$

3. a.
$\boxed{A}\quad x + 2y - z = 0$
$\boxed{B}\quad 2x + z = 4$
$\boxed{C}\quad 2x - y + 2z = 5$

Eliminate z from equations $\boxed{A}$ and $\boxed{B}$.

$\boxed{A}\quad x + 2y - z = 0$
$\boxed{B}\quad 2x + z = 4$
$3x + 2y = 4 \quad \boxed{D}$

Eliminate z from equations $\boxed{A}$ and $\boxed{C}$.

$\boxed{A}\quad x + 2y - z = 0 \xrightarrow{\text{Multiply by 2}} 2x + 4y - 2z = 0$
$\boxed{C}\quad 2x - y + 2z = 5 \qquad\qquad 2x - y + 2z = 5$
$4x + 3y = 5 \quad \boxed{E}$

Solve the systems of equations $\boxed{D}$ and $\boxed{E}$.

$\boxed{D}\quad 3x + 2y = 4 \xrightarrow{\text{Multiply by 4}} 12x + 8y = 16$
$\boxed{E}\quad 4x + 3y = 5 \xrightarrow{\text{Multiply by } -3} -12x - 9y = -15$
$-y = 1$
$y = -1$

Back substitute.

$\boxed{D}\quad 3x + 2y = 4$
$3x + 2(-1) = 4$
$3x = 6$
$x = 2$

$\boxed{B}\quad 2x + z = 4$
$2(2) + z = 4$
$4 + z = 4$
$z = 0$

$\{(2, -1, 0)\}$

b. Augmented matrix: $\left[\begin{array}{ccc|c} 1 & 2 & -1 & 0 \\ 2 & 0 & 1 & 4 \\ 2 & -1 & 2 & 5 \end{array}\right]$

$$\begin{array}{l} -2R_1 + R_2 \to R_2 \longrightarrow \\ -2R_1 + R_3 \to R_3 \longrightarrow \end{array} \left[\begin{array}{ccc|c} 1 & 2 & -1 & 0 \\ 0 & -4 & 3 & 4 \\ 0 & -5 & 4 & 5 \end{array}\right]$$

$$R_2 - R_3 \to R_2 \longrightarrow \left[\begin{array}{ccc|c} 1 & 2 & -1 & 0 \\ 0 & 1 & -1 & -1 \\ 0 & -5 & 4 & 5 \end{array}\right]$$

$$5R_2 + R_3 \to R_3 \longrightarrow \left[\begin{array}{ccc|c} 1 & 2 & -1 & 0 \\ 0 & 1 & -1 & -1 \\ 0 & 0 & -1 & 0 \end{array}\right]$$

$$-R_3 \to R_3 \longrightarrow \left[\begin{array}{ccc|c} 1 & 2 & -1 & 0 \\ 0 & 1 & -1 & -1 \\ 0 & 0 & 1 & 0 \end{array}\right]$$

$z = 0$

$y - z = -1$

$y = -1$

$x + 2y - z = 0$

$x + 2(-1) - 0 = 0$

$x - 2 = 0$

$x = 2$

$\{(2, -1, 0)\}$

c. Augmented matrix: $\left[\begin{array}{ccc|c} 1 & 2 & -1 & 0 \\ 2 & 0 & 1 & 4 \\ 2 & -1 & 2 & 5 \end{array}\right]$

$$\begin{array}{l} -2R_1 + R_2 \to R_2 \longrightarrow \\ -2R_1 + R_3 \to R_3 \longrightarrow \end{array} \left[\begin{array}{ccc|c} 1 & 2 & -1 & 0 \\ 0 & -4 & 3 & 4 \\ 0 & -5 & 4 & 5 \end{array}\right]$$

$$R_2 - R_3 \to R_2 \longrightarrow \left[\begin{array}{ccc|c} 1 & 2 & -1 & 0 \\ 0 & 1 & -1 & -1 \\ 0 & -5 & 4 & 5 \end{array}\right]$$

$$5R_2 + R_3 \to R_3 \longrightarrow \left[\begin{array}{ccc|c} 1 & 2 & -1 & 0 \\ 0 & 1 & -1 & -1 \\ 0 & 0 & -1 & 0 \end{array}\right]$$

$$-R_3 \to R_3 \longrightarrow \left[\begin{array}{ccc|c} 1 & 2 & -1 & 0 \\ 0 & 1 & -1 & -1 \\ 0 & 0 & 1 & 0 \end{array}\right]$$

$$-2R_2 + R_1 \to R_1 \longrightarrow \left[\begin{array}{ccc|c} 1 & 0 & 1 & 2 \\ 0 & 1 & -1 & -1 \\ 0 & 0 & 1 & 0 \end{array}\right]$$

$$\begin{array}{l} -R_3 + R_1 \to R_1 \longrightarrow \\ R_3 + R_2 \to R_2 \longrightarrow \end{array} \left[\begin{array}{ccc|c} 1 & 0 & 0 & 2 \\ 0 & 1 & 0 & -1 \\ 0 & 0 & 1 & 0 \end{array}\right]$$

$\{(2, -1, 0)\}$

d. Set up matrix $[A|I_3]$: $\left[\begin{array}{ccc|ccc} 1 & 2 & -1 & 1 & 0 & 0 \\ 2 & 0 & 1 & 0 & 1 & 0 \\ 2 & -1 & 2 & 0 & 0 & 1 \end{array}\right]$

$$\Rightarrow \begin{array}{l} -2R_1 + R_2 \to R_2 \longrightarrow \\ -2R_1 + R_3 \to R_3 \longrightarrow \end{array} \left[\begin{array}{ccc|ccc} 1 & 2 & -1 & 1 & 0 & 0 \\ 0 & -4 & 3 & -2 & 1 & 0 \\ 0 & -5 & 4 & -2 & 0 & 1 \end{array}\right]$$

$$-R_3 + R_2 \to R_2 \longrightarrow \left[\begin{array}{ccc|ccc} 1 & 2 & -1 & 1 & 0 & 0 \\ 0 & 1 & -1 & 0 & 1 & -1 \\ 0 & -5 & 4 & -2 & 0 & 1 \end{array}\right]$$

$$\Rightarrow \begin{array}{l} -2R_2 + R_1 \to R_1 \longrightarrow \\ 5R_2 + R_3 \to R_3 \longrightarrow \end{array} \left[\begin{array}{ccc|ccc} 1 & 0 & 1 & 1 & -2 & 2 \\ 0 & 1 & -1 & 0 & 1 & -1 \\ 0 & 0 & -1 & -2 & 5 & -4 \end{array}\right]$$

$$-R_3 \to R_3 \longrightarrow \left[\begin{array}{ccc|ccc} 1 & 0 & 1 & 1 & -2 & 2 \\ 0 & 1 & -1 & 0 & 1 & -1 \\ 0 & 0 & 1 & 2 & -5 & 4 \end{array}\right]$$

$$\Rightarrow \begin{array}{l} -R_3 + R_1 \to R_1 \longrightarrow \\ R_3 + R_2 \to R_2 \longrightarrow \end{array} \left[\begin{array}{ccc|ccc} 1 & 0 & 0 & -1 & 3 & -2 \\ 0 & 1 & 0 & 2 & -4 & 3 \\ 0 & 0 & 1 & 2 & -5 & 4 \end{array}\right]$$

$$\begin{bmatrix} x \\ y \\ z \end{bmatrix} = \begin{bmatrix} -1 & 3 & -2 \\ 2 & -4 & 3 \\ 2 & -5 & 4 \end{bmatrix} \begin{bmatrix} 0 \\ 4 \\ 5 \end{bmatrix} = \begin{bmatrix} -1(0) + 3(4) + (-2)(5) \\ 2(0) + (-4)(4) + 3(5) \\ 2(0) + (-5)(4) + 4(5) \end{bmatrix} = \begin{bmatrix} 2 \\ -1 \\ 0 \end{bmatrix}$$

$\{(2, -1, 0)\}$

e. Expand cofactors about the elements in the second row.

$$D=\begin{vmatrix}1 & 2 & -1\\ 2 & 0 & 1\\ 2 & -1 & 2\end{vmatrix}=2(-1)^{2+1}\begin{vmatrix}2 & -1\\ -1 & 2\end{vmatrix}+0(-1)^{2+2}\begin{vmatrix}1 & -1\\ 2 & 2\end{vmatrix}+1(-1)^{2+3}\begin{vmatrix}1 & 2\\ 2 & -1\end{vmatrix}$$

$$=2(-1)[(2)(2)-(-1)(-1)]+1(-1)[(1)(-1)-(2)(2)]=-2(3)-1(-5)=-6+5=-1$$

Expand cofactors about the elements in the third column.

$$D_x=\begin{vmatrix}0 & 2 & -1\\ 4 & 0 & 1\\ 5 & -1 & 2\end{vmatrix}=4(-1)^{2+1}\begin{vmatrix}2 & -1\\ -1 & 2\end{vmatrix}+0(-1)^{2+2}\begin{vmatrix}0 & -1\\ 5 & 2\end{vmatrix}+1(-1)^{2+3}\begin{vmatrix}0 & 2\\ 5 & -1\end{vmatrix}$$

$$=4(-1)[(2)(2)-(-1)(-1)]+1(-1)[(0)(-1)-(2)(5)]=-4(3)-1(-10)=-12+10=-2$$

Expand cofactors about the elements in the first row.

$$D_y=\begin{vmatrix}1 & 0 & -1\\ 2 & 4 & 1\\ 2 & 5 & 2\end{vmatrix}=1(-1)^{1+1}\begin{vmatrix}4 & 1\\ 5 & 2\end{vmatrix}+0(-1)^{1+2}\begin{vmatrix}2 & 1\\ 2 & 2\end{vmatrix}-1(-1)^{1+3}\begin{vmatrix}2 & 4\\ 2 & 5\end{vmatrix}$$

$$=1(1)[(4)(2)-(1)(5)]-1(1)[(2)(5)-(4)(2)]=1(3)-1(2)=3-2=1$$

Expand cofactors about the elements in the first row.

$$D_z=\begin{vmatrix}1 & 2 & 0\\ 2 & 0 & 4\\ 2 & -1 & 5\end{vmatrix}=1(-1)^{1+1}\begin{vmatrix}0 & 4\\ -1 & 5\end{vmatrix}+2(-1)^{1+2}\begin{vmatrix}2 & 4\\ 2 & 5\end{vmatrix}+0(-1)^{1+3}\begin{vmatrix}2 & 0\\ 2 & -1\end{vmatrix}$$

$$=1(1)[(0)(5)-(4)(-1)]+2(-1)[(2)(5)-(4)(2)]=1(4)-2(2)=4-4=0$$

$x=\dfrac{D_x}{D}=\dfrac{-2}{-1}=2$ $\quad y=\dfrac{D_y}{D}=\dfrac{1}{-1}=-1$ $\quad z=\dfrac{D_z}{D}=\dfrac{0}{-1}=0$ $\quad \{(2,-1,0)\}$

5. a. $\begin{vmatrix}1.5 & -2\\ -3 & 4\end{vmatrix}=(1.5)(4)-(-2)(-3)=6-6=0$ **b.** No

c. Augmented matrix: $\left[\begin{array}{cc|c}1.5 & -2 & 3\\ -3 & 4 & 12\end{array}\right]$

$2R_1\to R_1 \longrightarrow \left[\begin{array}{cc|c}3 & -4 & 6\\ -3 & 4 & 12\end{array}\right]$

$R_1+R_2\to R_2 \longrightarrow \left[\begin{array}{cc|c}3 & -4 & 6\\ 0 & 0 & 18\end{array}\right]$ Contradiction: { }

7. a. Expand cofactors about the elements in the first column.

$$\begin{vmatrix}1 & -3 & 7\\ -2 & 5 & -11\\ 1 & -5 & 13\end{vmatrix}=1(-1)^{1+1}\begin{vmatrix}5 & -11\\ -5 & 13\end{vmatrix}+(-2)(-1)^{2+1}\begin{vmatrix}-3 & 7\\ -5 & 13\end{vmatrix}+1(-1)^{3+1}\begin{vmatrix}-3 & 7\\ 5 & -11\end{vmatrix}$$

$$=1(1)[(5)(13)-(-11)(-5)]-2(-1)[(-3)(13)-(7)(-5)]+1(1)[(-3)(-11)-(7)(5)]$$

$$=1(10)+2(-4)+1(-2)=10-8-2=0$$

b. No

c. Augmented matrix: $\left[\begin{array}{rrr|r} 1 & -3 & 7 & 1 \\ -2 & 5 & -11 & -3 \\ 1 & -5 & 13 & -1 \end{array}\right]$

$$\begin{array}{l} 2R_1+R_2 \to R_2 \longrightarrow \\ -R_1+R_3 \to R_3 \longrightarrow \end{array} \left[\begin{array}{rrr|r} 1 & -3 & 7 & 1 \\ 0 & -1 & 3 & -1 \\ 0 & -2 & 6 & -2 \end{array}\right]$$

$$-R_2 \to R_2 \longrightarrow \left[\begin{array}{rrr|r} 1 & -3 & 7 & 1 \\ 0 & 1 & -3 & 1 \\ 0 & -2 & 6 & -2 \end{array}\right]$$

$$\begin{array}{l} 3R_2+R_1 \to R_1 \longrightarrow \\ \\ 2R_2+R_3 \to R_3 \longrightarrow \end{array} \left[\begin{array}{rrr|r} 1 & 0 & -2 & 4 \\ 0 & 1 & -3 & 1 \\ 0 & 0 & 0 & 0 \end{array}\right]$$

Identity:

$x-2z=4$ $\quad$ $y-3z=1$

$x=2z+4$ $\quad$ $y=3z+1$

$\{(2z+4, 3z+1, z) \mid z \text{ is any real number}\}$

Chapter 6 Review Exercises

1. $x-2y+3z=-1$

$y+4z=-11$

$z=-2$

Augmented matrix: $\left[\begin{array}{rrr|r} 1 & -2 & 3 & -1 \\ 0 & 1 & 4 & -11 \\ 0 & 0 & 1 & -2 \end{array}\right]$

$$2R_2+R_1 \to R_1 \longrightarrow \left[\begin{array}{rrr|r} 1 & 0 & 11 & -23 \\ 0 & 1 & 4 & -11 \\ 0 & 0 & 1 & -2 \end{array}\right]$$

$$\begin{array}{l} -11R_3+R_1 \to R_1 \longrightarrow \\ -4R_3+R_2 \to R_2 \longrightarrow \end{array} \left[\begin{array}{rrr|r} 1 & 0 & 0 & -1 \\ 0 & 1 & 0 & -3 \\ 0 & 0 & 1 & -2 \end{array}\right]$$

$\{(-1, -3, -2)\}$

3. $\left[\begin{array}{rr|r} 2 & -3 & 1 \\ 5 & 6 & -4 \end{array}\right] \xrightarrow[\text{Multiply row 1 by } \frac{1}{2}]{\frac{1}{2}R_1 \Leftrightarrow R_1} \left[\begin{array}{rr|r} 1 & -\frac{3}{2} & \frac{1}{2} \\ 5 & 6 & -4 \end{array}\right]$

5. Augmented matrix: $\left[\begin{array}{rr|r} -2 & 1 & -16 \\ 1 & -2 & 17 \end{array}\right]$

$$R_1 \Leftrightarrow R_2 \longrightarrow \left[\begin{array}{rr|r} 1 & -2 & 17 \\ -2 & 1 & -16 \end{array}\right]$$

$$2R_1+R_2 \to R_2 \longrightarrow \left[\begin{array}{rr|r} 1 & -2 & 17 \\ 0 & -3 & 18 \end{array}\right]$$

$$-\tfrac{1}{3}R_2 \to R_2 \longrightarrow \left[\begin{array}{rr|r} 1 & -2 & 17 \\ 0 & 1 & -6 \end{array}\right]$$

$$2R_2+R_1 \to R_1 \longrightarrow \left[\begin{array}{rr|r} 1 & 0 & 5 \\ 0 & 1 & -6 \end{array}\right]$$

$\{(5, -6)\}$

7. Augmented matrix: $\left[\begin{array}{rrr|r} 2 & -5 & 18 & 44 \\ 1 & -3 & 11 & 27 \\ 1 & -2 & 11 & 29 \end{array}\right]$

$$R_1 \Leftrightarrow R_2 \longrightarrow \left[\begin{array}{rrr|r} 1 & -3 & 11 & 27 \\ 2 & -5 & 18 & 44 \\ 1 & -2 & 11 & 29 \end{array}\right]$$

$$\left[\begin{array}{ccc|c} 1 & -3 & 11 & 27 \\ 0 & 1 & -4 & -10 \\ 0 & 1 & 0 & 2 \end{array}\right] \begin{array}{l} \\ -2R_1+R_2 \to R_2 \longrightarrow \\ -R_1+R_3 \to R_3 \longrightarrow \end{array}$$

$$R_2 \Leftrightarrow R_3 \longrightarrow \left[\begin{array}{ccc|c} 1 & -3 & 11 & 27 \\ 0 & 1 & 0 & 2 \\ 0 & 1 & -4 & -10 \end{array}\right]$$

$$\begin{array}{l} 3R_2+R_1 \to R_1 \longrightarrow \\ \\ -R_2+R_3 \to R_3 \longrightarrow \end{array} \left[\begin{array}{ccc|c} 1 & 0 & 11 & 33 \\ 0 & 1 & 0 & 2 \\ 0 & 0 & -4 & -12 \end{array}\right]$$

$$-\tfrac{1}{4}R_3 \to R_3 \longrightarrow \left[\begin{array}{ccc|c} 1 & 0 & 11 & 33 \\ 0 & 1 & 0 & 2 \\ 0 & 0 & 1 & 3 \end{array}\right]$$

$$-11R_3+R_1 \to R_1 \longrightarrow \left[\begin{array}{ccc|c} 1 & 0 & 0 & 0 \\ 0 & 1 & 0 & 2 \\ 0 & 0 & 1 & 3 \end{array}\right]$$

$\{(-1, 7, 1)\}$

9. Let x represent the amount Lily borrowed from her friend. Let y represent the amount she borrowed from the credit union. Let z represent the amount she borrowed from the bank.

$$\begin{array}{ll} x+y+z=10{,}000 & \to x+y+z=10{,}000 \\ 0.05y+0.075z=500 & \to 50y+75z=500{,}000 \\ x=z-1000 & \to x-z=-1000 \end{array}$$

Augmented matrix: $\left[\begin{array}{ccc|c} 1 & 1 & 1 & 10{,}000 \\ 0 & 50 & 75 & 500{,}000 \\ 1 & 0 & -1 & -1000 \end{array}\right]$

$$-R_1+R_3 \to R_3 \longrightarrow \left[\begin{array}{ccc|c} 1 & 1 & 1 & 10{,}000 \\ 0 & 50 & 75 & 500{,}000 \\ 0 & -1 & -2 & -11{,}000 \end{array}\right]$$

$$R_2 \Leftrightarrow R_3 \longrightarrow \left[\begin{array}{ccc|c} 1 & 1 & 1 & 10{,}000 \\ 0 & -1 & -2 & -11{,}000 \\ 0 & 50 & 75 & 500{,}000 \end{array}\right]$$

$$-R_2 \to R_2 \longrightarrow \left[\begin{array}{ccc|c} 1 & 1 & 1 & 10{,}000 \\ 0 & 1 & 2 & 11{,}000 \\ 0 & 50 & 75 & 500{,}000 \end{array}\right]$$

$$\begin{array}{l} -R_2+R_1 \to R_1 \longrightarrow \\ \\ -50R_2+R_3 \to R_3 \longrightarrow \end{array} \left[\begin{array}{ccc|c} 1 & 0 & -1 & -1000 \\ 0 & 1 & 2 & 11{,}000 \\ 0 & 0 & -25 & -50{,}000 \end{array}\right]$$

$$-\tfrac{1}{25}R_3 \to R_3 \longrightarrow \left[\begin{array}{ccc|c} 1 & 0 & -1 & -1000 \\ 0 & 1 & 2 & 11{,}000 \\ 0 & 0 & 1 & 2000 \end{array}\right]$$

$$\begin{array}{l} R_3+R_1 \to R_1 \longrightarrow \\ -2R_3+R_2 \to R_2 \longrightarrow \\ \\ \end{array} \left[\begin{array}{ccc|c} 1 & 0 & 0 & 1000 \\ 0 & 1 & 0 & 7000 \\ 0 & 0 & 1 & 2000 \end{array}\right]$$

Lily borrowed \$1000 from her friend, \$7000 from the credit union, and \$2000 from the bank.

11. $\left[\begin{array}{cc|c} 1 & -2 & 6 \\ 0 & 0 & 1 \end{array}\right]$

2nd row: $0=1$ is a contradiction.

$\{\ \}$

13. $\left[\begin{array}{ccc|c} 1 & 0 & -3 & 0 \\ 0 & 1 & 2 & 1 \\ 0 & 0 & 0 & 0 \end{array}\right]$

$x-3z=0 \to x=3z$

$y+2z=1 \to y=-2z+1$

$\{(3z, -2z+1, z) \mid z \text{ is any real number}\}$

15. $-(2x-y)=8-y \to -2x+2y=8$

$y=x-6 \quad\quad \to -x+y=-6$

Augmented matrix: $\left[\begin{array}{cc|c} -2 & 2 & 8 \\ -1 & 1 & -6 \end{array}\right]$

$$-\tfrac{1}{2}R_1 \to R_1 \longrightarrow \left[\begin{array}{cc|c} 1 & -1 & -4 \\ -1 & 1 & -6 \end{array}\right]$$

$$R_1+R_2 \to R_2 \longrightarrow \left[\begin{array}{cc|c} 1 & 1 & 4 \\ 0 & 0 & -10 \end{array}\right]$$

$\{\ \}$

17. Augmented matrix: $\left[\begin{array}{ccc|c} 1 & 0 & -3 & 5 \\ -2 & 1 & 10 & -7 \\ 1 & 1 & 1 & 8 \end{array}\right]$

$$\begin{array}{l} \\ 2R_1 + R_2 \to R_2 \longrightarrow \\ -R_1 + R_3 \to R_3 \longrightarrow \end{array} \left[\begin{array}{ccc|c} 1 & 0 & -3 & 5 \\ 0 & 1 & 4 & 3 \\ 0 & 1 & 4 & 3 \end{array}\right]$$

$$\begin{array}{l} \\ \\ -R_2 + R_3 \to R_3 \longrightarrow \end{array} \left[\begin{array}{ccc|c} 1 & 0 & -3 & 5 \\ 0 & 1 & 4 & 3 \\ 0 & 0 & 0 & 0 \end{array}\right]$$

$x - 3z = 5 \ \to x = 3z + 5$

$y + 4z = 3 \ \to y = -4z + 3$

$\{(3z+5, -4z+3, z) \mid z \text{ is any real number}\}$

19. $5y = x + 2z + 1 \qquad \to -x + 5y - 2z = 1$

$2(x - 5y) + 4z = -2 \ \to 2x - 10y + 4z = -2$

$3(x + 2z) = 15y - 3 \ \to 3x - 15y + 6z = -3$

Augmented matrix: $\left[\begin{array}{ccc|c} -1 & 5 & -2 & 1 \\ 2 & -10 & 4 & -2 \\ 3 & -15 & 6 & -3 \end{array}\right]$

$$-R_1 \Leftrightarrow R_1 \longrightarrow \left[\begin{array}{ccc|c} 1 & -5 & 2 & -1 \\ 2 & -10 & 4 & -2 \\ 3 & -15 & 6 & -3 \end{array}\right]$$

$$\begin{array}{l} \\ -2R_1 + R_2 \to R_2 \longrightarrow \\ -3R_1 + R_3 \to R_3 \longrightarrow \end{array} \left[\begin{array}{ccc|c} 1 & -5 & 2 & -1 \\ 0 & 0 & 0 & 0 \\ 0 & 0 & 0 & 0 \end{array}\right]$$

$\{(x, y, z) \mid x - 5y + 2z = -1\}$

21. $x_1 + 163 = x_2 + 142 \ \to x_1 - x_2 = -21$

$x_2 + 164 = x_3 + 131 \ \to x_2 - x_3 = -33$

$x_3 + 132 = x_1 + 186 \ \to x_3 - x_1 = 54$

Augmented matrix: $\left[\begin{array}{ccc|c} 1 & -1 & 0 & -21 \\ 0 & 1 & -1 & -33 \\ -1 & 0 & 1 & 54 \end{array}\right]$

$$\begin{array}{l} \\ \\ R_1 + R_3 \to R_3 \longrightarrow \end{array} \left[\begin{array}{ccc|c} 1 & -1 & 0 & -21 \\ 0 & 1 & -1 & -33 \\ 0 & -1 & 1 & 33 \end{array}\right]$$

$$\begin{array}{l} R_2 + R_1 \to R_1 \longrightarrow \\ \\ R_2 + R_3 \to R_3 \longrightarrow \end{array} \left[\begin{array}{ccc|c} 1 & 0 & -1 & -54 \\ 0 & 1 & -1 & -33 \\ 0 & 0 & 0 & 0 \end{array}\right]$$

a. $x_3 = 130$:

$x_1 - x_3 = -54$

$x_1 - 130 = -54$

$x_1 = 76$ vehicles per hour

$x_2 - x_3 = -33$

$x_2 - 130 = -33$

$x_2 = 97$ vehicles per hour

b. $x_3 = 100$:

$x_1 - x_3 = -54$

$x_1 - 100 = -54$

$x_1 = 46$ vehicles per hour

$x_2 - x_3 = -33$

$x_2 - 100 = -33$

$x_2 = 67$ vehicles per hour

$x_3 = 150$:

$x_1 - x_3 = -54$

$x_1 - 150 = -54$

$x_1 = 96$ vehicles per hour

$x_2 - x_3 = -33$

$x_2 - 150 = -33$

$x_2 = 117$ vehicles per hour

$46 \le x_1 \le 96$ vehicles per hour

$67 \le x_2 \le 117$ vehicles per hour

23. **a.** The matrix has 3 rows and 2 columns. 3×2

b. None of these

25. **a.** The matrix has 2 rows and 1 column. 2×1

b. Column matrix

27. a_{21} represents the element in the 2nd row, 1st column: 4

29. $x = 6, y = 3, z = 8$

31. $3A = 3\begin{bmatrix} -4 & 1 \\ 6 & -2 \\ 1 & 3 \end{bmatrix} = \begin{bmatrix} -12 & 3 \\ 18 & -6 \\ 3 & 9 \end{bmatrix}$

33. Matrix A is a 3×2 matrix, whereas matrix B is a 2×3 matrix.
The matrices are of different orders and therefore cannot be added.
The operation is not possible.

35. $2A-C=2\begin{bmatrix}-4 & 1\\ 6 & -2\\ 1 & 3\end{bmatrix}-\begin{bmatrix}\pi & 4\\ -3 & 1\\ 0 & 5\end{bmatrix}=\begin{bmatrix}-8 & 2\\ 12 & -4\\ 2 & 6\end{bmatrix}-\begin{bmatrix}\pi & 4\\ -3 & 1\\ 0 & 5\end{bmatrix}=\begin{bmatrix}-8-\pi & -2\\ 15 & -5\\ 2 & 1\end{bmatrix}$

37. $AB=\begin{bmatrix}-4 & 1\\ 6 & -2\\ 1 & 3\end{bmatrix}\begin{bmatrix}2 & 3 & -7\\ 1 & 5 & -6\end{bmatrix}=\begin{bmatrix}-4(2)+1(1) & -4(3)+1(5) & -4(-7)+1(-6)\\ 6(2)+(-2)(1) & 6(3)+(-2)(5) & 6(-7)+(-2)(-6)\\ 1(2)+3(1) & 1(3)+3(5) & 1(-7)+3(-6)\end{bmatrix}=\begin{bmatrix}-7 & -7 & 22\\ 10 & 8 & -30\\ 5 & 18 & -25\end{bmatrix}$

39. The number of columns of A does not equal the number of rows of C. Therefore, the product AC is undefined. The operation is not possible.

41. $A^2=\begin{bmatrix}2 & 6\\ -1 & 4\end{bmatrix}\begin{bmatrix}2 & 6\\ -1 & 4\end{bmatrix}=\begin{bmatrix}2(2)+6(-1) & 2(6)+6(4)\\ -1(2)+4(-1) & -1(6)+4(4)\end{bmatrix}=\begin{bmatrix}-2 & 36\\ -6 & 10\end{bmatrix}$

43. $BC=\begin{bmatrix}1\\ -3\end{bmatrix}\begin{bmatrix}2 & 7\end{bmatrix}=\begin{bmatrix}2 & 7\\ -6 & -21\end{bmatrix}$

45. **a.** $P-M=\begin{bmatrix}\$1365 & \$1222\\ \$1157 & \$1040\end{bmatrix}-\begin{bmatrix}\$1050 & \$940\\ \$890 & \$800\end{bmatrix}=\begin{bmatrix}\$315 & \$282\\ \$267 & \$240\end{bmatrix}$

This represents the profit that the store clears for each model.

b. $F=1.06P=1.06\cdot\begin{bmatrix}\$1365 & \$1222\\ \$1157 & \$1040\end{bmatrix}=\begin{bmatrix}\$1446.90 & \$1295.32\\ \$1226.42 & \$1102.40\end{bmatrix}$

47. **a.** $A=\begin{bmatrix}1 & 3 & 5\\ 1 & 4 & 2\end{bmatrix}$

b. $\begin{bmatrix}1 & 3 & 5\\ 1 & 4 & 2\end{bmatrix}+\begin{bmatrix}-3 & -3 & -3\\ -1 & -1 & -1\end{bmatrix}=\begin{bmatrix}-2 & 0 & 2\\ 0 & 3 & 1\end{bmatrix}$

c. $\begin{bmatrix}-1 & 0\\ 0 & 1\end{bmatrix}\cdot\begin{bmatrix}1 & 3 & 5\\ 1 & 4 & 2\end{bmatrix}=\begin{bmatrix}-1(1)+(0)(1) & -1(3)+(0)(4) & -1(5)+(0)(2)\\ (0)(1)+1(1) & (0)(3)+1(4) & (0)(5)+1(2)\end{bmatrix}=\begin{bmatrix}-1 & -3 & -5\\ 1 & 4 & 2\end{bmatrix}$

This matrix represents the reflection of the triangle across the y-axis.

d. $\begin{bmatrix}1 & 0\\ 0 & -1\end{bmatrix}\cdot\begin{bmatrix}1 & 3 & 5\\ 1 & 4 & 2\end{bmatrix}=\begin{bmatrix}1(1)+(0)(1) & 1(3)+(0)(4) & 1(5)+(0)(2)\\ (0)(1)+(-1)(1) & (0)(3)+(-1)(4) & (0)(5)+(-1)(2)\end{bmatrix}=\begin{bmatrix}1 & 3 & 5\\ -1 & -4 & -2\end{bmatrix}$

This matrix represents the reflection of the triangle across the x-axis.

49. $AB=\begin{bmatrix}4 & 1\\ 3 & 2\end{bmatrix}\begin{bmatrix}2 & -3\\ -1 & 4\end{bmatrix}=\begin{bmatrix}4(2)+1(-1) & 4(-3)+1(4)\\ 3(2)+2(-1) & 3(-3)+2(4)\end{bmatrix}=\begin{bmatrix}7 & -8\\ 4 & -1\end{bmatrix}$ No

51. $A^{-1}=\frac{1}{ad-bc}\begin{bmatrix} d & -b \\ -c & a \end{bmatrix}=\frac{1}{5(2)-(-2)(1)}\begin{bmatrix} 2 & 2 \\ -1 & 5 \end{bmatrix}=\frac{1}{12}\begin{bmatrix} 2 & 2 \\ -1 & 5 \end{bmatrix}=\begin{bmatrix} \frac{1}{6} & \frac{1}{6} \\ -\frac{1}{12} & \frac{5}{12} \end{bmatrix}$

53. $A^{-1}=\frac{1}{ad-bc}\begin{bmatrix} d & -b \\ -c & a \end{bmatrix}=\frac{1}{2(24)-(3)(16)}\begin{bmatrix} 24 & -3 \\ -16 & 2 \end{bmatrix}=\frac{1}{0}\begin{bmatrix} 24 & -3 \\ -16 & 2 \end{bmatrix}$ undefined

The matrix is singular.

55. Set up matrix $[A\,|\,I_3]$: $\left[\begin{array}{ccc|ccc} -5 & 4 & 1 & 1 & 0 & 0 \\ 15 & -12 & -4 & 0 & 1 & 0 \\ 4 & -3 & -1 & 0 & 0 & 1 \end{array}\right] \Rightarrow \begin{array}{l} -R_3-R_1\to R_1\longrightarrow \end{array} \left[\begin{array}{ccc|ccc} 1 & -1 & 0 & -1 & 0 & -1 \\ 15 & -12 & -4 & 0 & 1 & 0 \\ 4 & -3 & -1 & 0 & 0 & 1 \end{array}\right]$

$\begin{array}{r} \\ -15R_1+R_2\to R_2\longrightarrow \\ -4R_1+R_3\to R_3\longrightarrow \end{array}\left[\begin{array}{ccc|ccc} 1 & -1 & 0 & -1 & 0 & -1 \\ 0 & 3 & -4 & 15 & 1 & 15 \\ 0 & 1 & -1 & 4 & 0 & 5 \end{array}\right] \Rightarrow R_2\Leftrightarrow R_3\longrightarrow \left[\begin{array}{ccc|ccc} 1 & -1 & 0 & -1 & 0 & -1 \\ 0 & 1 & -1 & 4 & 0 & 5 \\ 0 & 3 & -4 & 15 & 1 & 15 \end{array}\right]$

$\begin{array}{r} R_2+R_1\to R_1\longrightarrow \\ \\ -3R_2+R_3\to R_3\longrightarrow \end{array}\left[\begin{array}{ccc|ccc} 1 & 0 & -1 & 3 & 0 & 4 \\ 0 & 1 & -1 & 4 & 0 & 5 \\ 0 & 0 & -1 & 3 & 1 & 0 \end{array}\right] \Rightarrow \begin{array}{r} R_2+R_1\to R_1\longrightarrow \\ \\ -R_3\to R_3\longrightarrow \end{array}\left[\begin{array}{ccc|ccc} 1 & 0 & -1 & 3 & 0 & 4 \\ 0 & 1 & -1 & 4 & 0 & 5 \\ 0 & 0 & 1 & -3 & -1 & 0 \end{array}\right]$

$\begin{array}{r} R_3+R_1\to R_1\longrightarrow \\ R_3+R_2\to R_2\longrightarrow \\ \end{array}\left[\begin{array}{ccc|ccc} 1 & 0 & 0 & 0 & -1 & 4 \\ 0 & 1 & 0 & 1 & -1 & 5 \\ 0 & 0 & 1 & -3 & -1 & 0 \end{array}\right] \qquad A^{-1}=\begin{bmatrix} 0 & -1 & 4 \\ 1 & -1 & 5 \\ -3 & -1 & 0 \end{bmatrix}$

57. $\begin{bmatrix} -3 & 7 & 0 \\ 4 & 0 & 2 \\ 2 & -1 & 5 \end{bmatrix}\begin{bmatrix} x \\ y \\ z \end{bmatrix}=\begin{bmatrix} 6 \\ -3 \\ 13 \end{bmatrix}$

59. $\begin{bmatrix} x \\ y \end{bmatrix}=\begin{bmatrix} \frac{1}{6} & \frac{1}{6} \\ -\frac{1}{12} & \frac{5}{12} \end{bmatrix}\begin{bmatrix} 26 \\ -2 \end{bmatrix}$

$\begin{bmatrix} x \\ y \end{bmatrix}=\begin{bmatrix} \frac{1}{6}(26)+\frac{1}{6}(-2) \\ -\frac{1}{12}(26)+\frac{5}{12}(-2) \end{bmatrix}=\begin{bmatrix} 4 \\ -3 \end{bmatrix}$

$\{(4,-3)\}$

61. $\begin{bmatrix} x \\ y \\ z \end{bmatrix}=\begin{bmatrix} 0 & -1 & 4 \\ 1 & -1 & 5 \\ -3 & -1 & 0 \end{bmatrix}\begin{bmatrix} 6 \\ -21 \\ -5 \end{bmatrix}$

$\begin{bmatrix} x \\ y \\ z \end{bmatrix}=\begin{bmatrix} (0)(6)+(-1)(-21)+4(-5) \\ 1(6)+(-1)(-21)+5(-5) \\ -3(6)+(-1)(-21)+(0)(-5) \end{bmatrix}=\begin{bmatrix} 1 \\ 2 \\ 3 \end{bmatrix}$

$\{(1,2,3)\}$

63. a. $M_{31}=\begin{vmatrix} 4 & -2 \\ 0 & 6 \end{vmatrix}$

$=(4)(6)-(-2)(0)$

$=24$

b. $(-1)^{i+j}M_{ij}=(-1)^{3+1}M_{31}$

$=(-1)^4(24)$

$=24$

65. a. $M_{23}=\begin{vmatrix}-5 & 4\\ 8 & -9\end{vmatrix}$
$=(-5)(-9)-(4)(8)$
$=13$

b. $(-1)^{i+j}M_{ij}=(-1)^{2+3}M_{23}$
$=(-1)^{5}(13)$
$=-13$

67. $|B|=ad-bc$
$=(3)(27)-(x)(x)=81-x^2$

69. Expand cofactors about the elements in the first row.

$$\begin{aligned}|D|&=4(-1)^{1+1}[(8)(3)-(-2)(5)]\\&\quad+(-1)(-1)^{1+2}[(6)(3)-(-2)(1)]\\&\quad+0(-1)^{1+3}[(6)(5)-(8)(1)]\\&=4(34)+1(20)\\&=136+20\\&=156\end{aligned}$$

71. Expand cofactors about the elements in the third column. Only one cofactor is nonzero. Expand cofactors of the resulting determinant about the elements in the first row.

$$\begin{aligned}|F|&=3(-1)^{1+3}\begin{vmatrix}1 & 1 & 5\\ 4 & 0 & -2\\ 0 & -3 & 6\end{vmatrix}\\&=3(1)\{1(-1)^{1+1}[(0)(6)-(-2)(-3)]\\&\quad+1(-1)^{1+2}[(4)(6)-(-2)(0)]\\&\quad+5(-1)^{1+3}[(4)(-3)-(0)(0)]\}\\&=3[1(1)(-6)+1(-1)(24)+5(1)(-12)]\\&=3(-6-24-60)\\&=3(-90)\\&=-270\end{aligned}$$

73. $9x=3y+5 \quad \rightarrow 9x-3y=5$
$-2(x+3y)=4 \rightarrow -2x-6y=4$
$a_1=9, b_1=-3, c_1=5, a_2=-2, b_2=-6, c_2=4$

$$\begin{aligned}D&=\begin{vmatrix}a_1 & b_1\\ a_2 & b_2\end{vmatrix}\\&=\begin{vmatrix}9 & -3\\ -2 & -6\end{vmatrix}\\&=(9)(-6)-(-3)(-2)\\&=-54-6\\&=-60\end{aligned}$$

$$\begin{aligned}D_x&=\begin{vmatrix}c_1 & b_1\\ c_2 & b_2\end{vmatrix}\\&=\begin{vmatrix}5 & -3\\ 4 & -6\end{vmatrix}\\&=(5)(-6)-(-3)(4)\\&=-30+12\\&=-18\end{aligned}$$

$$\begin{aligned}D_y&=\begin{vmatrix}a_1 & c_1\\ a_2 & c_2\end{vmatrix}\\&=\begin{vmatrix}9 & 5\\ -2 & 4\end{vmatrix}\\&=(9)(4)-(5)(-2)\\&=36+10\\&=46\end{aligned}$$

$x=\frac{D_x}{D}=\frac{-18}{-60}=\frac{3}{10}$

$y=\frac{D_y}{D}=\frac{46}{-60}=-\frac{23}{30}$

$\left\{\left(\frac{3}{10},-\frac{23}{30}\right)\right\}$

75. Use a graphing utility to find the determinants of the matrices.

$D=\begin{vmatrix}3 & -2 & 1\\ 5 & 3 & 6\\ -2 & 0 & 5\end{vmatrix}=125$

$D_x=\begin{vmatrix}4 & -2 & 1\\ 1 & 3 & 6\\ 7 & 0 & 5\end{vmatrix}=-35$

$$D_y = \begin{vmatrix} 3 & 4 & 1 \\ 5 & 1 & 6 \\ -2 & 7 & 5 \end{vmatrix} = -222$$

$$D_z = \begin{vmatrix} 3 & -2 & 4 \\ 5 & 3 & 1 \\ -2 & 0 & 7 \end{vmatrix} = 161$$

$$x = \frac{D_x}{D} = \frac{-35}{125} = -\frac{7}{25}$$

$$y = \frac{D_y}{D} = \frac{-222}{125} = -\frac{222}{125}$$

$$z = \frac{D_z}{D} = \frac{161}{125}$$

$$\left\{\left(-\frac{7}{25}, -\frac{222}{125}, \frac{161}{125}\right)\right\}$$

77. Use a graphing utility to find the determinants of the matrices.

$$D = \begin{vmatrix} -6 & 7 & 0 \\ 2 & 5 & 1 \\ 3 & 0 & 2 \end{vmatrix} = -67$$

$$D_y = \begin{vmatrix} -6 & 8 & 0 \\ 2 & -3 & 1 \\ 3 & 11 & 2 \end{vmatrix} = 94$$

$$y = \frac{D_y}{D} = \frac{94}{-67} = -\frac{94}{67}$$

Chapter 6 Test

1. $$\left[\begin{array}{ccc|c} 3 & 1 & 4 & -2 \\ 1 & 5 & -3 & 1 \\ 0 & 4 & 2 & 6 \end{array}\right] \xrightarrow[\text{Interchange rows 1 and 2.}]{R_1 \Leftrightarrow R_2} \left[\begin{array}{ccc|c} 1 & 5 & -3 & 1 \\ 3 & 1 & 4 & -2 \\ 0 & 4 & 2 & 6 \end{array}\right]$$

3. $$\left[\begin{array}{ccc|c} 3 & 1 & 4 & -2 \\ 1 & 5 & -3 & 1 \\ 0 & 4 & 2 & 6 \end{array}\right] \xrightarrow[\text{Multiply row 3 by } \frac{1}{4}.]{\frac{1}{4}R_3 \Leftrightarrow R_3} \left[\begin{array}{ccc|c} 3 & 1 & 4 & -2 \\ 1 & 5 & -3 & 1 \\ 0 & 1 & \frac{1}{2} & \frac{3}{2} \end{array}\right]$$

5. Augmented matrix: $\left[\begin{array}{cc|c} 1 & 4 & 2 \\ 0 & 1 & 3 \end{array}\right]$

$$-4R_2 + R_1 \to R_1 \longrightarrow \left[\begin{array}{cc|c} 1 & 0 & -10 \\ 0 & 1 & 3 \end{array}\right]$$

$\{(-10, 3)\}$

7. $x - 3z = 0 \to x = 3z$

$y + 2z = 5 \to y = -2z + 5$

$\{(3z, -2z + 5, z) \mid z \text{ is any real number}\}$

9. a. $$M_{31} = \begin{vmatrix} 3 & 0 \\ 5 & -4 \end{vmatrix} = (3)(-4) - (0)(5) = -12 - 0 = -12$$

b. $$(-1)^{i+j} M_{ij} = (-1)^{3+1} M_{31} = (-1)^4(-12) = -12$$

11. Expand cofactors about the elements in the first row.

$$|B| = -3(-1)^{1+1}\begin{vmatrix} -2 & 3 \\ 5 & 0 \end{vmatrix} + 4(-1)^{1+2}\begin{vmatrix} 1 & 3 \\ 6 & 0 \end{vmatrix}$$

$$+7(-1)^{1+3}\begin{vmatrix} 1 & -2 \\ 6 & 5 \end{vmatrix}$$

$$= -3(1)[(-2)(0)-(3)(5)]$$
$$+4(-1)[(1)(0)-(3)(6)]$$
$$+7(1)[(1)(5)-(-2)(6)]$$
$$= -3(-15)-4(-18)+7(17)$$
$$= 45+72+119$$
$$= 236$$

13. $-3(x+y) = 3y-12 \rightarrow 3x+6y=12$

$-3x = 4y-6 \quad \rightarrow 3x+4y=6$

Augmented matrix: $\left[\begin{array}{cc|c} 3 & 6 & 12 \\ 3 & 4 & 6 \end{array}\right]$

$\frac{1}{3}R_1 \rightarrow R_1 \longrightarrow \left[\begin{array}{cc|c} 1 & 2 & 4 \\ 3 & 4 & 6 \end{array}\right]$

$-3R_1 + R_2 \rightarrow R_2 \longrightarrow \left[\begin{array}{cc|c} 1 & 2 & 4 \\ 0 & -2 & -6 \end{array}\right]$

$-\frac{1}{2}R_2 \rightarrow R_2 \longrightarrow \left[\begin{array}{cc|c} 1 & 2 & 4 \\ 0 & 1 & 3 \end{array}\right]$

$-2R_2 + R_1 \rightarrow R_1 \longrightarrow \left[\begin{array}{cc|c} 1 & 0 & -2 \\ 0 & 1 & 3 \end{array}\right]$

$\{(-2, 3)\}$

15. $x-2y=5z+4 \quad \rightarrow x-2y-5z=4$

$6y+18z=2x-8 \quad \rightarrow -2x+6y+18z=-8$

$-3x+8y+20z=-18 \rightarrow -3x+8y+20z=-18$

Augmented matrix: $\left[\begin{array}{ccc|c} 1 & -2 & -5 & 4 \\ -2 & 6 & 18 & -8 \\ -3 & 8 & 20 & -18 \end{array}\right]$

$$\begin{array}{l} \\ 2R_1+R_2 \rightarrow R_2 \longrightarrow \\ 3R_1+R_3 \rightarrow R_3 \longrightarrow \end{array} \left[\begin{array}{ccc|c} 1 & -2 & -5 & 4 \\ 0 & 2 & 8 & 0 \\ 0 & 2 & 5 & -6 \end{array}\right]$$

$\frac{1}{2}R_2 \rightarrow R_2 \longrightarrow \left[\begin{array}{ccc|c} 1 & -2 & -5 & 4 \\ 0 & 1 & 4 & 0 \\ 0 & 2 & 5 & -6 \end{array}\right]$

$$\begin{array}{l} 2R_2+R_1 \rightarrow R_1 \longrightarrow \\ \\ -2R_2+R_3 \rightarrow R_3 \longrightarrow \end{array} \left[\begin{array}{ccc|c} 1 & 0 & 3 & 4 \\ 0 & 1 & 4 & 0 \\ 0 & 0 & -3 & -6 \end{array}\right]$$

$-\frac{1}{3}R_3 \rightarrow R_3 \longrightarrow \left[\begin{array}{ccc|c} 1 & 0 & 3 & 4 \\ 0 & 1 & 4 & 0 \\ 0 & 0 & 1 & 2 \end{array}\right]$

$$\begin{array}{l} -3R_3+R_1 \rightarrow R_1 \longrightarrow \\ -4R_3+R_2 \rightarrow R_2 \longrightarrow \\ \\ \end{array} \left[\begin{array}{ccc|c} 1 & 0 & 0 & -2 \\ 0 & 1 & 0 & -8 \\ 0 & 0 & 1 & 2 \end{array}\right]$$

$\{(-2, -8, 2)\}$

17. $a_1 = 3, b_1 = -5, c_1 = 7, a_2 = 11, b_2 = 2, c_2 = 8$

$$D = \begin{vmatrix} a_1 & b_1 \\ a_2 & b_2 \end{vmatrix} = \begin{vmatrix} 3 & -5 \\ 11 & 2 \end{vmatrix}$$
$$= (3)(2)-(-5)(11) = 6+55 = 61$$

$$D_x = \begin{vmatrix} c_1 & b_1 \\ c_2 & b_2 \end{vmatrix} = \begin{vmatrix} 7 & -5 \\ 8 & 2 \end{vmatrix}$$
$$= (7)(2)-(-5)(8) = 14+40 = 54$$

$$D_y = \begin{vmatrix} a_1 & c_1 \\ a_2 & c_2 \end{vmatrix} = \begin{vmatrix} 3 & 7 \\ 11 & 8 \end{vmatrix}$$
$$= (3)(8)-(7)(11) = 24-77 = -53$$

$$x = \frac{D_x}{D} = \frac{54}{61}$$

$$y = \frac{D_y}{D} = \frac{-53}{61} = -\frac{53}{40}$$

$$\left\{\left(\frac{54}{61}, -\frac{53}{61}\right)\right\}$$

19. $x=-9$, $y=2$, and $z=6$

21. Matrix A is a 2×3 matrix, whereas matrix B is a 3×2 matrix. The matrices are of different orders and therefore cannot be added. The operation is not possible.

23. $$BA=\begin{bmatrix}1 & 9\\ 0 & -1\\ 3 & 5\end{bmatrix}\begin{bmatrix}4 & 1 & -3\\ 2 & 4 & 6\end{bmatrix}=\begin{bmatrix}1(4)+9(2) & 1(1)+9(4) & 1(-3)+9(6)\\ (0)(4)+(-1)(2) & (0)(1)+(-1)(4) & (0)(-3)+(-1)(6)\\ 3(4)+5(2) & 3(1)+5(4) & 3(-3)+5(6)\end{bmatrix}=\begin{bmatrix}22 & 37 & 51\\ -2 & -4 & -6\\ 22 & 23 & 21\end{bmatrix}$$

25. $$A^{-1}=\frac{1}{ad-bc}\begin{bmatrix}d & -b\\ -c & a\end{bmatrix}=\frac{1}{3(4)-(2)(5)}\begin{bmatrix}4 & -2\\ -5 & 3\end{bmatrix}=\frac{1}{2}\begin{bmatrix}4 & -2\\ -5 & 3\end{bmatrix}=\begin{bmatrix}2 & -1\\ -\frac{5}{2} & \frac{3}{2}\end{bmatrix}$$

27. Set up matrix $[A\,|\,I_3]$: $$\left[\begin{array}{ccc|ccc}3 & -1 & -1 & 1 & 0 & 0\\ 2 & -1 & 1 & 0 & 1 & 0\\ -5 & 2 & 1 & 0 & 0 & 1\end{array}\right]$$

$$-R_2+R_1\to R_1 \longrightarrow \left[\begin{array}{ccc|ccc}1 & 0 & -2 & 1 & -1 & 0\\ 2 & -1 & 1 & 0 & 1 & 0\\ -5 & 2 & 1 & 0 & 0 & 1\end{array}\right]$$

$$\begin{array}{l}-2R_1+R_2\to R_2 \longrightarrow\\ 5R_1+R_3\to R_3 \longrightarrow\end{array}\left[\begin{array}{ccc|ccc}1 & 0 & -2 & 1 & -1 & 0\\ 0 & -1 & 5 & -2 & 3 & 0\\ 0 & 2 & -9 & 5 & -5 & 1\end{array}\right]$$

$$-R_2\to R_2 \longrightarrow \left[\begin{array}{ccc|ccc}1 & 0 & -2 & 1 & -1 & 0\\ 0 & 1 & -5 & 2 & -3 & 0\\ 0 & 2 & -9 & 5 & -5 & 1\end{array}\right]$$

$$-2R_2+R_3\to R_3 \longrightarrow \left[\begin{array}{ccc|ccc}1 & 0 & -2 & 1 & -1 & 0\\ 0 & 1 & -5 & 2 & -3 & 0\\ 0 & 0 & 1 & 1 & 1 & 1\end{array}\right]$$

$$\begin{array}{l}2R_3+R_1\to R_1 \longrightarrow\\ 5R_3+R_2\to R_2 \longrightarrow\end{array}\left[\begin{array}{ccc|ccc}1 & 0 & 0 & 3 & 1 & 2\\ 0 & 1 & 0 & 7 & 2 & 5\\ 0 & 0 & 1 & 1 & 1 & 1\end{array}\right]$$

$$A^{-1}=\begin{bmatrix}3 & 1 & 2\\ 7 & 2 & 5\\ 1 & 1 & 1\end{bmatrix}$$

29. $$\begin{bmatrix}x\\ y\\ z\end{bmatrix}=\begin{bmatrix}3 & 1 & 2\\ 7 & 2 & 5\\ 1 & 1 & 1\end{bmatrix}\begin{bmatrix}8\\ 0\\ -11\end{bmatrix}$$
$$=\begin{bmatrix}3(8)+1(0)+2(-11)\\ 7(8)+2(0)+5(-11)\\ 1(8)+1(0)+1(-11)\end{bmatrix}=\begin{bmatrix}2\\ 1\\ -3\end{bmatrix}$$

$\{(2,1,-3)\}$

31. $$CN=\begin{bmatrix}400 & 500 & 320\\ 550 & 780 & 480\end{bmatrix}\begin{bmatrix}6\\ 3\\ 5\end{bmatrix}$$
$$=\begin{bmatrix}400(6)+500(3)+320(5)\\ 550(6)+780(3)+480(5)\end{bmatrix}=\begin{bmatrix}5500\\ 8040\end{bmatrix}$$

This represents the total number of calories burned by two individuals with different weights after biking 6 hr, running 3 hr, and walking 5 hr. For example, the element 5500 in the first row tells us that 5500 cal would be burned by a 120-lb individual who biked 6 hr, ran 3 hr, and walked 5 hr in a given week.

Chapter 6 Cumulative Review Exercises

1. $\dfrac{f(x+h)-f(x)}{h}=\dfrac{(x+h)^3-x^3}{h}$

$=\dfrac{x^3+3x^2h+3xh^2+h^3-x^3}{h}$

$=\dfrac{3x^2h+3xh^2+h^3}{h}$

$=3x^2+3xh+h^2$

3. $f(x)=\sqrt{2x-3}$

$f\left(\dfrac{7}{2}\right)=\sqrt{2\left(\dfrac{7}{2}\right)-3}=\sqrt{7-3}=\sqrt{4}=2$

$f(2)=\sqrt{2(2)-3}=\sqrt{4-3}=\sqrt{1}=1$

Average rate of change $=\dfrac{f(x_2)-f(x_1)}{x_2-x_1}$

$=\dfrac{f(2)-f\left(\dfrac{7}{2}\right)}{2-\dfrac{7}{2}}=\dfrac{1-2}{-\dfrac{3}{2}}=-1\left(-\dfrac{2}{3}\right)=\dfrac{2}{3}$

5. $\dfrac{3+4i}{2-3i}=\dfrac{3+4i}{2-3i}\cdot\dfrac{2+3i}{2+3i}$

$=\dfrac{6+9i+8i+12i^2}{4-9i^2}$

$=\dfrac{-6+17i}{13}$

$=-\dfrac{6}{13}+\dfrac{17}{13}i$

7. The circle is written in standard form: $(x-h)^2+(y-k)^2=r^2$.

So $h=-\dfrac{5}{3}$, $k=0$, and $r=\sqrt{11}$.

Center: $\left(-\dfrac{5}{3},0\right)$; radius: $\sqrt{11}$

9. a. The graph of f is the graph of $y=e^x$ shifted 2 units to the right and 1 unit upward.

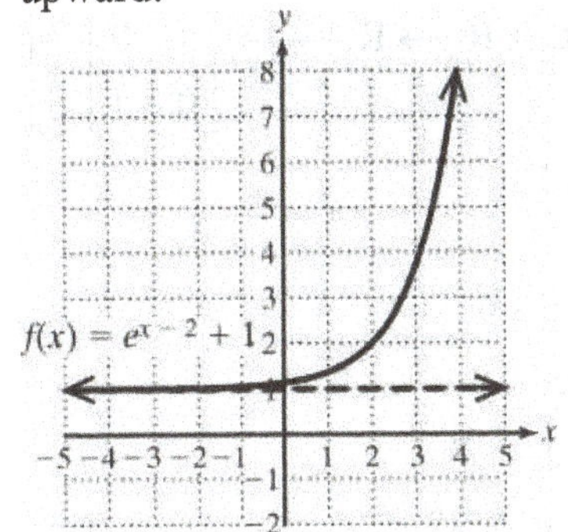

b. $(-\infty,\infty)$

c. $(1,\infty)$

11. a. The graph of h is the graph of $y=\sqrt{x}$ reflected across the y-axis.

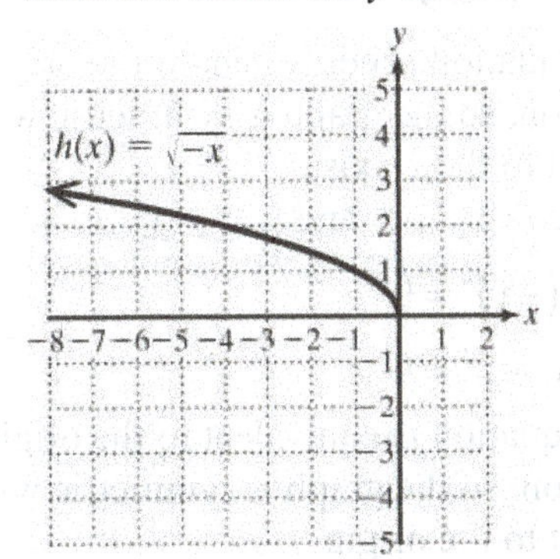

b. $(-\infty,0]$

c. $(1,\infty)$

13. $3(x-y)=y-3\rightarrow 3x-4y=-3$

$-2x+\dfrac{1}{2}y=-11\rightarrow -4x+y=-22$

Augmented matrix: $\left[\begin{array}{cc|c}3&-4&-3\\-4&1&-22\end{array}\right]$

$-R_2+(-R_1)\rightarrow R_1\longrightarrow\left[\begin{array}{cc|c}1&3&25\\-4&1&-22\end{array}\right]$

$$4R_1 + R_2 \to R_2 \longrightarrow \left[\begin{array}{cc|c} 1 & 3 & 25 \\ 0 & 13 & 78 \end{array}\right]$$

$$\frac{1}{13}R_2 \to R_2 \longrightarrow \left[\begin{array}{cc|c} 1 & 3 & 25 \\ 0 & 1 & 6 \end{array}\right]$$

$$-3R_2 + R_1 \to R_1 \longrightarrow \left[\begin{array}{cc|c} 1 & 0 & 7 \\ 0 & 1 & 6 \end{array}\right]$$

$\{(7, 6)\}$

15. $|x| + y^2 = 9$

Replace x by $-x$.

$|-x| + y^2 = 9$

$|x| + y^2 = 9$

This equation *is* equivalent to the original equation, so the graph is symmetric with respect to the y-axis.

Replace y by $-y$.

$|x| + (-y)^2 = 9$

$|x| + y^2 = 9$

This equation *is* equivalent to the original equation, so the graph is symmetric with respect to the x-axis.

Replace x by $-x$ and y by $-y$.

$|-x| + (-y)^2 = 9$

$|x| + y^2 = 9$

This equation *is* equivalent to the original equation, so the graph is symmetric with respect to the origin.

17. $AB = \begin{bmatrix} 2 & -3 \\ 1 & 6 \end{bmatrix}\begin{bmatrix} 1 & 4 \\ 0 & 7 \end{bmatrix}$

$= \begin{bmatrix} 2(1)+(-3)(0) & 2(4)+(-3)(7) \\ 1(1)+6(0) & 1(4)+6(7) \end{bmatrix}$

$= \begin{bmatrix} 2 & -13 \\ 1 & 46 \end{bmatrix}$

19. $4A - B = 4\begin{bmatrix} 2 & -3 \\ 1 & 6 \end{bmatrix} - \begin{bmatrix} 1 & 4 \\ 0 & 7 \end{bmatrix}$

$= \begin{bmatrix} 8 & -12 \\ 4 & 24 \end{bmatrix} - \begin{bmatrix} 1 & 4 \\ 0 & 7 \end{bmatrix}$

$= \begin{bmatrix} 7 & -16 \\ 4 & 17 \end{bmatrix}$

21. $A \cup B = [x \mid x > -2]$

23. $A \cap C = \{x \mid x \geq 5\}$

25. $|-x + 4| = |x - 4|$

$-x + 4 = x - 4$ or $-x + 4 = -(x - 4)$

$-2x = -8$ or $-x + 4 = -x + 4$

$x = 4$ or $0 = 0$

$\mathbb{R}$

27.

$\log_2(3x - 4) = \log_2(x + 1) + 1$

$\log_2(3x - 4) - \log_2(x + 1) = 1$

$\log_2\left(\frac{3x - 4}{x + 1}\right) = 1$

$\frac{3x - 4}{x + 1} = 2^1$

$\frac{3x - 4}{x + 1} = 2$

$3x - 4 = 2x + 2$

$x = 6$

$\{6\}$

29. The denominator is 0 at $x = 3$ and $x = -3$. r has vertical asymptotes at $x = 3$ and $x = -3$.
The degree of the numerator is 1.
The degree of the denominator is 2.
Since $n < m$, the line $y = 0$ is a horizontal asymptote of r.

Chapter 7 Analytic Geometry

Section 7.1 The Ellipse

1. $r^2 = 49$

$r = \sqrt{49} = 7$

Center: $(-4, 0)$; Radius: 7

3. $r^2 = 17$

$r = \sqrt{17}$

Center: $(0, 0)$; Radius: $\sqrt{17}$

5. $x^2 + y^2 + 10x - 4y + 22 = 0$

$(x^2 + 10x \quad) + (y^2 - 4y \quad) = -22$

$\left[\frac{1}{2}(10)\right]^2 = 25 \quad \left[\frac{1}{2}(-4)\right]^2 = 4$

$(x^2 + 10x + 25) + (y^2 - 4y + 4) = -22 + 25 + 4$

$(x+5)^2 + (y-2)^2 = 7$

7. conic

9. vertices

11. major

13. $\frac{x^2}{a^2} + \frac{y^2}{b^2} = 1$; $\frac{x^2}{b^2} + \frac{y^2}{a^2} = 1$

15. $(a, 0)$; $(-a, 0)$; $(0, b)$; $(0, -b)$

17. $(h, k+a)$; $(h, k-a)$; $(h+b, k)$; $(h-b, k)$

19. **a.** $d_1 = \sqrt{[3-(-4)]^2 + \left(\frac{12}{5} - 0\right)^2}$

$= \sqrt{7^2 + \left(\frac{12}{5}\right)^2}$

$= \sqrt{49 + \frac{144}{25}}$

$= \sqrt{\frac{1369}{25}} = \frac{37}{5}$

$d_2 = \sqrt{(4-3)^2 + \left(0 - \frac{12}{5}\right)^2}$

$= \sqrt{1^2 + \left(-\frac{12}{5}\right)^2} = \sqrt{1 + \frac{144}{25}}$

$= \sqrt{\frac{169}{25}} = \frac{13}{5}$

$d_3 = \sqrt{(-4-0)^2 + [0-(-3)]^2}$

$= \sqrt{(-4)^2 + 3^2} = \sqrt{16+9} = \sqrt{25} = 5$

$d_4 = \sqrt{(0-4)^2 + (-3-0)^2}$

$= \sqrt{(-4)^2 + (-3)^2} = \sqrt{16+9} = \sqrt{25} = 5$

b. $d_1 + d_2 = \frac{37}{5} + \frac{13}{5} = \frac{50}{5} = 10$

c. $d_3 + d_4 = 5 + 5 = 10$

d. They are the same.

e. The sums equal the length of the major axis.

21. **a.** Since $5 > 2$, $a^2 = 5$ and $b^2 = 2$. The major axis is vertical.

b. Since $5 > 2$, $a^2 = 5$ and $b^2 = 2$. The major axis is horizontal.

23. $\frac{x^2}{100} + \frac{y^2}{25} = 1$

Since $100 > 25$, $a^2 = 100$ and $b^2 = 25$.
Since the greater number in the denominator is found in the x^2 term, the ellipse is elongated horizontally.

a. Center: $(0, 0)$

b. $a = \sqrt{100} = 10$

c. $b = \sqrt{25} = 5$

d. Vertices: $(10, 0)$, $(-10, 0)$

e. Endpoints of minor axis: $(0,5)$, $(0,-5)$

f. $c^2 = a^2 - b^2 = 100 - 25 = 75$

$c = \sqrt{75} = 5\sqrt{3}$

Foci: $\left(5\sqrt{3}, 0\right)$, $\left(-5\sqrt{3}, 0\right)$

g. $2a = 2(10) = 20$

h. $2b = 2(5) = 10$

i. 

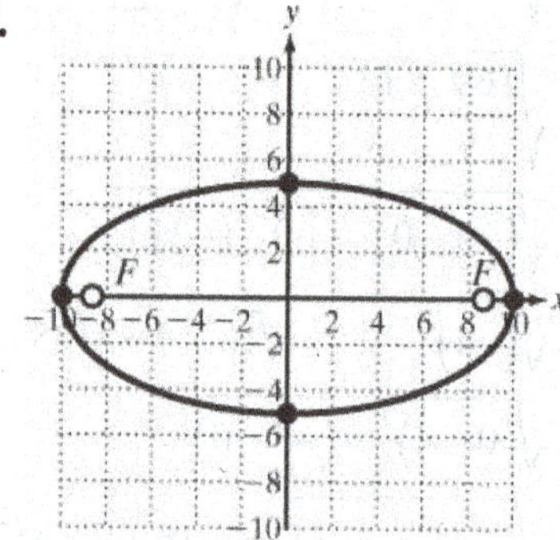

25. $\dfrac{x^2}{25} + \dfrac{y^2}{100} = 1$

Since $100 > 25$, $a^2 = 100$ and $b^2 = 25$.
Since the greater number in the denominator is found in the y^2 term, the ellipse is elongated vertically.

a. Center: $(0,0)$

b. $a = \sqrt{100} = 10$

c. $b = \sqrt{25} = 5$

d. Vertices: $(0,10)$, $(0,-10)$

e. Endpoints of minor axis: $(5,0)$, $(-5,0)$

f. $c^2 = a^2 - b^2 = 100 - 25 = 75$

$c = \sqrt{75} = 5\sqrt{3}$

Foci: $\left(0, 5\sqrt{3}\right)$, $\left(0, -5\sqrt{3}\right)$

g. $2a = 2(10) = 20$

h. $2b = 2(5) = 10$

i. 

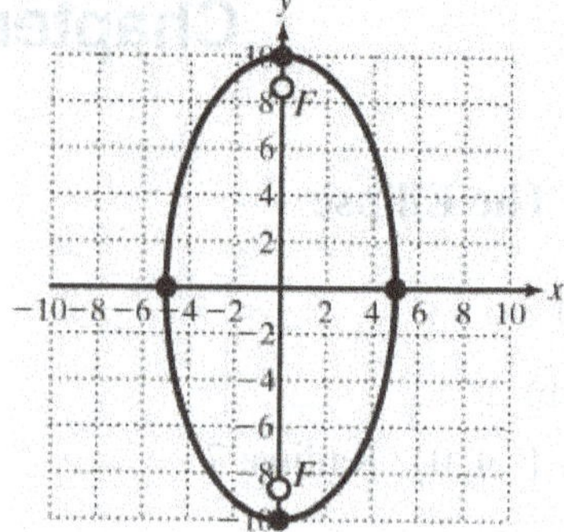

27. $4x^2 + 25y^2 = 100$

$\dfrac{4x^2}{100} + \dfrac{25y^2}{100} = \dfrac{100}{100}$

$\dfrac{x^2}{25} + \dfrac{y^2}{4} = 1$

Since $25 > 4$, $a^2 = 25$ and $b^2 = 4$.
Since the greater number in the denominator is found in the x^2 term, the ellipse is elongated horizontally.

a. Center: $(0,0)$

b. $a = \sqrt{25} = 5$

c. $b = \sqrt{4} = 2$

d. Vertices: $(5,0)$, $(-5,0)$

e. Endpoints of minor axis: $(0,2)$, $(0,-2)$

f. $c^2 = a^2 - b^2 = 25 - 4 = 21$

$c = \sqrt{21}$

Foci: $\left(\sqrt{21}, 0\right)$, $\left(-\sqrt{21}, 0\right)$

g. $2a = 2(5) = 10$

h. $2b = 2(2) = 4$

i.

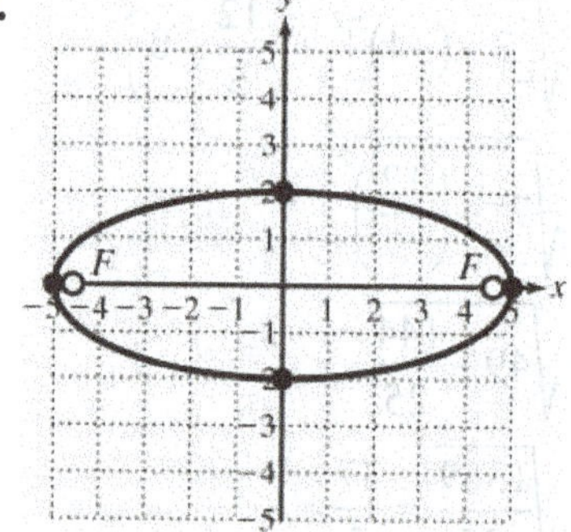

29. $-36x^2 - 4y^2 = -36$

$$\frac{-36x^2}{-36} + \frac{-4y^2}{-36} = \frac{-36}{-36}$$

$$\frac{x^2}{1} + \frac{y^2}{9} = 1$$

Since $9 > 1$, $a^2 = 9$ and $b^2 = 1$.
Since the greater number in the denominator is found in the y^2 term, the ellipse is elongated vertically.

a. Center: $(0, 0)$

b. $a = \sqrt{9} = 3$

c. $b = \sqrt{1} = 1$

d. Vertices: $(0, 3)$, $(0, -3)$

e. Endpoints of minor axis: $(1, 0)$, $(-1, 0)$

f. $c^2 = a^2 - b^2 = 9 - 1 = 8$

$c = \sqrt{8} = 2\sqrt{2}$

Foci: $(0, 2\sqrt{2})$, $(0, -2\sqrt{2})$

g. $2a = 2(3) = 6$

h. $2b = 2(1) = 2$

i.

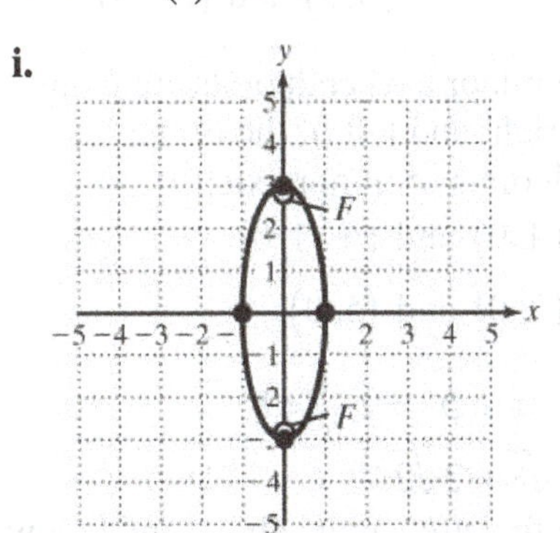

31. $\frac{x^2}{12} + \frac{y^2}{5} = 1$

Since $12 > 5$, $a^2 = 12$ and $b^2 = 5$.
Since the greater number in the denominator is found in the x^2 term, the ellipse is elongated horizontally.

a. Center: $(0, 0)$

b. $a = \sqrt{12} = 2\sqrt{3}$

c. $b = \sqrt{5}$

d. Vertices: $(2\sqrt{3}, 0)$, $(-2\sqrt{3}, 0)$

e. Endpoints of minor axis: $(0, \sqrt{5})$, $(0, -\sqrt{5})$

f. $c^2 = a^2 - b^2 = 12 - 5 = 7$

$c = \sqrt{7}$

Foci: $(\sqrt{7}, 0)$, $(-\sqrt{7}, 0)$

g. $2a = 2(2\sqrt{3}) = 4\sqrt{3}$

h. $2b = 2(\sqrt{5}) = 2\sqrt{5}$

i.

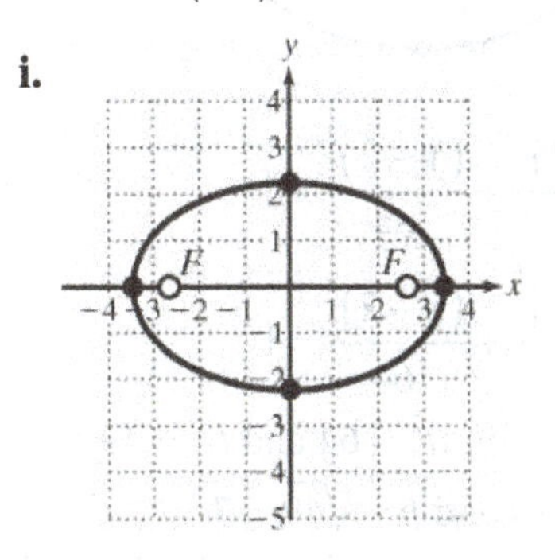

33. $\frac{(x-1)^2}{25} + \frac{(y+6)^2}{16} = 1$

Since $25 > 16$, $a^2 = 25$ and $b^2 = 16$.
$a = \sqrt{25} = 5$ and $b = \sqrt{16} = 4$.
Since the greater number in the denominator is found in the x^2 term, the ellipse is elongated horizontally.

a. Center: $(1, -6)$

b. The vertices are a units to the right and left of the center.
Vertices: $(1+5, -6)$ and $(1-5, -6)$
$(6, -6)$ and $(-4, -6)$

c. The minor axis endpoints are b units above and below the center.
Endpoints of minor axis:
$(1, -6+4)$ and $(1, -6-4)$
$(1, -2)$ and $(1, -10)$

d. $c^2 = a^2 - b^2 = 25 - 16 = 9$

$c = \sqrt{9} = 3$

The foci are c units to the right and left of the center.

Foci: $(1+3, -6)$ and $(1-3, -6)$

$(4, -6)$ and $(-2, -6)$

e.

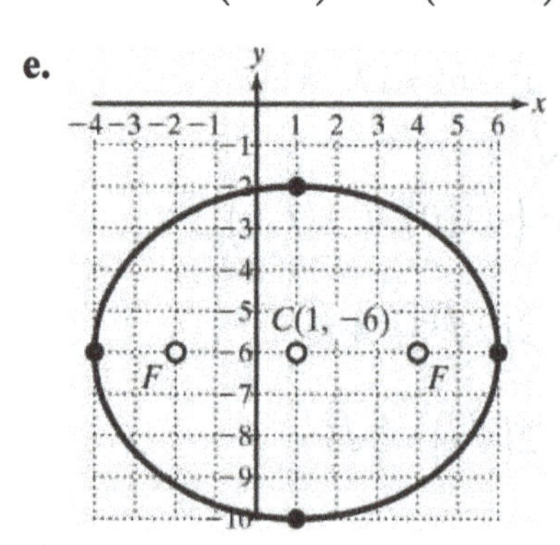

35. $\dfrac{(x+4)^2}{49} + \dfrac{(y-2)^2}{64} = 1$

$\dfrac{[x-(-4)]^2}{49} + \dfrac{(y-2)^2}{64} = 1$

Since $64 > 49$, $a^2 = 64$ and $b^2 = 49$.

$a = \sqrt{64} = 8$ and $b = \sqrt{49} = 7$

Since the greater number in the denominator is found in the y^2 term, the ellipse is elongated vertically.

a. Center: $(-4, 2)$

b. The vertices are a units above and below the center.

Vertices: $(-4, 2+8)$ and $(-4, 2-8)$

$(-4, 10)$ and $(-4, -6)$

c. The minor axis endpoints are b units to the right and left of the center.

Endpoints of minor axis:

$(-4+7, 2)$ and $(-4-7, 2)$

$(3, 2)$ and $(-11, 2)$

d. $c^2 = a^2 - b^2 = 64 - 49 - 15$

$c = \sqrt{15}$

The foci are c units above and below the center.

Foci: $\left(-4, 2+\sqrt{15}\right)$ and $\left(-4, 2-\sqrt{15}\right)$

e.

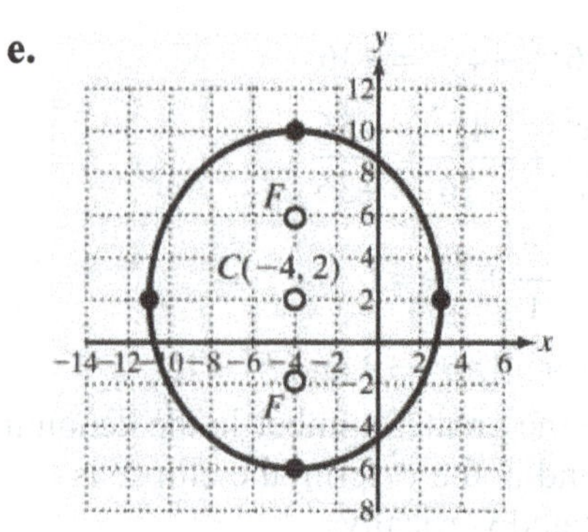

37. $(x-6)^2 + \dfrac{y^2}{9} = 1$

$\dfrac{(x-6)^2}{1} + \dfrac{(y-0)^2}{9} = 1$

Since $9 > 1$, $a^2 = 9$ and $b^2 = 1$.

$a = \sqrt{9} = 3$ and $b = \sqrt{1} = 1$.

Since the greater number in the denominator is found in the y^2 term, the ellipse is elongated vertically.

a. Center: $(6, 0)$

b. The vertices are a units above and below the center.

Vertices: $(6, 0+3)$ and $(6, 0-3)$

$(6, 3)$ and $(6, -3)$

c. The minor axis endpoints are b units to the right and left of the center.

Endpoints of minor axis:

$(6+1, 0)$ and $(6-1, 0)$

$(7, 0)$ and $(5, 0)$

d. $c^2 = a^2 - b^2 = 9 - 1 = 8$

$c = \sqrt{8} = 2\sqrt{2}$

The foci are c units above and below the center.

Foci: $\left(6, 0+2\sqrt{2}\right)$ and $\left(6, 0-2\sqrt{2}\right)$

$\left(6, 2\sqrt{2}\right)$ and $\left(6, -2\sqrt{2}\right)$

e.

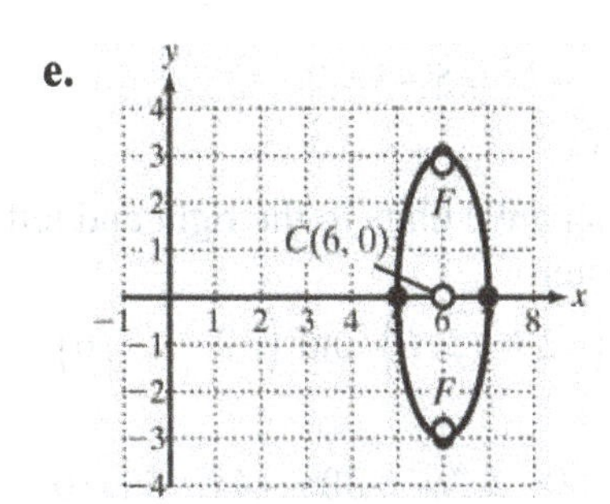

39. $$x^2+9(y+1)^2=81$$
$$(x-0)^2+9[y-(-1)]^2=81$$
$$\frac{(x-0)^2}{81}+\frac{9[y-(-1)]^2}{81}=\frac{81}{81}$$
$$\frac{(x-0)^2}{81}+\frac{[y-(-1)]^2}{9}=1$$

Since $81>9$, $a^2=81$ and $b^2=9$.

$a=\sqrt{81}=9$ and $b=\sqrt{9}=3$.

Since the greater number in the denominator is found in the x^2 term, the ellipse is elongated horizontally.

a. Center: $(0,-1)$

b. The vertices are a units to the right and left of the center.
Vertices: $(0+9,-1)$ and $(0-9,-1)$
$(9,-1)$ and $(-9,-1)$

c. The minor axis endpoints are b units above and below the center.
Endpoints of minor axis:
$(0,-1+3)$ and $(0,-1-3)$
$(0,2)$ and $(0,-4)$

d. $c^2=a^2-b^2=81-9=72$
$c=\sqrt{72}=6\sqrt{2}$
The foci are c units to the right and left of the center.
Foci: $(0+6\sqrt{2},-1)$ and $(0-6\sqrt{2},-1)$
$(6\sqrt{2},-1)$ and $(-6\sqrt{2},-1)$

e.

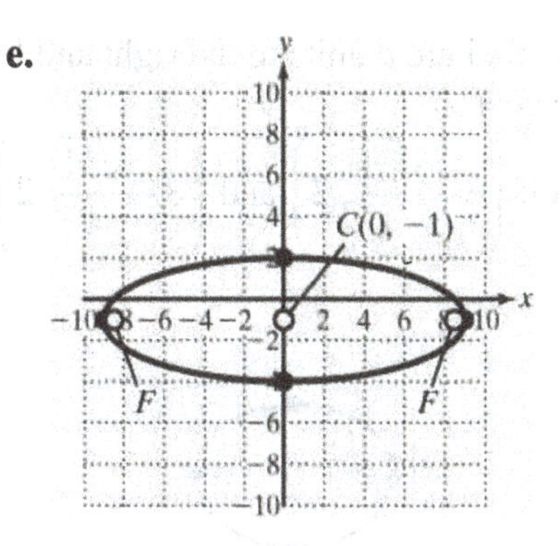

41. $$\frac{4(x-3)^2}{25}+\frac{16(y-2)^2}{49}=1$$
$$\frac{(x-3)^2}{\frac{25}{4}}+\frac{(y-2)^2}{\frac{49}{16}}=1$$

Since $\frac{25}{4}>\frac{49}{16}$, $a^2=\frac{25}{4}$ and $b^2=\frac{49}{16}$.

$a=\sqrt{\frac{25}{4}}=\frac{5}{2}$ and $b=\sqrt{\frac{49}{16}}=\frac{7}{4}$.

Since the greater number in the denominator is found in the x^2 term, the ellipse is elongated horizontally.

a. Center: $(3,2)$

b. The vertices are a units to the right and left of the center.
Vertices: $\left(3+\frac{5}{2},2\right)$ and $\left(3-\frac{5}{2},2\right)$
$\left(\frac{11}{2},2\right)$ and $\left(\frac{1}{2},2\right)$

c. The minor axis endpoints are b units above and below the center.
Endpoints of minor axis:
$\left(3,2+\frac{7}{4}\right)$ and $\left(3,2-\frac{7}{4}\right)$
$\left(3,\frac{15}{4}\right)$ and $\left(3,\frac{1}{4}\right)$

d. $c^2=a^2-b^2=\frac{25}{4}-\frac{49}{16}=\frac{100}{16}-\frac{49}{16}=\frac{51}{16}$
$c=\sqrt{\frac{51}{16}}=\frac{\sqrt{51}}{4}$

The foci are c units to the right and left of the center.

Foci: $\left(3+\frac{\sqrt{51}}{4}, 2\right)$ and $\left(3-\frac{\sqrt{51}}{4}, 2\right)$

e.

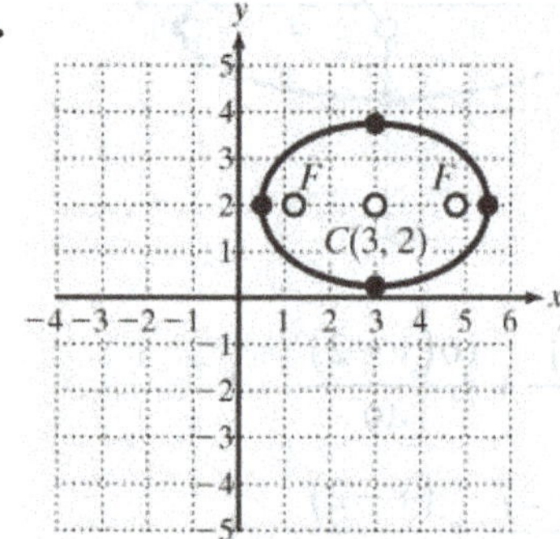

43. a. $3x^2+5y^2+12x-60y+177=0$

$3x^2+12x+5y^2-60y=-177$

$3\left(x^2+4x\quad\right)+5\left(y^2-12y\quad\right)=-177$

$\left[\frac{1}{2}(4)\right]^2=4 \quad \left[\frac{1}{2}(-12)\right]^2=36$

$3\left(x^2+4x+4\right)+5\left(y^2-12y+36\right)$

$=-177+12+180$

$3(x+2)^2+5(y-6)^2=15$

$\frac{3(x+2)^2}{15}+\frac{5(y-6)^2}{15}=\frac{15}{15}$

$\frac{(x+2)^2}{5}+\frac{(y-6)^2}{3}=1$

Since $5>3$, $a^2=5$ and $b^2=3$.

$a=\sqrt{5}$ and $b=\sqrt{3}$.

Since the greater number in the denominator is found in the x^2 term, the ellipse is elongated horizontally.

b. Center: $(-2, 6)$

The vertices are a units to the right and left of the center.

Vertices: $\left(-2+\sqrt{5}, 6\right)$ and $\left(-2-\sqrt{5}, 6\right)$

The minor axis endpoints are b units above and below the center.

Endpoints of minor axis:

$\left(-2, 6+\sqrt{3}\right)$ and $\left(-2, 6-\sqrt{3}\right)$

$c^2=a^2-b^2=5-3=2$

$c=\sqrt{2}$

The foci are c units to the right and left of the center.

Foci: $\left(-2+\sqrt{2}, 6\right)$ and $\left(-2-\sqrt{2}, 6\right)$

45. a. $3x^2+2y^2-30x-4y+59=0$

$3x^2-30x+2y^2-4y=-59$

$3\left(x^2-10x\quad\right)+2\left(y^2-2y\quad\right)=-59$

$\left[\frac{1}{2}(-10)\right]^2=25 \quad \left[\frac{1}{2}(-2)\right]^2=1$

$3\left(x^2-10x+25\right)+2\left(y^2-2y+1\right)$

$=-59+75+2$

$3(x-5)^2+2(y-1)^2=18$

$\frac{3(x-5)^2}{18}+\frac{2(y-1)^2}{18}=\frac{18}{18}$

$\frac{(x-5)^2}{6}+\frac{(y-1)^2}{9}=1$

Since $9>6$, $a^2=9$ and $b^2=6$.

$a=\sqrt{9}=3$ and $b=\sqrt{6}$.

Since the greater number in the denominator is found in the y^2 term, the ellipse is elongated vertically.

b. Center: $(5, 1)$

The vertices are a units above and below the center.

Vertices: $(5, 1+3)$ and $(5, 1-3)$

$(5, 4)$ and $(5, -2)$

The minor axis endpoints are b units to the right and left of the center.

Endpoints of minor axis:

$\left(5+\sqrt{6}, 1\right)$ and $\left(5-\sqrt{6}, 1\right)$

$c^2=a^2-b^2=9-6=3$

$c=\sqrt{3}$

The foci are c units above and below the center.

Foci: $\left(5, 1+\sqrt{3}\right)$ and $\left(5, 1-\sqrt{3}\right)$

47. a. $4x^2+y^2+14y+45=0$

$$4x^2+y^2+14y=-45$$

$$4x^2+\left(y^2+14y\quad\right)=-45$$

$$\boxed{\left[\frac{1}{2}(14)\right]^2=49}$$

$$4x^2+\left(y^2+14y+49\right)=-45+49$$

$$4x^2+(y+7)^2=4$$

$$\frac{4x^2}{4}+\frac{(y+7)^2}{4}=\frac{4}{4}$$

$$\frac{x^2}{1}+\frac{(y+7)^2}{4}=1$$

Since $4>1$, $a^2=4$ and $b^2=1$.
$a=\sqrt{4}=2$ and $b=\sqrt{1}=1$.
Since the greater number in the denominator is found in the y^2 term, the ellipse is elongated vertically.

b. Center: $(0,-7)$
The vertices are a units above and below the center.
Vertices: $(0,-7+2)$ and $(0,-7-2)$
$(0,-5)$ and $(0,-9)$
The minor axis endpoints are b units to the right and left of the center.
Endpoints of minor axis:
$(0+1,-7)$ and $(0-1,-7)$
$(1,-7)$ and $(-1,-7)$
$c^2=a^2-b^2=4-1=3$
$c=\sqrt{3}$
The foci are c units above and below the center.
Foci: $\left(0,-7+\sqrt{3}\right)$ and $\left(0,-7-\sqrt{3}\right)$

49. a. $36x^2+100y^2-180x+800y+925=0$

$$36x^2-180x+100y^2+800y=-925$$

$$36\left(x^2-5x\quad\right)+100\left(y^2+8y\quad\right)=-925$$

$$\boxed{\left[\frac{1}{2}(-5)\right]^2=\frac{25}{4}\quad\left[\frac{1}{2}(8)\right]^2=16}$$

$$36\left(x^2-5x+\frac{25}{4}\right)+100\left(y^2+8y+16\right)$$
$$=-925+225+1600$$

$$36\left(x-\frac{5}{2}\right)^2+100(y+4)^2=900$$

$$\frac{36\left(x-\frac{5}{2}\right)^2}{900}+\frac{100(y+4)^2}{900}=\frac{900}{900}$$

$$\frac{\left(x-\frac{5}{2}\right)^2}{25}+\frac{(y+4)^2}{9}=1$$

Since $25>9$, $a^2=25$ and $b^2=9$.
$a=\sqrt{25}=5$ and $b=\sqrt{9}=3$.
Since the greater number in the denominator is found in the x^2 term, the ellipse is elongated horizontally.

b. Center: $\left(\frac{5}{2},-4\right)$
The vertices are a units to the right and left of the center.
Vertices: $\left(\frac{5}{2}+5,-4\right)$ and $\left(\frac{5}{2}-5,-4\right)$
$\left(\frac{15}{2},-4\right)$ and $\left(-\frac{5}{2},-4\right)$
The minor axis endpoints are b units above and below the center.
Endpoints of minor axis:
$\left(\frac{5}{2},-4+3\right)$ and $\left(\frac{5}{2},-4-3\right)$
$\left(\frac{5}{2},-1\right)$ and $\left(\frac{5}{2},-7\right)$
$c^2=a^2-b^2=25-9=16$
$c=\sqrt{16}=4$
The foci are c units to the right and left of the center.
Foci: $\left(\frac{5}{2}+4,-4\right)$ and $\left(\frac{5}{2}-4,-4\right)$
$\left(\frac{13}{2},-4\right)$ and $\left(-\frac{5}{2},-4\right)$

51. The vertices have the same y-coordinate, so the major axis is horizontal and the ellipse is elongated horizontally. The standard form is

$$\frac{(x-h)^2}{a^2}+\frac{(y-k)^2}{b^2}=1$$

The center is midway between the vertices.

Center is $(0,0)$: $h=0$ and $k=0$

Vertices are 4 units from the center: $a=4$

Foci are 3 units from the center: $c=3$

$$b^2=a^2-c^2=(4)^2-(3)^2=16-9=7$$

$$b=\sqrt{7}$$

$$\frac{(x-0)^2}{(4)^2}+\frac{(y-0)^2}{(\sqrt{7})^2}=1$$

$$\frac{x^2}{16}+\frac{y^2}{7}=1$$

53. The foci have the same x-coordinate, so the major axis is vertical and the ellipse is elongated vertically. The standard form is

$$\frac{(x-h)^2}{b^2}+\frac{(y-k)^2}{a^2}=1$$

The center is midway between the foci.

Center is $(0,0)$: $h=0$ and $k=0$

Endpoints of the minor axis are $\sqrt{17}$ units from the center: $b=\sqrt{17}$

Foci are 9 units from the center: $c=9$

$$a^2=b^2+c^2=(\sqrt{17})^2+(9)^2=17+81=98$$

$$a=\sqrt{98}$$

$$\frac{(x-0)^2}{(\sqrt{17})^2}+\frac{(y-0)^2}{(\sqrt{98})^2}=1$$

$$\frac{x^2}{17}+\frac{y^2}{98}=1$$

55. The major axis is parallel to the x-axis, so the major axis is horizontal and the ellipse is elongated horizontally. The standard form is

$$\frac{(x-h)^2}{a^2}+\frac{(y-k)^2}{b^2}=1$$

Center is $(2,3)$: $h=2$ and $k=3$

Length of major axis is 14 units: $a=\frac{14}{2}=7$

Length of minor axis is 10 units: $b=\frac{10}{2}=5$

$$\frac{(x-2)^2}{(7)^2}+\frac{(y-3)^2}{(5)^2}=1$$

$$\frac{(x-2)^2}{49}+\frac{(y-3)^2}{25}=1$$

57. The foci have the same y-coordinate, so the major axis is horizontal and the ellipse is elongated horizontally. The standard form is

$$\frac{(x-h)^2}{a^2}+\frac{(y-k)^2}{b^2}=1$$

The center is midway between the foci.

Center is $(4,1)$: $h=4$ and $k=1$

Length of minor axis is 6 units: $b=\frac{6}{2}=3$

Foci are 4 units from the center: $c=4$

$$a^2=b^2+c^2=(3)^2+(4)^2=9+16=25$$

$$a=\sqrt{25}=5$$

$$\frac{(x-4)^2}{(5)^2}+\frac{(y-1)^2}{(3)^2}=1$$

$$\frac{(x-4)^2}{25}+\frac{(y-1)^2}{9}=1$$

59. The vertices have the same x-coordinate, so the major axis is vertical and the ellipse is elongated vertically. The standard form is

$$\frac{(x-h)^2}{b^2}+\frac{(y-k)^2}{a^2}=1$$

The center is midway between the foci.

Center is $(0,0)$: $h=0$ and $k=0$

Vertices are 5 units from the center: $a=5$

Substitute these values and the coordinates of the point $\left(\frac{16}{5},3\right)$ into the equation.

$$\frac{\left(\frac{16}{5}-0\right)^2}{b^2}+\frac{(3-0)^2}{(5)^2}=1$$

$$\frac{\frac{256}{25}}{b^2}+\frac{9}{25}=1$$

$$25b^2\cdot\left(\frac{256}{25b^2}+\frac{9}{25}\right)=25b^2\cdot 1$$

$$256-9b^2=25b^2$$

$$256=16b^2$$

$$16=b^2$$

$$4=b$$

$$\frac{(x-0)^2}{(4)^2}+\frac{(y-0)^2}{(5)^2}=1$$

$$\frac{x^2}{16}+\frac{y^2}{25}=1$$

61. The vertices have the same x-coordinate, so the major axis is vertical and the ellipse is elongated vertically. The standard form is

$$\frac{(x-h)^2}{b^2}+\frac{(y-k)^2}{a^2}=1$$

The center is midway between the vertices.

Center is $(3,0)$: $h=3$ and $k=0$

Vertices are 4 units from the center: $a=4$

Foci are $\sqrt{11}$ units from the center: $c=\sqrt{11}$

$$b^2=a^2-c^2=(4)^2-\left(\sqrt{11}\right)^2=16-11=5$$

$$b=\sqrt{5}$$

$$\frac{(x-3)^2}{\left(\sqrt{5}\right)^2}+\frac{(y-0)^2}{(4)^2}=1$$

$$\frac{(x-3)^2}{5}+\frac{y^2}{16}=1$$

63. The vertices have the same y-coordinate, so the major axis is horizontal and the ellipse is elongated horizontally. The standard form is

$$\frac{(x-h)^2}{a^2}+\frac{(y-k)^2}{b^2}=1$$

The center is midway between the vertices.

Center is $(-5,1)$: $h=-5$ and $k=1$

Vertices are 7 units from the center: $a=7$

Foci are $\sqrt{33}$ units from the center: $c=\sqrt{33}$

$$b^2=a^2-c^2=(7)^2-\left(\sqrt{33}\right)^2=49-33=16$$

$$b=\sqrt{16}=4$$

$$\frac{\left[x-(-5)\right]^2}{(7)^2}+\frac{(y-1)^2}{(4)^2}=1$$

$$\frac{(x+5)^2}{49}+\frac{(y-1)^2}{16}=1$$

65. b

67. a

69. $a^2=169$ $\qquad$ $b^2=25$

$$a=13$$

$$c^2=a^2-b^2=169-25=144$$

$$c=\sqrt{144}=12$$

$$e=\frac{c}{a}=\frac{12}{13}$$

71. $a^2=225$ $\qquad$ $b^2=144$

$$a=15$$

$$c^2=a^2-b^2=225-144=81$$

$$c=\sqrt{81}=9$$

$$e=\frac{c}{a}=\frac{9}{15}=\frac{3}{5}$$

73. $a^2=12$ $\qquad$ $b^2=6$

$$a=\sqrt{12}=2\sqrt{3}$$

$$c^2=a^2-b^2=12-6=6$$

$$c=\sqrt{6}$$

$$e=\frac{c}{a}=\frac{\sqrt{6}}{2\sqrt{3}}\cdot\frac{\sqrt{3}}{\sqrt{3}}=\frac{\sqrt{18}}{6}=\frac{3\sqrt{2}}{6}=\frac{\sqrt{2}}{2}$$

75. a. The eccentricity of the ellipse in Exercise 69 is closest to 1, so it is the most elongated.

b. The eccentricity of the ellipse in Exercise 71 is closest to 0, so it is the most circular.

77. The Earth's orbit is more circular, and the orbit for Halley's Comet is very elongated.

79. $e = \frac{c}{a} = 0.0549$

$$c = 0.0549a$$
$$a - c = 363{,}300$$
$$a = 363{,}300 + c$$
$$a = 363{,}300 + 0.0549a$$
$$0.9451a = 363{,}300$$
$$a = 384{,}404$$
$$c = a - 363{,}300 = 384{,}404 = 21{,}104$$
$$a + c = 384{,}404 + 21{,}104 \approx 405{,}500 \text{ km}$$

81. The major axis is vertical, so the ellipse is elongated horizontally. The standard form is

$$\frac{(x-h)^2}{b^2} + \frac{(y-k)^2}{a^2} = 1$$

Center is $(0, 0)$: $h = 0$ and $k = 0$

Length of major axis is 34 units: $a = \frac{34}{2} = 17$

$$e = \frac{c}{a} = \frac{15}{17}$$
$$c = \frac{15}{17}a = \frac{15}{17}(17) = 15$$
$$b^2 = a^2 - c^2 = (17)^2 - (15)^2 = 289 - 225 = 64$$
$$b = \sqrt{64} = 8$$
$$\frac{(x-0)^2}{(8)^2} + \frac{(y-0)^2}{(17)^2} = 1$$
$$\frac{x^2}{64} + \frac{y^2}{289} = 1$$

83. The foci have the same y-coordinate, so the major axis is horizontal and the ellipse is elongated horizontally. The standard form is

$$\frac{(x-h)^2}{a^2} + \frac{(y-k)^2}{b^2} = 1$$

The center is midway between the foci.
Center is $(4, -1)$: $h = 4$ and $k = -1$

Foci are 4 units from the center: $c = 4$

$$e = \frac{c}{a} = \frac{4}{5}$$
$$c = \frac{4}{5}a$$
$$4 = \frac{4}{5}a$$
$$5 = a$$
$$b^2 = a^2 - c^2 = (5)^2 - (4)^2 = 25 - 16 = 9$$
$$b = \sqrt{9} = 3$$
$$\frac{(x-4)^2}{(5)^2} + \frac{[y-(-1)]^2}{(3)^2} = 1$$
$$\frac{(x-4)^2}{25} + \frac{(y+1)^2}{9} = 1$$

85. Length of major axis is 90 in. $a = \frac{90}{2} = 45$

Length of minor axis is 30 in. $b = \frac{30}{2} = 15$

$$c^2 = a^2 - b^2 = (45)^2 - (15)^2 = 2025 - 225 = 1800$$
$$c = \sqrt{1800} = 30\sqrt{2}$$

The foci should be located $\pm 30\sqrt{2}$ in. (approximately 42.4 in.) from the center along the major axis.

87. Length of half of major axis is 22.3 cm.
$a = 22.3$

Length of minor axis is 29 cm. $b = \frac{29}{2} = 14.5$

$$c^2 = a^2 - b^2 = (22.3)^2 - (14.5)^2 = 287.04$$
$$c = \sqrt{287.04} \approx 16.94$$

The kidney stone should be located approximately 16.94 cm from the center along the major axis.

89. **a.** Make a table of values.

x	y	$y=2\sqrt{1-\frac{x^2}{9}}$	Ordered pair
−3	0	$y=2\sqrt{1-\frac{(-3)^2}{9}}=0$	$(-3,0)$
0	2	$y=2\sqrt{1-\frac{(0)^2}{9}}=2$	$(0,2)$
3	0	$y=2\sqrt{1-\frac{(3)^2}{9}}=0$	$(3,0)$

These points are on the top semiellipse.

b. Make a table of values.

x	y	$y=-2\sqrt{1-\frac{x^2}{9}}$	Ordered pair
−3	0	$y=-2\sqrt{1-\frac{(-3)^2}{9}}=0$	$(-3,0)$
0	−2	$y=-2\sqrt{1-\frac{(0)^2}{9}}=-2$	$(0,-2)$
3	0	$y=-2\sqrt{1-\frac{(3)^2}{9}}=0$	$(3,0)$

These points are on the bottom semiellipse.

c. Make a table of values.

x	y	$x=3\sqrt{1-\frac{y^2}{4}}$	Ordered pair
0	−2	$x=3\sqrt{1-\frac{(-2)^2}{4}}=0$	$(0,-2)$
3	0	$x=3\sqrt{1-\frac{(0)^2}{4}}=3$	$(3,0)$
0	2	$x=3\sqrt{1-\frac{(2)^2}{4}}=0$	$(0,2)$

These points are on the right semiellipse.

d. Make a table of values.

x	y	$x=-3\sqrt{1-\frac{y^2}{4}}$	Ordered pair
0	−2	$x=-3\sqrt{1-\frac{(-2)^2}{4}}=0$	$(0,-2)$
−3	0	$x=-3\sqrt{1-\frac{(0)^2}{4}}=-3$	$(-3,0)$
0	2	$x=-3\sqrt{1-\frac{(2)^2}{4}}=0$	$(0,2)$

These points are on the left semiellipse.

91. $\boxed{A}$ $\frac{x^2}{25}+\frac{y^2}{9}=1 \rightarrow 9x^2+25y^2=225$

$\boxed{B}$ $3x+5y=15 \rightarrow y=-\frac{3}{5}x+3$

$\boxed{A}$
$$9x^2+25y^2=225$$
$$9x^2+25\left(-\frac{3}{5}x+3\right)^2=225$$
$$9x^2+25\left(\frac{9}{25}x^2-\frac{18}{5}x+9\right)=225$$
$$9x^2+9x^2-90x+225=225$$
$$18x^2-90x=0$$
$$x^2-5x=0$$
$$x(x-5)=0$$
$$x=0 \text{ or } x=5$$

$\boxed{B}$ $y=-\frac{3}{5}x+3=-\frac{3}{5}(0)+3=3$

The solution is $(0,3)$.

$\boxed{B}$ $y=-\frac{3}{5}x+3=-\frac{3}{5}(5)+3=-3+3=0$

The solution is $(5,0)$.

$\{(0,3),(5,0)\}$

93. $\boxed{A}$ $\dfrac{x^2}{4}+\dfrac{y^2}{16}=1 \rightarrow 4x^2+y^2=16$

$\boxed{B}$ $y=-x^2+4 \rightarrow x^2=-y+4$

$\boxed{A}$
$$\begin{aligned} 4x^2+y^2&=16\\ 4(-y+4)+y^2&=16\\ -4y+16+y^2&=16\\ y^2-4y&=0\\ y(y-4)&=0\\ y=0 \text{ or } y&=4 \end{aligned}$$

$\boxed{B}$
$$\begin{aligned} x^2&=-y+4\\ x^2&=-(0)+4\\ x^2&=4\\ x&=\pm 2 \end{aligned}$$

The solutions are $(2,0)$ and $(-2,0)$.

$\boxed{B}$
$$\begin{aligned} x^2&=-y+4\\ x^2&=-(4)+4\\ x^2&=0\\ x&=0 \end{aligned}$$

The solution is $(0,4)$.

$\{(0,4),(-2,0),(2,0)\}$

95. a. Since $8>4$, $a^2=8$ and $b^2=4$.
$$\begin{aligned} a&=\sqrt{8}=2\sqrt{2}\\ b&=\sqrt{4}=2\\ A&=\pi ab\\ &=\pi\left(2\sqrt{2}\right)(2)\\ &=4\pi\sqrt{2} \text{ square units} \end{aligned}$$

b.
$$\begin{aligned} P&\approx \pi\sqrt{2\left(a^2+b^2\right)}\\ &\approx \pi\sqrt{2(8+4)}\\ &\approx \pi\sqrt{24}\\ &\approx 2\pi\sqrt{6} \text{ units} \end{aligned}$$

97. $\dfrac{x^2}{a^2}+\dfrac{y^2}{b^2}=1$

Substitute (c, y) into the equation.

$\dfrac{c^2}{a^2}+\dfrac{y^2}{b^2}=1$

Substitute a^2-b^2 for c^2.
$$\begin{aligned} \frac{a^2-b^2}{a^2}+\frac{y^2}{b^2}&=1\\ \frac{y^2}{b^2}&=1-\frac{a^2-b^2}{a^2}\\ \frac{y^2}{b^2}&=\frac{a^2-a^2+b^2}{a^2}\\ \frac{y^2}{b^2}&=\frac{b^2}{a^2}\\ y^2&=\frac{b^4}{a^2}\\ y&=\frac{b^2}{a} \end{aligned}$$

Length of latus rectum is $2y=\dfrac{2b^2}{a}$.

99. By the reflective property of an ellipse, any shot passing through one focus is reflected through the other focus.

101. The first equation represents an ellipse centered at the origin, whereas the second equation represents a circle with radius 2.

103. The eccentricity of an ellipse is a measure of elongation. It is computed by finding the ratio of c and a; that is, eccentricity is the ratio of the distance between a focus and the center to the distance between a vertex and the center.

105. a. $(a+c)+(a-c)=a+c+a-c=2a$

b. $\sqrt{(x+c)^2+y^2}+\sqrt{(x-c)^2+y^2}=2a$

c. $\sqrt{(x+c)^2+y^2}=2a-\sqrt{(x-c)^2+y^2}$

$$(x+c)^2+y^2=4a^2-4a\sqrt{(x-c)^2+y^2}+(x-c)^2+y^2$$

$$4a\sqrt{(x-c)^2+y^2}=4a^2+(x-c)^2-(x+c)^2$$

$$4a\sqrt{(x-c)^2+y^2}=4a^2+x^2-2xc+c^2-\left(x^2+2xc+c^2\right)$$

$$4a\sqrt{(x-c)^2+y^2}=4a^2-4xc$$

$$a\sqrt{(x-c)^2+y^2}=a^2-xc$$

d. $\left(a\sqrt{(x-c)^2+y^2}\right)^2=\left(a^2-xc\right)^2$

$$a^2\left[(x-c)^2+y^2\right]=a^4-2a^2xc+c^2x^2$$

$$a^2\left[x^2-2xc+c^2+y^2\right]=a^4-2a^2xc+c^2x^2$$

$$a^2x^2-2a^2xc+a^2c^2+a^2y^2=a^4-2a^2xc+c^2x^2$$

$$a^2x^2-c^2x^2+a^2y^2=a^4-a^2c^2$$

$$\left(a^2-c^2\right)x^2+a^2y^2=a^2\left(a^2-c^2\right)$$

e. $b^2x^2+a^2y^2=a^2b^2$

$$\frac{b^2x^2}{a^2b^2}+\frac{a^2y^2}{a^2b^2}=\frac{a^2b^2}{a^2b^2}$$

$$\frac{x^2}{a^2}+\frac{y^2}{b^2}=1$$

107. $\frac{x^2}{25}+\frac{y^2}{100}=1$

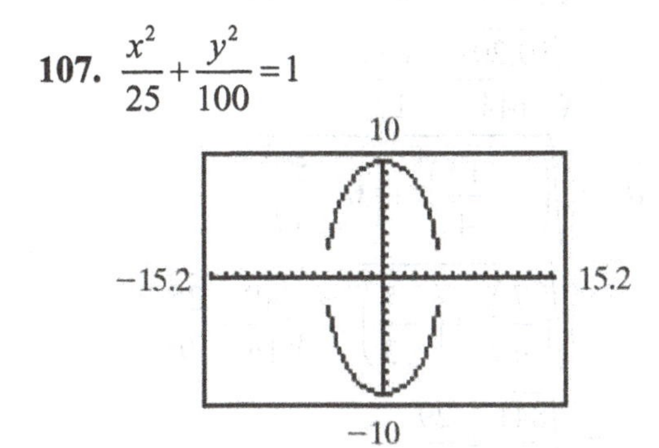

109. $\frac{4(x-3)^2}{25}+\frac{16(y-2)^2}{49}=1$

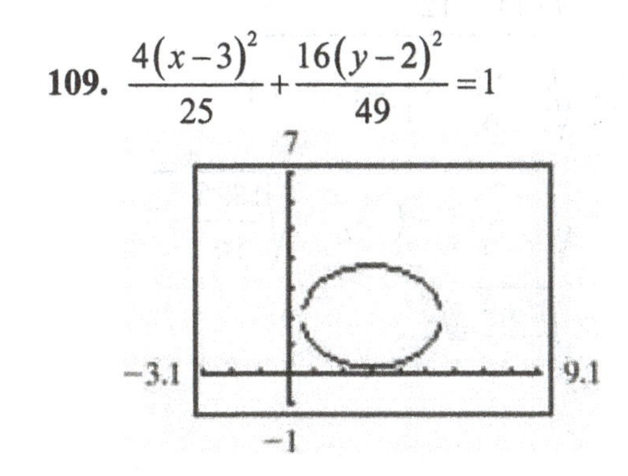

111.

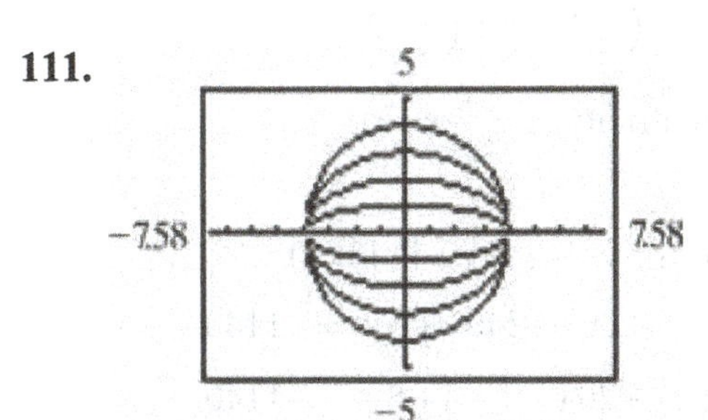

The eccentricity varies for different values of b. For values of b closer to 4, the ellipse appears more circular. For values of b closer to 0, the ellipse is more elongated. When $b=0$, the curve is a circle rather than an ellipse.

Section 7.2 The Hyperbola

1. $\dfrac{(x-3)^2}{169}+\dfrac{(y+10)^2}{144}=1$

$\dfrac{(x-3)^2}{169}+\dfrac{[y-(-10)]^2}{144}=1$

Center: $(3,-10)$

Since $169>144$, $a^2=169$ and $b^2=144$.

$a=\sqrt{169}=13$ and $b=\sqrt{144}=12$.

Since the greater number in the denominator is found in the x^2 term, the ellipse is elongated horizontally.

The vertices are a units to the right and left of the center.

Vertices: $(3+13,-10)$ and $(3-13,-10)$

$(16,-10)$ and $(-10,-10)$

$c^2=a^2-b^2=169-144=25$

$c=\sqrt{25}=5$

The foci are c units to the right and left of the center.

Foci: $(3+5,-10)$ and $(3-5,-10)$

$(8,-10)$ and $(-2,-10)$

Endpoints of minor axis:

$(3,-10+12)$ and $(3,-10-12)$

$(3,2)$ and $(3,-22)$

Eccentricity: $e=\dfrac{c}{a}=\dfrac{5}{13}$

3. $12x^2+5y^2+240x+1140=0$

$12x^2+240x+5y^2=-1140$

$12(x^2+20x\qquad)+5y^2=-1140$

$\boxed{\left[\frac{1}{2}(20)\right]^2=100}$

$12(x^2+20x+100)+5y^2=-1140+1200$

$12(x+10)^2+5y^2=60$

$\dfrac{12(x+10)^2}{60}+\dfrac{5y^2}{60}=\dfrac{60}{60}$

$\dfrac{(x+10)^2}{5}+\dfrac{y^2}{12}=1$

5. $(x+8)^2+\left(y-\dfrac{3}{2}\right)^2=20$

$[x-(-8)]^2+\left(y-\dfrac{3}{2}\right)^2=(2\sqrt{5})^2$

Center: $\left(-8,\dfrac{3}{2}\right)$; Radius: $2\sqrt{5}$

7. hyperbola; foci

9. transverse

11. horizontal; $(a,0)$; $(-a,0)$;

$y=\dfrac{b}{a}x$; $y=-\dfrac{b}{a}x$

13. conjugate

15. $(h,k+a)$; $(h,k-a)$

17. a. $d_1=\sqrt{\left[\dfrac{13}{4}-(-5)\right]^2+\left(\dfrac{5}{3}-0\right)^2}$

$=\sqrt{\left(\dfrac{33}{4}\right)^2+\left(\dfrac{5}{3}\right)^2}=\sqrt{\dfrac{1089}{16}+\dfrac{25}{9}}$

$=\sqrt{\dfrac{10{,}201}{144}}=\dfrac{101}{12}$

$d_2=\sqrt{\left(5-\dfrac{13}{4}\right)^2+\left(0-\dfrac{5}{3}\right)^2}$

$=\sqrt{\left(\dfrac{7}{4}\right)^2+\left(-\dfrac{5}{3}\right)^2}=\sqrt{\dfrac{49}{16}+\dfrac{25}{9}}$

$=\sqrt{\dfrac{841}{144}}=\dfrac{29}{12}$

$d_3=\sqrt{\left(-\dfrac{15}{4}-5\right)^2+(-3-0)^2}$

$=\sqrt{\left(-\dfrac{35}{4}\right)^2+(-3)^2}=\sqrt{\dfrac{1225}{16}+9}$

$=\sqrt{\dfrac{1369}{16}}=\dfrac{37}{4}$

$$d_4 = \sqrt{\left[-5-\left(-\frac{15}{4}\right)\right]^2 + \left[0-(-3)\right]^2}$$

$$= \sqrt{\left(-\frac{5}{4}\right)^2 + (3)^2} = \sqrt{\frac{25}{16}+9} = \sqrt{\frac{169}{16}} = \frac{13}{4}$$

b. $d_1 - d_2 = \frac{101}{12} - \frac{29}{12} = \frac{72}{12} = 6$

c. $d_3 - d_4 = \frac{37}{4} - \frac{13}{4} = \frac{24}{4} = 6$

d. They are the same.

e. The difference in distances equals the length of the transverse axis.

19. The x^2 term has the positive coefficient, indicating that the transverse axis and foci are on the x-axis.

21. The y^2 term has the positive coefficient, indicating that the transverse axis and foci are on the y-axis.

23. $\frac{x^2}{16} - \frac{y^2}{25} = 1$

The x^2 term has the positive coefficient, so the transverse axis is horizontal. The denominator of the x^2 term is a^2.

$a^2 = 16$, so $a = 4$.

$b^2 = 25$, so $b = 5$.

$c^2 = a^2 + b^2 = 16 + 25 = 41$

$c = \sqrt{41}$

a. Center: $(0, 0)$

b. Vertices: $(4, 0)$ and $(-4, 0)$

c. Foci: $(\sqrt{41}, 0)$ and $(-\sqrt{41}, 0)$

d. Asymptotes:

$y = \frac{b}{a}x$ $\quad$ $y = -\frac{b}{a}x$

$y = \frac{5}{4}x$ $\quad$ $y = -\frac{5}{4}x$

e.

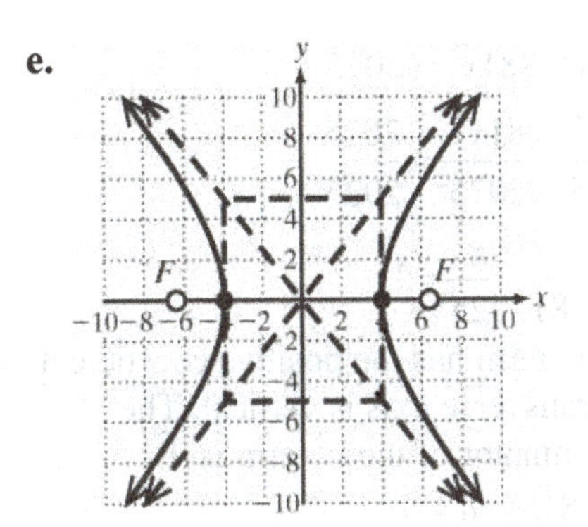

25. $\frac{y^2}{4} - \frac{x^2}{36} = 1$

The y^2 term has the positive coefficient, so the transverse axis is vertical. The denominator of the y^2 term is a^2.

$a^2 = 4$, so $a = 2$.

$b^2 = 36$, so $b = 6$.

$c^2 = a^2 + b^2 = 4 + 36 = 40$

$c = \sqrt{40} = 2\sqrt{10}$

a. Center: $(0, 0)$

b. Vertices: $(0, 2)$ and $(0, -2)$

c. Foci: $(0, 2\sqrt{10})$ and $(0, -2\sqrt{10})$

d. Asymptotes:

$y = \frac{a}{b}x$ $\quad$ $y = -\frac{a}{b}x$

$y = \frac{2}{6}x$ $\quad$ $y = -\frac{2}{6}x$

$y = \frac{1}{3}x$ $\quad$ $y = -\frac{1}{3}x$

e.

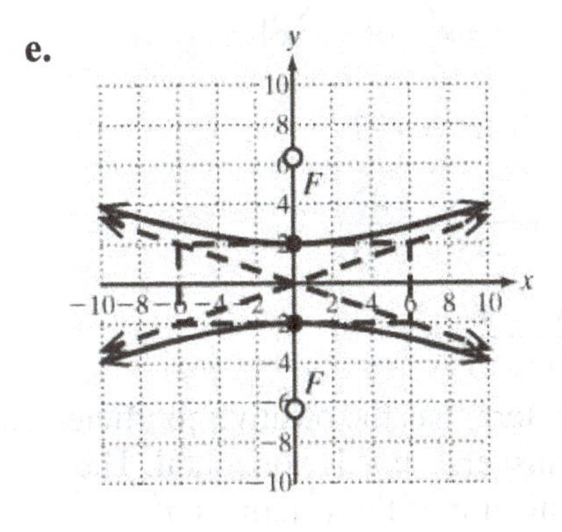

27. $25y^2 - 81x^2 = 2025$

$\frac{25y^2}{2025} - \frac{81x^2}{2025} = \frac{2025}{2025}$

$\frac{y^2}{81} - \frac{x^2}{25} = 1$

The y^2 term has the positive coefficient, so the transverse axis is vertical. The denominator of the y^2 term is a^2.

$a^2 = 81$, so $a = 9$.

$b^2 = 25$, so $b = 5$.

$c^2 = a^2 + b^2 = 81 + 25 = 106$

$c = \sqrt{106}$

a. Center: $(0, 0)$

b. Vertices: $(0, 9)$ and $(0, -9)$

c. Foci: $(0, \sqrt{106})$ and $(0, -\sqrt{106})$

d. Asymptotes:

$y = \frac{a}{b}x$ $\quad$ $y = -\frac{a}{b}x$

$y = \frac{9}{5}x$ $\quad$ $y = -\frac{9}{5}x$

e.

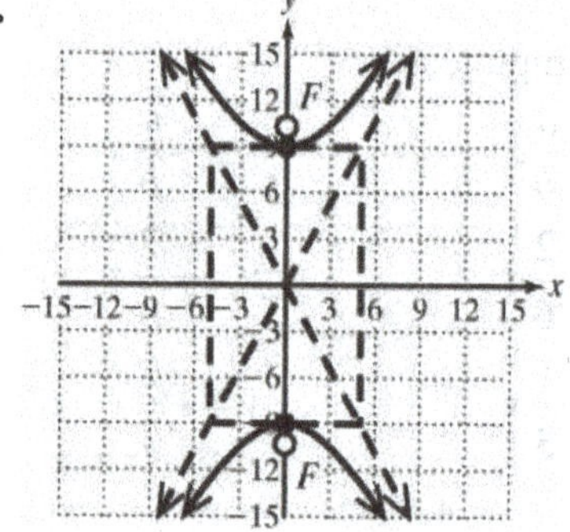

29. $-5x^2 + 7y^2 = -35$

$\frac{-5x^2}{-35} + \frac{7y^2}{-35} = \frac{-35}{-35}$

$\frac{x^2}{7} - \frac{y^2}{5} = 1$

The x^2 term has the positive coefficient, so the transverse axis is horizontal. The denominator of the x^2 term is a^2.

$a^2 = 7$, so $a = \sqrt{7}$.

$b^2 = 5$, so $b = \sqrt{5}$.

$c^2 = a^2 + b^2 = 7 + 5 = 12$

$c = \sqrt{12} = 2\sqrt{3}$

a. Center: $(0, 0)$

b. Vertices: $(\sqrt{7}, 0)$ and $(-\sqrt{7}, 0)$

c. Foci: $(2\sqrt{3}, 0)$ and $(-2\sqrt{3}, 0)$

d. Asymptotes:

$y = \frac{b}{a}x$ $\quad$ $y = -\frac{b}{a}x$

$y = \frac{\sqrt{5}}{\sqrt{7}}x \cdot \frac{\sqrt{7}}{\sqrt{7}}$ $\quad$ $y = -\frac{\sqrt{5}}{\sqrt{7}}x \cdot \frac{\sqrt{7}}{\sqrt{7}}$

$y = \frac{\sqrt{35}}{7}x$ $\quad$ $y = -\frac{\sqrt{35}}{7}x$

e.

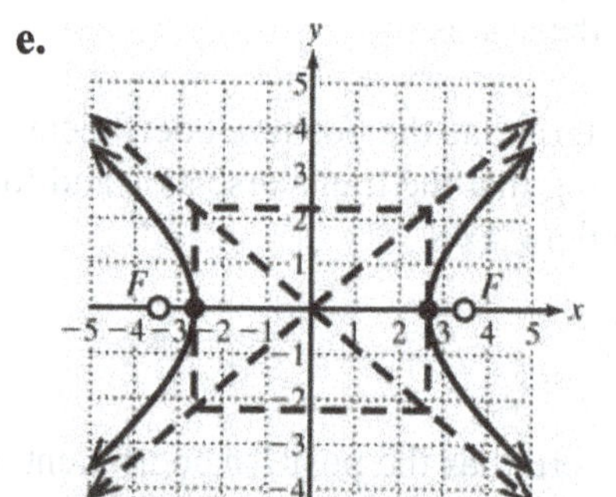

31. $\frac{4x^2}{25} - \frac{16y^2}{49} = 1$

$\frac{x^2}{\frac{25}{4}} - \frac{y^2}{\frac{49}{16}} = 1$

The x^2 term has the positive coefficient, so the transverse axis is horizontal. The denominator of the x^2 term is a^2.

$a^2 = \frac{25}{4}$, so $a = \frac{5}{2}$.

$b^2 = \frac{49}{16}$, so $b = \frac{7}{4}$

$c^2 = a^2 + b^2 = \frac{25}{4} + \frac{49}{16} = \frac{100}{16} + \frac{49}{16} = \frac{149}{16}$

$c = \sqrt{\frac{149}{16}} = \frac{\sqrt{149}}{4}$

a. Center: $(0, 0)$

b. Vertices: $\left(\frac{5}{2},0\right)$ and $\left(-\frac{5}{2},0\right)$

c. Foci: $\left(\frac{\sqrt{149}}{4},0\right)$ and $\left(-\frac{\sqrt{149}}{4},0\right)$

d. Asymptotes:

$$y=\frac{b}{a}x \qquad y=-\frac{b}{a}x$$

$$y=\frac{\frac{7}{4}}{\frac{5}{2}}x \qquad y=-\frac{\frac{7}{4}}{\frac{5}{2}}x$$

$$y=\frac{7}{4}\cdot\frac{2}{5}x \qquad y=-\frac{7}{4}\cdot\frac{2}{5}x$$

$$y=\frac{7}{10}x \qquad y=-\frac{7}{10}x$$

e.

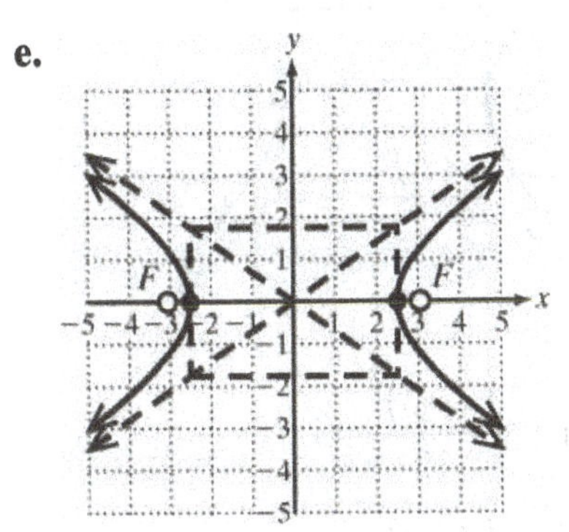

33. $$\frac{(x-4)^2}{9}-\frac{(y+2)^2}{16}=1$$

$$\frac{(x-4)^2}{9}-\frac{[y-(-2)]^2}{16}=1$$

The x^2 term has the positive coefficient, so the transverse axis is horizontal. The denominator of the x^2 term is a^2.

$a^2=9$, so $a=3$.

$b^2=16$, so $b=4$.

$c^2=a^2+b^2=9+16=25$

$c=\sqrt{25}=5$

a. Center: $(4,-2)$

b. The vertices are a units to the right and left of the center.

Vertices: $(4+3,-2)$ and $(4-3,-2)$

$(7,-2)$ and $(1,-2)$

c. The foci are c units to the right and left of the center.

Foci: $(4+5,-2)$ and $(4-5,-2)$

$(9,-2)$ and $(-1,-2)$

d. Asymptotes:

$$y-k=\frac{b}{a}(x-h) \qquad y-k=-\frac{b}{a}(x-h)$$

$$y+2=\frac{4}{3}(x-4) \qquad y+2=-\frac{4}{3}(x-4)$$

$$y+2=\frac{4}{3}x-\frac{16}{3} \qquad y+2=-\frac{4}{3}x+\frac{16}{3}$$

$$y=\frac{4}{3}x-\frac{22}{3} \qquad y=-\frac{4}{3}x+\frac{10}{3}$$

e.

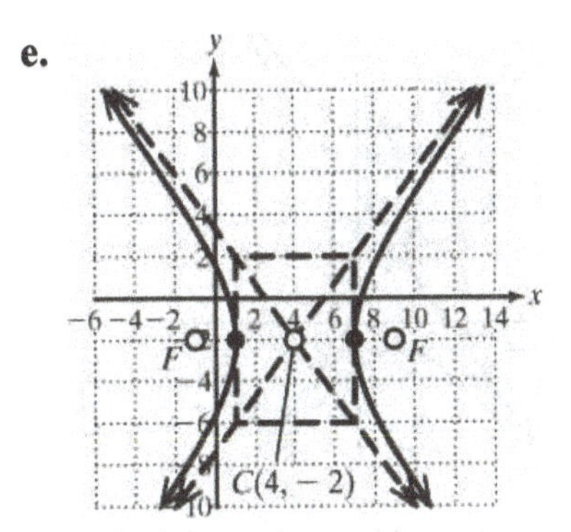

35. $$\frac{(y-5)^2}{49}-\frac{(x+3)^2}{25}=1$$

$$\frac{(y-5)^2}{49}-\frac{[x-(-3)]^2}{25}=1$$

The y^2 term has the positive coefficient, so the transverse axis is vertical. The denominator of the y^2 term is a^2.

$a^2=49$, so $a=7$.

$b^2=25$, so $b=5$.

$c^2=a^2+b^2=49+25=74$

$c=\sqrt{74}$

a. Center: $(-3,5)$

b. The vertices are a units above and below the center.

Vertices: $(-3,5+7)$ and $(-3,5-7)$

$(-3,12)$ and $(-3,-2)$

c. The foci are c units above and below the center.
Foci: $\left(-3, 5+\sqrt{74}\right)$ and $\left(-3, 5-\sqrt{74}\right)$

d. Asymptotes:

$$y-k=\frac{a}{b}(x-h) \qquad y-k=-\frac{a}{b}(x-h)$$
$$y-5=\frac{7}{5}(x+3) \qquad y-5=-\frac{7}{5}(x+3)$$
$$y-5=\frac{7}{5}x+\frac{21}{5} \qquad y-5=-\frac{7}{5}x-\frac{21}{5}$$
$$y=\frac{7}{5}x+\frac{46}{5} \qquad y=-\frac{7}{5}x+\frac{4}{5}$$

e.

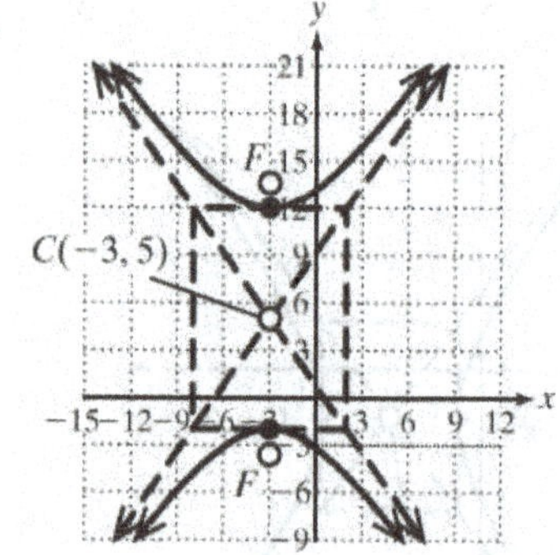

37. $100(y-7)^2-81(x+4)^2=-8100$

$$\frac{100(y-7)^2}{-8100}-\frac{81(x+4)^2}{-8100}=\frac{-8100}{-8100}$$
$$\frac{[x-(-4)]^2}{100}-\frac{(y-7)^2}{81}=1$$

The x^2 term has the positive coefficient, so the transverse axis is horizontal. The denominator of the x^2 term is a^2.
$a^2=100$, so $a=10$.
$b^2=81$, so $b=9$.
$c^2=a^2+b^2=100+81=181$
$c=\sqrt{181}$

a. Center: $(-4, 7)$

b. The vertices are a units to the right and left of the center.
Vertices: $(-4+10, 7)$ and $(-4-10, 7)$
$(6, 7)$ and $(-14, 7)$

c. The foci are c units to the right and left of the center.
Foci: $\left(-4+\sqrt{181}, 7\right)$ and $\left(-4-\sqrt{181}, 7\right)$

d. Asymptotes:

$$y-k=\frac{b}{a}(x-h) \qquad y-k=-\frac{b}{a}(x-h)$$
$$y-7=\frac{9}{10}(x+4) \qquad y-7=-\frac{9}{10}(x+4)$$
$$y-7=\frac{9}{10}x+\frac{18}{5} \qquad y-7=-\frac{9}{10}x-\frac{18}{5}$$
$$y=\frac{9}{10}x+\frac{53}{5} \qquad y=-\frac{9}{10}x+\frac{17}{5}$$

e.

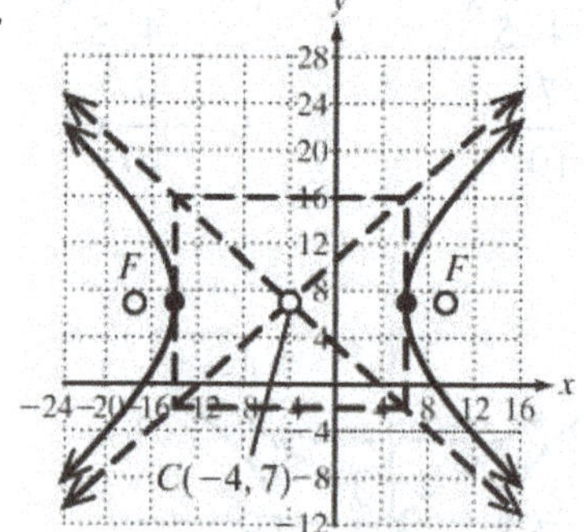

39. $y^2-\frac{(x-3)^2}{12}=1$

$$\frac{(y-0)^2}{1}-\frac{(x-3)^2}{12}=1$$

The y^2 term has the positive coefficient, so the transverse axis is vertical. The denominator of the y^2 term is a^2.
$a^2=1$, so $a=1$.
$b^2=12$, so $b=2\sqrt{3}$.
$c^2=a^2+b^2=1+12=13$
$c=\sqrt{13}$

a. Center: $(3, 0)$

b. The vertices are a units above and below the center.
Vertices: $(3, 0+1)$ and $(3, 0-1)$
$(3, 1)$ and $(3, -1)$

c. The foci are c units above and below the center.

Foci: $\left(3, 0+\sqrt{13}\right)$ and $\left(3, 0-\sqrt{13}\right)$

$\left(3, 0\sqrt{13}\right)$ and $\left(3, -\sqrt{13}\right)$

d. Asymptotes:

$$y-k=\frac{a}{b}(x-h)$$
$$y-0=\frac{1}{2\sqrt{3}}(x-3)$$
$$y=\frac{1}{2\sqrt{3}}x-\frac{3}{2\sqrt{3}}$$
$$y=\left(\frac{1}{2\sqrt{3}}x-\frac{3}{2\sqrt{3}}\right)\cdot\frac{\sqrt{3}}{\sqrt{3}}$$
$$y=\frac{\sqrt{3}}{6}x-\frac{\sqrt{3}}{2}$$
$$y-k=-\frac{a}{b}(x-h)$$
$$y-0=-\frac{1}{2\sqrt{3}}(x-3)$$
$$y=-\frac{1}{2\sqrt{3}}x+\frac{3}{2\sqrt{3}}$$
$$y=\left(-\frac{1}{2\sqrt{3}}x+\frac{3}{2\sqrt{3}}\right)\cdot\frac{\sqrt{3}}{\sqrt{3}}$$
$$y=-\frac{\sqrt{3}}{6}x+\frac{\sqrt{3}}{2}$$

e.

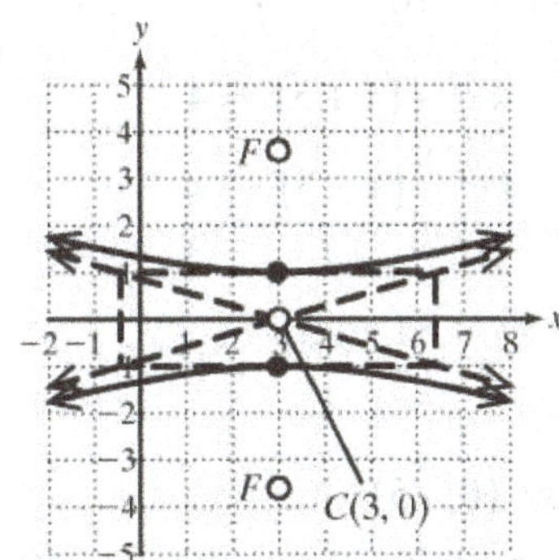

41. $$x^2-\frac{(y-4)^2}{8}=1$$
$$\frac{(x-0)^2}{1}-\frac{(y-4)^2}{8}=1$$

The x^2 term has the positive coefficient, so the transverse axis is horizontal. The denominator of the x^2 term is a^2.

$a^2=1$, so $a=1$.

$b^2=8$, so $b=2\sqrt{2}$.

$c^2=a^2+b^2=1+8=9$

$c=\sqrt{9}=3$

a. Center: $(0, 4)$

b. The vertices are a units to the right and left of the center.

Vertices: $(0+1, 4)$ and $(0-1, 4)$

$(1, 4)$ and $(-1, 4)$

c. The foci are c units to the right and left of the center.

Foci: $(0+3, 4)$ and $(0-3, 4)$

$(3, 4)$ and $(-3, 4)$

d. Asymptotes:

$y-k=\frac{b}{a}(x-h)$ $\quad$ $y-k=-\frac{b}{a}(x-h)$

$y-4=\frac{2\sqrt{2}}{1}(x-0)$ $\quad$ $y-4=-\frac{2\sqrt{2}}{1}(x-0)$

$y-4=2\sqrt{2}x$ $\quad$ $y-4=-2\sqrt{2}x$

$y=2\sqrt{2}x+4$ $\quad$ $y=-2\sqrt{2}x+4$

e.

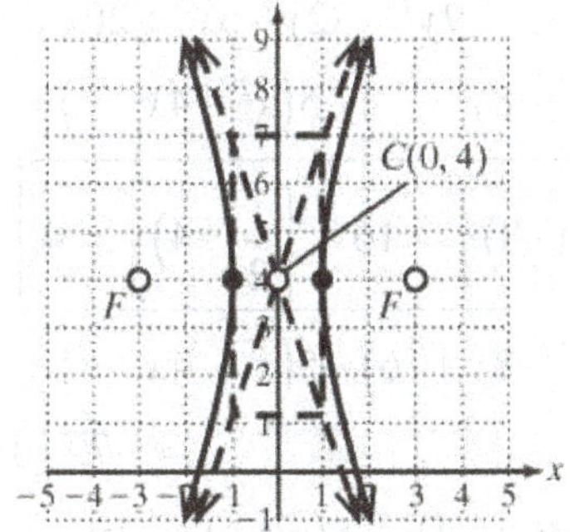

43. a. $7x^2-5y^2+42x+10y+23=0$

$7x^2+42x-5y^2+10y=-23$

$7(x^2+6x\quad)-5(y^2-2y\quad)=-23$

$\left[\frac{1}{2}(6)\right]^2=9 \quad \left[\frac{1}{2}(-2)\right]^2=1$

$7(x^2+6x+9)-5(y^2-2y+1)=-23+63-5$

$7(x+3)^2-5(y-1)^2=35$

$\frac{7(x+3)^2}{35}-\frac{5(y-1)^2}{35}=\frac{35}{35}$

$\frac{(x+3)^2}{5}-\frac{(y-1)^2}{7}=1$

b. Center: $(-3,1)$

The x^2 term has the positive coefficient, so the transverse axis is horizontal. The denominator of the x^2 term is a^2.

$a^2=5$, so $a=\sqrt{5}$.

$b^2=7$ so $b=\sqrt{7}$.

$c^2=a^2+b^2=5+7=12$

$c=\sqrt{12}=2\sqrt{3}$

The vertices are a units to the right and left of the center.

Vertices: $\left(-3+\sqrt{5},1\right)$ and $\left(-3-\sqrt{5},1\right)$

The foci are c units to the right and left of the center.

Foci: $\left(-3+2\sqrt{3},1\right)$ and $\left(-3-2\sqrt{3},1\right)$

45. a. $-5x^2+9y^2+20x-72y+79=0$

$9y^2-72y-5x^2+20x=-79$

$9(y^2-8y\quad)-5(x^2-4x\quad)=-79$

$\left[\frac{1}{2}(-8)\right]^2=16 \quad \left[\frac{1}{2}(-4)\right]^2=4$

$9(y^2-8y+16)-5(x^2-4x+4)$

$=-79+144-20$

$9(y-4)^2-5(x-2)^2=45$

$\frac{9(y-4)^2}{45}-\frac{5(x-2)^2}{45}=\frac{45}{45}$

$\frac{(y-4)^2}{5}-\frac{(x-2)^2}{9}=1$

b. Center: $(2,4)$

The y^2 term has the positive coefficient, so the transverse axis is vertical. The denominator of the y^2 term is a^2.

$a^2=5$, so $a=\sqrt{5}$.

$b^2=9$ so $b=3$.

$c^2=a^2+b^2=5+9=14$

$c=\sqrt{14}$

The vertices are a units above and below the center.

Vertices: $\left(2,4+\sqrt{5}\right)$ and $\left(2,4-\sqrt{5}\right)$

The foci are c units above and below the center.

Foci: $\left(2,4+\sqrt{14}\right)$ and $\left(2,4-\sqrt{14}\right)$

47. a. $4x^2-y^2-10y-29=0$

$4x^2-y^2-10y=29$

$4x^2-(y^2+10y\quad)=29$

$\left[\frac{1}{2}(10)\right]^2=25$

$4x^2-(y^2+10y+25)=29-25$

$4x^2-(y+5)^2=4$

$\frac{4x^2}{4}-\frac{(y+5)^2}{4}=\frac{4}{4}$

$x^2-\frac{(y+5)^2}{4}=1$

b. Center: $(0,-5)$

The x^2 term has the positive coefficient, so the transverse axis is horizontal. The denominator of the x^2 term is a^2.

$a^2=1$, so $a=1$.

$b^2=4$ so $b=2$.

$c^2=a^2+b^2=1+4=5$

$c=\sqrt{5}$

The vertices are a units to the right and left of the center.

Vertices: $(0+1,-5)$ and $(0-1,-5)$

$(1,-5)$ and $(-1,-5)$

The foci are c units to the right and left of the center.

Foci: $\left(0+\sqrt{5},-5\right)$ and $\left(0-\sqrt{5},-5\right)$

$\left(\sqrt{5},-5\right)$ and $\left(-\sqrt{5},-5\right)$

49. a. $-36x^2+64y^2+108x+256y-401=0$

$64y^2+256y-36x^2+108x=401$

$64\left(y^2+4y\quad\right)-36\left(x^2-3x\quad\right)=401$

$$\boxed{\left[\frac{1}{2}(4)\right]^2=4 \quad \left[\frac{1}{2}(-3)\right]^2=\frac{9}{4}}$$

$64\left(y^2+4y+4\right)-36\left(x^2-3x+\frac{9}{4}\right)$

$=401+256-81$

$64(y+2)^2-36\left(x-\frac{3}{2}\right)^2=576$

$$\frac{64(y+2)^2}{576}-\frac{36\left(x-\frac{3}{2}\right)^2}{576}=\frac{576}{576}$$

$$\frac{(y+2)^2}{9}-\frac{\left(x-\frac{3}{2}\right)^2}{16}=1$$

b. Center: $\left(\frac{3}{2},-2\right)$

The y^2 term has the positive coefficient, so the transverse axis is vertical. The denominator of the y^2 term is a^2.

$a^2=9$, so $a=3$.

$b^2=16$ so $b=4$.

$c^2=a^2+b^2=9+16=25$

$c=\sqrt{25}=5$

The vertices are a units above and below the center.

Vertices: $\left(\frac{3}{2},-2+3\right)$ and $\left(\frac{3}{2},-2-3\right)$

$\left(\frac{3}{2},1\right)$ and $\left(\frac{3}{2},-5\right)$

The foci are c units above and below the center.

Foci: $\left(\frac{3}{2},-2+5\right)$ and $\left(\frac{3}{2},-2-5\right)$

$\left(\frac{3}{2},3\right)$ and $\left(\frac{3}{2},-7\right)$

51. The foci have the same y-coordinate, so they are aligned horizontally and the transverse axis of the hyperbola is horizontal.

The standard form is

$$\frac{(x-h)^2}{a^2}-\frac{(y-k)^2}{b^2}=1$$

The center is midway between the vertices.

Center is $(0,0)$: $h=0$ and $k=0$

Vertices are 12 units from the center: $a=12$

Foci are 13 units from the center: $c=13$

$b^2=c^2-a^2=(13)^2-(12)^2=169-144=25$

$b=\sqrt{25}=5$

$$\frac{(x-0)^2}{(12)^2}-\frac{(y-0)^2}{(5)^2}=1$$

$$\frac{x^2}{144}-\frac{y^2}{25}=1$$

53. The vertices have the same x-coordinate, so they are aligned vertically and the transverse axis of the hyperbola is vertical.

The standard form is

$$\frac{(y-k)^2}{a^2}-\frac{(x-h)^2}{b^2}=1$$

The center is midway between the vertices.

Center is $(0,0)$: $h=0$ and $k=0$

Vertices are 12 units from the center: $a=12$

Asymptotes are $y=\pm\frac{4}{3}x=\pm\frac{a}{b}x$:

$\frac{4}{3}=\frac{12}{b}$

$4b=36$

$b=9$

$$\frac{(y-0)^2}{(12)^2}-\frac{(x-0)^2}{(9)^2}=1$$

$$\frac{y^2}{144}-\frac{x^2}{81}=1$$

55. The vertices have the same y-coordinate, so they are aligned horizontally and the transverse axis of the hyperbola is horizontal.
The standard form is

$$\frac{(x-h)^2}{a^2}-\frac{(y-k)^2}{b^2}=1$$

The center is midway between the vertices.
Center is $(2,-3)$: $h=2$ and $k=-3$
Vertices are 5 units from the center: $a=5$
Slope of asymptotes is $\pm\frac{7}{5}$.

$$\frac{7}{5}=\frac{b}{a}$$

$$\frac{7}{5}=\frac{b}{5}$$

$$b=7$$

$$\frac{(x-2)^2}{(5)^2}-\frac{[y-(-3)]^2}{(7)^2}=1$$

$$\frac{(x-2)^2}{25}-\frac{(y+3)^2}{49}=1$$

57. The vertices have the same x-coordinate, so they are aligned vertically and the transverse axis of the hyperbola is vertical.
The standard form is

$$\frac{(y-k)^2}{a^2}-\frac{(x-h)^2}{b^2}=1$$

The center is midway between the vertices.
Center is $(-3,4)$: $h=-3$ and $k=4$
Vertices are 4 units from the center: $a=4$
Foci are $\sqrt{21}$ units from the center: $c=\sqrt{21}$

$$b^2=c^2-a^2=\left(\sqrt{21}\right)^2-(4)^2=21-16=5$$

$$b=\sqrt{5}$$

$$\frac{(y-4)^2}{(4)^2}-\frac{[x-(-3)]^2}{\left(\sqrt{5}\right)^2}=1$$

$$\frac{(y-4)^2}{16}-\frac{(x+3)^2}{5}=1$$

59. The transverse axis is horizontal.
The standard form is

$$\frac{(x-h)^2}{a^2}-\frac{(y-k)^2}{b^2}=1$$

Width of reference rectangle is $8-(-6)=14$

$$2a=14$$

$$a=7$$

Height of reference rectangle is $7-(-3)=10$

$$2b=10$$

$$b=5$$

Center: $(-1,2)$

$$\frac{[x-(-1)]^2}{(7)^2}-\frac{(y-2)^2}{(5)^2}=1$$

$$\frac{(x+1)^2}{49}-\frac{(y-2)^2}{25}=1$$

61. a. Equation 1: $\frac{x^2}{144}-\frac{y^2}{81}=1$

The x^2 term has the positive coefficient.
The denominator of the x^2 term is a^2.
$a^2=144$, so $a=12$.
$b^2=81$
$c^2=a^2+b^2=144+81=225$
$c=\sqrt{225}=15$
$e=\frac{c}{a}=\frac{15}{12}=\frac{5}{4}$

Equation 2: $\frac{x^2}{81}-\frac{y^2}{144}=1$

The x^2 term has the positive coefficient.
The denominator of the x^2 term is a^2.
$a^2=81$, so $a=9$.
$b^2=144$
$c^2=a^2+b^2=81+144=225$
$c=\sqrt{225}=15$
$e=\frac{c}{a}=\frac{15}{9}=\frac{5}{3}$

b. As the eccentricity of a hyperbola increases, the branches open wider. Graph B represents Equation 1. Graph A represents Equation 2.

63. $\dfrac{\left(y-\frac{2}{3}\right)^2}{1600}-\dfrac{\left(x+\frac{7}{4}\right)^2}{81}=1$

The y^2 term has the positive coefficient. The denominator of the y^2 term is a^2.

$a^2=1600$, so $a=40$.

$b^2=81$

$c^2=a^2+b^2=1600+81=1681$

$c=\sqrt{1681}=41$

$e=\dfrac{c}{a}=\dfrac{41}{40}$

65. Since the transverse axis of length 24 units is horizontal, $2a=24$, so $a=12$.

Asymptotes are $y=\pm\dfrac{3}{4}x=\pm\dfrac{b}{a}x$:

$\dfrac{3}{4}=\dfrac{b}{12}$

$4b=36$

$b=9$

$c^2=a^2+b^2=(12)^2+(9)^2=144+81=225$

$c=\sqrt{225}=15$

$e=\dfrac{c}{a}=\dfrac{15}{12}=\dfrac{5}{4}$

67. The vertices have the same y-coordinate, so they are aligned horizontally and the transverse axis of the hyperbola is horizontal. The standard form is

$\dfrac{(x-h)^2}{a^2}-\dfrac{(y-k)^2}{b^2}=1$

The center is midway between the vertices.

Center is $(3,-1)$: $h=3$ and $k=-1$

Vertices are 4 units from the center: $a=4$

Eccentricity is $e=\dfrac{5}{4}=\dfrac{c}{a}$:

$\dfrac{5}{4}=\dfrac{c}{4}$

$c=5$

$b^2=c^2-a^2=(5)^2-(4)^2=25-16=9$

$b=\sqrt{9}=3$

$\dfrac{(x-3)^2}{(4)^2}-\dfrac{[y-(-1)]^2}{(3)^2}=1$

$\dfrac{(x-3)^2}{16}-\dfrac{(y+1)^2}{9}=1$

69. $d_1-d_2=(186\text{ mi/msec})(4\text{ msec})=744\text{ mi}$

$d_1-d_2=2a$

$744=2a$

$a=372\text{ mi}$

$c=500\text{ mi}$

$b^2=c^2+a^2=(500)^2-(372)^2$

$=138{,}384-250{,}000=111{,}616$

$b\approx 334\text{ mi}$

$\dfrac{x^2}{(372)^2}-\dfrac{y^2}{(334)^2}=1$

71. a. $\dfrac{x^2}{(1191.2)^2}-\dfrac{y^2}{(30.9)^2}=1$

$a=1191.2$ and $b=30.9$.

$c^2=a^2+b^2=(1191.2)^2+(30.9)^2$

$=1{,}419{,}912.25$

$c\approx 1191.6$

$1191.6-1191.2=0.4\text{ AU}$

b. $(0.4\text{ AU})(93{,}000{,}000\text{ mi/AU})$

$=37{,}200{,}000\text{ mi}$

73. The coefficients of both terms are positive. The equation represents an ellipse. Since the greater number in the denominator is found in the x^2 term, the ellipse is elongated horizontally. Graph B

75. The coefficient of one term is negative. The equation represents a hyperbola. The x^2 term has the positive coefficient, so the transverse axis is horizontal. Graph D

77. The ellipse is elongated vertically. The standard form is

$$\frac{(x-h)^2}{b^2}+\frac{(y-k)^2}{a^2}=1$$

Center is $(-1,0)$: $h=-1$ and $k=0$

Length of major axis is 10 units: $a=\frac{10}{2}=5$

Length of minor axis is 8 units: $b=\frac{8}{2}=4$

$$\frac{[x-(-1)]^2}{(4)^2}+\frac{(y-0)^2}{(5)^2}=1$$

$$\frac{(x+1)^2}{16}+\frac{y^2}{25}=1$$

79. The transverse axis of the hyperbola is vertical. The standard form is

$$\frac{(y-h)^2}{a^2}-\frac{(x-k)^2}{b^2}=1$$

Center is $(2,-3)$: $h=2$ and $k=-3$

Vertices are 3 units from the center: $a=3$

Slope of asymptotes is $\pm\frac{3}{4}$

$$\frac{3}{4}=\frac{a}{b}$$

$$\frac{3}{4}=\frac{3}{b}$$

$$b=4$$

$$\frac{[y-(-3)]^2}{(3)^2}-\frac{(x-2)^2}{(4)^2}=1$$

$$\frac{(y+3)^2}{9}-\frac{(x-2)^2}{16}=1$$

81.

$$\begin{array}{ll} \boxed{A}\ 9x^2-4y^2=36 \xrightarrow{\text{Multiply by 2.}} & 18x^2-8y^2=72 \\ \boxed{B}\ -13x^2+8y^2=8 & -13x^2+8y^2=8 \\ & \overline{5x^2 \qquad\quad =80} \\ & x^2=16 \\ & x=\pm 4 \end{array}$$

$x=4$: $\boxed{A}$

$$\begin{aligned} 9x^2-4y^2&=36 \\ 9(4)^2-4y^2&=36 \\ 144-4y^2&=36 \\ -4y^2&=-108 \\ y^2&=27 \\ y&=\pm 3\sqrt{3} \end{aligned}$$

$x=-4$: $\boxed{A}$

$$\begin{aligned} 9x^2-4y^2&=36 \\ 9(-4)^2-4y^2&=36 \\ 144-4y^2&=36 \\ -4y^2&=-108 \\ y^2&=27 \\ y&=\pm 3\sqrt{3} \end{aligned}$$

$$\{(4,3\sqrt{3}),(4,-3\sqrt{3}),(-4,3\sqrt{3}),(-4,-3\sqrt{3})\}$$

83. The transverse axis is horizontal if the coefficient of the x^2 term is positive. The transverse axis is vertical if the coefficient of the y^2 term is positive.

85. If the asymptotes are perpendicular, then $a=b$.

$$c^2=a^2+b^2=a^2+a^2=2a^2$$

$$c=\sqrt{2a^2}=a\sqrt{2}$$

87. $\frac{x^2}{a^2}-\frac{y^2}{b^2}=1$

$\frac{c^2}{a^2}-\frac{y^2}{b^2}=1$

$y=b\sqrt{\frac{c^2}{a^2}-1}=b\sqrt{\frac{c^2-a^2}{a^2}}=\frac{b}{a}\sqrt{c^2-a^2}$

Recall that $c^2=a^2+b^2$ or equivalently $b^2=c^2-a^2$ and $b>0$. Therefore, $y=\frac{b}{a}\sqrt{b^2}$ or $y=\frac{b^2}{a}$. The length of a latus rectum is $2y=\frac{2b^2}{a}$.

89.

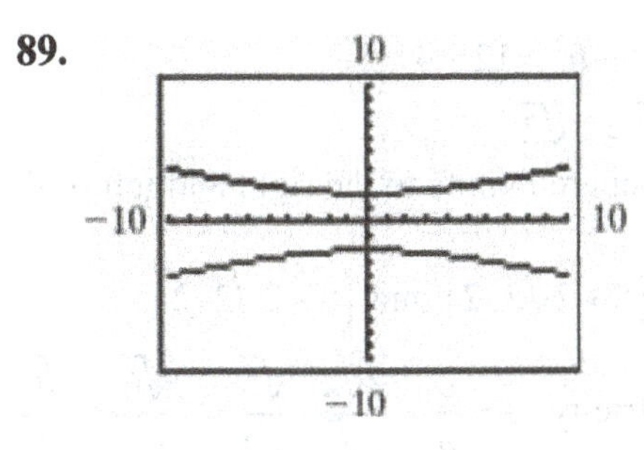

91.

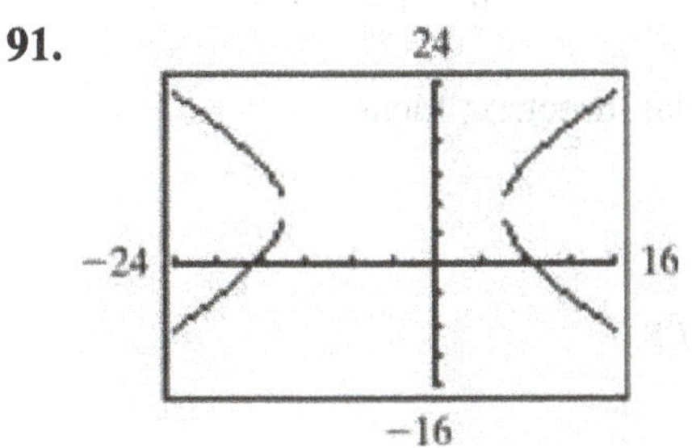

Section 7.3 The Parabola

1. $\frac{x^2}{9}-\frac{(y+7)^2}{16}=1$

Hyperbola

The x^2 term has the positive coefficient, so the transverse axis is horizontal. The denominator of the x^2 term is a^2.

Center: $(0,-7)$

$a^2=9$, so $a=3$.

$b^2=16$, so $b=4$.

$c^2=a^2+b^2=9+16=25$

$c=\sqrt{25}=5$

Vertices: $(0+3,-7)$ and $(0-3,-7)$

$(3,-7)$ and $(-3,-7)$

The foci are c units to the right and left of the center.

Foci: $(0+5-7)$ and $(0-5,-7)$

$(5,-7)$ and $(-5,-7)$

Asymptotes:

$y-k=\frac{b}{a}(x-h)$ $\qquad$ $y-k=-\frac{b}{a}(x-h)$

$y+7=\frac{4}{3}(x-0)$ $\qquad$ $y+7=-\frac{4}{3}(x-0)$

$y=\frac{4}{3}x-7$ $\qquad$ $y=-\frac{4}{3}x-7$

Eccentricity: $e=\frac{c}{a}=\frac{5}{3}$

3. $(x+3)^2+(y+4)^2=18$

Circle

$r^2=18$

$r=\sqrt{18}=3\sqrt{2}$

Center: $(-3,-4)$; Radius: $3\sqrt{2}$

5. $\frac{(x-8)^2}{12}+\frac{(y-2)^2}{4}=1$

Ellipse

Center: $(8,2)$

Since $12>4$, $a^2=12$ and $b^2=4$.

$a=\sqrt{12}=2\sqrt{3}$ and $b=\sqrt{4}=2$.

Since the greater number in the denominator is found in the x^2 term, the ellipse is elongated horizontally.

The vertices are a units to the right and left of the center.

Vertices: $(8+2\sqrt{3},2)$ and $(8-2\sqrt{3},2)$

The minor axis endpoints are b units above and below the center.

Endpoints of minor axis:

$(8,2+2)$ and $(8,2-2)$

$(8,4)$ and $(8,0)$

$c^2 = a^2 - b^2 = 12 - 4 = 8$
$c = \sqrt{8} = 2\sqrt{2}$
The foci are c units to the right and left of the center.
Foci: $\left(8+2\sqrt{2}, 2\right)$ and $\left(8-2\sqrt{2}, 2\right)$

Eccentricity: $e = \dfrac{c}{a} = \dfrac{2\sqrt{2}}{2\sqrt{3}} \cdot \dfrac{\sqrt{3}}{\sqrt{3}} = \dfrac{2\sqrt{6}}{6} = \dfrac{\sqrt{6}}{3}$

7. parabola; directrix; focus

9. vertex

11. $x^2 = 4py$

13. left

15. latus rectum

17. (h, k); $(h, k+p)$; $y = k - p$

19. vertically

21. a. $d_1 = \sqrt{(2-0)^2 + (1-1)^2} = \sqrt{2^2 + 0^2} = 2$
$d_2 = \sqrt{(2-2)^2 + (-1-1)^2}$
$= \sqrt{0^2 + (-2)^2} = 2$
$d_3 = \sqrt{\left[0-(-4)\right]^2 + (1-4)^2}$
$= \sqrt{4^2 + (-3)^2} = \sqrt{16+9} = \sqrt{25} = 5$
$d_4 = \sqrt{\left[-4-(-4)\right]^2 + \left[4-(-1)\right]^2}$
$= \sqrt{0^2 + 5^2} = 5$

b. They are the same.

c. They are the same.

23. $x^2 = 24y$

a. The equation is in the form $x^2 = 4py$.
$4p = 24$
$p = 6$

b. The focus is on the y-axis. Since $p > 0$, the parabola opens upward. The focus is $|p|$ units above the vertex: $(0, 6)$.

c. The directrix is the horizontal line $|p|$ units below the vertex: $y = -6$.

25. $y^2 = 36x$

a. The equation is in the form $y^2 = 4px$.
$4p = 36$
$p = 9$

b. The focus is on the x-axis. Since $p > 0$, the parabola opens to the right. The focus is $|p|$ units to the right of the vertex: $(9, 0)$.

c. The directrix is the vertical line $|p|$ units to the left of the vertex: $x = -9$.

27. $x^2 = -5y$

a. The equation is in the form $x^2 = 4py$.
$4p = -5$
$p = -\dfrac{5}{4}$

b. The focus is on the y-axis. Since $p < 0$, the parabola opens downward. The focus is $|p|$ units below the vertex: $\left(0, -\dfrac{5}{4}\right)$.

c. The directrix is the horizontal line $|p|$ units above the vertex: $y = \dfrac{5}{4}$.

29. $-x = y^2$
$y^2 = -x$

a. The equation is in the form $y^2 = 4px$.
$4p = -1$
$p = -\dfrac{1}{4}$

b. The focus is on the x-axis. Since $p < 0$, the parabola opens to the left. The focus is $|p|$ units to the left of the vertex: $\left(-\dfrac{1}{4}, 0\right)$.

c. The directrix is the vertical line $|p|$ units to the right of the vertex: $x = \frac{1}{4}$.

31. $x^2 = 25.2y$

The equation is in the form $x^2 = 4py$.

$4p = 25.2$

$p = \frac{25.2}{4} = 6.3$

a. The focus is on the y-axis. Since $p > 0$, the parabola opens upward. The focus is $|p|$ units above the vertex: $(0, 6.3)$.
Place the receiver 6.3 in. above the center of the dish.

b. The directrix is the horizontal line $|p|$ units below the vertex: $y = -6.3$.

33. $x^2 = -4y$

a. The equation is in the form $x^2 = 4py$.
The focus is on the y-axis and the parabola opens upward or downward.

$4p = -4$

$p = -1$

Since $p < 0$, the parabola opens downward.
The vertex is $(0, 0)$.
The focus is $(0, p)$, which is $(0, -1)$.
The focal diameter is $4|p|$, which is $4|-1| = 4$.

b. The endpoints of the latus rectum are $\frac{4}{2} = 2$ units to the right and to the left of the focus: $(-2, -1)$ and $(2, -1)$.

c.

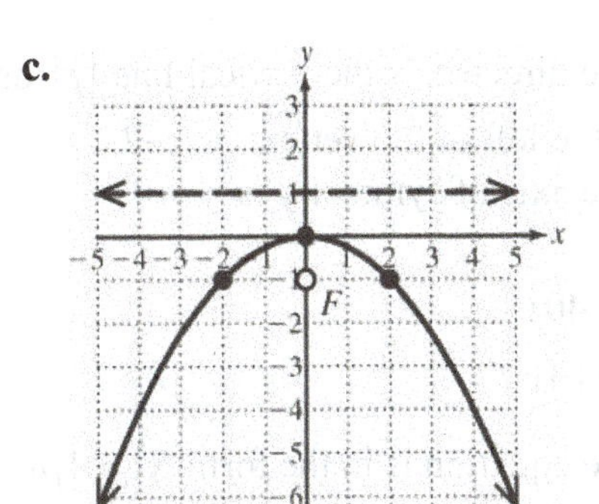

d. The directrix is the horizontal line $|p|$ units above the vertex: $y = 1$.
The axis of symmetry is $x = 0$.

35. $10y^2 = 80x$

$y^2 = 8x$

a. The equation is in the form $y^2 = 4px$.
The focus is on the x-axis and the parabola opens to the right or to the left.

$4p = 8$

$p = 2$

Since $p > 0$, the parabola opens to the right.
The vertex is $(0, 0)$.
The focus is $(p, 0)$, which is $(2, 0)$.
The focal diameter is $4|p|$, which is $4|2| = 8$.

b. The endpoints of the latus rectum are $\frac{8}{2} = 4$ units above and below the focus: $(2, 4)$ and $(2, -4)$.

c.

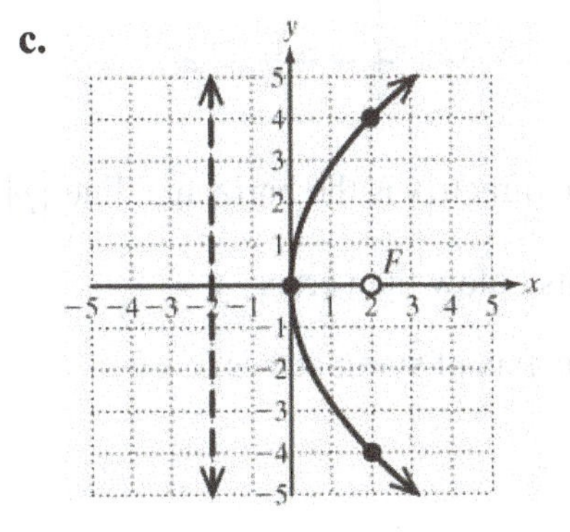

d. The directrix is the vertical line $|p|$ units to the left of the vertex: $x=-2$.
The axis of symmetry is $y=0$.

37. $4x^2=40y$

$x^2=10y$

a. The equation is in the form $x^2=4py$.
The focus is on the *y*-axis and the parabola opens upward or downward.

$4p=10$

$p=\frac{10}{4}=\frac{5}{2}$

Since $p>0$, the parabola opens upward.
The vertex is $(0, 0)$.

The focus is $(0, p)$, which is $\left(0, \frac{5}{2}\right)$.

The focal diameter is $4|p|$, which is $4\left|\frac{5}{2}\right|=10$.

b. The endpoints of the latus rectum are $\frac{10}{2}=5$ units to the right and to the left of the focus: $\left(5, \frac{5}{2}\right)$ and $\left(-5, \frac{5}{2}\right)$.

c.

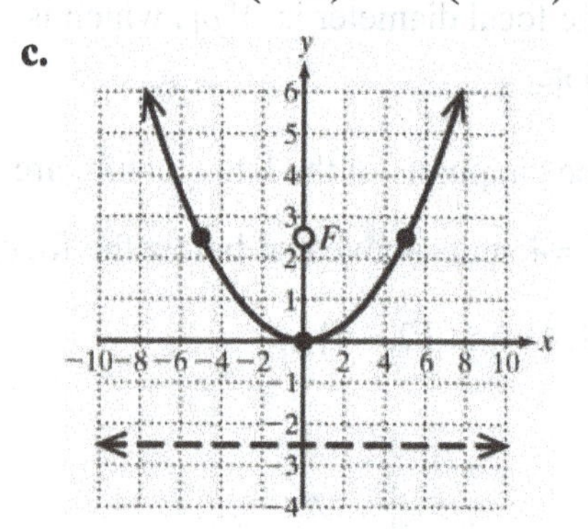

d. The directrix is the horizontal line $|p|$ units below the vertex: $y=-\frac{5}{2}$.

The axis of symmetry is $x=0$.

39. $y^2=-x$

a. The equation is in the form $y^2=4px$.
The focus is on the *x*-axis and the parabola opens to the right or to the left.

$4p=-1$

$p=-\frac{1}{4}$

Since $p<0$, the parabola opens to the left.
The vertex is $(0, 0)$.

The focus is $(p, 0)$, which is $\left(-\frac{1}{4}, 0\right)$.

The focal diameter is $4|p|$, which is $4\left|-\frac{1}{4}\right|=1$.

b. The endpoints of the latus rectum are $\frac{1}{2}$ units above and below the focus: $\left(-\frac{1}{4}, \frac{1}{2}\right)$ and $\left(-\frac{1}{4}, -\frac{1}{2}\right)$

c.

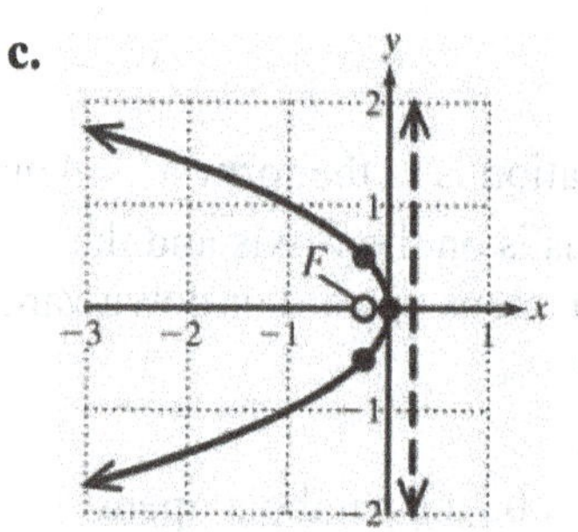

d. The directrix is the vertical line $|p|$ units to the right of the vertex: $x=\frac{1}{4}$.

The axis of symmetry is $y=0$.

41. $(y+1)^2=-12(x-4)$

a. The equation is in the form $(y-k)^2=4p(x-h)$, where $h=4$ and $k=-1$. The vertex is $(4, -1)$.

$4p=-12$

$p=-3$

The *x* term is linear; the parabola opens in the *x* direction.
Since $p<0$, the parabola opens to the left.
The focus is $(h+p, k)$, which is $(1, -1)$.

The focal diameter is $4|p|$, which is $4|-3| = 12$.

b. The endpoints of the latus rectum are $\frac{12}{2} = 6$ units above and below the focus: $(1, 5)$ and $(1, -7)$.

c.

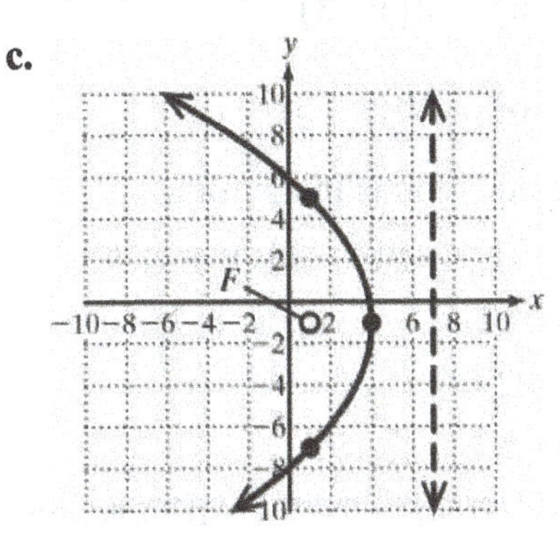

d. The directrix is the vertical line $|p|$ units to the right of the vertex: $x = 7$.
The axis of symmetry is $y = -1$.

43. $(x-1)^2 = -4(y+5)$

a. The equation is in the form $(x-h)^2 = 4p(y-k)$, where $h = 1$ and $k = -5$. The vertex is $(1, -5)$.

$4p = -4$

$p = -1$

The y term is linear; the parabola opens in the y direction.
Since $p < 0$, the parabola opens downward.
The focus is $(h, k+p)$, which is $(1, -6)$.
The focal diameter is $4|p|$, which is $4|-1| = 4$.

b. The endpoints of the latus rectum are $\frac{4}{2} = 2$ units to the right and to the left of the focus: $(3, -6)$ and $(-1, -6)$.

c.

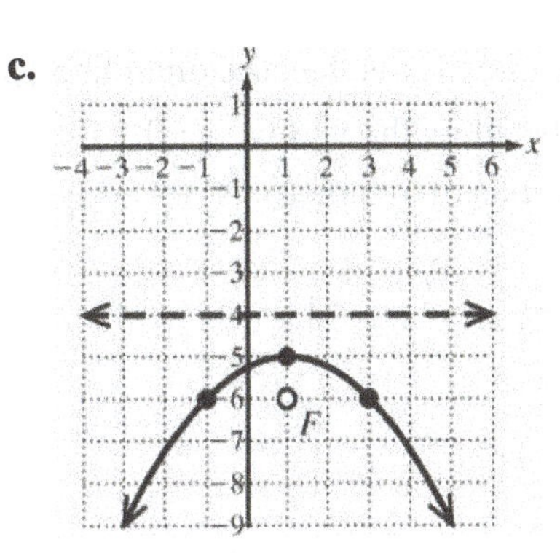

d. The directrix is the horizontal line $|p|$ units above the vertex: $y = -4$.
The axis of symmetry is $x = 1$.

45. $(x+3)^2 = 2\left(y - \frac{3}{2}\right)$

a. The equation is in the form $(x-h)^2 = 4p(y-k)$, where $h = -3$ and $k = \frac{3}{2}$. The vertex is $\left(-3, \frac{3}{2}\right)$.

$4p = 2$

$p = \frac{2}{4} = \frac{1}{2}$

The y term is linear; the parabola opens in the y direction.
Since $p > 0$, the parabola opens upward.
The focus is $(h, k+p)$, which is $(-3, 2)$.
The focal diameter is $4|p|$, which is $4\left|\frac{1}{2}\right| = 2$.

b. The endpoints of the latus rectum are $\frac{2}{2} = 1$ unit to the right and to the left of the focus: $(-2, 2)$ and $(-4, 2)$.

c.

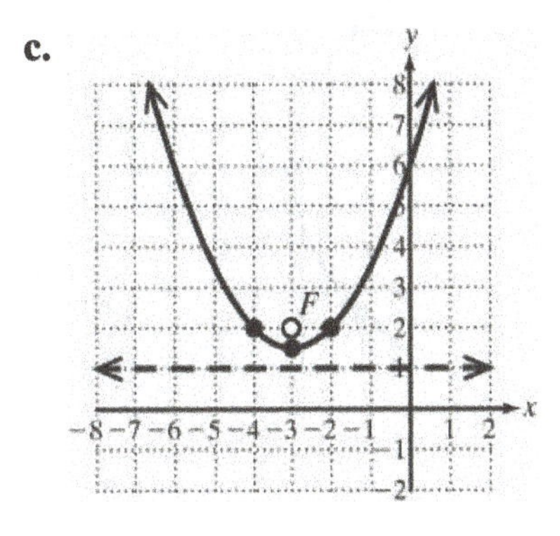

d. The directrix is the horizontal line $|p|$ units below the vertex: $y = 1$.
The axis of symmetry is $x = -3$.

47. $-(y-3)^2 = 7\left(x - \frac{1}{4}\right)$

$(y-3)^2 = -7\left(x - \frac{1}{4}\right)$

a. The equation is in the form $(y-k)^2 = 4p(x-h)$, where $h = \frac{1}{4}$ and $k = 3$. The vertex is $\left(\frac{1}{4}, 3\right)$.

$4p = -7$

$p = -\frac{7}{4}$

The x term is linear; the parabola opens in the x direction.
Since $p < 0$, the parabola opens to the left.
The focus is $(h+p, k)$, which is $\left(-\frac{3}{2}, 3\right)$.
The focal diameter is $4|p|$, which is $4\left|-\frac{7}{4}\right| = 7$.

b. The endpoints of the latus rectum are $\frac{7}{2}$ units above and below the focus: $\left(-\frac{3}{2}, \frac{13}{2}\right)$ and $\left(-\frac{3}{2}, -\frac{1}{2}\right)$.

c.

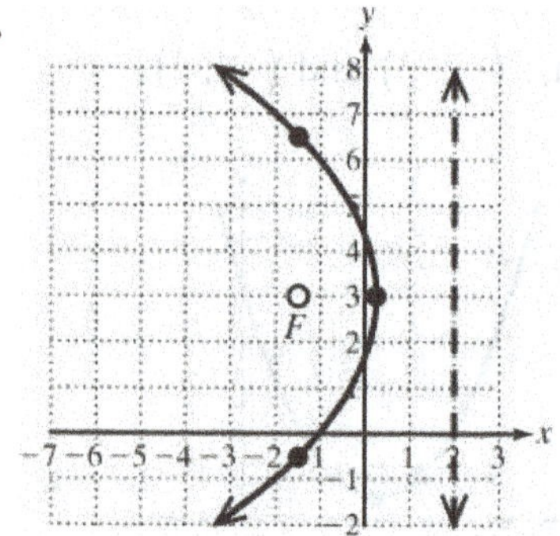

d. The directrix is the vertical line $|p|$ units to the right of the vertex: $x = 2$.
The axis of symmetry is $y = 3$.

49. $2(y-3) = \frac{1}{10}(x+6)^2$

$\frac{1}{10}(x+6)^2 = 2(y-3)$

$(x+6)^2 = 20(y-3)$

a. The equation is in the form $(x-h)^2 = 4p(y-k)$, where $h = -6$ and $k = 3$. The vertex is $(-6, 3)$.

$4p = 20$

$p = 5$

The y term is linear; the parabola opens in the y direction.
Since $p > 0$, the parabola opens upward.
The focus is $(h, k+p)$, which is $(-6, 8)$.
The focal diameter is $4|p|$, which is $4|5| = 20$.

b. The endpoints of the latus rectum are $\frac{20}{2} = 10$ units to the right and to the left of the focus: $(4, 8)$ and $(-16, 8)$.

c.

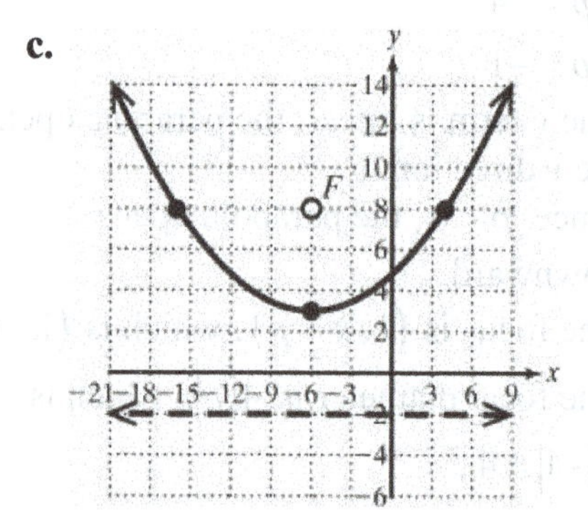

d. The directrix is the horizontal line $|p|$ units below the vertex: $y = -2$.
The axis of symmetry is $x = -6$.

51. a. $x^2-6x-4y+5=0$

$$x^2-6x=4y-5$$

$$\left(x^2-6x\quad\right)=4y-5$$

$$\boxed{\left[\frac{1}{2}(-6)\right]^2=9}$$

$$\left(x^2-6x+9\right)=4y-5+9$$

$$(x-3)^2=4y+4$$

$$(x-3)^2=4(y+1)$$

b. The equation is in the form $(x-h)^2=4p(y-k)$, where $h=3$ and $k=-1$. The vertex is $(3,-1)$.

$4p=4$

$p=1$

The y term is linear; the parabola opens in the y direction.
Since $p>0$, the parabola opens upward.
The focus is $(h,k+p)$, which is $(3,0)$.
The directrix is the horizontal line $|p|$ units below the vertex: $y=-2$.

53. a. $y^2+4y+8x+52=0$

$$y^2+4y=-8x-52$$

$$\left(y^2+4y\quad\right)=-8x-52$$

$$\boxed{\left[\frac{1}{2}(4)\right]^2=4}$$

$$\left(y^2+4y+4\right)=-8x-52+4$$

$$(y+2)^2=-8x-48$$

$$(y+2)^2=-8(x+6)$$

b. The equation is in the form $(y-k)^2=4p(x-h)$, where $h=-6$ and $k=-2$. The vertex is $(-6,-2)$.

$4p=-8$

$p=-2$

The x term is linear; the parabola opens in the x direction.
Since $p<0$, the parabola opens to the left.
The focus is $(h+p,k)$, which is $(-8,-2)$.
The directrix is the vertical line $|p|$ units to the right of the vertex: $x=-4$.

55. a. $4x^2-28x+24y+73=0$

$$4x^2-28x=-24y-73$$

$$4\left(x^2-7x\quad\right)=-24y-73$$

$$\boxed{\left[\frac{1}{2}(-7)\right]^2=\frac{49}{4}}$$

$$4\left(x^2-7x+\frac{49}{4}\right)=-24y-73+49$$

$$4\left(x-\frac{7}{2}\right)^2=-24y-24$$

$$4\left(x-\frac{7}{2}\right)^2=-24(y+1)$$

$$\left(x-\frac{7}{2}\right)^2=-6(y+1)$$

b. The equation is in the form $(x-h)^2=4p(y-k)$, where $h=\frac{7}{2}$ and $k=-1$. The vertex is $\left(\frac{7}{2},-1\right)$.

$4p=-6$

$p=-\frac{6}{4}=-\frac{3}{2}$

The y term is linear; the parabola opens in the y direction.
Since $p<0$, the parabola opens downward.
The focus is $(h,k+p)$, which is $\left(\frac{7}{2},-\frac{5}{2}\right)$.
The directrix is the horizontal line $|p|$ units above the vertex: $y=\frac{1}{2}$.

57. a. $16y^2 + 24y - 16x + 57 = 0$

$$16y^2 + 24y = 16x - 57$$

$$16\left(y^2 + \frac{3}{2}y \qquad\right) = 16x - 57$$

$$\boxed{\left[\frac{1}{2}\left(\frac{3}{2}\right)\right]^2 = \frac{9}{16}}$$

$$16\left(y^2 + \frac{3}{2}y + \frac{9}{16}\right) = 16x - 57 + 9$$

$$16\left(y + \frac{3}{4}\right)^2 = 16x - 48$$

$$16\left(y + \frac{3}{4}\right)^2 = 16(x - 3)$$

$$\left(y + \frac{3}{4}\right)^2 = (x - 3)$$

b. The equation is in the form $(y-k)^2 = 4p(x-h)$, where $h = 3$ and $k = -\frac{3}{4}$. The vertex is $\left(3, -\frac{3}{4}\right)$.

$4p = 1$

$p = \frac{1}{4}$

The x term is linear; the parabola opens in the x direction.

Since $p > 0$, the parabola opens to the right.

The focus is $(h + p, k)$, which is $\left(\frac{13}{4}, -\frac{3}{4}\right)$.

The directrix is the vertical line $|p|$ units to the left of the vertex: $x = \frac{11}{4}$.

59. Directrix: $y = 4$; Focus: $(1, 10)$

The vertex is halfway between the directrix and the focus: $(1, 7)$.

$p = \frac{10-4}{2} = \frac{6}{2} = 3$

61. Vertex: $(-3, 0)$; Focus: $(-7, 0)$

The vertex is halfway between the directrix and the focus, so the directrix is $x = 1$

$p = x - h = -7 - (-3) = -4$

63. $|p| = 3$

$4|p| = 4(3) = 12$

65. No; Focal length is a distance and gives no information regarding the orientation of a parabola.

67. The directrix is a vertical line $x = 4$ and the vertex is 4 units to the left of the directrix. The parabola must open away from the directrix and toward the focus which in this case is to the left. The standard equation is $(y-k)^2 = 4p(x-h)$.

The distance between the directrix and vertex is 4. Therefore, $|p| = 4$. Because the parabola opens to the left, $p < 0$. So, $p = -4$.

$$(y-k)^2 = 4p(x-h)$$

$$(y-0)^2 = 4(-4)(x-0)$$

$$y^2 = -16x$$

69. The focus and vertex are on the axis of symmetry $x = 2$. The vertex is 3 units below the focus.

The parabola must open towards the focus and away from the directrix which in this case is upward. The standard equation is $(x-h)^2 = 4p(y-k)$.

The distance between the focus and vertex is 3. Therefore, $|p| = 3$. Because the parabola opens upward, $p > 0$. So, $p = 3$.

$$(x-h)^2 = 4p(y-k)$$

$$(x-2)^2 = 4(3)(y-1)$$

$$(x-2)^2 = 12(y-1)$$

71. The directrix is a horizontal line $y=0$ and the focus is 2 units below the directrix.
The vertex is halfway between the directrix and focus. Therefore, the vertex is $\frac{2}{2}=1$ unit above the focus at $(-6,-1)$. This indicates that $h=-6$ and $k=-1$.
The parabola must open away from the directrix and toward the focus which in this case is downward.
The standard equation is
$(x-h)^2=4p(y-k)$.
The distance between the directrix and focus is 2. Therefore, $|p|=\frac{2}{2}=1$. Because the parabola opens downward, $p<0$. So, $p=-1$.

$$(x-h)^2=4p(y-k)$$
$$[x-(-6)]^2=4(-1)[y-(-1)]$$
$$(x+6)^2=-4(y+1)$$

73. Since the point $(6, 5)$ is above and to the right of the vertex $(2, 3)$, the parabola opens upward or to the right.
Standard equation for a parabola opening upward:

$$(x-h)^2=4p(y-k)$$
$$(6-2)^2=4p(5-3)$$
$$(4)^2=4p(2)$$
$$16=8p$$
$$2=p$$
$$(x-h)^2=4p(y-k)$$
$$(x-2)^2=4(2)(y-3)$$
$$(x-2)^2=8(y-3)$$

Standard equation for a parabola opening to the right:

$$(y-k)^2=4p(x-h)$$
$$(5-3)^2=4p(6-2)$$
$$(2)^2=4p(4)$$
$$4=16p$$
$$\frac{1}{4}=p$$
$$(y-k)^2=4p(x-h)$$
$$(y-3)^2=4\left(\frac{1}{4}\right)(x-2)$$
$$(y-3)^2=(x-2)$$

75. a. $p=57.6$

$$(x-h)^2=4p(y-k)$$
$$(x-0)^2=4(57.6)(y-0)$$
$$x^2=230.4y \text{ for } -1.2\le x\le 1.2$$

b.

$$x^2=230.4y$$
$$(1.2)^2=230.4y$$
$$1.44=230.4y$$
$$y=\frac{1.44}{230.4}=0.00625 \text{ m or } 6.25 \text{ mm}$$

77. Solve the second equation for y:

$$2x+y=9$$
$$y=-2x+9$$

Substitute into the first equation:

$$(y+3)^2=4(x-4)$$
$$(-2x+9+3)^2=4(x-4)$$
$$(-2x+12)^2=4(x-4)$$
$$4x^2-48x+144=4x-16$$
$$4x^2-52x+160=0$$
$$4(x^2-13x+40)=0$$
$$(x-5)(x-8)=0$$
$$x=5 \text{ or } x=8$$

Substitute the solutions back into the second equation:

$$y=-2x+9=-2(5)+9=-1$$
$$y=-2x+9=-2(8)+9=-7$$
$$\{(5,-1),(8,-7)\}$$

79. If the y term is linear, then the parabola opens vertically. If the x term is linear, then the parabola opens horizontally.

81.

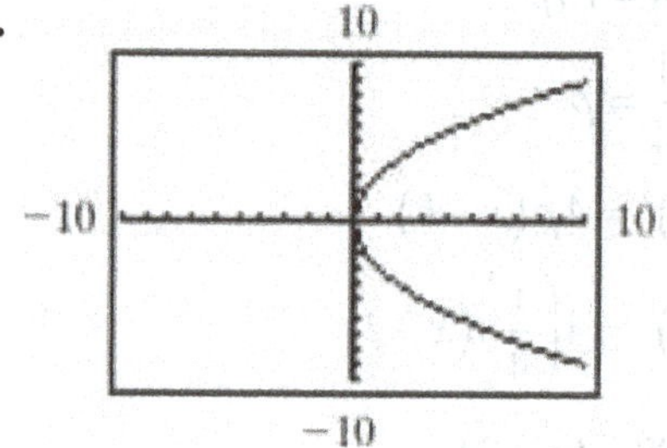

83.

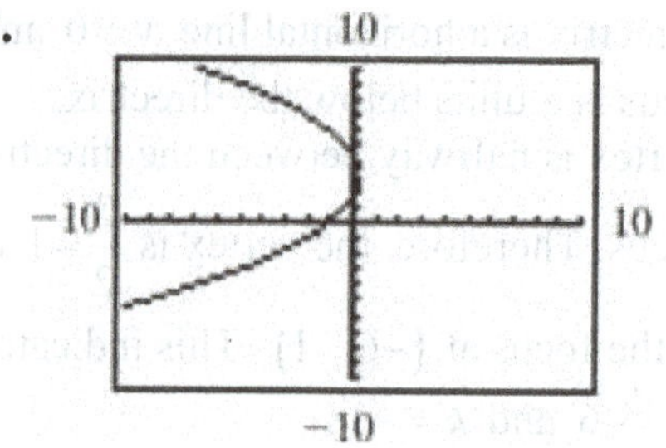

85. The greater the value of $|p|$ (the focal length), the wider the parabola.

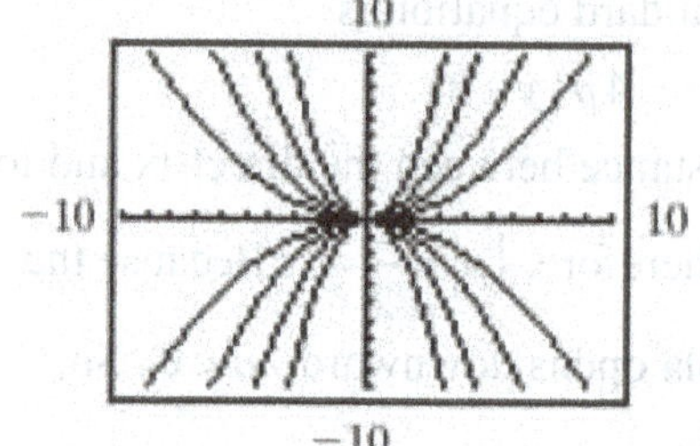

Problem Recognition Exercises: Comparing Equations of Conic Sections and the General Equation

1. $\dfrac{(x-2)^2}{16}-\dfrac{(y+2)^2}{9}=1$

$\dfrac{(x-2)^2}{16}-\dfrac{[y-(-2)]^2}{9}=1$

Hyperbola

Center: $(2,-2)$

The x^2 term has the positive coefficient, so the transverse axis is horizontal. The denominator of the x^2 term is a^2.

$a^2=16$, so $a=4$.

$b^2=9$, so $b=3$.

$c^2=a^2+b^2=16+9=25$

$c=\sqrt{25}=5$

The vertices are a units to the right and left of the center.

Vertices: $(2+4,-2)$ and $(2-4,-2)$

$(6,-2)$ and $(-2,-2)$

The foci are c units to the right and left of the center.

Foci: $(2+5,-2)$ and $(2-5,-2)$

$(7,-2)$ and $(-3,-2)$

Asymptotes:

$y-k=\dfrac{b}{a}(x-h)$ $\quad$ $y-k=-\dfrac{b}{a}(x-h)$

$y+2=\dfrac{3}{4}(x-2)$ $\quad$ $y+2=-\dfrac{3}{4}(x-2)$

$y+2=\dfrac{3}{4}x-\dfrac{3}{2}$ $\quad$ $y+2=-\dfrac{3}{4}x+\dfrac{3}{2}$

$y=\dfrac{3}{4}x-\dfrac{7}{2}$ $\quad$ $y=-\dfrac{3}{4}x-\dfrac{1}{2}$

Eccentricity: $e=\dfrac{c}{a}=\dfrac{5}{4}$

3. $(y-5)^2=-(x+2)$

Parabola

The equation is in the form $(y-k)^2=4p(x-h)$, where $h=-2$ and $k=5$.

Vertex: $(-2,5)$

$4p=-1$

$p=-\dfrac{1}{4}$

The x term is linear; the parabola opens in the x direction.

Since $p<0$, the parabola opens to the left.
The focus is $(h+p,k)$.

Focus: $\left(-\frac{9}{4},5\right)$

The directrix is the vertical line $|p|$ units to the right of the vertex.

Directrix: $x=-\frac{7}{4}$

Axis of symmetry: $y=5$

5. $16(x+1)^2+y^2=16$

$$\frac{16(x+1)^2}{16}+\frac{y^2}{16}=\frac{16}{16}$$

$$\frac{(x+1)^2}{1}+\frac{y^2}{16}=1$$

Ellipse

Center: $(-1,0)$

Since $16>1$, $a^2=16$ and $b^2=1$.

$a=\sqrt{16}=4$ and $b=\sqrt{1}=1$.

Since the greater number in the denominator is found in the y^2 term, the ellipse is elongated vertically.

The vertices are a units above and below the center.

Vertices: $(-1,0+4)$ and $(-1,0-4)$

$(-1,4)$ and $(-1,-4)$

The minor axis endpoints are b units to the right and left of the center.

Endpoints of minor axis:

$(-1+1,0)$ and $(-1-1,0)$

$(0,0)$ and $(-2,0)$

$c^2=a^2-b^2=16-1=15$

$c=\sqrt{15}$

The foci are c units above and below the center.

Foci: $\left(-1,0+\sqrt{15}\right)$ and $\left(-1,0-\sqrt{15}\right)$

$\left(-1,\sqrt{15}\right)$ and $\left(-1,-\sqrt{15}\right)$

Eccentricity: $e=\frac{c}{a}=\frac{\sqrt{15}}{4}$

7. $-(x+1)^2-(y+6)^2=-16$

$(x+1)^2+(y+6)^2=16$

Circle

The equation is in the form $(x-h)^2+(y-k)^2=r^2$, where $h=-1$, $k=-6$, and $r=4$.

Center : $(-1,-6)$; Radius: 4

9. A and C have opposite signs. Hyperbola

$$9x^2-16y^2-36x-64y-172=0$$

$$9x^2-36x-16y^2-64y=172$$

$$9\left(x^2-4x\quad\right)-16\left(y^2+4y\quad\right)=172$$

$$\left[\frac{1}{2}(-4)\right]^2=4 \qquad \left[\frac{1}{2}(4)\right]^2=4$$

$$9\left(x^2-4x+4\right)-16\left(y^2+4y+4\right)=172+36-64$$

$$9(x-2)^2-16(y+2)^2=144$$

$$\frac{9(x-2)^2}{144}-\frac{16(y+2)^2}{144}=\frac{144}{144}$$

$$\frac{(x-2)^2}{16}-\frac{(y+2)^2}{9}=1$$

11. A is zero. Parabola

$y^2-10y+x+27=0$

$y^2-10y+x=-27$

$y^2-10y\qquad=-x-27$

$$\left[\frac{1}{2}(-10)\right]^2=25$$

$y^2-10y+25=-x-27+25$

$(y-5)^2=-(x+2)$

13. A and C have the same sign. Ellipse

$$16x^2+y^2+32x=0$$
$$16x^2+32x+y^2=0$$
$$16\left(x^2+2x\quad\right)+y^2=0$$
$$\boxed{\left[\frac{1}{2}(2)\right]^2=1}$$
$$16\left(x^2+2x+1\right)+y^2=0+16$$
$$16\left(x^2+1\right)^2+y^2=16$$
$$\frac{16\left(x^2+1\right)^2}{16}+\frac{y^2}{16}=\frac{16}{16}$$
$$\left(x^2+1\right)^2+\frac{y^2}{16}=1$$

15. $A=C$. Circle

$$x^2+y^2+2x+12y+21=0$$
$$x^2+2x+y^2+12y=-21$$
$$\left(x^2+2x\quad\right)+\left(y^2+12y\quad\right)=-21$$
$$\boxed{\left[\frac{1}{2}(2)\right]^2=1\quad\left[\frac{1}{2}(12)\right]^2=36}$$
$$\left(x^2+2x+1\right)+\left(y^2+12y+36\right)=-21+1+36$$
$$(x+1)^2+(y+6)^2=16$$

17.

$$x^2+y^2-4x+2y+5=0$$
$$x^2-4x+y^2+2y=-5$$
$$\left(x^2-4x\quad\right)+\left(y^2+2y\quad\right)=-5$$
$$\boxed{\left[\frac{1}{2}(-4)\right]^2=4\quad\left[\frac{1}{2}(2)\right]^2=1}$$
$$\left(x^2-4x+4\right)+\left(y^2+2y+1\right)=-5+4+1$$
$$(x-2)^2+(y+1)^2=0$$

The only real numbers x and y that would make the sum of two squares equal to 0 are $x=2$ and $y=-1$. The graph is a single point: $(2,-1)$.

19.

$$4x^2-y^2-32x-4y+60=0$$
$$4x^2-32x-y^2-4y=-60$$
$$4\left(x^2-8x\quad\right)-\left(y^2+4y\quad\right)=-60$$
$$\boxed{\left[\frac{1}{2}(-8)\right]^2=16\quad\left[\frac{1}{2}(4)\right]^2=4}$$
$$4\left(x^2-8x+16\right)-\left(y^2+4y+4\right)=-60+64-4$$
$$4(x-4)^2-(y+2)^2=0$$
$$\frac{4(x-4)^2}{4}-\frac{(y+2)^2}{4}=\frac{0}{4}$$
$$(x-4)^2-\frac{(y+2)^2}{4}=0$$

Solve for y.

$$\frac{(y+2)^2}{4}=(x-4)^2$$
$$(y+2)^2=4(x-4)^2$$
$$y+2=\sqrt{4(x-4)^2}$$
$$y=-2\pm 2(x-4)$$
$$y=-2-2(x-4)\text{ or }y=-2+2(x-4)$$
$$y=-2-2x+8\quad\text{ or }y=-2+2x-8$$
$$y=-2x+6\quad\text{ or }y=2x-10$$

The graph is a pair of intersecting lines: $y=-2x+6$ and $y=2x-10$.

21.

$$9x^2+4y^2-16y+52=0$$
$$9x^2+4y^2-16y=-52$$
$$9x^2+4\left(y^2-4y\quad\right)=-52$$
$$\boxed{\left[\frac{1}{2}(-4)\right]^2=4}$$
$$9x^2+4\left(y^2-4y+4\right)=-52+16$$
$$9x^2+4(y-2)^2=-36$$
$$\frac{9x^2}{36}+\frac{4(y-2)^2}{36}=\frac{-36}{36}$$
$$\frac{x^2}{4}+\frac{(y-2)^2}{9}=-1$$

No solution. There are no real numbers x and y that would make the sum of two squares equal to -1.

Chapter 7 Review Exercises

1. Each equation represents an ellipse with a vertical major axis of length 30 units and horizontal minor axis of length 20 units. However, the first equation represents an ellipse centered at (0, 0), whereas the second equation represents an ellipse centered at $(1, -7)$.

3. $15x^2 + 9y^2 = 135$

$$\frac{15x^2}{135} + \frac{9y^2}{135} = \frac{135}{135}$$

$$\frac{x^2}{9} + \frac{y^2}{15} = 1$$

Since $15 > 9$, $a^2 = 15$ and $b^2 = 9$.

$a = \sqrt{15}$ and $b = \sqrt{9} = 3$.

Since the greater number in the denominator is found in the y^2 term, the ellipse is elongated vertically.

a. Center: $(0, 0)$

b. Vertices: $(0, \sqrt{15})$ and $(0, -\sqrt{15})$

c. Endpoints of minor axis: $(3, 0)$ and $(-3, 0)$

d. $c^2 = a^2 - b^2 = 15 - 9 = 6$

$c = \sqrt{6}$

Foci: $(0, \sqrt{6})$ and $(0, -\sqrt{6})$

e. $e = \frac{c}{a} = \frac{\sqrt{6}}{\sqrt{15}} \cdot \frac{\sqrt{15}}{\sqrt{15}} = \frac{\sqrt{90}}{15} = \frac{3\sqrt{10}}{15} = \frac{\sqrt{10}}{5}$

f.

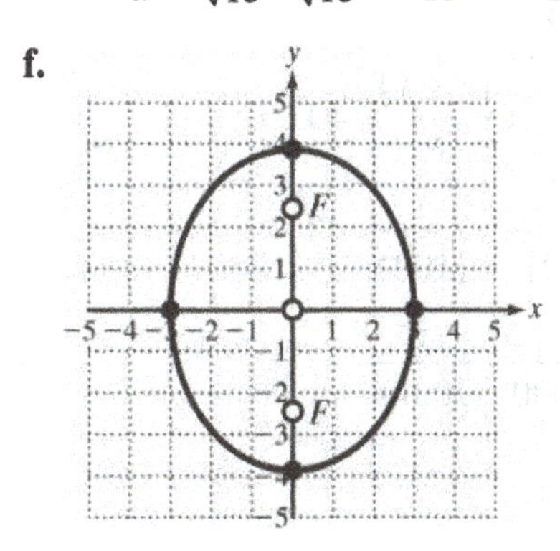

5. $\frac{(x-1)^2}{16} + \frac{(y-2)^2}{9} = 1$

Since $16 > 9$, $a^2 = 16$ and $b^2 = 9$.

$a = \sqrt{16} = 4$ and $b = \sqrt{9} = 3$.

Since the greater number in the denominator is found in the x^2 term, the ellipse is elongated horizontally.

a. Center: $(1, 2)$

b. The vertices are a units to the right and left of the center.

Vertices: $(1+4, 2)$ and $(1-4, 2)$

$(5, 2)$ and $(-3, 2)$

c. The minor axis endpoints are b units above and below the center.

Endpoints of minor axis:

$(1, 2+3)$ and $(1, 2-3)$

$(1, 5)$ and $(1, -1)$

d. $c^2 = a^2 - b^2 = 16 - 9 = 7$

$c = \sqrt{7}$

The foci are c units to the right and left of the center.

Foci: $(1+\sqrt{7}, 2)$ and $(1-\sqrt{7}, 2)$

e. $e = \frac{c}{a} = \frac{\sqrt{7}}{4}$

f.

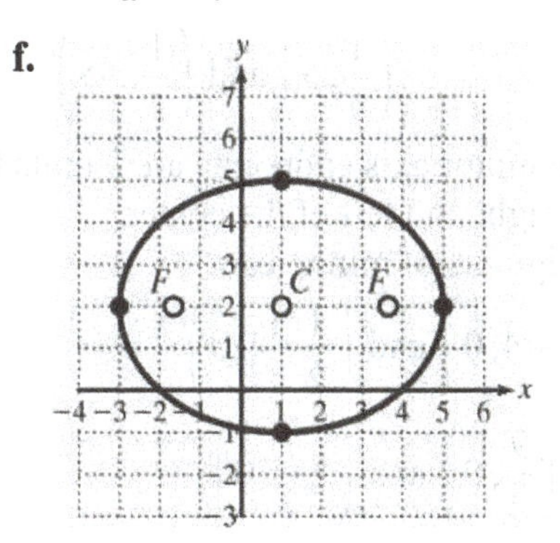

7. a. $100x^2 + 64y^2 - 100x - 1575 = 0$

$$100x^2 - 100x + 64y^2 = 1575$$

$$100\left(x^2 - x \quad\right) + 64y^2 = 1575$$

$$\boxed{\left[\frac{1}{2}(-1)\right]^2 = \frac{1}{4}}$$

$$100\left(x^2 - x + \frac{1}{4}\right) + 64y^2 = 1575 + 25$$

$$100\left(x - \frac{1}{2}\right)^2 + 64y^2 = 1600$$

$$\frac{100\left(x - \frac{1}{2}\right)^2}{1600} + \frac{64y^2}{1600} = \frac{1600}{1600}$$

$$\frac{\left(x - \frac{1}{2}\right)^2}{16} + \frac{y^2}{25} = 1$$

Since $25 > 16$, $a^2 = 25$ and $b^2 = 16$.

$a = \sqrt{25} = 5$ and $b = \sqrt{16} = 4$.

Since the greater number in the denominator is found in the y^2 term, the ellipse is elongated vertically.

b. Center: $\left(\frac{1}{2}, 0\right)$

The vertices are a units above and below the center.

Vertices: $\left(\frac{1}{2}, 0+5\right)$ and $\left(\frac{1}{2}, 0-5\right)$

$\left(\frac{1}{2}, 5\right)$ and $\left(\frac{1}{2}, -5\right)$

The minor axis endpoints are b units to the right and left of the center.

Endpoints of minor axis:

$\left(\frac{1}{2}+4, 0\right)$ and $\left(\frac{1}{2}-4, 0\right)$

$\left(\frac{9}{2}, 0\right)$ and $\left(-\frac{7}{2}, 0\right)$

$c^2 = a^2 - b^2 = 25 - 16 = 9$

$c = \sqrt{9} = 3$

The foci are c units above and below the center.

Foci: $\left(\frac{1}{2}, 0+3\right)$ and $\left(\frac{1}{2}, 0-3\right)$

$\left(\frac{1}{2}, 3\right)$ and $\left(\frac{1}{2}, -3\right)$

9. The foci have the same x-coordinate, so the major axis is vertical and the ellipse is elongated vertically. The standard form is

$$\frac{(x-h)^2}{b^2} + \frac{(y-k)^2}{a^2} = 1$$

The center is midway between the foci.

Center is $(3, 1)$: $h = 3$ and $k = 1$

Endpoints of the minor axis are 3 units from the center: $b = 3$

Foci are 4 units from the center: $c = 4$

$a^2 = b^2 + c^2 = (3)^2 + (4)^2 = 9 + 16 = 25$

$a = \sqrt{25} = 5$

$$\frac{(x-3)^2}{(3)^2} + \frac{(y-1)^2}{(5)^2} = 1$$

$$\frac{(x-3)^2}{9} + \frac{(y-1)^2}{25} = 1$$

11. The second ellipse with eccentricity $\frac{5}{7}$ is more elongated.

13. Length of major axis is 400 ft.

$a = \frac{400}{2} = 200$

Length of half of minor axis is 100 ft.

$b = 100$

$$\frac{(x-h)^2}{a^2} + \frac{(y-k)^2}{b^2} = 1$$

$$\frac{x^2}{(200)^2} + \frac{y^2}{(100)^2} = 1$$

$$\frac{x^2}{40{,}000} + \frac{y^2}{10{,}000} = 1$$

At $x = 50$:

$$\frac{(50)^2}{40{,}000} + \frac{y^2}{10{,}000} = 1$$

$$\frac{2500}{40{,}000} + \frac{y^2}{10{,}000} = 1$$

$$2500 + 4y^2 = 40{,}000$$
$$4y^2 = 37{,}500$$
$$y^2 = 9375$$
$$y = \pm\sqrt{9375}$$
$$\approx \pm 97$$

The height 50 ft from the center is approximately 97 ft.

15. The x^2 term has the positive coefficient, indicating that the transverse axis is horizontal.

17. $\frac{x^2}{9} - \frac{y^2}{4} = 1$

The x^2 term has the positive coefficient, so the transverse axis is horizontal. The denominator of the x^2 term is a^2.

$a^2 = 9$, so $a = 3$.

$b^2 = 4$, so $b = 2$.

$c^2 = a^2 + b^2 = 9 + 4 = 13$

$c = \sqrt{13}$

a. Center: $(0, 0)$

b. Vertices: $(3, 0)$ and $(-3, 0)$

c. Foci: $(\sqrt{13}, 0)$ and $(-\sqrt{13}, 0)$

d. Asymptotes:

$y = \frac{b}{a}x \qquad y = -\frac{b}{a}x$

$y = \frac{2}{3}x \qquad y = -\frac{2}{3}x$

e.

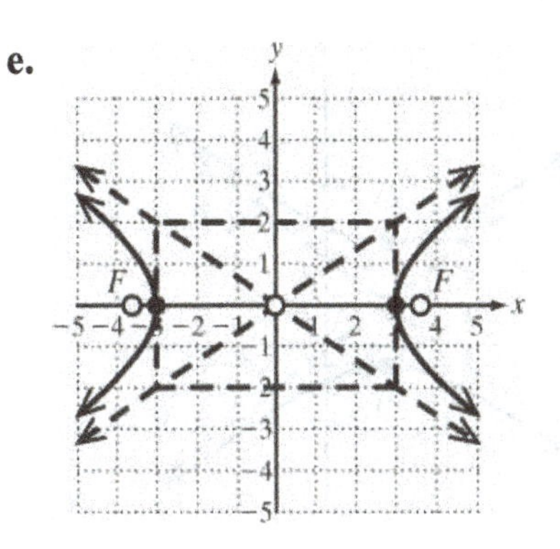

19. $-\frac{(x+3)^2}{16} + \frac{(y-2)^2}{9} = 1$

$$\frac{(y-2)^2}{9} - \frac{[x-(-3)]^2}{16} = 1$$

The y^2 term has the positive coefficient, so the transverse axis is vertical. The denominator of the y^2 term is a^2.

$a^2 = 9$, so $a = 3$.

$b^2 = 16$, so $b = 4$.

$c^2 = a^2 + b^2 = 9 + 16 = 25$

$c = \sqrt{25} = 5$

a. Center: $(-3, 2)$

b. The vertices are a units above and below the center.

Vertices: $(-3, 2+3)$ and $(-3, 2-3)$

$(-3, 5)$ and $(-3, -1)$

c. The foci are c units above and below the center.

Foci: $(-3, 2+5)$ and $(-3, 2-5)$

$(-3, 7)$ and $(-3, -3)$

d. Asymptotes:

$y - k = \frac{a}{b}(x-h) \qquad y - k = -\frac{a}{b}(x-h)$

$y - 2 = \frac{3}{4}(x+3) \qquad y - 2 = -\frac{3}{4}(x+3)$

$y - 2 = \frac{3}{4}x + \frac{9}{4} \qquad y - 2 = -\frac{3}{4}x - \frac{9}{4}$

$y = \frac{3}{4}x + \frac{17}{4} \qquad y = -\frac{3}{4}x - \frac{1}{4}$

e.

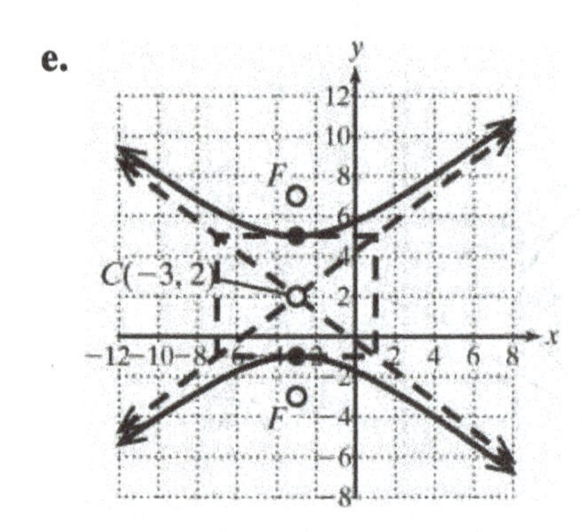

21. a.

$$11x^2 - 7y^2 + 44x - 56y - 145 = 0$$
$$11x^2 + 44x - 7y^2 - 56y = 145$$
$$11(x^2 + 4x \quad) - 7(y^2 + 8y \quad) = 145$$

$$\left[\frac{1}{2}(4)\right]^2 = 4 \quad \left[\frac{1}{2}(8)\right]^2 = 16$$

$$11(x^2 + 4x + 4) - 7(y^2 + 8y + 16)$$
$$= 145 + 44 - 112$$
$$11(x+2)^2 - 7(y+4)^2 = 77$$
$$\frac{11(x+2)^2}{77} - \frac{7(y+4)^2}{77} = \frac{77}{77}$$
$$\frac{(x+2)^2}{7} - \frac{(y+4)^2}{11} = 1$$

b. Center: $(-2, -4)$

The x^2 term has the positive coefficient, so the transverse axis is horizontal. The denominator of the x^2 term is a^2.

$a^2 = 7$, so $a = \sqrt{7}$.

$b^2 = 11$ so $b = \sqrt{11}$.

$c^2 = a^2 + b^2 = 7 + 11 = 18$

$c = \sqrt{18} = 3\sqrt{2}$

The vertices are a units to the right and left of the center.

Vertices:

$\left(-2+\sqrt{7}, -4\right)$ and $\left(-2-\sqrt{7}, -4\right)$

The foci are c units to the right and left of the center.

Foci: $\left(-2+3\sqrt{2}, -4\right)$ and $\left(-2-3\sqrt{2}, -4\right)$

23. The foci have the same y-coordinate, so they are aligned horizontally and the transverse axis of the hyperbola is horizontal.

The standard form is $\frac{(x-h)^2}{a^2} - \frac{(y-k)^2}{b^2} = 1$.

The center is midway between the vertices.

Center is $(0, 0)$: $h = 0$ and $k = 0$

Vertices are 4 units from the center: $a = 4$

Foci are 5 units from the center: $c = 5$

$b^2 = c^2 - a^2 = (5)^2 - (4)^2 = 25 - 16 = 9$

$b = \sqrt{9} = 3$

$$\frac{(x-0)^2}{(4)^2} - \frac{(y-0)^2}{(3)^2} = 1$$
$$\frac{x^2}{16} - \frac{y^2}{9} = 1$$

25. The transverse axis is horizontal.

The standard form is $\frac{(x-h)^2}{a^2} - \frac{(y-k)^2}{b^2} = 1$.

Width of reference rectangle is

$19 - (-15) = 34$

$2a = 34$

$a = 17$

Height of reference rectangle is

$15 - (-1) = 16$

$2b = 16$

$b = 8$

Center: $(2, 7)$

$$\frac{(x-2)^2}{(17)^2} - \frac{(y-7)^2}{(8)^2} = 1$$
$$\frac{(x-3)^2}{289} - \frac{(y-3)^2}{64} = 1$$

27. $\boxed{A}\ 6x^2 - 2y^2 = -12 \qquad\qquad 6x^2 - 2y^2 = -12$

$\boxed{B}\ 2x^2 + y^2 = 11 \xrightarrow{\text{Multiply by 2.}} \underline{4x^2 + 2y^2 = 22}$

$$10x^2 = 10$$
$$x^2 = 1$$
$$x = \pm 1$$

$x = 1$: $\boxed{B}\ 2x^2 + y^2 = 11$

$$2(1)^2 + y^2 = 11$$
$$2 + y^2 = 11$$
$$y^2 = 9$$
$$y = \pm 3$$

$x = -1$: $\boxed{B}\ 2x^2 + y^2 = 11$

$$2(-1)^2 + y^2 = 11$$
$$2 + y^2 = 11$$
$$y^2 = 9$$
$$y = \pm 3$$

$\{(1, 3), (1, -3), (-1, 3), (-1, -3)\}$

29. $x^2 = -2y$

The equation is in the form $x^2 = 4py$.

The focus is on the y-axis and the parabola opens upward or downward.

a. $4p = -2$

$$p = -\frac{2}{4} = -\frac{1}{2}$$

b. The vertex is $(0, 0)$.

c. The focus is $(0, p)$, which is $\left(0, -\frac{1}{2}\right)$.

d. The focal diameter is $4|p|$, which is $4\left|-\frac{1}{2}\right| = 2$.

e. The endpoints of the latus rectum are $\frac{2}{2} = 1$ unit to the right and to the left of the focus: $\left(1, -\frac{1}{2}\right)$ and $\left(-1, -\frac{1}{2}\right)$.

f. Since $p < 0$, the parabola opens downward. The directrix is the horizontal line $|p|$ units above the vertex: $y = \frac{1}{2}$.

g. The axis of symmetry is $x = 0$.

h.

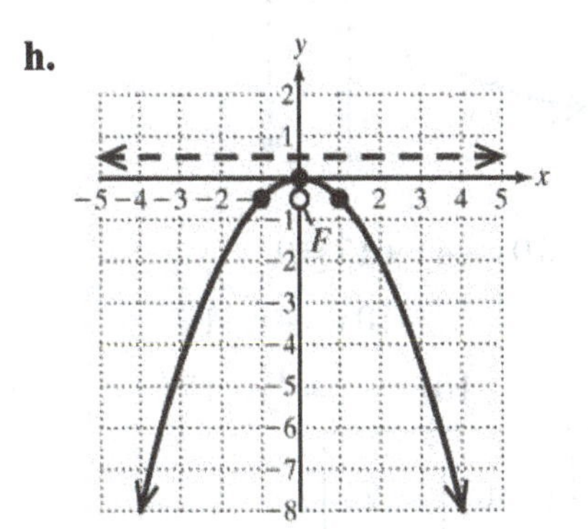

31. $(y + 2)^2 = -4(x - 3)$

The equation is in the form $(y - k)^2 = 4p(x - h)$, where $h = 3$ and $k = -2$.

The x term is linear; the parabola opens in the x direction.

a. $4p = -4$

$$p = -1$$

b. The vertex is $(h, k) = (3, -2)$.

c. The focus is $(h + p, k)$, which is $(2, -2)$.

d. The focal diameter is $4|p|$, which is $4|-1| = 4$.

e. The endpoints of the latus rectum are $\frac{4}{2}=2$ units above and below the focus: $(2,0)$ and $(2,-4)$.

f. Since $p<0$, the parabola opens to the left. The directrix is the vertical line $|p|$ units to the right of the vertex: $x=4$.

g. The axis of symmetry is $y=-2$.

h.

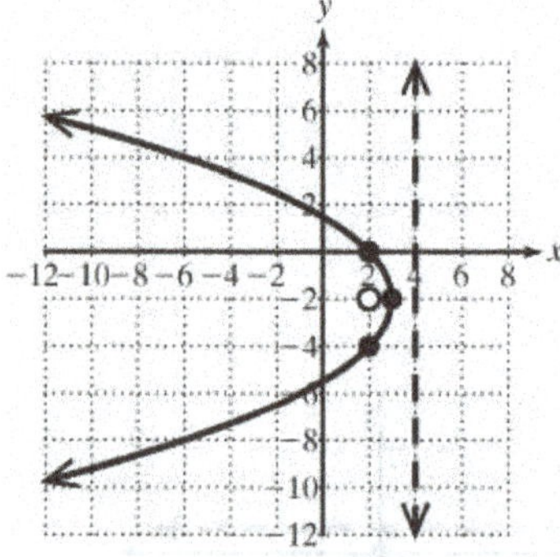

33. a.
$$x^2-10x-4y+17=0$$
$$x^2-10x=4y-17$$
$$x^2-10x \qquad =4y-17$$
$$\boxed{\left[\frac{1}{2}(-10)\right]^2=25}$$
$$x^2-10x+25=4y-17+25$$
$$(x-5)^2=4y+8$$
$$(x-5)^2=4(y+2)$$

b. The equation is in the form $(x-h)^2=4p(y-k)$, where $h=5$ and $k=-2$. The vertex is $(5,-2)$.
$$4p=4$$
$$p=1$$
The y term is linear; the parabola opens in the y direction.
Since $p>0$, the parabola opens upward.
The focus is $(h,k+p)$, which is $(5,-1)$.
The directrix is the horizontal line $|p|$ units below the vertex: $y=-3$.

35. Directrix: $y=3$; Focus: $(4,1)$
The vertex is halfway between the directrix and the focus: $(4,2)$.
$$p=\frac{1-3}{2}=\frac{-2}{2}=-1$$

37. $4|p|=12$
$$|p|=3$$

39. The focus and vertex are on the axis of symmetry $x=-4$. The vertex is 8 units above the focus.
The parabola must open towards the focus and away from the directrix which in this case is downward. The standard equation is $(x-h)^2=4p(y-k)$.
The distance between the focus and vertex is 8. Therefore, $|p|=8$. Because the parabola opens downward, $p<0$. So, $p=-8$.
$$(x-h)^2=4p(y-k)$$
$$[x-(-4)]^2=4(-8)(y-7)$$
$$(x+4)^2=-32(y-7)$$

41. Solve the second equation for y:
$$x-y=7$$
$$y=x-7$$
Substitute into the first equation:
$$(y+2)^2=3(x-5)$$
$$(x-7+2)^2=3(x-5)$$
$$(x-5)^2=3(x-5)$$
$$x^2-10x+25=3x-15$$
$$x^2-13x+40=0$$
$$(x-5)(x-8)=0$$
$$x=5 \text{ or } x=8$$
Substitute the solutions back into the second equation:
$$y=x-7=5-7=-2$$
$$y=x-7=8-7=1$$
$$\{(5,-2),(8,1)\}$$

Chapter 7 Test

1. Each equation represents an ellipse centered at the origin with a major axis of length 24 units and minor axis of length 12 units. However, the ellipse represented by the first equation has foci on the y-axis, whereas the second equation represents an ellipse with foci on the x-axis.

3. $\dfrac{(x-h)^2}{a^2}+\dfrac{(y-k)^2}{b^2}=1$

5. $\dfrac{(y-k)^2}{a^2}-\dfrac{(x-h)^2}{b^2}=1$

7. $(x-h)^2=4p(y-k)$

9. $\dfrac{(x-2)^2}{9}+\dfrac{y^2}{16}=1$

a. Ellipse

b.

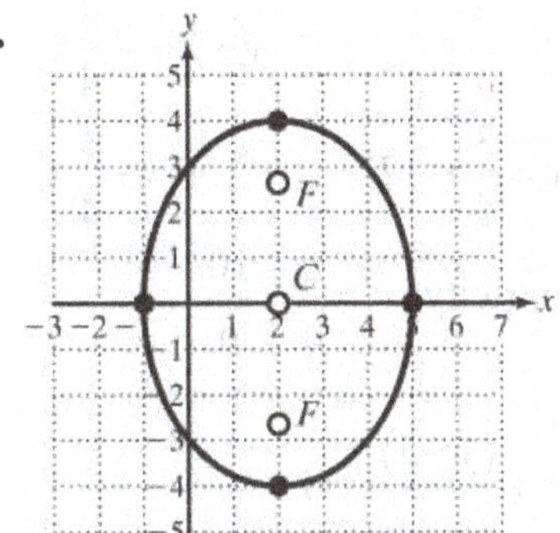

c. Center: $(2, 0)$

Since $16>9$, $a^2=16$ and $b^2=9$.

$a=\sqrt{16}=4$ and $b=\sqrt{9}=3$.

Since the greater number in the denominator is found in the y^2 term, the ellipse is elongated vertically.

The vertices are a units above and below the center.

Vertices: $(2, 0+4)$ and $(2, 0-4)$

$(2, 4)$ and $(2, -4)$

The minor axis endpoints are b units to the right and left of the center.

Endpoints of minor axis:

$(2+3, 0)$ and $(2-3, 0)$

$(5, 0)$ and $(-1, 0)$

$c^2=a^2-b^2=16-9=7$

$c=\sqrt{7}$

The foci are c units above and below the center.

Foci: $\left(2, 0+\sqrt{7}\right)$ and $\left(2, 0-\sqrt{7}\right)$

$\left(2, \sqrt{7}\right)$ and $\left(2, -\sqrt{7}\right)$

Eccentricity: $e=\dfrac{c}{a}=\dfrac{\sqrt{7}}{4}$

11. $\dfrac{(x-2)^2}{16}+y=0$

$\dfrac{(x-2)^2}{16}=-y$

$(x-2)^2=-16y$

$(x-2)^2=-16(y-0)$

a. Parabola

b.

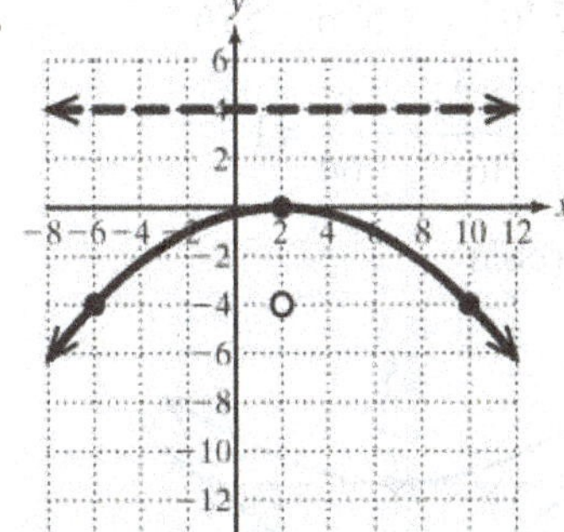

c. The equation is in the form $(x-h)^2=4p(y-k)$, where $h=2$ and $k=0$. The vertex is $(2, 0)$.

$4p=-16$

$p=-4$

The y term is linear; the parabola opens in the y direction.

Since $p < 0$, the parabola opens downward.

The focus is $(h, k+p)$, which is $(2, -4)$.

The focal diameter is $4|p|$, which is $4|-4| = 16$.

The endpoints of the latus rectum are $\frac{16}{2} = 8$ units to the right and to the left of the focus: $(10, -4)$ and $(-6, -4)$.

The directrix is the horizontal line $|p|$ units above the vertex: $y = 4$.

The axis of symmetry is $x = 2$.

13. $-9x^2 + 16y^2 + 64y - 512 = 0$

$$16y^2 + 64y - 9x^2 = 512$$

$$16(y^2 + 4y \quad) - 9x^2 = 512$$

$$\boxed{\left[\frac{1}{2}(4)\right]^2 = 4}$$

$$16(y^2 + 4y + 4) - 9x^2 = 512 + 64$$

$$16(y+2)^2 - 9x^2 = 576$$

$$\frac{16(y+2)^2}{576} - \frac{9x^2}{576} = \frac{576}{576}$$

$$\frac{(y+2)^2}{36} - \frac{x^2}{64} = 1$$

a. Hyperbola

b.

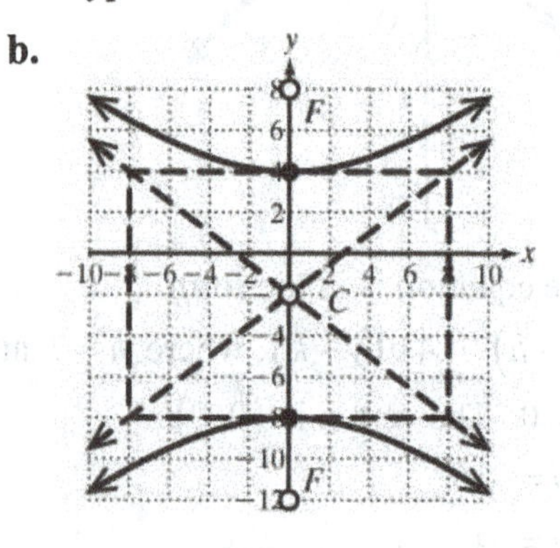

c. The y^2 term has the positive coefficient, so the transverse axis is vertical. The denominator of the y^2 term is a^2.

Center: $(0, -2)$

$a^2 = 36$, so $a = 6$.

$b^2 = 64$, so $b = 8$.

$c^2 = a^2 + b^2 = 36 + 64 = 100$

$c = \sqrt{100} = 10$

Vertices: $(0, -2+6)$ and $(0, -2-6)$

$(0, 4)$ and $(0, -8)$

The foci are c units to the right and left of the center.

Foci: $(0, -2+10)$ and $(0, -2-10)$

$(0, 8)$ and $(0, -12)$

Asymptotes:

$$y - k = \frac{a}{b}(x-h) \qquad y - k = -\frac{a}{b}(x-h)$$

$$y + 2 = \frac{6}{8}(x-0) \qquad y + 2 = -\frac{6}{8}(x-0)$$

$$y + 2 = \frac{3}{4}x \qquad y + 2 = -\frac{3}{4}x$$

$$y = \frac{3}{4}x - 2 \qquad y = -\frac{3}{4}x - 2$$

Eccentricity: $e = \frac{c}{a} = \frac{10}{6} = \frac{5}{3}$

15. $\frac{(x+4)^2}{4} + \frac{(y-1)^2}{4} = 1$

$$(x+4)^2 + (y-1)^2 = 4$$

a. Circle

b.

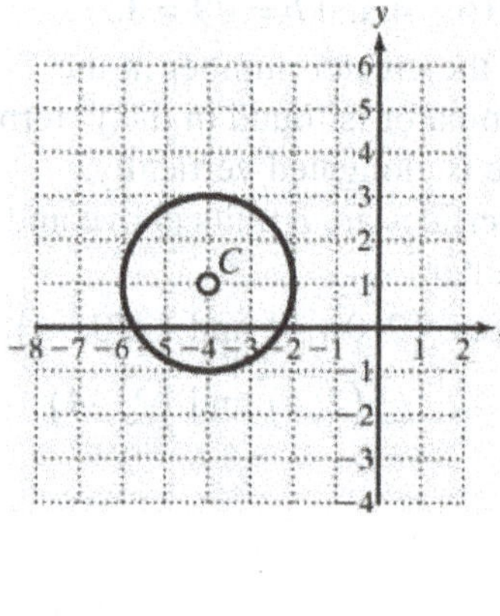

c. $r^2 = 4$

$r = \sqrt{4} = 2$

Center: $(-4, 1)$; Radius: 2

17. a. Length of major axis is 24 ft. $a = \frac{24}{2} = 12$

Length of half of minor axis is 6 ft. $b = 6$

$$\frac{(x-h)^2}{a^2} + \frac{(y-k)^2}{b^2} = 1$$

$$\frac{x^2}{(12)^2} + \frac{y^2}{(6)^2} = 1$$

$$\frac{x^2}{144} + \frac{y^2}{36} = 1$$

At $x = 12 - 6 = 6$:

$$\frac{(6)^2}{144} + \frac{y^2}{36} = 1$$

$$\frac{36}{144} + \frac{y^2}{36} = 1$$

$$\frac{1}{4} + \frac{y^2}{36} = 1$$

$$9 + y^2 = 36$$

$$y^2 = 27$$

$$y = \pm\sqrt{27} \approx \pm 5.2$$

$8 + 5.2 = 13.2$

The height 6 ft from edge of the tunnel is approximately 13.2 ft.

b. At $x = 12 - 3 = 9$:

$$\frac{(9)^2}{144} + \frac{y^2}{36} = 1$$

$$\frac{81}{144} + \frac{y^2}{36} = 1$$

$$\frac{81}{4} + y^2 = 36$$

$$y^2 = \frac{63}{4}$$

$$y = \pm\sqrt{\frac{63}{4}} \approx \pm 4.0$$

$8 + 4.0 = 12.0$

Yes; The height of the opening is approximately 12 ft at a point 3 ft from the edge.

c. Solve for y:

$$\frac{x^2}{144} + \frac{y^2}{36} = 1$$

$$\frac{y^2}{36} = 1 - \frac{x^2}{144}$$

$$y^2 = 36\left(1 - \frac{x^2}{144}\right)$$

$$y = \sqrt{36\left(1 - \frac{x^2}{144}\right)} = 6\sqrt{1 - \frac{x^2}{144}}$$

The function must include the height of the rectangular opening.

$h(x) = y + 8$

$$h(x) = 6\sqrt{1 - \frac{x^2}{144}} + 8$$

19. $d_1 - d_2 = (1100 \text{ ft/sec})(6 \text{ sec}) = 6600 \text{ ft}$

$d_1 - d_2 = 2a$

$6600 = 2a$

$a = 3300 \text{ ft}$

$c = \frac{7920 \text{ ft}}{2} = 3960 \text{ ft}$

$b^2 = c^2 + a^2 = (3960)^2 - (3300)^2$

$= 15{,}681{,}600 - 10{,}890{,}000 = 4{,}791{,}600$

$b \approx 2189 \text{ ft}$

$$\frac{x^2}{(3300)^2} - \frac{y^2}{(2189)^2} = 1$$

21. The directrix is a horizontal line $y = -2$ and the focus is 8 units above the directrix. The vertex is halfway between the directrix and focus. Therefore, the vertex is $\frac{8}{2} = 4$ units below the focus at $(1, 2)$. This indicates that $h = 1$ and $k = 2$.

The parabola must open away from the directrix and toward the focus which in this case is upward. The standard equation is $(x-h)^2 = 4p(y-k)$.

The distance between the directrix and focus is 8. Therefore, $|p| = \frac{8}{2} = 4$. Because the parabola opens upward, $p > 0$. So, $p = 4$.

$$(x-h)^2 = 4p(y-k)$$
$$(x-1)^2 = 4(4)(y-2)$$
$$(x-1)^2 = 16(y-2)$$

23. The vertices have the same x-coordinate, so they are aligned vertically and the transverse axis of the hyperbola is vertical. The standard form is

$$\frac{(y-k)^2}{a^2} - \frac{(x-h)^2}{b^2} = 1$$

The center is midway between the vertices.
Center is $(4, 3)$: $h = 4$ and $k = 3$
Vertices are 4 units from the center: $a = 4$
Foci are 6 units from the center: $c = \sqrt{6}$

$$b^2 = c^2 - a^2 = \left(\sqrt{6}\right)^2 - (4)^2 = 36 - 16 = 20$$
$$b = \sqrt{20} = 2\sqrt{5}$$

$$\frac{(y-3)^2}{(4)^2} - \frac{(x-4)^2}{\left(2\sqrt{5}\right)^2} = 1$$
$$\frac{(y-3)^2}{16} - \frac{(x-4)^2}{20} = 1$$

25. $2a = 26$ $\quad$ $2b = 10$

$a = 13$ $\quad$ $b = 5$

$$c^2 = a^2 - b^2$$
$$= (13)^2 - (5)^2$$
$$= 169 - 25$$
$$= 144$$
$$c = \sqrt{144}$$
$$= 12$$
$$e = \frac{c}{a} = \frac{12}{13}$$

27. Directrix: $x = 2$; Focus: $(-4, 0)$

The distance between the directrix and focus is 6. Therefore, $|p| = \frac{6}{2} = 3$.

Focal length: 3

Focal diameter: $4|p| = 4(3) = 12$

29. $\boxed{A}$ $3x^2 - 4y^2 = -13$
$\boxed{B}$ $5x^2 + 2y^2 = 13 \xrightarrow{\text{Multiply by 2.}} 10x^2 + 4y^2 = 26$

$$\begin{aligned} 3x^2 - 4y^2 &= -13 \\ 10x^2 + 4y^2 &= 26 \\ \hline 13x^2 &= 13 \\ x^2 &= 1 \\ x &= \pm 1 \end{aligned}$$

$x = 1$: $\boxed{B}$
$$5x^2 + 2y^2 = 13$$
$$5(1)^2 + 2y^2 = 13$$
$$5 + 2y^2 = 13$$
$$2y^2 = 8$$
$$y^2 = 4$$
$$y = \pm 2$$

$x = -1$: $\boxed{B}$
$$5x^2 + 2y^2 = 13$$
$$5(-1)^2 + 2y^2 = 13$$
$$5 + 2y^2 = 13$$
$$2y^2 = 8$$
$$y^2 = 4$$
$$y = \pm 2$$

$\{(1, 2), (1, -2), (-1, 2), (-1, -2)\}$

Chapter 7 Cumulative Review Exercises

1.
$$\begin{array}{r} 3x^2-2x-9 \\ x^2+0x+3\overline{)3x^4-2x^3+0x^2-5x+1} \\ -\left(3x^4+0x^3+9x^2\right) \\ \hline -2x^3-9x^2-5x \\ -\left(-2x^3+0x^2-6x\right) \\ \hline -9x^2+x+1 \\ -\left(9x^2+0x-27\right) \\ \hline x+28 \end{array}$$

$$3x^2-2x-9+\frac{x+28}{x^2+3}$$

3.
$$\begin{array}{r|rrrrr} 3 & 2 & -7 & 8 & -17 & 6 \\ & & 6 & -3 & 15 & -6 \\ \hline & 2 & -1 & 5 & -2 & \lfloor 0 \end{array}$$

By the factor theorem, since $f(3)=0$, $x-3$ is a factor of $f(x)$.

5.
$$\left(\frac{-2x^2y^{-1}}{z^3}\right)^{-4}\left(\frac{4x^{-5}}{y^7}\right)^2$$
$$=\left(\frac{z^3}{-2x^2y^{-1}}\right)^4\left(\frac{4x^{-5}}{y^7}\right)^2$$
$$=\left[(-2)^{-1}x^{-2}yz^3\right]^4\left(4x^{-5}y^{-7}\right)^2$$
$$=(-2)^{-4}x^{-8}y^4z^{12}(4)^2x^{-10}y^{-14}$$
$$=(-2)^{-4}(4)^2x^{-18}y^{-10}z^{12}$$
$$=\frac{16z^{12}}{16x^{18}y^{10}}$$
$$=\frac{z^{12}}{x^{18}y^{10}}$$

7.
$$f(x)=\sqrt[3]{x-7}$$
$$y=\sqrt[3]{x-7}$$
$$x=\sqrt[3]{y-7}$$
$$x^3=y-7$$
$$x^3+7=y$$
$$f^{-1}(x)=x^3+7$$

9.
$$f(x)=3|x|+5x^3$$
$$f(-x)=3|-x|+5(-x)^3$$
$$=3|x|-5x^3$$
$$-f(x)=-\left(3|x|+5x^3\right)$$
$$=-3|x|-5x^3$$
$$f(-x)\neq f(x)$$
$$f(-x)\neq -f(x)$$

The function is neither even nor odd.

11.
$$2-3\{4-[x+2(x-1)-3]+4\}>0$$
$$2-3\{4-[x+2x-2-3]+4\}>0$$
$$2-3\{4-[3x-5]+4\}>0$$
$$2-3\{4-3x+5+4\}>0$$
$$2-3\{-3x+13\}>0$$
$$2+9x-39>0$$
$$9x-37>0$$
$$9x>37$$
$$x>\frac{37}{9}$$

$$\left(\frac{37}{9},\infty\right)$$

13. $2x^2-8x+3\ge 0$

Find the real zeros of the related equation.

$2x^2-8x+3=0$

$$x=\frac{-(-8)\pm\sqrt{(-8)^2-4(2)(3)}}{2(2)}=\frac{8\pm\sqrt{40}}{4}=\frac{8\pm 2\sqrt{10}}{4}=\frac{4\pm\sqrt{10}}{2}$$

The boundary points are $\frac{4-\sqrt{10}}{2}$ and $\frac{4+\sqrt{10}}{2}$.

Sign of $\left[x-\left(\frac{4-\sqrt{10}}{2}\right)\right]$:	−	+	+
Sign of $\left[x-\left(\frac{4+\sqrt{10}}{2}\right)\right]$:	−	−	+
Sign of $\left[x-\left(\frac{4-\sqrt{10}}{2}\right)\right]\left[x-\left(\frac{4+\sqrt{10}}{2}\right)\right]$:	+	−	+

$\frac{4-\sqrt{10}}{2}$ $\quad$ $\frac{4+\sqrt{10}}{2}$

The solution set is $\left(-\infty,\frac{4-\sqrt{10}}{2}\right]\cup\left[\frac{4+\sqrt{10}}{2},\infty\right)$.

15. $e^{2x}-8e^x+15=0$

Let $u=e^x$.

$u^2-8u+15=0$

$(u-3)(u-5)=0$

$u=3$ or $u=5$

$e^x=3$ or $e^x=5$

$\ln e^x=\ln 3$ $\quad$ $\ln e^x=\ln 5$

$x=\ln 3$ $\quad$ $x=\ln 5$

$\{\ln 3,\ln 5\}$

17. $2x-y<4\xrightarrow[\text{equation}]{\text{related}}2x-y=4$

<u>x-intercept</u>:

$2x-(0)=4$

$2x=4$

$x=2$

<u>y-intercept</u>:

$2(0)-y=4$

$-y=4$

$y=-4$

The inequality is strict, so draw the line as a dashed line.

<u>Test (0, 0)</u>: $2x-y<4$

$2(0)-(0)\overset{?}{<}4$

$0\overset{?}{<}4$ ✓ true

The points on the side of the line containing (0, 0) are solutions.

$3x+y\le 1\xrightarrow[\text{equation}]{\text{related}}3x+y=1$

<u>x-intercept</u>:

$3x+(0)=1$

$3x=1$

$x=\frac{1}{3}$

<u>y-intercept</u>:

$3(0)+y=1$

$y=1$

The inequality symbol $\le$ allows for equality, so draw the line as a solid line.

<u>Test (0, 0)</u>: $3x+y\le 1$

$3(0)+(0)\overset{?}{\le}1$

$0\overset{?}{\le}1$ ✓ true

The points on the side of the line containing (0, 0) are solutions.

Find the point of intersection.

$$\begin{aligned} 2x - y &= 4 \\ 3x + y &= 1 \\ \hline 5x \quad &= 5 \\ x &= 1 \end{aligned}$$

$$\begin{aligned} 2x - y &= 4 \\ 2(1) - y &= 4 \\ -y &= 2 \\ y &= -2 \end{aligned}$$

Graph $(1, -2)$ as an open dot. It is not a solution to the first inequality.

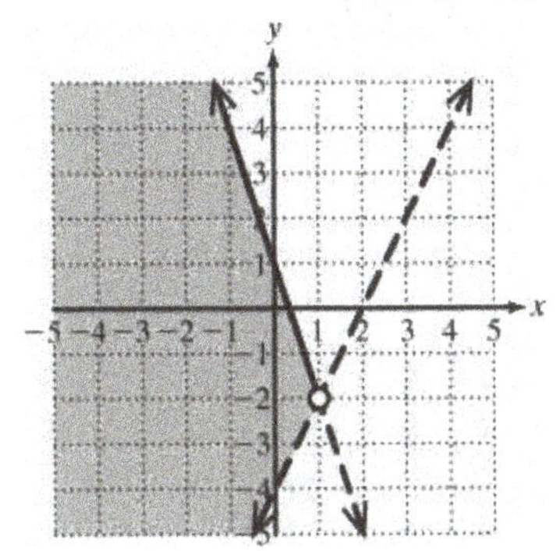

19. $(f \circ g)(x) = f(g(x)) = \dfrac{1}{g(x) - 2} = \dfrac{1}{\sqrt{x-1} - 2}$

$$\begin{aligned} \sqrt{x-1} - 2 &\neq 0 \\ \sqrt{x-1} &\neq 2 \\ x - 1 &\neq 4 \\ x &\neq 5 \end{aligned}$$

$$\begin{aligned} x - 1 &\geq 0 \\ x &\geq 1 \end{aligned}$$

Domain: $[1, 5) \cup (5, \infty)$

21. $\dfrac{x^2}{4} + \dfrac{(y-1)^2}{9} = 1$

Center: $(0, 1)$

Since $9 > 4$, $a^2 = 9$ and $b^2 = 4$.

$a = \sqrt{9} = 3$ and $b = \sqrt{4} = 2$.

Since the greater number in the denominator is found in the y^2 term, the ellipse is elongated vertically.

The vertices are a units above and below the center.

Vertices: $(0, 1+3)$ and $(0, 1-3)$

$(0, 4)$ and $(0, -2)$

The minor axis endpoints are b units to the right and left of the center.

Endpoints of minor axis:

$(0+2, 1)$ and $(0-2, 1)$

$(2, 1)$ and $(-2, 1)$

$c^2 = a^2 - b^2 = 9 - 4 = 5$

$c = \sqrt{5}$

The foci are c units above and below the center.

Foci: $\left(0, 1+\sqrt{5}\right)$ and $\left(0, 1-\sqrt{5}\right)$

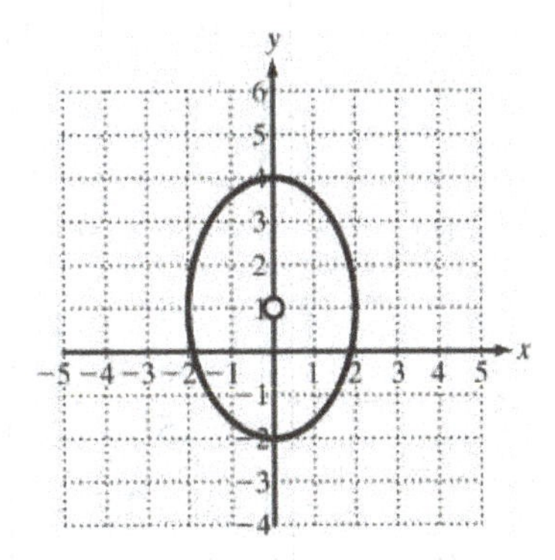

23. $x^2 = 4(y-2)$

The equation is in the form $(x-h)^2 = 4p(y-k)$, where $h = 0$ and $k = 2$. The vertex is $(0, 2)$.

$$\begin{aligned} 4p &= 4 \\ p &= 1 \end{aligned}$$

The y term is linear; the parabola opens in the y direction.

Since $p > 0$, the parabola opens upward.

The focus is $(h, k+p)$, which is $(0, 3)$.

The focal diameter is $4|p|$, which is $4|1| = 4$.

The directrix is the horizontal line $|p|$ units below the vertex: $y = 1$.

The axis of symmetry is $x = 0$.

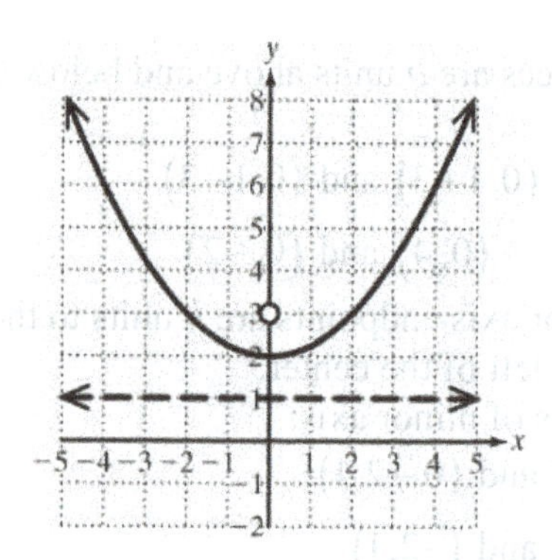

25. Augmented matrix: $\left[\begin{array}{rrr|r} 3 & -4 & 2 & 5 \\ 0 & 5 & -3 & -12 \\ 7 & 0 & 2 & 1 \end{array}\right]$

$$-2R_1+R_3 \to R_1 \longrightarrow \left[\begin{array}{rrr|r} 1 & 8 & -2 & -9 \\ 0 & 5 & -3 & -12 \\ 7 & 0 & 2 & 1 \end{array}\right]$$

$$-7R_1+R_3 \to R_3 \longrightarrow \left[\begin{array}{rrr|r} 1 & 8 & -2 & -9 \\ 0 & 5 & -3 & -12 \\ 0 & -56 & 16 & 64 \end{array}\right]$$

$$R_2 \Leftrightarrow R_3 \longrightarrow \left[\begin{array}{rrr|r} 1 & 8 & -2 & -9 \\ 0 & -56 & 16 & 64 \\ 0 & 5 & -3 & -12 \end{array}\right]$$

$$-R_2-11R_3 \to R_2 \longrightarrow \left[\begin{array}{rrr|r} 1 & 8 & -2 & -9 \\ 0 & 1 & 17 & 68 \\ 0 & 5 & -3 & -12 \end{array}\right]$$

$$\begin{array}{l} -8R_2+R_1 \to R_1 \longrightarrow \\ \\ -5R_2+R_3 \to R_3 \longrightarrow \end{array} \left[\begin{array}{rrr|r} 1 & 0 & -138 & -553 \\ 0 & 1 & 17 & 68 \\ 0 & 0 & -88 & -352 \end{array}\right]$$

$$-\frac{1}{88}R_3 \to R_3 \longrightarrow \left[\begin{array}{rrr|r} 1 & 0 & -138 & -553 \\ 0 & 1 & 17 & 68 \\ 0 & 0 & 1 & 4 \end{array}\right]$$

$$\begin{array}{l} 138R_3+R_1 \to R_1 \longrightarrow \\ -17R_3+R_2 \to R_2 \longrightarrow \\ \end{array} \left[\begin{array}{rrr|r} 1 & 0 & 0 & -1 \\ 0 & 1 & 0 & 0 \\ 0 & 0 & 1 & 4 \end{array}\right]$$

$\{(-1, 0, 4)\}$

27. $2AB = 2\begin{bmatrix} 2 & 4 \\ -1 & 6 \end{bmatrix} \cdot \begin{bmatrix} -3 & 8 \\ 0 & 4 \end{bmatrix}$

$$= 2\begin{bmatrix} 2(-3)+4(0) & 2(8)+4(4) \\ -1(-3)+6(0) & -1(8)+6(4) \end{bmatrix}$$

$$= 2\begin{bmatrix} -6 & 32 \\ 3 & 16 \end{bmatrix}$$

$$= \begin{bmatrix} -12 & 64 \\ 6 & 32 \end{bmatrix}$$

29. $|B| = ad - bc$

$$= (-3)(4) - (0)(8)$$

$$= -12 - 0$$

$$= -12$$

Chapter 8 Sequences, Series, Induction, and Probability

Section 8.1 Sequences and Series

1. sequence; finite

3. alternating

5. $n!$; 1

7. index; summation; lower; limit

9. $a_n = 2n^2 + 3$

$a_1 = 2(1)^2 + 3 = 2 + 3 = 5$

$a_2 = 2(2)^2 + 3 = 8 + 3 = 11$

$a_3 = 2(3)^2 + 3 = 18 + 3 = 21$

$a_4 = 2(4)^2 + 3 = 32 + 3 = 35$

$5, 11, 21, 35$

11. $c_n = 12\left(-\frac{1}{2}\right)^n$

$c_1 = 12\left(-\frac{1}{2}\right)^1 = 12\left(-\frac{1}{2}\right) = -6$

$c_2 = 12\left(-\frac{1}{2}\right)^2 = 12\left(\frac{1}{4}\right) = 3$

$c_3 = 12\left(-\frac{1}{2}\right)^3 = 12\left(-\frac{1}{8}\right) = -\frac{3}{2}$

$c_4 = 12\left(-\frac{1}{2}\right)^4 = 12\left(\frac{1}{16}\right) = \frac{3}{4}$

$-6, 3, -\frac{3}{2}, \frac{3}{4}$

13. $a_n = \sqrt{n+3}$

$a_1 = \sqrt{1+3} = \sqrt{4} = 2$

$a_2 = \sqrt{2+3} = \sqrt{5}$

$a_3 = \sqrt{3+3} = \sqrt{6}$

$a_4 = \sqrt{4+3} = \sqrt{7}$

$2, \sqrt{5}, \sqrt{6}, \sqrt{7}$

15. $a_n = e^{3\ln n}$

$a_1 = e^{3\ln 1} = e^0 = 1$

$a_2 = e^{3\ln 2} = e^{\ln 8} = 8$

$a_3 = e^{3\ln 3} = e^{\ln 27} = 27$

$a_4 = e^{3\ln 4} = e^{\ln 64} = 64$

$1, 8, 27, 64$

17. The odd-numbered terms are negative, and the even-numbered terms are positive.

19. $a_n = 2^n - 1$

$a_{10} = 2^{10} - 1 = 1024 - 1 = 1023$

21. $c_n = 5n - 4$

$c_{157} = 5(157) - 4 = 785 - 4 = 781$

23. c

25. a

27. $b_1 = -3;\ b_n = 3b_{n-1} + 4$

$b_1 = -3$

$b_2 = 3b_1 + 4 = 3(-3) + 4 = -9 + 4 = -5$

$b_3 = 3b_2 + 4 = 3(-5) + 4 = -15 + 4 = -11$

$b_4 = 3b_3 + 4 = 3(-11) + 4 = -33 + 4 = -29$

$b_5 = 3b_4 + 4 = 3(-29) + 4 = -87 + 4 = -83$

$-3, -5, -11, -29, -83$

29. $c_1 = 4;\ c_n = \frac{1}{2}c_{n-1} + 4$

$c_1 = 4$

$c_2 = \frac{1}{2}c_1 + 4 = \frac{1}{2}(4) + 4 = 2 + 4 = 6$

$c_3 = \frac{1}{2}c_2 + 4 = \frac{1}{2}(6) + 4 = 3 + 4 = 7$

$c_4 = \frac{1}{2}c_3 + 4 = \frac{1}{2}(7) + 4 = \frac{7}{2} + 4 = \frac{15}{2}$

$c_5 = \frac{1}{2}c_4 + 4 = \frac{1}{2}\left(\frac{15}{2}\right) + 4 = \frac{15}{4} + 4 = \frac{31}{4}$

$4, 6, 7, \frac{15}{2}, \frac{31}{4}$

31. $a_1 = 5; a_n = \dfrac{1}{a_{n-1}}$

$a_1 = 5$

$a_2 = \dfrac{1}{a_1} = \dfrac{1}{5}$

$a_3 = \dfrac{1}{a_2} = \dfrac{1}{\frac{1}{5}} = 5$

$a_4 = \dfrac{1}{a_3} = \dfrac{1}{5}$

$a_5 = \dfrac{1}{a_4} = \dfrac{1}{\frac{1}{5}} = 5$

$5, \dfrac{1}{5}, 5, \dfrac{1}{5}, 5$

33. a. 1, 1, 2, 3, 5, 8, 13, 21, 34, 55

b. $F_1 = 1, F_2 = 1, F_3 = 2$

35. $7! = 7 \cdot 6 \cdot 5 \cdot 4 \cdot 3 \cdot 2 \cdot 1 = 5040$

37. $0! = 1$

39. $\dfrac{8!}{6!} = \dfrac{8 \cdot 7 \cdot \cancel{6!}}{\cancel{6!}} = 56$

41. $\dfrac{9!}{5! \cdot 4!} = \dfrac{9 \cdot \cancel{8}^{2} \cdot 7 \cdot \cancel{6} \cdot \cancel{5!}}{\cancel{5!} \cdot \cancel{4} \cdot \cancel{3} \cdot \cancel{2} \cdot 1} = 126$

43. $\dfrac{(n-1)!}{(n+1)!} = \dfrac{\cancel{(n-1)!}}{(n+1)(n)\cancel{(n-1)!}} = \dfrac{1}{n(n+1)}$

45. $\dfrac{(2n)!}{(2n+1)!} = \dfrac{\cancel{(2n)!}}{(2n+1)\cancel{(2n)!}} = \dfrac{1}{2n+1}$

47. $a_n = \dfrac{2^n}{(n+2)!}$

$a_5 = \dfrac{2^5}{(5+2)!} = \dfrac{32}{7!} = \dfrac{32}{7 \cdot 6 \cdot 5 \cdot 4 \cdot 3 \cdot 2 \cdot 1} = \dfrac{2}{315}$

49. $c_n = \dfrac{(3n)!}{5n}$

$c_3 = \dfrac{(3 \cdot 3)!}{5(3)} = \dfrac{9!}{15} = \dfrac{9 \cdot 8 \cdot 7 \cdot 6 \cdot \cancel{5} \cdot 4 \cdot \cancel{3} \cdot 2 \cdot 1}{\cancel{15}}$

$= 24{,}192$

51. $\sum_{i=1}^{6}(3i-4)$

$= [3(1)-4]+[3(2)-4]+[3(3)-4]$

$+[3(4)-4]+[3(5)-4]+[3(6)-4]$

$= -1+2+5+8+11+14 = 39$

53. $\sum_{j=1}^{5}(-2j^2) = [-2(1)^2]+[-2(2)^2]+[-2(3)^2]$

$+[-2(4)^2]+[-2(5)^2]$

$= -2-8-18-32-50 = -110$

55. $\sum_{j=2}^{5}\left(\dfrac{1}{2}\right)^j = \left(\dfrac{1}{2}\right)^2+\left(\dfrac{1}{2}\right)^3+\left(\dfrac{1}{2}\right)^4+\left(\dfrac{1}{2}\right)^5$

$= \dfrac{1}{4}+\dfrac{1}{8}+\dfrac{1}{16}+\dfrac{1}{32} = \dfrac{15}{32}$

57. $\sum_{j=1}^{20} 6 = \underbrace{6+6+6+\cdots+6}_{\text{20 terms}} = 20(6) = 120$

59. $\sum_{k=3}^{7}(-1)^k(6k)$

$= [(-1)^3(6 \cdot 3)]+[(-1)^4(6 \cdot 4)]$

$+[(-1)^5(6 \cdot 5)]+[(-1)^6(6 \cdot 6)]$

$+[(-1)^7(6 \cdot 7)]$

$= -18+24-30+36-42 = -30$

61. $\sum_{m=1}^{3}\dfrac{m+1}{m} = \left(\dfrac{1+1}{1}\right)+\left(\dfrac{2+1}{2}\right)+\left(\dfrac{3+1}{3}\right)$

$= 2+\dfrac{3}{2}+\dfrac{4}{3} = \dfrac{29}{6}$

63. $\sum_{k=1}^{3}(k+2)(k+3)$

$=[(1+2)(1+3)]+[(2+2)(2+3)]$
$+[(3+2)(3+3)]$

$=12+20+30=62$

65. a. If n is even:

$$\sum_{i=1}^{n}(-1)^i=(-1)^1+(-1)^2+\cdots+(-1)^{n-1}+(-1)^n$$
$$=\underbrace{-1+1+\cdots+-1+1}_{\text{equal number of }-1\text{s and 1s}}=0$$

b. If n is odd:

$$\sum_{i=1}^{n}(-1)^i=(-1)^1+(-1)^2+\cdots+(-1)^{n-2}+(-1)^{n-1}+(-1)^n$$
$$=\underbrace{-1+1+\cdots-1+1}_{\text{equal number of }-1\text{s and 1s}}-1=-1$$

67. $\sum_{i=1}^{5}(n^2-2n)=[(1)^2-2(1)]+[(2)^2-2(2)]$
$+[(3)^2-2(3)]+[(4)^2-2(4)]$
$+[(5)^2-2(5)]$

$=-1+0+3+8+15=25$

69. $a_n=2^n$

71. $a_n=\dfrac{n+1}{3n}$

73. $a_n=(-1)^n(n^2)$

75. The series consists of n terms where a formula for the nth term is given.

$$\sum_{i=1}^{n}\frac{i^2}{i+1}$$

77. This is the sum of the integers 1 through 5.

$$\sum_{i=1}^{n}i$$

79. This is the sum of four eights. $\sum_{i=1}^{4}8$

81. The numerator is always 1. The denominator is 3 raised to the power of the term number: 3^i. The odd-numbered terms are positive. This can be generated by the factor $(-1)^{i+1}$.

$$\sum_{i=1}^{4}(-1)^{i+1}\frac{1}{3^i}$$

83. Each term is c raised to two more than the power of the term number: c^{i+2}.

$$\sum_{i=1}^{18}c^{i+2}$$

85. The numerator is x raised to the power of the term number: x^i. The denominator is factorial of the term number: $i!$.

$$\sum_{i=1}^{4}\frac{x^i}{i!}$$

87. $\sum_{i=1}^{8}i^2=\sum_{j=0}^{7}(j+1)^2=\sum_{k=2}^{9}(k-1)^2$

89. Expanding the series, there are n terms and each term is c. Therefore, the sum is cn.

$$\sum_{i=1}^{n}c=c+c+c+\cdots+c=cn$$

91. $\sum_{i=1}^{50}(i^2+3i)=\sum_{i=1}^{50}i^2+\sum_{i=1}^{50}(3i)=\sum_{i=1}^{50}i^2+3\sum_{i=1}^{50}i$

$=42{,}925+3(1275)=46{,}750$

93. $\sum_{i=1}^{50}(5i+4)=\sum_{i=1}^{50}5i+\sum_{i=1}^{50}4=5\sum_{i=1}^{50}i+\sum_{i=1}^{50}4$

$=5(1275)+4(50)=6575$

95. True

97. False

99. $a_n=1.03a_{n-1}+1000;\ n\ge 2$

101. $a_n=\dfrac{1}{2}n(n-1)$

$a_{12}=\dfrac{1}{2}(12)(12-1)=6(11)=66$

If 12 people are present at the meeting, then there will be 66 handshakes.

103. **a.** $\{31, 10, 17, 17, 11, 21\}$

b. $\frac{1}{6}\sum_{n=1}^{6} a_n = \frac{1}{6}(31+10+17+17+11+21)$

$= \frac{1}{6}(107) \approx 17.8$

The mean (average) number of tropical storms and hurricanes in the Atlantic was 17.8 during this period.

105. $\sum_{i=1}^{n}(a_i - \bar{a}) = \sum_{i=1}^{n} a_i - \sum_{i=1}^{n} \bar{a} = \sum_{i=1}^{n} a_i - n\bar{a}$

$= \sum_{i=1}^{n} a_i - n \cdot \frac{1}{n}\sum_{i=1}^{n} a_i = \sum_{i=1}^{n} a_i - \sum_{i=1}^{n} a_i$

$= 0$

107. A sequence is an ordered list of terms. A series is the sum of the terms of a sequence.

109. Write 1206! as (1206)(1205)(1204!) and then simplify the fraction before multiplying. The result is the product (1206)(1205) = 1,453,230.

111. $a_n = \frac{1}{2}\left(a_{n-1} + \frac{x}{a_{n-1}}\right)$

$a_1 = x = 2$

$a_2 = \frac{1}{2}\left(2 + \frac{2}{2}\right) = \frac{1}{2}(3) = \frac{3}{2} = 1.5$

$a_3 = \frac{1}{2}\left(\frac{3}{2} + \frac{2}{\frac{3}{2}}\right) = \frac{17}{12} \approx 1.4167$

$a_4 = \frac{1}{2}\left(\frac{17}{12} + \frac{2}{\frac{17}{12}}\right) = \frac{577}{408} \approx 1.4142$

$2, 1.5, 1.4167, 1.4142$

113.

```
Expr:12*(-1/2)^
Variable:n
start:1
end:4
step:1
Paste
```

```
seq(12*(-1/2)^n,
n,1,4,1)▸Frac
  {-6 3 -3/2 3/4}
```

115.

```
 7
 Σ ((-1)^n(6n))
n=3
                 -30
```

```
Σ((-1)^n*(6n),n,
3,7)
                 -30
```

117.

```
10
 Σ (1/n!)
n=1
      1.718281801
```

```
50
 Σ (1/n!)
n=1
      1.718281828
```

```
Σ(1/n!,n,1,10)
      1.718281801
Σ(1/n!,n,1,50)
      1.718281828
```

Section 8.2 Arithmetic Sequences and Series

1. $a_1 = 2;\ a_n = (a_{n-1})^2$

$a_1 = 2$

$a_2 = (a_1)^2 = (2)^2 = 4$

$a_3 = (a_2)^2 = (4)^2 = 16$

$a_4 = (a_3)^2 = (16)^2 = 256$

3. The numerator is 5 times the term number: $5n$. The denominator is 3 raised to the power of the term number: 3^n. $a_n = \frac{5n}{3^n}$

5. $\sum_{i=1}^{5}(2i^2) = 2(1)^2 + 2(2)^2 + 2(3)^2 + 2(4)^2 + 2(5)^2$

$= 2 + 8 + 18 + 32 + 50 = 110$

7. arithmetic

9. $a_1 + (n-1)d$

11. partial

13. $a_2 - a_1 = 19 - 15 = 4$

$a_3 - a_2 = 23 - 19 = 4$

$a_4 - a_3 = 27 - 23 = 4$

Yes; $d = 4$

15. $a_2 - a_1 = -2 - 9 = -11$

$a_3 - a_2 = -13 - (-2) = -11$

$a_4 - a_3 = -24 - (-13) = -11$

Yes; $d = -11$

17. $a_2 - a_1 = 22 - 18 = 4$

$a_3 - a_2 = 27 - 22 = 5$

$a_4 - a_3 = 33 - 27 = 6$

No

19. $a_2 - a_1 = \frac{14}{3} - 4 = \frac{14}{3} - \frac{12}{3} = \frac{2}{3}$

$a_3 - a_2 = \frac{16}{3} - \frac{14}{3} = \frac{2}{3}$

$a_4 - a_3 = 6 - \frac{16}{3} = \frac{18}{3} - \frac{16}{3} = \frac{2}{3}$

Yes; $d = \frac{2}{3}$

21. $3, 13, 23, 33, 43$ (differences: $+10$, $+10$, $+10$, $+10$)

23. $4, 2, 0, -2, -4$ (differences: -2, -2, -2, -2)

25. a. $a_n = a_1 + (n-1)d$

$a_n = -12 + (n-1)(5)$

$a_n = -12 + 5n - 5$

$a_n = 5n - 17$

b. $a_{20} = 5(20) - 17 = 100 - 17 = 83$

27. a. $a_n = a_1 + (n-1)d$

$a_n = \frac{1}{2} + (n-1)\left(\frac{1}{3}\right)$

$a_n = \frac{1}{2} + \frac{1}{3}n - \frac{1}{3}$

$a_n = \frac{1}{3}n + \frac{1}{6}$

b. $a_{10} = \frac{1}{3}(10) + \frac{1}{6} = \frac{10}{3} + \frac{1}{6} = \frac{21}{6} = \frac{7}{2}$

29. $a_n = a_1 + (n-1)d$

$a_n = 8 + 22 + (n-1)(22)$

$a_n = 30 + 22n - 22$

$a_n = 22n + 8$

31. a. $a_2 - a_1 = 50 - 34 = 16$

$a_3 - a_2 = 66 - 50 = 16$

$a_4 - a_3 = 82 - 66 = 16$

Yes

b. $a_n = a_1 + (n-1)d$

$a_n = 34 + (n-1)(16)$

$a_n = 34 + 16n - 16$

$a_n = 16n + 18$

c. $a_{10} = 16(10) + 18 = 160 + 18 = 178$

33. $v_{n+1} - v_n = [v_0 + a(n+1)] - (v_0 + an)$

$= v_0 + an + a - v_0 - an$

$= a$

The difference between two consecutive terms is the constant a. Therefore, the sequence is arithmetic.

35. a. $d = \frac{a_{10} - a_0}{10} = \frac{3133 - 2583}{10} = 55$

$a_n = a_1 + (n-1)d$

$a_n = 2583 + 55 + (n-1)(55)$

$a_n = 2583 + 55 + 55n - 55$

$a_n = 55n + 2583$, where a_n is represented in 1000s, and $n \geq 0$.

b. $a_5 = 55(5) + 2583 = 275 + 2583 = 2858$

Approximately 2,858,000 people were employed in nursing and residential care facilities in the year 2005.

37. $a_n = a_1 + (n-1)d$

$68 = -2 + (15-1)d$

$70 = 14d$

$d = 5$

$a_n = -2 + (n-1)(5)$

$a_n = -2 + 5n - 5$

$a_n = 5n - 7$

$a_8 = 5(8) - 7 = 40 - 7 = 33$

39. $a_n = a_1 + (n-1)d$

$-265 = 50 + (22-1)d$

$-315 = 21d$

$d = -15$

$a_n = 50 + (n-1)(-15)$

$a_n = 50 - 15n + 15$

$a_n = -15n + 65$

$a_{35} = -15(35) + 65 = -525 + 65 = -460$

41. $a_n = a_1 + (n-1)d$

$320 = 8 + (n-1)(6)$

$320 = 8 + 6n - 6$

$320 = 2 + 6n$

$318 = 6n$

$n = 53$

43. $a_n = a_1 + (n-1)d$

$-3.4 = 11 + (n-1)(-0.3)$

$-3.4 = 11 - 0.3n + 0.3$

$-3.4 = 11.3 - 0.3n$

$-14.7 = -0.3n$

$n = 49$

45. $a_n = a_1 + (n-1)d$ $\qquad$ $a_n = a_1 + (n-1)d$

$148 = a_1 + (14-1)d$ $\qquad$ $316 = a_1 + (35-1)d$

$148 = a_1 + 13d$ $\qquad$ $316 = a_1 + 34d$

$a_1 = 148 - 13d$ $\qquad$ $a_1 = 316 - 34d$

$148 - 13d = 316 - 34d$

$21d = 168$

$d = 8$

$a_1 = 148 - 13d = 148 - 13(8) = 148 - 104 = 44$

47. $S_n = \frac{n}{2}(a_1 + a_n)$

$S_{12} = \frac{12}{2}(4 + 37) = 6(41) = 246$

49. $a_n = a_1 + (n-1)d$

$a_n = 1 + (n-1)(5)$

$a_n = 1 + 5n - 5$

$a_n = 5n - 4$

$a_{40} = 5(40) - 4 = 200 - 4 = 196$

$S_n = \frac{n}{2}(a_1 + a_n)$

$S_{40} = \frac{40}{2}(1 + 196) = 20(197) = 3940$

51. $a_n = a_1 + (n-1)d$

$-30.5 = 5 + (n-1)(-0.5)$

$-30.5 = 5 - 0.5n + 0.5$

$-30.5 = 5.5 - 0.5n$

$0.5n = 36$

$n = 72$

$S_n = \frac{n}{2}(a_1 + a_n)$

$S_{72} = \frac{72}{2}[5 + (-30.5)] = 36(-25.5) = -918$

53. $a_n = a_1 + (n-1)d$

$-6 = a_1 + (3-1)(-3)$

$-6 = a_1 + (2)(-3)$

$-6 = a_1 - 6$

$a_1 = 0$

$a_2 = a_1 - 3 = 0 - 3 = -3$

55. $\sum_{j=1}^{18}(j+6)$
$=(1+6)+(2+6)+(3+6)+\cdots+(18+6)$
$=7+8+9+\cdots+24$
$a_1=7,\ a_{18}=24$
$S_n=\frac{n}{2}(a_1+a_n)$
$S_{18}=\frac{18}{2}(7+24)=9(31)=279$

57. $\sum_{i=1}^{50}(2i+6)=[2(1)+6]+[2(2)+6]$
$+[2(3)+6]+\cdots+[2(50)+6]$
$=8+10+12+\cdots+106$
$a_1=8,\ a_{50}=106$
$S_n=\frac{n}{2}(a_1+a_n)$
$S_{50}=\frac{50}{2}(8+106)=25(114)=2850$

59. $\sum_{k=1}^{162}\left(3-\frac{1}{2}k\right)=\left[3-\frac{1}{2}(1)\right]+\left[3-\frac{1}{2}(2)\right]$
$+\left[3-\frac{1}{2}(3)\right]+\cdots+\left[3-\frac{1}{2}(162)\right]$
$=\frac{5}{2}+2+\frac{3}{2}+\cdots+(-78)$
$a_1=\frac{5}{2},\ a_{162}=-78$
$S_n=\frac{n}{2}(a_1+a_n)$
$S_{162}=\frac{162}{2}\left(\frac{5}{2}-78\right)=81\left(-\frac{151}{2}\right)=-6115.5$

61. $a_n=a_1+(n-1)d$
$49=-1+(n-1)(5)$
$49=-1+5n-5$
$5n=55$
$n=11$
$S_n=\frac{n}{2}(a_1+a_n)$
$S_{11}=\frac{11}{2}(-1+49)=\frac{11}{2}(48)=264$

63. $a_n=a_1+(n-1)d$
$-39=-7+(n-1)(-4)$
$-39=-7-4n+4$
$4n=36$
$n=9$
$S_n=\frac{n}{2}(a_1+a_n)$
$S_9=\frac{9}{2}(-7-39)=\frac{9}{2}(-46)=-207$

65. a. Job 1:
$a_n=a_1+(n-1)d$
$a_n=64{,}000+(n-1)(3200)$
$a_n=64{,}000+3200n-3200$
$a_n=3200n+60{,}800$
$a_5=3200(5)+60{,}800=16{,}000+60{,}800$
$=76{,}800$
$S_n=\frac{n}{2}(a_1+a_n)$
$S_5=\frac{5}{2}(64{,}000+76{,}800)$
$=\frac{5}{2}(140{,}800)=\$352{,}000$
Job 2:
$a_n=a_1+(n-1)d$
$a_n=60{,}000+(n-1)(5000)$
$a_n=60{,}000+5000n-5000$
$a_n=5000n+55{,}000$
$a_5=5000(5)+55{,}000=25{,}000+55{,}000$
$=80{,}000$
$S_n=\frac{n}{2}(a_1+a_n)$
$S_5=\frac{5}{2}(60{,}000+80{,}000)$
$=\frac{5}{2}(140{,}000)=\$350{,}000$

b. Job 1:
$a_{10}=3200(10)+60{,}800=32{,}000+60{,}800$
$=92{,}800$

$S_n = \frac{n}{2}(a_1 + a_n)$

$S_{10} = \frac{10}{2}(64{,}000 + 92{,}800)$

$= 5(156{,}800) = \$784{,}000$

Job 2:

$a_{10} = 5000(10) + 55{,}000 = 50{,}000 + 55{,}000$

$= 105{,}000$

$S_n = \frac{n}{2}(a_1 + a_n)$

$S_{10} = \frac{10}{2}(60{,}000 + 105{,}000)$

$= 5(165{,}000) = \$825{,}000$

67. a. $d_n = d_1 + (n-1)d$

$d_n = 16 + (n-1)(32)$

$d_n = 16 + 32n - 32$

$d_n = 32n - 16$

b. $d_8 = 32(8) - 16 = 256 - 16 = 240$ ft

c. $S_n = \frac{n}{2}(a_1 + a_n)$

$S_8 = \frac{8}{2}(16 + 240) = 4(256) = 1024$ ft

69. a. $a_n = a_1 + (n-1)d$

$a_n = 15 + (n-1)(-1)$

$a_n = 15 - n + 1$

$a_n = -n + 16$

$a_{15} = -(15) + 16 = 1$ cup

b. $S_n = \frac{n}{2}(a_1 + a_n)$

$S_{15} = \frac{15}{2}(15 + 1) = \frac{15}{2}(16) = 120$ cups

71. a. 6, 4, 2, 0

b. $b_n = b_1 + (n-1)d$

$b_n = 6 + (n-1)(-2)$

$b_n = 6 - 2n + 2$

$b_n = -2n + 8$

c. $b_{30} = -2(30) + 8 = -60 + 8 = -52$

d. $S_n = \frac{n}{2}(a_1 + a_n)$

$S_{30} = \frac{30}{2}[6 + (-52)] = 15(-46) = -690$

e. $b_n = -2n + 8$

$-180 = -2n + 8$

$2n = 188$

$n = 94$

f. $b_{88} = -2(88) + 8 = -176 + 8 = -168$

$b_{20} = -2(20) + 8 = -40 + 8 = -32$

$b_{88} - b_{20} = -168 - (-32) = -136$

73. $256 - (-20) + 1 = 277$ integers

$a_n = a_1 + (n-1)d$

$256 = -20 + (n-1)(1)$

$256 = -20 + n - 1$

$256 = n - 21$

$n = 277$

$S_n = \frac{n}{2}(a_1 + a_n)$

$S_{277} = \frac{277}{2}(-20 + 256) = \frac{277}{2}(236) = 32{,}686$

75. $a_n = a_1 + (n-1)d$

$a_{50} = 5 + (50-1)(5) = 5 + 49(5) = 250$

$S_n = \frac{n}{2}(a_1 + a_n)$

$S_{50} = \frac{50}{2}(5 + 250) = 25(255) = 6375$

77. $a_n = a_1 + (n-1)d$

$99 = 42 + (n-1)(3)$

$99 = 42 + 3n - 3$

$99 = 3n + 39$

$3n = 60$

$n = 20$

$S_n = \frac{n}{2}(a_1 + a_n)$

$S_{20} = \frac{20}{2}(42 + 99) = 10(141) = 1410$

79. a. Substitute $a_1 = 1$ and $a_n = n$.

$$S_n = \frac{n}{2}(a_1 + a_n) = \frac{n}{2}(1+n)$$

b. $S_{100} = \frac{100}{2}(1+100) = 50(101) = 5050$

c. $S_{1000} = \frac{1000}{2}(1+1000)$

$= 500(1001) = 500{,}500$

81. $a_n = a_1 + (n-1)d$

$28 = 4 + (5-1)d$

$28 = 4 + 4d$

$4d = 24$

$d = 6$

$a_2 = a_1 + (2-1)(6) = 4 + 6 = 10$

$a_3 = a_1 + (3-1)(6) = 4 + 12 = 16$

$a_4 = a_1 + (4-1)(6) = 4 + 18 = 22$

The three arithmetic means between 4 and 28 are 10, 16, and 22.

83. The terms in the given sequence $a_1, a_2, a_3, \ldots$ differ by d units. Therefore, every other term would differ by $2d$ units. Thus, $a_1, a_3, a_5, \ldots$ is an arithmetic sequence with common difference $2d$.

85. $a_n = n^2 + 2n - \left[(n-1)^2 + 2(n-1)\right]$

$= n^2 + 2n - \left[(n-1)(n-1) + 2(n-1)\right]$

$= n^2 + 2n - (n-1+2)(n-1)$

$= n^2 + 2n - (n+1)(n-1)$

$= n^2 + 2n - (n^2 - 1)$

$= 2n + 1$

The sequence is $\{3, 5, 7, \ldots, (2n+1), \ldots\}$.

87.

```
Expr:5n-17
Variable:n
start:1
end:5
step:1
Paste
```

```
seq(5n-17,n,1,5,
1)
 {-12 -7 -2 3 8}
```

89.

```
 50
 Σ (2n+6)
n=1
                2850
```

```
Σ(2n+6,n,1,50)
              2850
```

Section 8.3 Geometric Sequences and Series

1. $a_n = a_1 + (n-1)d$

$29 = 5 + (5-1)d$

$24 = 4d$

$d = 6$

$a_n = 5 + (n-1)(6)$

$a_n = 5 + 6n - 6$

$a_n = 6n - 1$

$a_{46} = 6(46) - 1 = 276 - 1 = 275$

3. $\sum_{i=1}^{82}(5-7i) = [5-7(1)] + [5-7(2)]$

$+ [5-7(3)] + \cdots + [5-7(82)]$

$= (-2) + (-9) + (-16) + \cdots + (-569)$

$a_1 = -2,\ a_{82} = -569$

$S_n = \frac{n}{2}(a_1 + a_n)$

$S_{82} = \frac{82}{2}(-2-569) = 41(-571) = -23{,}411$

5. $a_n = a_1 + (n-1)d$

$622 = 28 + (n-1)(11)$

$622 = 28 + 11n - 11$

$622 = 17 + 11n$

$11n = 605$

$n = 55$

$S_n = \frac{n}{2}(a_1 + a_n)$

$S_{55} = \frac{55}{2}(28 + 622) = \frac{55}{2}(650) = 17{,}875$

7. geometric; ratio

9. finite

11. 0

13. annuity

15. $\frac{a_2}{a_1} = \frac{18}{6} = 3$

$\frac{a_3}{a_2} = \frac{54}{18} = 3$

$\frac{a_4}{a_3} = \frac{162}{54} = 3$

Yes; $r = 3$

17. $\frac{a_2}{a_1} = \frac{\frac{7}{2}}{-7} = -\frac{1}{2}$

$\frac{a_3}{a_2} = \frac{-\frac{7}{4}}{\frac{7}{2}} = -\frac{1}{2}$

$\frac{a_4}{a_3} = \frac{\frac{7}{8}}{-\frac{7}{4}} = -\frac{1}{2}$

Yes; $r = -\frac{1}{2}$

19. $\frac{a_2}{a_1} = \frac{12}{3} = 4$

$\frac{a_3}{a_2} = \frac{60}{12} = 5$

$\frac{a_4}{a_3} = \frac{360}{60} = 6$

No

21. $\frac{a_2}{a_1} = \frac{5}{\sqrt{5}} = \sqrt{5}$

$\frac{a_3}{a_2} = \frac{5\sqrt{5}}{5} = \sqrt{5}$

$\frac{a_4}{a_3} = \frac{25}{5\sqrt{5}} = \sqrt{5}$

Yes; $r = \sqrt{5}$

23. $\frac{a_2}{a_1} = \frac{\frac{4}{t}}{2} = \frac{2}{t}$

$\frac{a_3}{a_2} = \frac{\frac{8}{t^2}}{\frac{4}{t}} = \frac{8}{t^2} \cdot \frac{t}{4} = \frac{2}{t}$

$\frac{a_4}{a_3} = \frac{\frac{16}{t^3}}{\frac{8}{t^2}} = \frac{16}{t^3} \cdot \frac{t^2}{8} = \frac{2}{t}$

Yes; $r = \frac{2}{t}$

25. $a_1 = 7$

$a_2 = 7(2) = 14$

$a_3 = 14(2) = 28$

$a_4 = 28(2) = 56$

$a_5 = 56(2) = 112$

7, 14, 28, 56, 112

27. $a_1 = 24$

$a_2 = 24\left(-\frac{2}{3}\right) = -16$

$a_3 = -16\left(-\frac{2}{3}\right) = \frac{32}{3}$

$a_4 = \frac{32}{3}\left(-\frac{2}{3}\right) = -\frac{64}{9}$

$a_5 = -\frac{64}{9}\left(-\frac{2}{3}\right) = \frac{128}{27}$

$24, -16, \frac{32}{3}, -\frac{64}{9}, \frac{128}{27}$

29. $a_1 = 36$

$a_2 = \frac{1}{2}(36) = 18$

$a_3 = \frac{1}{2}(18) = 9$

$a_4 = \frac{1}{2}(9) = \frac{9}{2}$

$a_5 = \frac{1}{2}\left(\frac{9}{2}\right) = \frac{9}{4}$

$36, 18, 9, \frac{9}{2}, \frac{9}{4}$

31. $r = \frac{a_2}{a_1} = \frac{10}{5} = 2$

$a_n = a_1 r^{n-1}$

$a_n = 5(2)^{n-1}$

33. $r = \frac{a_2}{a_1} = \frac{-1}{-2} = \frac{1}{2}$

$a_n = a_1 r^{n-1}$

$a_n = -2\left(\frac{1}{2}\right)^{n-1}$

35. $r = \frac{a_2}{a_1} = \frac{4}{\frac{16}{3}} = 4 \cdot \frac{3}{16} = \frac{3}{4}$

$a_n = a_1 r^{n-1}$

$a_n = \frac{16}{3}\left(\frac{3}{4}\right)^{n-1}$

37. a. $a_1 = 100{,}000(0.85)$

$a_n = a_1 r^{n-1}$

$a_n = 100{,}000(0.85)(0.85)^{n-1}$

$a_n = 100{,}000(0.85)^n$

b. $a_5 = 100{,}000(0.85)^5 \approx \$44{,}000$

39. $a_1 = 24(1.3)$

$a_n = a_1 r^{n-1}$

$a_n = 24(1.3)(1.3)^{n-1}$

$a_n = 24(1.3)^n$

$a_{10} = 24(1.3)^{10} \approx 331$ cases

41. $r = \frac{a_2}{a_1} = \frac{-8}{12} = -\frac{2}{3}$

$a_n = a_1 r^{n-1}$

$a_n = 12\left(-\frac{2}{3}\right)^{n-1}$

$a_6 = 12\left(-\frac{2}{3}\right)^{6-1} = 12\left(-\frac{2}{3}\right)^5 = -\frac{128}{81}$

43. $a_n = a_1 r^{n-1}$

$16 = 2r^{4-1}$

$8 = r^3$

$r = 2$

$a_n = 2(2)^{n-1} = 2^n$

$a_{12} = 2^{12} = 4096$

45. $a_n = a_1 r^{n-1}$

$-6 = a_1\left(\frac{1}{2}\right)^{2-1}$

$-6 = a_1\left(\frac{1}{2}\right)$

$a_1 = -12$

$a_7 = -12\left(\frac{1}{2}\right)^{7-1} = -12\left(\frac{1}{2}\right)^6 = -\frac{3}{16}$

47. $a_n = a_1 r^{n-1}$

$-\frac{16}{9} = a_1\left(-\frac{2}{3}\right)^{5-1}$

$-\frac{16}{9} = a_1\left(-\frac{2}{3}\right)^4$

$a_1 = -\frac{16}{9}\left(\frac{81}{16}\right) = -9$

49. $r=\frac{a_3}{a_2}=\frac{75}{15}=5$

$a_n=a_1r^{n-1}$

$15=a_1(5)^{2-1}$

$15=a_1(5)^1$

$a_1=3$

51. $a_n=a_1r^{n-1}$

$18=a_1r^{2-1}$

$18=a_1r$

$a_1=\frac{18}{r}$

$a_n=a_1r^{n-1}$

$144=a_1r^{5-1}$

$144=a_1r^4$

$144=a_1r^4$

$144=\frac{18}{r}r^4$

$144=18r^3$

$8=r^3$

$r=2$

$a_1=\frac{18}{r}=\frac{18}{2}=9$

53. $a_n=a_1r^{n-1}$

$72=a_1r^{3-1}$

$72=a_1r^2$

$a_1=\frac{72}{r^2}$

$a_n=a_1r^{n-1}$

$-\frac{243}{8}=a_1r^{6-1}$

$-\frac{243}{8}=a_1r^5$

$-\frac{243}{8}=a_1r^5$

$-\frac{243}{8}=\frac{72}{r^2}r^5$

$-\frac{243}{8}=72r^3$

$-\frac{27}{64}=r^3$

$r=-\frac{3}{4}$

$a_1=\frac{72}{r^2}=\frac{72}{\left(-\frac{3}{4}\right)^2}=\frac{72}{\frac{9}{16}}=128$

55. $a_1=3, n=10, r=2$

$S_n=\frac{a_1(1-r^n)}{1-r}$

$S_{10}=\frac{3\left[1-(2)^{10}\right]}{1-2}=\frac{3(1-1024)}{-1}=3069$

57. $a_1=6, n=7, r=\frac{2}{3}$

$S_n=\frac{a_1(1-r^n)}{1-r}$

$S_7=\frac{6\left[1-\left(\frac{2}{3}\right)^7\right]}{1-\frac{2}{3}}=\frac{6\left(1-\frac{128}{2187}\right)}{\frac{1}{3}}=\frac{4118}{243}$

59. $a_1=15, n=6, r=\frac{a_2}{a_1}=\frac{5}{15}=\frac{1}{3}$

$S_n=\frac{a_1(1-r^n)}{1-r}$

$S_6=\frac{15\left[1-\left(\frac{1}{3}\right)^6\right]}{1-\frac{1}{3}}=\frac{15\left(1-\frac{1}{729}\right)}{\frac{2}{3}}=\frac{1820}{81}$

61. $r=\frac{a_2}{a_1}=\frac{6}{2}=3$

$a_n=2(3)^{n-1}$

$13{,}122=2(3)^{n-1}$

$6561=(3)^{n-1}$

$(3)^8=(3)^{n-1}$

$8=n-1$

$n=9$

$S_n=\frac{a_1(1-r^n)}{1-r}$

$S_9=\frac{2\left[1-(3)^9\right]}{1-3}=\frac{2(1-19{,}683)}{-2}=19{,}682$

63. $r=\dfrac{a_2}{a_1}=\dfrac{\frac{2}{3}}{1}=\dfrac{2}{3}$

$a_n=1\left(\dfrac{2}{3}\right)^{n-1}$

$\dfrac{32}{243}=\left(\dfrac{2}{3}\right)^{n-1}$

$\left(\dfrac{2}{3}\right)^5=(3)^{n-1}$

$5=n-1$

$n=6$

$S_n=\dfrac{a_1(1-r^n)}{1-r}$

$S_6=\dfrac{1\left[1-\left(\frac{2}{3}\right)^6\right]}{1-\frac{2}{3}}=\dfrac{1-\frac{64}{729}}{\frac{1}{3}}=\dfrac{665}{243}$

65. $a_1=1, r=\dfrac{a_2}{a_1}=\dfrac{\frac{1}{5}}{1}=\dfrac{1}{5}$

$|r|=\left|\dfrac{1}{5}\right|<1$

$S=\dfrac{a_1}{1-r}=\dfrac{1}{1-\frac{1}{5}}=\dfrac{1}{\frac{4}{5}}=\dfrac{5}{4}$

67. $a_1=-2, r=\dfrac{a_2}{a_1}=\dfrac{-\frac{1}{2}}{-2}=\dfrac{1}{4}$

$|r|=\left|\dfrac{1}{4}\right|<1$

$S=\dfrac{a_1}{1-r}=\dfrac{-2}{1-\frac{1}{4}}=\dfrac{-2}{\frac{3}{4}}=-\dfrac{8}{3}$

69. $a_1=2, r=\dfrac{a_2}{a_1}=\dfrac{8}{2}=4$

$|r|=|4|>1$; The sum does not exist.

71. $a_1=1, r=\dfrac{2}{3}$

$|r|=\left|\dfrac{2}{3}\right|<1$

$S=\dfrac{a_1}{1-r}=\dfrac{1}{1-\frac{2}{3}}=\dfrac{1}{\frac{1}{3}}=3$

73. $a_1=1, r=\dfrac{3}{2}$

$|r|=\left|\dfrac{3}{2}\right|>1$; The sum does not exist.

75. $\displaystyle\sum_{i=1}^{12}4(2)^i=\sum_{i=1}^{12}4\cdot 2(2)^{i-1}=\sum_{i=1}^{12}8(2)^{i-1}$

$a_1=8, n=12, r=2$

$S_n=\dfrac{a_1(1-r^n)}{1-r}$

$S_{12}=\dfrac{8\left[1-(2)^{12}\right]}{1-2}=\dfrac{8(1-4096)}{-1}=32{,}760$

77. $\displaystyle\sum_{n=3}^{\infty}4\left(\frac{1}{2}\right)^{n-1}=\sum_{n=1}^{\infty}4\left(\frac{1}{2}\right)^{n-1}-4\left(\frac{1}{2}\right)^{1-1}-4\left(\frac{1}{2}\right)^{2-1}$

$\displaystyle=\sum_{n=1}^{\infty}4\left(\frac{1}{2}\right)^{n-1}-4-2$

$\displaystyle=\sum_{n=1}^{\infty}4\left(\frac{1}{2}\right)^{n-1}-6$

$a_1=4, r=\dfrac{1}{2}$

$|r|=\left|\dfrac{1}{2}\right|<1$

$S=\dfrac{a_1}{1-r}=\dfrac{4}{1-\frac{1}{2}}=\dfrac{4}{\frac{1}{2}}=8$

$\displaystyle\sum_{n=3}^{\infty}4\left(\frac{1}{2}\right)^{n-1}=\sum_{n=1}^{\infty}4\left(\frac{1}{2}\right)^{n-1}-6=8-6=2$

79. $0.6\overline{4}=0.6444\ldots$

$=\dfrac{6}{10}+\underbrace{\left[\dfrac{4}{100}+\dfrac{4}{1000}+\dfrac{4}{10{,}000}+\ldots\right]}_{\text{Infinite geometric series}}$

$a_1 = \frac{4}{100} = \frac{1}{25}, r = \frac{1}{10}$

$S = \frac{a_1}{1-r} = \frac{\frac{1}{25}}{1-\frac{1}{10}} = \frac{\frac{1}{25}}{\frac{9}{10}} = \frac{1}{25} \cdot \frac{10}{9} = \frac{2}{45}$

$0.6\overline{4} = \frac{6}{10} + \frac{2}{45} = \frac{29}{45}$

81. $0.\overline{81} = 0.818181\ldots$

$= \underbrace{\left[\frac{81}{100} + \frac{81}{10{,}000} + \frac{81}{1{,}000{,}000} + \ldots\right]}_{\text{Infinite geometric series}}$

$a_1 = \frac{81}{100}, r = \frac{1}{100}$

$S = \frac{a_1}{1-r} = \frac{\frac{81}{100}}{1-\frac{1}{100}} = \frac{\frac{81}{100}}{\frac{99}{100}} = \frac{81}{100} \cdot \frac{100}{99} = \frac{9}{11}$

$0.\overline{81} = \frac{9}{11}$

83. $3.4\overline{25} = 3.4252525\ldots$

$= 3 + \frac{4}{10} + \underbrace{\left[\frac{25}{1000} + \frac{25}{100{,}000} + \ldots\right]}_{\text{Infinite geometric series}}$

$a_1 = \frac{25}{1000} = \frac{1}{40}, r = \frac{1}{100}$

$S = \frac{a_1}{1-r} = \frac{\frac{1}{40}}{1-\frac{1}{100}} = \frac{\frac{1}{40}}{\frac{99}{100}} = \frac{1}{40} \cdot \frac{100}{99} = \frac{5}{198}$

$3.4\overline{25} = 3 + \frac{4}{10} + \frac{5}{198} = \frac{3391}{990}$

85. a. $500{,}000(\$300) = \$150{,}000{,}000$

b. $a_1 = 150{,}000{,}000, r = 0.68$

$S = \frac{a_1}{1-r} = \frac{150{,}000{,}000}{1-0.68} = \frac{150{,}000{,}000}{0.32}$
$= \$468{,}750{,}000$

87. $a_1 = 11{,}520, r = 0.64$

$S = \frac{a_1}{1-r} = \frac{11{,}520}{1-0.64} = \frac{11{,}520}{0.36} = \$32{,}000$

89. $A = \frac{P\left[\left(1+\frac{r}{n}\right)^{nt} - 1\right]}{\frac{r}{n}}$

$A = \frac{200\left[\left(1+\frac{0.05}{12}\right)^{(12)(30)} - 1\right]}{\frac{0.05}{12}} \approx \$166{,}451.73$

91. a. $P = \$100, n = 12, r = 6\%, t = 20$ yr

$A = \frac{P\left[\left(1+\frac{r}{n}\right)^{nt} - 1\right]}{\frac{r}{n}}$

$A = \frac{100\left[\left(1+\frac{0.06}{12}\right)^{(12)(20)} - 1\right]}{\frac{0.06}{12}}$

$A \approx \$46{,}204.09$

b. $P = \$200, n = 12, r = 6\%, t = 20$ yr

$A = \frac{P\left[\left(1+\frac{r}{n}\right)^{nt} - 1\right]}{\frac{r}{n}}$

$A = \frac{200\left[\left(1+\frac{0.06}{12}\right)^{(12)(20)} - 1\right]}{\frac{0.06}{12}}$

$A \approx \$92{,}408.18$

$\frac{92{,}408.18}{46{,}204.09} = 2$

The value of the annuity doubles.

c. $P = \$100, n = 12, r = 6\%, t = 40$ yr

$A = \frac{P\left[\left(1+\frac{r}{n}\right)^{nt} - 1\right]}{\frac{r}{n}}$

$$A=\frac{100\left[\left(1+\frac{0.06}{12}\right)^{(12)(40)}-1\right]}{\frac{0.06}{12}}$$

$A\approx \$199{,}149.07$

$\frac{199{,}149.07}{46{,}204.09}\approx 4.3$

The value of the annuity more than doubles.

93. **a.** $a_1=6>0, r=0.4<1$; Decreasing

b. $a_1=3>0, r=1.4>1$; Increasing

95. Decreasing geometric sequence. Graph b

97. Alternating decreasing geometric sequence. Graph d

99. $a_1=24$ in., $r=98\%$

$S=\frac{a_1}{1-r}=\frac{24}{1-0.98}=\frac{24}{0.02}$

$=1200$ in. or 100 ft

101. $a_1=4$ ft, $r=\frac{1}{2}$

$S=\frac{a_1}{1-r}=\frac{4}{1-\frac{1}{2}}=\frac{4}{\frac{1}{2}}=8$ ft

Total vertical distance: $4+S=4+8=12$ ft

103. **a.** $a_1=0.01, r=2$

$a_n=a_1r^{n-1}$

$a_n=0.01(2)^{n-1}$ (dollars)

b. Day 10:

$a_{10}=0.01(2)^{10-1}=0.01(2)^9=\$5.12$

Day 20:

$a_{20}=0.01(2)^{20-1}=0.01(2)^{19}=\$5242.88$

Day 30:

$a_{30}=0.01(2)^{30-1}=0.01(2)^{29}$

$=\$5{,}368{,}709.12$

c. $S_n=\frac{a_1(1-r^n)}{1-r}$

$S_{30}=\frac{0.01\left[1-(2)^{30}\right]}{1-2}$

$=\frac{0.01(1-1{,}073{,}741{,}824)}{-1}$

$=\$10{,}737{,}418.23$

105. **a.** Arithmetic

$a_1=60{,}000, d=3000, n=20$

$a_n=a_1+(n-1)d$

$a_n=60{,}000+(n-1)(3000)$

$a_n=60{,}000+3000n-3000$

$a_n=3000n+57{,}000$

$a_{20}=3000(20)+57{,}000=60{,}000+57{,}000$

$=117{,}000$

$S_n=\frac{a_1(1-r^n)}{1-r}$

$S_{20}=\frac{20}{2}(60{,}000+117{,}000)$

$=10(177{,}000)=\$1{,}770{,}000$

b. Geometric

$a_1=56{,}000, r=1.06, n=20$

$S_n=\frac{a_1(1-r^n)}{1-r}$

$S_{20}=\frac{56{,}000\left[1-(1.06)^{20}\right]}{1-1.06}\approx \$2{,}059{,}993$

c. $\$2{,}059{,}993-\$1{,}770{,}000=\$289{,}993$

107. **a.** $a_1=2, r=2$

$a_n=a_1r^{n-1}$

$a_n=2(2)^{n-1}=2^n$

b. $a_{10}=2^{10}=1024$

109. In an arithmetic sequence, the difference between a term and its predecessor is a fixed constant. In a geometric sequence, the ratio between a term and its predecessor is a fixed constant.

111. Each term of the sequence $a_1, a_2, a_3, \ldots$ after the first term is obtained by multiplying the preceding term by r. Therefore, every other term is obtained by multiplying by r^2. Therefore, the sequence $a_1, a_3, a_5, \ldots$ is geometric with common ratio r^2.

113. The sequence $\frac{1}{a_1}, \frac{1}{a_2}, \frac{1}{a_3}, \frac{1}{a_4}, \ldots$ can be written as $\frac{1}{a_1}, \frac{1}{a_1 r}, \frac{1}{a_1 r^2}, \frac{1}{a_1 r^3}, \ldots$ or equivalently as

$$\frac{1}{a_1}, \frac{1}{a_1}\left(\frac{1}{r}\right), \frac{1}{a_1}\left(\frac{1}{r}\right)^2, \frac{1}{a_1}\left(\frac{1}{r}\right)^3, \ldots.$$

Therefore, the sequence is geometric with common ratio $\frac{1}{r}$.

115. Each term of the geometric sequence $a_1, a_2, a_3, \ldots$ can be written in the form $a_1 r^{n-1}$.
Therefore, $\log a_1, \log a_2, \log a_3, \ldots$

$$= \log a_1, \log a_1 r, \log a_1 r^2, \ldots, \log a_1 r^{n-1}, \log a_1 r^n, \ldots.$$

$$\begin{aligned} a_{n+1} - a_n &= \log a_1 r^n - \log a_1 r^{n-1} \\ &= \left(\log a_1 + \log r^n\right) - \left(\log a_1 + \log r^{n-1}\right) \\ &= \left(\log a_1 + n \log r\right) \\ &\quad - \left[\log a_1 + (n-1)\log r\right] \\ &= \log r \end{aligned}$$

The common difference is $\log r$.

117. $\ln 1, \ln 2, \ln 4, \ln 8, \ldots$

$$\begin{aligned} &= \ln 1, \ln\left[1(2)\right], \ln\left[1(2)^2\right], \ln\left[1(2)^3\right], \\ &\quad \ldots, \ln\left[1(2)^{n-1}\right], \ln\left[1(2)^n\right], \ldots \\ &= \ln 1, \ln\left[1(2)\right], 2\ln\left[1(2)\right], 3\ln\left[1(2)\right], \\ &\quad \ldots, (n-1)\ln\left[1(2)\right], n\ln\left[1(2)\right], \ldots \\ &= 0, \ln 2, 2\ln 2, 3\ln 2, \\ &\quad \ldots, (n-1)\ln 2, n\ln 2, \ldots \end{aligned}$$

Arithmetic; $d = \ln 2$

119.

121.

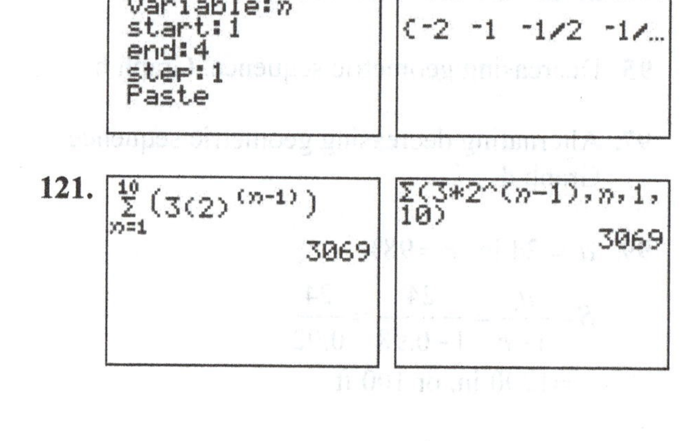

Problem Recognition Exercises: Comparing Arithmetic and Geometric Sequences and Series

1. Arithmetic

$$a_n = \frac{2n}{3} + \frac{1}{4}$$

$$a_n = \frac{1}{4} + \left(\frac{2}{3}\right)n$$

$$d = \frac{2}{3}$$

3. Geometric; $r = -1$ **5.** Neither

7. Geometric; $r = \sqrt{5}$ **9.** Arithmetic; $d = \sqrt{2}$

11. $\sum_{i=1}^{1506} 5 = 1506(5) = 7530$

13. $\sum_{i=1}^{716} (-1)^{i+1} = \sum_{i=1}^{716} (-1)^2 (-1)^{i-1} = \sum_{i=1}^{716} (-1)^{i-1}$

$a_1 = 1, n = 716, r = -1$

$$S_n = \frac{a_1(1-r^n)}{1-r}$$

$$S_{716} = \frac{1\left[1-(-1)^{716}\right]}{1-(-1)} = \frac{0}{2} = 0$$

15. $\sum_{n=1}^{3} \frac{n-2}{n+1} = \frac{1-2}{1+1} + \frac{2-2}{2+1} + \frac{3-2}{3+1}$

$$= \frac{-1}{2} + \frac{0}{3} + \frac{1}{4} = -\frac{1}{4}$$

17. $a_1 = -3, n = 4, r = \frac{1}{3}$

$$S_n = \frac{a_1(1-r^n)}{1-r}$$

$$S_4 = \frac{-3\left[1-\left(\frac{1}{3}\right)^4\right]}{1-\frac{1}{3}} = \frac{-3\left(1-\frac{1}{81}\right)}{\frac{2}{3}} = -\frac{40}{9}$$

19. $a_1 = 6, r = \frac{5}{3}$

$|r| = \left|\frac{5}{3}\right| > 1$; The sum does not exist.

21. $\sum_{i=1}^{27}(-6i-4) = -6\sum_{i=1}^{27} i + \sum_{i=1}^{27}(-4)$

$$= -6\left[\frac{27}{2}(1+27)\right] + (27)(-4)$$

$$= -6(378) - 108 = -2376$$

23. $\sum_{n=1}^{\infty} 8\left(\frac{1}{2}\right)^n = \sum_{n=1}^{\infty} 8\left(\frac{1}{2}\right)\left(\frac{1}{2}\right)^{n-1} = \sum_{n=1}^{\infty} 4\left(\frac{1}{2}\right)^{n-1}$

$a_1 = 4, r = \frac{1}{2}$

$|r| = \left|\frac{1}{2}\right| < 1$

$$S = \frac{a_1}{1-r} = \frac{4}{1-\frac{1}{2}} = \frac{4}{\frac{1}{2}} = 8$$

25. $a_1 = 36, r = \frac{a_2}{a_1} = \frac{30}{36} = \frac{5}{6}$

$|r| = \left|\frac{5}{6}\right| < 1$

$$S = \frac{a_1}{1-r} = \frac{36}{1-\frac{5}{6}} = \frac{36}{\frac{1}{6}} = 216$$

27. $a_n = a_1 + (n-1)d$

$363 = 3 + (n-1)(8)$

$363 = 3 + 8n - 8$

$363 = 8n - 5$

$8n = 368$

$n = 46$

$S_n = \frac{n}{2}(a_1 + a_n)$

$S_{46} = \frac{46}{2}(3+363) = 23(366) = 8418$

Section 8.4 Mathematical Induction

1. $\sum_{i=1}^{67}(3-8i) = \sum_{i=1}^{67} 3 - 8\sum_{i=1}^{67} i$

$$= (67)(3) - 8\left[\frac{67}{2}(1+67)\right]$$

$$= 201 - 8(2278) = -18{,}023$$

3. $a_1 = 5, n = 5, r = \frac{3}{4}$

$$S_n = \frac{a_1(1-r^n)}{1-r}$$

$$S_5 = \frac{5\left[1-\left(\frac{3}{4}\right)^5\right]}{1-\frac{3}{4}} = \frac{5\left(1-\frac{243}{1024}\right)}{\frac{1}{4}} = \frac{3905}{256}$$

5. induction; S_1; S_{k+1}

7. **1.** S_1 is true because $2 = 2(1)^2$.

2. Assume that $S_k = 2k^2$.

Show that $S_{k+1} = 2(k+1)^2$.

$$\begin{aligned} S_{k+1} &= [2+6+10+\cdots+(4k-2)] \\ &\quad + [4(k+1)-2] \\ &= 2k^2 + (4k+2) \text{ Inductive hypothesis} \\ &= 2(k^2+2k+1) \\ &= 2(k+1)^2 \text{ as desired.} \end{aligned}$$

The statement is true for all positive integers.

9. **1.** S_1 is true because $5 = \frac{1}{2}[3(1)+7]$.

2. Assume that $S_k = \frac{k}{2}(3k+7)$. Show that

$$S_{k+1} = \left(\frac{k+1}{2}\right)[3(k+1)+7] = \frac{(k+1)(3k+10)}{2}.$$

$$\begin{aligned} S_{k+1} &= [5+8+11+\cdots+(3k+2)] \\ &\quad + [3(k+1)+2] \\ &= \frac{k}{2}(3k+7)+(3k+5) \text{ Inductive hypothesis} \\ &= \frac{1}{2}(3k^2+7k)+3k+5 \\ &= \frac{3k^2+7k+6k+10}{2} \\ &= \frac{3k^2+13k+10}{2} \\ &= \frac{(k+1)(3k+10)}{2} \text{ as desired.} \end{aligned}$$

The statement is true for all positive integers.

11. **1.** S_1 is true because $8 = -2(1)(1-5)$.

2. Assume that $S_k = -2k(k-5)$. Show that

$$\begin{aligned} S_{k+1} &= -2(k+1)[(k+1)-5] \\ &= -2(k+1)(k-4) = -2k^2+6k+8 \end{aligned}$$

or equivalently,

$S_{k+1} = -2(k+1)(k-4)$.

$$\begin{aligned} S_{k+1} &= [8+4+0+\cdots+(-4k+12)] \\ &\quad + [-4(k+1)+12] \\ &= -2k(k-5)+(-4k+8) \text{ Inductive hypothesis} \\ &= -2k^2+10k-4k+8 \\ &= -2k^2+6k+8 \text{ as desired.} \end{aligned}$$

The statement is true for all positive integers.

13. **1.** S_1 is true because $1 = 2^1 - 1$.

2. Assume that $S_k = 2^k - 1$.

Show that $S_{k+1} = 2^{k+1} - 1$.

$$\begin{aligned} S_{k+1} &= [1+2+2^2+2^3+\cdots+2^{k-1}] + 2^{(k+1)-1} \\ &= (2^k-1)+2^k \text{ Inductive hypothesis} \\ &= 2\cdot 2^k - 1 \\ &= 2^{k+1}-1 \text{ as desired.} \end{aligned}$$

The statement is true for all positive integers.

15. **1.** S_1 is true because $\frac{3}{4} = 1 - \left(\frac{1}{4}\right)^1$.

2. Assume that $S_k = 1 - \left(\frac{1}{4}\right)^k$.

Show that $S_{k+1} = 1 - \left(\frac{1}{4}\right)^{k+1}$.

$$\begin{aligned} S_{k+1} &= \left[\frac{3}{4}+\frac{3}{16}+\frac{3}{64}+\cdots+\frac{3}{4^k}\right]+\frac{3}{4^{k+1}} \\ &= 1-\left(\frac{1}{4}\right)^k+\frac{3}{4^{k+1}} \text{ Inductive hypothesis} \\ &= 1-\frac{4}{4^{k+1}}+\frac{3}{4^{k+1}} \\ &= 1-\frac{1}{4^{k+1}} \\ &= 1-\left(\frac{1}{4}\right)^{k+1} \text{ as desired.} \end{aligned}$$

The statement is true for all positive integers.

17. 1. S_1 is true because $1\cdot 2=\frac{1(1+1)(1+2)}{3}$.

2. Assume that $S_k=\frac{k(k+1)(k+2)}{3}$.

Show that

$$S_{k+1}=\frac{(k+1)[(k+1)+1][(k+1)+2]}{3}=\frac{(k+1)(k+2)(k+3)}{3}$$

$$S_{k+1}=[1\cdot 2+2\cdot 3+3\cdot 4+\cdots+k(k+1)]+(k+1)[(k+1)+1]$$

$$=\frac{k(k+1)(k+2)}{3}+(k+1)(k+2) \quad \text{Inductive hypothesis}$$

$$=\frac{k(k+1)(k+2)}{3}+\frac{3(k+1)(k+2)}{3}$$

$$=\frac{(k+1)(k+2)(k+3)}{3} \text{ as desired.}$$

The statement is true for all positive integers.

19. 1. S_1 is true because $1-\frac{1}{2}=\frac{1}{1+1}$.

2. Assume that $S_k=\frac{1}{k+1}$.

Show that $S_{k+1}=\frac{1}{(k+1)+1}=\frac{1}{k+2}$.

$$S_{k+1}=\left(1-\frac{1}{2}\right)\left(1-\frac{1}{3}\right)\left(1-\frac{1}{4}\right)\cdots\left(1-\frac{1}{k+1}\right)\left[1-\frac{1}{(k+1)+1}\right]$$

$$=\left(\frac{1}{k+1}\right)\left(1-\frac{1}{k+2}\right) \quad \text{Inductive hypothesis}$$

$$=\left(\frac{1}{k+1}\right)\left(\frac{k+2}{k+2}-\frac{1}{k+2}\right)$$

$$=\left(\frac{1}{k+1}\right)\left(\frac{k+1}{k+2}\right)$$

$$=\left(\frac{1}{k+2}\right) \text{ as desired.}$$

The statement is true for all positive integers.

21. 1. S_1 is true because $\sum_{i=1}^{1}1=1$.

2. Assume that S_k is true; that is, $\sum_{i=1}^{k}1=k$.

Show that S_{k+1} is true: $\sum_{i=1}^{k+1}1=(k+1)$.

$$\sum_{i=1}^{k+1}1=\left(\sum_{i=1}^{k}1\right)+1=k+1 \text{ as desired.}$$

The statement is true for all positive integers.

23. 1. S_1 is true because

$$\sum_{i=1}^{1}i^2=(1)^2=\frac{1(1+1)[2(1)+1]}{6}.$$

2. Assume that S_k is true; that is,

$$\sum_{i=1}^{k}i^2=\frac{k(k+1)(2k+1)}{6}.$$

Show that S_{k+1} is true:

$$\sum_{i=1}^{k+1}i^2=\frac{(k+1)[(k+1)+1][2(k+1)+1]}{6}=\frac{(k+1)(k+2)(2k+3)}{6}$$

$$\sum_{i=1}^{k+1}i^2=1\sum_{i=1}^{k}i^2+(k+1)^2$$

$$=\frac{k(k+1)(2k+1)}{6}+(k+1)^2 \quad \text{Inductive hypothesis}$$

$$=\frac{k(k+1)(2k+1)}{6}+\frac{6(k+1)^2}{6}$$

$$=\frac{k(k+1)(2k+1)+6(k+1)(k+1)}{6}$$

$$=\frac{(k+1)(2k^2+k)+(k+1)(6k+6)}{6}$$

$$=\frac{(k+1)(2k^2+k+6k+6)}{6}$$

$$=\frac{(k+1)(2k^2+7k+6)}{6}$$

$$=\frac{(k+1)(k+2)(2k+3)}{6} \text{ as desired.}$$

The statement is true for all positive integers.

25. Let S_n be the statement: 2 is a factor of $5^n - 3$.

1. S_1 is true because 2 is a factor of $(5)^1 - 3 = 2$.

2. Assume that S_k is true; that is, assume that 2 is a factor of $5^k - 3$. This implies that $5^k - 3 = 2a$ and that $5^k = 2a + 3$ for some positive integer a.

Show that S_{k+1} is true; that is, show that 2 is a factor of $5^{k+1} - 3$.

$5^{k+1} - 3 = 5 \cdot 5^k - 3$

$= 5(2a+3) - 3$ Replace 5^k with $2a+3$.

$= 10a + 15 - 3$

$= 10a + 12$

$= 2(5a+6)$

Therefore, 2 is a factor of $5^{k+1} - 3$ as desired. The statement is true for all natural numbers.

27. Let S_n be the statement: $4^n - 1$ is divisible by 3. This is equivalent to saying that 3 is a factor of $4^n - 1$.

1. S_1 is true because 3 is a factor of $4^1 - 1 = 3$.

2. Assume that S_k is true; that is, assume that 3 is a factor of $4^k - 1$. This implies that $4^k - 1 = 3a$ and that $4^k = 3a + 1$ for some positive integer a.

Show that S_{k+1} is true; that is, show that 3 is a factor of $4^{k+1} - 1$.

$4^{k+1} - 1 = 4 \cdot 4^k - 1$

$= 4(3a+1) - 1$ Replace 4^k by $3a+1$.

$= 12a + 4 - 1$

$= 12a + 3$

$= 3(4a+1)$

Therefore, 3 is a factor of $4^{k+1} - 1$ as desired. The statement is true for all natural numbers.

29. $n! > 3^n$

Try $n=1$:	Try $n=2$:
$1! \overset{?}{>} 3^1$	$2! \overset{?}{>} 3^2$
$1 \overset{?}{>} 3$ No	$2 \overset{?}{>} 9$ No
Try $n=3$:	Try $n=4$:
$3! \overset{?}{>} 3^3$	$4! \overset{?}{>} 3^4$
$6 \overset{?}{>} 27$ No	$24 \overset{?}{>} 81$ No
Try $n=5$:	Try $n=6$:
$5! \overset{?}{>} 3^5$	$6! \overset{?}{>} 3^6$
$120 \overset{?}{>} 243$ No	$720 \overset{?}{>} 729$ No

Try $n=7$:

$7! \overset{?}{>} 3^7$

$5040 \overset{?}{>} 2187$ Yes

$n = 7$

31. $3n < 2^n$

Try $n=1$:	Try $n=2$:
$3(1) \overset{?}{<} 2^1$	$3(2) \overset{?}{<} 2^2$
$3 \overset{?}{<} 2$ No	$6 \overset{?}{<} 4$ No
Try $n=3$:	Try $n=4$:
$3(3) \overset{?}{<} 2^3$	$3(4) \overset{?}{<} 2^4$
$9 \overset{?}{<} 8$ No	$12 \overset{?}{<} 16$ Yes

$n = 4$

33. Let S_n be the statement: $n! > 3^n$ for $n \ge 7$.

1. S_7 is true because $7! = 5040$ and $3^7 = 2187$. Therefore, $7! > 3^7$.

2. Assume that $k! > 3^k$ for a natural number $k \ge 7$. Show that $(k+1)! > 3^{k+1}$.

$(k+1)! = (k+1) \cdot k!$

$> (k+1)(3^k)$ Inductive hypothesis

$> 3 \cdot 3^k$ Since $k \ge 7$, then $(k+1) > 3$.

$= 3^{k+1}$

Therefore, $(k+1)! > 3^{k+1}$ as desired. The statement is true for all natural numbers $n \ge 7$.

35. Let S_n be the statement: $3n < 2^n$ for $n \ge 4$.

1. S_4 is true because $3(4) = 12$ and $2^4 = 16$. Therefore, $3(4) < 2^4$.

2. Assume that $3k < 2^k$ for a natural number $k \ge 4$. Show that $3(k+1) < 2^{k+1}$.

$$3(k+1)=3k+3 \quad \text{Since } k \ge 4, \text{ then } 3k>3.$$
$$<3^k+3k$$
$$<2^k+2^k \quad \text{Inductive hypothesis}$$
$$=2(2^k)$$
$$=2^{k+1}$$

Therefore, $3(k+1)<2^{k+1}$ as desired.

The statement is true for all natural numbers $n \ge 4$.

37. Let S_n be the statement: $(xy)^n = x^n y^n$.

1. S_1 is true because $(xy)^1 = xy = x^1 y^1$.

2. Assume that S_k is true: $(xy)^k = x^k y^k$.

Show that S_{k+1} is true; that is, show that $(xy)^{k+1} = x^{k+1} y^{k+1}$.

S_{k+1} follows by multiplying S_k by (xy).

$(xy)^{k+1} = (xy)^k (xy) = (x^k y^k)(xy) = x^{k+1} y^{k+1}$

as desired.

The statement is true for positive integers n and real numbers x and y.

39. Let S_n be the statement: If $x>1$, then $x^n > x^{n-1}$.

1. S_1 is true because $x^1>1$ for $x>1$ and since $x^0=1$, we have $x^1>x^0$ for $x>1$.

2. Assume that S_k is true: $x^k > x^{k-1}$ for $x>1$. Show that S_{k+1} is true: $x^{k+1} > x^{(k+1)-1}$, or equivalently, $x^{k+1}>x^k$.

$$x^{k+1}=x^k \cdot x$$
$$>x^{k-1}\cdot x \quad \text{Inductive hypothesis}$$
$$=x^k x^{-1}\cdot x$$
$$=x^k \quad \text{as desired.}$$

The statement is true for positive integers n and real numbers x and y, provided that $x>0$.

41. The statement is false for $n=11$.

43. The principle of mathematical induction has us test the truth of a statement for $n=1$. The extended principle of mathematical induction has us test the truth of a statement for the first allowable value of *n*. In each case, the proof is concluded by showing that the truth of a statement for any other positive integer after the first allowable value of *n* follows directly from its predecessor.

45. Let S_n be the statement that n^2-n is even.

1. S_1 is true because $(1)^2-(1)=0$, which is even.

2. Assume that S_k is true; that is, assume that k^2-k is even. This implies that $k^2-k=2a$ for some natural number *a*, and that $k=k^2-2a$.

Show that S_{k+1} is true; that is, show that $(k+1)^2-(k+1)$ is even.

$$(k+1)^2-(k+1)$$
$$=k^2+2k+1-k-1$$
$$=k^2+k$$
$$=k^2+(k^2-2a) \quad \text{Inductive hypothesis}$$
$$=2k^2-2a$$
$$=2(k^2-a), \text{ which is an even integer.}$$

47. **1.** S_3 is true because $F_1+F_2+F_3=1+1+2=4$ and $F_{3+2}-1=F_5-1=5-1=4$.

2. Assume that S_k is true; that is, assume that $F_1+F_2+\cdots+F_k=F_{k+2}-1$.

Show that S_{k+1} is true; that is, show that $F_1+F_2+\cdots+F_k+F_{k+1}=F_{[(k+1)+2]}-1=F_{k+3}-1$

$$F_1+F_2+\cdots+F_k+F_{k+1}$$
$$=(F_1+F_2+\cdots+F_k)+F_{k+1}$$
$$=(F_{k+2}-1)+F_{k+1} \quad \text{Inductive hypothesis}$$
$$=(F_{k+1}+F_{k+2})-1 \quad \text{Replace } F_{k+1}+F_{k+2} \text{ by } F_{k+3}.$$
$$=F_{k+3}-1 \text{ as desired.}$$

Section 8.5 The Binomial Theorem

1. $\sum_{n=0}^{2}(x+y)^n=(x+y)^0+(x+y)^1+(x+y)^2$
$=1+x+y+x^2+2xy+y^2$

3. $\sum_{i=1}^{52}(3i+8)=3\sum_{i=1}^{52}i+\sum_{i=1}^{52}8=3\left[\frac{52(53)}{2}\right]+52(8)$
$=3(1378)+416=4550$

5. $a_1=3, n=4, r=\frac{1}{2}$

$$S_n=\frac{a_1(1-r^n)}{1-r}$$

$$S_4=\frac{3\left[1-\left(\frac{1}{2}\right)^4\right]}{1-\frac{1}{2}}=\frac{3\left(1-\frac{1}{16}\right)}{\frac{1}{2}}=\frac{45}{8}$$

7. $a_1=3, r=\frac{1}{2}$

$$|r|=\left|\frac{1}{2}\right|<1$$

$$S=\frac{a_1}{1-r}=\frac{3}{1-\frac{1}{2}}=\frac{3}{\frac{1}{2}}=6$$

9. $a_1=x^3, n=6, r=-x$

$$S_n=\frac{a_1(1-r^n)}{1-r}$$

$$S_6=\frac{x^3\left[1-(-x)^6\right]}{1-(-x)}=\frac{x^3(1-x^6)}{1+x}$$

$$=\frac{-x^3(x^6-1)}{x+1}=\frac{-x^3(x^3-1)(x^3+1)}{x+1}$$

$$=\frac{-x^3(x^3-1)\overset{1}{\cancel{(x+1)}}(x^2-x+1)}{\cancel{x+1}}$$

$$=(x^3-x^6)(x^2-x+1)$$

$$=x^3-x^4+x^5-x^6+x^7-x^8$$

11. binomial

13. n

15. $\binom{n}{k-1}a^{n-(k-1)}b^{k-1}$

17. a. See Figure 8-9 on page 753.

b. $(x-y)^4=x^4-4x^3y+6x^2y^2-4xy^3+y^4$

19. a. $\binom{4}{0}=\frac{4!}{0!\cdot(4-0)!}=\frac{4!}{0!\cdot 4!}=\frac{4!}{1\cdot 4!}=1$

b. $\binom{4}{1}=\frac{4!}{1!\cdot(4-1)!}=\frac{4!}{1!\cdot 3!}=\frac{4\cdot 3!}{1\cdot 3!}=4$

c. $\binom{4}{2}=\frac{4!}{2!\cdot(4-2)!}=\frac{4!}{2!\cdot 2!}=\frac{4\cdot 3\cdot 2!}{2\cdot 1\cdot 2!}=6$

d. $\binom{4}{3}=\frac{4!}{3!\cdot(4-3)!}=\frac{4!}{3!\cdot 1!}=\frac{4\cdot 3!}{3!\cdot 1}=4$

e. $\binom{4}{4}=\frac{4!}{4!\cdot(4-4)!}=\frac{4!}{4!\cdot 0!}=\frac{4!}{4!\cdot 1}=1$

21. $\binom{13}{3}=\frac{13!}{3!\cdot(13-3)!}=\frac{13!}{3!\cdot 10!}$

$$=\frac{13\cdot\overset{2}{\cancel{12}}\cdot 11\cdot\overset{1}{\cancel{10!}}}{\cancel{3}\cdot\cancel{2}\cdot 1\cdot\cancel{10!}}=286$$

23. $\binom{11}{5}=\frac{11!}{5!\cdot(11-5)!}=\frac{11!}{5!\cdot 6!}$

$$=\frac{11\cdot\overset{2}{\cancel{10}}\cdot\overset{3}{\cancel{9}}\cdot\overset{1}{\cancel{8}}\cdot 7\cdot\overset{1}{\cancel{6!}}}{\cancel{5}\cdot\cancel{4}\cdot\cancel{3}\cdot\cancel{2}\cdot 1\cdot\cancel{6!}}=462$$

25. The expression is in the form $(a+b)^5$ with $a=3x$ and $b=1$.

$$\begin{aligned}(3x+1)^5 &= \binom{5}{0}(3x)^5+\binom{5}{1}(3x)^4(1)+\binom{5}{2}(3x)^3(1)^2+\binom{5}{3}(3x)^2(1)^3+\binom{5}{4}(3x)(1)^4+\binom{5}{5}(1)^5\\ &= 1(243x^5)+5(81x^4)(1)+10(27x^3)(1)+10(9x^2)(1)+5(3x)(1)+1(1)\\ &= 243x^5+405x^4+270x^3+90x^2+15x+1\end{aligned}$$

27. The expression is in the form $(a+b)^3$ with $a=7x$ and $b=3$.

$$\begin{aligned}(7x+3)^3 &= \binom{3}{0}(7x)^3+\binom{3}{1}(7x)^2(3)+\binom{3}{2}(7x)(3)^2+\binom{3}{3}(3)^3\\ &= 1(343x^3)+3(49x^2)(3)+3(7x)(9)+1(27)\\ &= 343x^3+441x^2+189x+27\end{aligned}$$

29. The expression is in the form $(a+b)^4$ with $a=2x$ and $b=-5$.

$$\begin{aligned}(2x-5)^4 &= \binom{4}{0}(2x)^4+\binom{4}{1}(2x)^3(-5)+\binom{4}{2}(2x)^2(-5)^2+\binom{4}{3}(2x)(-5)^3+\binom{4}{4}(-5)^4\\ &= 1(16x^4)+4(8x^3)(-5)+6(4x^2)(25)+4(2x)(-125)+1(625)\\ &= 16x^4-160x^3+600x^2-1000x+625\end{aligned}$$

31. The expression is in the form $(a+b)^5$ with $a=2x^3$ and $b=-y$.

$$\begin{aligned}(2x^3-y)^5 &= \binom{5}{0}(2x^3)^5+\binom{5}{1}(2x^3)^4(-y)+\binom{5}{2}(2x^3)^3(-y)^2+\binom{5}{3}(2x^3)^2(-y)^3+\binom{5}{4}(2x^3)(-y)^4+\binom{5}{5}(-y)^5\\ &= 1(32x^{15})+5(16x^{12})(-y)+10(8x^9)(y^2)+10(4x^6)(-y^3)+5(2x^3)(y^4)+1(-y^5)\\ &= 32x^{15}-80x^{12}y+80x^9y^2-40x^6y^3+10x^3y^4-y^5\end{aligned}$$

33. The expression is in the form $(a+b)^6$ with $a=p^2$ and $b=-w^4$.

$$\begin{aligned}(p^2-w^4)^6 &= \binom{6}{0}(p^2)^6+\binom{6}{1}(p^2)^5(-w^4)+\binom{6}{2}(p^2)^4(-w^4)^2+\binom{6}{3}(p^2)^3(-w^4)^3+\binom{6}{4}(p^2)^2(-w^4)^4\\ &\quad+\binom{6}{5}(p^2)(-w^4)^5+\binom{6}{6}(-w^4)^6\\ &= 1(p^{12})+6(p^{10})(-w^4)+15(p^8)(w^8)+20(p^6)(-w^{12})+15(p^4)(w^{16})+6(p^2)(-w^{20})+1(w^{24})\\ &= p^{12}-6p^{10}w^4+15p^8w^8-20p^6w^{12}+15p^4w^{16}-6p^2w^{20}+w^{24}\end{aligned}$$

35. The expression is in the form $(a+b)^4$ with $a=0.2$ and $b=0.1k$.

$$\begin{aligned}(0.2+0.1k)^4 &= \binom{4}{0}(0.2)^4+\binom{4}{1}(0.2)^3(0.1k)+\binom{4}{2}(0.2)^2(0.1k)^2+\binom{4}{3}(0.2)(0.1k)^3+\binom{4}{4}(0.1k)^4\\ &= 1(0.0016)+4(0.008)(0.1k)+6(0.04)(0.01k^2)+4(0.2)(0.001k^3)+1(0.0001k^4)\\ &= 0.0016+0.0032k+0.0024k^2+0.0008k^3+0.0001k^4\end{aligned}$$

37. The expression is in the form $(a+b)^3$ with $a=\frac{c}{2}$ and $b=-d$.

$$\left(\frac{c}{2}-d\right)^3=\binom{3}{0}\left(\frac{c}{2}\right)^3+\binom{3}{1}\left(\frac{c}{2}\right)^2(-d)+\binom{3}{2}\left(\frac{c}{2}\right)(-d)^2+\binom{3}{3}(-d)^3$$

$$=1\left(\frac{c^3}{8}\right)+3\left(\frac{c^2}{4}\right)(-d)+3\left(\frac{c}{2}\right)(d^2)+1(-d^3)=\frac{1}{8}c^3-\frac{3}{4}c^2d+\frac{3}{2}cd^2-d^3$$

39. $n=10, k=7, a=m, b=n$

$$\binom{n}{k-1}a^{n-(k-1)}b^{k-1}=\binom{10}{7-1}m^{10-(7-1)}n^{7-1}$$

$$=\binom{10}{6}m^4n^6=210m^4n^6$$

41. $n=8, k=4, a=c, b=-d$

$$\binom{n}{k-1}a^{n-(k-1)}b^{k-1}=\binom{8}{4-1}c^{8-(4-1)}(-d)^{4-1}$$

$$=\binom{8}{3}c^5(-d)^3=-56c^5d^3$$

43. $n=15, k=10, a=u^2, b=2v^4$

$$\binom{n}{k-1}a^{n-(k-1)}b^{k-1}=\binom{15}{10-1}(u^2)^{15-(10-1)}(2v^4)^{10-1}$$

$$=\binom{15}{9}(u^2)^6(2v^4)^9$$

$$=5005u^{12}\cdot 512v^{36}$$

$$=2{,}562{,}560u^{12}v^{36}$$

45. $n=9, k=4, a=\sqrt{3}x^2, b=y^3$

$$\binom{n}{k-1}a^{n-(k-1)}b^{k-1}=\binom{9}{4-1}(\sqrt{3}x^2)^{9-(4-1)}(y^3)^{4-1}$$

$$=\binom{9}{3}(\sqrt{3}x^2)^6(y^3)^3$$

$$=84\cdot 27x^{12}y^9=2268x^{12}y^9$$

47. $n=12, k=7, a=h^4, b=-1$

$$\binom{n}{k-1}a^{n-(k-1)}b^{k-1}=\binom{12}{7-1}(h^4)^{12-(7-1)}(-1)^{7-1}$$

$$=\binom{12}{6}(h^4)^6(-1)^6=924h^{24}$$

49. $n=8, a=p^4, b=3q$

$$\binom{n}{k-1}a^{n-(k-1)}b^{k-1}=\binom{8}{k-1}(p^4)^{8-(k-1)}(3q)^{k-1}$$

$$p^{12}=(p^4)^{8-(k-1)}$$

$$p^{12}=(p^4)^{9-k}$$

$$p^{12}=p^{36-4k}$$

$$12=36-4k$$

$$4k=24$$

$$k=6$$

$$\binom{8}{6-1}(p^4)^{8-(6-1)}(3q)^{6-1}=\binom{8}{5}(p^4)^3(3q)^5$$

$$=56p^{12}\cdot 243q^5$$

$$=13{,}608p^{12}q^5$$

51. The expression is in the form $(a+b)^4$ with $a=e^x$ and $b=-e^{-x}$.

$$(e^x-e^{-x})^4=\binom{4}{0}(e^x)^4+\binom{4}{1}(e^x)^3(-e^{-x})+\binom{4}{2}(e^x)^2(-e^{-x})^2+\binom{4}{3}(e^x)(-e^{-x})^3+\binom{4}{4}(-e^{-x})^4$$

$$=1(e^{4x})+4(e^{3x})(-e^{-x})+6(e^{2x})(e^{-2x})+4(e^x)(-e^{-3x})+1(e^{-4x})=e^{4x}-4e^{2x}+6-\frac{4}{e^{2x}}+\frac{1}{e^{4x}}$$

53. The expression is in the form $(a+b)^3$ with $a=x+y$ and $b=-z$.

$$\begin{aligned}(x+y-z)^3 &= \binom{3}{0}(x+y)^3+\binom{3}{1}(x+y)^2(-z)+\binom{3}{2}(x+y)(-z)^2+\binom{3}{3}(-z)^3\\ &=1(x^3+3x^2y+3xy^2+y^3)+3(x^2+2xy+y^2)(-z)+3(x+y)(z^2)+1(-z^3)\\ &=x^3+3x^2y+3xy^2+y^3-3x^2z-6xyz-3y^2z+3xz^2+3yz^2-z^3\end{aligned}$$

55. The expression is in the form $(a+b)^4$ with $a=1$ and $b=0.01$.

$$\begin{aligned}(1+0.01)^4 &= \binom{4}{0}(1)^4+\binom{4}{1}(1)^3(0.01)+\binom{4}{2}(1)^2(0.01)^2+\binom{4}{3}(1)(0.01)^3+\binom{4}{4}(0.01)^4\\ &=1(1)+4(1)(0.01)+6(1)(0.0001)+4(1)(0.000001)+1(0.00000001)\\ &=1+0.04+0.0006+0.000004+0.00000001=1.04060401\end{aligned}$$

57. The expression is the difference of two expressions in the form $(a+b)^3$. The first with $a=x$ and $b=y$. The second with $a=x$ and $b=-y$.

$$\begin{aligned}(x+y)^3-(x-y)^3 &= \binom{3}{0}(x)^3+\binom{3}{1}(x)^2(y)+\binom{3}{2}(x)(y)^2+\binom{3}{3}(y)^3\\ &\quad -\left[\binom{3}{0}(x)^3+\binom{3}{1}(x)^2(-y)+\binom{3}{2}(x)(-y)^2+\binom{3}{3}(-y)^3\right]=2\binom{3}{1}(x)^2(y)+2\binom{3}{3}(y)^3\\ &=6x^2y+2y^3\end{aligned}$$

59.

$$\begin{aligned}\frac{f(x+h)-f(x)}{h} &= \frac{2(x+h)^3+4-(2x^3+4)}{h}\\ &=\frac{2\left[\binom{3}{0}(x)^3+\binom{3}{1}(x)^2(h)+\binom{3}{2}(x)(h)^2+\binom{3}{3}(h)^3\right]+4-2x^3-4}{h}\\ &=\frac{2(x^3+3x^2h+3xh^2+h^3)-2x^3}{h}=\frac{2x^3+6x^2h+6xh^2+2h^3-2x^3}{h}=\frac{6x^2h+6xh^2+2h^3}{h}\\ &=6x^2+6xh+2h^2\end{aligned}$$

61.

$$\begin{aligned}&\frac{f(x+h)-f(x)}{h}\\ &=\frac{(x+h)^4-5(x+h)^2+1-(x^4-5x^2+1)}{h}\\ &=\frac{\binom{4}{0}(x)^4+\binom{4}{1}(x)^3(h)+\binom{4}{2}(x)^2(h)^2+\binom{4}{3}(x)(h)^3+\binom{4}{4}(h)^4-5(x^2+2xh+h^2)+1-x^4+5x^2-1}{h}\\ &=\frac{x^4+4x^3h+6x^2h^2+4xh^3+h^4-5x^2-10xh-5h^2-x^4+5x^2}{h}=\frac{4x^3h+6x^2h^2+4xh^3+h^4-10xh-5h^2}{h}\\ &=4x^3+6x^2h+4xh^2+h^3-10x-5h\end{aligned}$$

63. The expression is in the form $(a+b)^3$ with $a=2$ and $b=i$.

$$(2+i)^3=\binom{3}{0}(2)^3+\binom{3}{1}(2)^2(i)+\binom{3}{2}(2)(i)^2+\binom{3}{3}(i)^3$$
$$=1(8)+3(4)i+3(2)(i^2)+1(i^3)=8+12i+6(-1)+(-i)=2-11i$$

65. The expression is in the form $(a+b)^4$ with $a=5$ and $b=-2i$.

$$(5-2i)^4=\binom{4}{0}(5)^4+\binom{4}{1}(5)^3(-2i)+\binom{4}{2}(5)^2(-2i)^2+\binom{4}{3}(5)(-2i)^3+\binom{4}{4}(-2i)^4$$
$$=1(625)+4(125)(-2i)+6(25)(4i^2)+4(5)(-8i^3)+1(16i^4)$$
$$=625-1000i-600+160i+16=41-840i$$

67. $\binom{n}{r}=\dfrac{n!}{r!\cdot(n-r)!}$ and

$$\binom{n}{n-r}=\frac{n!}{(n-r)!\cdot[n-(n-r)]!}=\frac{n!}{(n-r)!\cdot r!}.$$

By the commutative property of multiplication, $\dfrac{n!}{r!\cdot(n-r)!}=\dfrac{n!}{(n-r)!\cdot r!}$.

69. a. 1, 2, 4, 8, 16, 32, 64, 128, 256

b. $a_n=2^{n-1}$

71. In this expression $n=3$ and $r=5$. The value $(n-r)!$ is $(3-5)!=(-2)!$, which is undefined.

73. a. $8!\approx\sqrt{2\pi\cdot 8}\left(\dfrac{8}{e}\right)^8\approx 39{,}902$

b. $8!=40{,}320$

c. $\left|\dfrac{40{,}320-39{,}902}{40{,}320}\right|\approx 0.01=1.0\%$ error

75.
$$(1+0.4)^{3/2}\approx 1+\left(\frac{3}{2}\right)(0.4)+\frac{\left(\frac{3}{2}\right)\left(\frac{3}{2}-1\right)(0.4)^2}{2!}+\frac{\left(\frac{3}{2}\right)\left(\frac{3}{2}-1\right)\left(\frac{3}{2}-2\right)(0.4)^3}{3!}$$
$$=1.656$$

Section 8.6 Principles of Counting

1.
$$\binom{11}{5}=\frac{11!}{5!\cdot(11-5)!}=\frac{11!}{5!\cdot 6!}$$
$$=\frac{11\cdot \overset{2}{\cancel{10}}\cdot \overset{3}{\cancel{9}}\cdot \overset{1}{\cancel{8}}\cdot 7\cdot \overset{1}{\cancel{6!}}}{\cancel{5}\cdot\cancel{4}\cdot\cancel{3}\cdot\cancel{2}\cdot 1\cdot\cancel{6!}}=462$$

3. Arithmetic

5. Neither

7. principle; counting, $m\cdot n$

9. permutation

11. $\dfrac{n!}{r!}$

13. $\dfrac{n!}{(n-r)!}$

15. 10

17. 8

19. 2

21. $4\cdot 3\cdot 6\cdot 4=288$

23. a. $26^3\cdot 10^3=17{,}576{,}000$

b. $26\cdot 25\cdot 24\cdot 10\cdot 9\cdot 8=11{,}232{,}000$

25. a. $2\cdot 26^3=35{,}152$

b. $2\cdot 25\cdot 24\cdot 23=27{,}600$

27. $5^{10} \cdot 2^4 = 156{,}250{,}000$

29. $2^8 = 256$

31. a. $7! = 5040$

b. $\dfrac{11!}{4!\cdot 4!\cdot 2!} = 34{,}650$

33. $5! - 1 = 120 - 1 = 119$

35. $4! = 24$

37. ${}_6P_4 = \dfrac{6!}{(6-4)!} = \dfrac{6!}{2!} = 360$

39. ${}_{12}P_2 = \dfrac{12!}{(12-2)!} = \dfrac{12!}{10!} = 132$

41. ${}_9P_9 = \dfrac{9!}{(9-9)!} = \dfrac{9!}{0!} = 9! = 362{,}880$

43. ${}_{20}P_3 = \dfrac{20!}{(20-3)!} = \dfrac{20!}{17!} = 6840$

There are 6840 ways to select 3 distinct items in a specific order from a group of 20 items.

45. ${}_6C_4 = \dfrac{6!}{4!\cdot(6-4)!} = \dfrac{6!}{4!\cdot 2!} = 15$

47. ${}_{12}C_2 = \dfrac{12!}{2!\cdot(12-2)!} = \dfrac{12!}{2!\cdot 10!} = 66$

49. ${}_9C_9 = \dfrac{9!}{9!\cdot(9-9)!} = \dfrac{9!}{9!\cdot 0!} = 1$

51. ${}_{20}C_3 = \dfrac{20!}{3!\cdot(20-3)!} = \dfrac{20!}{3!\cdot 17!} = 1140$

There are 1140 ways to select 3 distinct items in no specific order from a group of 20 items.

53. a. AB, BA, AC, CA, BC, CB

b. AB, AC, BC (Note: The order within the individual combinations does not matter. That is, AB or BA represents the same group of two elements.)

55. ${}_9P_3 = \dfrac{9!}{(9-3)!} = \dfrac{9!}{6!} = 504$

57. a. ${}_{24}C_5 = \dfrac{24!}{5!\cdot(24-5)!} = \dfrac{24!}{5!\cdot 19!} = 42{,}504$

b. ${}_{24}P_5 = \dfrac{24!}{(24-5)!} = \dfrac{24!}{19!} = 5{,}100{,}480$

59. ${}_{16}C_2 = \dfrac{16!}{2!\cdot(16-2)!} = \dfrac{16!}{2!\cdot 14!} = 120$

61. ${}_{47}C_5 = \dfrac{47!}{5!\cdot(47-5)!} = \dfrac{47!}{5!\cdot 42!} = 1{,}533{,}939$

63. $({}_9C_3)\cdot({}_7C_3) = 84\cdot 35 = 2940$

65. $10^4 = 10{,}000$

67. ${}_{12}C_5 = \dfrac{12!}{5!\cdot(12-5)!} = \dfrac{12!}{5!\cdot 7!} = 792$

69. ${}_5P_5 = \dfrac{5!}{(5-5)!} = \dfrac{5!}{0!} = 5! = 120$

71. a. ${}_4C_3 = \dfrac{4!}{3!\cdot(4-3)!} = \dfrac{4!}{3!\cdot 1!} = 4$

b. ${}_{16}C_3 = \dfrac{16!}{3!\cdot(16-3)!} = \dfrac{16!}{3!\cdot 13!} = 560$

c. $({}_{16}C_2)\cdot({}_4C_1) = 120\cdot 4 = 480$

73. a. { }, {a}, {b}, {c}, {d}, {a, b}, {a, c}, {a, d}, {b, c}, {b, d}, {c, d}, {a, b, c}, {a, b, d}, {a, c, d}, {b, c, d}, {a, b, c, d}

b. ${}_4C_0 = 1$, ${}_4C_1 = 4$, ${}_4C_2 = 6$, ${}_4C_3 = 4$, and ${}_4C_4 = 1$

75. $40^3 = 64,000$

77. Let s represent the number of sides.
$_sC_2 - s = {_4C_2} - 4 = 6 - 4 = 2$

79. Let s represent the number of sides.
$_sC_2 - s = {_6C_2} - 6 = 15 - 6 = 9$

81. $6^3 = 216$

83. $5 \cdot 6 \cdot 6 = 180$

85. The code must end in a 0, 2, or 4.
$5 \cdot 6 \cdot 3 = 90$

87. a. $a_n = 2(2)^{n-1} = 2^n$
$a_3 = 2^3 = 8$
$a_4 = 2^4 = 16$
$a_5 = 2^5 = 32$

b. $2^3 = 8$; BBB, BBG, BGB, BGG, GBB, GBG, GGB, GGG

c. $2^4 = 16$

89. $_{160}C_4 = \dfrac{160!}{4! \cdot (160-4)!} = \dfrac{160!}{4! \cdot 156!}$

$= \dfrac{160 \cdot 159 \cdot 158 \cdot 157 \cdot \cancel{156!}^{1}}{4! \cdot \cancel{156!}} = 26,294,360$

91. $2 \cdot 4 \cdot 3 \cdot 2 \cdot 1 = 48$

93. Using the fundamental principle of counting, we have $8 \cdot 7 \cdot 6 = 336$. There are 8 horses that can cross the finish line first. Once the first horse finishes, there are 7 horses remaining that can come in second. Then there are 6 horses that are available for third place. Alternatively, the number of first-, second-, and third-place ordered arrangements can be found by taking the number of permutations of 8 horses taken 3 at a time, $_8P_3$.

95. $\left[({_3P_3}) \cdot ({_4P_4}) \cdot ({_2P_2})\right] \cdot ({_3P_3}) = [6 \cdot 24 \cdot 2] \cdot 6$
$= 1728$

Section 8.7 Introduction to Probability

1. $_{30}C_5 = \dfrac{30!}{5! \cdot (30-5)!} = \dfrac{30!}{5! \cdot 25!} = 142,506$

3. $A \cup B = \{1, 2, 3, 4, 5, 6, 7, 8\}$

5. $A \cap B = \{5\}$

7. experiment

9 event

11. $\dfrac{n(E)}{n(S)}$

13. complement; 1

15. $P(A) + P(B) - P(A \cap B)$

17. independent

19. $0 \le P(E) \le 1$, so the values in c, d, g, and h can represent the probability of an event.

21. c

23. e

25. f

27. E: 1, 2, 3, or 4
$\dfrac{n(E)}{n(S)} = \dfrac{4}{10} = \dfrac{2}{5}$

29. E: 4, 5, 6, 7, 8, 9, or 10
$\dfrac{n(E)}{n(S)} = \dfrac{7}{10}$

31. E: none
$\dfrac{n(E)}{n(S)} = \dfrac{0}{10} = 0$

33. E: 1, 2, 3, 4, 5, 6, 7, 8, 9, or 10
$\dfrac{n(E)}{n(S)} = \dfrac{10}{10} = 1$

35. $n(S)=6+8+16=30$

a. $\frac{n(E)}{n(S)}=\frac{16}{30}=\frac{8}{15}$

b. $\frac{n(E)}{n(S)}=\frac{6}{30}=\frac{1}{5}$

c. $\frac{n(E)}{n(S)}=\frac{0}{30}=0$

37. a. $\frac{n(E)}{n(S)}=\frac{18}{38}=\frac{9}{19}$

b. $\frac{n(E)}{n(S)}=\frac{7}{38}$

c. $\frac{n(E)}{n(S)}=\frac{37}{38}$

39. $P(\overline{E})=1-P(E)=1-0.842=0.158$

41. $P(\overline{E})=1-P(E)=1-0.016=0.984$

For Exercises 43–45, refer to Figure 8-11 on page 774.

43. a. $\frac{n(E)}{n(S)}=\frac{3}{36}=\frac{1}{12}$ **b.** $\frac{n(E)}{n(S)}=\frac{33}{36}=\frac{11}{12}$

45. a. $\frac{n(E)}{n(S)}=\frac{6}{36}=\frac{1}{6}$ **b.** $\frac{n(E)}{n(S)}=\frac{3}{36}=\frac{1}{12}$

47. $P(E)=1-(0.68+0.22)=1-0.9=0.1$

49. a. $\frac{n(E)}{n(S)}=\frac{{}_{18}C_9}{{}_{34}C_9}=\frac{48{,}620}{52{,}451{,}256}\approx 0.00093$

b. $\frac{n(E)}{n(S)}=\frac{{}_{16}C_9}{{}_{34}C_9}=\frac{11{,}440}{52{,}451{,}256}\approx 0.00022$

c. The events from parts (a) and (b) are not complementary events. There are many other cases to consider regarding the number of male and female jurors: for example, 4 male, 5 female, etc.

51. $\frac{n(E)}{n(S)}=\frac{1}{{}_{39}C_5}=\frac{1}{575{,}575}$

53. a.

	Y	y
Y	YY	Yy
y	yY	yy

b. $\frac{n(E)}{n(S)}=\frac{1}{4}$

c. $\frac{n(E)}{n(S)}=\frac{3}{4}$

55. $P(E)=\frac{222}{4624}\approx 0.048$

57. Sample size: $4+8+7+31=50$

a. $P(E)=\frac{7}{50}=0.14$

b. $P(E)=\frac{31}{50}=0.62$

For Exercises 59–67, refer to Figure 8-12 on page 776.

59. $P(J\cup Q)=P(J)+P(Q)=\frac{4}{52}+\frac{4}{52}=\frac{8}{52}=\frac{2}{13}$

61. $P(J\cup D)=P(J)+P(D)-P(J\cap D)$

$=\frac{4}{52}+\frac{13}{52}-\frac{1}{52}=\frac{16}{52}=\frac{4}{13}$

63. $P(J\cup Q\cup K)=P(J)+P(Q)+P(K)$

$=\frac{4}{52}+\frac{4}{52}+\frac{4}{52}=\frac{12}{52}=\frac{3}{13}$

65. $P(J\cup Q\cup K\cup R)=P(J\cup Q\cup K)+P(R)$

$-P[(J\cup Q\cup K)\cap R]$

$=\frac{12}{52}+\frac{26}{52}-\frac{6}{52}=\frac{32}{52}=\frac{8}{13}$

67. $P(H \cup C \cup S) = P(H) + P(C) + P(S)$

$= \frac{13}{52} + \frac{13}{52} + \frac{13}{52} = \frac{39}{52} = \frac{3}{4}$

69.

	Normal Cholesterol	Elevated Cholesterol	Total
≤ 30	14	4	18
31-60	52	28	80
≥ 61	22	80	102
Total	88	[112]	200

$\frac{112}{200} = \frac{14}{25}$

71.

	Normal Cholesterol	Elevated Cholesterol	Total
≤ 30	14	4	[18]
31-60	52	28	[80]
≥ 61	22	80	102
Total	88	112	200

$\frac{18}{200} + \frac{80}{200} = \frac{98}{200} = \frac{49}{100}$

73.

	Normal Cholesterol	Elevated Cholesterol	Total
≤ 30	[14]	4	18
31-60	[52]	28	80
≥ 61	[[22]]	[80]	102
Total	88	112	200

$\frac{88}{200} + \frac{102}{200} - \frac{22}{200} = \frac{168}{200} = \frac{21}{25}$

75.

	Normal Cholesterol	Elevated Cholesterol	Total
≤ 30	14	[4]	18
31-60	[52]	[[28]]	80
≥ 61	22	[80]	102
Total	88	112	200

$\frac{80}{200} + \frac{112}{200} - \frac{28}{200} = \frac{164}{200} = \frac{41}{50}$

77. $P(5 \text{ and heads}) = P(5) \cdot P(\text{heads})$

$= \frac{1}{6} \cdot \frac{1}{2} = \frac{1}{12}$

79. $P(2 \text{ survive 5 yr})$

$= P(1 \text{ survives 5 yr}) \cdot P(1 \text{ survives 5 yr})$

$= (0.88)(0.88)$

$= 0.7744$

81. a. $P(\text{once}) = \frac{1}{127} \approx 0.007874$

b. $P(\text{meet twice})$

$= P(\text{meet once}) \cdot P(\text{meet once})$

$= \frac{1}{127} \cdot \frac{1}{127}$

≈ 0.000062

83. $\left(\frac{1}{2}\right)^5 \cdot \left(\frac{1}{4}\right)^5 = \frac{1}{32} \cdot \frac{1}{1024}$

$= \frac{1}{32{,}768}$

≈ 0.0000305

85. a. $P(\text{Rh}^+) = 0.374 + 0.357 + 0.085 + 0.034$

$= 0.85$

b. $P(\text{Rh}^+ \text{ and Rh}^+ \text{ and Rh}^+)$

$= P(\text{Rh}^+) \cdot P(\text{Rh}^+) \cdot P(\text{Rh}^+)$

$= (0.85)^3 \approx 0.614$

87. a. O^+ **b.** AB^-

89. Blood must be B or O, and Rh^+ or Rh^-.
$0.374 + 0.066 + 0.085 + 0.015 = 0.54$

91. a. Blood must be O and Rh^-. 0.066

b. O^- blood is absent all three antigens and will not introduce a new antigen to the recipient's blood.

93. $\left(\frac{1}{11}\right)^3 = \frac{1}{1331}$

95. a. Area of entire target:

$$\pi r^2 = \pi\left(\frac{61.0}{2}\right)^2 = 930.25\pi \text{ cm}^2$$

Area of gold region:

$$\pi r^2 = \pi\left(\frac{12.2}{2}\right)^2 = 37.21\pi \text{ cm}^2$$

$$\frac{\text{gold area}}{\text{target area}} = \frac{37.21\pi}{930.25\pi} = 0.04$$

b. Area of blue/red/gold region:

$$\pi r^2 = \pi\left(\frac{36.6}{2}\right)^2 = 334.89\pi \text{ cm}^2$$

Area of red/gold region:

$$\pi r^2 = \pi\left(\frac{24.4}{2}\right)^2 = 148.84\pi \text{ cm}^2$$

Area of blue region:

$$334.89\pi - 148.84\pi = 186.05 \text{ cm}^2$$

$$\frac{\text{blue area}}{\text{target area}} = \frac{186.05\pi}{930.25\pi} = 0.20$$

c. Area of blue/red/gold region:

$334.89\pi \text{ cm}^2$

$$\frac{\text{blue/red/gold area}}{\text{target area}} = \frac{334.89\pi}{930.25\pi} = 0.36$$

97. a.

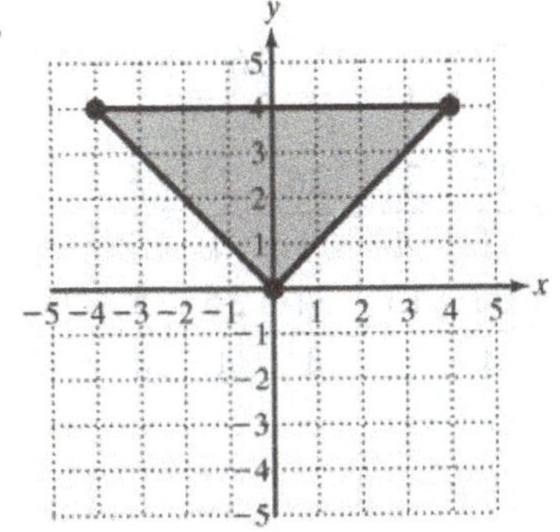

b. Area of entire target: $10 \times 10 = 100 \text{ units}^2$

Area of shaded region:

$$\frac{1}{2}bh = \frac{1}{2}(8)(4) = 16 \text{ units}^2$$

$$\frac{\text{shaded area}}{\text{target area}} = \frac{16}{100} = 0.16$$

99. Observe the game being played by other players. Approximate the probability by dividing the number of times a player wins to the number of games played.

101. Answers will vary.

103. a. $P(\text{two comedies})$

$$= P(\text{comedy}) \cdot P(\text{comedy})$$

$$= \frac{5}{15} \cdot \frac{5}{15} = \frac{1}{9}$$

b. No

c. $P(\text{two comedies})$

$$= P(\text{comedy}) \cdot P(\text{comedy})$$

$$= \frac{5}{15} \cdot \frac{4}{14} = \frac{2}{21}$$

105. a. $\frac{{}_{13}C_5}{{}_{52}C_5} = \frac{1287}{2{,}598{,}960} \approx 0.000495$

b. $4\left(\frac{{}_{13}C_5}{{}_{52}C_5}\right) = 4\left(\frac{1287}{2{,}598{,}960}\right) \approx 0.001981$

107. $P(4) = {}_{10}C_4(0.32)^4(0.68)^{(10-4)}$

$$= 210(0.32)^4(0.68)^6 \approx 0.2177$$

109. a. $1 - 0.01 = 0.99$

b. $(0.99)^5 \approx 0.951$

c. All five employees test negative.

d. $1 - 0.951 = 0.049$

e. The lab must test the individual blood samples separately to determine which individual (or individuals) is positive.

f. Roughly 95% of the time, the mixture will test negative, and the lab determines the results for five employees for the cost of doing one test. On the other hand, only 5% of the time does the mixture test positive making it necessary to test the individual samples. In such a case, the lab tests five samples for the cost of running six samples. But again, this occurs infrequently.

Chapter 8 Review Exercises

1. $\dfrac{8!}{3!\cdot 5!}=\dfrac{8\cdot 7\cdot \overset{1}{\cancel{6}}\cdot \overset{1}{\cancel{5!}}}{\cancel{3}\cdot \cancel{2}\cdot 1\cdot \cancel{5!}}=56$

3. $a_n=\dfrac{(n+1)!}{n!}$

$a_1=\dfrac{(1+1)!}{1!}=\dfrac{2!}{1!}=\dfrac{2\cdot \overset{1}{\cancel{1!}}}{\cancel{1!}}=2$

$a_2=\dfrac{(2+1)!}{2!}=\dfrac{3!}{2!}=\dfrac{3\cdot \overset{1}{\cancel{2!}}}{\cancel{2!}}=3$

$a_3=\dfrac{(3+1)!}{3!}=\dfrac{4!}{3!}=\dfrac{4\cdot \overset{1}{\cancel{3!}}}{\cancel{3!}}=4$

$a_4=\dfrac{(4+1)!}{4!}=\dfrac{5!}{4!}=\dfrac{5\cdot \overset{1}{\cancel{4!}}}{\cancel{4!}}=5$

$a_5=\dfrac{(5+1)!}{5!}=\dfrac{6!}{5!}=\dfrac{6\cdot \overset{1}{\cancel{5!}}}{\cancel{5!}}=6$

2, 3, 4, 5, 6

5. The odd-numbered terms will be positive, and the even-numbered terms will be negative.

7. a. $a_n=2n+3$

$a_1=2(1)+3=2+3=5$

$a_2=2(2)+3=4+3=7$

$a_3=2(3)+3=6+3=9$

$a_4=2(4)+3=8+3=11$

$a_5=2(5)+3=10+3=13$

5, 7, 9, 11, 13

b. $\sum_{i=1}^{5}(2i+3)=5+7+9+11+13=45$

c. $S_n=\dfrac{n}{2}(a_1+a_n)$

$S_5=\dfrac{5}{2}(a_1+a_5)=\dfrac{5}{2}(5+13)=\dfrac{5}{2}(18)=45$

9. $\sum_{i=3}^{8}2i$

$=2(3)+2(4)+2(5)+2(6)+2(7)+2(8)$

$=6+8+10+12+14+16$

$=66$

11. $\sum_{i=1}^{60}(-1)^{i+1}=(-1)^2+(-1)^3+\cdots+(-1)^{60}+(-1)^{61}$

$=\underbrace{1+(-1)+\cdots+1+(-1)}_{\text{equal number of }(-1)\text{s and 1s}}$

$=0$

13. The numerator is always 2 more than the term number: $i+2$. The denominator is 2 raised to the power of the term number: 2^i.

$\sum_{i=1}^{5}\dfrac{i+2}{2^i}$

15. $\sum_{i=1}^{10}i^3=\sum_{j=0}^{9}(j+1)^3=\sum_{k=2}^{11}(k-1)^3$

17. $a_n=2^n;\ 1\le n\le 12$

19. $a_2-a_1=\dfrac{21}{4}-4=\dfrac{21}{4}-\dfrac{16}{4}=\dfrac{5}{4}$

$a_3-a_2=\dfrac{13}{2}-\dfrac{21}{4}=\dfrac{26}{4}-\dfrac{21}{4}=\dfrac{5}{4}$

$a_4-a_3=\dfrac{31}{4}-\dfrac{13}{2}=\dfrac{31}{4}-\dfrac{26}{4}=\dfrac{5}{4}$

Yes; $d=\dfrac{5}{4}$

21. 4, 12, 20, 28, 36 (+8, +8, +8, +8)

23. $a_n = a_1 + (n-1)d$

$239 = 15 + (57-1)d$

$239 = 15 + 56d$

$56d = 224$

$d = 4$

$a_n = 15 + (n-1)(4)$

$a_n = 15 + 4n - 4$

$a_n = 4n + 11$

$a_{23} = 4(23) + 11 = 92 + 11 = 103$

25. $a_n = a_1 + (n-1)d$

$122 = 11 + (n-1)(3)$

$122 = 11 + 3n - 3$

$122 = 8 + 3n$

$114 = 3n$

$n = 38$

27. $a_n = a_1 + (n-1)d$

$a_n = -1 + (n-1)(-8)$

$a_n = -1 - 8n + 8$

$a_n = -8n + 7$

$a_{35} = -8(35) + 7 = -280 + 7 = -273$

$S_n = \frac{n}{2}(a_1 + a_n)$

$S_{35} = \frac{35}{2}(-1 - 273) = \frac{35}{2}(-274) = -4795$

29. $\sum_{n=1}^{36}(2-5n) = [2-5(1)] + [2-5(2)] + [2-5(3)] + \cdots + [2-5(36)]$

$= (-3) + (-8) + (-13) + \cdots + (-178)$

$a_1 = -3, a_{36} = -178$

$S_n = \frac{n}{2}(a_1 + a_n)$

$S_{36} = \frac{36}{2}(-3 - 178) = 18(-181) = -3258$

31. $a_n = a_1 + (n-1)d$

$a_n = 50 + (n-1)(10)$

$a_n = 50 + 10n - 10$

$a_n = 10n + 40$

$S_n = \frac{n}{2}(a_1 + a_n)$

$3960 = \frac{n}{2}(50 + 10n + 40)$

$3960 = \frac{n}{2}(10n + 90)$

$3960 = 5n^2 + 45n$

$5n^2 + 45n - 3960 = 0$

$n^2 + 9n - 792 = 0$

$(n-24)(n+33) = 0$

$n = 24$ or $n = -33$

24 months

33. $\frac{a_2}{a_1} = \frac{9}{3} = 3$

$\frac{a_3}{a_2} = \frac{36}{9} = 4$

$\frac{a_4}{a_3} = \frac{180}{36} = 5$

No

35. $a_1 = 120$

$a_2 = 120\left(\frac{2}{3}\right) = 80$

$a_3 = 80\left(\frac{2}{3}\right) = \frac{160}{3}$

$a_4 = \frac{160}{3}\left(\frac{2}{3}\right) = \frac{320}{9}$

$a_5 = \frac{320}{9}\left(\frac{2}{3}\right) = \frac{640}{27}$

$120, 80, \frac{160}{3}, \frac{320}{9}, \frac{640}{27}$

37. $r = \frac{a_2}{a_1} = \frac{12}{4} = 3$

$a_n = a_1 r^{n-1}$

$a_n = 4(3)^{n-1}$

$a_6 = 4(3)^{6-1} = 4(3)^5 = 972$

39. $a_n = a_1 r^{n-1}$

$\frac{1}{64} = a_1\left(-\frac{1}{4}\right)^{7-1}$

$\frac{1}{64} = a_1\left(-\frac{1}{4}\right)^{6}$

$a_1 = \frac{1}{64}(4096) = 64$

41. $a_1 = 5, n = 7, r = 3$

$S_n = \frac{a_1(1-r^n)}{1-r}$

$S_7 = \frac{5\left[1-(3)^7\right]}{1-3} = \frac{5(1-2187)}{-2} = 5465$

43. $a_1 = 5, r = \frac{5}{6}$

$|r| = \left|\frac{5}{6}\right| < 1$

$S = \frac{a_1}{1-r} = \frac{5}{1-\frac{5}{6}} = \frac{5}{\frac{1}{6}} = 30$

45. $\sum_{n=3}^{\infty} 6\left(\frac{1}{2}\right)^{n-1} = \sum_{n=1}^{\infty} 6\left(\frac{1}{2}\right)^{n-1} - 6\left(\frac{1}{2}\right)^{1-1} - 6\left(\frac{1}{2}\right)^{2-1}$

$= \sum_{n=1}^{\infty} 6\left(\frac{1}{2}\right)^{n-1} - 6 - 3$

$= \sum_{n=1}^{\infty} 6\left(\frac{1}{2}\right)^{n-1} - 9$

$a_1 = 6, r = \frac{1}{2}$

$|r| = \left|\frac{1}{2}\right| < 1$

$S = \frac{a_1}{1-r} = \frac{6}{1-\frac{1}{2}} = \frac{6}{\frac{1}{2}} = 12$

$\sum_{n=3}^{\infty} 6\left(\frac{1}{2}\right)^{n-1} = \sum_{n=1}^{\infty} 6\left(\frac{1}{2}\right)^{n-1} - 9 = 12 - 9 = 3$

47. $0.8\overline{7} = 0.8777\ldots$

$= \frac{8}{10} + \underbrace{\left[\frac{7}{100} + \frac{7}{1000} + \frac{7}{10{,}000} + \ldots\right]}_{\text{Infinite geometric series}}$

$a_1 = \frac{7}{100}, r = \frac{1}{10}$

$S = \frac{a_1}{1-r} = \frac{\frac{7}{100}}{1-\frac{1}{10}} = \frac{\frac{7}{100}}{\frac{9}{10}} = \frac{7}{100} \cdot \frac{10}{9} = \frac{7}{90}$

$0.8\overline{7} = \frac{8}{10} + \frac{7}{90} = \frac{79}{90}$

49. $A = \frac{P\left[\left(1+\frac{r}{n}\right)^{nt} - 1\right]}{\frac{r}{n}}$

$A = \frac{150\left[\left(1+\frac{0.04}{12}\right)^{(12)(16)} - 1\right]}{\frac{0.04}{12}} \approx \$40{,}250.86$

51. a. $P = \$100, n = 24, r = 5.5\%,$
$t = 62 - 28 = 34$ yr

$A = \frac{P\left[\left(1+\frac{r}{n}\right)^{nt} - 1\right]}{\frac{r}{n}}$

$A = \frac{100\left[\left(1+\frac{0.055}{24}\right)^{(24)(34)} - 1\right]}{\frac{0.055}{24}}$

$A \approx \$238{,}884.21$

b. $P = \$100, n = 24, r = 5.5\%,$
$t = 65 - 28 = 37$ yr

$$A = \frac{P\left[\left(1+\frac{r}{n}\right)^{nt} - 1\right]}{\frac{r}{n}}$$

$$A = \frac{100\left[\left(1+\frac{0.055}{24}\right)^{(24)(37)} - 1\right]}{\frac{0.055}{24}}$$

$A \approx \$289,503.57$

53. **1.** S_1 is true because $3 = 1\left[2(1)+1\right]$.

2. Assume that $S_k = k(2k+1)$. Show that $S_{k+1} = (k+1)\left[2(k+1)+1\right] = (k+1)(2k+3)$.

$$\begin{aligned} S_{k+1} &= \left[3+7+11+\cdots+(4k-1)\right] \\ &\quad +\left[4(k+1)-1\right] \\ &= k(2k+1)+(4k+3) \quad \text{Inductive hypothesis} \\ &= 2k^2+k+4k+3 \\ &= 2k^2+5k+3 \\ &= (k+1)(2k+3) \text{ as desired.} \end{aligned}$$

The statement is true for all positive integers.

55. **1.** S_1 is true because $1 = \frac{1}{3}\left(4^1-1\right)$.

2. Assume that $S_k = \frac{1}{3}\left(4^k-1\right)$.

Show that $S_{k+1} = \frac{1}{3}\left(4^{k+1}-1\right)$.

$$\begin{aligned} S_{k+1} &= \left[1+4+16+\cdots+4^{k-1}\right]+4^{(k+1)-1} \\ &= \frac{1}{3}\left(4^k-1\right)+4^k \quad \text{Inductive hypothesis} \\ &= \frac{1}{3}4^k-\frac{1}{3}+4^k \\ &= \frac{4}{3}4^k-\frac{1}{3} \\ &= \frac{1}{3}\left(4\cdot 4^k-1\right) \\ &= \frac{1}{3}\left(4^{k+1}-1\right) \text{ as desired.} \end{aligned}$$

The statement is true for all positive integers.

57. Let S_n be the statement: $4^n < (n+2)!$ for $n \ge 2$.

1. S_2 is true because $4^2 = 16$ and $(2+2)! = 24$. Therefore, $4^2 < (2+2)!$.

2. Assume that $4^k < (k+2)!$ for an integer $k \ge 2$.

Show that $4^{k+1} < \left[(k+1)+2\right]!$, or equivalently, $4^{k+1} < (k+3)!$.

$$\begin{aligned} 4^{k+1} &= 4\cdot 4^k \\ &< 4\cdot(k+2)! \quad \text{Inductive hypothesis} \\ &< (k+3)(k+2)! \quad \text{Since } k \ge 2, \text{ then } (k+3) > 4. \\ &= (k+3)! \text{ as desired.} \end{aligned}$$

The statement is true for all integers $n \ge 2$.

59. **a.** $\binom{3}{0} = \frac{3!}{0!\cdot(3-0)!} = \frac{3!}{0!\cdot 3!} = \frac{3!}{1\cdot 3!} = 1$

b. $\binom{3}{1} = \frac{3!}{1!\cdot(3-1)!} = \frac{3!}{1!\cdot 2!} = \frac{3\cdot 2!}{1\cdot 2!} = 3$

c. $\binom{3}{2} = \frac{3!}{2!\cdot(3-2)!} = \frac{3!}{2!\cdot 1!} = \frac{3\cdot 2!}{2!\cdot 1} = 3$

d. $\binom{3}{3} = \frac{3!}{3!\cdot(3-3)!} = \frac{3!}{3!\cdot 0!} = \frac{3!}{3!\cdot 1} = 1$

The values of $\binom{3}{0}$, $\binom{3}{1}$, $\binom{3}{2}$, and $\binom{3}{3}$ match the entries in the fourth row of Pascal's triangle.

61. The expression is in the form $(a+b)^5$ with $a=2x$ and $b=-3$.

$$(2x-3)^5=\binom{5}{0}(2x)^5+\binom{5}{1}(2x)^4(-3)+\binom{5}{2}(2x)^3(-3)^2+\binom{5}{3}(2x)^2(-3)^3+\binom{5}{4}(2x)(-3)^4+\binom{5}{5}(-3)^5$$
$$=1(32x^5)+5(16x^4)(-3)+10(8x^3)(9)+10(4x^2)(-27)+5(2x)(81)+1(-243)$$
$$=32x^5-240x^4+720x^3-1080x^2+810x-243$$

63. The expression is in the form $(a+b)^6$ with $a=t^5$ and $b=u^3$.

$$(t^5+u^3)^6=\binom{6}{0}(t^5)^6+\binom{6}{1}(t^5)^5(u^3)+\binom{6}{2}(t^5)^4(u^3)^2+\binom{6}{3}(t^5)^3(u^3)^3+\binom{6}{4}(t^5)^2(u^3)^4+\binom{6}{5}(t^5)(u^3)^5+\binom{6}{6}(u^3)^6$$
$$=t^{30}+6t^{25}u^3+15t^{20}u^6+20t^{15}u^9+15t^{10}u^{12}+6t^5u^{15}+u^{18}$$

65. $n=6, k=5, a=5x, b=4y$

$$\binom{n}{k-1}a^{n-(k-1)}b^{k-1}=\binom{6}{5-1}(5x)^{6-(5-1)}(4y)^{5-1}=\binom{6}{4}(5x)^2(4y)^4=15(25x^2)(256y^4)=96{,}000x^2y^4$$

67. $n=9, a=2c^2, b=-d^5$

$$\binom{n}{k-1}a^{n-(k-1)}b^{k-1}=\binom{9}{k-1}(2c^2)^{9-(k-1)}(-d^5)^{k-1}$$
$$d^{25}=(d^5)^{k-1}$$
$$d^{25}=d^{5(k-1)}$$
$$25=5(k-1)$$
$$5=k-1$$
$$k=6$$
$$\binom{9}{6-1}(2c^2)^{9-(6-1)}(-d^5)^{6-1}=\binom{9}{5}(2c^2)^4(-d^5)^5=126(16c^8)(-d^{25})=-2016c^8d^{25}$$

69. The expression is in the form $(a+b)^4$ with $a=3$ and $b=2i$.

$$(3+2i)^4=\binom{4}{0}(3)^4+\binom{4}{1}(3)^3(2i)+\binom{4}{2}(3)^2(2i)^2+\binom{4}{3}(3)(2i)^3+\binom{4}{4}(2i)^4$$
$$=1(81)+4(27)(2i)+6(9)(4i^2)+4(3)(8i^3)+1(16i^4)=81+216i-216-96i+16=-119+120i$$

71. $7!=5040$

73. $5!-1=120-1=119$

75. a. $5^3=125$

b. The code must end in a 4, 6, or 8.
$5\cdot5\cdot3=75$

c. The code must end in a 5 and neither of the first two digits cannot be a 5.
$4\cdot3\cdot1=75$

77. $_{21}C_4 = \dfrac{21!}{4!\cdot(21-4)!} = \dfrac{21!}{4!\cdot 17!} = 5985$

There are 5985 ways to select 4 distinct items in no specific order from a group of 21 items.

79. $_{90}C_{15} = \dfrac{90!}{15!\cdot(90-15)!} = \dfrac{90!}{15!\cdot 75!} \approx 4.58\times10^{16}$

81. $_{40}P_3 = \dfrac{40!}{(40-3)!} = \dfrac{40!}{37!}$

$= \dfrac{40\cdot39\cdot38\cdot\cancel{37!}}{\cancel{37!}} = 59{,}280$

83. $({}_{10}C_2)\cdot({}_6C_3)\cdot({}_7C_2) = 45\cdot20\cdot21 = 18{,}900$

85. Not likely

87. $P(\bar{E}) = 1 - P(E) = 1 - 0.73 = 0.27$

89. Refer to Figure 8-11 on page 774.

a. $\dfrac{n(E)}{n(S)} = \dfrac{5}{36}$

b. $\dfrac{n(E)}{n(S)} = \dfrac{6}{36} = \dfrac{1}{6}$

c. The numbers form a sum of 5 or 10.

$\dfrac{n(E)}{n(S)} = \dfrac{7}{36}$

91. a. $\dfrac{n(E)}{n(S)} = \dfrac{_4C_3}{_{15}C_3} = \dfrac{4}{455} \approx 0.00879$

b. $\dfrac{n(E)}{n(S)} = \dfrac{_{11}C_3}{_{15}C_3} = \dfrac{165}{455} \approx 0.36264$

c. The events from parts (a) and (b) are not complementary events. There are many other cases to consider regarding the number of defective and good lightbulbs: for example, 2 good, 1 defective, etc.

93. $P(10 \text{ survive to } 51) = [P(1 \text{ survives to } 51)]^{10}$

$= (0.9959)^3 \approx 0.9597$

95.

	Normal BP	Elevated BP	Total
Smokers	42	28	70
Nonsmokers	80	10	90
Total	122	38	160

$\dfrac{90}{160} = \dfrac{9}{16}$

97.

	Normal BP	Elevated BP	Total
Smokers	42	28	70
Nonsmokers	80	10	90
Total	122	38	160

$\dfrac{122}{160} + \dfrac{70}{160} - \dfrac{42}{160} = \dfrac{150}{160} = \dfrac{15}{16}$

99.

	Normal BP	Elevated BP	Total
Smokers	42	28	70
Nonsmokers	80	10	90
Total	122	38	160

$\dfrac{90}{160} + \dfrac{122}{160} - \dfrac{80}{160} = \dfrac{132}{160} = \dfrac{33}{40}$

For Exercises 101–103, refer to Figure 8-12 on page 776.

101. $P(R\cup5) = P(R) + P(5) - P(R\cap5)$

$= \dfrac{26}{52} + \dfrac{4}{52} - \dfrac{2}{52} = \dfrac{28}{52} = \dfrac{7}{13}$

103. $P(\text{two kings}) = P(\text{king})\cdot P(\text{king})$

$= \dfrac{4}{52}\cdot\dfrac{4}{52} = \dfrac{1}{169}$

Chapter 8 Test

1. $a_1 = -2,\ a_2 = 3,\ a_n = a_{n-2} + a_{n-1}$, for $n \geq 3$

$a_1 = -2$

$a_2 = 3$

$a_3 = a_1 + a_2 = -2 + 3 = 1$

$a_4 = a_2 + a_3 = 3 + 1 = 4$

$a_5 = a_3 + a_4 = 1 + 4 = 5$

$a_6 = a_4 + a_5 = 4 + 5 = 9$

$-2, 3, 1, 4, 5, 9$

3. a. $a_2 - a_1 = 0.00139 - 0.139 = -0.13761$

$a_3 - a_2 = 0.0000139 - 0.00139$

$= -0.0013761$

Not arithmetic

$\frac{a_2}{a_1} = \frac{0.00139}{0.139} = 0.01$

$\frac{a_3}{a_2} = \frac{0.0000139}{0.00139} = 0.01$

Geometric

b. $r = 0.01$

c. $a_n = a_1 r^{n-1} = 0.139(0.01)^{n-1}$

5. a. $a_2 - a_1 = \frac{6}{25} - \frac{3}{5} = -\frac{9}{25}$

$a_3 - a_2 = \frac{9}{125} - \frac{6}{25} = -\frac{21}{125}$

$a_4 - a_3 = \frac{12}{625} - \frac{9}{125} = -\frac{33}{625}$

Not arithmetic

$\frac{a_2}{a_1} = \frac{\frac{6}{25}}{\frac{3}{5}} = \frac{2}{5}$

$\frac{a_3}{a_2} = \frac{\frac{9}{125}}{\frac{6}{25}} = \frac{3}{10}$

$\frac{a_4}{a_3} = \frac{\frac{12}{625}}{\frac{9}{125}} = \frac{4}{15}$

Not geometric

b. Not applicable **c.** $a_n = \frac{3n}{5^n}$

7. $a_n = a_1 + (n-1)d$

$67 = 10 + (20-1)d$

$57 = 19d$

$d = 3$

$a_n = 10 + (n-1)(3)$

$a_n = 10 + 3n - 3$

$a_n = 3n + 7$

$a_2 = 3n + 7 = 3(2) + 7 = 6 + 7 = 13$

$a_3 = 3n + 7 = 3(3) + 7 = 9 + 7 = 16$

$a_4 = 3n + 7 = 3(4) + 7 = 12 + 7 = 19$

$a_5 = 3n + 7 = 3(5) + 7 = 15 + 7 = 22$

10, 13, 16, 19, 22

9. $a_1 = -3, r = 2$

$a_n = a_1 r^{n-1}$

$a_n = -3(2)^{n-1}$

$a_6 = -3(2)^{6-1} = -3(2)^5 = -96$

11. $r = \frac{a_2}{a_1} = \frac{8}{16} = \frac{1}{2}$

$a_n = 16\left(\frac{1}{2}\right)^{n-1}$

$\frac{1}{16} = 16\left(\frac{1}{2}\right)^{n-1}$

$\left(\frac{1}{16}\right)^2 = \left(\frac{1}{2}\right)^{n-1}$

$\left(\frac{1}{2}\right)^8 = \left(\frac{1}{2}\right)^{n-1}$

$8 = n - 1$

$n = 9$

13. $a_n = a_1 r^{n-1}$

$20 = a_1 r^{3-1}$

$20 = a_1 r^2$

$a_1 = \frac{20}{r^2}$

$a_n = a_1 r^{n-1}$

$640 = a_1 r^{8-1}$

$640 = a_1 r^7$

$640 = a_1 r^7$

$640 = \frac{20}{r^2} r^7$

$640 = 20r^5$

$32 = r^5$

$r = 2$

$a_1 = \frac{20}{r^2} = \frac{20}{(2)^2} = 5$

15. $\sum_{k=1}^{54}(3k+7) = [3(1)+7]+[3(2)+7]$

$+[3(3)+7]+\cdots+[3(54)+7]$

$=10+13+16+\cdots+169$

$a_1 = 10,\ a_{54} = 169$

$S_n = \frac{n}{2}(a_1 + a_n)$

$S_{54} = \frac{54}{2}(10+169) = 27(179) = 4833$

17. $\sum_{k=1}^{4} k! = 1!+2!+3!+4! = 1+2+6+24 = 33$

19. a. $50{,}000(\$15) = \$750{,}000$

b. $a_1 = 750{,}000,\ r = 0.65$

$S = \frac{a_1}{1-r} = \frac{750{,}000}{1-0.65} = \frac{750{,}000}{0.35}$

$= \$2{,}142{,}857$

21. $a_n = (3.50+4.00)n^2 = 7.50n^2$

23. 1. S_1 is true because $6 = 1[2(1)+4]$.

2. Assume that $S_k = k(2k+4)$. Show that

$S_{k+1} = (k+1)[2(k+1)+4] = (k+1)(2k+6)$.

$S_{k+1} = [6+10+14+\cdots+(4k+2)]+[4(k+1)+2]$

$= k(2k+4)+(4k+6)$ Inductive hypothesis

$= 2k^2+4k+4k+6$

$= 2k^2+8k+6$

$= (k+1)(2k+6)$ as desired.

The statement is true for all positive integers.

25. Let S_n be the statement: 2 is a factor of $7^n - 5$.

1. S_1 is true because 2 is a factor of $7^1 - 5 = 2$.

2. Assume that S_k is true; that is, assume that 2 is a factor of $7^k - 5$. This implies that $7^k - 5 = 2a$ and that $7^k = 2a+5$ for some positive integer a.

Show that S_{k+1} is true; that is, show that 2 is a factor of $7^{k+1} - 5$.

$7^{k+1} - 5 = 7 \cdot 7^k - 5$

$= 7(2a+5) - 5$ Replace 7^k by $2a+5$.

$= 14a + 35 - 5$

$= 14a + 30$

$= 2(7a+15)$

Therefore, 2 is a factor of $7^{k+1} - 5$ as desired. The statement is true for all positive integers.

27. The expression is in the form $(a+b)^5$ with $a = 4c^2$ and $b = -t^4$.

$$(4c^2 - t^4)^5 = \binom{5}{0}(4c^2)^5 + \binom{5}{1}(4c^2)^4(-t^4) + \binom{5}{2}(4c^2)^3(-t^4)^2 + \binom{5}{3}(4c^2)^2(-t^4)^3 + \binom{5}{4}(4c^2)(-t^4)^4 + \binom{5}{5}(-t^4)^5$$

$$= 1(1024c^{10}) + 5(256c^8)(-t^4) + 10(64c^6)(t^8) + 10(16c^4)(-t^{12}) + 5(4c^2)(t^{16}) + 1(-t^{20})$$

$$= 1024c^{10} - 1280c^8t^4 + 640c^6t^8 - 160c^4t^{12} + 20c^2t^{16} - t^{20}$$

29. $n = 7, a = 3x, b = y^2$

$$\binom{n}{k-1}a^{n-(k-1)}b^{k-1} = \binom{7}{k-1}(3x)^{7-(k-1)}(y^2)^{k-1}$$

$$y^6 = (y^2)^{k-1}$$
$$y^6 = y^{2(k-1)}$$
$$6 = 2(k-1)$$
$$3 = k-1$$
$$k = 4$$

$$\binom{7}{4-1}(3x)^{7-(4-1)}(y^2)^{4-1} = \binom{7}{3}(3x)^4(y^2)^3$$
$$= 35(81x^4)(y^6)$$
$$= 2835x^4y^6$$

31. A permutation of n items taken r at a time is an arrangement of r items taken from a group of n items in a specific order. A combination of n items taken r at a time is a group of r items taken from a group of n items in no particular order.

33. $4 \cdot 5 \cdot 3 = 60$

35. **a.** ${}_{56}C_{12} \approx 5.584 \times 10^{11}$

b. $({}_{30}C_6) \cdot ({}_{26}C_6) \approx 1.367 \times 10^{11}$

c. $\dfrac{n(E)}{n(S)} = \dfrac{({}_{30}C_6) \cdot ({}_{26}C_6)}{{}_{56}C_{12}} \approx 0.245$

d. $\dfrac{n(E)}{n(S)} = \dfrac{{}_{26}C_{12}}{{}_{56}C_{12}} = \dfrac{11{,}440}{52{,}451{,}256} \approx 0.0000173$

37. $2^6 \cdot 3^4 = 64 \cdot 81 = 5184$

39. $\dfrac{36{,}000}{300{,}000{,}000} = 0.00012$

41. Refer to Figure 8-11 on page 774.

$$\frac{n(E)}{n(S)} = \frac{16}{36} = \frac{4}{9}$$

43. $(0.319)^3 \approx 0.0325$

45.

	Cash	Credit Card	Check	Total
Male	19	24	13	56
Female	14	30	20	64
Total	33	54	33	120

$$\frac{56}{120} + \frac{33}{120} - \frac{13}{120} = \frac{76}{120} = \frac{19}{30}$$

47.

	Cash	Credit Card	Check	Total
Male	19	24	13	56
Female	14	30	20	64
Total	33	54	33	120

$$\frac{33}{120} + \frac{64}{120} - \frac{14}{120} = \frac{83}{120}$$

Chapter 8 Cumulative Review Exercises

1. The inequality is strict. False

3. Since all elements in B are also in A the intersection of A and B is B.

5. **a.** $|t-5|$ or $|5-t|$

b. for $t < 5, |t-5| = -(t-5) = 5-t$

7. $-27^{-4/3} = -\left(3^3\right)^{-4/3} = -3^{-4} = -\frac{1}{3^4} = -\frac{1}{81}$

9. $\frac{4-x^{-2}}{6-3x^{-1}} = \frac{4-\frac{1}{x^2}}{6-\frac{3}{x}} = \frac{x^2\cdot\left(4-\frac{1}{x^2}\right)}{x^2\cdot\left(6-\frac{3}{x}\right)} = \frac{4x^2-1}{6x^2-3x}$

$= \frac{(2x+1)\cancel{(2x-1)}}{3x\cancel{(2x-1)}} = \frac{2x+1}{3x}$

11. $\sqrt{2x^2-x-14}+4=x$

$\sqrt{2x^2-x-14}=x-4$

$\left(\sqrt{2x^2-x-14}\right)^2=(x-4)^2$

$2x^2-x-14=x^2-8x+16$

$x^2+7x-30=0$

$(x+10)(x-3)=0$

$x=-10$ or $x=3$

Check: $x=-10$

$\sqrt{2x^2-x-14}+4=x$

$\sqrt{2(-10)^2-(-10)-14}+4 \stackrel{?}{=} -10$

$\sqrt{200+10-14}+4 \stackrel{?}{=} -10$

$\sqrt{196}+4 \stackrel{?}{=} -10$

$18 \stackrel{?}{=} -10$ false

Check: $x=3$

$\sqrt{2x^2-x-14}+4=x$

$\sqrt{2(3)^2-(3)-14}+4 \stackrel{?}{=} 3$

$\sqrt{18-3-14}+4 \stackrel{?}{=} 3$

$\sqrt{1}+4 \stackrel{?}{=} 3$

$5 \stackrel{?}{=} 3$ false

$\{\ \}$; The values 3 and -10 do not check.

13. $|3x-5|=|2x+1|$

$3x-5=2x+1$ or $3x-5=-(2x+1)$

$x=6$ or $5x=4$

or $x=\frac{4}{5}$

$\left\{6, \frac{4}{5}\right\}$

15. $0 \le |x+7|-6$

$6 \le |x+7|$

$|x+7| \ge 6$

$x+7 \le -6$ or $x+7 \ge 6$

$x \le -13$ or $x \ge -1$

$(-\infty, -13] \cup [-1, \infty)$

17. $\log_4(2x+7)=2+\log_4 x$

$\log_4(2x+7)-\log_4 x=2$

$\log_4\left(\frac{2x+7}{x}\right)=2$

$\frac{2x+7}{x}=4^2$

$2x+7=16x$

$14x=7$

$x=\frac{1}{2}$

$\left\{\frac{1}{2}\right\}$

19. $f(x)=2x^3-5x^2-28x+15$

a. $\frac{\text{Factors of } 15}{\text{Factors of } 1}$

$= \frac{\pm1, \pm3, \pm5, \pm15}{\pm1, \pm2}$

$= \pm1, \pm3, \pm5, \pm15, \pm\frac{1}{2}, \pm\frac{3}{2}, \pm\frac{5}{2}, \pm\frac{15}{2}$

$$\begin{array}{r|rrrr} 5 & 2 & -5 & -28 & 15 \\ & & 10 & 25 & -15 \\ \hline & 2 & 5 & -3 & \lfloor 0 \end{array}$$

$f(x)=(x-5)(2x^2+5x-3)$

$f(x)=(x-5)(x+3)(2x-1)$

The zeros are $-3, 5, \frac{1}{2}$.

b. x-intercepts: $(-3,0), (5,0), \left(\frac{1}{2},0\right)$

c. $f(0)=2(0)^3-5(0)^2-28(0)+15=15$

y-intercept: $(0,15)$

d.

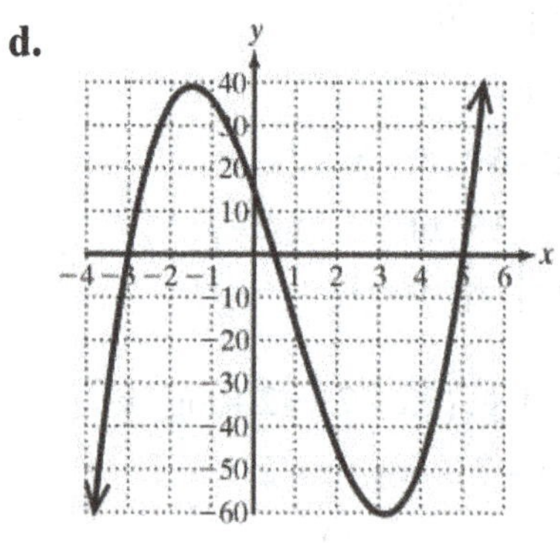

e. $(-\infty,-3)\cup\left(\frac{1}{2},5\right)$

21. a. $x^2+y^2-8x-2y+1=0$

$(x^2-8x \quad)+(y^2-2y \quad)=-1$

$$\left[\frac{1}{2}(-8)\right]^2=16 \qquad \left[\frac{1}{2}(-2)\right]^2=1$$

$(x^2-8x+16)+(y^2-2y+1)=-1+16+1$

$(x-4)^2+(y-1)^2=16$

b. Center: $(4,1)$; Radius: 4

c.

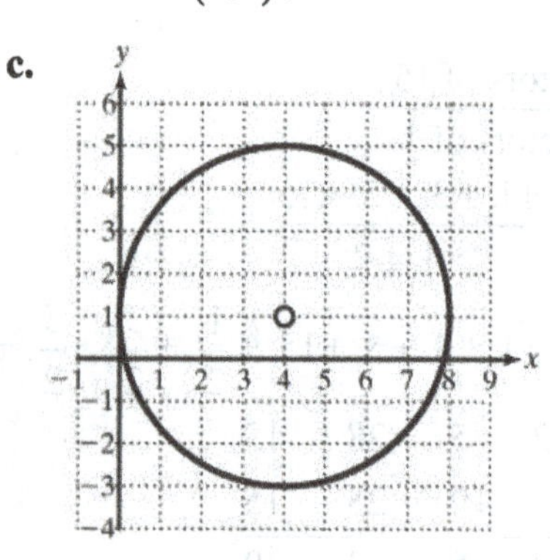

23. $\dfrac{(x+1)^2}{4}+\dfrac{(y-2)^2}{9}=1$

Center: $(-1,2)$

Since $9>4$, $a^2=9$ and $b^2=4$.

$a=\sqrt{9}=3$ and $b=\sqrt{4}=2$.

Since the greater number in the denominator is found in the y^2 term, the ellipse is elongated vertically.

The vertices are a units above and below the center.

Vertices: $(-1,2+3)$ and $(-1,2-3)$

$(-1,5)$ and $(-1,-1)$

The minor axis endpoints are b units to the right and left of the center.

Endpoints of minor axis:

$(-1+2,2)$ and $(-1-2,2)$

$(1,2)$ and $(-3,2)$

$c^2=a^2-b^2=9-4=5$

$c=\sqrt{5}$

The foci are c units above and below the center.

Foci: $\left(-1,2+\sqrt{5}\right)$ and $\left(-1,2-\sqrt{5}\right)$

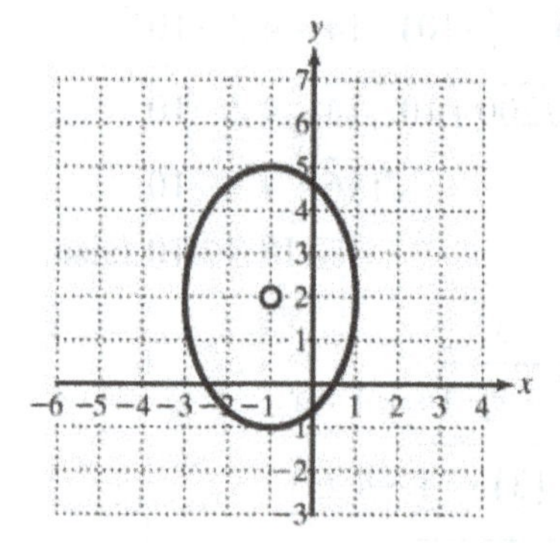

25.

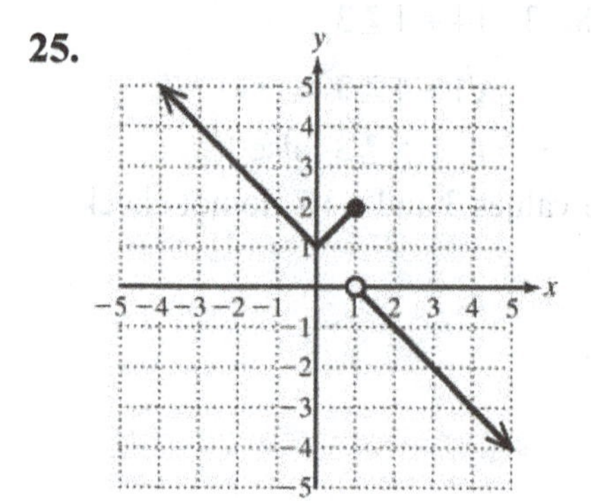

27. a. $\frac{f(x+h)-f(x)}{h}=\frac{4(x+h)^3-3(x+h)-(4x^3-3x)}{h}=\frac{4(x^3+3x^2h+3xh^2+h^3)-3x-3h-4x^3+3x}{h}$

$=\frac{4x^3+12x^2h+12xh^2+4h^3-3x-3h-4x^3+3x}{h}=\frac{12x^2h+12xh^2+4h^3-3h}{h}$

$=12x^2+12xh+4h^2-3$

b. $f(x)=4x^3-3x$

$f(1)=4(1)^3-3(1)=4-3=1$

$f(3)=4(3)^3-3(3)=108-9=99$

Average rate of change $=\frac{f(x_2)-f(x_1)}{x_2-x_1}=\frac{f(3)-f(1)}{3-1}=\frac{99-1}{2}=\frac{98}{2}=49$

c. $f(x)=4x^3-3x$

$f(-x)=4(-x)^3-3(-x)$

$f(-x)=-4x^3+3x$

$-f(x)=-(4x^3-3x)=-4x^3+3x$

$f(-x)=-f(x)$ ⇒ The function is odd.

29. a.

$\underline{2+i}$	1	−1	−7	15
		$2+i$	$1+3i$	-15
	1	$1+i$	$-6+3i$	0

Since $f(2+i)=0$, $2+i$ is a zero of $f(x)$.

b.

$\underline{-3}$	1	−1	−7	15
		−3	12	−15
	1	−4	5	0

Since $f(-3)=0$, $(x+3)$ is a factor of $f(x)$.

31. a. $P(t)=P_0e^{kt}$

$P(t)=320{,}000e^{kt}$

$360{,}800=320{,}000e^{k(12)}$

$\frac{360{,}800}{320{,}000}=e^{k(12)}$

$\ln\left(\frac{451}{400}\right)=12k$

$k=\frac{\ln\left(\frac{451}{400}\right)}{12}\approx 0.01$

$P(t)=320{,}000e^{0.01t}$

b. $P(15)=320{,}000e^{0.01(15)}\approx 371{,}800$

c $400{,}000=320{,}000e^{0.01t}$

$\frac{400{,}000}{320{,}000}=e^{0.01t}$

$\frac{5}{4}=e^{0.01t}$

$\ln\left(\frac{5}{4}\right)=0.01t$

$t=\frac{\ln\left(\frac{5}{4}\right)}{0.01}\approx 22.3$ yr

33. $\log_2\frac{1}{16}=\log_2(2^{-4})=-4$

35. Augmented matrix: $\left[\begin{array}{cc|c} 2 & -5 & 13 \\ -3 & 2 & -3 \end{array}\right]$

$-R_1 - R_2 \to R_1 \longrightarrow \left[\begin{array}{cc|c} 1 & 3 & -10 \\ -3 & 2 & -3 \end{array}\right]$

$3R_1 + R_2 \to R_2 \longrightarrow \left[\begin{array}{cc|c} 1 & 3 & -10 \\ 0 & 11 & -33 \end{array}\right]$

$\frac{1}{11}R_2 \to R_2 \longrightarrow \left[\begin{array}{cc|c} 1 & 3 & -10 \\ 0 & 1 & -3 \end{array}\right]$

$-3R_2 + R_1 \to R_1 \longrightarrow \left[\begin{array}{cc|c} 1 & 0 & -1 \\ 0 & 1 & -3 \end{array}\right]$

$\{(-1,-3)\}$

37. $\boxed{A}$ $2x - y = 6$

$\boxed{B}$ $x^2 + y = 9 \to y = -x^2 + 9$

$\boxed{A}$ $2x - y = 6$

$2x - (-x^2 + 9) = 6$

$2x + x^2 - 9 = 6$

$x^2 + 2x - 15 = 0$

$(x+5)(x-3) = 0$

$x = -5$ or $x = 3$

$\boxed{B}$ $y = -x^2 + 9 = -(-5)^2 + 9 = -16$

The solution is $(-5,-16)$.

$\boxed{B}$ $y = -x^2 + 9 = -(3)^2 + 9 = 0$

The solution is $(3,0)$.

$\{(-5,-16),(3,0)\}$

39. $13x - 2y = 11$

$5x + 3y = 6$

$a_1 = 13, b_1 = -2, c_1 = 11, a_2 = 5, b_2 = 3, c_2 = 6$

$D = \begin{vmatrix} a_1 & b_1 \\ a_2 & b_2 \end{vmatrix} = \begin{vmatrix} 13 & -2 \\ 5 & 3 \end{vmatrix}$

$= (13)(3) - (-2)(5) = 39 + 10 = 49$

$D_x = \begin{vmatrix} c_1 & b_1 \\ c_2 & b_2 \end{vmatrix} = \begin{vmatrix} 11 & -2 \\ 6 & 3 \end{vmatrix} = (11)(3) - (-2)(6)$

$= 33 + 12 = 45$

$D_y = \begin{vmatrix} a_1 & c_1 \\ a_2 & c_2 \end{vmatrix} = \begin{vmatrix} 13 & 11 \\ 5 & 6 \end{vmatrix} = (13)(6) - (11)(5)$

$= 78 - 55 = 23$

$x = \frac{D_x}{D} = \frac{45}{49} =, y = \frac{D_y}{D} = \frac{23}{49}$

$\left\{\left(\frac{45}{49}, \frac{23}{49}\right)\right\}$

41. $-5A + 2B = -5\begin{bmatrix} 4 & -3 \\ 5 & 9 \end{bmatrix} + 2\begin{bmatrix} -1 & 6 \\ 3 & 7 \end{bmatrix}$

$= \begin{bmatrix} -20 & 15 \\ -25 & -45 \end{bmatrix} + \begin{bmatrix} -2 & 12 \\ 6 & 14 \end{bmatrix}$

$= \begin{bmatrix} -22 & 27 \\ -19 & -31 \end{bmatrix}$

43. $|A| = \begin{vmatrix} 4 & -3 \\ 5 & 9 \end{vmatrix} = (4)(9) - (-3)(5) = 36 + 15 = 51$

45. Matrix C is a 3×3 matrix, whereas matrix D is a 2×3 matrix. The number of columns in the first matrix is not equal to the number of rows in the second matrix. The operation is not possible.

47. Set up matrix $[A \mid I_3]$: $\left[\begin{array}{ccc|ccc} 1 & 2 & -1 & 1 & 0 & 0 \\ 0 & 1 & 3 & 0 & 1 & 0 \\ 1 & 0 & 2 & 0 & 0 & 1 \end{array}\right]$

$-R_1 + R_3 \to R_3 \longrightarrow \left[\begin{array}{ccc|ccc} 1 & 2 & -1 & 1 & 0 & 0 \\ 0 & 1 & 3 & 0 & 1 & 0 \\ 0 & -2 & 3 & -1 & 0 & 1 \end{array}\right]$

$$\begin{array}{r} -2R_2+R_1\to R_1\longrightarrow \\ \\ 2R_2+R_3\to R_3\longrightarrow \end{array}\left[\begin{array}{ccc|ccc} 1 & 0 & -7 & 1 & -2 & 0 \\ 0 & 1 & 3 & 0 & 1 & 0 \\ 0 & 0 & 9 & -1 & 2 & 1 \end{array}\right]$$

$$\frac{1}{9}R_3\to R_3\longrightarrow \left[\begin{array}{ccc|ccc} 1 & 0 & -7 & 1 & -2 & 0 \\ 0 & 1 & 3 & 0 & 1 & 0 \\ 0 & 0 & 1 & -\frac{1}{9} & \frac{2}{9} & \frac{1}{9} \end{array}\right]$$

$$\begin{array}{r} 7R_3+R_1\to R_1\longrightarrow \\ -3R_3+R_2\to R_2\longrightarrow \\ \\ \end{array}\left[\begin{array}{ccc|ccc} 1 & 0 & 0 & \frac{2}{9} & -\frac{4}{9} & \frac{7}{9} \\ 0 & 1 & 0 & \frac{1}{3} & \frac{1}{3} & -\frac{1}{3} \\ 0 & 0 & 1 & -\frac{1}{9} & \frac{2}{9} & \frac{1}{9} \end{array}\right] \qquad A^{-1}=\begin{bmatrix} \frac{2}{9} & -\frac{4}{9} & \frac{7}{9} \\ \frac{1}{3} & \frac{1}{3} & -\frac{1}{3} \\ -\frac{1}{9} & \frac{2}{9} & \frac{1}{9} \end{bmatrix}$$

49. $(y+2)^2=-8(x-1)$

a. The equation is in the form $(y-k)^2=4p(x-h)$, where $h=1$ and $k=-2$. The vertex is $(1,-2)$.

b. $4p=-8$

$p=-2$

The focus is $(h+p,k)$, which is $(-1,-2)$.

c. The directrix is the vertical line $|p|$ units to the right of the vertex: $x=3$.

51. $a_n=a_1+(n-1)d$

$a_n=6.9+(n-1)(0.3)$

$a_n=6.9+0.3n-0.3$

$a_n=0.3n+6.6$

$a_{500}=0.3(500)+6.6=150+6.6=156.6$

53. $a_1=6, n=8, r=2$

$$S_n=\frac{a_1(1-r^n)}{1-r}$$

$$S_8=\frac{6\left[1-(2)^8\right]}{1-2}=\frac{6(1-256)}{-1}=1530$$

55. **1.** S_1 is true because $8=2(1)(1+3)$.

2. Assume that $S_k=2k(k+3)$. Show that $S_{k+1}=2(k+1)\left[(k+1)+3\right]=2(k+1)(k+4)$.

$$\begin{aligned} S_{k+1}&=\left[8+12+16+20+\cdots+(4k+4)\right] \\ &\quad+\left[4(k+1)+4\right] \\ &=2k(k+3)+(4k+8) \quad \text{Inductive hypothesis} \\ &=2k^2+6k+4k+8 \\ &=2k^2+10k+8 \\ &=2(k+1)(k+4) \text{ as desired.} \end{aligned}$$

The statement is true for all positive integers.

57. $$\frac{(3n-1)!}{2!(3n+1)!}=\frac{\overset{1}{\cancel{(3n-1)!}}}{2!(3n+1)(3n)\cancel{(3n-1)!}}=\frac{1}{2(3n+1)(3n)}$$

59. $$\frac{{}_8C_2\cdot{}_5C_2}{{}_{13}C_4}\approx 0.3916$$